Heinz-Wolfram Kasemir:
His Collected Works

Vladislav Mazur and Lothar H. Ruhnke, Editors

American Geophysical Union

Library of Congress Cataloging-in-Publication Data

Kasemir, Heinz W.
  [Works. Selections]
  The collected writings of Heinz Kasemir / Vladislav Mazur and Lothar Ruhnke, editors.
      pages cm
  ISBN 978-0-87590-737-6
  1. Atmospheric electricity.   I. Mazur, Vladislav.   II. Ruhnke, Lothar H.   III. Title.
  QC961.2.K372 2012
  551.56'3—dc23

                                                                    2012038540

  Book doi: 10.1029/SP066

Heinz-Wolfram Kasemir (1913–2007)

CONTENTS

PREFACE

Historically, the science of atmospheric electricity has evolved, largely, based on field observations and measurements taken during fair weather and thunderstorms. In its early stages, experimentalists were the primary developers of this field, reporting and interpreting, to the best of their abilities, the different manifestations of the electrical processes found in the atmosphere. Meanwhile, as a branch of physics, this new field required a sound knowledge of theoretical physics and mathematics, in order to correctly interpret observations that were frequently obtained with sensors of limited capabilities. In the late 1940s, the first critical studies of the relations of observed variables to the laws of physics were undertaken by Heinz-Wolfram Kasemir, and he continued to advance such studies throughout the rest of his scientific career. Heinz-Wolfram Kasemir (Heinz, as we, the Editors, called him) was a physicist by education, and also a talented and tireless experimentalist and innovative designer of scientific instruments.

Many of the physical concepts presented by Kasemir in his manuscripts on atmospheric electricity and lightning physics contradicted the prevailing contemporary interpretations of the physics of atmospheric electrical processes. Not surprisingly, therefore, many of the papers Kasemir submitted to peer-reviewed publications in the United States had serious difficulties with reviewers, and the majority of the papers were rejected. This could have been because either the reviewers' abilities were simply not equal to understanding the new physical concepts developed by Kasemir, or the reviewers were biased toward the prevailing interpretations of that time. In the face of such constantly frustrating, exhausting fights for acceptance of his papers in scientific journals, Kasemir finally stopped submitting them to peer-reviewed journals. As his close associates and friends, we knew of his feelings on this issue. Most of Kasemir's publications, therefore, are either in technical reports, or in the proceedings of scientific conferences, making them difficult or impossible for interested researchers to access.

Eduard Bazelyan, a well-known Russian physicist in the field of spark discharges, recently shared with us an interesting story related to Kasemir: "None of us thought about the possibility of starting and developing a lightning flash without any contact with a high-voltage electrode in the laboratory, until we started working on our book in the late 1990s. The possibility of the simultaneous development of positive and negative leaders in the volume between high-voltage electrodes appeared to us to be a brilliant idea, and was supported by our laboratory experiment. We were so proud of ourselves for coming up with it. Our euphoria, however, lasted only for a couple of weeks until, by chance, we came across a paper written by Kasemir a half-century earlier, where he had described a similar idea. After this understandably-great disappointment, we found satisfaction in proceeding with numerical calculations to develop this idea, which Kasemir could not perform because of the lack of powerful computers. So, Kasemir's idea provided a quantitative foundation, and his name received the well-deserved recognition and respect of our research community. It is simply a pity that it took so long."

Heinz's early, fundamental papers were published in German, and were not translated into English until now, for this book. Most of the work he conducted in the U.S. was presented in conference proceedings and technical reports, with only a few papers published in scientific journals. The main reasons for publishing this collection of Kasemir's papers and presentations are that Kasemir's ideas were far ahead of their time, that many of his publications are not readily accessible, and that it is important to make them available to researchers currently pursuing a better understanding of atmospheric electricity and lightning physics, a field making rapid advances at the moment. It would be a huge loss to the research community if most of Kasemir's scientific legacy were to remain unavailable. Since most of the papers were not reviewed by his peers, this book is a rare opportunity to experience the real, "uncensored" thinking of a prominent scientist in the field of atmospheric electricity. Kasemir's papers are not easy reading, but a persistent reader will find great pleasure in discovering in them clear ideas expressed in very precise language.

In the late 1990s, we approached Heinz with the suggestion of publishing a collection of his work. He was very receptive to this idea, and we started, with his assistance, the preliminary assembly of his publications. The project stalled, however, because he insisted on revisiting his old papers, to critically evaluate their contents for possible additional commentaries. We told him that we would wait until he finished this process. Sadly, however, it became obvious to us very soon that Heinz would not be able to proceed with these revisions: the unmistakable signs of dementia had already become very noticeable by this time.

In 2010, we submitted a proposal to the National Science Foundation, asking for funding of the publication of a collection of the scientific papers by Heinz-Wolfram Kasemir. This was not a typical NSF proposal, so we solicited the co-participation of the Book Department of the American Geophysical Union (AGU) as our partner in publishing it. We are most grateful for the consideration, support and promotion of our idea for this book by Dr. Bradley F. Small, Program Director of Physical and Dynamic Meteorology in the Division of Atmospheric and Geospace Sciences at the NSF. Also, we could not have had

Preface

a better assistance in our project than that provided by Ms. Colleen Matan from the Book Department of the AGU, whose suggestions were always on-target and highly appreciated, and who has tirelessly advanced this project as an example for future projects of this type by the AGU. The excellent German-to-English translation of Kasemir's early papers was done by Apex Translation Inc. Our friend, Tom Warner, generously contributed a photograph from his collection of exceptionally beautiful images of thunderstorms for the cover of this book.

The lightning literature includes several books that are essentially reference books, which review the published results of the lightning research community (e.g., two volumes of *Lightning* edited by R. H. Golde, 1977; *Lightning Discharge* by M. Uman, 1987; and *Lightning: Physics and Effects* by V. Rakov and M. Uman, 2003). The first book to directly address the various issues of the physical concepts of lightning was "*Lightning Physics and Lightning Protection*" by E. M. Bazelyan and Yu. P. Raizer, 2000. There is still, however, a void in the literature on issues that address the physical interpretations of many lightning observations and measurements. The publication of this collection of Kasemir's papers is intended, to some degree, to fill this void.

The reproduction of some of the earlier papers was a challenging task, even with the use of modern technology; so, the clarity of some images and formulas in those papers is not as good as we would like it to be, and reading these papers may require some effort and patience. In addition, some of the copies we had were undated and we were unable to track down any identifying information, and they are noted as "no date (n.d.)."

This collection of the work of Heinz-Wolfram Kasemir consists of 55 manuscripts, which comprise journal articles, conference presentations and technical reports, and is organized into five topics: Fair Weather Electricity, Global Circuit, Thunderstorm Electricity, Lightning Physics, and Measurement Techniques. Twelve early papers by Kasemir, published in German (and now translated into English), are included in the collection. In the Table of Contents, for each paper listed, the comments by the Editors serve to emphasize the important points of each manuscript, and to connect it to current issues in the fields of atmospheric electricity and lightning research.

The project was supported by the National Science Foundation, Federal Award ID Number 1138919.

Introduction

INTRODUCTION
PERSONAL REMARKS BY LOTHAR H. RUHNKE

This volume contains all of the significant papers of Heinz-Wolfram Kasemir. His accomplishments in the field of atmospheric electricity research can be best understood by considering his life's time line.

After he finished his courses toward an advanced degree in physics (Diplom Physiker), World War II broke out, and he was assigned to work at the Research Institute in Oberpfaffenhofen, Bavaria, with Professor Hans Israel, who was the prominent authority on atmospheric electricity in Germany. After the war, he stayed for 7 more years with Professor Israel, who became the head of the German Meteorological Service in Aachen, Germany. These years were, in Kasemir's own words, the most productive years of his research. He remembered these years after the war as a time of severe financial limitations at German research institutions, but under the leadership of Israel, research was free of red tape and restrictions by management. The fundamental problem in atmospheric electricity at the time was, as defined by Israel, the preservation of the fair-weather electrical current resulting from thunderstorm activity around the world. Toward a solution of this problem, a project was established to measure the electric field and current density at a site on one of Europe's highest mountains: Jungfraujoch. After Kasemir developed sensors suitable for measurements in the extremely harsh environment, he collected a considerable amount of valuable data. But the result of several years of effort was disappointing, as measuring the so-called Carnegie curve was not clearly established. Kasemir shifted his effort at that time to theories of the electrical state of the atmosphere. Of interest in his first publication after the war was a question about a pet problem of Israel's: At what altitude does the current from thunderstorms flow to the fair-weather areas? This is also the first paper he gave me to read when I first met him. I was disappointed in the logic shown in the paper. He agreed with me and told me that this paper should have never been written but that he wanted to please Israel. As a result of his own disappointment, Kasemir carefully reconsidered the problem of global current flow, which resulted in defining it successfully in his doctoral dissertation. From then on, Kasemir only considered writing papers with a solid foundation in physics or with a careful analysis of measurements, regardless of whether his colleagues or the scientific community liked it or not.

In 1954, Kasemir and his wife Poldi and four children immigrated to the United States under the project "Paper Clip." He settled near Fort Monmouth, New Jersey, to work for the U.S. Army Signal Corps. Basically, he was a "loner," meaning he preferred to work alone, in contrast to the prevailing trend of the time in research to form teams in order to encourage cooperation among researchers. The laboratory he set up was in a fenced-in area with a 200 foot high radar tower and a large building where he was the only occupant. Scientifically, he was in lull for a while, being occupied with setting up a research facility and adjusting, with his family, to the living environment in the United States. He presented a significant paper at the First International Conference on Atmospheric Electricity in 1954 on measuring conduction current in the atmosphere. Up to that time, the mixture of conduction and displacement currents plagued current sensors, which prevented proving, through experiments, Ohm's law in the atmosphere.

In 1957, I immigrated to the United States from Germany and was assigned to a secret project at Fort Monmouth. I was told that I had to wait 6 months for my clearance and could not enter any facility except the library and cafeteria. I asked for a reassignment and, after asking me to name my home town, a U.S. Army officer assigned me to a physicist who had been born and raised in the same town (Tilsit, Germany): Dr. Heinz-Wolfram Kasemir. His first sentence to me was "I do not want and I do not need an associate, so go away." But that changed rapidly after finding out my background in VLF propagation, in the frequency range of 3 Hz to 20 kHz, where lightning signals were most often the only source of the VLF radiation from which to obtain propagation data. He asked me what I knew about lightning, and I told him that I had always assumed lightning to be a Dirac impulse. He smiled and welcomed me to the job. The research environment was perfect: we both had complete freedom to choose our tasks and work independently. Within 6 months, we had prepared a paper that dealt with potential and current sensors as an active two-terminal circuit, the area where my engineering skills came in handy. I welcomed the many detailed discussions we had on everything we worked on. He was a wonderful teacher, as well as a good listener, and we both profited from our discussions, a practice he had not engaged in much before that period of his life. His research shifted slowly from fair-weather electricity to thunderstorm physics and then to lightning physics. In that period, we both had a chance to do research on the ice cap of northern Greenland; Kasemir tried to record the "Carnegie curve" there, while I measured the effect of aerosol on conductivity in very clean air. Although digital technology and numerical methods for solving complex physics problems were highly advanced at that time, Kasemir, very fluent in

Introduction

handling partial differential equations, was not impressed and to the end of his productive life used computers mainly to plot his analytical solutions.

In 1961, I left the government job and went into private industry. Our research association, however, stayed alive, as I had contract support from him and interacted in many ways with his research. New in this period of time was his belief that charges in thunderstorms are much higher than those shown in the literature and that it might be possible to influence the amount of charge and high electric fields in thunderstorms by seeding the storm clouds with radar chaff fibers. Laboratory experiments required high-voltage generators to study corona emission and self-charge of chaff fibers. I designed and constructed a pair of 500 kV Van de Graaff machines by using kerosene instead of a belt to pump charges to a high-voltage electrode. These units were comparatively compact, which allowed them to be used in his research building. Kasemir's experiments and theoretical treatment of chaff fibers in high electric fields were a trailblazing effort for in-cloud experiments, as well as in the understanding of the corona charging of rockets and in-cloud lightning channels.

Around that time, the Atmospheric Physics Branch of the U.S. Army Signal Corp was dissolved, and most members, including Kasemir, were transferred to the Department of Commerce Research Laboratory in Boulder, Colorado. I changed jobs and also became a member of this Boulder laboratory, but I was assigned to the Mauna Loa Observatory in Hawaii for a 2 year term as director. Kasemir continued his chaff-seeding project in Boulder and concentrated also on electric field measurements with airplanes. Toward that goal, he developed the cylindrical field mill and flew near and under many thunderstorms. His project, now conducted with a team of associates, was successful, as his many reports and presentations at conferences clearly show.

In 1970, I was back in Boulder, united again with Kasemir and his group, continuing research discussions and independent fieldwork and laboratory work. Of special interest was the Apollo 12 incident, when in light rain and without natural lightning, the Apollo spacecraft initiated a lighting strike following launch. The incident aroused many lightning researchers, as well as Kasemir. For the next few years, Kasemir's work concentrated on the thunderstorm and lightning problems of the Kennedy Space Center. He defined the need for an electric field–measuring network there and worked on the problem of deriving cloud electric field values from electric fields on the ground, as well as the conditions for triggering lightning discharges from a rocket in flight.

Kasemir retired from his Department of Commerce position in 1978 and devoted all of his time to research in atmospheric electricity by forming a nonprofit, one-man corporation. He received some contract support from the Kennedy Space Center and from me, after I had accepted a Branch Head position at the Naval Research Laboratory in Washington, D. C. Of special interest at the time was the technology to trigger at will lightning channels using special rockets connected in flight to the ground with a wire. This gave researchers the chance to observe lightning channels in detail through experiments. Kasemir pioneered the theory of influence charges on lightning channels and found that influence charges are proportional to the environmental potentials. Also significant here is the fact that he found, from theory, a small decrease of the charge on an upward leader during its growth due to the screening effect of the leader charge on the nearby electric field. Kasemir also constructed sensors for the electric field that can be used to reconstruct lightning channel geometries from collected data. Personally significant for me is the fact that we were coauthors on his last paper, as we were on my very first paper when I met him 30 years earlier.

DOCUMENT AND COMMENTARY INDEX

SECTION 1. FAIR WEATHER ELECTRICITY

1. Atmospheric Electrical Diurnal Variations and Mass Exchange in the High Mountain Range of the Alps. The Atmospheric Electrical Conditions at Jungfraujoch (3472 m) (Part I). I. H. Israël, H. W. Kasemir and K. Wienert, Archiv für Meteorologie, Geophysik und Bioklimatologie. Serie A: Band III, 5. Heft, Springer Verlag, Wien, 1951.

    Editors' remarks: The paper describes in detail measurements of atmospheric electric variables at a high mountain top. The reason for this project came from the belief that in areas of low convective activity and low aerosol contamination, the daily variation of electric field and current are of global character and vary with universal time, as demonstrated with the data of the Carnegie cruises. Convective activity over land varies with local time and does not show a global daily variation. This paper gives first preliminary results of the expedition.

2. Atmospheric Electrical Diurnal Variations and Mass Exchange in the High Mountain Ranges of the Alps, The Atmospheric Electrical Conditions at Jungfraujoch (3472 m)(Part II). H. Israël, H. W. Kasemir and K. Wienert, Archiv für Meteorologie, Geophysik und Bioklimatologie. Serie A, Band 8, 1.-2. Heft, 1955, Springer Verlag, Wien, 1955.

    Editors' remarks: This research supplements the previous paper on atmospheric electric measurements at a high mountain top. The Jungfraujoch laboratory located at 2600 m altitude was not high enough during the period of measurements to be above the exchange layer, and therefore, the typical Carnegie curve could not be found. As a last effort, it was decided to make measurements in the winter, in very harsh experimental conditions. Simultaneously, a second station at another mountain station (Sonnblick), about 400 km away, was also used to make measurements. The results showed clearly that during the winter, high mountain stations are free from local conductivity changes and produce clearly the same universal time daily variation of currents and electric fields as those in the oceanic recordings. The paper also shows calibration constants for electric field data, which can be applied to the old data presented in the previous publication.

3. Atmospheric Electric Measurements in the Arctic and Antarctic, H. W. Kasemir, Pure and Applied Geophysics (PAGEOPH), Vol.100, 1972/VIII, pp.70-80, Birkhäuser Verlag, Basel, 1972.

    Editors' remarks: The paper is Kasemir's final effort to find locations, other than on oceans or on high mountain tops, to record a worldwide daily variation. This was a project to make simultaneous recordings of atmospheric electric variables in the Arctic and Antarctic. The data show clearly a high correlation of daily variations of electric field between the two stations.

4. Studies regarding the Atmospheric Potential Gradient IV: Examples for the Behavior of Atmospheric Electrical Elements in Fog, H. Israël and H.W. Kasemir, Archiv für Meteorologie, Geophysik und Bioklimatologie. Serie A, Band 5, Heft 1, Springer Verlag, Wien, 1952.

    Editors' remarks: Kasemir analyzes the various forms of fog, in relation to current and electric field. Of interest is the fact that shape and convective activity influence currents and electric fields in fog. Significant is his finding that conductivity is decreased inside nonraining clouds and fog by a factor of 3, regardless of the shape of the clouds or fog. The data on fog also point to the importance of convection that generates electric fields.

5. Studies regarding the Atmospheric Potential Gradient V, Regarding the Current Theory of the Atmospheric Electrical Field (Part I), H.W. Kasemir, Archiv für Meteorologie, Geophysik und Bioklimatologie. Serie A, Band 3, Heft 1 und 2, Springer Verlag, Wien, 1950.

    Editors' remarks: Kasemir considers the factor of the current flow in a system that is conductive, e.g., the atmosphere. This represents a fundamental change in the theory of atmospheric electricity, which up to this time was guided by electrostatic theory. Although extremely important in thunderstorm research, the new theory is applied to conditions in fair weather. In essence, Kasemir's theory shows that for fast changes of a source function (like a step function), the electrostatic treatment is applicable, but, depending on the relaxation time ε/λ, the current after some time is guided by conductivity rather than by the dielectric constant. The realization of this fact has profound effects on understanding of both the electric fields and the space charges in the atmosphere. The conditions for the global circuit as well as the conditions in and near clouds are analyzed. The current flow theory also applies to the thunderstorm conditions. For example, the electric field from a lightning flash is determined at first by the laws of electrostatics, but within about 20 seconds the electric field is determined by the atmospheric conductivity. In this example, the current flow theory explains the nature of the so-called recovery time in the electric field records of lightning flashes.

Document and Commentary Index

6. Studies regarding the Atmospheric Potential Gradient V, Regarding the Current Theory of the Atmospheric Electrical Field (Part II), H.W. Kasemir, Archiv für Meteorologie, Geophysik und Bioklimatologie. Serie A, Band 5, Heft 1, Springer Verlag, Wien, 1952.

   Editors' remarks: In this paper, Kasemir uses his current theory to derive potential and electric fields in and near any rotational ellipsoid, considering conductivities inside and outside the ellipsoid. This is of special interest for probing experimentally and theoretically fields and potentials near clouds and fog of various dimensions. His theory is also useable to calculate form factors of conductors in an electric field.

7. On the Theory of The Atmospheric Electric Circuit Flow, IV, H. W. Kasemir, U.S. Army Electronics Research and Development Laboratories, Fort Monmouth, N.J.,USA, Tech. Report # 2724, 1963.

   Editors' remarks: The paper is the final work on atmospheric electric current flow theory. It discusses in detail the physical meaning of the mathematical manipulations. What is missing is a consideration of the Maxwell equations in relation to the influence of curl E on the results. A lengthy public discussion with R. Boström ended with a statement that for Kasemir's applications in atmospheric electricity, curl E is of no concern.

8. Regarding the Shielding Effect of Buildings on the Fluctuations of the Atmospheric Electrical Field, H. Israël and H. W. Kasemir, Extrait des Annales de Geophysique, Tome 7, fascicule 1, 1951.

   Editors' remarks: The paper deals with an application of the current flow theory. The penetration of atmospheric electric field fluctuations into a building is not a problem in fair weather but is possibly significant for lightning frequencies. The paper is of interest mainly because of the skill shown in handling a complex problem with precise physics.

9. How to Measure and Analyze Solar-Terrestrial Weather Relationships, H. W. Kasemir, Colorado Scientific Research Corporation, Berthoud, Colorado, Copy of original manuscript, no date (n.d.).

   Editors' remarks: Kasemir analyzes the effect of currents in the ionosphere on the global atmospheric electric circuit. The current flow theory, potential distributions, charge distribution, and electric fields in the space between ground and ionosphere are discussed in detail.

10. Analytic Solutions of the Problem of Down-Mapping of Electric Fields and Currents of Solar Events to the Earth Surface, H. W. Kasemir, Weather and Climate Responses to Solar Variations, Ed. Billy M. McCormac, Colorado Associated University Press, 1983.

    Editors' remarks: Kasemir expects near-surface effects of solar events mainly at very high latitudes where Earth magnetic field lines are fairly vertical. He calculates vertical currents along the magnetic field lines that attenuate by 4 orders of magnitude on the way down to the surface. Horizontal electric fields are "down-mapped" without any attenuation until close to the surface.

11. The Exchange Generator. Archiv für Meteorologie, Geophysik und Bioklimatologie. Serie A: Band 9, 3. Heft, Springer Verlag, Wien, 1956.

    Editors' remarks: Using his own data taken in 1953, Kasemir discusses here the so-called sunrise effect. His measurements of current density and electric field show consistently a decrease in conductivity on calm, sunny days during sunrise, with an increase in current density at the surface occurring at the same time. This is consistent with a generator effect produced by an upward flux of the surface space charge layer. Kasemir's data clearly show a mixture of the conduction and convection currents.

SECTION 2. GLOBAL CIRCUIT

1. At What Altitude Does the Global Atmospheric Electricity Circuit Connect? H. Israël and H.W. Kasemir, Extrait des Annales de Geophysique, Tome 5, fascicule 4, 1949.

   Editors' remarks: The main question in the atmospheric electricity research at the time was to find a physical explanation of how thunderstorms produce a worldwide effect in fair-weather areas, as demonstrated by the data from the "Carnegie cruises." The accepted logic was to assume a path of current flow from thunderstorms to the fair-weather areas at some altitude (electrosphere). As this paper shows, it was known already that the conductivity of the atmosphere increases exponentially with altitude, and any charge in a thunderstorm head will decay with a current flow upward and never sideways, as long as the assumption of the exponential conductivity is valid. Kasemir reluctantly participated in the work on this paper that is based on the ideas of H. Israel but later regretted his participation, because of the nonphysical approach in solving the problem. This paper is included in the book for historical reasons, because it left a lasting impression on Kasemir, because he vowed never again to write a paper without a rigorous physical treatment of the problem, regardless of the insistence of his boss or nonacceptance by his peers. The real question for the problem at hand would have been to ask how thunderstorms are able to produce fair-weather electric fields.

Document and Commentary Index

2. The Current Efficiency of the Storm Generator with Regard to the Atmospheric Electrical Vertical Current of the Fair Weather Areas, H. W. Kasemir, Berichte des Deutschen Wetterdienstes in der UA-Zone, Nr. 38, Bad Kissingen, 1952.
   Editors' remarks: Here Kasemir fully answers the question of how currents from thunderstorms can spread over the whole globe to produce the electric field in fair-weather areas. The theory of electric current flow and the assumption of an exponential increase of conductivity with altitude lead to the surface of the Earth becoming the equalizing path for the global effect. The conductivity at very high altitude has no effect on the transfer of charges flowing from the thunderstorms to the fair-weather areas. This leads to the negative total charge of the Earth's surface and with it to a substantial potential of the Earth against infinity. For a positive charge in a thundercloud at 10 km altitude, the effectiveness of the charge transport to fair-weather areas is about 90 percent. Holzer and Saxon discussed the same problem using the traditional view of a spherical condenser model in their paper published in the same year. A more detailed analysis of the physics of the problem is given in the next Kasemir paper in this section.

3. The Storm as a Generator in the Atmospheric Electrical Circuit, H. W. Kasemir, Zeitschrift für GEOPHYSIK, Sonderdruck Jahrg. 25, Heft 2, 1959.
   Editors' remarks: The paper consists of two parts: part 1 is the dissertation of Kasemir from 1954, and part 2 is a more detailed discussion of the physics of the global circuit with a detailed and comprehensive mathematical treatise on the problem area. An appendix gives the experienced physicist a chance to verify Kasemir's conclusions in detail. This work is the culmination of Kasemir's ideas on the thunderstorm generator and its implication for the global circuit. The connection of the thunderstorm's generation of electricity to the fair-weather electric fields and currents required a shift of thinking from electrostatic theory to the current flow theory. The results of his research are often at odds with widely accepted intuitions as to the mechanism of behavior of electrical fields and currents in the atmosphere. Even today, more than 50 years after his work, the contents of this paper will be a valuable source of knowledge for any aspiring scientist in the field of the atmospheric electricity research.

4. Theoretical Problems of the Global Atmospheric Electric Circuit, H. W. Kasemir,. (Proc.Conference on Electrical Processses in Atmospheres, Ed. H. Dolezalek and R. Reiter, Steinkopf Verlag, Darmstadt, 1977.
   Editors' remarks: The paper covers many aspects of the global circuit research. The spherical capacitor model of the global circuit is challenged, as well as the assumption of the ionosphere being an equalizing layer of high conductivity. The conditions for variability of current profiles in fair weather are discussed, as well as sources for convection currents. The paper contains a large number of formulas that apply to the global circuit.

5. Current Budget of the Atmospheric Electric Global Circuit, H. W. Kasemir, J. Geophys. Res., Vol. 99, No. D5, pp.10,701-10,708. 1994.
   Editors' remarks: The paper covers the same subject as his early paper in 1952 (paper 2 in this section) but specifies in more detail how thunderstorms produce the fair-weather electric fields. It is clear from comments printed in *JGR* that the general public has not fully accepted the Kasemir model even after 42 years from its introduction.

6. The Atmospheric Electric Ring Current in the Higher Atmosphere, H. W. Kasemir, Pure and Applied Geophysics (PAGEOPH), Vol.84,1971/I, pp.76-88, Birkhäuser Verlag, Basel, 1971.
   Editors' remarks: The paper is a critique of the spherical condenser model of the global circuit. Kasemir considers atmospheric electric currents that flow in thunderclouds upward, toward the ionosphere. Magnetic effects produce the tensor behavior of air conductivity, and would induce higher vertical current densities near the poles and fairly low currents near the equator region. Also, Kasemir predicts an east-west ring current at ionosphere heights. Therefore, it is not possible for these effects to produce the fairly uniform surface current density found in fair-weather regions.

SECTION 3. THUNDERSTORM ELECTRICITY

1. The Thundercloud, H.W. Kasemir, Problems in Atmospheric and Space Electricity, Ed. S. Coroniti, Elsevier Publishing Co., Amsterdam, London, New York, 1965.
   Editors' remarks: Kasemir compares data and ideas of several leading scientists on the mechanisms of charge generation in thunderclouds, particularly the charge amount and its location. Using his previous work on current flow, Kasemir concludes that the amount of charge in a typical thunderstorm is by a factor of 10 higher than that traditionally accepted. He also discusses here the physics of the recovery curve in electric field records of lightning flashes. A lengthy but interesting debate on the subject of the paper follows.

Document and Commentary Index

Document and Commentary Index

SECTION 4. LIGHTNING PHYSICS

1. Qualitative Overview of Potential, Field and Charge Conditions in the Case of a Lightning Discharge in the Storm Cloud, H. Kasemir, Das Gewitter: Ergbnisse und Probleme der modernen Gewitterforschung, ed. H. Israel, Akademische Verlagsgesellschaft Geest & Portig,. Leipzig, 1950.

   Editors' remarks: The paper is Kasemir's first attempt to qualitatively describe the relation of cloud charges to potentials and associated charges on lightning channels. He uses a simple but very illustrative way to get information on basic variables like the potentials of a floating conductor, the enhancement of electric field at the tips of a conductor in a cloud electric field, and charges and possible currents along the conductor. It starts with the case of a simple conductor in a constant electric field and proceeds to the case of a conductor in the electric field in a thunderstorm with a tripole charge structure. The conditions for leader propagating in both directions or in one direction are considered. The paper shows what type of potential distribution produces cloud-to-ground discharges and what potential distribution produces intracloud discharges. The paper does not have mathematical equations and is easy to read. The main emphasis is on understanding basic lightning processes. Kasemir also comments on the life cycle of a thunderstorm and its relevance to the type of lightning.

2. A Contribution to the Electrostatic Theory of a Lightning Discharge, H.W. Kasemir, J. Geophys. Res., Vol.65, No. 7, July 1960.

   Editors' remarks: As in the previous paper, Kasemir discusses here in detail his model of thunderstorm charges, the resulting cloud electric fields, electrostatic potentials, and their relation to the charges on lightning channels. This paper is highly cited yet not well accepted by some scientists, because of the contrast to the Schonland model of lightning development.

3. Static Discharge and Triggered Lightning. H.W. Kasemir, Proc. 8th International Aerospace and Ground Conference on Lightning and Static Electricity, Fort Worth, Texas. June 1983,

   Editors' remarks: This is an excellent review of the misconception of the Schonland model of unipolar charges on the leader channel. In contrast, Kasemir outlines again the solid physical model of a bipolar but essentially uncharged leader channel. Also, the bipolar characteristic of aircraft-initiated discharges is discussed in detail.

4. The Field Equations of the Radiating Dipole, H.W. Kasemir, Atmosphärische Elektrizität, Teil II, Felder, Ladungen und Ströme p.409, H.Israel ed., Akademische Verlagsgesellschaft Geest & Portig, Leipzig, 1961.

   Editors' remarks: Starting with the Maxwell equations, Kasemir defines the need for the vector potential and proceeds, with almost pedantic details, to the differential equations needed to solve the radiation pattern of an elementary dipole. Although the field equation of the radiating dipole is often quoted, its derivation, in elegant logic, is hard to find in the literature.

5. The Electric Field of Lightning Discharges as an Observational Tool in Thunderstorm Research, H. W. Kasemir, U.S. Army Electronics and Development Laboratories, Forth Monmouth, New Jersey, NOAA, no date (n.d.).

   Editors' remarks: Here Kasemir points out the unique relationship between cloud charges, with their effect on the potential distribution, and the charge distribution on lightning channels, which determines the type of lightning flash.

6. Lightning Hazards to Rockets during Launch I, H. W. Kasemir, Technical Report ERL 143-APCL 11, US Dept. of Commerce Publication, 1969.

   Editors' remarks: The report was written in response to the Apollo 12 lightning incident. It explains, in simple language, that although natural lightning may not develop in shower clouds, rockets in flight in the same conditions may trigger a lightning discharge. The ambient electric field required for triggering lightning by a rocket is 1/10 the strength necessary for initiating natural lightning.

7. Lightning Hazards to Rockets during Launch II, H. W. Kasemir, Technical Report ERL 144-APCL 12, US Dept. of Commerce Publication. 1970.

   Editors' remarks: The report addresses designs of a warning system for safe rocket launch operations. Three approaches are discussed: (1) using aircraft to determine the maximum electric field inside clouds; (2) inferring cloud charge distributions from a network of surface electric field sensors, and determining the electric fields aloft from such charge distribution; and (3) using small test rockets to sense corona onset conditions, and from them determining safe conditions for rocket launch. Over the next few years, all three methods were tested at Kennedy Space Center.

Document and Commentary Index

Document and Commentary Index

17. Electrostatic fields of Ground-Triggered Lightning, L. H. Ruhnke, H. W. Kasemir, Proc. Int. Conf. on Atm. Electricity, Uppsala, Sweden, 1988.
   Editors' remarks: A model of upward triggered lightning was used to calculate base currents, charges on the leader, and propagation velocities utilizing data on electric fields measured at several stations a few kilometers away from the rocket triggering site. The results show the feasibility of reconstructing leader channel variables from surface electric field measurements.

SECTION 5. MEASUREMENT TECHNIQUES

1. An Apparatus for Simultaneous Registration of Potential Gradient and Air-Earth Current (Description and First Results), H. W. Kasemir, J. Atm. and Terr. Physics, Vol. 2,pp.32-37, 1951, Pergamon Press Ltd., London, 1951.
   Editors' remarks: Kasemir addresses the need for measuring simultaneously electric fields and current density at remote field sites, and for developing a system that would be reliable in a harsh environment at mountain tops. This is the first attempt to measure, using proper input time constants, a conduction current density and to exclude displacements currents. First results from the Nebelhorn station show a remarkable similarity to the data from the Carnegie cruises.

2. Measurement of the Air-Earth Current Density, H. W. Kasemir, Proc.First Intentional Conference on Atmospheric Electricity, AFCRC-TR-55-206, U.S. Dept. of Commerce, Office of Technical Services, Washington, D.C., 1954.
   Editors' remarks: For a long time, measurements of air-Earth currents' density posed problems because of the simultaneous presence of conduction and displacement currents, which made averaging over 1 hour necessary. By applying his theory, Kasemir showed that it is feasible to separate, with any time resolution, conduction currents from Maxwell currents by using input time constants equal to the relaxation time of air.

3. Antenna Problems of Measurements of the Air-Earth Current, H.W. Kasemir and L. H. Ruhnke, Proc. Second Int. Conf. on Atmospheric Electricity, Recent Advances in Atmospheric Electricity, Pergamon Press, New York, 1958.
   Editors' remarks: The paper discusses current and electric field sensors as seen by electrical engineers. Both types of sensors are active two-terminals and equally represent current or potential sources. Equivalent circuit diagrams are discussed for the general case and for applications as current as well as potential probes. Attention is given to errors made by using too low effective heights, errors due to long cables with piezoelectric effects, and the need to keep the ratio of conduction current to convection current high.

4. A Radiosonde for Measuring the Air-Earth Current Density, H.W. Kasemir,
   USASRDL Technical Report 2125, Fort Monmouth, New Jersey, 1960.
   Editors' remarks: The report describes in detail an instrument to record current density with an ascending balloon. It also contains calculations of effective areas of dipole antennas and discusses the errors posed by altitude variations of air relaxation times.

5. Field Component Meter, H.-W. Kasemir, Tellus, Vol. 3, November, 1951.
   Editors' remarks: The paper describes the first version of Kasemir's design of a cylindrical field mill to be used on an aircraft. However, at the time, no aircraft in Germany was available for conducting atmospheric research. It took another 20 years for Kasemir to put his design into practice and to construct a cylindrical field mill for airplanes to measure electric fields in and around thunderstorms.

6. The Cylindrical Field Mill, H. W. Kasemir, Meteorologische Rundschau, 25. Jahrgang, 2. Heft, 1972.
   Editors' remarks: This is a description of the version of Kasemir's cylindrical field mill that was built in 1964. Kasemir also comments on the interpretation of records of three vector components of electric field obtained during flying near charged clouds.

7. Electric Field Measurements from Airplanes, H. W. Kasemir, Proc. Fourth Symposium on Meteorological Observations and Instrumentation, Denver, CO, published by American Meteorological Society, Boston, April, 1978.
   Editors' remarks: The paper discusses usage of radioactive probes as well as Kasemir's cylindrical field mill system for measuring ambient electric field from airplanes. The emphasis is on the major role played by form factors, the location of sensors, and proper calibration procedures in the separation of external field from fields due to self-charge on aircraft.

Document and Commentary Index

8. Evaluation of Mighty Mouse Trigger Rocket Flights, H. W. Kasemir, Progress Report on Kennedy Space Center Contract No. CC-88025, Part D., 1971.
   Editors' remarks: The report discusses the statistical data showing that small rockets trigger lightning at much reduced cloud electric fields, with the effect of reducing natural lightning for several seconds thereafter.

9. Airborne Warning Systems for Natural and Aircraft-Initiated Lightning, L. W. Parker and H. W. Kasemir, IEEE Transactions on Electromagnetic Compatibility, VOL. EMC-24, No.2, May 1982.
   Editors' remarks: The authors discuss various aircraft warning systems for lightning danger to aircraft. Two types of systems are differentiated: one based on the use of electric fields from charges in the cloud and another based on using electromagnetic or optical sensing of existing lightning strikes. Because of the high incidence of aircraft-triggered lightning in fields that do not produce natural lightning, the authors favor warning devices that use electric field sensing (field mills, radioactive potential sensors, and corona probes).

10. Theoretical and Experimental Determination of Field, Charge, and Current on an Aircraft hit by Natural or Triggered Lightning, H. W. Kasemir, International Aerospace and Ground Conference on Lightning and Static Electricity, NASA, 1984.
    Editors' remarks: Kasemir discusses electric fields and current waveforms on an aircraft to differentiate between triggered and natural lightning discharges. Using his theories, Kasemir predicts various waveforms measured on an aircraft, which are produced by natural and triggered lightning. His analysis takes into account the airplane form factors, its self-charge from precipitation impact, and the electric field and current produced by the lightning discharge.

11. Ranging and Azimuthal Problems of an Airborne Crossed Loop Used as a Single-Station Lightning Locator, L. W. Parker and H. W. Kasemir, Proc. 10th International Aerospace and Ground conference on Lightning and Static Electricity. Paris, France, 1985.
    Editors' remarks: The authors use data from a crossed loop location system to evaluate a commercial lightning locating system for errors in range and in direction to lightning. Their analysis shows considerable range errors, making it unreliable to predict distance to the radiation source. Errors in direction are found to be fair for distances of about 20 km, but direction accuracies deteriorate for short distances of less than 3 km.

12. Predicted Aircraft Field Concentration Factors and their Relation to Triggered Lightning, L. W. Parker, H. W. Kasemir, Proc. International Aerospace and Ground Conference on Lightning and Static Electricity, Orlando, FL, 1984.
    Editors' remarks: Form factors as they apply to the enhancement of the ambient electric field on various parts of an aircraft in flight are discussed through detailed theory and model calculations. The analysis shows form factors up to 130, making it feasible to trigger lightning in fairly low thunderstorm fields.

SECTION 1. FAIR WEATHER ELECTRICITY

1. Atmospheric Electrical Diurnal Variations and Mass Exchange in the High Mountain Range of the Alps. The Atmospheric Electrical Conditions at Jungfraujoch (3472 m) (Part I). I. H. Israël, H. W. Kasemir and K. Wienert, Archiv für Meteorologie, Geophysik und Bioklimatologie. Serie A: Band III, 5. Heft, Springer Verlag, Wien, 1951.

2. Atmospheric Electrical Diurnal Variations and Mass Exchange in the High Mountain Ranges of the Alps, The Atmospheric Electrical Conditions at Jungfraujoch (3472 m)(Part II). H. Israël, H. W. Kasemir and K. Wienert, Archiv für Meteorologie, Geophysik und Bioklimatologie. Serie A, Band 8, 1.-2. Heft, 1955, Springer Verlag, Wien, 1955.

3. Atmospheric Electric Measurements in the Arctic and Antarctic, H. W. Kasemir, Pure and Applied Geophysics (PA-GEOPH), Vol.100, 1972/VIII, pp.70-80, Birkhäuser Verlag, Basel, 1972.

4. Studies regarding the Atmospheric Potential Gradient IV: Examples for the Behavior of Atmospheric Electrical Elements in Fog, H. Israël and H.W. Kasemir, Archiv für Meteorologie, Geophysik und Bioklimatologie. Serie A, Band 5, Heft 1, Springer Verlag, Wien, 1952.

5. Studies regarding the Atmospheric Potential Gradient V, Regarding the Current Theory of the Atmospheric Electrical Field (Part I), H.W. Kasemir, Archiv für Meteorologie, Geophysik und Bioklimatologie. Serie A, Band 3, Heft 1 und 2, Springer Verlag, Wien, 1950.

6. Studies regarding the Atmospheric Potential Gradient V, Regarding the Current Theory of the Atmospheric Electrical Field (Part II), H.W. Kasemir, Archiv für Meteorologie, Geophysik und Bioklimatologie. Serie A, Band 5, Heft 1, Springer Verlag, Wien, 1952.

7. On the Theory of The Atmospheric Electric Circuit Flow, IV, H. W. Kasemir, U.S. Army Electronics Research and Development Laboratories, Fort Monmouth, N.J.,USA, Tech. Report # 2724, 1963.

8. Regarding the Shielding Effect of Buildings on the Fluctuations of the Atmospheric Electrical Field, H. Israël and H. W. Kasemir, Extrait des Annales de Geophysique, Tome 7, fascicule 1, 1951.

9. How to Measure and Analyze Solar-Terrestrial Weather Relationships, H. W. Kasemir, Colorado Scientific Research Corporation, Berthoud, Colorado, Copy of original manuscript, no date (n.d.).

10. Analytic Solutions of the Problem of Down-Mapping of Electric Fields and Currents of Solar Events to the Earth Surface, H. W. Kasemir, Weather and Climate Responses to Solar Variations, Ed. Billy M. McCormac, Colorado Associated University Press, 1983.

11. The Exchange Generator. Archiv für Meteorologie, Geophysik und Bioklimatologie. Serie A: Band 9, 3. Heft, Springer Verlag, Wien, 1956.

Special print from

Archive for Meteorology, Geophysics and Bioclimatology

Series A: Meteorology and Geophysics, Volume III, Issue 5, 1951

Published by
Dr. W. Mörikofer, Davos and Prof. Dr. F. Steinhauser, Vienna
Springer-Verlag in Vienna

551.594.1(494)

(Atmospheric Electrical Research Center of the Observatory in Friedrichshafen in Buchau a. F., National Weather Service Württemberg-Hohenzollern.)

Atmospheric Electrical Diurnal Variations and Mass Exchange in the High Mountain Range of the Alps.

The Atmospheric Electrical Conditions at Jungfraujoch (3472 m) I.

By

H. Israël, H. W. Kasemir and K. Wienert

H. ISRAËL, H. W. KASEMIR AND K. WIENERT:

Summary. The present paper gives an account of researches on atmospheric electricity carried out at Jungfraujoch (Switzerland). During two periods of several weeks each the potential gradient and the vertical air-earth current have been recorded. In the first place the diurnal variations on fine days at high pressure weather situation are elaborated. The daily variations of the three elements: potential gradient, vertical current, and conductivity, show that in summer and autumn two different types of the conditions of atmospheric electricity can be discerned: In summer the well-known type of the lowland stations is predominating also at this high altitude station, the type where the potential gradient is very nearly inverse to conductivity and vertical current. In autumn on the contrary the potential gradient and the vertical current are essentially parallel since the conductivity shows a slight diurnal variation in contrary sense. Thus, in summer prevails at the altitude of Jungfraujoch in the Alps the *continental type*, whilst in autumn the conditions approach to the *oceanic type*. It results from a rigorous analysis that in summer the diurnal variation of the *columnar resistance* on Jungfraujoch is in close correlation to the local air resistance (the reciprocal figure of conductivity) at the measuring point, whereas in autumn this is no longer the case.

To explain these results we must assume as follows: In midsummer the diurnal variation of vertical mass exchange produces alterations of the aerosol up to the altitude of the crest of the Alps, which is perceptible from the fact that diurnal variations of atmospheric electricity start essentially from the lower layers. In autumn this influence has almost disappeared in the summit altitude of the Alps, as the daily variation of exchange does not advance to this altitude. The remaining variations of the conductivity point at an advective exchange of air. The behaviour of vapour pressure in summer and autumn at Jungfraujoch gives support to this interpretation.

Résumé. L'article traite des résultats de recherches d'électricité atmosphérique au Jungfraujoch (Suisse) et de leur interprétation. Deux périodes de mesures de plusieurs semaines chacune ont été consacrées à l'enregistrement du gradient de potentiel et du courant vertical. On décrit tout d'abord la variation diurne lors des jours sereins de temps anticyclonique. On distingue au Jungfraujoch en été et en automne respectivement deux types très différents de la marche diurne du gradient de potentiel, du courant vertical et de la conductibilité. En été le type bien connu des stations de plaine pour lequel le gradient de potentiel varie à peu près inversement à la conductibilité et au courant vertical s'observe aussi au Jungfraujoch. Par contre en automne le gradient de potentiel et le courant vertical varient en général dans le même sens, tandis que la conductibilité varie, elle, en sens contraire, bien que faiblement. Il y a donc à l'altitude du Jungfraujoch en été le «type continental», alors qu'en automne les conditions se rapprochent du «type océanique». Une analyse plus détaillée montre qu'en été la variation diurne de la «résistance tubulaire» (columnar resistance) au-dessus du Jungfraujoch est en liaison étroite avec la résistance locale de l'air au point de mesure (inverse de la conductibilité), tandis qu'en automne ce n'est plus le cas.

Pour interpréter ces résultats il faut admettre ce qui suit: En été la variation diurne de l'échange vertical turbulent modifie les aérosols jusqu'au

Atmospheric Electrical Diurnal Variations and Mass Exchange in the High
Mountains

niveau de la crête alpine, ce que l'on reconnaît au fait que les variations diurnes des phénomènes d'électricité atmosphérique sont largement conditionnés par les couches basses. En automne par contre cet effet a disparu sur les Alpes, car la variation de l'échange turbulent n'a plus d'influence à ce niveau. La variation de la conductibilité qui subsiste suggère un échange d'air par advection. L'allure de la pression de vapeur d'eau au Jungfraujoch en été et en automne confirme cette hypothèse.

I. Definition of the Problem.

In one of our papers appearing recently [1], a depiction is given how the method of perception and work in the field of atmospheric electricity has changed over the last decade. Instead of a long prevailing isolationist notion of these things, which – in the meteorological influences on the atmospheric electrical phenomena – only saw a turbidity of its overall image that was to be abstracted as possible by weather events, the recognition has grown that both distribution areas – the atmospheric electricity and meteorology – are tightly intermeshed with each other and that therefore the atmospheric electrical phenomena cannot be understood as an abstract of the meteorological phenomena. – Therefore, the familiar connections of the atmospheric electricity to other (geophysical and cosmic) occurrences are not assessed inferiorly. Moreover, particularly the encounter of these various types of influences leads to the multifarious image of the atmospheric electricity that we know.

The following image is largely offered in this case:

1. The roots of the atmospheric electrical phenomena in general must be sought in global meteorological events or – perhaps more generally – in the atmospheric condensation occurrences and their electrical shades [1, 2, 3].

2. The decrease of altitude of the atmospheric electrical field strengths closely affiliated with the increase of conductivity with altitude originating through cosmic ultra-radiation [4]. The influence of the polarity on the field and vertical current that is still controversial [5, 6] points to cosmic-terrestrial effects just like the correlation to occurrences in the ionosphere [7] and to the sun spots [8].

3. The diurnal variation of the atmospheric electrical field that passes by on high sea and in the Arctic or the Antarctic according to universal time, i.e. its extreme values traverse the entire world, is connected over the mainland to the time of day (regional time), therefore, it may depend on the meteorological elements of the vertical exchange of mass and its diurnal variations [8, 9]. The atmospheric mass dependence of these variations [1] and the difference of the average values and the character of the diurnal variations (whole-day oscillation and half-day oscillation) subject to the degree of purity of the air [10] suggest this as well.

The duty of the atmospheric electrical research is to recognize these connections individually and to achieve better understanding in its interaction with the complex image of the atmospheric electrical phenomena.

24*

H. ISRAËL, H. W. KASEMIR AND K. WIENERT:

It goes without saying that with this aspiration of focusing on the atmospheric electrical-meteorological correlations as particularly important, a contribution for the meteorological-aerological research may also be expected, as this would have been inconceivable 10 or 15 years ago.

As already presented earlier [1], the central point of all of the considerations is based on the difference of the atmospheric electrical behavior over land and sea. Because the diurnal variability of the meteorological elements is considerable over land, although it is low at sea, one may assume that the vertical mass exchange, which is really the cause of these variations, fluctuates significantly over land, while fluctuating only insignificantly at sea throughout the course of a day. Because this exchange further modifies the suspended contents of the atmosphere and therefore its conductivity, the globally-homogenous character of the atmospheric electrical field curve, which can be viewed as the primary phenomenon,

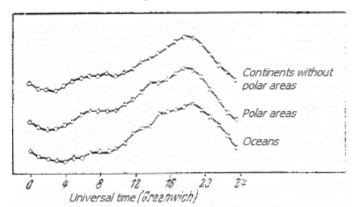

Figure 1. Universal time curves of the atmospheric potential gradient for non-polar mainland stations, polar areas and oceans in a relative display according to N. A. Paramonoff.

only occurs at sea and in the polar areas, while it is completely covered over the mainland through its dependence on the exchange. The fact that this *universal time rate* is also kept hidden within the diurnal variation on the mainland, as one must expect according to this notion, is demonstrated in the result of an analysis by N. A. Paramonoff [11] in Figure 1. It summarizes the results of 60 non-polar mainland stations according to the individual geographical longitudinal ranges, gives these individual areas the same significance and averages the curves indicated according to Greenwich Time. Because this process encompasses the entire earth, in this process, the influences bound to regional time must be stressed. The result is the uppermost curve in Figure 1. The other two curves comparatively represent the familiar universal time curves of the potential gradient in the polar areas and over the oceans.

This working hypothesis, namely that the vertical mass exchange in the atmosphere necessitates the emergence of the atmospheric electrical variations bound to regional time through its changes and is therefore capable of providing the key for interpreting the meteorological atmospheric electrical correlations, simultaneously indicates the path to its closer analysis. If one intends to record this exchange influence, one must first seek an abstraction from it and examine the atmospheric electrical behavior at those measurement sites that are largely detracted from its influence.

Atmospheric Electrical Diurnal Variations and Mass Exchange in the High
Mountains

One route for this is the acquisition of the atmospheric electrical diurnal variations at high sea, as the results of the Carnegie journey showed. The direct discovery of the universal time period of the potential gradient is, as already mentioned, only possible over the oceans due to the nearly complete absence of temporal exchange variations.

Another possible is to analyze the atmospheric electrical ratios at higher altitudes. As we have already presented in detail in one of our earlier work [12], one can differentiate two layers of various behavior in the electrical system of the atmosphere – a "ground layer" that is subject to regional exchange influences, and an "upper layer" that is not dependent on exchange. As a juncture between these two layers, at its time an altitude of approx. 2 km was assumed, an assumption, however, that as we will see generally does not correspond to the facts.

These consideration lead to the plan to establish atmospheric electrical registrations at a mountain observatory situated as high as possible, because it was possible to expect that at altitudes of 3000+ meters, as is available in the European mountain observatories, the conductivity-altering influence of the exchange is practically gone.

The intention of the following is to report about the initial experiences of this type gained during the summer and fall of 1950 at Jungfraujoch (Switzerland).

II. Work Method.

The idea that one can differentiate a "lower layer" and an "upper layer" of the atmosphere with regard to atmospheric electricity requires some supplementation for a more in-depth observation:

The ratios in a simple electrical circuit can be understood if at least two of the three variables connected to each other in Ohm's law are known. For our atmospheric electrical observation, this means that the measurement or registration of the potential gradient does not suffice alone for statements regarding the entire event. Only the addition of at least one of the two other "*basic ohmic variables*" – conductivity or vertical current – of the atmosphere is capable of providing more detailed information*.

Therefore, it is not possible to derive a "lower" or an "upper layer" from the behavior of the potential gradient alone.

The resistor W_H of a vertical columnar resistance with a cross-section of 1 cm^2 sitting on the ground initially increases rapidly with altitude and then continues to decelerate (see Figure 2a). The potential gradient that is measured at any point can be understood,

---

* The fact that the statements regarding the global character of the atmospheric electrical circuit process familiar from the Carnegie measurements of the potential gradient over the ocean have been able to be deduced, was only possible because one of the three basic ohmic variables – conductivity – is virtually constant over the oceans.

H. ISRAËL, H. W. KASEMIR AND K. WIENERT:

as can be seen from the equivalent circuit diagram of Figure 2b, as the voltage drop E_w on a partial element w of the entire columnar resistance W:

$$E_w = w \cdot \frac{V}{W}, \tag{1}$$

Where V represents the total potential difference maintained between the ground and compensating layer in the upper atmosphere (details will follow [1]).

V and W are not directly measureable, although their ratio V/W that represents the vertical current density i can. With that, (1) transfers to

$$E_w = w \cdot i \tag{2}$$

in which the three atmospheric electrical variables directly available for the measurement are connected to each other. Please see [2] for more details regarding the analysis of the atmospheric electrical measurement series based on these relationships. We dealt with details regarding the stationary and non-stationary ratios

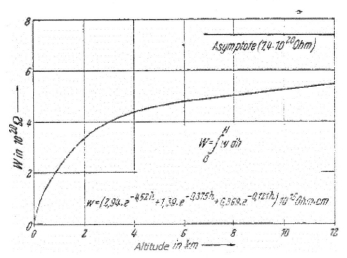

Figure 2a. Resistance of an atmospheric column with a cross-section of 1 cm2 sitting on the ground according to O. H. Gish [13].

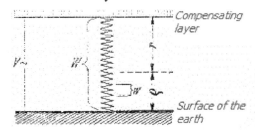

Figure 2b. Equivalent circuit diagram of the atmosphere.

("turn-on processes") in one of our [14] papers that appeared recently.

At this point W_H, which is given by the relationship,

$$W_H = \int_0^H \frac{dh}{(\lambda_h{}^+ + \lambda_h{}^-)} \tag{3}$$

$$\left.\begin{array}{l} \lambda_h{}^+ = \text{positive} \\ \lambda_h{}^- = \text{negative} \end{array}\right\} \text{ Polar conductivity at the altitude } h$$

is significantly altered by the exchange in the lower layers of the atmosphere. Because half of the total resistance W of a vertical columnar resistance is omitted at the lowest two to three kilometers of altitude, the exchange influence near the surface must heavily effect the atmospheric electrical behavior of the entire atmosphere. Therefore, E_w or i are also influenced at higher altitudes by changes to w, thus they are indirectly controlled by the surface-level exchange effect. In other words, in the free atmosphere the "upper layer" *cannot* be characterized in that the universal time dominates in it.

The ratios are different at the mountain stations – if they are high enough that they protrude beyond the layers, in which the conductivity is changed by the diurnal variation of the exchange, diurnal constancy of the remainder of the columnar resistance lying over the station and therefore the emergence of the universal time period must be expected*.

Therefore, we need to look for another more general criteria for the limitation of the "bottom layer" against the "upper layer" and go back to a formula already provided in an earlier paper [15], which – under the requirement of the stationary rations (see [14]) – emerges directly from equation (1) or (2):

If the atmospheric electrical potential gradient changes at the site of measurement without a change of vertical current, then these are locally-conditioned phenomena; in contrast, if both variables change simultaneously, the cause for this is not the site of measurement.

Thus, when analyzing altitudes of the atmospheric electrical ratios, regardless of whether in the open atmosphere or on mountain peaks, we must analyze whether the field and vertical current change in a similar manner or not. With regard to our notions surrounding the vertical mass exchange and altitude of its diurnal variations, we may expect to find a gradual transition to the synchronization of the field and vertical current with an ever higher ascent over the surface and to be able to segregate our layers.

At first glance, one could object to the fact that this procedure seems unnecessarily complicated and that the measurement of the conductivity could provide the same result because it is determined by the ratio of the vertical current and field according to equation (2). With regard to the segregation of the "lower layer" and "upper layer", this may be true; in fact one only needs to find the altitude, at which the conductivity no longer produces a diurnal variation. On the other hand, for the knowledge of the atmospheric electrical processes, the determination of the material constant of the conductivity alone can suffice just a little as the already criticized limitation to the potential gradient alone.

It is only natural that research of this type depends first and foremost on the gathering of the *relative changeability of the atmospheric electrical elements*. The average absolute level of field strength, etc. at the measurement site of concern is only of secondary interest in this connection. This offers a great benefit in this respect because, as a result, the most strived for "reduction to the free plane", which in the case of measurements on exposed mountain peaks is very difficult to achieve, is not necessary for this.

---

* However, this presupposes that the relevant massif is expanded in such a manner that the neighboring parties of the free atmosphere lying over the plane with their exchange reactions will not react, in other words, that the field lines will remain vertical. Otherwise, the assumption of a diurnally constant columnar resistance is not permissible because the field lines will experience a variable deviation from the vertical course throughout the day in this case. This may be true in general with exposed mountain peaks [14].

H. ISRAËL, H. W. KASEMIR AND K. WIENERT:

The abandonment of "absolute values" hereby expressed may seem unusual. However, this is merely the strict consequence of known experience that such a "reduction to undisturbed ratios" does not initially provide insight for the true atmospheric electrical state of an inhomogeneous electrical field in an ion atmosphere and, as H. Benndorf emphasized quite some time ago [16], does not enable a satisfactory solution.

As already indicated, all aforementioned considerations relate to stationary ratios. Therefore, in this case we have provisionally only taken those observations of temporally variable processes into account that may be understood as a juxtaposition of the stationary individual states. Because the "relaxation time" for the atmospheric electrical events is normally 15 to 20 minutes, all fluctuation periods of a respective and shorter duration must therefore be perceived as non-stationary and initially excluded.

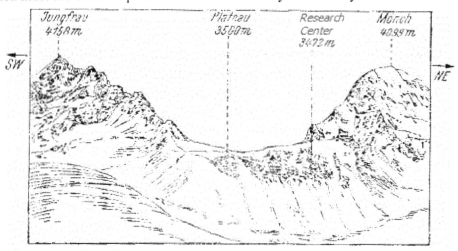

Figure 3. The location of the High Alpine Research Station Jungfraujoch (view from SE)

The main emphasis of our observations presented in the following will therefore be on the slow fluctuations, thus, placed primarily on the diurnal variations.

With that, the requirements and the functional mission of our *"atmospheric electrical mountain peak program"* are depicted in their significant traits.

After brief pre-analysis trip to Nebelhorn (2224 m) near Oberstdorf (August 1949) and to Zugspitze (2960 m, March 1950), Jungfraujoch (3472 m) was chosen as the first site of measurement for longer analyses. The location of the station on the slope of the Sphinx Summit between Jungfrau and Mönch can be recognized in Figure 3, which provides a view from the southeast.

The atmospheric electrical potential gradient and the vertical current were sufficient for the registration. The device used for the simultaneous recording of field and vertical current [17], which we developed in Buchau and recently described in detail, served as the measuring device. It was mounted in the wooden shanty on the deck of the tower of the research center.

Atmospheric Electrical Diurnal Variations and Mass Exchange in the High Mountains.

The radioactive collector (polonium compound of the usual strength) was attached to the southwest wall, the power supply for recording the vertical current was attached to the southeast wall of the wooden structure facing the Great Aletsch Glacier. A six-color recorder manufactured by Whico (Düsseldorf) served as the measuring instrument. A measuring circuit of the recorder was connected to a depletion layer photocell manufactured by B. Lange (Berlin) that was mounted to the roof of the shanty; this record of the "brightness of the location" proved to be very practical for the subsequent verification of the lack of the clouds and especially fog at the location of measurement. The required meteorological data (temperature, moisture, duration of sunshine, etc.) were determined with the usual meteorologically logical devices.

III. Results.
a) Duration of the Test.

In the case of the atmospheric electrical research at Jungfraujoch, unfortunately it was not possible to conduct an uninterrupted series of registrations over a period of one year as would have been desirable. In order to gain an overview of the changes of the atmospheric electrical diurnal variations of a season regardless, the total accessible period was divided into three measuring periods respectively of 1½ to 2 months. These were divided temporally in such a manner that the typical ratios in the middle of summer and the middle of winter as well as in the transitional period of the equinox could be gathered. The first measurement period comprised the period from the end of May to the end of July and the second from the beginning of October to the end of November 1950. To gather the ratios in the middle of winter, finally, a third measurement period was conducted during February/March 1951.

If some results of both initial measurement periods are reported on prior to the conclusion of the entire measurement process, this is due to the fact that they may be able to deliver some important results that pertain to the working hypothesis presented in the introduction and seem worthy of a separate illustration.

b) Selection of the Material.

The question concerned in this case necessitates a strict selection of the material. A broad abstraction of all effective influences in the region and the atmospheric electricity must be pursued if it is intended that the large-scale exchange effect should occur. Therefore, all incidental registrations in particular must be separated in the case of cloud cover over the station as an electrical generator effect can likely in general be attributed to the clouds [18].

This condition heavily restricts the likelihood of obtaining useful diurnal series in the case of the High Alpine location of the station. Therefore, one will generally have the best prospects for success for such atmospheric electrical analyses in the high mountains if one limits the research to stabile high-pressure weather conditions from the onset.

H. ISRAËL, H. W. KASEMIR AND K. WIENERT:

This circumstance even subsequently justifies the replacement of an interrelated annual registration through selected measurement periods, which became necessary for our research at Jungfraujoch, because they provided nearly the same scope of results as an uninterrupted registration would have provided.

Table 1. *Hourly averages of the potential gradient at Jungfraujoch on the sought out, disturbance-free days in the summer of 1950. The information is provided in volts (collector voltage, unreduced).*

Date	0—1	1—2	2—3	3—4	4—5	5—6	6—7	7—8	8—9	9—10	10—11	11—12
5-30	85	83	70	70	65	60	65	73	65	60	73	83
6-5	138	115	130	150	133	145	158	168	158	120	118	135
6-29	160	150	150	153	148	150	148	145	153	145	160	145
6-30	113	98	100	93	95	113	140	163	175	168	213	198
7-4	143	125	88	68	75	55	93	110	88	115	123	98
7-7	125	160	120	118	110	125	148	168	140	155	188	200
7-8	173	190	170	155	153	165	158	150	153	153	125	120
8-9	115	105	90	95	98	115	120	115	115	123	113	105
7-16	158	170	150	130	128	115	123	120	118	108	100	103
7-17	148	153	133	130	120	138	138	160	145	130	148	150
7-18	113	150	125	220	203	145	138	128	118	108	120	123
7-19	150	158	143	123	120	128	125	125	120	138	133	123
7-21	123	138	143	135	123	110	110	128	140	125	125	125
7-25	145	135	133	125	108	93	78	80	93	98	110	108
7-30	63	50	45	65	95	98	85	113	128	125	135	155
Average:	130	132	119	122	118	117	122	130	127	111	132	131

Date	12—13	13—14	14—15	15—16	16—17	17—18	18—19	19—20	20—21	21—22	22—23	23—24
5-30	73	70	85	105	130	163	140	140	160	150	113	105
6-5	135	118	115	120	110	130	190	183	190	140	113	105
6-29	155	163	165	163	163	138	138	135	125	128	125	128
6-30	180	188	193	165	163	155	175	200	212	195	190	185
7-4	100	93	110	118	108	113	138	168	193	170	163	118
7-7	215	180	150	123	140	160	200	202	190	180	193	170
7-8	120	135	145	138	163	170	163	170	188	148	125	123
8-9	98	90	98	120	133	133	155	180	188	188	200	208
7-16	118	115	140	175	215	190	220	208	220	205	148	150
7-17	175	165	153	165	183	235	212	205	205	210	200	145
7-18	115	113	120	140	168	140	145	138	138	125	125	133
7-19	158	148	200	193	185	183	185	175	175	155	173	190
7-21	125	103	118	133	133	133	170	210	220	220	237	212
7-25	110	103	110	135	150	143	150	175	178	170	185	178
7-30	145	150	163	148	165	165	170	153	153	145	145	150
Average:	135	129	138	145	154	157	169	175	184	169	162	153

Atmospheric Electrical Diurnal Variations and Mass Exchange in the High Mountains.

As demonstrated in Tables 1 to 7, the months of July and October of 1950 provided beneficial weather at Jungfraujoch in a very special way for our purposes. During the first measurement period, 15 complete diurnal series (eleven of these attributed to the month of July) fulfilling all of the requirements for the absence of disturbances in the region were achieved. The fall period was not quite so beneficial. It provided us with nine complete diurnal series.

Table 2. *Hourly averages of the atmospheric electrical vertical current at Jungfraujoch on the sought out, disturbance-free days in the summer of 1950. The information is provided in 10^{-17} amps/cm^2 of the used grid (unreduced).*

Date	0—1	1—2	2—3	3—4	4—5	5—6	6—7	7—8	8—9	9—10	10—11	11—12
5-30	538	538	510	510	510	781	510	557	490	491	500	538
6-5	708	699	737	803	717	708	708	708	670	595	623	547
6-29	860	793	727	793	783	699	661	651	823	623	651	566
6-30	547	510	538	481	510	547	595	623	670	651	736	690
7-4	557	520	415	312	340	293	415	481	415	453	472	378
7-7	350	520	557	548	491	481	491	595	575	595	680	736
7-8	615	680	737	700	680	708	690	670	642	605	529	481
8-9	615	605	529	520	510	548	614	585	548	520	481	425
7-16	605	623	660	614	585	548	520	481	453	453	425	415
7-17	425	435	397	368	453	406	425	406	406	340	284	284
7-18	453	547	454	585	700	575	575	547	529	453	387	359
7-19	453	444	435	435	425	435	444	481	481	500	444	378
7-21	425	462	462	453	453	435	406	406	415	415	378	359
7-25	529	510	481	453	415	387	368	368	368	378	387	387
7-30	482	463	435	435	453	462	444	500	557	520	491	500
Average:	545	555	538	534	536	534	525	538	523	506	498	496

Date	12—13	13—14	14—15	15—16	16—17	17—18	18—19	19—20	20—21	21—22	22—23	23—24
5-30	491	453	453	462	510	614	548	435	519	614	520	490
6-5	500	472	463	510	500	557	623	605	642	575	490	453
6-29	529	520	472	453	444	406	425	453	490	566	585	623
6-30	727	680	623	529	538	491	491	642	745	755	736	755
7-4	359	340	368	425	406	368	463	595	698	632	605	472
7-7	690	472	350	284	236	264	481	481	415	254	368	547
7-8	453	472	453	435	472	481	529	510	557	642	642	632
8-9	378	350	350	359	387	387	387	415	490	557	585	660
7-16	415	406	435	378	321	293	396	425	444	462	472	462
7-17	293	312	331	350	330	349	221	378	378	368	415	481
7-18	359	368	378	387	415	396	425	453	462	472	472	481
7-19	387	368	368	340	293	274	246	245	292	330	378	472
7-21	340	302	284	264	274	284	321	321	359	284	208	259
7-25	397	378	359	378	350	330	378	453	510	566	632	660
7-30	500	500	463	406	350	415	462	472	481	481	406	406
Average:	455	426	410	398	388	394	426	459	499	505	502	524

12

H. ISRAËL, H. W. KASEMIR AND K. WIENERT:

To adapt the results of this period to the scope of that of the summer period, ten partial elements of diurnal series fulfilling the requirements were added to these nine series. The results were derived nearly exclusively from the month of October.

Table 3. *Hourly averages of the (total) conductivity of the atmosphere at Jungfraujoch on the sought out, disturbance-free days in the summer of 1950 – calculated from the values of the Tables 1 and 2. The information is provided in 10^{-4} electrostatic units (unreduced values).*

Date	0—1	1—2	2—3	3—4	4—5	5—6	6—7	7—8	8—9	9—10	10—11	11—12
5-30	57	59	65	65	71	72	71	69	68	74	62	59
6-5	47	55	51	48	49	42	41	38	38	45	47	37
6-29	48	48	44	47	48	42	41	41	37	38	37	35
6-30	43	47	48	47	48	44	38	35	34	35	31	31
7-4	35	37	43	42	41	48	40	39	43	35	35	35
7-7	25	29	42	42	40	35	30	32	37	35	33	33
7-8	32	32	39	40	40	39	39	40	38	36	38	36
8-9	48	52	53	49	47	43	46	46	43	38	39	36
7-16	31	34	39	43	41	43	38	36	35	38	38	36
7-17	26	26	27	25	27	26	28	25	25	23	17	17
7-18	36	33	33	24	31	36	38	38	40	38	29	27
7-19	27	25	28	32	32	31	32	35	36	33	30	28
7-21	31	30	29	30	33	35	33	29	27	30	27	26
7-25	33	34	33	33	35	38	43	42	36	35	32	32
7-30	69	84	87	60	43	43	47	40	39	38	33	29
Average:	38	42	44	42	42	41	40	39	38	38	38	33

Date	12—13	13—14	14—15	15—16	16—17	17—18	18—19	19—20	20—21	21—22	22—23	23—24
5-30	61	58	48	40	35	34	35	28	30	35	42	42
6-5	33	36	36	38	41	38	30	30	31	37	39	39
6-29	31	29	26	25	25	27	28	30	35	40	42	44
6-30	36	33	29	29	30	28	25	29	32	35	35	37
7-4	32	33	30	33	34	30	30	32	33	33	33	35
7-7	29	24	21	21	15	15	18	21	20	14	17	29
7-8	34	32	28	29	26	25	30	27	27	39	46	46
8-9	35	35	32	27	27	27	23	24	24	27	27	29
7-16	32	32	28	19	14	14	16	18	18	20	29	28
7-17	15	17	19	19	16	13	14	17	17	16	19	30
7-18	28	30	28	25	22	25	26	30	30	34	34	33
7-19	22	22	17	16	14	14	13	13	15	19	20	23
7-21	24	26	22	18	19	19	17	14	15	12	8	15
7-25	32	33	30	25	21	21	23	23	26	30	31	34
7-30	31	30	26	25	19	23	24	28	29	30	25	24
Average:	32	31	28	26	24	23	23	24	25	28	30	33

Atmospheric Electrical Diurnal Variations and Mass Exchange in the High Mountains.

Table 4. *Hourly averages of the potential gradient at Jungfraujoch on the sought out, disturbance-free days in the fall of 1950, supplemented through disturbance-free partial diurnal series. The information is provided in volts (collector voltage, unreduced).*

Date	0—1	1—2	2—3	3—4	4—5	5—6	6—7	7—8	8—9	9—10	10—11	11—12
10-5	—	—	—	—	—	—	—	—	—	—	—	—
10-6	73	65	80	80	90	98	95	85	65	80	75	70
10-7	118	120	118	105	125	150	143	133	135	128	128	135
10-8	108	100	93	95	95	105	110	118	118	110	103	115
10-12	—	—	—	—	—	—	—	—	—	—	—	—
10-13	60	55	63	65	60	70	75	88	88	100	98	93
10-14	65	63	60	55	55	65	80	—	—	—	—	60
10-15	78	80	68	63	73	78	75	88	83	88	85	85
10-16	75	78	75	90	75	65	68	60	78	78	95	85
10-17	—	—	—	—	—	—	—	—	—	—	—	—
10-19	—	—	—	—	—	—	—	—	—	113	103	98
10-20	70	70	70	75	73	83	80	88	110	103	100	110
10-21	133	118	108	103	100	100	90	90	85	80	78	90
10-22	63	60	70	63	83	95	98	78	73	65	65	73
10-23	75	73	78	68	63	68	68	78	90	93	78	75
10-24	65	60	65	68	68	70	68	70	80	68	63	65
10-25	—	—	—	—	—	—	—	60	65	63	53	60
10-31	—	—	—	—	—	—	—	—	88	93	103	
11-1	138	123	128	73	73	70	108	100	98	110	108	108
Average:	86	82	83	77	80	86	89	89	91	91	88	89

Date	12—13	13—14	14—15	15—16	16—17	17—18	18—19	19—20	20—21	21—22	22—23	23—24
10-5	—	70	98	108	104	100	95	100	98	93	93	83
10-6	75	110	125	142	182	193	148	128	138	138	138	125
10-7	143	148	170	188	193	198	198	175	180	170	135	120
10-8	120	135	123	120	133	145	105	83	125	143	—	—
10-12	—	—	—	—	—	—	—	—	90	75	65	65
10-13	98	135	95	105	120	135	140	143	123	95	90	83
10-14	80	93	90	90	90	103	108	110	98	78	85	78
10-15	90	95	83	88	95	98	105	125	113	98	83	73
10-16	90	—	—	—	—	—	—	—	—	—	—	—
10-17	105	113	118	163	173	195	165	160	185	—	—	—
10-19	98	93	83	88	105	135	128	123	120	115	100	75
10-20	135	148	160	168	175	185	200	195	188	178	170	158
10-21	103	148	138	133	140	148	148	95	128	130	108	75
10-22	98	118	120	140	125	115	113	110	105	93	83	78
10-23	80	85	113	123	118	123	105	110	88	83	85	78
10-24	73	78	78	78	113	93	95	103	—	—	—	—
10-25	65	85	120	135	90	148	153	140	170	143	123	95
10-31	115	145	155	170	148	140	148	163	135	128	150	125
11-1	125	125	125	135	145	148	138	133	125	118	100	83
Average:	101	113	117	128	132	142	135	129	131	118	107	93

H. ISRAËL, H. W. KASEMIR AND K. WIENERT:

Table 5. *Hourly averages of the atmospheric electrical vertical current at Jungfraujoch on the sought out, disturbance-free days in the fall of 1950, supplemented through disturbance-free partial diurnal series. The information is provided in 10^{-17} amps/cm² of the used grid (unreduced).*

Date	0—1	1—2	2—3	3—4	4—5	5—6	6—7	7—8	8—9	9—10	10—11	11—12
10-5	—	—	—	—	—	—	—	—	—	—	—	—
10-6	359	331	359	349	387	307	415	406	415	379	379	387
10-7	576	557	529	500	529	614	642	632	614	557	510	491
10-8	520	453	406	406	425	453	491	520	520	500	462	453
10-12	—	—	—	—	—	—	—	—	—	—	—	—
10-13	350	312	312	331	331	321	331	397	397	406	406	378
10-14	378	350	321	312	293	321	378	—	—	—	—	274
10-15	293	302	274	274	302	331	340	378	368	378	359	368
10-16	368	368	368	406	387	321	312	350	359	350	378	350
10-17	—	—	—	—	—	—	—	—	—	—	—	—
10-19	—	—	—	—	—	—	—	—	—	510	500	453
10-20	387	378	368	378	387	406	397	406	453	444	406	387
10-21	548	520	491	462	453	453	444	425	415	368	358	350
10-22	359	330	340	330	359	397	406	387	359	321	302	302
10-23	302	293	302	302	292	292	283	274	283	302	264	264
10-24	302	293	302	302	312	321	321	340	340	330	283	283
10-25	—	—	—	—	—	—	—	312	312	293	283	293
10-31	—	—	—	—	—	—	—	—	—	255	293	302
11-1	444	444	444	330	283	368	350	368	368	368	378	378
Average:	399	380	370	360	364	384	393	399	400	385	372	357

Date	12—13	13—14	14—15	15—16	16—17	17—18	18—19	19—20	20—21	21—22	22—23	23—24
10-5	—	302	359	415	453	462	453	472	462	425	444	406
10-6	406	481	557	595	660	755	698	623	623	623	651	632
10-7	491	481	519	566	642	670	680	680	736	736	700	605
10-8	425	415	462	453	472	520	462	406	491	548	—	—
10-12	—	—	—	—	—	—	—	435	406	359	359	
10-13	397	387	406	435	500	548	605	651	605	529	482	453
10-14	302	302	293	312	312	340	387	455	425	350	321	283
10-15	359	359	340	349	368	397	425	472	481	472	406	397
10-16	368	—	—	—	—	—	—	—	—	—	—	—
10-17	283	283	293	312	520	651	605	576	670	—	—	—
10-19	387	425	387	396	453	566	632	613	613	595	556	472
10-20	359	397	444	500	529	613	680	680	680	651	622	595
10-21	387	435	481	538	566	614	614	538	557	566	520	425
10-22	321	368	387	444	453	444	435	406	387	359	340	331
10-23	255	264	340	406	406	425	415	397	368	321	330	321
10-24	283	302	321	340	378	378	397	406	—	—	—	—
10-25	302	330	415	510	444	500	556	576	642	623	597	444
10-31	321	378	435	519	529	547	556	538	491	472	453	435
11-1	415	444	453	481	510	520	500	463	463	425	397	321
Average:	357	374	405	445	482	526	535	544	537	507	475	432

15

Atmospheric Electrical Diurnal Variations and Mass Exchange in the High Mountains.

Table 6. *Hourly averages of the (total) conductivity of the atmosphere at Jungfraujoch on the sought out, disturbance-free days in the fall of 1950 – calculated from the values of the Tables 4 and 5. The information is provided in 10^{-4} electrostatic units (unreduced values).*

Date	0—1	1—2	2—3	3—4	4—5	5—6	6—7	7—8	8—9	9—10	10—11	11—12
10-5	—	—	—	—	—	—	—	—	—	—	—	—
10-6	45	48	40	39	39	37	39	43	44	45	47	50
10-7	44	42	41	43	38	37	41	43	41	39	30	33
10-8	44	41	39	38	40	39	40	40	40	41	41	36
10-12	—	—	—	—	—	—	—	—	—	—	—	—
10-13	54	51	45	46	49	41	40	41	41	37	37	38
10-14	52	50	48	51	48	41	43	—	—	—	—	41
10-15	34	34	37	39	37	38	41	39	40	39	38	39
10-16	44	43	44	41	47	45	42	39	42	41	36	37
10-17	—	—	—	—	—	—	—	—	—	—	—	—
10-19	—	—	—	—	—	—	—	—	—	41	44	42
10-20	50	49	47	45	48	44	45	42	37	39	36	32
10-21	35	40	41	41	41	41	44	42	44	41	41	35
10-22	52	50	44	48	39	37	37	45	45	44	42	38
10-23	36	36	35	40	42	39	38	32	28	29	31	32
10-24	42	44	42	40	42	41	43	44	38	44	42	39
10-25	—	—	—	—	—	—	—	47	43	42	48	44
10-31	—	—	—	—	—	—	—	—	—	26	29	27
11-1	29	32	31	41	35	47	29	33	34	30	32	32
Average:	43	43	41	43	42	41	40	41	40	39	39	38

Date	12—13	13—14	14—15	15—16	16—17	17—18	18—19	19—20	20—21	21—22	22—23	23—24
10-5	—	39	33	35	38	42	41	48	43	42	43	44
10-6	49	39	40	38	33	35	43	44	41	41	43	44
10-7	31	30	28	27	30	31	31	35	37	39	47	45
10-8	32	28	34	34	32	32	39	43	35	35	—	—
10-12	—	—	—	—	—	—	—	—	44	49	50	50
10-13	37	26	38	37	37	36	39	41	44	49	48	49
10-14	34	30	30	31	31	30	32	35	39	40	34	34
10-15	36	34	37	36	35	37	36	34	38	44	44	48
10-16	37	—	—	—	—	—	—	—	—	—	—	—
10-17	24	23	22	17	27	30	33	33	33	—	—	—
10-19	36	42	42	41	39	38	45	45	46	47	50	56
10-20	24	24	25	27	27	30	31	31	33	33	33	34
10-21	34	27	32	26	36	37	37	41	39	39	44	51
10-22	30	28	29	29	33	35	35	33	33	35	37	38
10-23	29	28	27	30	31	31	36	33	34	35	35	37
10-24	35	35	37	39	30	39	37	36	—	—	—	—
10-25	42	35	31	34	44	31	33	37	34	39	40	43
10-31	25	23	25	28	32	35	34	30	33	33	27	31
11-1	30	32	33	32	32	32	33	31	33	33	36	35
Average:	33	31	32	32	33	34	36	37	38	40	40	43

H. ISRAËL, H. W. KASEMIR AND K. WIENERT:

Table 7. *Hourly averages of the steam pressure at Jungfraujoch during both measurement periods in the summer and fall of 1950 – calculated from the hourly averages of the same days or partial days, on which disturbance-free atmospheric electrical registrations were achieved (see Table 1 to 3 or 4 to 6).*

Time	0—1	1—2	2—3	3—4	4—5	5—6	6—7	7—8	8—9	9—10	10—11	11—12
Summer	2.47	2.53	2.65	2.58	2.58	2.65	2.72	2.80	2.89	2.94	3.23	3.51
Fall	2.59	2.50	2.63	2.59	2.58	2.48	2.42	2.33	2.28	2.22	2.21	2.18

Time	12—13	13—14	14—15	15—16	16—17	17—18	18—19	19—20	20—21	21—22	22—23	23—24
Summer	3.76	3.95	4.27	4.48	4.83	4.55	4.27	3.91	3.76	3.58	3.36	3.08
Fall	2.28	2.31	2.55	2.68	2.70	2.64	2.58	2.49	2.45	2.60	2.55	2.61

c) Individual Results.

Figures 4 and 5 reflect the individual registrations as examples. The thick solid curve indicates the potential difference of the radioactive collector against the earth; the thin solid curve is the record of the vertical current and the dotted curve reflects the registration of the photocell.

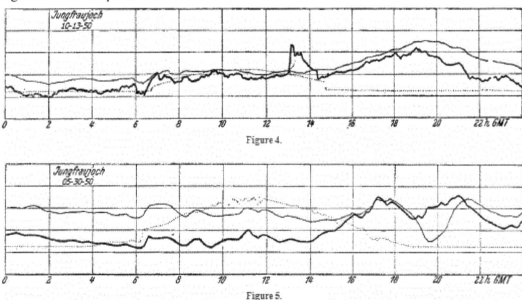

Figure 4.

Figure 5.

Figures 4 and 5. Registration strips with the record of the potential gradient, vertical current and local brightness at Jungfraujoch. Thick curve: potential gradient; thin curve: vertical current; dotted curve: local brightness. Upper diagram: May 30, 1950; lower diagram: October 13, 1950. The values are respectively given as a percentage of the diurnal average.

The jumps in the dotted curve in the late morning and afternoon are a result of the fact that the horizon is not clear. The sun is blocked by the mountain slopes both after rising and before setting. An image of the actual contour of the horizon at Jungfraujoch can be found in Figure 5 from the research of W. Mörikofer and U. Chorus [19].

17

Atmospheric Electrical Diurnal Variations and Mass Exchange in the High Mountains.

The hourly averages of the registrations of potential gradient and vertical current as well as the conductivity values calculated from these are consistently provided in an unreduced, relative measurement in Tables 1 to 6.

As a relative measurement for the field strength, Tables 1 and 4 provide the recorded, unreduced voltage values of the radioactive collector, which was located at a distance of approx. 80 cm from the southwest wall of the observation cabin – approx. 2 meters above the small platform there.

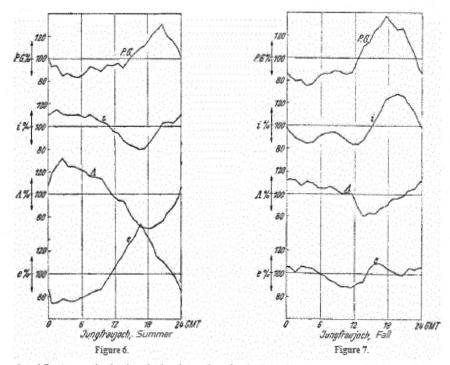

Figure 6.   Figure 7.

Figures 6 and 7. Atmospheric electrical ratios at Jungfraujoch in the summer and fall of 1950. Uppermost curve: diurnal variation of the potential gradients; second curve: diurnal variation of the vertical current density; third curve: diurnal variation of the conductivity; lowest curve: diurnal variation of the steam pressure. All curves are marked as percentages of the average. Left: summer 1950; right: fall 1950.

Tables 2 and 5 contain the values of the vertical current, which merged into the 1 square meter fine-mesh net mounted vertically at a distance of approx. 30 cm from the southeast wall of the cabin, in a unit of measurement of 10^{-17} amps/cm^2.

Finally, Tables 3 and 6 provide the hourly values of the conductivity, which are calculated from field strength and vertical current. Naturally, they are also provided in a relative measurement, although they are converted to an electrostatic measurement and provided in 10^{-4} electrostatic units to enhance the ability to compare with other results.

In Table 7, the steam pressure values are compiled for the same measurement periods as in the previous tables. All time-related information is related to Greenwich Time.

The averages of these tables are presented graphically in Figures 6 and 7.

Arch. Met. Geoph. Biokl. A. Bd. III, H. 5.   25

H. ISRAËL, H. W. KASEMIR AND K. WIENERT:

The potential gradient practically only features the 24-hour variations in the summer as well as in the fall with the lowest value in the early hours of the morning and the highest value in the late afternoon – as it has been recognized by all high-altitude stations.

The reaction of the vertical current is interesting. While it runs more or less laterally reversed to the potential gradients in the summer, a widely parallel variation exists between the two elements in the fall. This indicates that a fundamental change of the atmospheric electrical state apparently emerges at Jungfraujoch from the summer to the fall. The two conductivity curves provide us with more detailed information in this regard.

At Jungfraujoch, the conductivity in the summer as well as in the fall displays a distinct diurnal variation with the highest values at night and the smallest values in the afternoon. This is precisely the same behavior that is familiar to the stations in the lowlands. The amplitude of the variation, however, strongly decreases from the summer to the fall; it drops from 60 % to approx. 30% of the average.

This behavior of the conductivity is perhaps the most interesting and simultaneously the most surprising result of the Jungfraujoch analyses, for the interpretation that immediately comes to mind for this is that the exchange effect must apparently play a decisive role as well in the configuration of the atmospheric electrical ratios at the high-altitude station at Jungfraujoch.

This is surprising for the following reason – the vertical mass exchange as such is naturally a phenomenon, which is *not* limited to the surface-level layers of the atmosphere, but rather comprises the entire *troposphere* and only vanishes once it is in the lower stratosphere. However, as we have previously illustrated, the *atmospheric electrical behavior of a modifying element* in this process is not the exchange itself, but rather its diurnal variation and the aerosol change caused by this. Because this occurs through the vertical transport of suspensions, which derived from the lowest atmospheric layers, it is only necessary to expect that an atmospheric electrical exchange rhythm can only be found up to a certain altitude. For this variation can only penetrate up to a limited altitude due to its limited period of variation.

As to the question regarding how high this area extends, in which the diurnal variations caused by the vertical transport from below become noticeable, the analysis from J. Reger [20] can provide specific information regarding the diurnal steam pressure variation in the free atmosphere over Lindenberg, thus, over the North German Plain, "the influence of the midday exchange with the surface layers on clear summer days appears to be over with a maximum altitude of 2500 meters", while "the midday moisture current in the winter does not reach the altitude of 500 meters". According to this, one should expect that the sphere of influence is already exceeded at the altitude of Jungfraujoch with its 3500 meters. However, as our results demonstrate, this is not the case.

Atmospheric Electrical Diurnal Variations and Mass Exchange in the High Mountains.

To allow this to become even clearer, the diurnal steam pressure variations are entered in the respectively lowest curves of Figure 2 and 3. They show that a very distinct variation of the steam pressure appears in the summer, while it is nearly gone in the fall and most likely entire gone in the winter. That confirms the assumption that we have a penetration of the diurnal convection to an altitude above Jungfraujoch during the summer in the high mountains, while this altitude is no longer achieved in the fall.

d) Columnar Resistance and Conductivity.

The different behavior of the vertical current at Jungfraujoch during the summer and fall becomes understandable based on the previous results. During the summer, this is apparently decisively controlled by the aerosol change and we find a similar variation of current and conductivity. In contrast, during the fall, the influence diminishes and then allows the other atmospheric electrical mechanism to emerge, according to which the current is controlled by the potential difference between the surface of the earth and the "atmospheric electrical compensating layer" at an altitude of 60 to 70 km [4]; as a result, we can observe a similar variation of current and potential gradient.

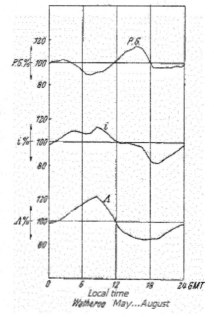

Figure 8. The atmospheric electrical ratios in Watheroo (Western Australia) in the months from May to August.

Thus, interestingly at Jungfraujoch we find two completely different types of atmospheric electrical behavior depending on the season, which we can identify as the *continental* and *oceanic type* by analogy to the ratios familiar in low-lying locations, whereby the expressions "continental" and "oceanic" in the present case are intended to be understood not as the geographic position, but rather the influence of the subsurface.

We can explain the materialization of these two types in the following manner for example:

According to equation (1) and (2), the field strength E, the conductivity λ, the total potential difference V between the surface and the compensating layer, the atmospheric total resistance W and the vertical current density i are connected through the relationships:

$$E \cdot \lambda = i,$$

$$E = w \cdot \frac{V}{W} \quad (w = 1/\lambda).$$

If w varies very heavily, this impresses its stamp on the variation of i, as can be recognized, for example in Figure 8, which represents the atmospheric electrical ratios during the months of May to August in Watheroo (Western Australia) [21] (*continental type*).

25*

20

H. ISRAËL, H. W. KASEMIR AND K. WIENERT:

However, if w in nearly constant, a parallel variation of E and i will result, as was demonstrated in Carnegie's measurements [22] on the oceans (oceanic type, see Figure 9.).

Depending on the regional, geographic and climatic conditions, mixed types can be found at the various measurement sites, which primarily approximate the one case or the other.

One can get an even better look into these conditions according to a suggestion by G. R. Wait [23] through the observation of the so-called "columnar resistance" and its connection to the changes in conductivity at the

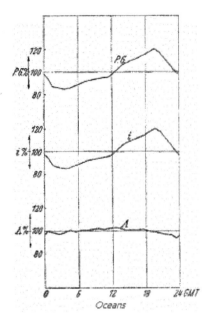

Figure 9. The atmospheric electrical conditions over the oceans.

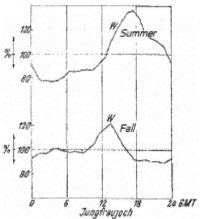

Figure 10. Percentage of diurnal variation of the columnar resistance over Jungfraujoch in the summer and fall.

measurement site. If we turn this analysis procedure to the results at Jungfraujoch, the following is revealed:

Figure 10 demonstrates the variation of the columnar resistance for both measurement periods; this represents the resistance of a vertical atmospheric column with a cross-section of 1 cm^2, which extends from the measurement site to the compensating layer. The determination occurs in the familiar manner through the combination of the regional measurements with the values determined for the oceans through the formation of the quotients i_0/i_s (i_0 or i_s = vertical current density over the ocean or at the station). The averages provided by W. C. Parkinson and O. W. Torreson [24] were taken as a basis for the i_0–values.

The diurnal variation of the columnar resistance demonstrates an exceptionally inverse course to the conductivity in the summer. The opposite course is less heavily pronounced in the fall period.

The columnar resistance is comprised of the diurnal constant element lying in the upper layer and the element allocated to the lower layer, which changes inversely to the exchange-related conductivity fluctuation.

Atmospheric Electrical Diurnal Variations and Mass Exchange in the High Mountains.

If one identifies the temporally constant portion with r, the variable portion with $Q = h \cdot w$, whereby h stands for the altitude of the border between the lower and upper layer and w stands for the specific resistance applied at the same altitude in the initial approximation within the lower layer ($w = 1/\lambda$), then W applies for the total resistance.

$$W = r + \varrho = r + h \cdot w. \quad (4)$$

Because now:

$$W/W_o = i_o/i$$

(index 0 stands for the oceanic value), the following relationship is revealed:

$$\frac{i_o}{i} = \frac{1}{W_o} \cdot (r + h \cdot w). \quad (6)$$

This is the equation of a straight line with w as an independent and i_0/i as a dependent variable.

Thus, if one applies i_0/i dependent upon the reciprocal conductivity, a straight lines results, which - due to its location - allows for certain statements regarding the two resistance portions. This occurred for both measurement periods in Figure 11. If one extrapolates the straight lines up to the intersection with the ordinate axis, the variable of the ordinate section provides a measurement for the portion of the temporally constant

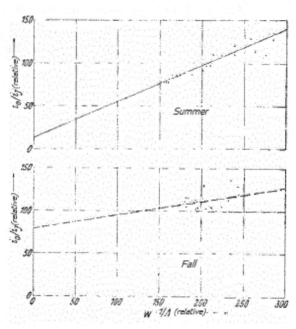

Figure 11. The columnar resistance of the atmosphere over Jungfraujoch dependent upon the conductivity at the station for the summer and fall (depiction in a relative dimension).

contribution to the total resistance. One can see from the comparison of both depictions that, the relationship of W to the change in conductivity on the surface during the summer at Jungfraujoch is much closer than in the fall. This confirms our assumption expressed above that during the summer in the high mountains of Jungfraujoch the diurnal variation of W must primarily originate from the exchange, while this is only the case at a minimal degree in the fall – provided that such a slanted position of the curve may even be derived from the heavy dispersion of the point cloud in the lower part of Figure 11.

Principally, one can derive a determination of the altitude of the lower layer, as G. R. Wait successfully shows us [23], from this. Although, this point would be especially interesting for us, this will not be possible in the present case. This is because, on the hand we are only dealing with unreduced relative values in the case of individual values and on the other hand, such a method of observation and division of the columnar resistance requires large, flat areas, over which a vertical course of the field lines may be assumed.

H. ISRAËL, H. W. KASEMIR AND K. WIENERT:

Therefore, it strictly cannot be applied in an orographic field skewed in such a way, as is the case in the high mountain. As a result, it is necessary for us to limit ourselves at this point to the qualitative statements above.

c) Conclusions.

Through this observation, we can also find confirmation for the fact that significant communication with the atmosphere stemming from low locations must occur in the highest locations of the Alps on clear summer days with high pressure. With regard to our initially handled opinion of a lower and upper layer of the atmosphere to be defined atmospherically-electrically, this means that the border between these two areas in the high mountains will be raised sharply and obliterated.

This result should not surprise because the orographic conditions of the high mountains must result in an effective and significantly higher reaching vertical transport of atmosphere from low areas through the diurnal circulation of the mountain and valley wind as would be the case over the free plane.

Something else will be added. In the same way that an interaction of an *advective* exchange of atmosphere must be assumed [20] as with the interpretation of the diurnal steam pressure variation in the free atmosphere above the plane, this advectively conditional change of the atmospheric characteristics can and will play an additional and decisive role for the formation of the atmospheric electrical conditions in the high mountains. The result, for example of the fall period, during which a significant variation of the columnar resistance still occurs without being closely bound to the conductivity variation at the measurement location as is the case during the summer, speaks for this. Today, this change of resistance is apparently on supported to a small degree by the vertical exchange and its aerosol changing fluctuations, but rather is caused primarily through (potentially advectively conditional) variations of the atmospheric additives.

If we summarize, we can obtain the following image for example. The results of the atmospheric electrical analyses at Jungfraujoch confirm in general our opinions and expectations through the inner connections between the diurnal variations of the atmospheric electrical elements and those of the vertical mass exchange. At the same time, they provide a visible image of the connections, which are extremely involved in particular in high mountain locations, between the exchange, orographically conditioned circulation (mountain and valley wind) and the large-scale circulation above the massif. As a result, at least the possibility was created for approaching the complex meteorological-aerological problem of the atmospheric circulation and of the atmospheric exchange in the horizontal and vertical direction.

Atmospheric Electrical Diurnal Variations and Mass Exchange in the High Mountains.

The reported results are initial random advances in this direction and, to some extent, can create the foundation for a program in this direction through the emerging connections. In one of our [1] papers mentioned in the introduction, at the end in Figure 14 was a schematic equivalent circuit diagram was provided for understanding the exchange effect on the atmospheric electrical elements at various altitudes of the stations. According to that, the following was revealed as a particular necessity *the simultaneous collaboration at various altitudes. In connection with this and the above results, the natural continuation of this type of analyses lies in its expansion to multiple simultaneously functioning stations at various points of the massif and at various altitudes of the same*.* Thus, hand in hand, suitable research of the atmospheric electrical nature must function in the open atmosphere over wide even areas, for such research allows the exchange and advection events to appear above an orographically uniform base in an atmospheric electrical manner and then to clearly put forth the special high mountain influence on the atmospheric circulation in close connection with the research at the mountain stations.

Without providing the significant statements regarding these connections, the fact remains without a doubt that from this consciously close coupling of the atmospheric electrical and meteorological-aerological analyses we may anticipate a further approximation toward the clarification of the inner causal connections of the atmospheric circulation.

The execution of these analyses under the difficult post-war conditions was only possible through the sympathetic support of numerous Swiss and German agencies. Our special thanks go to Prof. Dr. A. von Muralt, Bern, Prof. Dr. J. Lugeon, Zurich, Dr. W. Mörikofer, Davos, of the Swiss Atmospheric Electrical Commission, the Association for the Promotion of Science in Germany as well as the German Ministry for Education and Cultural Affairs and the German Ministry of Finance of Württemberg-Hohenzollern.

Appendix.

In the present paper, only the results of our atmospheric electrical analyses conducted at Jungfraujoch connected to the atmospheric mass exchange. All remaining results will be handled in a second report after the conclusion of the third measurement period planned for the middle of winter. The comparisons to earlier atmospheric electrical high mountain analyses remain reserved to this as well. At this point, we would only like to make a brief note regarding a particularity in the above reported numerical material.

---

* In the meantime, we have made an initial start with the construction of a parallel station at Sonnblick Summit (3106 meters) in the Hohe Tauern Mountains. We will report on these results at another point.

H. ISRAËL, H. W. KASEMIR AND K. WIENERT:

The numerical values of the Tables 1 to 6 are, we already mentioned, unreduced, direct measurement values (or the quotient in the case of the conductivity). Because the observations of this research are only tied to the relative variability, going into the numerical absolute altitude of the elements was not necessary. Nonetheless, the unusually high the vertical current values and the conductivities derived from them and the potential gradient will strike the attention of the reader.

Without going deeper into this point, the individual discussion of which is also reserved for the second publication, we would however like to note that these evidently high vertical current values must be treated as real according to all previous tests. The used measuring equipment, which features a modification of Wilson's earth plate method [17], has previously been in operation at multiple locations of the flatland and the high mountains for longer and shorter periods. The average vertical current values found at flatland stations revealed the adequately familiar level. In Buchau a. F. resulted in approx. $2 \cdot 10^{-16}$ amps/cm^2, in Uppsala approx. $4 \cdot 10^{-16}$ amps/cm^2, and from that in Davos: a somewhat increased value of approx. $6 \cdot 10^{-16}$ amps/cm^2. In contrast, in the case of all of our previous high mountain measurement locations (Nebelhorn, Zugspitze, Jungfraujoch and Sonnblick) we consistently achieved the same high average level of approx. 30 to $50 \cdot 10^{-16}$ amps/cm^2.

Bibliography.

1. Israël, H.: Zur Entwicklung der luftelektrischen Grundanschauungen. Arch. Met. Geoph. Biokl. A 3, 1—16 (1950).
2. — und G. Lahmeyer: Studien über das atmosphärische Potentialgefälle, I: Das Auswahlprinzip der luftelektrisch „ungestörten Tage". Terr. Magn. 53, 373—386 (1948).
3. — Das Gewitter. Ergebnisse und Probleme der modernen Gewitterforschung. Probleme der kosmischen Physik, Bd. XXV, Leipzig 1950.
4. — und H. W. Kasemir: In welcher Höhe geht der weltweite luftelektrische Ausgleich vor sich? Ann. Géophys. 5, 313—324 (1949).
5. Scholz, J.: Polarlichtuntersuchungen auf Franz-Josephs-Land. Gerl. Beitr. Geophys. 44, 145—156 (1935).
6. Israël, H.: Extraterrestrische Einflüsse auf das luftelektrische Feld. Wiss. Arb. d. Dtsch. Met. Dienst. i. franz. Bes.-Geb. 1, 62—67 (1947).
7. — Die Unruhe des elektrischen Feldes (Beispiele für den Witterungseinfluß auf das Feldverhalten am Boden). Met. Zeitschr. 60, 56—62 (1943). Vgl. auch [6].
8. Brown, J. G.: The local variation of the earth electric field. Terr. Magn. 40, 413—424 (1935).
9. Israël, H.: Die Tagesvariation des elektrischen Widerstandes der Atmosphäre (Studien über das atmosphärische Potentialgefälle, II). Ann. Géophys. 5, 196—210 (1949).
10. — Die halbtägige Welle des luftelektrischen Potentialgefälles. Arch. Met. Geoph. Biokl. A 1, 247—251 (1948).
11. Paramonoff, N. A.: Über die Weltzeitperiode des luftelektrischen Potentialgradienten (russisch). Dokl. Acad. Nauk. S. S. S. R. (Akad. d. Wissensch. d. UdSSR.) 70, 37—38 (1950).
12. Israël, H.: Der Elektrizitätshaushalt der Erdatmosphäre. Naturwiss. 29, 700 bis 706 (1941).
13. Gish, O. H.: Evaluation and interpretation of the columnar resistance of the atmosphere. Terr. Magn. 49, 159—168 (1944).

Heinz-Wolfram Kasemir: His Collected Works

Atmospheric Electrical Diurnal Variations and Mass Exchange in the High Mountains.

14. KASEMIR, H. W.: Studien über das atmosphärische Potentialgefälle. IV: Zur Strömungstheorie des luftelektrischen Feldes. 1. Arch. Met. Geoph. Biokl. A 3, 84—97 (1950).

15. ISRAËL, H.: Gedanken und Vorschläge zur luftelektrischen Arbeit. Wissensch. Abhandl. Reichsamt. f. Wetterd. V, Nr. 12, Berlin 1939, 26 S.

16. BENNDORF, H.: Grundzüge einer Theorie des elektrischen Feldes der Erde. Wien. Ber. IIa, 136, 175—194 (1927).

17. KASEMIR, H. W.: An apparatus for simultaneous registrations of potentialgradient and air-earth current. Journ. Atm. Terr. Physics (1951, im Druck).

18. FRENKEL, J.: A theory of the fundamental phenomena of atmospheric electricity. Journ. Physics 8, 285—304 (1944).

19. MÖRIKOFER, W. und U. CHORUS: Meteorologische Untersuchungen auf dem Jungfraujoch während des Polarjahres 1932/33. Public. de la Station Centrale Suisse de Météorologie. Zürich, 1941.

20. REGER, J.: Der tägliche Gang der Feuchtigkeit über Lindenberg. Arb. Preuß. Aeron. Obs. Lindenberg. Wiss. Abh. 14, 44—61 (1922).

21. WAIT, G. R. and O. W. TORRESON: Atmospheric-Electric Results from Watheroo/Western Australia, for the Period 1924—1934. Terr. Magn. 46, 319—342 (1941).

22. BAUER, L. A. and W. F. G. SWANN: Results of Atmospheric-Electric Observations made Aboard the Galilee (1907—1908), and the Carnegie (1909—1916). Publ. Carnegie Inst. Wash. 3, Nr. 175, 361—422 (1917).

23. WAIT, G. R.: Electrical Resistance of a vertical colum of air over Watheroo (Western Australia) and over Huancayo (Peru). Terr. Magn. 47, 243—249 (1942).

24. PARKINSON, C. C. and O. W. TORRESON: The diurnal variation of the electric potential of the atmosphere over the Oceans. U. G. G. I., Compt. Rend. Assemblée Stockholm (Bull. Nr. 8) 340—345, Paris, 1931.

Special print from

Archive for Meteorology, Geophysics and Bioclimatology

Series A: Meteorology and Geophysics, Volume 8, Issue 1 – 2, 1955

Published by
Dr. W. Mörikofer, Davos and Prof. Dr. F. Steinhauser, Vienna
Springer-Verlag in Vienna

(Atmospheric Electrical Research Center of the Observatory in Friedrichshafen
in Buchau a. F., National Weather Service Württemberg-Hohenzollern.)

Atmospheric Electrical Diurnal Variations and Mass Exchange in the High Mountain Ranges of the Alps.

The Atmospheric Electrical Conditions at Jungfraujoch (3472 m) I.

By

H. Israël, H. W. Kasemir and K. Wienert

Summary. The investigations on atmospheric electricity of summer and autumn 1950 on the Jungfraujoch have been continued with a third recording-period in the winter. By these investigations the former results which are already reported in this review are completed to a full-year-research. In spite of bad weather, which strongly limited the extent of the results, the conclusions drawn from earlier investigations could be entirely verified. In this Alpine altitude the systematic variation of the conductivity, that has been found high in summer and scarcely remarkable in autumn, has completely disappeared in winter. The daily variations of the potential gradient and the air-earth-current are essentially parallel; it accounts for that in winter at the ridge of the Alps the oceanic type of atmospheric electricity is predominant. Obviously, the daily variations of the mass exchange do not reach these high levels in the winter months.

In autumn 1950 simultaneous records of the potential gradient and the air-earth-current at Jungfraujoch and at Sonnblick, 400 km distant, have been made during some days. The two elements are in good agreement both with regard to the mean daily variations as to particulars. Hence it follows that the results received at Jungfraujoch are representative for the summit of the Alps except for cases with local disturbances.

Furthermore an investigation has been carried out at Jungfraujoch on the electric field and the current during precipitation. It resulted that "dry snow", i. e. snow originating by sublimation only, always carries a positive charge and generates an electric field of positive direction. On the contrary, the solid precipitation particles which are formed by graupel formation ("wet snow", soft hail, granular snow) carry generally a negative charge and generate an electric field of negative direction. In summer the occurence of snow after soft hail or rain can be taken as a remarkable symptom for developing a thunderstorm.

Résumé. Les recherches d'électricité atmosphérique effectuées au Jungfraujoch pendant l'été et l'automne 1950 et dont il a été question précédemment dans cette revue ont été complétées par une série hivernale. Malgré un temps défavorable qui a limité les recherches, les résultats précédents ont pu être confirmés. La variation systématique de la conductibilité observée à haute altitude, forte en été et encore faiblement présente en automne, disparaît complètement en hiver. Le gradient de potentiel et le courant vertical présentent des marches diurnes parallèles, c'est-à-dire qu'en hiver, à l'altitude du faîte alpin qui n'est apparemment plus affecté par la variation diurne de l'échange de masse, s'établit le type océanique de l'état électrique de l'atmosphère.

En automne 1950 les enregistrements du gradient de potentiel et du courant vertical furent exécutés pendant quelques jours simultanément au Jungfraujoch et au Sonnblick, distant de 400 km. du premier; il y a bon accord entre les mesures des deux éléments, tant pour la variation diurne moyenne que pour les cas isolés. Il s'ensuit que les résultats obtenus au Jung-

H. ISRAËL, H. W. KASEMIR AND K. WIENERT:

fraujoch, pour autant qu'ils ne sont pas troublés localement, sont représentatifs de la crête des Alpes.

Une étude du champ et du courant au Jungfraujoch pendant des précipitations montre que la neige sèche, c'est à dire comme produit pur de sublimation, est toujours chargée positivement et engendre un champ positif (direction normale). Par contre les particules solides issues du grésil (neige mouillée, neige en grains, grésil) sont en général négativement chargées, dans des champs négatifs (direction inversée). Il est en particulier remarquable qu'en été la neige succédant au grésil ou à la pluie soit dans la règle le signe d'un orage en formation.

I. Introduction.

Some time ago, we reported the results of registrations of atmospheric electrical potential gradient and the vertical current at the High Alpine Research Station Jungfraujoch (3472 m elevation) in the summer and fall of 1950 in an initial paper of the same title [1]. The goal of these papers was the analysis of the atmospheric electrical ratios at a high altitude, from which we were able to assume, that it will no longer be gathered due to the diurnal variations of the atmospheric mass exchange and its consequences.

The results achieved in the first two measurement periods – June/July and October/November of 1950 – showed that the daily exchange variation up to the summit region of the Alps is noticeable in the high summer, thus it reaches significantly higher than in the free atmosphere, while this influence at the elevation of Jungfraujoch is reduced to the extent in the fall that the atmospheric electrical ratios now approximate the dominating oceanic type with an absence or heavy reduction of the exchange variation. To verify this determination and to complete the results for an annual overview, both measurement periods of the summer and fall were supplemented by one in the winter (in February/March 1951). Due to external reasons, the processing of these results was only possible at this time. The following is intended to report on the results.

Furthermore, we are only now able to report on the parallel registration of potential gradient and vertical current on the peak of the Sonnblick (elevation: 3100m; distance from Jungfraujoch: 400m) conducted during the fall of 1950.

II. Results.

a) Preliminary remarks.

As the third "high winter" measurement period, the timeframe of January/February 1951 was chosen, although the start had to be postponed until February due to inclement weather. Unfortunately, the weather proved to be less beneficial for atmospheric electrical research in the following time period as well due to the fact that, even in clear weather, snow drifts often occurred.

As a result, despite a two-month stay, the volume of the observational material not disturbed by weather-related influences and local effects lagged heavily behind that of the earlier measurement periods.

Measurement location and measuring device were the same for both of the earlier measurement periods. All associated details, therefore, can be taken from the first publication [1]. With regard to the technology, merely a single change is noteworthy insofar as to supplement the potential gradient registrations, namely a field meter registration was used intermittently.

b) Potential Gradient, Vertical Current and Conductivity in Winter.

After excluding the periods of inclement weather and all of the registrations that were disturbed by snow drifts, only the diurnal series or diurnal partial series compiled in Table 1 remained for the mentioned measurement period. The information regarding the potential gradient values deviates from the information in the first publication because they are now reduced and expressed in Volt/m through comparison measurements on the Jungfraujoch Plateau. The vertical current values are provided like earlier in 10^{-17} Amp/cm^2 of the used grid. The conductivity data must be derived from the (unreduced) current values and the (reduced) potential gradient values and multiplied with the earlier values – Table 3 and 6 of the first paper – by 1.48.

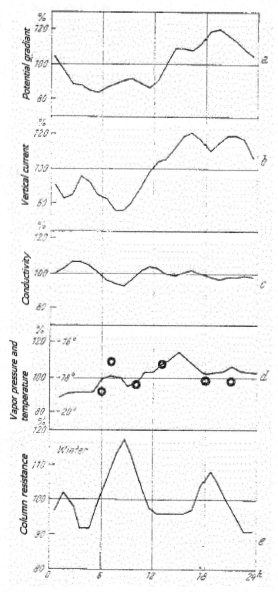

Figure 1. The atmospheric electrical ratios on Jungfraujoch in the winter (February/March 1951). Diurnal variation of the potential gradient a, of the vertical current density b, of the conductivity c, of the temperature (circles) and of the vapor pressure d, as well as of the column resistance e in percentages of the mean values.

Table 1. *Hourly values of the atmospheric electrical potential gradient, of the vertical current and the conductivity at Jungfraujoch on selected disturbance free days or portions of days in February/March of 1951.*

Potential gradient: Information in Volt/m, reduced to the Jungfraujoch Plateau[1]; vertical current: in units of 10^{-17} Amp/cm^2 of the used grid (not reduced); conductivity: information in 10^{-4} electrostatic units, calculated from the (unreduced) current values and the (reduced) potential gradient values[2]. .

		12—1 am	1—2 am	2—3 am	3—4 am	4—5 am	5—6 am	6—7 am	7—8 am	8—9 am
Potential Gradient	February 27									
	February 28	193	237		278	207	200	218	240	222
	March 1	307	304	237	211	189	163	166	163	133
	March 4								122	104
	March 5	170	137	137	137	130	148	181	196	189
	March 6	100	89	78	74	81	96	100	100	89
	March 7	115	104	100	96	93	89	74	81	104
	March 10								172	185
	Balanced average	177	166	152	149	145	143	148	152	154
	„ „ in %	104	97	89	88	85	84	87	89	91
Vertical Current	February 27									
	February 28	450	530		730	490	420	460	470	400
	March 1			820	640	490				
	March 4								270	250
	March 5	450	400	380	370	350	380	470	510	310
	March 6	300	290	280	270	270	290	310	310	290
	March 7	310	310	300	280	280	280	280	220	390
	March 10								240	170
	Balanced average	410	370	380	430	420	380	370	340	340
	„ „ in %	91	83	85	96	93	85	83	76	76
Conductivity	February 27									
	February 28	32	31		36	33	28	29	26	25
	March 1			53	46	41				
	March 4								30	32
	March 5	36	40	38	37	37	35	35	36	37
	March 6	42	45	48	49	44	42	43	43	45
	March 7	37	41	42	40	41	44	53	36	52
	March 10								17	13
	Balanced average	37	40	42	42	41	39	37	34	34
	„ „ in % after being smoothed twice ..	100	103	107	107	105	101	96	94	93

---

[1] The values of the summer and fall measurement periods in the first publication [1] are given in Volt. To reduce them to the Jungfraujoch Plateau, the figures (Table 1 and 4 in the paper [1]) must be multiplied by 1.48.

[2] To be able to compare the conductivity values with those provided earlier (see Table 3 and 6 of the first paper [1]), they must be multiplied by 1.48.

(This is the second part of Table 1 on the previous page)

9—10 am	10—11 am	11 am — 12 pm	12 — 1 pm	1 — 2 pm	2 — 3 pm	3 — 4 pm	4 — 5 pm	5 — 6 pm	6 — 7 pm	7 — 8 pm	8 — 9 pm	9 — 10 pm	10 — 11 pm	11 pm — 12 am	Average
		122	163	207	248	270	259	207	278	296	288	248	204	177	
244												274	270	300	
137	118	137	166	185	233										
93	96	115	104	129	159	187	155	193	196	230	222	218	196	189	
137	126	111	133	148	155	159	166	204	196	166	137	107	85	93	
130	159	185	211		262	181	200			170	144	118	107	118	
248	248	159													
163	140	170	163	140	133	133	137	140	152	163	192	218	207	200	
156	151	148	155	170	185	186	184	191	202	204	199	192	183	178	170
92	89	87	91	100	109	109	108	112	119	120	117	113	108	105	
		400	480	550	620	700	710	660	680	660	650	600	480	160	
360	390	390	440	550	420										
560	440	460	510	520	640										
280	280	320	300	340	400	460	430	490	490	580	590	530			
410	350	330	350	410	410	450	470	540	550	510	450	350	320	300	
390	490	630	650		740	650	730			480	410	360			
			650												
240	330	440	440	370	360	390	430	420	460	470	510	590	570	530	
360	400	440	470	480	500	530	540	520	490	510	530	530	520	480	447
81	89	98	104	106	112	119	121	117	110	115	119	119	117	106	
		45	41	36	34	35	37	43	34	30	30	33	33	35	
20												39	35	32	
56	50	46	42	39	37										
41	40	40	40	36	34	37	37	35	34	34	36	33			
41	38	41	36	37	36	38	39	36	38	42	44	45	52	45	
41	42	46	42		39	49	50			39	39	42			
		55													
20	32	35	37	36	37	40	43	41	41	39	36	37	37	36	
37	40	42	40	37	37	39	40	39	37	37	37	38	39	37	38,3
97	102	104	103	100	99	101	102	100	98	97	98	98	99	98	

In Figure 1, the diurnal variations are presented that result on average from the values of Table 1[3]. The curves are marked in percentages. The potential gradient and

[3] Considering the low volume of material, the value series of the daytime curves are smoothed as usual according to the formula $(a + 2b + c)/4$, namely the curves for the potential gradient and vertical current through a single application of the smoothing formula and the curve for the conductivity, which displayed strong, irregular fluctuations according to its origination through the formation of quotients, through two applications of the smoothing formula.

vertical current display equally-directed and homogenous diurnal variation, thus they behave the same in the winter as in the fall. The diurnal variation of conductivity is of particular interest. If one disregards the somewhat irregular curve, which is without a question the result of the low volume of material, their curve implies that the traits of the conductive variation typical for mainland stations at Jungfraujoch in the winter, i.e. high values during the night, low values during the day.

c) Annual Overview.

Now, together with the results of the first two measurement periods (see Table 1 to 6 and Figure 6/7 in paper [1]), we can finally obtain the following diagram for the behavior of the atmospheric electrical elements throughout the day during the various seasons:

Table 2. *The diurnal variations of potential gradient, vertical current and conductivity at Jungfraujoch during the various seasons.*

Season	Potential Gradient	Vertical Current	Conductivity
Summer	simple 24-hour fluctuation; min.: in the early morning; max.: in the later afternoon. Curve therefore opposing the vertical current.	simple 24-hour fluctuation; min.: afternoons; max.: in the first hours of the morning. Curve therefore opposing the potential gradient.	simple 24-hour fluctuation of significant amplitude; min.: afternoons; max.: in the first hours of the morning.
Fall	the same; Curve now in same direction as the vertical current.	simple 24-hour fluctuation, although now: min. and max. correlate temporally to the potential gradient; Thus, curve in same direction as the potential gradient	similar as above, however, only a weak daytime amplitude remains
Winter	the same; in same direction as the vertical current	the same; in same direction as the potential gradient	*no* diurnal variation

As already presented in earlier papers [1, 2, 3, 4], one can differentiate two borderline cases in the behavior of atmospheric electricity – a *"continental type"* and an *"oceanic type"*; decisive for the type structure in this process is the question, which of the two known control mechanisms provides the amplitude in the individual case. One can easily differentiate both types at first sight. In one case, the conductivity is subject to a strong meteorologically conditioned fluctuation, i.e. to the effects of the exchange variations, which results in a mostly opposing course of potential gradient and vertical current (*continental type*). In the other case, the conductivity remains nearly constant from a temporal standpoint, which causes parallel courses of potential gradient and vertical current, with an exchange that alters only little or not at all (*oceanic type*). In the first case, the local influences dominate, in the second case, it is the global influences. As a result of this, in the first case, the diurnal variations "*associated to the local time*" emerge to the front, in the other case, the diurnal variations are "*associated with global time*".

The Jungfraujoch analyses show that different types of variation dominate there during different seasons. Summer clearly demonstrates the continental type; during this time, the influence of the daily exchange variation reaches into the heights of the Alpine range. In contrast, during the fall and winter, the exchange variation no longer reaches this height; as result, the oceanic type applies there.

This concept derived from the results of the summer and fall measurement periods has been completely certified through the results of the winter measurements described above despite the significantly lower registration material of this period and, therefore, may be assumed as verified.

The diurnal variation of the *column resistance* for both measurement periods was also determined and presented in the first paper for the completion of this diagram (see Figure 10 in paper [1]). Figure 1e shows the same variable for the third measurement period.

While a laterally-reversed course of column resistance and conductivity resulted in the summer, the correlation between these two variables was much less developed already in the fall. It was completely absent in the winter as shown in Figure 1. While the conductivity remained practically constant, the column resistance still passes through a developed diurnal variation. Therefore, an assumption that was implied to its time [1, 5] prevails that a circadian, broad circulation over the massif manifests itself in the diurnal variation of the column resistance over the mountain range, which no longer comes from exchange variations – a new solid support.

If one individually compares the curves traits of the resistance courses, it becomes further evident that, aside from the decrease of amplitude from the summer course (depending on the exchange) to the fall and winter course (independent of the exchange), a displacement of the maximum occurs from 5 pm to 6 pm in the summer or from 1 pm to 2 pm in the fall as well as an additional displacement and simultaneous division into two maxima in the winter in the time periods from 8 am to 9 am and 6 pm to 7 pm. In the case of the relatively low amount of measurement material from the third measurement period, an attempt to interpret these determinations would be premature; if the findings could verified in a future analyses, then, as a result, an entirely new possibility for broad aerological observations would be enabled without a question.

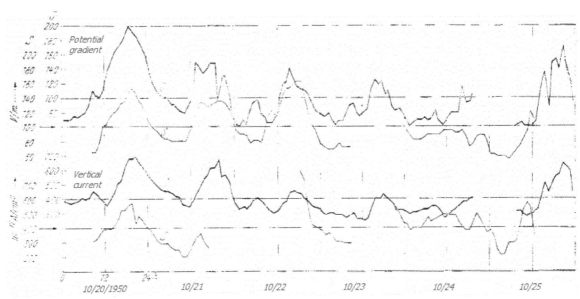

Figure 2. Parallel registrations of potential gradient and vertical current and at Jungfraujoch J and Sonnblick S in October 1950. Above: potential gradient, below: vertical current: thick curves: Jungfraujoch, thin curves: Sonnblick.

d) Comparative Registrations at Jungfraujoch and Sonnblick.

As indicated in the first paper, it formed one of our program points, namely that of achieving a verification of the achieved results for the analyses through parallel registrations at Jungfraujoch and Sonnblick (elevation 3100m) located approx. 400 km away in the Hohe Tauern mountains, and to simultaneously attempt a separation between regionally-bound and large-scale influences. The simultaneous analyses at two elevated stations were conducted during the second measurement period in the fall of 1950. The device at Sonnblick was the same as at Jungfraujoch; the collector and the grid were mounted to different walls of the house at Sonnblick. Unfortunately, registrations at Sonnblick were heavy challenged due to external difficulties so that we were able to attain only a relatively low amount of comparative material (five incomplete diurnal series).

In Table 3, the individual hourly averages of the potential gradient of the vertical current and the conductivity are compiled for the Sonnblick registrations and summarized in diurnal variations. The dimensions for the individual variables are the same as for the Jungfraujoch data. As a comparison, the averages of the Jungfraujoch registrations for the same days are respectively entered in the table. The results are graphically represented in Figure 2 and 3.

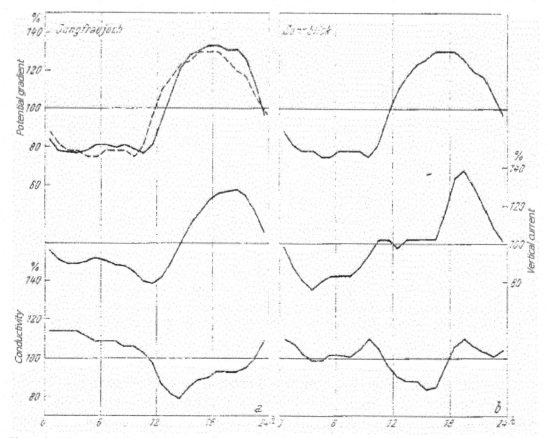

Figure 3. Diurnal variations of potential gradient, vertical current and conductivity after five-day simultaneous registrations at Jungfraujoch a and Sonnblick b in October 1950 (percentage depiction of the balanced diurnal series). As a better comparison, the hashed curve in Figure 3a above shows once again the Sonnblick curve of the potential gradient from Figure 3b above.

The result of the comparative registrations is still surprising despite the low volume of material. If a beneficial parallelism in the course of both elements emerges at both stations when ignoring the details apparently dependent on the region, then the conformity for the summarization of diurnal variations must therefore almost unexpectedly be identified as good.

The characteristic traits of the diurnal variations of potential gradient, vertical current and conductivity (parallel course of field and current, low conductive variations with a minimum in the early hours of the afternoon) are the same at both stations, Jungfraujoch and Sonnblick; therefore, they may certainly be viewed as typical for the Alpine ridge in the fall.

Table 3. *Hourly values of the atmospheric electrical potential gradient, of the vertical current and the conductivity at Sonnblick on five selected disturbance-free days or part of days in October 1950.*

Information is in the same units as the Jungfraujoch results. For comparison the average values of the registrations at Jungfraujoch in per cent is shown for the same days.

Potential Gradient

	am	am	am	am	am	am	am	am	am
October 20									65
October 21	104	98	94	88	84	84	81	81	81
October 22	94	91	91	88	78	81	81	78	94
October 23	75	72	68	68	72	78	72	72	72
October 24	94	88	81	81	81	85	88	88	88
October 25	62	62	59	59	59	55	62	68	75
Balanced average	88	81	78	78	75	75	78	78	78
" " in %	88	81	78	78	75	75	78	78	78
Jungfraujoch in %	84	78	77	77	78	81	81	80	81

Vertical Current

	am	am	am	am	am	am	am	am	am
October 20									306
October 21	394	350	306	292	292	292	263	248	248
October 22									
October 23	380	350	365	336	308	306	292	292	292
October 24	423	380	394	394	394	423	467	438	452
October 25	380	336	234	204	219	292	292	306	409
Balanced average	394	350	321	306	321	336	336	336	350
" " in %	98	87	80	76	80	83	83	83	87
Jungfraujoch in %	96	91	89	89	90	92	91	89	88

Conductivity

	am	am	am	am	am	am	am	am	am
October 20									26
October 21	21	20	18	18	19	19	18	17	17
October 22									
October 23	28	27	30	28	24	22	23	23	23
October 24	25	24	27	27	27	28	30	28	29
October 25	34	36	22	20	21	30	26	25	31
Balanced average	26	26	24	24	24	24	24	24	25
" " in %	110	108	102	99	99	102	102	101	105
Jungfraujoch in %	114	111	114	114	111	109	109	109	106

This result provides very important support to the result obtained at Jungfraujoch and guarantees its representative meaning for the Alpine ridge; because due to the fact that the conformity of the significant traits in the atmospheric electrical diagram proven in the fall may now be expected with particular certainty for the other seasons as well.

(This is the second part of Table 3 on the previous page)

9—10 am	10—11 am	11 am—12 pm	12—1 pm	1—2 pm	2—3 pm	3—4 pm	4—5 pm	5—6 pm	6—7 pm	7—8 pm	8—9 pm	9—10 pm	10—11 pm	11 pm—12 am	Average
65	78	101	104	114	120	123	133	136	149	152	149	136	130	117	
81	81	91	107	120	133	133	127	130	133	136	136	133	123	114	
114	123	117	123	143	153	153	159	162	162	149	127	104	97	84	
88	91	94	94	94	91	91	94	88	72	75	85	88	88	81	
84	94	120	110												
75	81	97	110	117	123	126	130	130	130	126	120	117	107	97	100
75	81	97	110	117	123	126	130	130	130	126	120	117	107	97	
79	77	81	94	109	121	127	130	133	133	131	131	126	113	97	
321	365	394	394	394	408	452	525	525	555	569	440	482	452	408	
204	204	234	292	336	365	306	262								
									790	730	540	423	408	394	
												540	496	438	
467	481	467	482	482	482	438	394	380	380	423	496	453	408	365	
525	570	540	350												
280	408	408	394	408	408	408	408	467	540	555	525	481	438	408	403
94	102	102	98	102	102	102	102	116	134	138	130	120	109	102	
85	80	79	82	89	99	110	116	122	126	127	128	125	117	106	
28	26	22	21	19	19	21	22	22	21	21	20	20	20	20	
14	14	14	15	15	15	13	12								
									27	27	24	23	24	26	
												27	28	24	
30	30	28	29	29	30	28	23	24	30	32	33	29	26	25	
35	34	25	18												
26	25	23	21	21	21	20	20	23	25	26	25	25	24	25	24
110	105	96	90	88	88	84	85	95	106	110	106	103	101	104	
106	103	98	87	82	79	85	89	90	93	93	93	95	100	109	

c) Precipitation Charge and Field Direction in Rain and Storm Clouds.

The primary goal of the atmospheric electrical measurements conducted at Jungfraujoch was to obtain registrations of undisturbed days. Nonetheless, measurements were also taken in disturbed conditions, i.e. in rain and storm clouds, to the extent this was possible from a technical standpoint. Throughout the three measurement periods, a sufficiently large volume of material was accrued in this process, which contains genuinely interesting results regarding precipitation charge and field strength in rain and storm clouds. Prior to a compilation and interpretation of the results, however, we must conducts some observations regarding the extent to which the values measured with the field and vertical registering device used are real.

I. Criticism of the Measuring Device

As already stated in the description of the device [1, 6], the potential gradient was registered using a radioactive collector with a subsequently switched electrometer tube and recorder. For this purpose, as a comparative registration we used a field mill in the third measurement period, the reading of which was sent to an ink recorder with a rapid paper feed for fine resolution, by which, however, a sphere of the recorder was also simultaneously controlled. Through a comparison of the field strength curves recorded in this manner on the same register strips, the accuracy of the field reading of the collector was also verified during precipitation. Only in the case of very high fields did the reading of the collector lag behind that of the field meter due to the saturation. In the case of strong precipitation, it would be conceivable that this would cause the collector reading to be falsified. As demonstrated by a raw rough estimate and the experimental findings, however, this is not the case. If we assume the transfer resistance of the collector at around 19^9 Ohm, a current of 10^{-7} amps would be necessary for a voltage drop of 100 volts. Because the collector assumes voltages of 1000+ volts with rain and storm clouds, an error reading of 100 volts would be just barely acceptable. At this point we have to estimate whether such a voltage of 100 volts can be administered to the collector system through the largest possible precipitation current. We can gather from the research by Simpson [7] that precipitation currents occur up to approx. 10^{-12} amps/cm^2. If we apply the effective collection area of the collector and its handle with 200 cm^2, we can calculate the greatest possible precipitation supplied for 2 to 10^{-10} amps. We can see that the seemingly still acceptable current of 10^{-7} amps will not be reached even remotely as well. Therefore, we can ignore the influence of the precipitation current on the field registration without hesitation.

For the measurement of the precipitation current with the aid of the vertical current grid, it is necessary to apply significantly more limitations. Principally, the conductivity and convection current is always measured by the grid. However, we can obtain the approximate convection current as soon as it significantly dominates the conductive current.

The fact that this is almost always the case with precipitation is clearly indicated from the precipitation registrations. The conductive current only dominates and the influence of the convection current can hardly be noticed in the registration when the snowfall (flurries) or the rainfall (drizzle) is very weak. With a precipitation that is becoming somewhat stronger (snow 0, rain 0), the convection current increases so rapidly that the conductive current is almost completely covered. This can be immediately recognized by the fact that the sensitivity of the measurement sphere for the vertical current in contrast to that for the potential gradient must be set back greatly to still keep the registration deflection of the measuring device on the paper strip. A second piece of evidence for the domination of the convection current is the differing polarity between the field and current reading. In the case, the convection current must compensate the conductive current, while the remainder controls the current reading of the instruments. However, because the precipitation current is normally 10 to 100 times stronger than the conductive current, one can also expect a fairly accurate reading in this process.

As a next concern toward a real reading of the precipitation current, we must invoke the Lenard effect during rain and static electricity during snow. The fact that these effects –

should they even be applicable – are emphasized by the precipitation charge, shows the conditions that both polarities sometimes occur subsequently in a heavy exchange. If one, however, assumes, e.g. a Lenard effect as the cause of the charge of the electrical grid, then the rain would always have to produce a charge with the same sign. Similar considerations would have to be made for the snow. In this case, the static electricity would also be somewhat dependent on the velocity, with which snowflakes drift toward or past the grid. In the case of flurries, the grid may not be charged with static electricity, whereas, the charge would have to become continually greater with increasing wind. However, we were not able to observe such an effect throughout the numerous registrations. With the absence of wind, the snowflakes falling slowly yielded partially greater convection currents as snowflakes blown against the grid during storm-like winds. Furthermore, we would have to consider that rain dripping from the grid as well as lumps of snow falling from the grid divert a certain amount of influence charge away. This effect of the precipitation collector must be negligibly small, however, because it is known that precipitation collectors have to be operated with a higher impedance than, e.g. radioactive collectors. A bleeder resistance of 10^{10} ohms, as it was used for the measurement of voltage with the grid, would depress every effect of the precipitation collector to virtually 0. Because lumps of snow fall in a much greater timeframes than exist, e.g. in the case of drop sequences of the precipitation collector, we can ignore this source of error.

A precise experimental analysis regarding the extent, to which the registration of current during precipitation reflects the convection current with sufficient precision, should still be conducted. For the analyses at Jungfraujoch, however, we can assume with enough certainty that at least the polarity of the precipitation current was accurately reflected.

2. Results.

The measurement results are arranged in Table 4 according to the following aspects. Column 1 contains the date and column 2 the beginning and end of the event. The meteorological data is provided in column 3a and b with the conventional sign. Sleet, small hail, rain and storms are marked with the following symbol in 3a: $\triangleq$, $\triangle$, $\bigcirc$ and $\boxtimes$, while only snow $\ast$ is listed in 3b. The fourth column displays the polarity of the field and the current*, while the fifth column contains the sensitivity of the sphere to the field measurement and that of the sphere to the measurement of vertical current below that. In this connection, the given figure represents the factor by which the values are increased compared to the normal sensitivity.

Upon briefly reviewing the table, we can immediately see that snow or blowing snow and snow drifts with a positive field and a positive current only occur in the fall and winter months of November, February and March. The only exception to this is established by the registration on November 9, 1950, in which a downpour of small hail with a negative field and current was registered. This abruptly leads us to the thought of investigating the polarity of the just the cases of small hail during the summer months. In this connection, we

---

*The "normal" course, i.e. field lines directed upward with a positive current supply to the surface of the earth, is characterized with + and vice versa with -.

notice that of the nine cases of small hail alone, there were seven cases with a negative field and current and two with an alternating positive and negative field and current. This suggests that small hail is negatively charged and produces a negative field. The deviations thereof can be explained through observation of the remaining fields of precipitation. A change of polarity of the field and current can be frequently observed if small hail turns into snowfall and vice versa. Based on this we can presume that we simply missed observing this transition into snow with regard to just the cases of small hail on July 13, 1950 between 3:50 and 5:30 pm and on July 20, 1950 from 3:40 to 7:30 pm. This becomes that much more probable as it was not even our intention at that time of analyzing the disturbed cases of the atmospheric electrical field, thus the information regarding the "small hail shower at 3:50 pm on July 13, 1950" should be more of an indication of the general character of the disturbed atmospheric electrical state and not be viewed as proof that this small hail shower that lasted until 5:30 pm consisted of small hail throughout. The same applies for the registration on July 20, 1950.

Furthermore, it is interesting that the note "wet snow" was added with the registration on July 28, 1950 during the snowfall, through which the negative polarity uncommon for snow would be explained in this case. All the cases of snow during the winter months could have only been dry snow. Therefore, it is possible to conclude that the snow with a negative polarity in the summer months was a wet snow.

The following conclusion, for example, can be made from all of these registrations: dry snow, i.e. snow as a pure sublimation product always carries a positive charge and produces a positive field. Now, if condensation sets in, then the transition of the snow to a wet snow, sleet, white frost or frost pellets is provided depending on the strength of the condensation, in which the precipitation now assumes a negative charge and produces a negative field. The change of polarity of the precipitation charge and the field indicates that a change has occurred from a dry to a wet snow for the respective registration (this also includes sleet, white frost or frost pellets). It is particularly noteworthy at this point that a storm was most frequently observed when a sequence of sleet and snow or rain and snow was noted. A typical example of a registration of just small hail, dry snowfall and a mixture of snowfall and small hail is reflected in Figure 4a – c.

In the case of snow drifts, the snow always consists of dry snow. Therefore, it is not surprising that a positive polarity of the field and convection current was always measured here as well. This correlates to the observations of Scholz in Franz-Josephs-Land [7]. The counter-polarity of precipitation charge and field primarily observed by J. Küttner was limited in these observations to temporary cases. For example, if the polarity of the precipitation during a small hail shower was negative, then it could occur that the field, which generally had a negative sign as well, could transition to positive values for approx. one half hour and then later return to negative values.

In conclusion, two registrations in Figures 4d and 4e reflected two strong cumulus clouds at the Jungfraujoch summit.

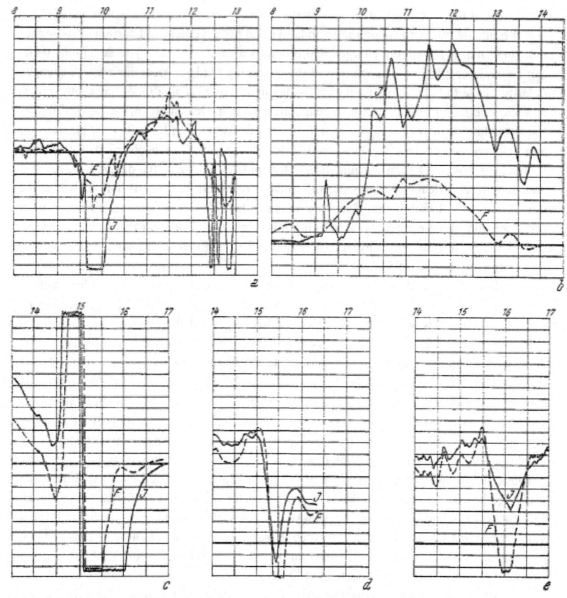

Figure 4.   Registrations of the atmospheric electrical potential gradient *F* and the vertical current *J* at Jungfraujoch. *a* Pure case of small hail on June 14, 1950 from 9:30 to 10:30 am and 12:15 to 1 pm. *b* Dry snowfall on June 1, 1950 from 10 am to 12:30 pm. *c* Mixed snowfall/small hail on June 17, 1950 - 2:30 to 3:05 pm snow and 3:05 to 3:30 pm small hail. *d* Stationary Cu at the Jungfraujoch summit on June 6, 1950 – 3:10 to 3:45 pm. *e* Stationary Cu at the Jungfraujoch summit on June 18, 1950 – 3:30 to 4:30 pm.

Table 4. *Polarity of the field and precipitation current in storm and rain clouds at Jungfraujoch.*

(The following symbols were used for the meteorological observation: snow*, sleet △, small hail △, storms 𝕂, rain O, blowing snow * ↑→, snow drifts * ↓→, snowstorm * St, rainfall ▽.)

Column 1: Date; Column 2: Beginning and End; Column 3: Sleet, Small Hail, Rain, Rainfall, Storms; Column 5: Sensitivity for the field above / for precipitation current below. The given figure represents the factor by which the values are increased compared to the normal sensitivity.

1	2		3 a	3 b	4	5
May 23, 1950	14³⁰	15⁰⁰	△	✻	— / +	3 / 4
May 23, 1950	19⁰⁰	23⁰⁰		✻	+	1 / 9
May 26, 1950	14³⁰	16⁰⁰	△	✻	— / —	3 / 70
May 26, 1950	21⁰⁰	23⁰⁰	△	✻	— —+ / —++	1 / 4
May 27, 1950	0⁰⁰	1³⁰		✻ ↑→	+ / +	3 / 9
May 31, 1950	16⁵⁰	17⁰⁰		✻ ⁰	+ / +	1 / 9
June 1, 1950	10⁰⁰	12³⁰		✻ ⁰	+ / +	1 / 9
June 1, 1950	13⁰⁰	16³⁰	△ △		— / —	1 / 9
June 2, 1950	14³⁰	17⁰⁰		✻	—	1 / 9
June 3, 1950	13⁰⁰	15³⁰	△ ⁰		·	1 / 9
June 12, 1950	12¹⁵	15⁰⁰	△ 𝕂	✻	—+—+— / —+—+—	17 / 70
June 13, 1950	11³⁰	12²⁵		✻	+ / +	17 / 70
June 13, 1950	15⁰⁰	15¹⁰		✻	+ / +	3 / 9
June 13, 1950	16¹⁵	16³⁰	△	✻	+— / +—	17 / 9
June 14, 1950	9³⁰	10³⁰	△		—	1 / 9
June 14, 1950	12³⁰	13⁰⁰	△ △		— / —	1 / 9
June 15, 1950	13⁴⁵	15¹⁰		✻ ⁰ — ✻ ¹	—	17 / 70
June 15, 1950	18¹⁹	20⁰⁰		✻ ↑→	— / —	3 / 4
June 16, 1950	6³⁰	6⁴⁵	△		— / —	3 / 9
June 16, 1950	8⁴⁵	10⁰⁰	△	✻	—	1 / 4
June 16, 1950	13²⁰	14⁰⁰	△		—	1 / 4
June 16, 1950	15⁵⁰	17³⁰	△	✻	+·+ +—+ / +—+—+	3 / 4
June 17, 1950	7⁵⁰	8³⁰	△		— / —	1 / 1
June 17, 1950	8³⁰	8⁴⁰	△		—	1 / 1
June 17, 1950	9¹⁵	11⁰⁰	△ 𝕂		—+—+ / —+—+	3 / 70

(Continuation see p.)

43

H. Israël, H. W. Kasemir and K. Wienert: Atmospheric Electrical Diurnal Variations

(Continuation from Table 4)

1	2		3 a	3 b	4	5
June 17, 1950	14^{30}	16^{10}	$\wedge$	✳		3 4
June 19, 1950	18^{30}	19^{30}	$\wedge$ ⬚	✳		1 1
June 25, 1950	13^{00}	15^{00}	$\triangle$ ⬚			1
July 3, 1950	12^{30}	13^{30}	$\wedge$ ⬚			1
July 10, 1950	8^{50}	10^{30}	$\wedge \vee$ ⬚			1 1
July 10, 1950	14^{30}	15^{30}	$\wedge \vee$ ⬚			1
July 13, 1950	15^{50}	17^{30}	$\wedge$			1
July 14, 1950	12^{50}	15^{00}	$\bigcirc \wedge$ ⬚			1
July 20, 1950	2^{50}	4^{15}	$\wedge$			1
July 20, 1950	15^{40}	19^{30}	$\triangle$			1
July 22, 1950	6^{20}	7^{00}		✳		1
July 22, 1950	15^{40}	24^{00}	$\bigcirc \triangle$ ⬚	✳		1
July 26, 1950	13^{00}	18^{00}	$\bigcirc$	✳		1
July 28, 1950	12^{00}	12^{20}	$\wedge \triangledown$			1
July 28, 1950	16^{15}	20^{00}		✳ wet		1
August 1, 1950	0^{45}	6^{00}	$\triangle$			1
October 9, 1950	6^{50}	7^{10}		✳		1
October 11, 1950	0^{00}	15^{00}		✳		1
October 12, 1950	6^{50}	7^{30}		✳		1
October 16, 1950	14^{00}	23^{00}	$\triangle$	✳		1
October 17, 1950	4^{00}	9^{00}		✳		1
October 26, 1950	10^{00}	24^{00}		✳		1
October 27, 1950	0^{00}	17^{00}		✳		3 70
October 28, 1950	2^{30}	9^{30}		✳		3 9
November 2, 1950	15^{20}	21^{30}		✳		1 9
November 3, 1950	0^{00}	24^{00}		✳ ✳		1 1
November 4, 1950	0^{00}	20^{00}		✳		1 70
November 7, 1950	16^{30}	24^{00}		✳ .Chryst.		1 1

(Continuation on p. 92)

H. Israël, H. W. Kasemir and K. Wienert: Atmospheric Electrical Diurnal Variations

(Continuation from Table 4)

1	2		3		4	5
			a	b		
November 8, 1950	0^{00}	16^{00}		⊛ ⊛ $\uparrow\rightarrow$	+ / +	1 / 70
November 9, 1950	23^{00}	24^{00}	∧		-- / --	1 / 1
November 11, 1950	7^{30}	24^{00}		⊛ St	+ / + .	1 / 9
November 12, 1950	0^{00}	24^{00}		⊛ St	+ / +	1 / 9
November 13, 1950	0^{00}	24^{00}		⊛ St	+ / +	1 / 9
November 14, 1950	0^{00}	24^{00}		⊛	-+ / +	1 / 9
November 15, 1950	0^{00}	24^{00}		⊛	+ / +	1 / 9
November 19, 1950	0^{00}	24^{00}		⊛ $\downarrow\rightarrow$ ⊛	-+ / -+	1 / 9
November 21, 1950 ., ...	0^{00}	16^{00}		⊛ $\downarrow\rightarrow$ ⊛	-+ --- -+ / + -+ +	1 / 4
February 24, 1950	5^{00}	17^{00}		⊛	+ / +	3 / 70
March 1, 1951	16^{00}	21^{00}		⊛ $\downarrow\rightarrow$	-+ / -+	1 / 1
March 2, 1951	11^{00}	17^{00}		⊛ $\downarrow\rightarrow$	+ / -+	3 / 9
March 11, 1951	5^{00}	24^{00}		⊛ St	-+ /	1 / 9
March 12, 1951	12^{00}	14^{00}		⊛ $\downarrow\rightarrow$	+ / -+	1 / 1
March 15, 1951	19^{30}	22^{00}		⊛ St	-+ / +	1 / 9

No precipitation from the clouds was observed. Nonetheless, the polarity of the field and current (purely conductive current in this case) demonstrates that the negative charge in the lower portion of the cloud had to have been measured. The regularity of these phenomena was evident. These cumulus formed on every warm summer day in the afternoon hours between 2 and 3 pm and normally dispersed, e.g. from 4:30 to 5 pm. In all of these cases, the field and current registrations, as they are found from Figure 4*d* and 4*e*, were extraordinarily similar.

In conclusion, at this time we would like to once again extent our most sincere gratitude to our colleagues and to the institutes in Switzerland, whose gracious support allowed us to conduct the research at Jungfraujoch. In the same manner, our colleagues in Austria as well as the Central Institute for Meteorology and the Aviation Weather Service in Salzburg, whose continual kind support enabled the parallel analyses at Sonnblick. Furthermore, the former Ministry of Education and Arts from Württemberg-Hohenzollern in Tübingen provided financial support for the acquisition of equipment and the German Research Community contributed significantly by paying for a calculation tool for the success of the research. We thank you once again as well!

H. Israël, H. W. Kasemir and K. Wienert: Atmospheric Electrical Diurnal
Variations

Bibliography.

1. ISRAËL, H., H. W. KASEMIR und K. WIENERT: Luftelektrische Tagesgänge und Massenaustausch im Hochgebirge der Alpen. Die luftelektrischen
Verhältnisse am Jungfraujoch (3470 m) I. Arch. Met. Geoph. Biokl. A 3,
357—381 (1951).

2. ISRAËL, H.: Zur Entwicklung der luftelektrischen Grundanschauungen.
Arch. Met. Geoph. Biokl. A 3, 1—16 (1950).

3. — Zum Tagesgang des luftelektrischen Potentialgefälles II. Potentialgefälle und Dampfdruck. Ber. d. Dtsch. Wetterd. US-Zone 6, Nr. 38,
409—411 (1952).

4. — The atmospheric electric field and its meteorological causes. "Thunderstorm electricity" edited by Horace R. Byers, S. 4—23. Chicago 1953.

5. — Luftelektrizität und Meteorologie. Ber. d. Dtsch. Wetterd. US-Zone 5,
Nr. 35, 217—223 (1952).

6. KASEMIR, H. W.: An apparatus for simultaneous registration of potential
gradient and air-earth current. J. Atm. Terr. Phys. 2, 32—37 (1951).

7. SCHOLZ, J.: Luftelektrische Messungen auf Franz-Josephs-Land während
des II. internationalen Polarjahres 1932—33. Transact. Arctic Inst. Leningrad 1935.

8. KÜTTNER, J.: Ist die Wolkenluft oder der Niederschlag Träger der Haupt-
ladungen im Gewitter? Meteor. Rundsch. 3, 145—147 (1950).

Atmospheric Electric Measurements in the
Arctic and Antarctic

by

HEINZ W. KASEMIR

Reprint from the Review
PURE AND APPLIED GEOPHYSICS (PAGEOPH)
Formerly 'Geofisica pura e applicata'

Vol. 100, 1972/VIII   BIRKHÄUSER VERLAG BASEL   Pages 70–80

Atmospheric Electric Measurements in the Arctic and Antarctic

By Heinz W. Kasemir[1])

Summary – During the International Geophysical Year, 1958, and extending into 1959, the atmospheric electric field, current, and conductivity were recorded at Thule, Greenland (78°N). During the International Year of the Quiet Sun, 1964, records of the atmospheric electric field were obtained at the Amundsen-Scott Station at the South Pole (90°S). The diurnal variation averaged over the year of the normalized current at Thule and the normalized field at the South Pole show a surprisingly good agreement. These two curves combined into one represent the world time variation of the air-earth current (or field) in the Polar regions. Compared with the oceanic diurnal field variation obtained at the Carnegie ship cruises, the Polar curve shows a very similar shape but a much reduced amplitude. The maximum and minimum in the Polar regions are 1.07 and 0.92. The corresponding values on the oceans are 1.20 and 0.85. The difference is greater than the measuring error or statistical scatter and has to be accepted as real. No conclusive explanation of the deviation of the two curves can be offered.

The diurnal variation of the Polar data averaged over a season displays very smooth and similar curves during Northern autumn and winter. The spring and summer curves show a much more detailed structure with several maxima and minima. It is somewhat unexpected that the summer curve with a variety of fine structure is the flattest curve of all seasons. The minimum never drops below 0.95, and the maximum does not exceed 1.06. If the data are broken down into hourly means averaged over one month and split into an Arctic and Antarctic part, the similarity between corresponding curves of the same month vanishes for the months of January to July. This may partly be due to the fact that the number of fair-weather days of the individual month is too small to obtain a representative statistical average. Usually averaging over seven or more days is necessary for the oceanic pattern to emerge. However, there is a strong possibility that another agent besides the worldwide thunderstorm activity modulates the global circuit. The seasonal differences, and especially the difference between Arctic and Antarctic pattern, point to such a conclusion.

Zusammenfassung – Während des internationalen geophysikalischen Jahres (IGY) 1958 und bis in das Jahr 1959 hinein wurden Registrierungen des luftelektrischen Feldes, des Vertikalstromes und der Leitfähigkeit durchgeführt in Thule, Grönland (78°N). Während des internationalen Jahres der Ruhigen Sonne (IQSY) 1964 wurde das luftelektrische Feld an der Amundsen-Scott Station am Südpol (90°S) registriert. Die normalisierte Tagesvariation des Stromes, gemittelt über das Jahr 1958, in Thule, und die normalisierte Tagesvariation des Feldes am Südpol, gemittelt über das Jahr 1964, zeigen eine überraschend gute Übereinstimmung. Diese zwei Tagesgänge sind zu einem gemittelten Tagesgang zusammengefasst, der den weltzeitlichen Tagesgang des Stromes oder des Feldes in den polaren Regionen repräsentiert. Im Vergleich zu dem Tagesgang des Feldes auf den Ozeanen, wie er während der Carnegie-Fahrten bestimmt wurde, zeigt der Tagesgang in polaren Gebieten einen sehr ähnlichen Verlauf, hat aber eine viel kleinere Amplitude. Die Werte für das Tagesmaximum und Minimum in polaren Gebieten sind 1.07 und 0.92. Die entsprechenden Werte auf dem Ozean sind 1.20 und 0.85. Der Unterschied ist so gross, dass er nicht durch Messungenauigkeit oder statistische

[1]) Dr. H. W. Kasemir, NOAA, APCL, R 31, Boulder, CO, 80302, USA.

Electric Measurements in Arctic and Antarctic

Streuung hätte hervorgerufen werden können. Er muss deshalb als real akzeptiert werden. Eine Erklärung für diesen Unterschied konnte nicht gefunden werden.

Der Tagesgang in polaren Gebieten gemittelt über die verschiedenen Jahreszeiten zeigt für die nördlichen Herbst und Wintermonate sehr glatte und ähnliche Kurven. Die Frühlings- und Sommer- kurven haben eine mehr detaillierte Struktur mit mehreren Maxima und Minima. Es ist etwas überraschend, dass die Sommerkurve mit einer grossen Variation in der Feinstruktur die flacheste Kurve von allen Jahreszeiten ist. Die Minima sind niemals kleiner als 0.95 und die Maxima über- schreiten nicht den Wert 1.06. Wenn die Daten weiter unterteilt werden in Tagesgänge gemittelt über Monate, dann verschwindet die Ähnlichkeit zwischen arktischen und antarktischen Gängen desselben Monates für die Monate Januar bis Juli. Das mag teilweise darauf zurückzuführen sein, dass die Anzahl der Schönwettertage für die einzelnen Monate zu klein ist, um statistisch repräsentativ zu sein. Eine Mittelung über mindestens 7 Tage ist notwendig, damit der weltweite Tagesgang zum Vorschein kommt. Es ist aber auch sehr gut möglich, dass andere Einflüsse als die weltweite Gewit- tertätigkeit den Tagesgang modulieren. Unterschiede im Tagesgang der Jahreszeiten und auch des vollen Jahres legen eine solche Erklärung nahe.

1. Introduction

It was one of the main achievements of the last half century in the field of atmos- pheric electricity that the worldwide thunderstorm activity was recognized as the driv- ing generator of the global atmospheric electric current. (For the history of this devel- opment see ISRAËL [3, 4]). The proof of this theory was based mainly on the close cor- relation of the average diurnal variation of the world thunderstorm activity and the average diurnal variation of the atmospheric electric field on the oceans (WHIPPLE and SCRASE [11][2]). Measurements supporting this theory were carried out by several scientists of the Carnegie Institute during a number of cruises from 1915 to 1929 (MAUCHLEY, [8]; PARKINSON and TORRESON, 1931). Since this time, the pattern of the average diurnal variation of the field on the oceans has been the yardstick for atmos- pheric electric measurements, which claim to reflect the global atmospheric electrical current flow. Even field or current records of a single day or sporadic balloon or air- plane soundings have been compared with the oceanic pattern and usually they agree with it. It may be well to remember that the oceanic field variation is a statistical average of about 130 days spread over several years. The chance to obtain this statistical curve on a single day or as the average of a few days is small. Under these circumstances, one would expect a number of reports that the oceanic field variation has not been found. The lack of such publications shows probably more a bias in selecting the material as a true state of affairs.

Furthermore, we should keep in mind that the world time variation of the field is obtained over the oceans. It does not follow that the same variation has to appear in the Polar areas or at other places which are considered to reflect the global pattern. We will see later in this report that the average field or current variation in the Arctic and Antarctic shows similarities as well as deviations from the oceanic picture.

In general, atmospheric field or current measurements on the continents do not follow the global pattern. This indicates an overriding local influence. The driving

[2]) Numbers in brackets refer to References, page 80.

49

force of the local effect is the austausch (turbulence) in the lower layer of the atmosphere. The effect on the atmospheric electric circuit is twofold. Positive space charge is carried by the austausch to higher altitudes against the force of the existing electric field. This constitutes the generator effect of the austausch (KASEMIR [6]). Aitken nuclei are distributed throughout the austausch region which lower the conductivity and increase the columnar resistance in the affected area. The austausch generator will affect the atmospheric electric field and current locally, i.e., in the region where the austausch occurs. Its global component is probably negligible. The same is not true for the variation of the columnar resistance, if larger areas – for instance, continents – are affected.

The increase of the columnar resistance over continents facing the sun would deflect a part of the air-earth current to the night side of the globe or to regions where the columnar resistance is constant during the whole day, for instance, the oceans or the polar regions. The rotation of the earth combined with this resistance modulation would produce a similar variation of the field and current in areas with constant columnar resistance as the worldwide thunderstorm activity (KASEMIR [5]). This indicates the need to have the thunderstorm activity determined by an independent parameter, for instance, lightning count in the extremely low frequency range (ELF) and to work out a detailed analysis of the columnar resistance variation.

If we assume that the austausch phenomena is the controlling factor of the local influence on the atmospheric electric parameters, it is easy to specify the requirements of a location at which global variations may be recorded. The austausch itself should be absent or at least the acting agents, space charge and Aitken nuclei should be absent or very much reduced. These conditions are almost ideally fulfilled at the poles. For this reason, an atmospheric electric station was set up at Thule, Greenland, 78°N, during the International Geophysical Year (IGY), 1958, and a field recorder at the South Pole Amundsen-Scott Station during the International Year of the Quiet Sun (IQSY), 1964.

The instruments and the results of the measurements of these two stations will be discussed in this report.

2. Atmospheric electric instruments used at Thule, Greenland, and Amundsen-Scott Station at the South Pole

The instruments used at Thule, Greenland, have been described in detail by J. KRIEG [7]. The measuring system consists of an antenna array for field, current, positive and negative conductivity feeding into a four-channel input switch which is connected to a picoamperemeter. The picoamperemeter acts as an electrometer amplifier and its output is connected to a four-channel potentiometric recorder. Input switch and recorder channel are switched synchronously so that the four atmospheric electric parameters are recorded in cyclic rotation with about 4-sec time interval from one measurement to the next. To have the sensors of the four atmospheric electric parameters as similar and comparable as possible, horizontal wire antenna were used throughout. The field antenna was about 10 m long and 1.5 m above the ground with a radioactive probe attached to its midpoint. The load resistor was 10^{12} ohm. The current antenna was of about the same height and length as the field antenna. The time constant of the input impedance was 600 sec. For the conductivity measurement an antenna arrangement was used similar to that of SCHERING [10]. It is, in essence, a large Gerdien tube with natural ventilation. The inner electrode is again a horizontal wire of 10 m length, which is connected to the input switch of the picoamperemeter. The outer electrode with a driving voltage of ± 250 volt is not a solid tube as by the Gerdien apparatus, but consists of large numbers of wire antennas symmetrical surrounding the central wire in a cylindrical fashion.

The system operated very satisfactorily over the whole year with a minimum of maintenance. Snowstorms, even blowing snow, can be recognized from the records and separated from the fair weather records. However, there are two error sources which could easily go undetected. These are the change of the ground level by snow cover and the icing up of the antenna and antenna insulators. Especially during the long polar night, these events may be overlooked.

The best method to detect – even afterwards – a malfunction of the instrument or to determine the so-called fairweather condition is the fulfillment of Ohm's Law. For the last decades it has been an open question how to define in an atmospheric electric sense fair weather and disturbed weather. The main characteristic of disturbed weather is precipitation and the strong and varying fields accompanying it. From the electrical point of view, the essential fact would be the convection current represented by the charged precipitation. It may be worthwhile to consider to base the separation of the two main electric conditions of the atmosphere on the presence or absence of convection current. The first state would indicate that the station is inside a generator or – expressed in the more meteorological terminology – in disturbed weather. In this case, Ohm's Law would not be fulfilled. In the second case, the station would be outside a generator or in fair weather. Ohm's Law should be fulfilled.

This definition of fair weather conditions has been applied to the Thule data and all times have been eliminated during which Ohm's Law was not fulfilled inside an accuracy of ± 15 percent. The term, 'Ohm's Law,' is used here in a somewhat modified sense. The relation between the air-earth current density i, the atmospheric electric field E, and the positive and negative conductivity λ_+ and λ_- defined in the textbooks as Ohm's Law is the following:

$$i = E(\lambda_+ + \lambda_-). \qquad (1)$$

However, after careful calibration of all atmospheric electric channels in Thule, the relation was obtained

$$i = E(\lambda_+ + \lambda_-)/2. \qquad (2)$$

This is probably due to the fact that the slightest air movement prevents the establishment of an electrode effect on the wire antenna. As no negative ions can emerge from the surface of the wire, the electrode effect would be a necessity to enable the positive ions to carry the full current. Without electrode effect the antenna will receive only about half of the current which is then given by

$$i = E \, \lambda_+ . \tag{3}$$

If the positive and negative conductivity is almost equal as was the case in Thule, Greenland, equation (2) can be reduced to equation (3). It is quite possible that this reasoning will apply to all current sensors built from nonporous solid material, for instance, the Wilson plate. This would explain the difference between theory and current measurement sometimes noted in the literature.

The selection principle for fair weather based on the fulfillment of Ohm's Law resulted in a reliable and homogeneous material extracted from the Thule records. Another advantage is that the selection can be done by a computer. As soon as the percentage of the permitted deviation from Ohm's Law is given, the simple yes or no answer for automatic selection can be programmed. The application of this method to the evaluation of the material compiled during the atmospheric electric ten year program can be highly recommended.

As could be expected from theoretical considerations and the experience of others (HOLZER [2], REITER [9], COBB [1]), the air-earth current is not as much influenced by local conductivity changes as the field. For this reason, the current records are used in the following discussion. At the South Pole, only a field recorder was set up, but – as will be seen later – the close agreement between the average current variation at Thule and the average field variation at the South Pole may justify the comparison between these two different atmospheric electric parameters. Furthermore, it can be assumed that at the Pole the diurnal variation of the austausch has completely vanished. Therefore, the conductivity would be constant or at least would not show any diurnal variation. In such a case field and current records would be equivalent.

The field at the South Pole was measured with a quadrant electrometer recorder. Each of the quadrant pairs was connected to a horizontal wire antenna with a radio-active probe at its midpoint. The vertical distance between the two antennas was 1 m with the lower antenna about 1 m above ground. Measured is the voltage difference between the two wires. This set-up has several advantages as compared with the single horizontal wire. If the field is constant with altitude in the first 2 m above ground level, difficulties in obtaining a good ground would be avoided. The circuit would act like one with a balanced floating input. If there is an electrode effect in the lowest meter above ground, the measurement would not be affected by it. The instrument itself has no electronics and the chart drive is powered by a hand-wound spring. This makes the whole system independent of electric power and its voltage and frequency changes. On the other hand, the recorder is a delicate instrument, needs careful handling, and cannot compete in accuracy, linearity, and overall roughness with the more commonly used potentiometric recorder.

3. Evaluated field and current records from the South Pole and Thule, Greenland

The discussion in this report is limited to the diurnal variation averaged over a month, season, and year of fair weather days extracted from the field records at the South Pole and the current records at Thule, Greenland. To compare the two different atmospheric electric parameters, they are represented as the deviation from the mean value, i.e., the hourly absolute value has been divided by the mean value of the month, season, or year, respectively. Only full days were accepted for evaluation with very few exceptions. If, for instance, the zero check was recorded for almost one hour on an otherwise perfect day, the beginning and the end of the break in the record was connected by a straight line and the hourly mean estimated using the so completed record. Sometimes, even on a single day, the record was similar to the oceanic pattern, but more often the shape of the curve was dissimilar to it. Usually it requires 7 to 10 day averaging for the oceanic pattern to emerge. At the South Pole, fair weather records were obtained on 105 days during 1964 with the month of December completely missing. At Thule, Greenland, 190 days of fair weather records were obtained with the month of November missing, there the records started on May 1958 and ended April 1959.

Figure 1 shows the normalized diurnal variation averaged over the full year, with the solid line representing the air-earth current density at Thule and the dashed line the electric field at the South Pole. The time is GMT or world time. It is remarkable that from 1000 to 2400 the curves are practically the same. The difference never exceeds one percent. In the forenoon, the South Pole curve shows a more pronounced dip with a minimum of 0.88 at 0230 in the morning. It recovers then faster than the Thule curve and passes through the overall average value 1.00 at about 0715. The

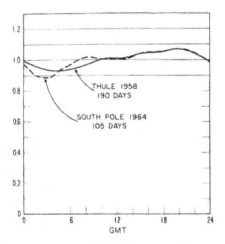

Figure 1

Normalized diurnal variation of the air-earth current density at Thule, Greenland, 1958 (solid line).
Normalized diurnal variation of the atmospheric electric field at the South Pole, 1964 (dashed line)

Thule curve has a less pronounced dip with a minimum of 0.93 at about 0400 and a crossing of the 1.00 line at 0920. The maximum difference between the two curves does not exceed five percent. Several explanations can be offered for the deviation of the two curves in the forenoon. The two stations were located almost on antipode points on the globe. One set of data was obtained in 1958, the year with maximum sunspot activity, the other set, 1964, during the year of the quiet sun. There is also still the possibility that the deviation is a result of comparing two different atmospheric electric parameters or that the superior selection methods and the much larger number of the Thule data is the cause of the deviation. However, the difference of five percent is relatively small, and it may be justified to combine the Thule and South Pole data according to their weight in days to one single curve representing the world time variations in the Polar regions.

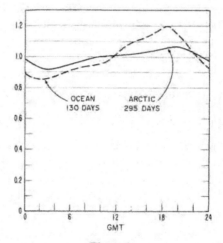

Figure 2
Normalized diurnal variation of the air-earth current density (field) in the Arctic and Antarctic, 1958 and 1964 (solid line). Normalized diurnal variation of the atmospheric electric field on the oceans (WHIPPLE and SCRASE [11]) (dashed line)

This curve is shown in Figure 2 as a solid line and compared to the normalized oceanic variation after Whipple and SCRASE [11], drawn as a dashed line. The similarity in shape is quite apparent. Both curves have their minimum at about 0330, crossing the 1.00 line at 0930 and 1130, respectively, and have their maxima at 1830 to 2030. The dissimilarity lies in the smooth appearance and the remarkably smaller swing of the Polar curve. The oceanic maximum and minimum is 1.20 and 0.85, whereas the Polar values are 1.07 and 0.92. The maximum difference between the two curves occurs at 1830, and it is 14 percent. This is a rather large deviation and its significance is emphasized by the fact that the two Polar curves differ less than one percent from each other in the afternoon, and only five percent in the forenoon. It

Vol. 100, 1972/VIII)   Electric Measurements in Arctic and Antarctic

is not too likely that the material being from different years will have caused the deviation. The two Polar curves, figure 1, are also from different years but agree much better with each other. An explanation based on the fact that one set of data was obtained on the ocean and the other set in the Polar regions would imply a damping effect there. However, such an effect is hard to visualize under the assumption that the earth is an equipotential surface.

Statistical results always show an inherent scatter. But here again the close agreement between the Polar curves does not point to a solution in that direction. Nevertheless, a more detailed statistical analysis is planned to determine the probability that the deviation is inside or outside a reasonable confidence level.

Again the question may be asked, could the difference be due to the fact that two different atmospheric electric parameters, field and current, are compared? The field at Thule has a larger swing than the current due to the fact that the conductivity is

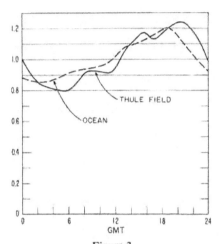

Figure 3
Normalized diurnal variation of the atmospheric electric field at Thule, Greenland (solid line). Normalized diurnal variation of the atmospheric electric field on the oceans
(WHIPPLE and SCRASE [11]) (dashed line)

not exactly constant during the day. The normalized field curve of Thule is shown in Figure 3 as the solid line and in comparison the oceanic curve as a dashed line. The three peaks reflecting the maximum thunderstorm activity on the three main land masses are very pronounced in the Thule curve. For demonstrating the correlation between the thunderstorm generator and the global circuit, one cannot ask for a better example. Even so, the Thule curve is more modulated and overshoots the oceanic curve sometimes by 15 percent. The overall impression is that of a better fit then between the oceanic field and the Arctic field curve. Because of the larger standard deviation of the field as compared to the current, it is quite possible that the two field curves agree with each other inside a reasonable scatter band. The discussion of the somewhat controversial conclusion arising from such a result will be postponed until the statistical analysis is carried out.

Figure 4 shows the Polar variation for the four seasons. During the months August, September, October and during the months November, December, and January, the curves are very smooth and quite similar to each other. The three peaks which reflect the maximum thunderstorm activity on the three main continents are almost absent. The August-to-October curve is shaped very symmetrically. The minimum of 0.89 occurs at 0500 in the morning. The curve then increases with a constant slope, crosses the 1.00 line after eight hours at 1300, reaches its maximum at 2100 and

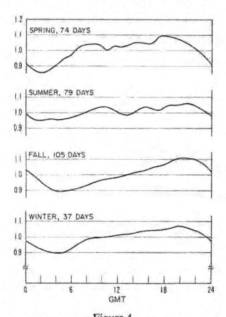

Figure 4

Normalized diurnal variation of the air-earth current density (field) in the Polar areas averaged over the four seasons. Spring: February to April; Summer: May to July; Fall: August to October; Winter: November to January

drops then steadily to the minimum with a 1.00 crossing at 0100. In the November-to-January curve, there is a very slight secondary maximum at about 0800 which probably reflects the thunderstorm activity in Australia. In the February-to-July curves, this peak becomes more pronounced and shifts to 1000. This could be interpreted by the fact that the thunderstorm activity in the forenoon of GMT is now produced by the Asiatic continent, which is larger than Australia and has its maximum thunderstorm activity about two hours later. The February-to-July curves lose the smooth appearance of the August-to-October curve and contain a number of maxima, which could reflect the peak thunderstorm activities of the different continents. However, a detailed analysis does not lead to a clear-cut picture. One gets the impression that besides the thunderstorm activity, another modulating influence exists. This influence is more pronounced in the February-to-July period than in that of August-to-January. The same conclusion can be drawn from Figure 5, which shows the diurnal variation averaged over each month on the left side of the current at Thule and on the right side of the field at the South Pole. From August to December, the curves of each station show a fair resemblance to the Polar pattern averaged over the year (Figure 2). From January to July, at least on one, more often on both stations, the similarity is either weak or lacking.

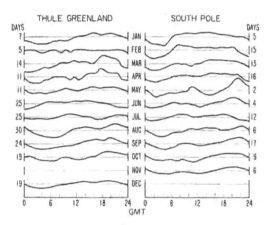

Figure 5

Normalized diurnal variation of the air-earth current density at Thule, Greenland, averaged over one month (left side) and of the atmospheric electric field at the South Pole averaged over one month (right side)

4. Conclusions

The similarity in the shape of the oceanic and the polar curves may be considered as a confirmation of the former. However, if we stop with this result the gain of new knowledge would be small because the validity of the oceanic field pattern is generally accepted. The more significant aspect, which is emphasized in this report, is the deviation between these two curves. The Polar curve is more flat with a much reduced maximum and minimum as compared with the oceanic curve. The close agreement between Arctic and Antarctic variation shows that the deviation from the oceanic pattern cannot be explained by the fact that the measurements were made in different years or that it is a mere statistical scatter. There seems to be a better fit of the oceanic field pattern with the Thule field pattern than there is with the current pattern. However, the conductivity in Thule was not constant during the day. Therefore, the atmospheric electric field there is not completely free from local influence. The ocean field pattern was obtained with constant conductivity; therefore, the better agreement between the two field curves is somewhat mysterious. A final solution to this problem can only be obtained by future measurement carried out simultaneously in the Polar

Heinz W. Kasemir

areas and on the oceans. It would be necessary to measure all three basic atmospheric electric elements, field, current, and conductivity, and to apply an objective selection principle of the so-called fair-weather days. The fulfillment of Ohm's Law inside a certain error limit is suggested for the determination of fair-weather records. The worldwide thunderstorm activity should be obtained by separate means, for instance ELF or VLF recording, again preferable in the Polar regions. Most of these suggestions have also been proposed by other scientists. The international atmospheric electric Ten-Year Program would be an excellent opportunity to realize them.

5. Acknowledgement

The author wishes to express his sincere thanks to persons and institutions who made it possible to carry out the reported research. These are Dr. LOTHAR RUHNKE who helped to set up and kept operating the atmospheric electric station at Thule, Greenland, and the U.S. Army Radio Propagation Agency, which housed and maintained this equipment at their Ionospheric Station at Thule, Greenland. The field measurements at the South Pole were carried out by Mr. SCHROEDER at the Amundsen-Scott Station, which was maintained by the Overseas Operations Division of the National Oceanic and Atmospheric Administration and the National Science Foundation.

REFERENCES

[1] W. E. COBB, *The atmospheric electric climate at Mauna Loa, Hawaii*. Presented at AMS-AGU Meeting, Washington, D.C. (1967). (In print).

[2] R. E. HOLZER, *Studies of the universal aspect of atmospheric electricity*. Contract No. AF 19(122) –254 (Final Report 1955).

[3] H. ISRAËL, *Atmosphärische Elektrizität*, Teil I: *Grundlagen, Ionen, Leitfähigkeit* (1957); *Atmosphärische Elektrizität*, Teil II: *Felder, Ladungen, Ströme* (1961). Akademische Verlagsgesellschaft Geest U. Portig KG. Leipzig.

[4] H. ISRAËL, *Atmospheric Electricity*, Volume I: *Fundamentals, Ions, Conductivity*. Israel Program for Scientific Translations, Jerusalem (1970).

[5] H. W. KASEMIR, *Zur Strömungstheorie des luftelektrischen Feldes I*. Arch. Meteor., Geophys., Biokl (A) *3* (1950), 84–97.

[6] H. W. KASEMIR, *Der Austauschgenerator*. Arch. Meteor., Geophys., Biokl. (A) *9* (1956), 357–370.

[7] J. J. KRIEG, *Measurement of electric field, air-earth current, and conductivity*. USASRDL Tech. Report 2085 (1959).

[8] S. J. MAUCHLY, *Studies in atmospheric electricity based on observations made on the Carnegie, 1915–1921*. Res. Rep. Terr. Magn. Carnegie Institution V, 383–424 (1926).

[9] R. REITER, *Felder, Ströme und Aerosole in der unteren Troposphäre*. (Steinkopff Verlag, Darmstadt, 1964).

[10] H. SCHERING, *Registrierungen des spezifischen Leitvermögens der atmosphärischen Luft*. Göttinger Nachrichten, 201–218 (1908).

[11] F. J. H. WHIPPLE and F. J. SCRASE, *Point discharge in the electric field of the earth*. Geophysical Memoirs *VII*, No. 68, (1936), 3–20.

Special print from

Archive for Meteorology, Geophysics and Bioclimatology

Series A: Meteorology and Geophysics, Volume V, Issue 1, 1952

Published by
Dr. W. Mörikofer, Davos and Prof. Dr. F. Steinhauser, Vienna
Springer-Verlag in Vienna

551.594.2:551.575

(From the Atmospheric Electrical Research Center of the Observatory in Friedrichshafen
in Buchau a. F., National Weather Service Württemberg-Hohenzollern.)

Studies regarding the Atmospheric Potential Gradient* IV:
Examples
for the Behavior of Atmospheric Electrical Elements in Fog.
By
H. Israël and H.W. Kasemir.

Summary. The work on atmospheric electricity to-day can be divided in two main areas of activities, one occupied with world-wide, the other with local meteorological influences on the formation of atmospheric electrical conditions. In this paper the influence of fog on atmospheric electrical elements is treated as a problem pertaining to the latter of the above mentioned activities. From a large number of records of the potential gradient and the air-earth-current at times of fog at two low country stations (Buchau and Uppsala) and two high mountain stations (Jungfraujoch and Sonnblick) relations can be found where several characteristical types of reaction are to distinguish. The different types are discussed here in detail and an attempt is made to explain their origin, as far as possible also theoretically. Special attention is given to the theoretically expected and experimentally confirmed behaviour of the potential gradient and the air-earth-current in vast flat fog layers in the plain and in the cumulus-interior of cloud-covered mountain-tops. While in the low country it is predominantly the potential gradient which reacts (increases), since the air-earth-current remains practically unchanged, inside the cumulus the potential gradient remains practically unchanged and the current decreases remarkably.

Both possibilities demonstrate that the conductivity inside the fog drops to one third of its value in an atmosphere free from fog.

Résumé. Les recherches d'électricité atmosphérique se poursuivent aujourd'hui dans deux domaines principaux, celui des phénomènes à l'échelle planétaire et celui des effets des conditions météorologiques locales. Le présent article traite un problème de la deuxiéme catégorie, soit l'influence du brouillard sur l'électricité atmosphérique. Se fondant sur l'examen d'un grand nombre d'enregistrements du champ et du courant vertical par brouillard en deux stations de plaine (Buchau et Uppsala) et deux stations de haute montagne (Jungfraujoch et Sonnblick), on peut établir des relations de la nature que voici. On peut distinguer plusieurs types caractéristiques de réaction par brouillard; on expose ceux-ci en détail et on explique leur origine, si possible théoriquement aussi. Le comportement du champ et du courant dans de vastes couches plates de brouillard en plaine et à l'intérieur des cumulus recouvrant des sommets montagneux, tel que la théorie le prévoit et que l'expérience le vérifie, est particulièrement intéressant: tandis qu'en plaine le champ surtout réagit (croît) et que le courant vertical reste pratiquement inchangé, à l'intérieur des cumulus le champ est pratiquement constant et le courant est fortement réduit. Dans les deux cas il s'avère que la conductibilité dans le brouillard tombe à environ un tiers de sa valeur dans l'air sans brouillard.

I. INTRODUCTION.

The atmospheric electrical state with its changes must be perceived as the result of the interaction of two spheres of influence, as is generally known – the one sphere of influence, with its worldwide influence, is largely responsible for the existence and the behavior of this state; the other sphere of influence, controlled by local meteorological influences, forms the individual phenomena in the atmospheric electrical state.

The atmospheric electrical research can be divided accordingly into two main directions that seek the discovery and causal interpretation of the effects coming from the one or the other sphere of influence. Due to the fact that the global interrelations can be considered as clarified to a large degree, the problem of the interrelations between meteorological and electrical events is the primary focus today. In the following, therefore, we will single out an essential issue from this complex of meteorological electrical interrelations, which has previously been given proportionally little attention, and observe the influence of fog on atmospheric electrical elements.

A practical interest is simultaneously conjoined with this problem – namely several attempts have been made to also use the atmospheric electrical measurements and experiences to forecast fog – a task that is still important and practical today despite the aviation technological advancement of blind takeoffs and blind landings. Such attempts are naturally presupposed to a systematic inspection and sufficiently precise knowledge of the impact of the atmospheric electrical circumstances due to fog.

As is generally known, fog and haze significantly affect the electrical conditions of a location by reducing the conductivity of the air. Accordingly, one should expect that the potential gradient is increased or that the vertical current is decreased or that both occur simultaneously in the case of fog and haze. Beyond these "regular" reactions, however, occasionally other effects are observed that are sometimes contradicting. With this approach, J. A. Chalmers and E. W. R. Little [1] determined, e.g. the reversal of the field and current direction for practically unchanged conductivity in haze. F. J. Scrase [2] frequently finds high current values of a reverse direction with simultaneously high potential gradient of a normal direction in dense fog.

These and similar observations show that multiples effects are apparently possible with the influence of haze and fog, which cannot be understood based on the reduction of conductivity alone and that various effects materialize depending on the preponderance of the one or the other influence. For this reason, in the case of our previous synchronous registrations, we have explored the question surrounding the atmospheric electrical field and the vertical current in Buchau and Uppsala* as well as at the mountain stations, Jungfraujoch and Sonnblick, and have handled a large number of cases involving fog regarding their atmospheric electrical effectiveness. The results are reported in the following.

II. The Types of Fog Influence on the Atmospheric Electrical Elements.
a) General Information.

From the registrations, a total of 31 cases of fog could be obtained, which were sufficiently secured through simultaneous meteorological observations. They are compiled in Table 1. During the compilation, it was demonstrated that, in the case of fog, one must differentiate various types in the behavior of the atmospheric electrical elements; generally, the following cases occur:

I. Atmospheric electrical field and vertical current have the same direction**.

II. Atmospheric electrical field and vertical current have different directions.

For the first type, we have further differentiated the following individual cases:

I, 1. Field and current simultaneously change their sign multiple times.

I, 2. Field and current maintain their normal direction; the field is increased and the current is unchanged.

I, 3. Direction, as in I, 2; the field is increased, the current is decreased;

I, 4. Direction, as in I, 2; the field and current are unchanged.

Type II, which resembles the above mentioned observations of J. A. Chalmers and E. W. R. Little [1] and of F. J. Scrase [2], is relatively seldom and is provisionally not further subdivided as a result. Examples for the various types are reflected in Figure 3 (Type I, 1), 1 (Type I, 2), 7/8 (Type I, 3), 2 (Type I, 4) and 5/6 (Type II).

---

* The registrations in Uppsala were conducted with the same equipment [3] at the Institutet för Högspänningsforskning. They are currently being processed and their results will be reported on later. The values from Uppsala used in Table 1 have been taken from the material of Prof. Dr. H. Norinder with his gracious approval, for which we would also like to express our thanks at this point.

** According to the common method of designating terminology in atmospheric electricity, the information regarding the "direction" of atmospheric electrical field and vertical current must be perceived in such way that the potential difference between a point in the atmosphere and the surface of the earth is normally positive and that the vertical current of the surface of the earth receives a positive charge in this case. At this point, the field and current are designated as "positive" and of the "same voltage". The case that field and current "have a various direction" indicates that another charge transport, presupposed in a mechanically-convective manner, must be added to the charge transport by electrically-charged carriers under the effect of the field.

H. ISRAËL and H.W. KASEMIR:

b) Fog and Clouds.

Table I indicates that the individual types occur with various frequencies and that, apart from their frequency, they appear to be dependent on the orographic position of the measurement site. Thus, it is immediately obvious that Type I, 2 apparently dominates in the flatland (Buchau, Uppsala), while Type I, 3 appears most frequently in the mountain range (Jungfraujoch, Sonnblick). This provides a starting point for interpretation:

In our [4] previous work, it is stated that the current density in the flat layers of fog must be constant within and outside of the area of fog, while the field strengths of the space filled with fog and that without fog react opposite to the conductivity. However, if one transitions from broadly extended layers of fog to those, for which the expansion of the sides and height are comparable (e.g. cumulus clouds), then the field increase will be small in the cloud space while a reduction of current applies simultaneously. This effect goes so far that, in the case of bodies, the height of which is several times that of its lateral expansion; the field strength in the inner and outer space remains practically unchanged while the current densities react like the conductivities. Thus, exactly the opposite is demonstrated as with the expanded layer of fog. This fact can be easily understood physically – the current cannot swerve out of the way of a broadly expanded layer of fog as easily as a cu-shaped due to the fact that the resistance of the long lateral detour is larger than the direct short path through the layer of fog with the low conductivity.

Now the fog dimensions of the station in the flatland and in the high mountains are quite different. In the flatland, the fog is generally broadly expanded horizontal layers of a low vertical thickness while in the mountains, the covering generally occurs through cumulus clouds, the lateral expansion of which is comparable to or smaller than its vertical reach. In the first instance, therefore, it is necessary to expect almost the same current density in the foggy area and in the area without fog while in the second instance, a main part of the current flows around the cumulus cloud, whereby the current density in the cloud is lowered naturally. The field strength in this process is the secondary partner and is attuned to $E = \dfrac{i}{\lambda}$ according to Ohm's law.

This concept, as seen in Table 1, is completely verified through experimental findings. Only case no. 27 that depicts Type I, 2 is dropped in the case of the high mountain measurements. However, according to the meteorological observation only the above listed explanation verifies this because this and only this registration deals with an expanded stratus cover that initially extends below the height of the peak in the south and the north, which however, ascended with time until it's southern and northern parts finally joined above Jungfraujoch, during which the measurement station was encapsulated. All other mountain registrations dealt more or less with high cumulus clouds that covered the measurement station.

Studies regarding the Atmospheric Potential Gradient IV

Table 1. *The Behavior of Atmospheric Electrical Field and Vertical Current in 31 Individual Cases of Fog.*

Meaning of the individual columns:

Column 1: Continual number.

Column 2: Location and date. B. = Buchau; U. = Uppsala; J. = Jungfraujoch; S. = Sonnblick.

Column 3-7: Behavior of field and vertical current according to the grouping provided in the text.

Column 8: Meteorological information.

Column 9: Relationship of the conductivity Λ_1 before or after the fog to Λ_2 during the fog.

No.	Location, date	I,1	I,2	I,3	I,4	II	Remark	Λ_1/Λ_2
1	B. 19. VIII. 1951		×					3,6 see Figure 1
2	B. 9. IX. 1951		×					2,9 see Figure 2c
3	B. 11. IX. 1951		×					2,8
4	B. 12. IX. 1951		×					2,5
5	B. 25. IX. 1951		×					3,4
6	B. 28. IX. 1951				×			
7	B. 5. X. 1951					×		see Figure 5c
8	B. 8. X. 1951				×			
9	B. 13. X. 1951		×					2,6
10	B. 19. X. 1951	×						see Figure 3
11	B. 21. X. 1951					×		see Figure 5b
12	B. 22. X. 1951		×					2,8
13	B. 24. X. 1951					×		see Figure 5a
14	B. 28. X. 1951				×			see Figure 2a
15	B. 30. X. 1951				×			
16	B. 31. X. 1951		×					2,5
17	U. 27. VIII. 1950					×		see Figure 6
18	U. 14. I. 1951		×					3,1
19	J. 30. V. 1950			×				
20	J. 13. VI. 1950			×				
21	J. 26. VI. 1950			×				
22	J. 6. VII. 1950			×				
23	J. 26. VII. 1950			×				
24	J. 9. X. 1950			×				
25	J. 17. X. 1950			×				
26	J. 24. X. 1950			×				3,63 see Figure 7
27	J. 29. X. 1950		×				St	
28	J. 12. III. 1951			×				
29	S. 16. X. 1950			×				3,53 see Figure 8
30	S. 17. X. 1950			×				
31	S. 25. X. 1950				×			

c) Individual Cases.

To achieve quantitative relationships, we will observe some typical individual cases and begin with that, which preferably occurs in the flatland

Type 1, 2.

Figure 1 reflects a typical example of such a case. The assessment is difficult in that the course of the field not influenced by the fog is not known with sufficient assuredness. One can assume with some level of assuredness according to our above represented experience that the vertical current is not noticeably influenced in the case of expanded flat layers of fog in a curve running parallel and correlating to the vertical current of the hashed path shown on the bottom in Figure 1.

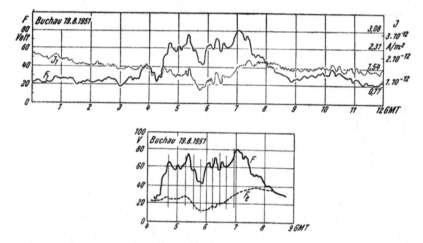

Fig.1 Above: Registration of atmospheric electric fields (solid curve) and vertical current (dotted lines) in Buchau a .F. on August 19, 1051 from o to 12 hr GMT. The labeling of the ordinate shows the potential of the collector against ground in volt (left side) meaning the relative value of the atmospheric electric field, and (on the right) the vertical current in A/m^2 ; the numbers on the right side inside the figure give the current values belonging to the horizontal lines. Below: Representation of the influence of fog on the atmospheric electric field: solid line (F) = field during fog; dotted line (Fe) = assumed field without fog (for an explanation, see text).

We establish the relations of fog affected field and non fog affected field values on ten evenly divided times during fog periods- as shown in the lower part of Figure 1 to determine the variation of conductivity due to fog. This will establish the relation of conductivity in and outside areas of fog. The ten ratios of conductivities Λ_1/Λ_2 are then averaged and this will establish the characteristic of this fog event. These averaged values are listed in the last column of Table I for all I.2-cases of the stations in the flatlands. The highest value is 5.2 (registration. on 19. August 1951) and the lowest value is 1.86 (registration on 31. October 1951). The total average for all processed data for type I.2 is 2.9. For the mean value we have $\Lambda_1/\Lambda_2 = 3$, i.e. during ground fog the conductivity is reduced to 1/3. It is of special interest to compare this value with the measurements of P. Pluvinage [5] who used a very different method for determining the lowering of conductivity in fog. If we assume an approximate mean value of air conductivity of $2 \cdot 10^{-4}$ esu , then we obtain from the eleven measured values by Pluvinage a mean value of 3.15 for Λ_1/Λ_2 with the extreme values of 1.7 and 5.4. The agreement with our results is therefore surprisingly good. If these results are limited to the I.2 cases, as up to now the only available data base with sufficient number of measurements, or if these results have more general significance, has to be decided from a more detailed examination.

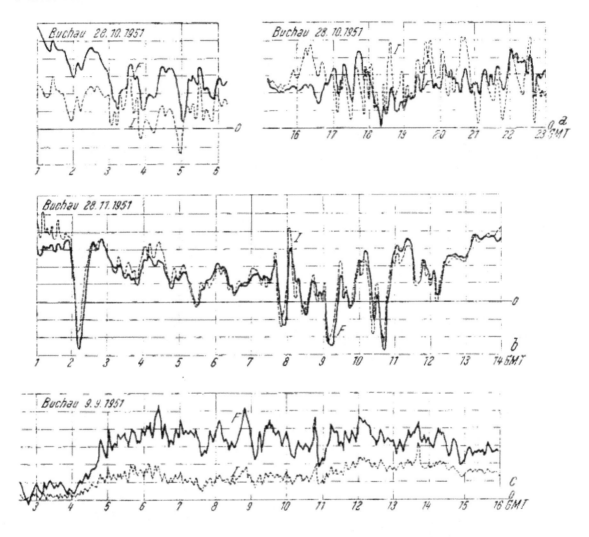

Figure 2. Registrations of the atmospheric electrical field (expanded curves) and the vertical current (hashed curves) in Buchau during fog and mist. *a* Registration of fog of Type 1, 4 on October 28, 1951, 1 am to 6 am and 3 pm to 11 pm GMT. *b* Registrations of haze on November 28, 1951, 1 am to 2 pm GMT. c Registration of fog and haze on September 9, 1951, 2:30 am to 4 pm GMT. The sensitivities have the following values in Figures 2, 3 and 5: Registration of the field: A deflection of the variables of the distance between two horizontal lines in the figure patter correlates to 20,3V of change of the potential difference between the collector and the earth. Registration of vertical current: The distance of two horizontal lines of the pattern correlates to a change of the vertical current by $0.46 \cdot 10^{-12}$ Amp/m^2.

Studies regarding the Atmospheric Potential Gradient IV

We will look now at the second most frequent flatland fog – **Type I, 4.**

The requirement that field and current remain unchanged for this type is merely related to the average value – because compared to a fair weather registration, the fog curves of this type demonstrate strong fluctuations – as can be seen in Figure 2a.

In the case of the previously observed registrations of Type I, 2, such strong fluctuations did not occur. However, it must be emphasized that both Types I, 2 and I, 4 cannot be sharply separated, but rather that continual transitions exist. This is valid – to mention that here – in general terms. Thus, by no means have we included all cases of fog in the compilation of Table 1, rather we have knowingly omitted all fog types of mixed conditions.

A secondary registration of Type I, 4 is represented in Figure 2a. Figure 2b shows such a registration during strong haze, which however could not be considered as fog. Figure 2c finally provides a registration of the fog, which turned into strong *haze** at around 8:30 am. However, it is noteworthy that all three examples from Figure 2 are so similar in their appearance that no considerable difference exists between a registration of Type I, 4 in fog and a registration during strong haze. This is most clearly demonstrated in Figure 2c, which depicts a registration of fog until 8:30 am and from then on haze. Figure 2c also depicts a mixed type, for which the I, 2 characteristic of the increased field is combined with the I, 4 characteristic of the increased disturbance.

With the registration of haze on November 28, 1951 (Figure 2b), where no further clouds were observed above the layer of haze, it is evident that current and field occasionally demonstrate strong deflections into the negative range – a phenomenon that seems highly noteworthy in a cloudless sky and without any precipitation, although atmospheric electricity is registered. This peculiarity guides to our Type **I, 1.**

Here, the similarity of fog and haze registration is even more noticeable (see Figure 3). Regarding the daily climate schedule, at 6:22 am GMT, there was fog* and no wind; throughout the late morning, the fog cleared and transitioned to haze and at 1:22 pm, it was mostly hazy*.

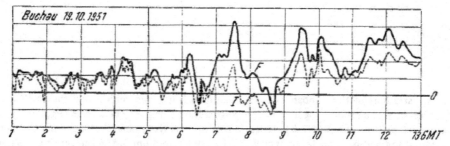

Figure 3. Registration of the atmospheric electrical field (solid curve) and the vertical current (hashed curve) in Buchau on October 19, 1951. Sensitivities as above in Figure 2.

---

* The term "haze" is used here and in the following if visibility is only slightly more than 1 km without providing a more detailed characteristic for the cloudy particles. In the case of "haze" directly before or after fog, accordingly, we may have been dealing with so-called "pre-condensation" or "haze" (=) in the case of November 28, 1951 (Figure 2b), in contrast, we had "dry haze" (∞).

Studies regarding the Atmospheric Potential Gradient IV

Because the lack of wind or only slight wind are a common characteristic of these registrations of Type I, 1 with their field dispersion in some half-hour lapses, slow thermally conditioned redistributions in the surface-level atmospheric layers can be assumed as the cause of these fluctuations in fog and mist.

The analyses on a smaller climatic scale at the Federsee basin (Buchau) by E. Huss [6] can provide confirming information in this direction, for which the distributions in a 10m thick layer of fog are made clearly visible through temperature and wind registrations.

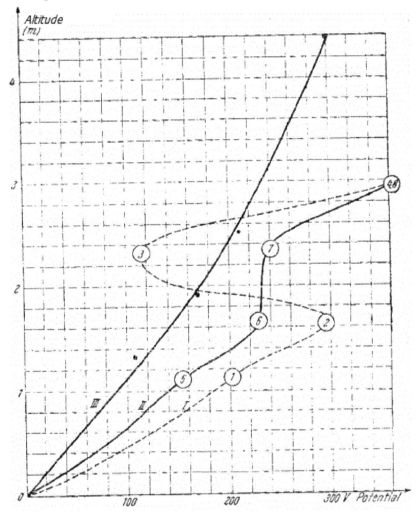

Figure 4. Measurements of the potential curve with the altitude in Buchau a. F. on June 10, 1951, 11 pm to 12 pm GMT. Curve I: Measurement at the Kanzach Channel in the pre-stage of fog. Curve II: Measurement at the same site somewhat later. Curve III: Measurement at Forsthaus Kappel, approx. 10 m above the bottom of the valley. At the Kanzach Channel curves I and II), the metrological values are marked through circles with listed digits (continual numbering), at Forsthaus Kappel by small crosses.

To allow such distributions to become electrically efficient, there is a requirement that the atmosphere was originally categorized in layers of greatly differing conductivity, for which a positive or negative space charge was formed for every change of conductivity. The potential curve with the altitude must demonstrate a respective reaction in such a case. Figure 4 provides an interesting illustration.

67

H. ISRAËL and H.W. KASEMIR:

Curve III, which was recorded outside of the area of ground fog (at Forsthaus Kappel), shows the normal increase of potential difference versus the ground with the altitude. In contrast, curve I allows a fully other curve to be seen in the area of the ground fog. (In this context, it must be stated that the height of 2.5 m, in which the potential difference shows its deep recess correlates to the latter upper border of the ground fog.) Curve II, which was recorded shortly after curve I, shows the deviating course much weaker – a sign that a distribution of the layers apparently occurred.

This example can serve as a sign that the space charge must be taken into account together with its thermal distributions in order to understand the effects of fog. More detailed information regarding the mechanisms indicated here must be expected from simultaneous registrations of the field in various altitudes, the conductive current and the meteorological elements.

A somewhat more detailed approach are required by the registrations of fog of **Type II,** of which four cases are contained in Table 1. As already indicated above, these cases can only be understood with the presence of a convective charge transport. Indeed, we have separated all secondary cases connected with precipitation of any kind for eliminating such convectively conditioned currents; however, we must assume that some convection current is active in the cases of Type II.

Three such cases are presented in Figure 5. On October 24, 1951 there was predominant wet fog of the density 0 under an extended stratus cover. During the whole night, the current was more or less negative with a normal positive field direction. There was no precipitation of any kind, however, the fact that there was "moisture" indicated that apparently a convective charge transport was associated with it, which affected atmospherically-electrically similar to normal precipitation. In the second case (Figure 5b), fog^0 and dew^1 was noted during the 7 am observation. It is logical to address the somewhat improbable occurrence of dew during fog as a result of "moisture" of the fog. Finally, in the third case (Figure 5c), there was clearly wet fog again.

In summary, one may subsequently say that fog type II resembling the type of precipitation with opposite field and current direction can only be generated by a non-electrically conditioned convection current, which can indeed be seen in the extremely weak precipitation of the moisture. The fact that on the other hand not every wet fog produces this "convective type", can be seen in Table 1.

It can be assumed that other registrations of fog also contain a convection current, which is however not strong enough to drown out the purely electrical current, and thus, to imprint the negative sign on the total current or to depress it to the value of 0.

The observation is noteworthy that this convection current occurring during fog apparently carries a negative charge to the earth, while a primarily positive charge is transported downward during rain. However, at this time we cannot go into further detail regarding this point.

H. ISRAËL and H.W. KASEMIR:

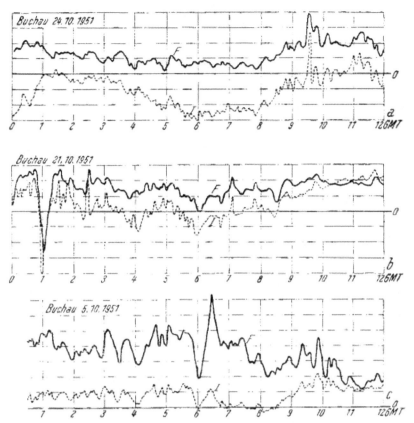

Figure 5. Registrations of the atmospheric electrical field (solid curves) and vertical current (hashed curves) in Buchau a. F. during fog, Type II. a October 24, 1951, 12 am to 12 pm GMT. b October 21, 1951, 12 am to 12 pm GMT. c October 5, 1951, 12:30 am to 12 pm GMT. Sensitivities as above Figure 2.

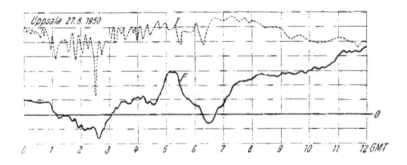

Figure 6. Registration of the atmospheric electrical field (solid curves) and vertical current (hashed curves) in Uppsala on August 27, 1950, 12 am to 12 pm GMT. Sensitivities: 5 V (field) and $0.41 \cdot 10^{-12}$ Amp/m^2 must be applied instead of the values given in Figure 2.

In addition, the opposite case occasionally occurs that the convection type II appears *without* the fog being wet. Figure 6 demonstrates such a registration from Uppsala. The field curve shows a very low, even partially negative field with a comparably high positive current. The meteorological observation indicates fog with visibility of 150 m; precipitation or "moisture" is not noted.

69

H. ISRAËL and H.W. KASEMIR:

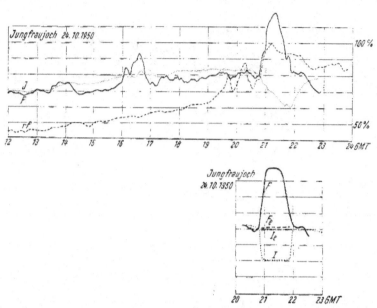

Figure 7. Above: Registration of the atmospheric electrical field (solid curve), the vertical current (dotted curve) and the relative humidity (hashed curve) at Jungfraujoch on October 24, 1950, 12 pm to 12 am GMT. Below: Schematic diagram of the influence of fog: F and J stand for the field and current during the fog, Fe and Je indicate the course of the field and current, as it may have dominated without fog..* Sensitivity:: The distance between two horizontal grid lines correlates to 24.6 V (for the registration of the field) or $9.45 \cdot 10^{-12}$ Amp/m^2 (for the registration of the vertical current)..

Nonetheless, following the course of the registration, it is possible to safely assume that convection current must have been present. **Type I, 3,** which occurs preferably in high mountains, has already been described and justified above. The attempt was also made for this type just as for Type I, 2 preferable in the flatlands to determine the conductive change. However, severe difficulties were encountered insofar as the path of the current in the fog is also influenced by this and therefore can no longer be used for the construction of the field curve undisturbed by the fog. In this case, we are instructed to supplement the undisturbed field as well as the undisturbed current during the fog from the curve sections before and after the fog. This is possible if the emergence of clouds lasts only a short time and the current and field curve of the remaining day demonstrates a proportionally balanced course. Some cases of this type are represented in Figures 7 and 8.

The beginning and end of the fog was visually observed on Sonnblick; on the Jungfraujoch they were determined from the simultaneous registration of relative humidity.

---

* In the case of the supplemented curves, the deformation of the recording through the inertia of the measuring equipment is eliminated by the consideration of the time constants determined from the right bottom partial depiction of Figure 8. Naturally, when immersing the measuring station into the clouds, field and current react immediately according to the bottom partial depictions 7 and 8, while the "fog values" from the registering device are only achieved after approx. ¼ to ½ an hour as a result of the high time constant.

H. ISRAËL and H.W. KASEMIR:

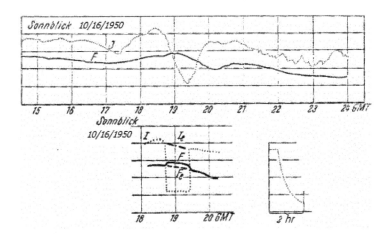

Figure 8. Analogous to Figure 7, for Sonnblick on October 16, 1950, 2:30 pm to 12 am GMT. On the bottom right is the abatement of a voltage imprinted on the vertical current measurement section depicted for demonstrating the time constants. Sensitivity: Instead of the figures provided in Figure 7, we get the following: 21 V (field) or $5 \cdot 10^{-12}$ Amp/m^2 (current).

The conductivity ratio between the atmosphere and cloud results in both cases for Jungfraujoch in Figures 7 and 8 in 3.63, and 3.53 for Sonnblick. The decision cannot be made provisionally whether this difference is real. At any rate, the fact remains surprising that the conductivity is decreased to approx. 1/3 in the same manner as in the fog of the level as well as in the cloud air.

Furthermore, according to the theoretical formula 30 of the previous work by Kasemir [4] one can calculate the form factor and, thus, the ratio of altitude to width of the clouds from the electrical registrations and obtain the value of 2/4 for the Jungfraujoch registration and 3.1 for the Sonnblick registration. Therefore, these had to have been cumulus clouds that had a height of 2.5 to 3 times their width. Unfortunately, observations were not able to be made in this regard to his time. Even if the assumed geometric ratio figures for the clouds actually lie within the realm of possibilities, we should not hide the fact that they must be recorded with a certain level of cautiousness because the requirements of the theory, which was established for level problems, are no longer met due to the deformation of the field and current lines in the high mountains. The application of the theory only therefore seems possible in approximation because the deformation during the fog and in the fog-free state may have been approximately the same and therefore become less significant for the formation of ratios conducted in this case.

H. ISRAËL and H.W. KASEMIR:

Bibliography.

1. CHALMERS, J. A. and E. W. R. LITTLE: Currents of atmospheric electricity. Terr. Magn. and Atm. Elec. 52, 239—260 (1947).
2. SCRASE, F. J.: The Air-Earth Current at Kew Observatory. Geophys. Memoirs, VII. Nr. 58, S. 22 (1937).
3. Sämtliche hier verwandte Messungen wurden mit der in Buchau neu entwickelten Feld-Vertikalstrom-Apparatur gewonnen. Beschreibung der Apparatur bei H. W. KASEMIR: An Apparatus for Simultaneous Registration of Potential-Gradient and Air-Earth-Current. Journ. Atm. Terr. Physics 2, 32—37 (1951).
4. KASEMIR, H. W.: Studien über das atmosphärische Potentialgefälle V. Zur Strömungstheorie des luftelektrischen Feldes II. Arch. Met. Geoph. Biokl. Ser. A, 5, 56—70 (1952).
5. PLUVINAGE, P.: Étude théorique et expérimentale de la conductibilité électrique dans les nuages non orageux. Ann. de Géophys. 2, 31—54, 160—178 (1946).
6. HUSS, E.: Kleinklimatische Studien im Federseegebiet (in Vorbereitung).

Special print from

Archive for Meteorology, Geophysics and Bioclimatology

Series A: Meteorology and Geophysics, Volume III, Issue 1 and 2, 1950

Published by
Dr. W. Mörikofer, Davos and Prof. Dr. F. Steinhauser, Vienna
Springer-Verlag in Vienna

531.594.11

(From the Atmospheric Electrical Research Center of the Observatory in Friedrichshafen in Buchau a. F.)

Studies regarding the Atmospheric Potential Gradient* IV:
Regarding the Current Theory of the Atmospheric Electrical Field I**.
By
H.W. Kasemir.

Summary. These past 25 years, since BENNDORF [1] gave his theory of the electric field of the earth, the conception has been endorsed that the meteorological phenomena and especially the world-wide thunderstorm activity must be considered as the origin of the electric field of current. In this article the theory of BENNDORF, which is very general as regards the generator of atmospheric electricity, is applied to the thunderstorm activity being the current's source. It may be emphasized that many of the equations laid down here have already been developed by BENNDORF for the case of the spherical condenser. Here the theoretical deductions are different: giving prominence to the field of flow, they proceed from the second equation of Maxwell and lead in a continuous development to the construction of an electro-technical equivalent circuit of the current in the spherical condenser. Proceeding from the technical notions the general physical deductions get much more distinctness. In addition, it can be shown that the space charges, which play the leading part in the electrostatic views on atmospheric electricity, have but little importance for the field of flow and have no influence on the current density or the field force. Thereby new prospects follow for the interpretation of the records of the electric field in a thunderstorm. The electro-technical equivalent circuit shows furthermore immediately that the world-time curve of the potential gradient on sea is not uninfluenced by the variable air-resistance on the mainland and that it reflects therefore only approximately the world-wide thunderstorm activity. Although this influence on the world-time daily variation is small — according to a rough valuation less than 17% — the desire arises for closer experimental and theoretical investigations about this problem.

---

* Currently, the following works from the Atmospheric Electrical Research Center Buchau a. F. appeared under this frame title:

1. ISRAËL, H. und G. LAHMEYER: Studien über das atmosphärische Potentialgefälle I: Das Auswahlprinzip der luftelektrisch „ungestörten Tage". Terr. Magn. **53**, 373—386 (1948).

2. ISRAËL, H.: Die Tagesvariation des elektrischen Widerstandes der Atmosphäre. (Studien über das atmosphärische Potentialgefälle II.) Ann. Géophys. **5**, 196—210 (1949).

3. ISRAËL, H.: Luftelektrische Tagesgänge und Luftkörper. (Studien über das atmosphärische Potentialgefälle III.) Journ. atmosph. terr. Phys. **1**, 26—31 (1950).

In seeking a development of the theoretical depictions, the present work provides the attempt of an advancing physical-mathematical penetration of the atmospheric electrical problems.

H. ISRAËL.

** Dedicated to Prof. Dr. H. Benndorf on the occasion of his 80th birthday.

H. W. Kasemir

Résumé. Depuis que BENNDORF [1] a donné, il y a 25 ans, sa théorie du champ électrique terrestre, l'opinion générale s'est de plus en plus affirmée que les phénomènes météorologiques et en particulier l'activité orageuse mondiale constituaient la source d'énergie du champ des courants électriques. Dans la présente étude, la théorie générale de BENNDORF relative au générateur de l'électricité atmosphérique est appliquée plus particulièrement à l'activité orageuse considérée comme source de courant. Il faut remarquer que beaucoup d'équations établies ici avaient déjà été développées par BENNDORF dans le cas d'un condensateur sphérique. La voie suivie est, il est vrai, différente, car en insistant sur la notion de champ de courant on part de la deuxième équation de Maxwell et on aboutit à une image électrotechnique du courant dans le condensateur sphérique. En utilisant les représentations de la technique, les principes physiques et les déductions gagnent beaucoup en clarté; il apparaît en outre que les charges électriques libres qui jouent un si grand rôle dans l'électrostatique atmosphérique sont très peu importants dans le champ de courant et n'exercent aucune influence sur la densité de courant ni sur l'intensité du champ. Cela ouvre quelques nouvelles perspectives sur l'interprétation d'un enregistrement du champ électrique lors d'un orage. De plus le schéma électrotechnique montre immédiatement que la courbe du gradient de potentiel sur mer en fonction du temps universel n'est pas indépendante de la résistance variable de l'air sur terre et ne reflète donc qu'approximativement l'activité orageuse mondiale. Bien que cet effet sur la variation diurne soit heureusement faible (probablement inférieur à 17%), il est à souhaiter que l'on poursuive sur ce point les recherches expérimentales et théoriques.

I. General Output Equations.

To take the character of the current field in the atmospheric electricity into account immediately at the onset, at this point we need to start with the electro-dynamical core equations, namely the equations of Maxwell and their supplemental axioms. The electrostatic field, which often forms the starting point for atmospheric electrical theoretical considerations, appears in our work only as a borderline case, as a temporal marginal condition so to speak. In respect of the use of vector calculus*, the calculation and dimension of the individual variables and therefore the selection of the practical measuring system, we will draw heavily on the depiction of A. Sommerfeld [2].

We will transform the second equation of Maxwell dominating our problem

$$\frac{\partial \vec{D}}{\partial t} + \vec{i} = \text{curl } \vec{H}$$

by forming divergence in the mode appropriate in this case and obtain our four output equations with the three supplemental axioms.

----

* Vectors are marked with an arrow above the letter.

75

H. W. Kasemir

$$\text{div} \left(\frac{\partial \vec{D}}{\partial t} + \vec{i} \right) = 0, \tag{1}$$

$$\text{div} \, \vec{D} = q, \tag{2}$$

$$\vec{i} = \lambda \, \vec{E}, \tag{3}$$

$$\vec{D} = \varepsilon \, \vec{E}. \tag{4}$$

In this case, the following meanings apply:

$\vec{D} \, (x, y, z, t)$ the electrical displacement $\left[\frac{C}{m^2}, \, C = \text{Coulomb} \right]$,

$\vec{i} \, (x, y, z, t)$ the current density $\left[\frac{A}{m^2} \right]$,

$\vec{E} \, (x, y, z, t)$ the electrical field strength $\left[\frac{V}{m} \right]$,

$q \, (x, y, z, t)$ the space charge $\left[\frac{C}{m^3} \right]$,

$\lambda \, (x, y, z)$ the conductivity $\left[\frac{S}{m}, \, S = \text{Siemens} = 1/\text{Ohm} \right]$,

ε the dielectric constant $\left[\frac{F}{m}, \, F = \text{Farad} \right]$,

x, y, z the space coordinates $[m]$,

t the time $[s]$.

$$\tag{5}$$

With the aid of the equations (3) and (4), we can introduce and then integrate a uniform unknown function in (1). We will do that here in detail for $\vec{i}$. From (3) and (4) we obtain

$$\vec{D} = \frac{\varepsilon}{\lambda} \, \vec{i},$$

as well as

$$\frac{\partial \vec{D}}{\partial t} = \frac{\varepsilon}{\lambda} \, \frac{\partial \vec{i}}{\partial t}, \tag{6}$$

and with (1)

$$\text{div} \left(\frac{\varepsilon}{\lambda} \, \frac{\partial \vec{i}}{\partial t} + \vec{i} \right) = 0. \tag{7}$$

Equation (7) can immediately be integrated using vector analysis and we obtain

$$\frac{\varepsilon}{\lambda} \, \frac{\partial \vec{i}}{\partial t} + \vec{i} = \vec{i}_s \tag{8}$$

with the marginal condition

$$\text{div} \, \vec{i}_s = 0. \tag{8a}$$

76

H. W. Kasemir

The temporarily still unknown vector $\vec{i}_s = \vec{i}_s(x, y, z, t)$ - quasi the constant of our unspecified integration – serves to adapt our equation (8) to the marginal conditions predetermined by the problem. It temporal dependence is given by the voltage fluctuations of the electrical source and the marginal condition $\operatorname{div} \vec{i}_s = 0$ points to the fact that we are dealing with the current vector of the quasi stationary state. Despite the provisional uncertainty of $\vec{i}_s$, the differential equation (8) can be solved in general and we obtain

$$\vec{i} = e^{-\frac{\lambda}{\varepsilon}t}\left(\vec{i}_a + \frac{\lambda}{\varepsilon}\int_0^t \vec{i}_s\, e^{\frac{\lambda}{\varepsilon}t}\, dt\right). \tag{9}$$

The vector $\vec{i}_a(x, y, z)$ newly occurring in this case is given by the initial state of the current, which we recognize immediately if we set $t = 0$ in equation (9). Then we obtain $\vec{i} = \vec{i}_a$.

In a similar way, we could also achieve the corresponding equations for $\vec{E}$ and q, though we can obtain these much quicker with the aid of (4) and (2). If we also designate the initial and stationary state in this case through the indices a or s, then (4) and (9) directly result in

$$\vec{E} = e^{-\frac{\lambda}{\varepsilon}t}\left(\vec{E}_a + \frac{\lambda}{\varepsilon}\int_0^t \vec{E}_s\, e^{\frac{\lambda}{\varepsilon}t}\, dt\right) \tag{10}$$

and the following through the forming of divergence of (10) with consideration for (2)

$$q = e^{-\frac{\lambda}{\varepsilon}t}\left(q_a + \frac{\lambda}{\varepsilon}\int_0^t q_s\, e^{\frac{\lambda}{\varepsilon}t}\, dt\right). \tag{11}$$

In this general form, the equations (9) and (11) still cannot be properly understood. Therefore, we need to introduce the additional condition that the variables marked with s change only slightly with time compared to the function $e^{\lambda t/\varepsilon}$, thus defining them directly as quasi stationary through this limitation. The integrals can then be solved according to the mean value theorem by placing the quasi stationary variables in front of the integral.

We obtain

$$\vec{i} = \vec{i}_a\, e^{-\frac{\lambda}{\varepsilon}t} + \vec{i}_s\left(1 - e^{-\frac{\lambda}{\varepsilon}t}\right), \tag{12}$$

$$\vec{E} = \vec{E}_a\, e^{-\frac{\lambda}{\varepsilon}t} + \vec{E}_s\left(1 - e^{-\frac{\lambda}{\varepsilon}t}\right), \tag{13}$$

$$q = q_a\, e^{-\frac{\lambda}{\varepsilon}t} + q_s\left(1 - e^{-\frac{\lambda}{\varepsilon}t}\right). \tag{14}$$

The equations clearly convey to us the meaning of the individual variables. The initial state is eliminated with the factor $e^{\lambda t/\varepsilon}$, while the final state is formed with the factor $1 - e^{-\lambda t/\varepsilon}$. $\vec{i} \to \vec{i}_s$ applies for $t \to \infty$. In this manner, our previous assumption that $\vec{i}_s$ represents the state of the stationary current is verified. For example, if $\vec{i}_s = 0$ now, equation (12) depicts the subsiding of the current field upon disengaging the source of power, if $\vec{i}_\alpha = 0$, then we obtain the formation of the current field when abruptly engaging the power source. If $\vec{i}_\alpha$ and $\vec{i}_s \neq 0$ and not identical to each other, then the transition of the initial state $\vec{i}_\alpha$ into the final state $\vec{i}_s$ upon an abrupt voltage jump of the power source is reflected through (12).

The time function $e^{-\lambda t/\varepsilon}$ with the time constant $T = \varepsilon/\lambda$ very much resembles the discharge of a condenser through resistance and we will benefit from that later when developing a technical equivalent circuit diagram. Due to the increase of conductivity with the altitude, the time constant T also becomes increasingly smaller with increasing altitude. On the surface of the earth we must calculate approx. $T = 400$ seconds in atmospheric electricity; at an altitude of 10 km, T is approx. 8 seconds and at 50 km of altitude $8 \cdot 10^{-3}$ seconds. The voltage variations of the generator that occur in a few 100 seconds are followed by the electrical elements current, field and space charge at an altitude of 10 km in a manner preserving the amplitudes and phases, thus quasi stationary, while we are required to deal with the non-stationary state on the surface of the earth. This results in the possibility of making a metrological decision based on simultaneous registration on the surface of the earth and at altitude, whether a turn-on process was registered on the surface of the earth or a slow voltage variation of the generator in the quasi stationary state. The decision cannot be made through the simultaneous registration of field, current and space charge in one location due to the fact that all three elements follow the same temporal law as equations (12) to (14) clearly indicate. However, we can reach our objective easier if we determine the time constant at the measurement site by measuring the conductivity, for they are not subject to the temporal variations of the atmospheric electrical elements as a material constant. Then we can immediately determine that field, current and space charge fluctuations occur in a quasi stationary manner if they occur in time periods that are large compared to the time constant. In another case, it is necessary to view them as turn-on processes.

II. The Current in the Spherical Condenser

Based on the example of the current in the spherical condenser, we would like to track how the initial and final state is derived from the marginal conditions. In this process, we arrive at the formulas that Benndorf [1] provided in his theory of the electrical field of the earth.

The following should apply (see Figure 1):

$r \, [m]$	the distance of the grid point from the middle point of the sphere,
$R \, [m]$	the inner radius of the spherical condenser (earth's radius),
$R + h \, [m]$	the outer radius (at which h is the altitude of the compensating layer above the earth),
$U \, [V]$	the voltage of the generator (in this case the voltage between both spherical shells),
$I \, [A]$	the total current flowing through the condenser.

H. W. Kasemir

The variables still used maintain their meaning according to (3). In this process, we still wish to assume that λ increases from its surface value λ_0 to an e function with the altitude, and can thus be represented through the equation

$$\lambda = \lambda_0\, e^{k\,(r-R)}.\qquad(15)$$

In this process, the value k results in

$$k = \frac{\ln 10}{10\,\mathrm{km}},\qquad(15a)$$

if λ should have 10 times its surface value at an altitude of 10 km.

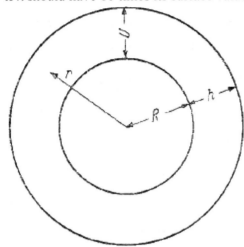

Figure 1. Spherical condenser.

Upon applying the voltage U to the time $t = 0$, no space charge has formed yet in the spherical condenser. It is $\mathrm{div}\,\vec{E} = 0$. Therefore, we can calculate the initial state of the field in a purely electrostatic manner and obtain

$$E_a = -\frac{U\,(R+h)\,R}{h\,r^2}.\qquad(16)$$

Due to the fact that our vectors $\vec{i}$ and $\vec{E}$ always lay in the r direction when there is current in the spherical condenser, it is sufficient in this case to calculate only with the values. For this reason, we have already eliminated the direction in equation (16). Due to the fact that the distance h of the condenser plate is still very small in our example compared to the radius of the sphere R, our r can vary by no more than 1%. As a result, we will set r = const = R and also disregard h against R. Physically, these approximations mean that we actually calculating a plate condenser of the area $4\,\pi\,R^2$ and ignoring the marginal disturbance. Equation (16) is then simplified to

$$E_a = -\frac{U}{h},\qquad(17)$$

and according to (3) we obtain the initial state of the current for

$$i_a = -\lambda\,\frac{U}{h}.\qquad(18)$$

The final state is subject to the condition $\mathrm{div}\,\vec{i_s} = 0$ according to equation (8a). Thus, we can transfer the electrostatic field of the initial state directly to the current field of the final state and obtain

$$i_s = -\frac{I_s}{4\,\pi\,R^2},\qquad(19)$$

and again according to (3)

$$E_s = -\frac{I_s}{\lambda\,4\,\pi\,R^2}.\qquad(20)$$

In order to create the connection between I and U, we form the following from (20)

H. W. Kasemir

$$U = -\int_{R}^{R+h} E_\delta \, dr = \frac{I}{4\pi R^2} \int_{R}^{R+h} \frac{dr}{\lambda} .$$ (21)

The integral can be easily solved with the definition of our λ according to (15) and we obtain the following if λ_h provides the value of the conductivity at the altitude h

$$U = \frac{I}{4\pi R^2 k} \left(\frac{1}{\lambda_0} - \frac{1}{\lambda_h} \right).$$ (22)

If we set the initial and final values into the equations (12) and (13), we will obtain the current and field curve at every altitude dependent upon the time. We will spare ourselves from explicitly writing the formulas and turn immediately to the graphical depiction of the turn-on process for i in Figure 2. The current curve in this case is i/i_s applied when turning on the generator for various altitudes. The time as well as the current scale is logarithmically selected in order to be able to graphically depict the numerous variables passing through the decimal powers. In this process, naturally we lose the usual shape of the e function that we expect according to the formulas. Nonetheless, the graphical image conveys very interesting details.

As a result of the atmospheric distribution of conductivity according to equation (15), the initial state surpasses the final state at an altitude of 60 km by more than 4.5 decimal powers. In this process, the final state arrives in approx. $5 \cdot 10^{-3}$ seconds. On the surface of the earth, the initial value of the current is only around 1/10 of the final value, therefore lower, whereat however the final value is only reached in about 10^3 seconds. This indicates that the initial and final value of the current must be equal at some altitude between the earth and the compensating layer, thus in this case the turn-on process is not even applied. In fact, we find this distinguished altitude in our example at 11.4 km above the surface of the earth. In this case, the current and field will momentarily follow every fluctuation of the generator however quick.

However, we wish to postpone further discussion of these interesting results and first develop the technical equivalent circuit diagram for the spherical condenser due to the fact that many of these results, which seem odd at the moment, will be easily comprehendible based on this circuit diagram.

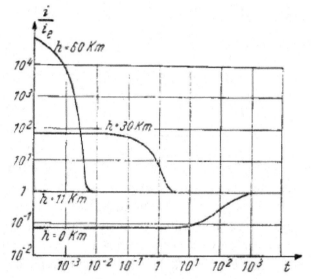

Figure 2. Turn-on process of the current in the spherical condenser.

H. W. Kasemir

III. The Equivalent Circuit Diagram of the Current in the Spherical Condenser

If we look at the dimension of our material constants, namely Siemens/meter for the conductivity and Farad/meter for the dielectric constant, we recognized in the latter the capacity of the unit volume.

With this, the components are given for our equivalent circuit, which are presented according to Figure 3 as a resistance condenser chain for every electrical conduit. If we ignore the low cross section variation of an electrical conduit with the altitude, which we have already done with the mathematical approach, then the values of $1/\lambda$ and ε directly provide the variables of the equivalent resistance and capacities. We must imagine a resistance condenser chain established for every m^2 of the earth's surface. Only two such chains are depicted in Figure 3.

The increase of conductivity with the altitude is indicated by a decrease of the resistance symbols in the circuit diagram. The capacity is equal at every altitude. We recognize the field strength in the voltage over a link in the chain, namely the voltage per unit of length. The current density i is the flow of current in a chain. When abruptly applying a cumulative voltage U to the chain, initially it distributes itself evenly across all capacities. This is our electrostatic initial state. In contrast, the resistance chain in the current state is decisive for the voltage division and because the individual resistances are not equal in size, this voltage division is something other than that given by the condenser chain. Thus, a charge reversal occurs until it has adapted to the resistance voltage division.

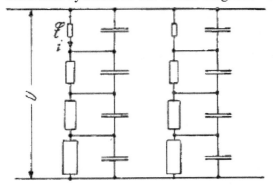

Figure 3. Equivalent circuit diagram for the current in the spherical condenser.

At first glance, we see that the voltage drop in the condenser, thus the electrostatic initial state, outweighs the voltage drop across the small resistance in the stationary current state at altitude, while the opposite is true on the surface of the earth, where the final state is larger than the initial state. Even without calculation, we can immediately state that the initial state and final state are equal across the link of the chain, the resistance value of which corresponds to the arithmetical average of all chain resistances. In the process, every chain link arrives at the final value according to its time constants defined by resistance and capacity. These statements seem to be almost too trivial from a technical perspective to be particularly emphasized. Their intent is primarily to serve us in shedding light on the previously defined calculation and also to gain trust for the further conclusions that we will achieve from the circuit diagram.

IV. The Space Charge.

The space charge is reflected in our circuit diagram through the net charge on two connected plates from respectively two adjacent condensers.

81

Studies regarding the atmospheric potential gradient IV.

The net charge is naturally equal to 0 on the plates of the same condenser. However, in this way it is dependent on the variation of the respectively adjacent resistances, i.e. the space charge is determined by the change of conductivity. If all resistances are equal, then no space charge is present regardless of however great the current and field strength is. Of course, the opposite statement also applies: An equally large space charge has no effect on current or field strength in a current field - be it in direct proximity or distant. This contradicts the mentality towards atmospheric electricity developed from electrostatics so much that the earnest conclusions are drawn very reluctantly even today. For the immediate consequence is namely that we cannot even register the charge of a precipitation-free cloud through a field measurement outside of the cloud or that these charges have absolutely no influence on current and field. Even in the case of a rain or storm cloud, we may not base the electrostatic field of the space charges on an observation alone, but rather we must first decide whether the electrostatic or the current field predominates. An extensive approach of the problems posed in that manner would be troublesome in this case, though. We will have to reserve them for a special representation. Nonetheless, we do want to at least discuss the current conditions in clouds to the extent that they are apt to clearly differentiate between the electrostatic and current field. We will assume that the clouds are free of precipitation, the space charge formation of which may only be ascribed to the altered conductivity.

V. The Clouds Free of Precipitation

If an extensive cloud layer lies above the earth, e.g. large-scale stratus cover, then the conductivity therein is heavily reduced and we will have to place greater resistances at this point in our circuit diagram. The result from this is a greater voltage drop across them and the respective capacities are charged higher only due to this. Nonetheless, the formation of this space charge has no effect on the increase of the field strength so that we could remove the capacities or the space charges without concern. The field strength would remain unaffected as a result. As well, a remote effect of this space charge caused by the increase of resistance will not occur if we assume that this local increase does not significantly change the total resistance of an air column between the earth and the compensating layer. As a result, it remains close under and above the cloud of the current and field curve as if the cloud was not even there. Numerous atmospheric electrical measurement flights in a glider by F. Rossmann and by the author verify this statement, for a noticeable increase of the field was only found as soon as the glider immerged into the cloud.

Now, if we exclude the above made condition that we are dealing with a broadly extended cloud layer, thus if we observe, for example, a high cumulus of nice weather, then the image will change somewhat. Through the local high resistance of the cloud, the current lines will be forced out, which will naturally lead to a compression of the current and field lines on the edges of the cloud. In this case, a precise calculation must first show what impact this marginal effect is capable of creating. At any rate, it cannot be significant as we merely need to think of the current line image of a sphere around which fluid flows, which represents the most extreme borderline case for our problem, whereby the resistance in the cloud is infinite and the conductivity is zero. In any case, this marginal disturbance will no longer be noticeable on the surface of the earth. That is also

Studies regarding the atmospheric potential gradient IV.

consistent with the metrological experience as it is not necessary to determine a significant impact of clouds free of precipitation on the ground registration of the atmospheric electrical field.

VI. The Storm Cloud.

In the case of rain and storm clouds, we are dealing with the formation of space charges that do not arise from the electrical current field, but rather are of a convective nature. To the extent we remain outside of the area of precipitation, however, we are dealing with a purely electrical current field if not with a current in the spherical condenser. The space charges of the storm cloud correlate to the surface charge of the upper shell of the spherical condenser that we expressed there with the total voltage U. Whether they act in an electrostatic manner or as generator voltage in the current field depends on whether we have to expect a stationary or non-stationary state. If we compare the time constant of our current system of approx. 7 minutes with the duration of a storm lasting approx. 1 hour, we will assume the stationary state and consider rapid redistributions of charge as disturbances, which will destabilize the stationary state more or less frequent, though only temporarily.

We will see how fast such a disturbance yields to the stationary state with the lightning field jump and the recovery curve. The charge redistribution through a lightning discharge occurs in such a short time that the electrostatic approach of the lightning field jump is absolutely justified. The recovery curve, however, falls into the area of the turn-on process, i.e. into the abatement of a disturbance with the final state 0. We cannot primarily ascribe the recovery curve to an effect of the generator activity of the storm as the lightning field jump would also dissipate if the generator activity of the storm machine were to be suddenly interrupted with the lightning strike. In any case, our observation of the recovery curve leads us to perceiving the so-called stationary storm field properly as a stationary current state.

With that we must also recognize the consequences emphasized above, namely that the charge quantities or even the charge arrangements in the storm cloud cannot be deduced from a field registration on the surface of the earth. The primary characteristic of the storm generator is it current production and not the stationary charges.

The charge concentration and with that the voltage in the storm cloud also depends on the load as well as on the question whether our storm generator works in the short circuit or in idle.

With that, we need to conclude the observation of the current field of a storm cloud. We will have to postpone the quantitative calculation and comprehensive discussion of this problem for another work.

VII. The Diurnal Variation of the Potential Gradient at Sea and on the Top of a High Mountain.

We will turn once again to the stationary current state in the spherical condenser and come to the diurnal variation of the atmospheric electrical variables and the universal time storm curve. In this connection, we can still simplify our circuit diagram significantly according to Figure 4 if we leave out the capacities altogether as they do not

Studies regarding the atmospheric potential gradient IV.

have any effect on current and voltage in the stationary state. Moreover, we will combine the entire atmospheric resistance between the earth and the compensating layer in two resistors connected in parallel W_M and W_L, of which the first represents the atmospheric resistance over the seas and the second represents that over all the land area*. W_M in this connection has been specified as a constant, while we adopt W_L as variable resistance due to the fact that it is subject to a daily fluctuation as a result of the exchange. Furthermore, we have additionally marked the storm generator with the voltage U_G in Figure 3, which is connected to it through the resistance W_G of the air column between the head of the storm and the compensating layer. Initially, we will not take the resistance W_B

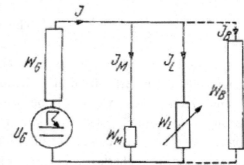

Figure 4. Equivalent circuit diagram of the atmospheric electrical circuit above the earth.

connected through the dotted line into account. It will help us later to understand the conditions during a mountain registration.

Now, if we clarify the scale of the individual resistors, we recognize immediately that W_G is larger than W_L and W_M by multiple decimal powers because the size of the resistors is inversely proportionate to the area covered by them. A quarter or three-fourths of the entire surface of the earth reacts to the expansion of a storm over an area of 10+ square kilometers.

This overbalance of W_G will not be neutralized even if we take into consideration that the head of the storm is located at an altitude of approx. 7 km and therefore approx. 80% of the resistance is bypassed, which we ascribe to an air column from the unit cross section and the altitude 'earth – compensating layer (columnar resistance)'. However, the consequence of this is that the net current I is determined by U_G and W_G.

$$I = \frac{U_G}{W_G}. \tag{23}$$

This net current I is then divided across the resistors W_M and W_G according to their size. This shows that the current I_M through the resistance W_M cannot be independent from the fluctuation of the resistance W_L. The formulas for the currents I_M and I_L can be easily read from Figure 4. It is

$$I_M = \frac{I}{1 + \frac{W_M}{W_L}}, \tag{24}$$

$$I_L = \frac{I}{1 + \frac{W_L}{W_M}}. \tag{25}$$

Now, to be able to quantitatively estimate the influence of W_L on I_M as well, we will assume that the sea and land areas behave 3:1 and the air column resistance on land can fluctuate from a single to a triple value of that at sea. That means $3\,W_M < W_L < 9\,W_M$ and we obtain a variation of I_M through W_L of no more than 17% if we insert the extreme values of W_L in equation (24). In doing so, we can engage additional considerations in

Studies regarding the atmospheric potential gradient IV.

that direction that this maximum value will probably not be reached. For one, not all continents have day and night simultaneously. Thus, the fluctuations of the air column resistance of the individual countries overlap with a phase displacement and level the total fluctuation. On the other hand, the exchange in the polar areas is largely prevented even over land, whereby the total fluctuation of W_L also becomes smaller. Therefore, we can state in conclusion that the diurnal variation of current and field at sea reflects the voltage fluctuations of our storm generator barring a slight error.

For a registration of the vertical current on land, we can gather from equation (25) that it is approximately proportional to I/W_L under the previous assumptions that in addition to the voltage fluctuation of the generator it is completely subjected to the variation of W_L. However, in this case, an in-depth approach to the problem will be necessary because the phase position of the approximate surroundings will be dominating for every measurement site, while continents located further away can only be made noticeable through a load variation of the generator.

In conclusion, we would still like to briefly turn to the registration on the top of a mountain. The more we lift ourselves from the surface-level atmospheric layers, the more we deprive ourselves of the influence of the exchange events. The conductivity not only becomes greater with an increasing altitude, but its diurnal variation becomes smaller in comparison. At an altitude of 6000 meters, we have already left 80% of the air column resistance below us and we can assume with great probability that the remaining 20% has hardly any diurnal variation. As a result, we find the same conditions here as at sea. If we now observe the previously neglected resistance W_B (Figure 4), which represents the atmospheric resistance between the top of the mountain and the compensating layer, then we obtain the current I_B flowing there for

$$I_B = I \, \frac{W_M}{W_B} \cdot \frac{1}{1 + \dfrac{W_M}{W_L}}. \tag{26}$$

Thus, except for the constant factor W_M/W_B for I_B, we find the same value as for I_M. The diurnal variation of current and field at sea and on a tall mountain must therefore be the same. In this case, however, the strengthened desire arises to calculate the spatial current field so as to determine to what extent the current line concentration at the top of a mountain remains uninfluenced by the fluctuation of the resistance of the surface-level atmospheric layers at sea level.

Bibliography.

1. BENNDORF, H.: Grundzüge einer Theorie des elektrischen Feldes der Erde, I. und II. Sitz. Ber. Akad. Wien. II a, 134, 281 (1925); 136, 175 (1927).
2. SOMMERFELD, A.: Vorlesungen über theoretische Physik, Bd. III. Wiesbaden 1948.
3. ISRAËL, H. und G. LAHMAYER: Studien über das atmosphärische Potentialgefälle I: Das Auswahlprinzip der luftelektrisch „ungestörten Tage". Terr. Magn. 53, 373 (1948).

Special print from

Archive for Meteorology, Geophysics and Bioclimatology

Series A: Meteorology and Geophysics, Volume V, Issue 1, 1952

Published by
Dr. W. Mörikofer, Davos and Prof. Dr. F. Steinhauser, Vienna
Springer-Verlag in Vienna

531.594.11

(From the Atmospheric Electrical Research Center of the Observatory in Friedrichshafen
in Buchau a. F.; National Weather Service Württemberg-Hohenzollern.)

Studies regarding the Atmospheric Potential Gradient V
Regarding the Current Theory of the Atmospheric Electrical Field II.
By
H.W. Kasemir.

H. W. Kasemir:

Summary. After the description of some problems of the atmospheric electricity in [1] (in view of the theory of current and described qualitatively), the single problems shall be treatened more in detail and quantitatively computed in the following papers, as far as this will be possible with moderate mathematical means.

In this paper clouds without precipitation, layers of haze, mist and fog will be described mathematically as ellipsoids of rotation with constant internal conductivity. Later on the field- and current conditions inside and outside the ellipsoids will be examined, if these bodies are brought into a homogeneous current field of vertical direction. It can be shown that the increase of the field inside is dependent on the conductivity and very much on the geometric form of the body. In the case of a layer infinitely extended in two directions (degeneration of the oblate rotation ellipsoid) the current density inside and outside the body is equal, since the potential gradients are inversely proportional to the conductivities. In the case of an infinitely long cylinder (degeneration of the prolate ellipsoid) the potential gradients are equal inside and outside while the current densities are proportional to the conductivities. Between these two extreme cases the field forces and the current densities may have any value according to the rate of the axes of the ellipsoid.

The structure of the potential gradient close to the disturbing body requires a special interest. Applying the results to the haze layer which lies over the continents (exchange layer), some valuable hints result for records of the potential gradient at the shore or over a lake.

Résumé. Alors que dans [1] on a décrit qualitativement quelques problèmes d'électricité atmosphérique du point de vue de la théorie de courant, il s'agit dans les études suivantes de traiter plus à fond les questions particulières et d'en préciser par le calcul l'aspect quantitatif pour autant que cela se peut sans surcharger l'appareil mathématique.

Dans le présent article on applique le calcul à des nuages ne donnant pas de précipitations et à des couches de brume ou de brouillard, considérés comme des ellipsoïdes de révolution à conductibilité interne constante. On étudie ensuite les conditions de champ et de courant à l'intérieur et à l'extérieur lorsqu'on place ces corps dans un champ de courant homogène et vertical. On constate que l'augmentation du champ interne dépend non seulement de la conductibilité, mais aussi fortement de la forme géométrique du corps. Dans le cas d'une couche illimitée à deux dimensions (ellipsoïde aplati de révolution dégénéré), il règne à l'intérieur et à l'extérieur la même densité de courant, tandis que les intensités de champ sont inversement proportionnelles aux conductibilités. Pour le cylindre infiniment long (ellipsoïde prolongé de révolution dégénéré), les intensités de champ sont les mêmes à l'intérieur qu'à l'extérieur, tandis que les densités de courant sont proportionnelles aux conductibilités. Entre ces deux cas extrêmes, les intensités de champ et les densités de courant peuvent prendre toutes les valeurs intermédiaires selon le rapport des axes de l'ellipsoïde.

La structure du champ au voisinage immédiat du corps perturbateur est particulièrement intéressante. Si l'on applique les résultats à la couche de brume reposant sur les continents (couche d'échange turbulent), on peut en tirer d'utiles indications pour un enregistrement du champ sur les côtes de la mer ou sur un lac interne.

H. W. Kasemir:

I. The Rotational Ellipsoid in the Homogenous Current Field

For the mathematical representation of the obstructions, such as clouds, layers of haze, etc., which become deposited in the undisturbed atmospheric electrical field F, we will choose the rotational ellipsoid. This can be defined by providing a single coordinate in the appropriately chosen elliptical coordinate system. The formulas for the potential, field and current distribution, etc. obtain a closed and relatively simple form in the process. In addition, the rotational ellipsoid with its devolutions, originating from the layer expanded infinitely over the flat rotational ellipsoid (disc) through the sphere and the elongated rotational ellipsoid up to an infinitely long cylinder, is capable of being altered in its form in such a manner that we can approach a large quantity of the naturally occurring obstructions with sufficient precision through an appropriately selected rotational ellipsoid.

The following designations and dimensions [equation (1)] apply for the occurring variables:

V	=	Potential [V]
E	=	Field strength (general) [V/m]
F	=	Field strength of the undisturbed primary field [V/m]
i	=	Current density [A/m^2]
λ	=	Conductivity [S/m = l/Ohm meter]
x, y, z	=	Cartesian coordinates [m], [m], [m]
r, ϑ, φ	=	Polar coordinates [m], [∢],[∢]
u, v, φ	=	Elliptical coordinates [m], [m], [∢]

Indexes:

1	=	Outer space
2	=	Inner space
s	=	Layer
f	=	Flat rotational ellipsoid
k	=	Sphere
g	=	Elongated rotational ellipsoid
z	=	Cylinder

We need to presuppose the correlation between Cartesian and polar coordinates as known and, in contrast, provide the equation of combination for the elliptical coordinates to the Cartesian. Because we are dealing with rotationally symmetric problems, we can simplify our formulas by always apply $z = 0$ and $\varphi = 0$ without limiting the general validity:

Thus, the following applies with the elliptical coordinates, u, ε for the flat rotational ellipsoid:

$$\frac{y^2}{u^2} + \frac{x^2}{u^2 - \varepsilon^2} = 1 \text{ für } \varepsilon \leqq u \leqq \infty,$$

$$\frac{y^2}{v^2} - \frac{x^2}{\varepsilon^2 - v^2} = 1 \text{ für } 0 \leqq v \leqq \varepsilon, \tag{2 a}$$

and for the elongated rotational ellipsoid,: whereat ε indicates the eccentricity of the confocal pencil of the ellipsoid and hyperboloid.

88

H. W. Kasemir:

$$\frac{x^2}{u^2} + \frac{y^2}{u^2 - \varepsilon^2} = 1 \ \text{für} \ \varepsilon \leqq u \leqq \infty,$$

$$\frac{x^2}{v^2} - \frac{y^2}{\varepsilon^2 - v^2} = 1 \ \text{für} \ 0 \leqq v \leqq \varepsilon, \tag{2 b}$$

Next, we will provide the equations of the potential distribution, which emerges, if we add a flat or elongated rotational ellipsoid or a sphere with the inner conductivity of λ_2 into a homogenous current field with the outer conductivity of λ_1*.

The following applies in the outer space:

$$V_{f1} = -F \frac{\sqrt{(u^2 - \varepsilon^2)(\varepsilon^2 - v^2)}}{\varepsilon} \cdot$$

$$\cdot \left(1 - \frac{(\lambda_2 - \lambda_1)\, p}{\lambda_1 + (\lambda_2 - \lambda_1)\, p} \cdot \frac{\frac{\varepsilon}{\sqrt{u^2 - \varepsilon^2}} - \operatorname{arctg} \frac{\varepsilon}{\sqrt{u^2 - \varepsilon^2}}}{\frac{\varepsilon}{\sqrt{u_0^2 - \varepsilon^2}} - \operatorname{arctg} \frac{\varepsilon}{\sqrt{u_0^2 - \varepsilon^2}}} \right), \tag{3}$$

$$V_{g1} = -F \frac{u\, v}{\varepsilon} \left(1 - \frac{(\lambda_2 - \lambda_1)\, q}{\lambda_1 + (\lambda_2 - \lambda_1)\, q} \cdot \frac{\operatorname{ArTg} \frac{\varepsilon}{u} - \frac{\varepsilon}{u}}{\operatorname{ArTg} \frac{\varepsilon}{u_0} - \frac{\varepsilon}{u_0}} \right), \tag{4}$$

$$V_{k1} = -F\, r \cos \vartheta \left(1 - \frac{\lambda_2 - \lambda_1}{\lambda_2 + 2\, \lambda_1} \cdot \frac{a^3}{r^3} \right) \tag{5}$$

and in the inner space

$$V_{f2} = -F \frac{\lambda_1}{\lambda_1 + (\lambda_2 - \lambda_1)\, q} \cdot \frac{\sqrt{(u^2 - \varepsilon^2)(\varepsilon^2 - v^2)}}{\varepsilon}, \tag{6}$$

$$V_{g2} = -F \frac{\lambda_1}{\lambda_1 + (\lambda_2 - \lambda_1)\, q} \cdot \frac{u\, v}{\varepsilon}, \tag{7}$$

$$V_{k2} = -F \frac{3\, \lambda_1}{\lambda_2 + 2\, \lambda_1}\, r \cos \vartheta \tag{8}$$

with

$$p = \frac{u_0^2 \sqrt{u_0^2 - \varepsilon^2}}{\varepsilon^3} \left(\frac{\varepsilon}{\sqrt{u_0^2 - \varepsilon^2}} - \operatorname{arctg} \frac{\varepsilon}{\sqrt{u_0^2 - \varepsilon^2}} \right) \tag{9}$$

and

$$q = \frac{u_0 (u_0^2 - \varepsilon^2)}{\varepsilon^3} \left(\operatorname{ArTg} \frac{\varepsilon}{u_0} - \frac{\varepsilon}{u_0} \right). \tag{10}$$

The factors p and q appearing in the equations (3), (4), (6) and (7) and defined by (9) and (10) are known as *deelectrification factors* and depend solely on the form of the selected basic ellipsoid u_0. Because deelectrification factors do not appear to be included for this purpose in our problem, we will name these variables simply from now on *form factors*.

* These equations are obtained through the simple transformation to the current field from the formulas, which F. Ollendorff provided for electrostatic problems in [2].

The analogous structure of the equations (3) to (8) is more clearly expressed if we introduce a mixed coordinate system. For this, we will take the x-coordinate from the Cartesian coordinate system and the u or the r-coordinate from the elliptical or the polar coordinate system. The introduction of this somewhat uncommon coordinate system is done for the following reason. For all aircraft ascent previously made known for the analysis of the electrical state of the free atmosphere, only the vertical component – in our coordinate system, the x-coordinate – of the atmospheric electrical field was measured. This is because of the fact that we only expect a vertically-directed field for so-called undisturbed atmospheric electrical relationships. Based on the same reasoning, we have also assumed our undisturbed field F in the x-direction. Once an obstruction is present in this vertically homogenous field F, the other field components naturally emerge. Nonetheless, we are primarily interested in the x-components for the undisturbed field as well because measurement results exist only for these. In the case of the heavy emphasis on the x-direction in our premise (F in the x-direction) as well as for the variables of interest, it is not surprising if the formulas as well as the graphic depictions are simple in the x, u or x, r-coordinate system, as we would otherwise be solely preoccupied with writing when using another coordinate system.

Therefore, if we transcribe the equations (3) to (8), we obtain

$$V_{f1} = -Fx\left(1 - \frac{(\lambda_2 - \lambda_1)\,p}{\lambda_1 + (\lambda_2 - \lambda_1)\,p} \cdot \frac{\dfrac{\varepsilon}{\sqrt{u^2 - \varepsilon^2}} - \operatorname{arctg}\dfrac{\varepsilon}{\sqrt{u^2 - \varepsilon^2}}}{\dfrac{\varepsilon}{\sqrt{u_0^2 - \varepsilon^2}} - \operatorname{arctg}\dfrac{\varepsilon}{\sqrt{u_0^2 - \varepsilon^2}}}\right), \quad (11)$$

$$V_{g1} = -Fx\left(1 - \frac{(\lambda_2 - \lambda_1)\,q}{\lambda_1 + (\lambda_2 - \lambda_1)\,q} \cdot \frac{\operatorname{ArTg}\dfrac{\varepsilon}{u} - \dfrac{\varepsilon}{u}}{\operatorname{ArTg}\dfrac{\varepsilon}{u_0} - \dfrac{\varepsilon}{u_0}}\right), \quad (12)$$

$$V_{k1} = -Fx\left(1 - \frac{(\lambda_2 - \lambda_1)\dfrac{1}{3}}{\lambda_1 + (\lambda_2 - \lambda_1)\dfrac{1}{3}} \cdot \frac{a^3}{r^3}\right) = -Fx\left(1 - \frac{\lambda_2 - \lambda_1}{\lambda_2 + 2\lambda_1} \cdot \frac{a^3}{r^3}\right), \quad (13)$$

$$V_{f2} = -\frac{\lambda_1}{\lambda_1 + (\lambda_2 - \lambda_1)\,p}\,Fx, \quad (14)$$

$$V_{g2} = -\frac{\lambda_1}{\lambda_1 + (\lambda_2 - \lambda_1)\,q}\,Fx, \quad (15)$$

$$V_{k2} = -\frac{\lambda_1}{\lambda_1 + (\lambda_2 - \lambda_1)\dfrac{1}{3}}\,Fx = -\frac{3\lambda_1}{\lambda_2 + 2\lambda_1}\,Fx. \quad (16)$$

Studies regarding the Atmospheric Potential Gradient V.

In this case, the similarity of the three last equations (14) to (16), which are only differentiated by the factor *p, q* or 1/3, is particularly apparent. With the transition to the sphere, *p* and *q* namely assume the value of 1/3. In Figure 1, this form factor is applied subject to the axis ratio b/a of the ellipsoid, in which *a* always signifies the large half-axis and *b* the small half-axis.

In the borderline case of the infinitely long cylinder, this form factor has the value of 0 and continually increases for elongated ellipsoids becoming shorter until it reaches the value of 1/3 for the sphere. When passing through sphere, the elongated ellipsoid becomes flat with the

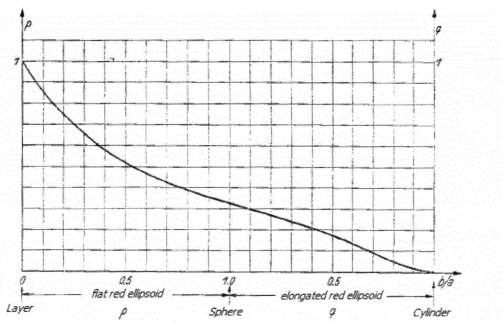

Figure 1. Form factor (deelectrification factor) of the flat and the elongated rotational ellipsoid subject to the axis ratio.

borderline case of the two-dimensionally, infinitely elongated layer. The form factor continues to increase in this process and assumes the value of 1 for the borderline case of the layer.

With the aid of the written equations (11) to (16), we can answer all desired questions. However, we do want to go into a lengthy discussion regarding the potential equations ourselves and we will now turn to the formation of the *x*-components of the field. To differentiate according to *x*, we need the auxiliary equations to be obtained from (2a) and (2b). For the flat rotational ellipsoid

$$\frac{\partial u}{\partial x} = \frac{x\,u\,(u^2 - \varepsilon^2)}{x^2\,\varepsilon^2 + (u^2 - \varepsilon^2)^2} \tag{17}$$

and for the elongated rotational ellipsoid

$$\frac{\partial u}{\partial x} = \frac{x\,u\,(u^2 - \varepsilon^2)}{u^4 - x^2\,\varepsilon^2}. \tag{18}$$

91

With that, the potential equations can be easily differentiated according to x and we obtain the following for the field strength in the x-direction

$$E_{f1} = F \left(1 - \frac{(\lambda_2 - \lambda_1)\, p}{\lambda_1 + (\lambda_2 - \lambda_1)\, p} \cdot \frac{\dfrac{\varepsilon}{\sqrt{u^2 - \varepsilon^2}} \dfrac{(u^2 - \varepsilon^2)^2}{x^2 \varepsilon^2 + (u^2 - \varepsilon^2)^2} - \operatorname{arctg} \dfrac{\varepsilon}{\sqrt{u^2 - \varepsilon^2}}}{\dfrac{\varepsilon}{\sqrt{u_0^2 - \varepsilon^2}} - \operatorname{arctg} \dfrac{\varepsilon}{\sqrt{u_0^2 - \varepsilon^2}}} \right), \tag{19}$$

$$E_{g1} = F \left(1 - \frac{(\lambda_2 - \lambda_1)\, q}{\lambda_1 + (\lambda_2 - \lambda_1)\, q} \cdot \frac{\operatorname{ArTg} \dfrac{\varepsilon}{u} - \dfrac{\varepsilon}{u} \dfrac{u^4}{u^4 - x^2 \varepsilon^2}}{\operatorname{ArTg} \dfrac{\varepsilon}{u_0} - \dfrac{\varepsilon}{u_0}} \right), \tag{20}$$

$$E_{k1} = F \left[1 - \frac{\lambda_2 - \lambda_1}{\lambda_2 + 2\lambda_1} \cdot \frac{a^3}{r^3} \left(1 - \frac{3\, x^2}{r^2} \right) \right], \tag{21}$$

$$E_{f2} = F \frac{\lambda_1}{\lambda_1 + (\lambda_2 - \lambda_1)\, p}, \tag{22}$$

$$E_{g2} = F \frac{\cdot\ \lambda_1}{\lambda_1 + (\lambda_2 - \lambda_1)\, q}, \tag{23}$$

$$E_{k2} = F \frac{3\,\lambda_1}{\lambda_2 + 2\lambda_1}. \tag{24}$$

[In equations (5), (13) and (21), $a = b$ logically indicates the spherical radius.]

We would like to continue to deal with the application of these equations (19) to (24) on the disturbance of the atmospheric electrical field through embedded objects.

II. Expanded Layers of Haze and Fog.

We will observe the expanded layers of mist and fog, which we depict through a flat rotational ellipsoid, as the first example of the application of our equations. Accordingly, equations (19) and (22) will be observed for this problem. It is a known fact and can also be readily seen from the potential and field equations (14) to (16) and (22) to (24) that the field strength is constant in the inside of the ellipsoid and has the same direction as the undisturbed outer field strength. Thus, it only differentiates itself from the latter through a constant factor, which depends upon the ratio of conductivity of the inner and outer space and the form factor. For flat rotational ellipsoids, the axis ratio b/a of which is smaller than 1/10, the form factor p is equal to 1 (Figure 1) with the exception of an error of no more than 15%. Thus, the following is approached

$$\frac{E_{f2}}{F} = \frac{\lambda_1}{\lambda_2}. \tag{25}$$

Studies regarding the Atmospheric Potential Gradient V.

This equation strictly applies for the infinitely expanded layer. It is often used for the observation of broad layers of fog and can be immediately accessed in a stationary current form the condition of continuity of the current. Because the current density in the fog layer and the outer space must be equally large ($i_1 = i_2$), the following results instantly with the aid of the relationship $i_1 = \lambda_1\,F$ and $i_1 = \lambda_2\,E_{s2}$

$$\frac{E_{s2}}{F} = \frac{\lambda_1}{\lambda_2}.$$

This, however, is our equation (25) for the borderline of the layer. Through the derivation of this formula on the indirect route over the flat rotational ellipsoid, in addition to the strictly applicable equation (22), we can achieve the possibility of defining the validity or the error of the formula (25) if we have an object of finite expansion in front of us, as is indeed always the case in reality. Later we will return to the influence of the form factor that was omitted here.

We want to maintain that, as the result, we may only deduce the ratio of the conductivity of clouds to the atmosphere from a field registration within and outside of a layer of fog or cloud if the lateral expansion of the cloud is at least 10 times as tall as its thickness. The error is then already 20%. If we are dealing with ground fog, we must even stipulate a ratio of lateral expansion to overall height as 20:1 due to the reflecting influence of the surface of the earth because, as we will see in a moment, only the upper half of the ellipsoid is available to us for representing the fog.

Furthermore, we will raise the question regarding the degree to which the introduced obstruction influences the outer field, therefore, e.g. in which distance does an approaching layer of fog become noticeable in a field registration. In this regard, we will initially limit ourselves to the observation of the $x = 0$-plane, i.e. the area that divides our ellipsoid in two equally sized half-shells. Because this $x = 0$-plane is additionally an equipotential plane, we can replace it with a conductive plane and thereby obtain a replica of the surface of the earth. Our fog layer then depicted by an upper ellipsoid half-shell lying on the surface of the earth. If we continue to take into account that $y = u$ in this case according to equation (2a), then equation (19) is simplified with $\sqrt{u_0^2 - \varepsilon^2} = b$ to

$$E_{f1} = F\left(1 - \frac{(\lambda_2 - \lambda_1)\,p}{\lambda_1 + (\lambda_2 - \lambda_1)\,p} \cdot \frac{\dfrac{\varepsilon}{\sqrt{y^2 - \varepsilon^2}} - \operatorname{arctg}\dfrac{\varepsilon}{\sqrt{y^2 - \varepsilon^2}}}{\dfrac{\varepsilon}{b} - \operatorname{arctg}\dfrac{\varepsilon}{b}}\right) \qquad (26)$$

for $x = 0$

If we limit ourselves to the flat ellipsoid with an axis ratio of $b/a < 1/10$, then the approximations $p = 1$ apply with an error of no more than 13%, $\varepsilon = a\,(1 - 1/200) = a$ with an

H. W. Kasemir:

error of approx. 0.5%, arctg ε/b = arctg 10 = 1.47 = $\pi/2$ with an error of 7%.

Additionally, it will demonstrate that the disturbance of the outer field only becomes noticeable in a very narrow boundary zone. Therefore, we will introduce the distance of $z = y - a$ from the edge of the ellipsoid instead of the distance y from the center of the ellipsoid. In the process, we will limit ourselves to the distances of z, for which the approximation is applicable

$$y^2 - \varepsilon^2 = 2\,a\,z \doteq b^2.$$

Then, we will obtain the approximation equation from equation (26)

$$\frac{E_{f1}}{F} = 1 + \frac{\lambda_1 - \lambda_2}{\lambda_2} \cdot \frac{\dfrac{1}{\sqrt{2z/a + b^2/a^2}} - \dfrac{\pi}{2}}{a/b - \pi/2}. \tag{27}$$

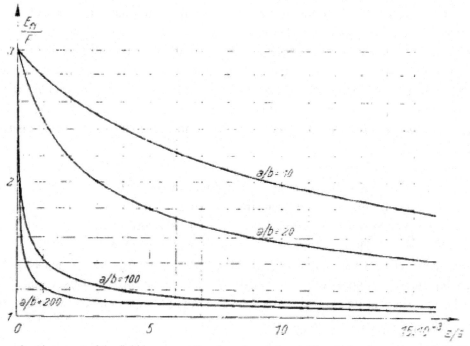

Figure 2. Abatement of the field increase at the edge of a rotational ellipsoid in a homogenous field. Parameters a/b − 10; 20; 100; 200.

In Figure 2, the field strength E_{f1}/F normalized to F and subject to the distance z/a normalized to a is applied for the axis ratios a/b = 10; 20; 100; 200.

In this process, we have selected the ratio $\lambda_1 = \lambda_2 = 3/1$ for the conductivity. When selecting another conductivity ratio, only the ordinate scale changes. We see that the field increase caused by the obstruction strives toward the value of the undisturbed field continually more rapid with a continually growing axis ratio of a/b if we distance ourselves from the edge of the ellipsoid. The field value of E_{f1}/F = 1.02, i.e. an abatement of the field increase to 10% of the maximum value, is reached for the a/b-values provided above at a distance of s = 0.125a; 0.05a; 0.004a; 0.0012a. A layer of fog, e.g. of 2.5m in thickness and approx. 1km in diameter would only effectuate a field increase of 10% of the maximum value with an approximation at 25m.

94

Studies regarding the Atmospheric Potential Gradient V.

The lateral long-distance effect of such flat layers of clouds, fog and mist is extremely remote. The field disturbed by an approaching patch of fog is practically only registered by a measuring device if the fog has reached the measurement site.

One problem also falling under the question regarding the long-distance effect would be the influence of the layer of mist lying over the mainland (exchange layer) on the field registration at sea. This question was answered in [1] with the aid of a simple equivalent circuit diagram to the effect that the influence of the atmospheric resistance, which fluctuate with the exchange events over the mainland, on the universal diurnal variation of the potential gradient is no more than 17% at sea. Based on the calculation conducted here, it is now clear that the main percentage of current, which is forced away by the increased resistance over the mainland, flows into the sea in a narrow boundary zone and that the surroundings around the oceans distant from the coast must therefore be practically undisturbed. However, there are two concerns regarding a quantitative calculation of this reaction with the formulas developed in this case. The precondition for the theoretical calculation pertaining to current was that the conductivity λ_1 in the outer space is constant and that the undisturbed field F is homogenous and the surface of the earth is an infinitely expanded plane. These conditions only apply to our problems if the observed objects are small compared to the radius of the earth. That is, however, no longer the case when transitioning to entire continents. The current forced away, which in our theoretical case provides access to an infinitely large plane for distribution, in reality is limited to the earth's sphere, i.e. to a finite area. In addition, the condition of the constant conductivity in the outer space does not apply because at altitude the reaction of the exchange layer strives toward the current line curve, in which the conductivity of the atmosphere has already increased significantly.

Despite the concerns that exist with the marginal effect of the layer of mist over the mainland against the quantitative evaluation of equation (27), we would still like to analyze an interesting thought process in this direction. If we ignore the $\pi/2$ in the numerator and denominator of the fraction in equation (27), which is allowed with a sufficiently large a/b, then we obtain

$$\frac{E_{f1}}{F} = 1 + \left(\frac{\lambda_1}{\lambda_2} - 1\right) \cdot \frac{b}{\sqrt{2\,z/a + b^2/a^2}}. \tag{28}$$

We can see that the disturbance, i.e. the second term on the right side of equation (28), is dependent upon the altitude of the exchange layer b and the ratio of conductivity λ_1/λ_2. Therefore, in the proximity of the coast, we have the beneficial possibility of analyzing these two variables that are so important for the atmospheric electrical diurnal variation through registrations. If we conduct this registration in the middle of a lake surrounded by land instead of on the coast of a mainland surrounded by ocean, then we may expect a significant increase of this reaction because the current forced away by the mainland is concentrated on the small area of the lake. The diurnal variation of a field registration on land and on a lake would therefore have to run directly inversely to each other.

H. W. Kasemir:

III. Clouds of a Random Lateral Expansion and Elevation

At this point, we will remove the limitation that the obstruction should have a large lateral expansion with regard to their thickness. In addition to the stratus clouds, we will then gather the middle-atmospheric and high cumuli up to the extreme case of the vertically ascending pillars of smoke. Our initial observation applies to the ratio of the inner to the undisturbed outer field strength or the inner and the outer current density subject to the form factor and the ratio of the conductivity. In this regard, we will normalize the excessive inner field to the undisturbed outer field in a similar manner as for equation (28) and obtain the following from equation (22) and (24)

$$\frac{E_{f2}}{F} = \frac{1}{p\dfrac{\lambda_2}{\lambda_1} + 1 - p}, \tag{29}$$

$$\frac{E_{g2}}{F} = \frac{1}{q\dfrac{\lambda_2}{\lambda_1} + 1 - q}, \tag{30}$$

$$\frac{E_{k2}}{F} = \frac{1}{\dfrac{1}{3}\dfrac{\lambda_2}{\lambda_1} + \dfrac{2}{3}}. \tag{31}$$

Through multiplication with λ_2/λ_1, we achieve the respective equations for i_2/i_v.

$$\frac{i_{f2}}{i_1} = \frac{1}{(1-p)\dfrac{\lambda_1}{\lambda_2} + p}, \tag{32}$$

$$\frac{i_{g2}}{i_1} = \frac{1}{(1-q)\dfrac{\lambda_1}{\lambda_2} + q}, \tag{33}$$

$$\frac{i_{k2}}{i_1} = \frac{1}{\dfrac{1}{3} + \dfrac{2}{3}\dfrac{\lambda_1}{\lambda_2}}. \tag{34}$$

In Figure 3, the normalized field and current value are depicted subject to λ_2/λ_1, namely for the layer, the flat rotational ellipsoid with $a/b = 3/1$, the sphere, the elongated rotational ellipsoid with $a/b = 3/1$ and the cylinder. Because we have always assumed λ_1 larger than λ_2, the field strength within the object is naturally always larger or at least identical to the field strength in the outer space. $1 \leq E_2/F$. In contrast, the current density in the cloud is always smaller or at least equal to the outer current density $i_2/i_1 \leq 1$. The equal sign for i_2/i_1 applies, as we have seen above, for the infinitely expanded layer. However, it is interesting that there is an analogous case for the field strength, namely the infinitely long cylinder.

Studies regarding the Atmospheric Potential Gradient V.

In this instance, the field strength in the inner and outer space is equal to and independent from how low the conductivity in the inner space of the cylinder is assumed. On the other hand, with a decreasing conductivity the current becomes lower to the same extent so that the relationship $i_2/\lambda_2 = E_2$ remains fulfilled.

If we briefly look at the ideas in electrical engineering for the purpose of finding a parallel case, this is displayed surprisingly simple in the diagram of two resistors switched subsequently or parallel to each other. The same current flows through two subsequently switched resistors of an electrical circuit, while the voltage drop through them is proportional to the resistance value. However, on resistors switched parallel to each other, the same voltage drop dominates while the current values react opposite to the resistors. The incidents, namely the sphere and the ellipsoid in the current field, can no longer be reproduced through such simple equivalent circuit diagrams comprised of two resistors alone. Here, the marginalization of the current lines in the areas of greater conductivity would have to be taken into account through cross resistors in the equivalent circuit diagram. Although it is possible to attribute these cases as well to the bridge-like equivalent circuit diagrams, at the moment we do not want to further preoccupy ourselves in this regard. We have namely already seen in Section II that, in the case of the current theory, the current density distribution plays a very significant role. However, this can no longer be reflected by a technical equivalent circuit diagram.

We will now return to the intermediate forms, i.e. to the sphere and the ellipsoid in Figure 3. Here it is noteworthy how fast the field increase recedes in the inner space of the object if we transition from the infinitely expanded layer to objects with similar length and height expansions.

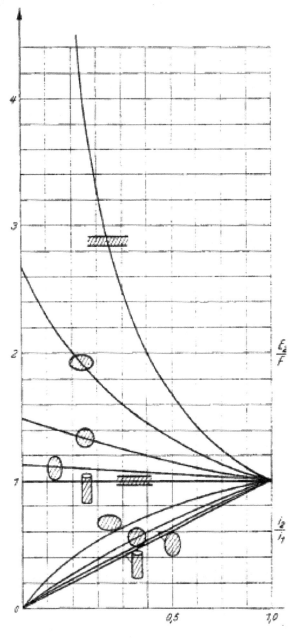

Figure 3. Ratio of outer and inner field strengths and current density subject to the ratio of outer and inner conductivity for the ellipsoid and the sphere in a homogenous current field.

H. W. Kasemir:

The maximum field increase for inner conductivity $\lambda_2 \to 0$ can only be 50% of the outer field for the sphere, while for the expanded rotational ellipsoid with the axis ratio $a/b = 3/1$, i.e. for a high cumulus, in the best scenario, a field increase materializes by 12%, while only a field increase of 8% remains for the average available conductivity ratio $\lambda_2/\lambda_1 = 1/3$. With no insight into the meaning of the form factor, it would be more difficult to understand why such relatively low field strengths are registered in contrast to the stratus clouds in high cumuli free of precipitation, though we must assume the same conductivity values for both types of clouds. Therefore, it is extremely important to enter the ratio of height and length expansion for field measurements. We can already very generally see from equation (29) that the field registration, i.e. the knowledge of E12 and F alone is not even enough to determine the conductivity. If for any reason it is not possible to determine the geometric dimensions of the cloud, we must acquire the aid of the registration of a second atmospheric electrical element, e.g. that of the current. Naturally, we can determine the conductivities directly from the relationship $\lambda_1 = i_1/F$ and $\lambda_2 = i_2/F_{f2}$, although we therefore have the possibility of calculating the form factor only from the electrical measurements. With the knowledge of the geometric dimensions of the cloud through self-observation or measurement, we could mutually inspect the obtained values and test theory derived at this point for its usability*.

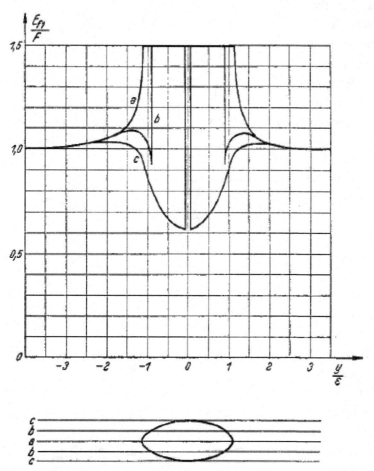

Figure 4. Atmospheric electrical field during a flight through a flat cumulus (rotational ellipsoid) in the horizontal sections *a, b, c*.

_____

* A first contribution for this is contained in the following work by H. Israël and H. W. Kasemir [4].

Studies regarding the Atmospheric Potential Gradient V.

This practical test of the conductivity or the accuracy limit of our theoretical formulas is also stipulated by the deviation of the theoretical assumption of the actual ratios. This is because the continual increase of conductivity with the altitude or the deposit of layers of mist in the atmospheric space, which effectuate a dramatic change of the conductivity of the outer space [3], is added to the difference in the geometric form between a rotational ellipsoid and a cloud. However, at this point we would like to refrain from expanding this theory according to these two directions before the conductivity of this simplified theory is limited not through experimental analysis.

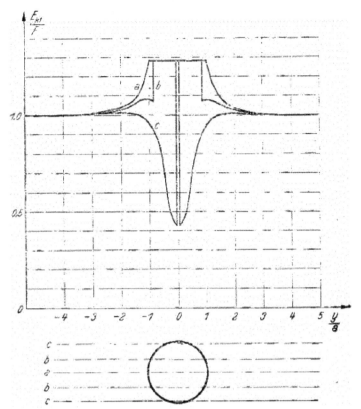

Figure 5. Atmospheric electrical field during a flight through a medium - height cumulus (sphere) in the horizontal sections *a, b, c*.

In conclusion, we would still like to turn to the question regarding what a field registration during cloud-level flights even looks like. If we imagine an aircraft equipped with a field registration device [5] flying through a cloud in various horizontal sections, then Figures 4, 5 and 6 reflect the theoretical field registration curve, namely for a flat, a medium-high and a high cumulus. The respective obstruction is mapped accordingly under the field curves with the flights paths. The field curves have the same characteristic traits with all three cases. If we are separated from the surface of the cloud by a greater distance than correlates, e.g. to its diameter, then a disturbance of the outer field by the cloud cannot be noticed. Only with a greater approximation to the cloud does

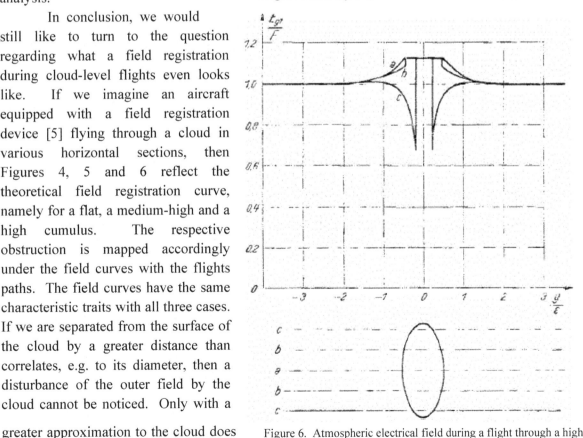

Figure 6. Atmospheric electrical field during a flight through a high cumulus (expanded rotational ellipsoid) in the horizontal sections *a, b, c*.

99

Heinz-Wolfram Kasemir: His Collected Works

H. W. Kasemir: Studies regarding the Atmospheric Potential Gradient V.

the field strength begins to increase more and more. If the flight path goes through the center of the cloud, this increase occurs steadily until the edge of the cloud. The maximum marginal value is remains constant within the cloud and once again transitions into the undisturbed field upon exiting in the same way as when flying into the cloud. However, if we were to fly through the cloud in a section that lies over or under the center of the cloud, the disturbed outer field demonstrates a peculiar variation. Shortly before we reach the edge of the cloud, the field strength decreases again. The preceding field increase is followed by a weakening of the field. It even goes so far near the head and the base of the cloud that the resulting field becomes even smaller as the undisturbed outer field. Only when we fly into the cloud does the field strength suddenly increase to its constant, high inner value. This is a phenomenon that is caused by our unilateral selection of the x-component of the field. The absolute value of the field strength increases steadily in every horizontal section upon approaching the cloud. The x-component, however, only does this on the vertical section that leads through the center of the cloud because it is identical to the absolute value of the field strength in this case. For section lines located lower or higher, however, the y-component contributes a significant portion at to the absolute value of the field strength on the edge of the cloud, which is viewed as a deceleration of the ascent or even as a weakening for the x-component.

If we may not expect such an abrupt jump of field strength – as it occurs in our theoretical curves – when flying into the cloud as a result of the unclear boundary of the cloud or due to the potential presence of inertia of the measuring device, a more or less abraded field jump would still have to be noticeable when flying through the head of the cumulus.

Bibliography.

1. KASEMIR, H. W.: Studien über das atmosphärische Potentialgefälle IV: Zur Strömungstheorie des luftelektrischen Feldes I. Arch. Met. Geoph. Biokl. Ser. A, 3, 84—97 (1950).

2. OLLENDORFF, F.: Potentialfelder der Elektrotechnik. Berlin: Springer-Verlag. 1932.

3. PLUVINAGE, P.: Étude théorique et expérimentale de la conductibilité électrique dans les nuages non orageux. Ann. de Géophys. 2, 31—54, 160—178 (1946).

4. ISRAËL, H. und H. W. KASEMIR: Studien über das atmosphärische Potentialgefälle VI: Beispiele für das Verhalten luftelektrischer Elemente bei Nebel. Arch. Met. Geoph. Biokl. Ser. A, 5, 71—85 (1952).

5. KASEMIR, H. W.: Die Feldkomponentenmühle. Ein Gerät zur Messung der drei Komponenten des luftelektrischen Feldes und der Flugzeugeigenladung bei Flugzeugaufstiegen. Tellus, 3, 240—247 (1951).

ON THE THEORY OF THE ATMOSPHERIC ELECTRIC CURRENT FLOW, IV

Heinz W. Kasemir

DA Task 1A0-11001-3-021-03

Abstract

Starting from the differential equation of the continuity of the current flow, a general solution is given for problems in which the conductivity is given as a function of space and time. A physical interpretation for the complete Maxwell current is obtained, and it is shown how the Maxwell current can be composed of two field vectors, namely, that of the electrostatic field and of the stationary current flow. Each vector is weighted by a different time function, which can be calculated from the time function of the current source.

A method is developed for the calculation of the stationary current flow field; and the eigen functions for cartesian, cylindrical, and polar coordinates are given. The mirror law for the current flow is determined in a medium where the conductivity increases with altitude according to an e-function. The mathematical formalism is explained, using the example of the field of a decaying point source.

U. S. ARMY ELECTRONICS RESEARCH AND DEVELOPMENT LABORATORIES
FORT MONMOUTH, NEW JERSEY

ON THE THEORY OF THE ATMOSPHERIC ELECTRIC CURRENT FLOW, IV

INTRODUCTION

Problems in atmospheric electricity are usually treated according to the electrostatic theory in spite of the fact that a three-dimensional current flow is being dealt with here. The reason for this unsatisfactory state is that the electrostatic theory is worked out in great detail in numerous text books of mathematics and physics,[1,2,3] but problems of the three-dimensional current-flow theory--especially with a variable conductivity or a convection current--are only briefly discussed, if at all. Therefore, the author started to publish some of his notes on the current-flow theory under the general title, "Zur Stroemungs-theorie des luftelektrischen Feldes I, II, III." [4,5,6] Also, two papers,[7,8] which deal with the calculation of the three-dimensional current flow under certain boundary conditions.

Preceding the author,[8] Holzer and Saxon[9] published their famous paper, "Distribution of Electrical Conduction Currents in the Vicinity of Thunderstorms," and in 1955 Tamura presented at the First International Conference on Atmospheric Electricity his excellent "Analysis of Electric Field after Lightning Discharges." [10] In both papers the calculation was carried out according to the current-flow theory. But in the majority of publications in the field of atmospheric electricity, the electrostatic theory is still used for calculations, and this quite often leads to erroneous conclusions.

It soon became obvious that the solution of a few single problems was not enough to introduce the application of the theory of current flow to atmospheric electric problems in general. A more systematic discussion of the methods of solution, with a strong accent on the physical meaning of the mathematical manipulations and the consequences of the introduced assumptions, is necessary. This discussion is given in this report.

DISCUSSION

The Differential Equation of the Three-Dimensional Current Flow

Table I gives a list of the symbols used, with their meanings and dimensions.

In the electrostatic theory, only the quantities on the left side of the list are present; while in the current-flow theory there are, in addition, the four quantities of the right side. It is seen that the dimensions of the quantities of a corresponding pair differ only by the addition of the time unit, s, to the dimension of the electrostatic quantity. This will prove very fortunate in the conversion of solutions of the electrostatic theory into these of the corresponding solution of the current-flow theory, as will be shown later. However, this "fortunate" relationship has a deeper physical meaning, which becomes much clearer if the dimension of the charge cb (coulomb) instead of the current A (ampere) is introduced into the dimension of the quantities.

Table I. Symbols Used, with Their Meanings and Dimensions

$Q \left[As \right]$ = charge $\qquad$ $I \left[A \right]$ = strength of current source

$q \left[\dfrac{As}{m^3} \right]$ = space charge density $\qquad$ $\omega \left[\dfrac{A}{m^3} \right]$ = space current source density

$\vec{D} \left[\dfrac{As}{m^2} \right]$ = dielectric displacement

$\vec{i} \left[\dfrac{A}{m^2} \right]$ = current density

$\varepsilon \left[\dfrac{As}{Vm} \right]$ = dielectric constant

$\lambda \left[\dfrac{A}{Vm} \right]$ = conductivity

$\vec{E} \left[\dfrac{V}{m} \right]$ = field strength

$\phi \left[V \right]$ = potential function

With the relation As = cb, the following are obtained:

$Q \quad \left[cb \right]$ $\qquad$ $I \quad \left[\dfrac{cb}{s} \right]$

$q \quad \left[\dfrac{cb}{m^3} \right]$ $\qquad$ $\omega \quad \left[\dfrac{cb}{sm^3} \right]$

$\vec{D} \quad \left[\dfrac{cb}{m^2} \right]$ $\qquad$ $\vec{i} \quad \left[\dfrac{cb}{sm^2} \right]$

$\varepsilon \quad \left[\dfrac{cb}{Vm} \right]$ $\qquad$ $\lambda \quad \left[\dfrac{cb}{sVm} \right]$

In this way the time has been removed from the electrostatic quantities, where it does not belong, and introduced into the current-flow theory, where one would expect it to be. If the time is now extended, e.g., one second to infinity, then the movement of all particles would freeze and everything would be static. The four specific quantities of the current-flow theory, I, ω, $\vec{i}$, and λ, would vanish because the time is in the denominator of the dimensions, and the electrostatic conditions are obtained. So, one could say that electrostatic is a special case of the current-flow theory.

In the current-flow theory, all the quantities listed above can be functions of space and time. The problems encountered are mostly of the following type: Given are the spatial distribution and the time function of the current source ω and the electrical properties of space ε and λ as functions of space and time. This includes also the boundary condition such as a perfect conducting earth surface, or sometimes an ionosphere of infinite conductivity. The problem is to calculate the potential function and the field and current distribution.

The dielectric constant ϵ is always a true constant in space and time, since the difference of ϵ in air and in a vacuum is negligible. The manner in which the conductivity λ is given determines the grade of difficulty of the problem, and to some extent also the method of solution. Therefore, the problems may be subdivided into the following six classes:

1. The conductivity is constant in space and time.

2. The conductivity is constant in space, but a function of time.

3. The conductivity is constant in time, but a function of space.

4. The conductivity is a function of space and time.

5. The conductivity is a tensor (influence of the earth's magnetic field in the ionosphere).

6. The assumption for the definition of the conductivity is no longer valid.

This is the case in outer space, where the mean free path of the moving particle is greater than either the antenna of the measuring instrument or the space under consideration.

Classes 1, 2, and 3 are special cases of class 4. Class 4, again, could be considered as a special case of 5. However, the solution of 5 as well as of 6 class problems requires a much larger mathematical effort. Therefore, this report is limited to a general solution of 1 to 4 class problems.

The solution to class 1 problems can be obtained easily if the corresponding electrostatic solution is known. It is necessary only to substitute for the electrostatic quantities, Q, q, $\vec{D}$, and ϵ, the corresponding current-flow quantities I, ω, $\vec{i}$, and λ of table I in the formulas of the potential function ϕ or of the electric field distribution $\vec{E}$. For instance, the potential function $\phi = \frac{Q}{4\pi\epsilon}\frac{1}{r}$ of a point charge leads to the potential function $\phi = \frac{I}{4\pi\lambda}\frac{1}{r}$ of a point current source.

The applicability of class 1 problems in atmospheric electricity seems to be very limited as, for instance, to the calculation of the effective altitude or area of field or current antennas, where the conductivity in the space considered may be assumed to be constant. But it is shown later that the electrostatic solution is needed in addition to that of the stationary current flow for the general solution of class 3 and 4 problems. Also, the important matching condition[11] for air-earth-current measurements can be obtained with the assumption that the conductivity is constant in space and time.

If the conductivity changes with time but is constant in space, this would result in mismatching, which is a problem belonging to class 2. An analysis of these conditions for air-earth-current measurement at the ground is given by Ruhnke.[12] At a larger scale the same problem is encountered by the air-earth-current radiosonde,[13] where the conductivity increases during the flight time from the low ground value by a factor 100 or more to that in the higher altitudes. Here the assumption is made that the conductivity can be considered as constant in the space occupied by the antennas of the radiosonde, but will increase in time with the ascent of the balloon.

The calculations given by the author[5,7,13] belong to class 3.

There is no example of class 4 problems in the literature. This type of problem will be encountered by the calculation of the austausch generator, where the air turbulence will change the conductivity in space and time and also cause a convection current which is a function of space and time.

The differential equation of the three-dimensional current flow, which has to be fulfilled by all solutions, is derived from the second Maxwell equation and is well known as the equation of the continuous current flow. It is written in the following form:

$$\text{div} \left(\frac{\partial \vec{D}}{\partial t} + \vec{i} \right) = \omega. \tag{1}$$

The intensity of the current production is usually given by a convection current $\vec{\zeta}$. In this case the space current source density ω is given by div $\vec{\zeta}$. If this is introduced into Eq. (1),

$$\text{div} \left(\frac{\partial \vec{D}}{\partial t} + \vec{i} \right) = \text{div} \, \vec{\zeta}. \tag{2}$$

From Eq. (2) follows

$$\frac{\partial \vec{D}}{\partial t} + \vec{i} = \vec{\zeta} + \vec{c}. \tag{3}$$

$\vec{c}$, which appears here as a kind of integration constant with the condition div $\vec{c} = 0$, is known as the complete Maxwell current. It is a current flow, which originates at the boundaries of the considered space. In thunderstorm problems, $\vec{c}$ would represent the fair-weather current between the ionosphere and the earth, which penetrates the thunderstorm areas as well as the fair-weather areas. However, the fair-weather current density is much smaller than the current density caused by the thunderstorm, and therefore it is usually neglected. This cannot be done in the case of the austausch generator, which works in the fair-weather region. Here, the convection current density of the austausch generator $\vec{\zeta}$ and the current $\vec{c}$ produced by the charged earth are of the same order of magnitude. (The conduction current density of $\vec{c}$ is the well-known air-earth-current density.) Therefore, in this case, both current sources have to be considered. But as the principle of superposition also holds in the current-flow theory, the current distribution of the two sources may be calculated separately, and the solutions

105

superimposed. The dielectric displacement $\vec{D}$ and the current density $\vec{I}$ are connected with the electric field $\vec{E}$ through the following two equations:

$$\epsilon \, \vec{E} = \vec{D}. \tag{4}$$

$$\lambda \, \vec{E} = \vec{I}. \tag{5}$$

(If λ is a tensor Λ as in class 5 problems, $\Lambda \cdot \vec{E} = \vec{I}$, and from here on the calculation of class 5 problems would branch off.) With Eqs. (4) and (5), $\vec{E}$ is substituted for $\vec{I}$ and $\vec{D}$ in (3), and

$$\frac{\partial \vec{E}}{\partial t} + \frac{\lambda}{\epsilon} \vec{E} = \frac{1}{\epsilon} (\vec{\zeta} + \vec{c}) \tag{6}$$

is obtained. The solution of this differential equation is

$$\vec{E} = e^{-\int \frac{\lambda dt}{\epsilon}} \left(\vec{E}_{(t=0)} + \frac{1}{\epsilon} \int (\vec{\zeta} + \vec{c}) \, e^{\int \frac{\lambda dt}{\epsilon}} \, dt \right). \tag{7}$$

This is the general solution for $\vec{E}$ for all class 1 to 4 problems. The integration constant $\vec{E}_{(t=0)}$ gives the field distribution for t=0. On account of the integration constant, the integral can be written without boundaries; and the letter t, which ordinarily would appear as the upper boundary of the integral, is used as the integration variable of the integral.

According to the above-mentioned principle of superposition, the field-vector $\vec{E}$ may be split up into two parts; namely, $\vec{E}_\zeta$, which results from the convection current $\vec{\zeta}$, and $\vec{E}_b$, which results from current emitted at the boundaries.

$$\vec{E}_\zeta = e^{-\int \frac{\lambda dt}{\epsilon}} \left(\vec{E}_{\zeta(t=0)} + \frac{1}{\epsilon} \int \vec{\zeta} \, e^{\int \frac{\lambda dt}{\epsilon}} \, dt \right). \tag{8}$$

and

$$\vec{E}_b = e^{-\int \frac{\lambda dt}{\epsilon}} \left(E_{b(t=0)} + \frac{1}{\epsilon} \int \vec{c} \, e^{\int \frac{\lambda dt}{\epsilon}} \, dt \right). \tag{9}$$

Equation (8) cannot be discussed without further knowledge of the convection current $\vec{\zeta}$. The analysis of Eq. (9), especially the physical meaning of the complete Maxwell current $\vec{c}$ is given below.

The Solution of Boundary Problems without Convection Currents

Set $\vec{\zeta} = 0$, and from Eq. (6) the electric field $\vec{E}_b$ due to the boundary conditions may be obtained. With the further omission of the subscript b, it is

$$\frac{\partial \vec{E}}{\partial t} + \frac{\lambda}{\epsilon} \vec{E} = \frac{\vec{c}}{\epsilon} \; . \tag{10}$$

The form and the physical meaning of the complete Maxwell current $\vec{c}$ will now be deduced from the three boundary values $\vec{c}_b$, $\vec{c}_e$, and $\vec{c}_s$ which $\vec{c}$ has to assume, first at the boundary $(\vec{c}_b)$, second, if $\lambda = 0$ $(\vec{c}_e)$, and third, if $\frac{\partial \vec{E}}{\partial t} = 0$ $(\vec{c}_s)$.

At the boundary of the current source, the field E_b is given by the field strength F_b and the time function T of the current source. Therefore,

$$\vec{E}_b = \vec{F}_b \cdot T. \tag{11}$$

$\vec{F}_b(x,y,z)$ is hereby a function of the space coordinates x, y, and z only, and $T_{(t)}$ is a pure time function. This implies that on all parts of the boundary of the current source the field changes with the same time function. From Eqs. (10) and (11) follows

$$\frac{\vec{c}_b}{\epsilon} = \vec{F}_b \frac{dT}{dt} + \frac{\lambda_b}{\epsilon} \vec{F}_b \, T. \tag{12}$$

If $\lambda = 0$, the electrostatic field distribution is obtained, which will be indicated by a subscript e at the Maxwell current $\vec{c}_e$ and the field vector $\vec{F}_e$. Furthermore, the finite propagation velocity of the electric field with the speed of light will be neglected, which means that the electric field $\vec{E}$ follows everywhere the time function T of the source momentarily. $\vec{E} = \vec{F}_e \cdot T$. Hereby excluded are all problems where the propagation velocity of electric signals becomes important, for instance, by the electromagnetic wave of a lightning discharge (sferics). From Eq. (10) follows

$$\frac{\vec{c}_e}{\epsilon} = \vec{F}_e \frac{dT}{dt}. \tag{13}$$

If the time function of the current source is a constant, the condition of the stationary current flow is obtained, which is indicated by the subscript s. With $\frac{\partial \vec{E}_s}{\partial t} = 0$ and $\vec{E}_s = \vec{F}_s \, T$, from Eq. (10) is obtained

$$\frac{\vec{c}_s}{\epsilon} = \frac{\lambda}{\epsilon} \vec{F}_s \, T. \tag{14}$$

Equations (12), (13), and (14) indicate the form which the Maxwell current $\vec{c}$ in general will assume. It may be inferred that $\vec{c}$ is composed in the following manner:

$$\vec{c} = \epsilon \vec{F}_e \frac{dT}{dt} + \lambda \vec{F}_s \, T. \tag{15}$$

107

Equation (15) fulfills all the required conditions. At the boundary, where $\vec{F}_{eb} = \vec{F}_{sb} = \vec{F}_b$, Eq. (15) changes to Eq. (12), and therefore meets the boundary conditions. Furthermore, div $\vec{c} = 0$, because div $\vec{F}_e = 0$ as well as div $\lambda \vec{F}_s = 0$. Therefore the complete Maxwell current $\vec{c}$ is given in the form of Eq. (15).

If Eq. (15) is introduced into Eq. (9), with the omission of the subscript b, with $\vec{E}'_{(t=0)} = \vec{F}_0$,

$$\vec{E} = \vec{F}_0 \cdot e^{-\int \frac{\lambda dt}{\varepsilon}} + \vec{F}_e \, e^{-\int \frac{\lambda dt}{\varepsilon}} \int \frac{\partial T}{\partial t} e^{\int \frac{\lambda dt}{\varepsilon}} \, dt$$

$$+ \vec{F}_s \, e^{-\int \frac{\lambda dt}{\varepsilon}} \int \frac{\lambda}{\varepsilon} T \, e^{\int \frac{\lambda dt}{\varepsilon}} \, dt. \tag{16}$$

The physical meaning of Eq. (16) becomes much clearer if the following abreviations are introduced for the time-function factors of $\vec{F}_0$, $\vec{F}_e$, and $\vec{F}_s$. It will be

$$T_e = e^{-\int \frac{\lambda dt}{\varepsilon}} \int \frac{dT}{dt} e^{\int \frac{\lambda dt}{\varepsilon}} \, dt. \tag{17}$$

$$T_s = e^{-\int \frac{\lambda dt}{\varepsilon}} \int \frac{\lambda}{\varepsilon} T e^{\int \frac{\lambda dt}{\varepsilon}} \, dt. \tag{18}$$

$$T_0 = e^{-\int \frac{\lambda dt}{\varepsilon}}. \tag{19}$$

By partial integration of Eq. (17),

$$T_e = T - e^{-\int \frac{\lambda dt}{\varepsilon}} \int \frac{\lambda}{\varepsilon} T e^{\int \frac{\lambda dt}{\varepsilon}}$$

is obtained, and with Eq. (18),

$$T_e = T - T_s, \quad T = T_e + T_s. \tag{20}$$

The sum of the time functions of the electrostatic field and the stationary current-flow field is the time function of the current source. These time-function factors are obtained by the prescribed integration process on the time function T of the source. They are different at each point in space, because λ is a function of space; and they are of different form for the electrostatic field $\vec{F}_e$, the stationary current-flow field $\vec{F}_s$, and the decaying initial field $\vec{F}_0$.

If Eqs. (17), (18), and (19) are introduced into (16),

$$\vec{E} = \vec{F}_O \, T_O + \vec{F}_e \, T_e + \vec{F}_s \, T_s. \tag{21}$$

The electric field $\vec{E}$ is composed of three vectors, each of them modulated by a different time function. $\vec{F}_O$ is the field distribution at the time $t = 0$ and decays with the time constant of the observation point. $\vec{F}_e$ is the electrostatic field, and the weighting function T_e is zero for a constant-current source, but becomes very large for fast time variations of the source. $\vec{F}_s$ is the field of the stationary current flow, and the factor T_s becomes one for a constant current source. It is pointed out that in general $\vec{F}_O$, $\vec{F}_e$, and $\vec{F}_s$ do not have the same directions, and the rules of vector addition have to be applied. After the decay of $\vec{F}_O$, the field-vector $\vec{E}$ is limited to the space between $\vec{F}_e$ and $\vec{F}_s$.

To illustrate the point, the field lines of a point source are drawn in Fig. 1 for the electrostatic case (broken lines), and for the stationary current-flow case in which the conductivity increases with altitude according to an e-function (solid lines). It is seen that for almost any point in space the field direction and amplitude--the amplitude is indicated by the spacing of the lines--is different in the two cases. If one arbitrary observation point is selected, a picture of the composition of the field-vector $\vec{E}$ is obtained, as given in Fig. 2. The weighting time functions will change with time, and accordingly the vector $\vec{E}$ will change its amplitude and shift its position inside the hatched space.

For a fast time variation, $\vec{E}$ is predominantly given by $\vec{F}_e$; and for slow time variation, $\vec{E}$ will move back to the position of $\vec{F}_s$. Hence it becomes obvious that for a complete solution the electrostatic field has to be calculated as well as the field of the stationary current flow. The electrostatic, the current flow, and the initial field $\vec{F}_e$, $\vec{F}_s$, and $\vec{F}_O$ are given by the gradient operation from their potential functions ϕ_e, ϕ_s, and ϕ_O, respectively.

$$\vec{F}_e = - \text{grad } \phi_e, \quad \vec{F}_s = - \text{grad } \phi_s, \quad \vec{F}_O = - \text{grad } \phi_O. \tag{22}$$

Hence, the electric field $\vec{E}$ is also given by the gradient of a potential function ϕ. $\vec{E} = - \text{grad } \phi$. Therefore, with respect to Eq. (21),

$$\text{grad } \phi = T_O \text{ grad } \phi_O + T_e \text{ grad } \phi_e + T_s \text{ grad } \phi_s. \tag{23}$$

Equations (21) and (23) are the general solutions to all class 1 to 4 problems in the absence of a convection current.

Solution of the Differential Equation of the Stationary Current Flow

If the current flow is stationary, derivatives with respect to time are zero, and from Eq. (2) is obtained the differential equation of the stationary current flow.

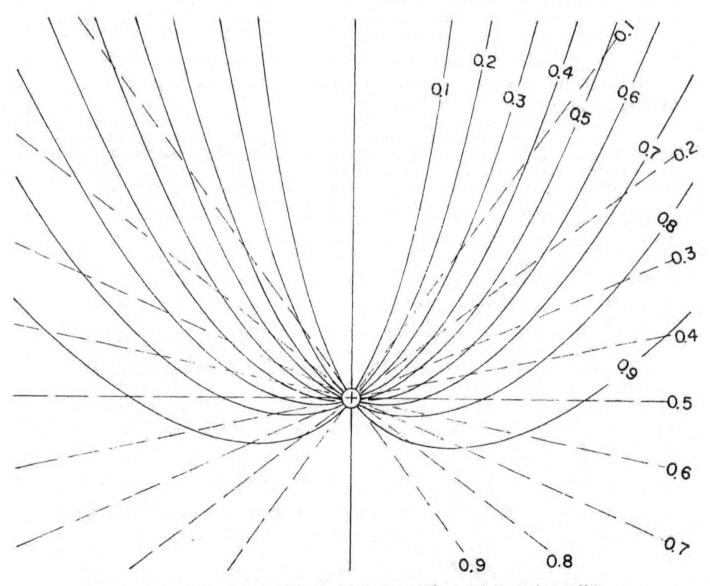

Fig. 1. Lines of the electric field or current flow of 1) a point source in a medium of constant conductivity (broken lines), and 2) a point source in a medium of conductivity, which increase with altitude (solid lines).

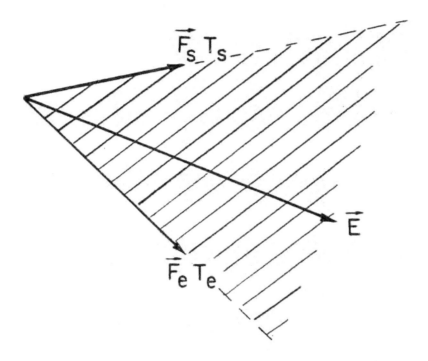

Fig. 2. Composition of the general field vector $\vec{E}$ by vector addition from the electrostatic field vector $\vec{F_e}$ weighted by the time function T_e and the stationary current flow field vector $\vec{F_s}$ weighted by the time function T_s.

$$\operatorname{div} \vec{i_s} = 0. \tag{24}$$

With $\vec{i_s} = \lambda \vec{F_s}$ and $\vec{F_s} = -\operatorname{grad} \phi_s$,

$$\operatorname{div} \lambda \operatorname{grad} \phi_s = 0. \tag{25}$$

The following self-explanatory operations are carried out on this equation:

$$\operatorname{div} \lambda \operatorname{grad} \phi = 0.$$

$$\lambda \operatorname{div} \operatorname{grad} \phi + \operatorname{grad} \lambda \operatorname{grad} \phi = 0.$$

Divide by $\lambda^{\frac{1}{2}}$, add and subtract $\phi \operatorname{div} \operatorname{grad} \lambda^{\frac{1}{2}}$.

$$\lambda^{\frac{1}{2}} \operatorname{div} \operatorname{grad} \phi + 2 \operatorname{grad} \lambda^{\frac{1}{2}} \operatorname{grad} \phi + \phi \operatorname{div} \operatorname{grad} \lambda^{\frac{1}{2}} - \phi \operatorname{div} \operatorname{grad} \lambda^{\frac{1}{2}} = 0.$$

$$\operatorname{div}(\lambda^{\frac{1}{2}} \phi) - \phi \operatorname{div} \operatorname{grad}(\lambda^{\frac{1}{2}} = 0.$$

$$\frac{\operatorname{div} \operatorname{grad} \lambda^{\frac{1}{2}} \phi}{\lambda^{\frac{1}{2}} \phi} - \frac{\operatorname{div} \operatorname{grad} \lambda^{\frac{1}{2}}}{\lambda^{\frac{1}{2}}} = 0. \tag{26}$$

Here are defined two new functions:

$$M = \lambda^{\frac{1}{2}} \phi \text{ and } N = \lambda^{\frac{1}{2}}. \tag{27}$$

Introducing these new functions into Eq. (24), the relatively simple equation is obtained:

$$\frac{\operatorname{div} \operatorname{grad} M}{M} = \frac{\operatorname{div} \operatorname{grad} N}{N}. \tag{28}$$

With this equation there are available all the solutions of the electrostatic theory div grad M = 0, if it is possible to present λ by a function N^2, of which div grad N = 0.

It is pointed out that different coordinate systems may be used for the representation of M and N because the operation div grad is invariant against a change of coordinates.

As an example, assume that it is desired to calculate the potential function of a point source imbedded in a medium where the conductivity can be represented or approximated in the space under consideration by a suitable piece of a parabola.

$$\lambda = \lambda_0 (mz - a)^2. \tag{29}$$

The parameters m, a, and λ_0 can be used to select the best-fitting piece of the parabola.

From Eq. (27) follows

$$N = \lambda^{\frac{1}{2}} = \lambda_0^{\frac{1}{2}} (mz - a), \tag{30}$$

and

$$\text{div grad } N = \frac{d^2N}{dz^2} = 0. \tag{31}$$

Introducing this result in Eq. (28),

$$\text{div grad } M = 0. \tag{32}$$

This equation means that any known electrostatic potential function or any sum of them may be chosen in any coordinate system which suits the purpose best. For a point source, polar coordinates would certainly be chosen, where the function M is given by

$$M = \frac{A}{r}. \tag{33}$$

A is here a constant given by the strength of the source I, and r is the distance of the observation point from the center of the point source. If the point source is located in the cartesian coordinates at the point x_0, y_0, z_0, then r is given by $r = \left[(x-x_0)^2 + (y-y_0)^2 + (z-z_0)^2\right]^{\frac{1}{2}}$. The potential function is now easily obtained by Eqs. (27), (29), and (33).

$$\phi = \frac{M}{\lambda^{\frac{1}{2}}} = \frac{A}{\lambda^{\frac{1}{2}} (mz - a) r}. \tag{34}$$

The determination of A remains as the last step. If the point source is inclosed by a small sphere of the radius r_0, which is chosen so small that the conductivity λ can be considered as constant with the value at the center of the sphere $\lambda = \lambda_0 (m z_0 - a)^2$, then, according to Eq. (34),

$$\phi = \frac{A}{\lambda_0^{\frac{1}{2}} (m z_0 - a) r}. \tag{35}$$

The field in the r direction on the little sphere $r = r_0$ is given by

$$F = -\frac{\partial \phi}{\partial r} = \frac{A}{\lambda_0^{\frac{1}{2}} (m z_0 - a)} \frac{1}{r_0^2},$$

and the current density according to Ohm's law $i = \lambda F$,

$$i_r = \frac{A \lambda_0 (m z_0 - a)^2}{\lambda_0^{\frac{1}{2}} (m z_0 - a) r_0^2} = \frac{A \lambda_0^{\frac{1}{2}} (m z_0 - a)}{r_0^2}.$$

Integration over the surface O of the little sphere gives the current strength I of the source.

$$I = \int_0 i_r \, dO = \frac{A \lambda_0^{\frac{1}{2}} (m z_0 - a)}{r_0^2} 4 \pi r_0^2,$$

113

or

$$A = \frac{I}{4 \pi \lambda_0^{\frac{1}{2}} (m z_0 - a)} . \tag{36}$$

If this result is introduced in Eq. (34), the final solution for the potential function ϕ_s of the problem is obtained.

$$\phi_s = \frac{1}{4 \pi \lambda_0 (m z_0 - a)} \cdot \frac{1}{(m z - a) \cdot r} . \tag{37}$$

Unfortunately, in atmospheric electric problems, the conductivity is usually represented by an e-function. For instance, in a cartesian coordinate system x, y, z, in the manner

$$\lambda = \lambda_0 e^{2 kz} . \tag{38}$$

The x,y plane with z = 0 represents hereby the earth's surface, and λ_0 the conductivity at the ground. This function (given in Eq. (38)) does not fulfill Laplace's equation, but it is

$$\text{div grad } \lambda^{\frac{1}{2}} = \lambda_0^{\frac{1}{2}} k^2 e^{kz} = k^2 N, \text{ and}$$

$$\frac{\text{div grad } N}{N} = k^2 . \tag{39}$$

Therefore, the function M is given as the solution of the differential equation

$$\text{div grad } M - k^2 M = 0. \tag{40}$$

To solve Eq. (40), the method of the eigen functions will be applied, but confined to problems of rotational symmetry. This means that, for instance, in polar coordinates r, Θ, ϕ, the function M is given by the product of a function $f_{(r)}$, which depends on r only, and another function $g_{(\Theta)}$, which depends only on Θ.

$$M = f_{(r)} g_{(\Theta)} . \tag{41}$$

This leads with Eq. (40) to the differential equation

$$\text{div grad } f \cdot g - k^2 f \cdot g = 0,$$

or, in polar coordinates,

$$g \left(\frac{d^2 f}{dr^2} + \frac{2}{r} \frac{df}{dr} \right) + \frac{f}{r^2 \sin \Theta} \frac{d}{d\Theta} \sin \Theta \frac{dg}{d\Theta} - k^2 fg = 0, \tag{42}$$

or

$$\frac{1}{k^2 f} \left(\frac{d^2 f}{dr^2} + \frac{2}{r} \frac{df}{dr} \right) + \frac{1}{r^2 k^2 g \sin \Theta} \frac{d}{d\Theta} \sin\Theta \frac{dg}{d\Theta} - 1 = 0. \tag{43}$$

114

If $\quad \dfrac{1}{g \sin \Theta} \dfrac{d}{d\Theta} \sin \Theta \dfrac{dg}{d\Theta} = - n\,(n+1)$ (44)

where $n = 0, 1, 2, 3, \ldots$, the spherical functions of the first kind P_n (Legendre's functions) are obtained as a solution for g.

$$g = P_{n(\cos \Theta)}.$$ (45)

If Eq. (44) is inserted in (43), the differential equation

$$\dfrac{d^2f}{dr^2} + \dfrac{2}{r} \dfrac{df}{dr} - \left(\dfrac{n\,(n+1)}{r^2} + k^2 \right) f = 0 \quad \text{is obtained for f.}$$ (46)

Equation 46 can be solved with the help of cylinder functions Z_p, with a noninteger parameter $p = n + \frac{1}{2}$. The solution is given by

$$f = (j\,k\,r)^{-\frac{1}{2}} Z_{n+\frac{1}{2}} (j\,k\,r).$$ (47)

Thereby, j is the imaginary unit $j = \sqrt{-1}$.

The complete solution for M follows from Eqs. (41), 45), and (47).

$$M_n = A_n\,P_n\,(j\,k\,r)^{-\frac{1}{2}} Z_{n+\frac{1}{2}} (j\,k\,r).$$ (48)

A_n is an arbitrary constant, which serves to meet the prescribed boundary condition. M may be presented in a more general way as a super-position of the M_n.

$$M = \sum_{n=0}^{\infty} M_n.$$ (49)

The cylinder functions Z_p of a noninteger parameter p can be expressed in given polynomials of the argument and sin and cos functions. For instance, for n = 0,

$$Z_{\frac{1}{2}} = \left(\dfrac{2}{\pi\,j\,k\,r} \right)^{\frac{1}{2}} \sin j\,k\,r.$$

With $P_0 = 1$, and $\sin j\,k\,r = j \sin h\,k\,r = \dfrac{j}{2} \left[\exp\,(kr) - \exp\,(-kr) \right]$,

$$M_0 = A_0 \left[\dfrac{\exp\,(kr)}{kr} - \dfrac{\exp\,(-kr)}{kr} \right].$$ (50)

The constant factor $(2\,\pi)^{-\frac{1}{2}}$ is hereby combined with A_0.

Each of the e-function terms in the bracket of Eq. (50) fulfills the differential equation. Hence, if the boundary condition requests a pole $(M_0 \to \infty)$ at the origin $r = 0$, only the second term is used. This will then lead to the potential function of a point source.

According to Eq. (27), the potential function is given by

$$\phi = M\,\lambda^{-\frac{1}{2}}.$$

Using only the second term in the bracket of Eq. (50), and letting A_0 take care of the sign and the constant factor $\lambda_0^{-\frac{1}{2}}$, the potential function of a point source is obtained.

$$\phi = A_0 \frac{\exp\left[-k(r + z)\right]}{kr} . \tag{51}$$

In a manner very similar as before, it is possible to calculate from Eq. (51) the field strength in the r direction on the surface of a very small sphere around the current source I, convert the field into the current density, and integrate over the surface. This gives the strength of the current source I, and the constant A_0 can be determined. It is

$$A_0 = \frac{I\,k}{4\,\pi\,\lambda_0} , \tag{52}$$

whereby λ_0 denotes the conductivity value at the current source. From Eqs. (51) and (52), the final solution is obtained:

$$\phi = \frac{I}{4\,\pi\,\lambda_0} \frac{\exp\left[-k(r + z)\right]}{r} . \tag{53}$$

For k = 0, Eq. (38) is $\lambda = \lambda_0 \exp(2\,kz) = \lambda_0$, a constant, and Eq. (53) simplifies to

$$\phi = \frac{I}{4\,\pi\,\lambda_0} \cdot \frac{1}{r} .$$

This is the right solution for the point source in a medium of constant conductivity.

The equation (53), which follows here as the simplest solution of Eqs. (48) and (49), is the key solution in the calculation of the electric field of thunderstorms of Holzer and Saxon,[9] in the calculation of the recovery curve of lightning discharges by Tamura,[10] and in the calculation of the thunderstorm generator by the author.[8]

It is mentioned here that a similar procedure to that used for polar coordinates led to the eigen functions of the differential equation, div grad M - k^2M = 0 in cylindrical coordinates z, R, ϕ. Again assuming rotational symmetry,

$$M_n = A_n Z_{0(n\,k\,R)} \exp\left[\pm(1 - n^2)^{\frac{1}{2}} k\,z\right] . \tag{54}$$

Z_0 is the cylinder function of the order zero. n is here not confined to an integer, but may assume any value. For n = 0,

$$M_0 = A_0 \exp(\pm kz).$$

116

It is seen that the root of the conductivity function $\lambda^{\frac{1}{2}} = \lambda_0^{\frac{1}{2}} e^{kz}$ is the eigen function of the order zero in cylindrical coordinates. In cartesian coordinates the eigen functions are given by

$$M_n = A_n \exp(\pm c_1 x) \exp(\pm c_2 y) \exp(\pm c_3 z). \tag{55}$$

c_1, c_2, and c_3 are arbitrary functions of n, restricted only by the equation

$$c_1^2 + c_2^2 + c_3^2 = k^2. \tag{56}$$

A final remark is made about the representation of λ in polar coordinates. It is seen from Eq. (48) and from Eq. (50) that a pure e-function is not an eigen function in the polar coordinate system. Therefore it is not possible to obtain a simple solution of the potential function ϕ for $\lambda = \lambda_0 \exp(2 kr)$. The best that can be done is to choose the following function for λ:

$$\lambda = \lambda_0 \frac{r_0^2 \exp\left[2k(r - r_0)\right]}{r^2} . \tag{57}$$

In this case,

$$N = \lambda^{\frac{1}{2}} = \lambda_0^{\frac{1}{2}} \exp(-k r_0) \cdot r_0 \cdot \frac{e^{kr}}{r} ,$$

which is an eigen function according to Eq. (50). A_0 is given by $A_0 = \lambda^{\frac{1}{2}} r_0 \exp(-k r_0)$. If r_0 denotes the earth's radius, and this discussion is confined to the space between the earth's surface and about 100-km altitude, then the r^2 in the denominator changes very little, but the e-function in the nominator increases in the requested way. Hence, the representation of λ by Eq. (57) is absolutely feasible. This will, then, bring again the full benefit of the simple solutions as outlined above.

The Mirror Law in the Current-Flow Theory

To introduce the earth's surface as an equipotential layer, the mirror law is applied in the electrostatic theory. As this is a powerful method which leads to simple solutions of boundary problems, its application in the current-flow theory will be discussed briefly.

Take a point source I and place it in a cylindrical coordinate system z, R, ϕ on the positive z axis at the point z = h. With regard to the coordinate system used for formula (53), the zero point has been shifted down the z axis by the distance h. As can be easily verified, this means that the potential function ϕ is now expressed in the following form:

$$\phi_n = \frac{I}{4 \pi \lambda_n} \frac{\exp\left[-k(r + z - h)\right]}{r} . \tag{58}$$

$\lambda_n = \lambda_0 \exp(2 kh)$ is here the conductivity at the altitude h, and r is given by $r = \left[R^2 + (z - h)^2\right]^{\frac{1}{2}}$.

117

To now introduce the plane z = 0 (earth surface) as an equipotential layer, place, analog to the electrostatic theory, a point source of the strength I* at the mirror point of I, i.e., at z = -h. The potential function ϕ* of this source will be

$$\phi* = \frac{I*}{4\pi\lambda*} \frac{\exp\left[-k(r* + z + h)\right]}{r*} , \qquad (59)$$

with r* = $\left[(z + h)^2 + R^2\right]^{\frac{1}{2}}$ and λ* = λ_0 exp (-2k h), the conductivity value at the point z = -h.

The superposition of ϕ_h and ϕ* results in a potential function ϕ, which will be zero for z = 0.

$$\phi = \phi_h + \phi* = \frac{I}{4\pi\lambda_h} \frac{\exp\left[-k(r + z - h)\right]}{r}$$

$$+ \frac{I*}{4\pi\lambda*} \frac{\exp\left[-k(r* + z + h)\right]}{r*} . \qquad (60)$$

For z = 0 is r = r*, $0 = \frac{I}{\lambda_h} \exp(kh) + \frac{I*}{\lambda*} \exp(-kh)$, or

$$I* = - \frac{\lambda* \exp(2 kh)}{\lambda_h} I. \qquad (61)$$

If Eq. (61) is inserted in Eq. (60), the final formula for the potential function ϕ is obtained:

$$\phi = \frac{I}{4\pi\lambda_h} \left[\frac{\exp(-kr)}{r} - \frac{\exp(-kr*)}{r*}\right] \exp -k(z-h) . \qquad (62)$$

For k = 0, Eq. (62) will change to the potential function ϕ_e of the electrostatic theory.

$$\phi_e = \frac{I}{4\pi\lambda_h} \left[\frac{1}{r} - \frac{1}{r*}\right]. \qquad (63)$$

With λ* exp (2 kh) = λ_0, from Eq. (61),

$$I* = - \frac{\lambda_0}{\lambda_h} I. \qquad (64)$$

In the electrostatic theory, the mirror source is placed at the mirror point -h and its strength is of the same amount but of opposite sign of the original source. It is seen here that in the current-flow theory the location and the sign reversal are retained, but that the amount of the mirror source is smaller than the original source in the same proportion as the conductivity at the ground λ_0 is smaller compared to the conductivity λ_h at the height of the original source.

Applied to the current flow of a thunderstorm, this result has far-reaching consequences, as pointed out by the author.[8] It means that only a part of the conduction current produced by the thunderstorm--namely, that given by I*--flows to the earth's surface in the immediate neighborhood of the storm, while the other part is drained off to the ionosphere and contributes to the air-earth current of the fair-weather areas. The difference between the lines of current flow of a point source in a medium of constant conductivity, $\lambda = \lambda_0$ (equivalent to the field lines of the electrostatic case), and those of a point source in a medium of increasing conductivity, $\lambda = \lambda_0 \exp(2 kz)$, can be easily recognized from Figs. 3a and 3b.[8] This difference also affects very drastically the amount of charges of the thunderstorm, as calculated from field measurements at the ground,[14, 15] the field reversal at the ground from a bipolar thunderstorm, the recovery curve of the lightning stroke,[10] and many other phenomena. But, as the object of this report is to outline the methods of calculation more than to discuss specific application, the mathematical aspects will be continued.

The Potential Function of a Decaying Current Source

This problem finds its application in the calculation of the recovery curve after a lightning discharge. Field records of a thunderstorm taken at the ground show that each lightning flash increases (or decreases) the so-called stationary field of the thunderstorm very suddenly, and that after the flash is completed the field returns to its preflash value in an e-function fashion, whereby the time constant of the decay is approximately ten seconds. The return is called the recovery curve.

This phenomenon is rather puzzling because the time constant of the air at the ground is about 600 seconds, and charges separated or produced by the lightning flash should therefore decay much slower. Tamura[10] showed in an excellent analysis that a much faster decrease of the field at the ground will result if, besides the decay of the lightning charges with the time constant in the cloud and the regenerating effect of the thunderstorm's charging mechanism, the screening effect of the space charge is taken into account, which builds up under the influence of the conductivity which increases with altitude. In other words, his calculations are based on the theory of current flow. The calculation given here may be considered as a part of Tamura's analysis, but presented from a different point of view, namely, to show a classical example of the full application of the current-flow theory.

The key problem is that of the decaying point charge. Assume that the charge $+Q_0$ is deposited by the lightning flash at a point at the altitude h. (If it is a cloud flash, in addition a negative charge $-Q_0$ would have to be placed at a higher altitude H and the resulting fields superimposed.) The charge decays with the time constant Θ' at this altitude. $\Theta' = \dfrac{\epsilon}{\lambda'}$. The charge Q left at the time t is then given by

$$Q = Q_0 e^{-\frac{\lambda' t}{\epsilon}} .$$

(65)

119

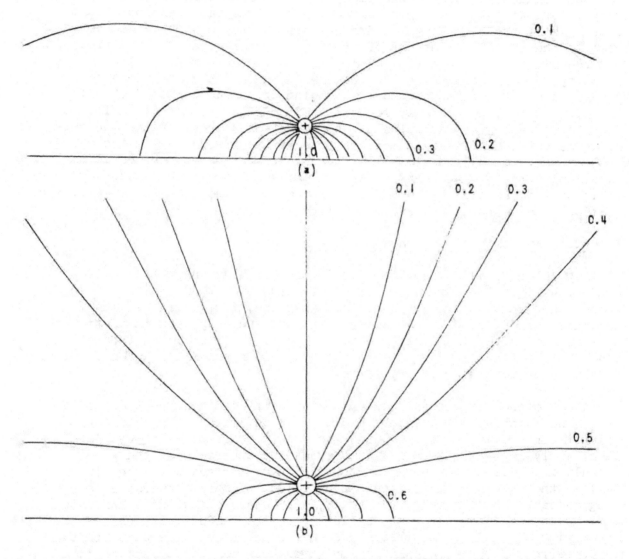

FIG. 3. FIELD AND CURRENT LINES OF A POINT SOURCE ABOVE A CONDUCTING PLANE IN A MEDIUM WITH
(a) CONSTANT CONDUCTIVITY
(b) CONDUCTIVITY INCREASING WITH ALTITUDE

The current output I is given by

$$I = -\frac{dQ}{dt} = \frac{\lambda'}{\epsilon} Q_o\, e^{-\frac{\lambda' t}{\epsilon}} \,. \tag{66}$$

Therefore, the time function T of the current source is given by the exponential function of Eq. (66).

$$T = e^{-\frac{\lambda' t}{\epsilon}} \,. \tag{67}$$

The conductivity λ depends here only on the space coordinates, but is constant with time. The weighting time functions T_e, T_s, and T_c given by Eqs. (17), (18), and (19) simplify to

$$T_e = e^{-\frac{\lambda t}{\epsilon}} \int \frac{dT}{dt}\, e^{\frac{\lambda t}{\epsilon}}\, dt. \tag{68}$$

$$T_s = e^{-\frac{\lambda t}{\epsilon}} \int T\, e^{\frac{\lambda t}{\epsilon}}\, dt. \tag{69}$$

$$T_o = e^{-\frac{\lambda t}{\epsilon}} \,. \tag{70}$$

With T given by Eq. (67), the integrals can be solved, and it becomes

$$T_e = -\frac{\lambda'}{\lambda - \lambda'}\, T. \tag{71}$$

$$T_s = \frac{\lambda}{\lambda - \lambda'}\, T. \tag{72}$$

The equation $T_e + T_s = T$ is fulfilled by Eqs. (71) and (72). The next step is to determine $\vec{F}_o$ from the initial condition. For $t = 0$, it is found that the field $\vec{E}$ shall be the electrostatic field $\vec{F}_e$, since the lightning flash occurs in such a short time that the space charges of the stationary current-flow field due to the conductivity variation have not accumulated. From Eqs. (21), (67), (70), (71), and (72) follows for $t = 0$,

$$\vec{F}_e = \vec{F}_o - \frac{\lambda'}{\lambda - \lambda'}\, \vec{F}_e + \frac{\lambda}{\lambda - \lambda'}\, \vec{F}_s,$$

or

$$\vec{F}_o = \frac{\lambda}{\lambda - \lambda'}\, \vec{F}_e + \frac{\lambda}{\lambda - \lambda'}\, \vec{F}_s. \tag{73}$$

If Eq. (73) is introduced into Eq. (21), obtained with regard to Eqs. (71) and (72) is

121

$$\vec{E} = \frac{\lambda}{\lambda - \lambda'} \left(T_0 - \frac{\lambda'}{\lambda} T\right) \vec{F}_e + \frac{\lambda}{\lambda - \lambda'} (T - T_0) \vec{F}_s. \tag{74}$$

Equation (74) is the final form of the field equation of the decaying point source. It is valid for every point in space. But it must be remembered that vector addition is required, because $\vec{F}_e$ and $\vec{F}_s$ do not have the same direction. Only at the ground is the direction of $\vec{E}$, $\vec{F}_e$, and $\vec{F}_s$ the same, namely, vertical to the ground surface. In this case the vector equation (74) simplifies to a scalar equation. Notice that the sum of the weighting time functions of $\vec{F}_e$ and $\vec{F}_s$ again is T, the time function of the current source.

REFERENCES

1. Courant, R., and D. Hilbert, Methoden der Mathematischen Physik I, II, Verlag Julius Springer, Berlin, 1937.

2. Smyth, W. R., Static and Dynamic Electricity, McGraw-Hill Book Co., New York, 1950.

3. Ollendorff, F., Potentialfelder der Elektrotechnik, Verlag Julius Springer, Berlin, 1932.

4. Kasemir, H. W., Zur Stroemungstheorie des Luftelektrischen Feldes I, Arch. Met. Geophys. Biokl., A 3, 85-97, 1950.

5. Kasemir, H. W., Zur Stroemungstheorie des Luftelektrischen Feldes II, Arch. Met. Geophys. Biokl., A 5, 56-70, 1952.

6. Kasemir, H. W., Zur Stroemungstheorie des Luftelektrischen Feldes III, Arch. Met. Geophys. Biokl., A 9, 357-370, 1956.

7. Kasemir, H. W., Die Stromausbeute des Gewittergenerators, Ber. d. Wetterd. d. US Zone 6, Nr. 38, 428-434, 1952.

8. Kasemir, H. W., Der Gewittergenerator im Luftelektrischen Stromkreis I, II, Z. f. Geophys., 25, 33-96, 1959.

9. Holzer, R. E., and D. S. Saxon, Distribution of Electrical Conduction Currents in the Vicinity of Thunderstorms, J. Geophys. Res., 57, 207-216, 1952.

10. Tamura, Y., An Analysis of Electric Field after Lightning Discharges, Proc. Conf. Atmospheric Electricity, 207-216, 1955.

11. Kasemir, H. W., Measurement of the Air-Earth Current Density, Proc. Conf. Atmospheric Electricity, 91-95, 1955.

12. Ruhnke, L. H., The Effect of Mismatching in the Measurement of the Air-Earth Current Density, USASRDL Technical Report 2232, Fort Monmouth, N. J., Oct 61.

13. Kasemir, H. W., A Radiosonde for Measuring the Air-Earth Current Density, USASRDL Technical Report 2125, Fort Monmouth, N. J., Jun 60.

14. Simpson, G. C., and G. D. Robinson, The Distribution of Electricity in Thunderclouds, Proc. Roy. Soc., A 177, 281-329, 1940

15. Malan, D. J., Les Descharges dans l Air et la Charge Inference Positive d'un Nuage Orageux, Ann. Geophys., 8, 385-401, 1952.

Extrait des ANNALES DE GÉOPHYSIQUE

Tome 7, fascicule 1. — Janvier-Mars 1951.

REGARDING THE SHIELDING EFFECT OF BUILDINGS
ON THE FLUCTUATIONS OF THE ATMOSPHERIC ELECTRICAL FIELD

by H. Israël and H. W. Kasemir

(Atmospheric Electrical Research Center of the Observatory Friedrichshafen in Buchau a. F.:
National Weather Service Württemberg-Hohenzollern).

RÉSUMÉ. — *Pour l'information des biologistes on calcule la réduction d'un champ électrique variable traversant les parois d'un bâtiment faiblement conducteur.*

SUMMARY. — *Screening of varying electrical field by poorly conducting building is calculated, interesting as it for biological purposes.*

I. INTRODUCTION

Recently, the question has often arisen, whether the fluctuations of the natural atmospheric electrical field is capable of penetrating the interior of a residential building.

The reason for this was by large a hypothesis emerging in bioclimatology according to which greater significance must be attributed to atmospheric electrical events and specifically the atmospheric electrical field with its variations in respect to bioclimatology as one previously assumed (1) (2).

This thought is not new, for such suppositions of causal correlations between the atmospheric electrical and the biological event have been frequently discussed in bioclimatology. On the other hand, all previous attempts of proving such effects as factually present have always more or less proceeded negatively.

One particular explanation for the negative failure of these attempts provides the following consideration: It is well known that so-called "meteorosensitivity" in enclosed spaces also occurs like in the open. Therefore, it must also be required of the climate factors, which one attempts to hold responsible for such effects, that they can also be effective in enclosed residential spaces. Moreover, they must be capable of noticeably influencing or modifying the artificial climate that we create in those living spaces.

Both of these conditions heavily constrain the area of factors coming into question and enable an assessment as to whether a weather or climate element under consideration can come into question for climate effects or not.

Applied to the elements of the atmospheric electrical events, this means the following: Both variations coming into bioclimatic consideration are fluctuations of the "aerosol" (ions and solid content of the atmosphere) and that of the atmospheric electrical field. The former are capable of penetrating into living spaces through windows and gaps around doors with the general exchange of air (1) (2) (3); however, the ion and core content of the room air is so heavily altered through human action (smoking, heating, etc.) and alone through inhabitation (4) that it is questionable, whether the significantly weaker variations penetrating from the outside are capable of playing an additional role. — The stationary atmospheric electrical field is already sufficiently shielded by the walls of a building. From a practical standpoint, the same applies for its gradual fluctuations as we gather them with the usual atmospheric electrical metrology. It is naturally different with very rapid fluctuations of a highly frequent type. These penetrate the walls of a house at a noticeable level; otherwise radio reception with an indoor antenna would be impossible! -- Thus, from a physical perspective, only the area of the rapid field fluctuations remains for the atmospheric electro-biological correlations!

For an exact definition of what is meant by "rapid" field fluctuations, up to this point, we do not have the fundamentals in the form of a precisely definable specification regarding the shielding effect of house walls on fluctuations of a varying frequency. The objective of the following observations is to provide clarification for this question.

2. THE ELECTRICAL SHIELDING EFFECT OF SPHERE-SHAPED SHIELDS.

For our case, we wish to assume a primary electrical field in the outer space of the most-simple form, namely a homogenous field of constant direction and strength. We will represent the shielding building through a hollow sphere, for which, however, we do not want to make any limiting assumptions with regard to its thickness. We will also reserve the possibility to provide arbitrary values of conductivity and dielectric constants to the outer space, intermediate space and interior space of the hollow sphere. We will omit the third spatial characteristic – namely the induction constant – and thus limit our case to a frequency range of 0 to a few MHz or to wave lengths greater than our building dimensions.

A hollow sphere with the inner radius a and the outer radius b (Figure 1) is situated concentrically to the coordinate origin in a three-dimensional space depicted through a polar coordinate system r, θ, φ. This hollow sphere is subject to the influence of an outer homogenous electrical field F_1. In this connection, the outer space, inner space and the intermediate space (space between the radii a and b) of the hollow sphere should have a varying conductivity and dielectric constant. We will indicate the variables of the outer space, such as

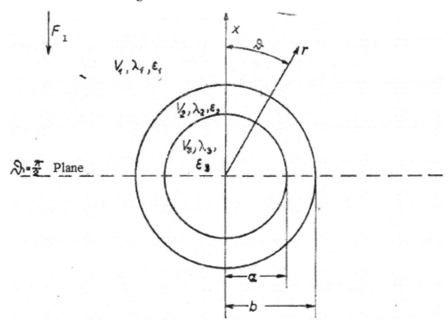

Figure 1. – Hollow sphere of finite conductivity in the homogenous field of a stationary current

Taking the coupling between rapidly changing electrical and magnetic fields through the induction constant μ into account, we can no longer ignore the velocity of propagation $c = 1/\sqrt{\varepsilon\mu}$ given through ε and μ. With that, we will briefly look at the range of retarded potential. However, as we will see later, due to the fact that an alternating field of frequency of $3\cdot10^5$ Hz or a wave length of 1000 m completely penetrates our shield, the range of ultra-short waves already lies outside of our problem. First, we will calculate the stationary current field and then later see that the solution also applies for alternating fields with a simple transition into the complex, insofar as we remain in the range of frequency given above and limit ourselves to sinusoidal fluctuations. We could liberate ourselves from the latter limitation if we use the Fourier representation for a random fluctuation.

Following these brief preliminary notes, we can dress our problem in the following mathematical formulation. :

potential, conductivity, dielectric constant, etc. through index 1, i.e. v_1, λ_1, ε_1, etc. and those of the intermediate space or inner space through the indexes 2 or 3. In general, we indicate the space ratio with k ($k = 1, 2, 3$).

We are searching for the shielding effect of the hollow sphere on the inner space and define it as the relationship of the field strength of the inner space F_3 to the field strength F_1 of the undisturbed outer field.

$$(1) \qquad S = \frac{F_3}{F_1}.$$

The potential, field and current distribution is regulated by the differential equation of the current theory [5], which in turn is a special case of the equation of Maxwell. Indeed, if we perceive the conductivity and the dielectric constant in the various spaces differently, yet within the same space, as constant, then the differential equation of the stationary current formally identical to Laplace's equation in potential theory and we have k for all spaces

$$(2) \qquad \text{div grad } v_k = 0.$$

If we rotate the $\theta = 0$ axis of our polar coordinate system in the direction of the field vector $\vec{F}_1$ of the undisturbed field in the outer space, then our problem possesses rotational symmetry and every potential that obeys the equation (2) can be applied in the form

$$(3) \qquad v_k = \sum_{n=0}^{\infty} \left(A_{kn} r^n + \frac{B_{kn}}{r^{n+1}} \right) P_n.$$

In this case, P_n stands for the Legendre spherical functions and A_{kn}, B_{kn} are the constant that serve to fulfill the prescribed boundary conditions.

As marginal conditions of our problem, the following are sought – the equality of the potentials and the continuity of the normal components of the current thickness on the spherical shells of our hollow sphere. Furthermore, the potential of the outer space should be transferred from the hollow sphere to the potential of the undisturbed homogenous primary field for large distances, while the potential of the inner space should remain finite for $r = 0$. Both of the last mentioned conditions imply that only coefficients for $n = 1$ in the molecular formula differ from 0 and, in addition, B_{31} disappears in the inner space. The latter statement is easy to understand because the r is in the denominator for B_{31}, thus, it provides an infinite point for $r = 0$ in this case, which is only necessary to avoid, if we determine $B_{31} = 0$. The first limitation, only A_{k1} and $B_{k1} \neq 0$ can only be justified in a very cumbersome manner. If we were to carry sum terms other than for $n = 1$, we would later be lead to the fact that the coefficients A_n, $B_n = 0$ are for $n \neq 1$ when applying the boundary conditions to the spherical shells. In this case, we need to limit ourselves merely to the terms with $n = 1$, and see whether we can obtain the solution to our present problem with these alone. Thus, a differentiation of the individual terms is invalid through the index n, which we will omit in the future. If we still take into account that $P_1 = \cos\theta$, the potentials in our 3 spaces take on the following form

$$(4) \qquad \begin{aligned} v_1 &= \left(A_1 r + \frac{B_1}{r^2} \right) \cos\theta \\ v_2 &= \left(A_2 r + \frac{B_2}{r^2} \right) \cos\theta \\ v_3 &= A_3 r \cos\theta. \end{aligned}$$

In this connection, it is still necessary to notice that $A1 = -F$ in the first equation of (4) if we specify the undisturbed potential of the outer space v_u in the following form

$$(5) \qquad v_u = -Fr \cos\theta.$$

Then, v_1 is transferred namely to v_u for capital r because the second term is transferred to v_1 for capital r toward o. v_u is the representation in coordinates of the known form $v_u = -Fx$ of the potential of a homogenous field. (E.g. plate condenser).

The equations (4) are pleasantly simple. We will determine the still unknown constants based on the boundary conditions on the spherical shells. Prior to that, we would like to take note that all potentials for the $\theta = \pi/2$ plane disappear. This plane is therefore equipotential plane and we can replace it with a conductive plane. Thus, we have also solved the problem of half of the hollow sphere, which sits on a conductive plane and is located in a homogenous current field. This diagram would reflect the relationships of the shielding effect of a house versus the electrical fields even better because, in this case, the conductive surface of the earth is also provided. The house would then be represented by the half of the hollow sphere.

The equations (4) continue to show that the following generalization can be easily conducted. Namely, if we not only have a spherical shell, but rather any number situated concentrically over each other and in every one a different conductivity, then the representation of the potential applies for every space limited by two spherical shells in the following form

$$(6) \qquad v_k = \left(A_k r + \frac{B_k}{r^2} \right) \cos\theta.$$

For the specification of the constants A_k and B_k reoccurring with every additional spherical shell, we will always obtain two new conditional equations at the new boundaries, so that there are always $2(k-1)$ conditional equations available for the $2(k-1)$ unknown constants.

With that, the path is then also pointed out for a problem, for which λ can no longer be dramatically changed, but rather is provided with continuity and, e.g. as a function of r.

According to this small digression, we can return to the specification of our constants from the boundary conditions for the spherical shells. If we take into account that the current density i_k is given through the equation

$$(7) \qquad \vec{i}_k = \lambda_k \vec{E}_k$$

with

$$(8) \qquad \vec{E}_k = -\text{grad } v_k.$$

126

then we can write our boundary conditions:

$$(9) \quad \begin{array}{ll} v_1 = v_2, & i_{r1} = i_{r2} \quad \text{für } r = b \\ v_2 = v_3, & i_{r2} = i_{r3} \quad \text{für } r = a. \end{array}$$

In this process, the normal components of the current density is the meant for the spherical shell under i_{rk}, which can be calculated according to (7) and (8) for

$$(10) \qquad i_{rk} = -\lambda_k \frac{\partial v_k}{\partial r}.$$

With the aid of (4) and considering that $A_1 = -F$, we can then obtain,

$$(11) \quad \begin{aligned} -Fb + \frac{B_1}{b^2} &= A_2 b - \frac{B_2}{b^2} \\ \lambda_1\left(-F - \frac{2B_1}{b^3}\right) &= \lambda_2\left(A_2 - \frac{2B_2}{b^3}\right) \\ A_2 a + \frac{B_2}{a^2} &= A_3 a \\ \lambda_2\left(A_2 - \frac{2B_2}{a^3}\right) &= \lambda_3 A_3. \end{aligned}$$

These are the 4 equations that serve the 4 still unknown constants in (4).

In this case, we can still make a simplification adapted to our problem by applying $\lambda_1 = \lambda_3$. The value of the atmospheric conductivity must be applied later for the outer and inner space. The solution to (11) then results in the following

$$(12) \quad \begin{aligned} B_1 &= -F\frac{(b^3 - a^3)(2\lambda_2 - \lambda_1)(\lambda_2 - \lambda_1)b^3}{2(b^3 - a^3)(\lambda_1^2 - \lambda_2^2) - \lambda_1\lambda_2(5b^3 - 4a^3)} \\ A_2 &= -F\frac{3\lambda_1(2\lambda_2 - \lambda_1)b^3}{2(b^3 - a^3)(\lambda_1^2 - \lambda_2^2) - \lambda_1\lambda_2(5b^3 - 4a^3)} \\ B_2 &= Fa^3\frac{\lambda_2 - \lambda_1}{2\lambda_2 + \lambda_1} \\ A_3 &= -F\frac{9\lambda_1\lambda_2 b^3}{2(b^3 - a^3)(\lambda_1^2 - \lambda_2^2) - \lambda_1\lambda_2(5b^3 - 4a^3)} \end{aligned}$$

Applying these constants in (4) provides the complete solution to our problem in the stationary current field. We would still like to briefly discuss a few passages in order to gain some insight into the equation.

For $\lambda_1 = \lambda_2$ we get $B_1 = B_2 = 0$ and $A_2 = A_3 = -F$, i.e. the original undisturbed field dominates everywhere, which is necessary to expect, if no difference exists between the conductivity of the spherical shell and the outer and inner space.

For $a = b$, we also get $A_3 = -F$, $B_1 = 0$. This means that the reaction of the hollow sphere, which has an infinitely thin wall thickness, on the outer space disappears and the same field dominates in the inside and outside. However, it is somewhat surprising that the values for A_2 and B_2 remain finite in the intermediate space, which is actually no longer even present. This ambiguity resolves itself, however, if we apply these values for A_2 and B_2 to the associated potential equation for v_2 and make the passage $a = r = b$. Then v_2 takes on the value of the potential of the undisturbed field on the sphere $r = b$, i.e. the potential of this infinitely thin

intermediate space is arranged without a break in the potential of the undisturbed field.

However, we are most interested by the shielding factor S. In this case, we benefit extensively from the fact that a homogenous field dominates from a great distance in the inside of the hollow sphere like in the outer space. Therefore, our shielding factor S is a constant for the entire inner space, which was absolutely not expected initially. It is simply defined for

$$(13) \quad S = \frac{A_3}{-F} = \frac{9\lambda_1\lambda_2 b^3}{2(b^3 - a^3)(\lambda_1^2 - \lambda_2^2) + \lambda_1\lambda_2(5b^3 + 4a^3)}$$

Following a simple transformation, from this we obtain

$$(14) \quad S = \frac{9\lambda_1\lambda_2}{(2\lambda_2 - \lambda_1)(2\lambda_1 - \lambda_2) - 2\frac{a^3}{b^3}(\lambda_2 - \lambda_1)^2}.$$

From (14) in particular, we can see that the shielding factor becomes smaller with an increasing difference of λ_1 and λ_2 as well as of a and b. That means that the inner field almost vanishes with an approximation of the conductivity of the walls of our shielding building to some metallic conductivity. The inner field also becomes weaker with a greater layer thickness. However, the path of the shielding factor is very dependent on the individual values of the given variables and, in general, cannot be discussed in detail very well.

Therefore, we will now turn to the general solution for alternating fields.

In this case, in addition to the pure conductive current, we must also consider the displacement current and expand our boundary conditions accordingly. Most frequently, the series connection of conductive current and displacement current opposes us in the physical-technical problems, e.g. in the case of the radiant antenna. At the point where the metallic line ended – and with that the conductive current as well – the displacement current takes over the closure of the electrical circuit through the non-conductive atmosphere. However, in this case, we must expect the juxtaposition of conductive current and displacement current in every unit of volume. Maxwell calls this cumulative current c and defines it through the equation

$$(15) \qquad \vec{c} = \vec{i} + \frac{\partial \vec{D}}{\partial t}$$

In this connection, D signifies the dielectric stimulation and its temporal d, i.e. the displacement current. We now have to apply that, which was previously applicable for the conductive current i, to the total current density. We can derive the current density c from a potential v due to the fact that the known relationships apply

$$\vec{i} = \lambda\vec{E}$$

(16)
$$\vec{D} = \varepsilon\vec{E}$$

$$\vec{E} = -\,\text{grad}\ v.$$

Our previous boundary conditions (9), therefore, extend to

(17)
$$v_1 = v_2, \quad c_{r1} = c_{r2} \quad \text{for}\ r = b$$
$$v_2 = v_3, \quad c_{r2} = c_{r3} \quad \text{for}\ r = a.$$

The meaning of c_{rk} in this case is the normal component of the total current density c for the observation of spherical shells, i.e. therefore, the current components in the r-direction. With the aid of (15) and (16), we obtain from this

(18)
$$\lambda_1\frac{\partial v_1}{\partial r} + \varepsilon_1\frac{\partial}{\partial t}\left(\frac{\partial v_1}{\partial r}\right) = \lambda_2\frac{\partial v_2}{\partial r} + \varepsilon_2\frac{\partial}{\partial t}\left(\frac{\partial v_2}{\partial r}\right) \text{ for } r = b$$
$$\lambda_2\frac{\partial v_2}{\partial r} + \varepsilon_2\frac{\partial}{\partial t}\left(\frac{\partial v_2}{\partial r}\right) = \lambda_3\frac{\partial v_3}{\partial r} + \varepsilon_3\frac{\partial}{\partial t}\left(\frac{\partial v_3}{\partial r}\right) \text{ for } r = a.$$

We have to supplement the approach for the potential (4) with the temporal law of fluctuation, thus multiply it with $e^{j\omega t}$.

(19)
$$v_1 = \left(-\,Fr + \frac{B_1}{r_2}\right)\cos\theta\ e^{j\omega t} \quad \text{for } b \leqslant r \leqslant \infty$$
$$v_2 = \left(A_2 r + \frac{B_2}{r^2}\right)\cos\theta\ e^{j\omega t} \quad \text{for } a \leqslant r \leqslant b$$
$$v_3 = A_3 r \cos\theta\ e^{j\omega t}. \quad \text{for } 0 \leqslant r \leqslant a.$$

Of course, we cannot expect that the potentials fluctuate in the same phase in the outer, inner and intermediate space. Therefore, we must anticipate that our constants A_k and B_k are of a complex nature, i.e. that they also contain a factor for the phase shift.

Now, if we arrive at the boundary conditions (17) and (18) with the approach (19), then we will obtain the following conditional equations for the constants A_k and B_k.

(20)
$$-\,Fb^3 + B_1 = A_2 b^3 + B_2$$
$$A_2 a^3 + B_2 = A_3 a^3$$
$$(\lambda_1 + j\omega\varepsilon_1)(Fb^3 + 2B_1)$$
$$= (\lambda_2 + j\omega\varepsilon_2)(-\,A_2 b^3 + 2B_2)$$
$$(\lambda_2 + j\omega\varepsilon_2)(A_2 a^3 - 2B_2) = (\lambda_3 + j\omega\varepsilon_3)\,A_3 a^3.$$

These equations are very similar to those from (11), with the exception that in this case the complex digits $\lambda_k + j\omega\varepsilon_k$ apply instead of the purely real values λ_k. In optics, the designation of 'complex dielectric constant' has been attributed to these complex values in the case of the theory of the absorbed media. We can also call it 'complex conductivity' with the same right. We take the authority to transfer this naming convention to

spatially-extensive current fields as well from the work [5], in which is demonstrated that we can regard the dielectric constant ε [Farad /m] as the capacity of the unit of volume according to the conductivity per unit of volume λ [Siemens /m].

If we designate the complex conductivities in the individual spaces with

(21)
$$L_1 = \lambda_1 + j\omega\varepsilon_1$$
$$L_2 = \lambda_2 + j\omega\varepsilon_2$$
$$L_3 = \lambda_3 + j\omega\varepsilon_3,$$

then we can simply assume the values listed in (12) for the constants A_k and B_k if we replace λ_k with the respective L_k. The fact that our now complex constants are transferred to the real constants from (12) for $w = 0$, i.e. the stationary current, is immediately apparent. The boundary observations conducted in conclusion on (12) will also remain intact in their complexity. At this point, we will turn to a somewhat more in-depth observation of the complex shielding factor, which will designate with s. Simplification $L_1 = L_3$ for

(22)
$$s = \frac{9L_1\,L_2}{(2L_2 + L_1)(2L_1 + L_2) - 2\dfrac{a^3}{b^3}(L_2 - L_1)^2}.$$

The phase shift between the alternating field of the outer and inner space is of little interest to us at the moment. However, to achieve the amplitude ratio of the field strength, we must establish the figure for s, which we designate with $|s|$ as usual. Considering the problem characterized in the beginning, we will also apply $\lambda_1 = \lambda_2$ and $\varepsilon_1 = \varepsilon_2 = \varepsilon_3 = \varepsilon$ in this case and obtain

(23)
$$|s| = \frac{9\sqrt{\lambda_1^2\lambda_2^2 - \varepsilon^2\omega^2(\lambda_1^2 + \lambda_2^2) + \varepsilon^4\omega^4}}{\sqrt{\left[(2\lambda_1 + \lambda_2)(2\lambda_2 + \lambda_1) - 9\varepsilon^2\omega^2 - 2\dfrac{a^3}{b^3}(\lambda_2 - \lambda_1)^2\right]^2 + 81\varepsilon^2\omega^2(\lambda_1 + \lambda_2)^2}}$$

If the formula for the shielding factor (14) was not understandable in a simple manner for the case of the stationary current, then we absolutely cannot expect to understand it based on this general formula (23). Therefore, for the discussion regarding (23), we would like to take its graphical depiction in Figure 2 as the basis and select the following values for the constants:

$$\lambda_1 = 2.10^{-14} \text{ Simens/m}$$
$$\lambda_2 = 2.10^{-4} \text{ Simens/m}$$
$$\varepsilon = 8,9 \; 10^{-12} \text{ Farad/m}$$
$$a = 3 \; m$$
$$b = 3,2 \; m.$$

In Figure 2, the figure of the shielding factor s is applied in a double-logarithmic scale dependent upon the fluctuation frequency f = w/2π of the

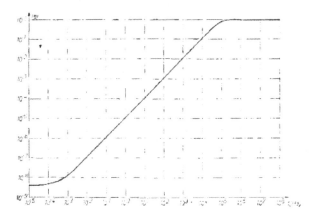

Figure 2. -- Shielding factor |s| of a hollow sphere in the homogenous alternating field as a function of the frequency.

alternating field. For extremely slow frequencies, such as the diurnal variation of the atmospheric field, we are dealing the stationary current state. The shielding factor has the constant value of $5 \cdot 10^{-9}$ from the frequency 0 to approx. 10^{-4} Hz. Thus, for example, if an atmospheric electrical field of 100 V/m dominates on the earth, then we may only expect a field strength of approx. $5 \cdot 10^{-7}$ in the room. However, if we go further to higher frequencies, then the inner field will increase. We arrive at a frequency range, in which the conductive current in the atmospheric spaces (outer and inner space) more readily accommodates the displacement current. In the wall, we must still expect almost exclusively conductive current. Following a small transitional range of 10^{-4} to 10^{-3} Hz, the shielding factor increases proportionally with the frequency to approx. $2 \cdot 10^4$. In this manner, we find the shielding factor at approx. $6 \cdot 10^{-4}$, e.g. for 50 Hz. Thus, if a power line carrying 220 V crosses over a small house with a clearance of roughly 5 m and we anticipate an average field strength of approx. 20 to 40 V/m, then we can assume that only one field of approx. 0.01 to 0.02 V/m penetrates into the house. In contrast, in a radio wave range of 10^5 Hz, we only have a shielding factor of 0.8 to 0.9. We should also expect that because radio reception within a house is not noticeably impacted by the walls. For higher frequencies, the shielding factor =

1. Now, the conductive current has accommodated the displacement current in the walls as well – and because we applied the dielectric constant as equal everywhere in our example, although the displacement current only depends on ε, there is no longer a reason for a shielding effect of the walls.

For the material constants we chose, simple and well-suited approximation formulas can be provided in the individual frequency ranges instead of the complicated formulas (23). The following applies

$$(24) \qquad |s| = \frac{9}{2\left(1 - \dfrac{a^3}{b^3}\right)} \frac{\lambda_1}{\lambda_2} \qquad \text{for } 0 \leqslant f \leqslant 10^{-4}$$

$$|s| = \frac{9}{2\left(1 - \dfrac{a^3}{b^3}\right)} \frac{\varepsilon \omega}{\lambda_2} \qquad \text{for } 10^{-3} < f < 2 \; 10^4$$

with $\omega = 2\pi f$.

Manuscrit reçu le 17 *janvier* 1951.

LITERATURE

[1] H. ISRAËL, Die luftelektrischen Elemente und ihre mögliche bioklimatische Bedeutung. *Dtsch. Mediz. Woch. schr.*, **75**, 202-205, 1950.
Bemerkung zur » Entgegnung « von R. REITER und J. KAMPIK auf obige Arbeit. *Dtsch. Mediz. Woch. schr.*, **75**, 1754, 1950.
Luftelektrizität und Bioklimatologie. *Experimental Medicine and Surgery* (im Druck).

[2] P. COURVOISIER, Die Schwankungen des elektrischen Feldes in der Atmosphäre und ihre Beziehungen zur Meteoropathologie. (Vortrag auf der Bioklimatikertagung in Freiburg/Breisgau im März 1950). *Experientia* (im Druck).
Ueber das Eindringen von Schwankungen der meteorologischen Elemente in Gebäude. *Arch. f. Meteorol., Geophys. u. Bioklimatolog.* (B) **2**, 161-169, 1950.

[3] K. EGLOFF, Ueber das Klima im Zimmer und seine Beziehungen zum Aussenklima, mit besonderer Berücksichtigung von Feuchtigkeit, Staub- und Ionengehalt der Luft. *Dissertation*, Zürich 1933 (1934 ?)

[4] G. R. WAIT, A cause for the decrease in the number of ions in air of occupied rooms. *Journ. of Industrial Hygiene*, **16**, 147-159, 1934.

[5] H. W. KASEMIR, Studien über das atmosphärische Potentialgefälle IV : Zur Strömungstheorie des luftelektrischen Feldes I. *Arch. f. Meteorol., Geophys. u. Bioklimatol.* (A) **3**, 84-97, 1950.

HOW TO MEASURE AND ANALYZE SOLAR-TERRESTRIAL WEATHER RELATIONSHIPS

Heinz W. Kasemir

Colorado Scientific Research Corporation
Berthoud, Colorado

I. GENERAL REMARKS

Solar and ionospheric events are observed in the iono- and magneto-sphere. On the other hand weather is a phenomenon of the tropo- and the stratosphere. Therefore, in a study of solar-terrestrial weather relationship it would be desirable to determine how much electric energy of the particles coming from the sun penetrates into the strato- and troposphere. To calculate the electric energy flow we define the following vectors:

$\overline{P}$ = Pointing vector of energy-flow.

$\overline{E}$ = Electric field

$\overline{D} = \varepsilon\overline{E}$ = dielectric displacement

$\overline{H}$ = Magnetic field   (1)

$\overline{B} = \mu\overline{H}$ Magnetic induction

$\overline{J} = \lambda\overline{E}$ = conduction current density

λ = Conductivity

All vectors are functions of space and time. Differentiation to time is indicated by a dot above the vector.

$$\dot{\overline{D}} = \partial\overline{D}/\partial t \qquad (1a)$$

The Pointing vector is given by

$$\overline{P} = \overline{E} \times \overline{H} \qquad (2)$$

and the divergence of $\overline{P}$ by

$$- \text{div } \overline{P} = \overline{H} \cdot \dot{\overline{B}} + \overline{E} \cdot \dot{\overline{D}} + \overline{E} \cdot \overline{J} \qquad (3)$$

If we can assume stationary conditions equation (3) would reduce to:

$$- \text{div } \overline{P} = \overline{E} \cdot \overline{J} \qquad (4)$$

The term $\overline{E} \cdot \overline{J}$ represents Joule's Heat generated per second and cubic meter. This term would also contain the electric wind if not only increase of random motion of ions but also directed motion has to be taken into account. If energetic charged particles penetrate into the strato- or troposphere they would appear in (3) or (4) as impressed currents or electromotive forces as part of $\overline{J}$ or $\overline{E}$.

With the relations given in (1) we would have to determine $\overline{E}$ and $\overline{H}$ - assuming that λ and ε are known - as functions of space and time to obtain the electric energy deposited in the strato- and troposphere by solar events. The determination of the space vector $\overline{E}$ for stationary conditions from measurements at the ground or from an airplane will be discussed in the next paragraph. This method would also apply to the determination of $\overline{H}$ from ground measurements and could be extended to include time dependent functions. Special attention should be given to the separation of the atmospheric electric and the earth magnetic field from the fields produced by ionospheric and solar events. A few problems of the atmospheric electric global circuit will be discussed at the end of this paper. Instrumentation such as field mills, long wire antennas, ground current probes, magnetometers, etc will not be discussed here. It shall be sufficient to point out here that the reliability of the analysis will increase with the number of stations and the quality of the data. An important point here is that data are recorded in such a way that they are easy to dupli-cate and distribute among the scientific community and that the format is com-patible with widely used computer processing equipment.

II. DATA ANALYSIS

The method suggested is based on two powerful physical theorems: the expansion of the potential function in harmonics of the chosen coordinate system and the least square fit method. The proper coordinates for our problem are the spherical coordinates r, θ, ϕ. Therefore, the potential function V is a function of r, θ, ϕ. Harmonics or eigenfunctions are solutions of the governing differential equation for a potential function given in the form

$$V_n^* = A_n R_n(r) \Theta_n(\theta) \, \Phi_n(\phi) \tag{5}$$

with

$$V = \sum_{n=0}^{\infty} A_n R_n(r) \, \Theta_n(\theta) \, \Phi_n(\phi) \tag{6}$$

Since this is only a brief outline of the method it may suffice to explain (6) by using the more familiar example of electrostatic potential theory (i.e. $\lambda = 0$ or $\lambda = $ const) and then switch back to our problem with the conductivity being an exponential function or a given tensor function.

In electrostatic theory the governing differential equation is Laplace's equation

$$\nabla^2 V = 0 \tag{7}$$

and (6) has the form

$$V = \sum_{n=0}^{\infty} (A_n r^n + B_n/r^{n+1}) \, Y_n(\theta,\phi) \tag{8}$$

Thereby the radial function $R_n(r)$ splits up in two parts

$$R_n(r) = A_n r^n + B_n/r^{n+1} \tag{8a}$$

and $\Theta_n \Phi_n$ are combined into Yn known as the spherical surface harmonics.

$$Y_n = \Theta_n \Phi_n \tag{8b}$$

In our case the governing differential equation is that of the continuity of current flow (Preservation of charge)

$$\nabla \bar{J} = 0 \tag{9}$$

and the solution for an exponential conductivity $\lambda = \lambda_0 \exp(r-a)$ and the relations

$$\bar{J} = \lambda\bar{E}; \quad \bar{E} = -\nabla V \tag{9a}$$

$$V = \sum_{n=0}^{\infty} (A_n R_{1n}(r) + B_n R_{2n}(r)) Y_n \tag{10}$$

Since λ is a function of r, only the radial functions R_{1n} and R_{2n} are different from those in equation (8).

The $A_n R_{1n} Y_n$ of (10) describe all current sources in the ionosphere or solar sources that are penetrating the ionosphere and the R_{1n} give the down mapping radial functions. The $B_n R_{2n} Y_n$ represent all current sources on the earth surface and their up mapping radial functions is R_{2n}. This includes for $n = 0$ the fair weather current and for $n \geq 1$ the reaction of the earth to the down coming ionospheric currents. For these cases the A_n and B_n are related and B_n for instance can be expressed by A_n from the condition that the earth is an equipotential surface.

If we have determined the A_n and B_n by the least square fit method we have not only the potential function, electric and magnetic field, current density, space charge density, electric energy, etc. but we will also obtain a good picture of the current sources and their spacial distribution. Each individual term in (10) represents a well defined current source. The most simple one is A_o. Since $n = o$, $Y_o = 1$ and $R_{10} = 1$ it follows from (8) as well as from (10) that $V = A_o = $ const.

In electrostatics the charge distribution required to produce a constant potential is well known. It is a net charge Q that is equally distributed over a spherical shell. Such a charge distribution would produce inside the shell with the radius a the constant potential

$$V_i = Q/4\pi\varepsilon a = Q/C$$

where C is the capacity of the spherical shell.

In current flow problems charges have to be replaced by current sources. Using equation (8) and (9) given by Kasemir 1977 for the analog current flow problem, we have the potential of a spherical shell with the net current I and exponential conductivity of a scale factor s

$$V_i = IR, \quad R = s/4\pi a^2 \lambda_i$$

For $V_i = 300,000$ V, $\lambda_i = 2\cdot 10^{-4}$ $1/\Omega m$, $a = 6\cdot 4\cdot 10^6 m$ and $s = 10^4 m/\ln(10)$ the current I is

$$I = 7\cdot 10^{12} A$$

This means that to add a constant potential of 300,000 V to the atmospheric electric circuit so that the earth potential is zero and the ionospheric potential $V_i = 300,000$ V would imply that in the ionosphere exists a spherical symmetric current source with a net current of $I = 7\cdot 10^{12} A$. Since there is no reason to assume the existance of such a current source, we set $A_o = 0$. The ramification in the atmospheric electric global circuit will be discussed later in more detail.

III. LEAST SQUARE FIT

The least square fit method fits a certain number N of measured data D_i to a theoretically assumed relationship as for instance expressed in equation (10).

134

In our case this could be measurements of the electric field at a number of stations distributed over the earth. Since we measure the field at the earth surface only the radial component is different from zero. From (10) we obtain by differentiation to r with

$$dR_{1n}/dr = R'_{1n} \quad ; \quad dR_{2n}/dr = R'_{2n}$$

$$E = -\sum_{n} (A_n R'_{1n} + B_n R'_{2n}) Y_n \tag{11}$$

Our interest is now centered on the determination of the A_n. As mentioned above the B_n can be determined from the A_n from the boundary condition at the earth surface by a relation of the type

$$B_n = a_n A_n \tag{12}$$

Substituting (12) in (11)

$$E = - \sum_{n} A_n (R'_{1n} + a_n R'_{2n}) Y_n \tag{13}$$

The expression in the brackets is for all stations the same since r = a = earth radius. Y_n is a function of latitude θ and longitude ϕ therefore it has to be calculated for each station i. We set

$$C_{ni} = (R'_{1n} + a_n R'_{2n}) Y_{in} \tag{14}$$

C_{ni} is a constant but a different constant for each multipole n and each station i. IF E_i is the theoretical field at the ith station, then

$$E_i = - \sum_{n} A_n C_{ni} \tag{15}$$

135

The statistic χ^2 is defined as the square of the difference between the measured field D_i and the theoretical field E_i summed over all stations i.

$$\chi^2 = \sum_i (D_i - E_i)^2 \tag{16}$$

The A_n are now adjusted in such a way that χ^2 is a minimum. The mathematical procedure for this adjustment is the following. χ^2 is partially differentiated to each A_n and the resulting equation set to zero. This gives n+1 equations linear in the A_n of the type for a specific n = m

$$\sum_i \cdot (D_i - E_i) \, C_{mi} = 0 \tag{17}$$

or rearranged and E_i replaced by (15)

$$\sum_i D_i \, C_{mi} = \sum_n A_n \sum_i C_{ni} C_{mi} \tag{18}$$

The C_{ni}, C_{mi} are known constants that depend only on the station coordinates and the expressions

$$\sum_i C_{ni} \, C_{mi} = C_{nm} \tag{19}$$

can be calculated once and for all. The expressions

$$\sum_i D_i C_{mi} = D_m \tag{20}$$

have to be recalculated for each new data set D_i. With expressions (19) and (20)

$$D_m = \sum_n A_n \, C_{nm} \tag{21}$$

or expanded

$$D_0 = A_0 \, C_{00} + A_1 \, C_{01} + A_2 \, C_{02} + \ldots$$

$$D_1 = A_0 \, C_{10} + A_1 \, C_{11} + A_2 \, C_{12} + \ldots$$

$$D_2 = A_0 C_{20} + A_1 \, C_{21} + A_{22} + \ldots$$

(21a)

.
.
.
.

There are standard computer programs available to solve a set of linear equations. Furthermore for the Y_n of order $n \geq 3$ recurrence equations can be used and it may be possible to derive recurrance equations for the R_{1n} and R_{2n}. The calculation of the A_n by multipole expansion and least square fit may appear formidable but with a certain know how even a desk top computer could be programmed to evaluate the data of, lets say, 100 stations. A similar procedure could also be used to evaluate the electric field records obtained from an airplane or the magnetic field records obtained at ground stations.

IV. COMMENTS ON INSTRUMENTS AND STATION NETWORKS.

The larger the number of stations the more detailed and reliable will be the analysis. It is not necessary that the stations are lined up along a certain latitude or longitude since the evaluation covers the whole globe. However, if we, for instance, expect the ionospheric current sources to be concentrated in the auroral zone, station in this zone would have the best signal to noise ratio and would contribute most to the reliability of the analysis. However, stations in the midlatitude or equatorial zone, that may not be strongly affected by auroral currents, are important to determine amplitude and time variation of the noise. Noise means here all other electrical effects besides the auroral currents that produce electric fields. The largest contributor will be the atmospheric electric field. Special attention should be given to separate the atmospheric electric field from the field produced by ionospheric or solar events.

V. PHYSICAL AND ATMOSPHERIC ELECTRIC CONVENTION OF THE SIGN OF THE FIELD AND THE ZERO POTENTIAL.

It is not always realized that some important definitions or conventions in atmospheric electricity are contrary to the generally accepted convention in physics. Since we are dealing here with a super-position of potential functions and electric fields from both disciplines it is necessary to convert the atmospheric electric into the physical conventions.

The problem can be best explained by a brief discussion of the atmospheric electric global circuit. The earth is charged negative and produces a current density i that flows from the earth to the ionosphere in the fair weather areas. The current flow in turn generates the atmospheric electric fair weather field E according to Ohm's law

$$\overline{I} = \lambda \overline{E}.; \quad \overline{E} = \overline{I} / \lambda \tag{22}$$

Since $\overline{I}$ is approximately constant and λ increase exponentially with altitude, $\overline{E}$ has to decrease according to (22). The second effect of the increase of λ will be the build up of a positive space charge which is proportional to the gradient of λ. The potential at the ground is zero and at ionospheric altitude about 300,000 V. The current flow and the potential difference between earth and the ionosphere is maintained by the world-wide thunderstorm activity which acts as the generator in the global circuit.

This model of the global circuit seems to be reasonable but it is incompatible with theoretical physics on two counts. The atmospheric electric field

138

is defined to be positive and the earth potential is assumed to be zero. This would necessitate a change in some well established physical relations and implies some unrealistic assumptions.

1) All relations containing E would need a sign change. For instance, the charge density q at the earth surface is given by $q = \varepsilon E$. Since the dielectric constant ε is positive, the relation would give for a positive field a positive surface charge and in consequence a positive net charge of the earth. However, even according to atmospheric electric textbooks the earth charge is negative.

2) The same discrepancy exist by the sign of the space charge density ρ. Here the physical relation is $\rho = \varepsilon \ \text{div} \ E$. Since E has only a radial component and decreases with altitude the calculated space charge density would be negative, however measurements and textbooks state that it is positive.

3) In physics the field is defined as the negative gradient of the potential function $E = - \ \text{grad} \ \phi$ however in atmospheric electric field and potential gradient have the same sign. $E = \text{grad} \ \phi$.

4) It is common practice to determine the potential of the ionosphere by measuring the radial field by balloons or airplane and integrating the field from the earth surface to peak altitude of the flight h. This would give the potential V_h.

$$V_h = \int_0^h E \ dz \qquad (23)$$

With the assumption that I is constant with altitude the current density I_h at h is then calculated by Ohm's law from the measured E_h and an assumed or measured conductivity λ_h.

$$I_h = E_h \lambda_h \quad \text{or}$$

$$\left.\begin{array}{c} \\ \\ \\ \end{array}\right\} \quad (24)$$

$$E_h = I_h/\lambda_h$$

With the relation of (24) and the assumed exponential increase of the conductivity and constancy of the current density $I = I_h$ a correction term V_c is calculated to extend V_h to the ionospheric potential V_i at altitude H

$$V_c = I_h \int_h^H \frac{dz}{\lambda} \qquad (25)$$

$$V_i = V_h + V_c \qquad (26)$$

Since both E and I are assumed to be positive, the potentials V_h, V_c, and consequently V_i are also positive. This leads to the curious result that a negative charged earth produce a positive potential function with the zero potential at the earth surface, the potential function increasing with increasing distance from the earth surface and obtaining a value of 300,000 V at ionospheric altitude. Every physicist would expect that a negative charged body would produce a negative potential function which has its maximum negative value at the surface of the body and decreases to zero as the distance goes to infinity.

It is easy to realize that with the physical sign convention $\overline{E} = -\text{grad } \theta$ we would obtain the expected negative potential function, but the problem of the zero potential at the earth surface would still remain. It is usually argued that it is always possible to add a constant potential to a solution of the differential equation (7) or (9) and therefore by choosing a suitable constant the potential zero at the earth surface can be enforced.

It has already been shown in section II that even a constant potential requires a cause, in our case a spherical symmetric current source in the ionosphere of $7 \cdot 10^{12}$A. This would be an unrealistic and unjustified assumption. If we add to this the confusion generated by the different sign convention of the atmospheric electric fair weather field, it becomes necessary not to rewrite theoretical physics but to recalculate the atmospheric electric global circuit. This has been done by Kasemir (1977) with the result that the earth potential is between -218 to -577 kV and the ionospheric potential between -22 and -58 μV. This calculation has been based on the physical sign convention $\bar{E} = -$ grad ϕ and the physical convention that in three dimensional space the potential function approaches zero as r the distance from the source goes to infinity.

It is obvious that by superimposing potential functions of ionospheric and atmospheric electric origin it makes a vast difference if we add to the horizontal potential difference in the order of 10 kV of the ionospheric dynamo or of the dawn-dusk potential pattern of ±60 kV potential difference an atmospheric electric potential of 300 kV or of -30 μV. For the 300 kV atmospheric electric potential, the ionospheric events introduce only minor variations which seems to be rather unlikely. In the second case of -30 μV atmospheric electric potential, the ionospheric potential distributions are dominant in the ionosphere which is a more acceptable result. Unfortunately, interesting papers such as Hays and Roble's "A Quasi-Static Model of Global Atmospheric Electricity 1". JGR, 84, A7, 3291, 1979, or Roble and Hays "A Quasi-Static Model of Global Atmospheric Electricity 2", JGR, 84, A12, 7247, 1979 are much reduced in their value because of the superposition of the ill defined atmospheric electric potential function.

VI. DETERMINATION OF DIFFERENCES IN THE IONOSPHERIC POTENTIAL.

The method described above of deducing the ionospheric potential from measurements of the radial field by airplanes or balloon soundings, integrating this field to flight altitude and then extrapolating it to ionospheric altitude has been applied to the determination of potential differences in the ionosphere by making these soundings simultaneously at two different locations on the globe. These differences in the order of kilovolts at stations 30^0 to 40^0 longitude apart have been attributed to ionospheric or solar influences. The error in this procedure lies in the calculation of the extrapolation using equation (24). For the mapping up of the fair weather field it is admissable to set the current density approximately constant with altitude. Then the integral of the correction term depends only on the conductivity λ. (see (24)). However a horizontal oriented current source in the ionosphere will not produce a vertical current density that is constant with altitude. Therefore, the correction term V_c given in (24) cannot be applied indiscriminately to the mapping up of the atmospheric fair weather field and the mapping down of ionospheric or solar events.

Beside the thunderstorm generator producing a world-wide contribution to the fair weather field, we have a more regional or continental generator, namely, the austausch generator. The field produced by this generator varies with local time and is usually strong enough to mask the world-wide variation caused by the thunderstorm generator. Even at the so-called global representative stations, for instance, on the oceans, in the arctic regions, at mountain tops, etc. it requires averaging over 7 to 10 days until the world-wide diurnal variation of the thunderstorm generator emerges. Therefore, we should expect that the potential at peak flight altitude obtained by integrating the vertical field component contains a contribution of the austausch generator that may be large or small according to local time, weather and location on the globe. It is much more likely that differences in the potential at peak flight altitude determined at two widely separated locations on the globe are caused by local generators in the ground layer than by the down mapping of ionospheric or solar events. At least without a determined effort to eliminate the local influence,conclusions with regard to ionospheric influences cannot be drawn.

SUMMARY

1) The electric energy of solar or ionospheric current sources deposited in the tropo and stratosphere is given by divergence of the Pointing vector $\overline{P}$.

$$\overline{P} = \overline{E} \times \overline{H}$$

$$- \text{div} \ \overline{P} = \overline{H} \cdot \dot{\overline{B}} + \overline{E} \cdot \dot{\overline{D}} + \overline{E} \cdot \overline{J}$$

or in the stationary case

$$- \text{div} \ \overline{P} = \overline{E} \cdot \overline{J}$$

2) The electric parameters to be determined are $\overline{H}$ and $\overline{E}$

3) $\overline{H}$ and $\overline{E}$ as functions of space can be obtained from measurement at the ground at a large number of stations distributed over the globe. The method suggested is expansion of the electric parameter in multipole functions and fitting the measured data by a least square fit method to the theoretical functions.

4) Special attention should be given to the separation of the field caused by ionospheric energy sources and the atmospheric electric field. It is important to realize clearly the ramifications caused by different sign convention of the atmospheric electric field and the zero potential of the earth in atmospheric electricity.

REFERENCES

Kasemir, H.W., "Theoretical Problems of the Global Atmospheric Electric Circuit." Electrical Processes in Atmospheres. Ed. H. Dolezalek, R. Reiter, Steinkopff Verlag Darmstadt 1977.

Hays, P.B., R. G. Roble, "A Quasi-Static Model of Global Atmospheric Electricity 1." JGR, 84, A7, 3291, 1979.

Roble, R.G., P.B.Hays, "A Quasi-Static Model of Global Atmospheric Electricity 2." JGR, 84, A12, 7247, 1979.

ANALYTIC SOLUTIONS OF THE PROBLEM OF DOWN-MAPPING OF ELECTRIC FIELDS AND CURRENTS OF SOLAR EVENTS TO THE EARTH SURFACE

Heinz W. Kasemir
Colorado Scientific Research Corporation

ABSTRACT

An analytic solution of the current continuity equation div I = 0 in the form of harmonic functions has been found that is valid in the ionosphere (anisotropic conductivity) as well as in the troposphere and stratosphere (isotropic conductivity). The Earth's magnetic field is assumed to be vertical (limited to polar caps). This solution is applicable to the calculation of field, current, potential, space charge, energy-density functions, etc., generated by solar and ionospheric generators. The equipotential and current flow lines of such a harmonic generator (shown in Figure 3) have two layers between 0- and 15-km and 70- and 90-km altitude with unexpected but quite interesting features. In the ionosphere the current flow is vertical in the direction of the magnetic field lines but it bends rather abruptly into a horizontal direction at and below the lower ionospheric boundary. In 70- to 90-km altitude the current flow lines are compressed, the density is even greater than at the source, and the largest electric-thermodynamic energy conversion occurs here. About 80% of the current output of the generator flows through this layer. In a study of solar-terrestrial manifestations this area deserves special consideration.

In the ground layer of 0- to 15-km altitude the equipotential lines show peculiarities similar to the current flow lines at the ionospheric boundary. They bend from the vertical to the horizontal direction. Over 90% of the equipotential lines generated at the source enter into and are compressed in this layer. This would provide a unique opportunity to monitor and study ionospheric potential distributions at or close to the ground. Unfortunately the atmospheric potential function is dominant in this region, and a perceptive knowledge of the problem is required to develop a measuring program that is capable of separating the ionospheric from the atmospheric electric potential function. The analytic solution presented here will be of considerable value in solving the problem and in avoiding the errors commonly made in extrapolating measurements made close to or at the ground to ionospheric altitude. In this respect the horizontal component of the field has received special attention since it is practically free from interference by the fair weather field. The importance of the horizontal component is borne out in the calculation by the result that this component is constant in good approximation from 15-km altitude up to the ionospheric source.

1. DEFINITION OF PROBLEM

In a cartesian coordinate system x, y, z, a cylindrical box of height h (z-axis), width w (x-axis), and of infinite length (y-axis) is defined (Figure 1). At the top of the box current is injected into the left half x = 0 to x = w/2 and ejected from the right half x = w/2 to x = w of the box. The Earth's magnetic field is assumed to be vertical, which limits the application of the solution to the polar caps. The conductivity L1 is scalar and exponential below the ionosphere and splits inside the ionosphere into the parallel and transverse (Pedersen) conductivity L1 and L2. L1 increases exponentially from the ground to source height h. L2 = L1 below the ionosphere. L2 has a maximum close to the ionospheric boundary and decreases exponentially with increasing altitude (Figure 2). The sharp decrease of L2 in the ionosphere is not realistic but is adopted here for greater simplicity of the calculation. The second maximum of L2 at about 125 km (Park, 1976) cannot be ignored if ionospheric currents are of primary concern.

The boundary conditions are as follows: 1) No current shall flow through the side walls, here Ix = 0; 2) The potential function V0 at the ground shall be constant and can be set to zero; and 3) The input and output current are of the same amount but of opposite polarity. The average input and output currents are given parameters.

The problem is to calculate the potential function V(x, z), and the vector functions of the electric field E(x, z) and the current density I(x, z) inside the box. The current vector I has three components Ix, Iz, and Iy. Iy is the Hall current. It is source free (div Iy = 0) and is zero below the ionosphere. The Hall current will not be discussed here.

2. SOLUTION OF THE PROBLEM

Symbols and functions:

x, y, z	Cartesian coordinates
Functions of x and z:	
V(x, z)	Potential function
E(x, z)	Electric field vector
Ex(x, z)	X-component of E
Ez(x, z)	Z-component of E
I(x, z)	Current density vector
Ix(x, z)	X-component of I
Iz(x, z)	Z-component of I
Functions of z:	
L1 = L0 exp (2 kz)	Scalar and parallel conductivity
L2 = L1/(1 + a exp [2 kz])²	Transverse conductivity
Arbitrary parameters:	
L0	Conductivity value at the ground
1/2 k	Scale height of conductivity

147

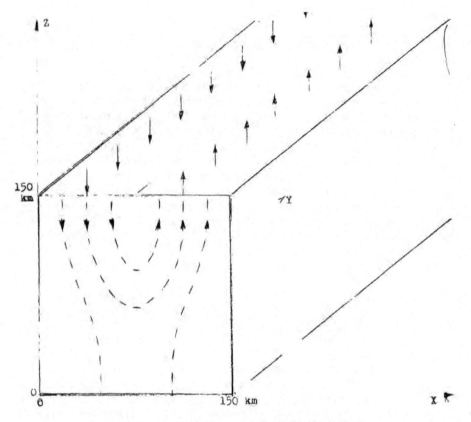

Figure 1. Current in- and outflow of cylindrical box.

b	Altitude of ionospheric boundary
$a = \exp(-2\,kb)$	Constant determined by 2 kb
h	Altitude of box or source
w	Width of box
$s = pi/w$	Separation constant of diff. equation
	also scale factor of x

The arguments of the functions are only shown if important; otherwise, they are omitted in the following calculation. For instance, $Ix(x, z) = Ix$. Partial differentiation symbols are replaced by ordinary differentiation symbols; for instance, $dIx(x, z)/dx = dIx/dx$ indicates the partial differentiation of Ix to x.

Since we restrict the problem to quasi-stationary conditions and do not assume generators to be located inside the box, the governing differential equation is

$$\text{div } I = dIx/dx + dIz/dz = 0. \tag{1}$$

With the relations

$$E = -\text{grad } V, \quad Ex = -dV/dx, \quad Ez = -dV/dz, \quad Ix = L2Ex, \quad Iz = L1Ez \tag{2}$$

148

we obtain from Equations (1) and (2) the differential equation

$$d(L2\ dV/dx)/dx\ +\ d(L1\ dV/dz)/dz\ =\ 0. \tag{3}$$

We set now

$$V\ =\ C\ f(z)\ g(x) \tag{4}$$

which leads to solutions given in eigenfunctions or harmonics. Furthermore, we will use the symbols ' and " to indicate single or double differentiation. For instance

$$df/dz\ =\ f',\ d(dg/dx)/dx\ =\ g''. \tag{5}$$

Substituting Equations (4) and (5) in Equation (3) and rearranging in two terms of which the first depends only on x and the second only on z, we obtain

$$g''/g\ +\ (L1\ f''\ +\ L1'\ f')/L2f\ =\ 0. \tag{6}$$

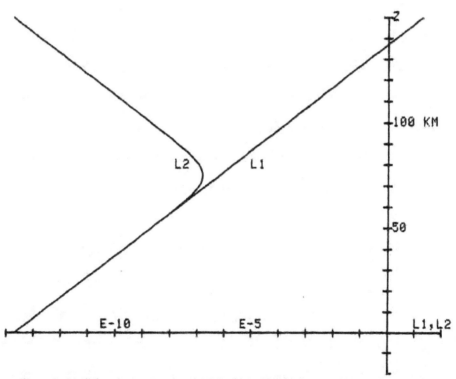

Figure 2. Parallel and transverse conductivity L1 and L2 from ground to 150-km altitude.

149

Using now the separation constant s^2 we can split Equation (6) into two second-order ordinary differential equations:

$$g'' = -s^2 g, \quad L1 f'' + L1' f' - s^2 L2 F = 0. \tag{7}$$

The solution for g is the well-known exponential function

$$g = \exp(+-isx) = \cos(sx). \tag{8}$$

We will use here only the real part of g which is the cosine function. However, keep in mind that using a complex g and s will extend the range of solutions.

The way to a solution of f is more involved and discussed in detail by Kasemir (1981). We obtain here two independent solutions f1 and f2:

$$f1 = (\exp[-2\ kz] + a)^m, \quad f2 = (\exp[-2\ kz] + a)^n \tag{9}$$

where

$$m = ([1 + \{s/k\}^2]^{1/2} + 1)/2, \quad n = -([1 + \{s/k\}^2]^{1/2} - 1)/2 \tag{10}$$

and finally from Equations (4), (8), (9), and (10) the two potential functions

$$V1 = C1\ f1\ g, \quad V2 = C2\ f2\ g. \tag{11}$$

V1 is a potential function which decreases with increasing z, in this case the source is at the ground, and V2 increases with increasing distance from the ground indicating that the source is in the ionosphere. V1 and V2 are the harmonics or eigenfunctions of the differential Equation (1) and we use them to fulfill the boundary conditions.

1. No current flow through the walls is automatically fulfilled by the harmonics since the horizontal currents have $\sin(sx)$ dependence. This function is zero for $x = 0$ and $x = sw$.

2. Potential function $V = 0$ for $z = 0$. This we achieve by superposition of the source function V2 with function V1 of the mirror image of the source at the ground. We set

$$V = V1 + V2 = (C1f1 + C2f2)\cos(sx) \tag{12}$$

and for $z = 0$

$$0 = (C1f1[0] + C2f2[0])\cos(sx), \text{ or } C1 = -C2f2(0)/f1(0). \tag{13}$$

Inserting Equation (13) in Equation (12) gives

$$V = C2(f2 - f2[0]f1/f1[0])\cos(sx). \tag{14}$$

With

$$F(z) = f2 - f2(0)f1/f1(0) \tag{15}$$

and dropping the now superfluous index 2 on C we obtain the required potential function V and the field and current components Ex, Ez, Ix, Iz by simple differentiation:

$$V = CF\cos(sx)$$
$$Ex = sCF\sin(sx), \quad Ix = L2sCF\sin(sx) \tag{16}$$
$$Ez = -CF'\cos(sx), \quad Iz = -L1CF'\cos(sx).$$

3. The last boundary condition that the net input and output current is given, determines the still arbitrary constant C. This easy calculation is not carried out here.

3. DISCUSSION OF SOLUTION

The solution of Equation (16) is one special case of an infinite series of harmonics of the differential Equation (1). If we consider s as a continuous variable we may expand any given potential or current density function at the source into a Fourier series or integral. The same method may also be used to eliminate the boundary condition 1, that no current shall pass the side walls. Other restrictions may be relaxed at the cost of more complex solutions, such as accommodating different conductivity profiles, extension to time-dependent problems, or different coordinate systems. For our discussion here we will use the simple solution (16).

In Figure 3 we see the equipotential and current flow lines of our problem. This picture is truly remarkable since it shows two unusual features, namely the concentration of the equipotential lines at the ground and of the current flow lines at the lower boundary of the ionosphere and not at the source as one would expect. In the ionosphere both equipotential and flow lines are vertical, i.e., they follow the magnetic field lines. This is a well-known fact but here it generates confidence in the validity of the solution. As soon as parallel and transverse conductivities have the same order of magnitude the current breaks away from its vertical path along the magnetic field lines and assumes a horizontal path across the magnetic lines. The flow lines concentrate— between 68- and 88-km altitude in our example—by about a factor 8 as compared with the vertical path. We will find the highest energy concentration in this region. About 80% of the net current flows here and only 10% is left for the whole area between ground and 68 km. Since the flow lines are drawn in steps of 10% no flow lines are shown in Figure 3 in this region. From the downward path of the equipotential lines we may infer that the horizontal flow will prevail all the way down to an altitude of about 12 km. This predominance of the horizontal over the vertical component of the current and the field in the altitude range from 12 to 80 km is a characteristic feature of ionospheric generators and in contrast to the fair weather field and current. There the horizontal component is either zero or extremely weak. This is an important result with respect to separating the fields or the currents of ionospheric and atmospheric generators, which at or very close to the ground is one of the major problems.

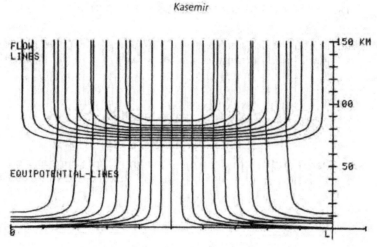

Figure 3. Equipotential and current-flow lines.

The equipotential lines are equally remarkable. Here the vertical trend reaches all the way down to about 10-km altitude. In this shallow layer from ground to 10 km we have practically the whole menu of the potential lines generated in the ionosphere. This would provide an ideal situation to obtain the potential distribution in the ionosphere by measurements carried out close to the ground. However, as mentioned above, we are in the domain of the fair weather field and may expect a signal-to-noise ratio of about 1:5 or more probably of 1:10, depending on the strength of the ionospheric generator.

Figure 4 shows the two components Ix and Iz normalized to the average current density at the source in logarithmic scale as a function of altitude. In the ionosphere the vertical current Iz is practically constant with altitude. In the same manner as it is produced at the source it flows along the magnetic lines without deviation, concentrating, or spreading. The horizontal component Ix is smaller than the vertical component at the source by about six powers of ten, but increases exponentially with decreasing altitude. At the lower boundary of the ionosphere Ix finally catches up with Iz, surpassing it by about a factor of 8. This produces the dramatic concentration and change in direction in the flow lines in Figure 3. Due to the logarithmic scale of Ix and Iz in Figure 4, the little overshoot of Ix over Iz is not very impressive. From here on downward to about 13-km altitude Ix is larger than Iz by about a factor of 10, but both components decrease exponentially at the same rate. At 13 km another reversal takes place. Ix decreases rapidly to zero at the ground, whereas Iz levels off to a fairly constant value in the last 10 km.

It should be mentioned here that the profiles of Ix and Iz given in Figure 4 are not drawn for the same footpoint, i.e., for the same x value. The profiles with maximum Ix and Iz values are drawn for Ix in the middle of the box $(x = w/2)$ and for Iz at the side wall $(x = 0)$. Ix follows the $sin(sx)$ and Iz the $cos(sx)$ function. The same is true for the profiles of Ex and Ez.

Figure 5 shows the field components Ex and Ez in logarithmic scale as a function of altitude. The most striking feature here is that Ex is practically constant from about 10-km altitude up to the ionospheric source. For the mid-atmosphere this is nothing new and many scientists have pointed out this remarkable fact. Mozer and Serlin (1969) have measured the horizontal field as a function of

DOWN-MAPPING OF ELECTRIC FIELDS AND CURRENTS

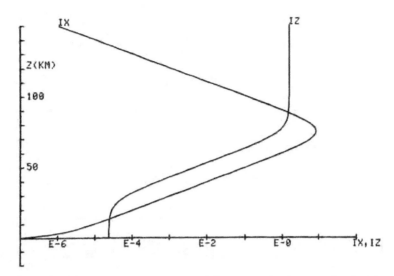

Figure 4. Profiles of the components Iz and Iz of the current density normalized to the average source current density.

altitude from a balloon from 10 to about 36 km and found it to be practically constant with a value of about 40 mV m⁻¹. What is new here is that according to Figure 5 the constant Ex can be extended high into the ionosphere all the way to the source as long as the transverse conductivity is much smaller than the parallel conductivity and assuming that the concept of conductivity is still valid. It is remarkable that even the drastic change of direction and concentration of the current flow at the ionospheric boundary does not interrupt nor disturb the constancy of Ex. It may be a coincidence, but worth mentioning, that Heppner (1972) published records of Ex measured from the OGO-6 satellite above the ionospheric F-layer on a polar dawn-dusk traverse which show horizontal field values of –40 to +60 mV m⁻¹, i.e., the same order of magnitude as Mozer and Serlin's Ex measurements at 20 to 36 km above ground. Mozer's balloons were launched from Fort Churchill in Canada during a magnetic bay. Simultaneous satellite and balloon measurements would be necessary to confirm the prediction that horizontal fields in the ionosphere up to or above the F-layer could be measured from balloons flying at or above 20-km altitude.

It is interesting to observe the exchange of field and current in the profiles of atmospheric electric and ionospheric generators. In atmospheric electricity the current Iz is constant and the field Ez decreases exponentially with altitude. Ionospheric generators produce a constant Ex field but the Ix current increases exponentially with altitude. In both cases field and current are related by Ohm's law, I = LE. The extrapolation formula used in atmospheric electricity to obtain the ionospheric potential is based on the assumption that the Iz current is constant with altitude. This method cannot be applied to ionospheric generators since this assumption is not valid here. Differences in the ionospheric potential caused by ionospheric generators cannot be determined

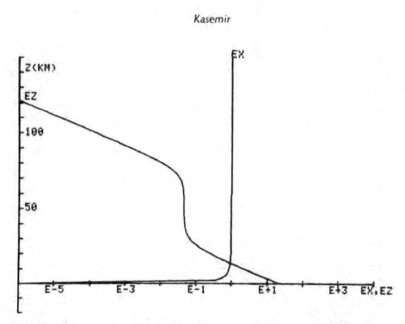

Figure 5. Profiles of the components Ex and Ez of the electric field normalized to Ex-max at the source.

by this method. It is much more probable that the differences obtained by measurements of the ionospheric potential at two different locations are caused by electric effects of eddy diffusion at the ground (austausch-generator), which depends on local weather and time and may have very well been different at the two locations.

Assuming that our model is in essence applicable (see arbitrary parameter), the analytic solution enables us from one measurement to calculate the numerical value of any other parameter or function of the circuit. Let's take the Mozer and Serlin value of $(Ex) = 40$ mV m^{-1}. The parentheses around Ex indicate that this is one measured value. We have to know where it was measured in reference to the coordinate system of the generator. Let's assume this value was obtained close enough to the center of the box where approximately $\sin(sx) = 1$ and at 30-km altitude. Then, from Equation (16),

$$(Ex) = sCF(30), \quad C = (Ex)/sF(30). \tag{17}$$

We have now to replace the C-value given by Equation (17) in any of the five equations given in Equation (16) to obtain the potential function and the Ex, Ez, Ix, and Iz components as functions of altitude and horizontal distance from the center of the source. We may even calculate Joule's heat $W = EI$, the potential distribution at the source, the Hall current, or any other parameter or function of the circuit. For instance, what kind of ground field can we expect which corresponds to the measured (Ex)? From Equations (16) and (17) we obtain for $z = 0$

$$Ez(0) = -(Ex)F'(0)\cos(sx)/sF(30). \tag{18}$$

154

DOWN-MAPPING OF ELECTRIC FIELDS AND CURRENTS

First of all, right under the center of the source Ez(0) = 0 since sx = pi/2 and cos(sx) = 0. This would be the worst place to choose for measuring the influence of ionospheric generators on the ground field. The best place would be at the side walls x = 0 or x = w, where the ground field is a maximum. Here the numerical value would be Ez(0) = +−1.2 V m⁻¹. There is a factor of 30 between the maximum ground field and (Ex). This can clearly be seen in Figure 5. Fields of the order of 1 V m⁻¹ are not difficult to measure. However, how to separate the ionospheric from the atmospheric electric field of about 100 V m⁻¹ is a question which should be addressed and solved first before launching a major experiment.

Another interesting but simple result can be obtained from Equations (16) and (17). We may ask what kind of current density would correspond to (Ex) = 40 mV m⁻¹ in mid-atmosphere where L1 = L2. The answer expressed in an analytical formula is

$$Ix = L1(Ex)Fsin(sx)/F(30), \quad Iz = -L1(Ex)F'cos(sx)/F(30). \tag{19}$$

The design of an experiment is quite evident from these equations. Construct a radiosonde which can measure all three components of the current density vector. Launch one balloon below the center of the ionospheric generator where sin(sx) = 1 and cos(sx) = 0 and another one at the side where sin(sx) = 0 and cos(sx) = 1. At the center you have maximum Ix and Iz = 0. The Iz which you will measure is of atmospheric electric origin and will serve as a good check on the proper operation of the instrument and in comparison with the Iz of the second balloon. The Ix is of ionospheric origin and should be free from any interference of the atmospheric electric Iz. At the side Ix = 0, which is a good check that your balloon is at the right spot. Iz is a superposition of the atmospheric electric and ionospheric Iz. The ionospheric Iz is probably too weak to show, but the comparison with the control Iz of the other flight will tell if this is so or not. The balloons will not stay over their launch pads but each will follow its own path as tracked by the receiver at the ground. This will probably furnish enough information to reconstruct from the sine-cosine dependence of the components the position of the generator with respect to the launch sites. There are many more relationships to check and information to get out of these two flight records, which are rather elementary but may not be so obvious to the reader at first glance. However, space limitation doesn't permit further discussion. The last remark will be just the statement that Ix will be 0.1 or 1 pA m⁻¹ at 20- or 30-km altitude, respectively, corresponding to a horizontal field Ex = 50 mV m⁻¹. This is important to know for designing the proper instrument.

There is no claim that this analytic model will fit all ionospheric or solar generators. Indeed, it will not fit a PCA event at all where electrons and protons penetrate into the mid-atmosphere. They would show up as an electric generator inside the considered space in contradiction to the assumption div I = 0 in Equation (1). This condition excludes generators inside the given boundaries. We are dealing here with electromagnetically driven currents which enter and depart at the boundary. This model may serve as a guideline on how to study these types of currents and design suitable experiments. Refinements and corrections will come from an interplay of theoretical and experimental investigations.

REFERENCES

Heppner, J. P.: 1972, in *Critical Problems of Magnetospheric Physics*, ed. E. R. Dyer, p. 107.

Kasemir, H. W.: 1981, Analytic model of aurora currents, final report Naval Research Laboratory, Washington, D.C., Contract N00014-81-C-2275.

Mozer, F. S. and Serlin, R.: 1969, *J. Geophys. Res.*, **74**, 4734.

Park, C. G.: 1976, *J. Geophys. Res.*, **81**, 168.

On the Electrodynamic Theory of the Atmospheric Electric Field: The Turbulence Generator.

H. W. KASEMIR

Meteorologisches Observatorium des

Deutschen Wetterdienstes

The records of the atmospheric electric field and the air-to-earth current in the fall of 1953 in Buchau show a remarkable increase during the sunrise on clear and cloudless days. This prompted a closer investigation of the so-called sunrise effect. Since a decrease of the conductivity takes place at the same time, the increase of the current can not be accounted by a variation of the conductivity. In order to explain this phenomenon satisfactorily, one must assume a local fair weather generator on land in addition to the world wide thunderstorm generator. The experimental results lead to the conclusion that this fair weather generator must be attributed to the turbulent mixing, which, with the help of the electrode effect, builds up a positive space charge within the turbulent mixing layer.

1. INTRODUCTION

During an extended period of good weather in late summer of 1953, at the atmospheric electric registration station in Buchau, a striking increase of the atmospheric electric field and the vertical electric current was observed at sunrise. This increase of field and current of up to at least twice the values preceding sunrise was especially conspicuous at times of calm and clear weather. Also there occurred frequently, in connection with the sunrise effect, a negative current, even though the field was positive. This started shortly after midnight and continued until sunrise. This opposition of sign between field and current indicates a negative convection current that exceeds the positive conduction current. The striking fact here in that the carriers of this negative convection current could not be observed. There were no signs of the slightest precipitation or fog. The density of haze or dew was not appreciably different on those days than on those having a positive current. Although the positive current trend at sunrise could be accounted for, in terms of the eddy exchange generator action, the negative current effect is obscure.

2. RECORDING ARRANGEMENTS

Before analyzing the results of measurements, a sketch of the recording arrangements will be given. The field and current recording apparatus described in reference [1] was modified for the registration of the vertical current [2]. This program was amplified by recording also sunrise and wind. The field strength, the current sunshine and wind were recorded in cyclical order every 20 s on the same recording strip.

The radioactive collector used in the electrical field measurements was mounted on the wall of the laboratory building at a distance of 4 m above ground level. The collecting net for the current registration was located 25 m from any building. This net was placed at the ground level over a pit. The wind vane was mounted on a mast of 12 m long and installed at a distance of 8 m from one side of the pit. The mast, which carries the photocell for registration sunshine, was 4 m long and located at a distance of 5 m on the opposite side of the pit.

Clear weather could be distinguished from cloudy weather by the smooth arcs on the registration curve, as contrasted with the agitated and small graphs on cloudy weather. Calm weather could be readily recognized on the wind register from the flat character of the curve, whereas, even with wind velocities of as small as 2 m s^{-1} and above, the wind vane would oscillate back and forth and give an agitated appearance to the registered curve.

3. OBSERVATIONS

The period of observation extended from the beginning of July 1953 until the middle of September 1953. After eliminating days with precipitation, fog, and other disturbances, there remained 48 days, which were divided into three groups as follows:

1. Clear and calm days with positive current (26 days),
2. Cloudy and windy days with positive current at sunrise (12 days),
3. Clear and calm days with negative current before sunrise (10 days).

There were no windy and cloudy sunrises when negative current was observed.

In each of these groups the registrations were synchronized on the time of sunrise and then averaged. The results are shown in Figs. 1, 2, and 3 with numbers corresponding to the groups above. The curve for the field strength is always the continuous curve and that for current is the dashed curve.

Legends of figures.

Fig. 1. Electric field and vertical current variation on cloudless and sunrises (positive current).

Fig. 2. Electric field and vertical current on windy and cloudy sunrises (positive current).

Fig. 3. Electric field and vertical current on cloudless and calm sunrises (negative current).

In Fig. 1, one sees clearly the sunrise effect in the field as well as in the current graph. At sunrise, a steady rise begins in both curves that level off about 2.5 hours later. In this time, the field increases 2.4 fold and the current increases to twice the value before sunrise. The steeper rise

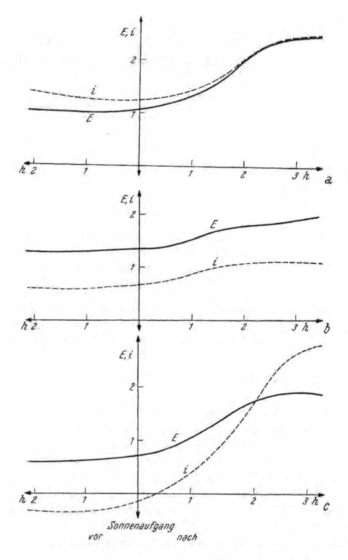

Figure.1 Traces of potential field E and the vertical current density i before and after sunrise:

(a) on sunrises with calm and clear sky conditions and with positive current densities,

(b) on sunrises with windy and cloudy conditions and with positive current densities,

(c) on sunrises with calm and clear sky conditions and with negative current densities.

in the field relative to that of the current indicates a small decrease in the air conductivity of about 20 %. About a 25% increase of conductivity occurs on windy and cloudy sunrises (Fig. 2). However, in this ease the conductivity prior to sunrise is only about 1/2 the value that is found an days without wind or clouds at sunrise. The increase in field and current is also noticeable in Fig. 2, even though it is not so pronounced as in Fig. 1. The current increases 1.8 fold and the field about 1.5 fold of the value before sunrise.

The sunrise effect in clear and calm weather is shown in Fig. 3: the current having an initially negative value up to and before sunrise, the field increase amounts to about threefold. To calculate in this case either the current or conductivity increase has no meaning, because there is a superposition of a conduction current and a convection current in the registered values of current. The calculation of conductivity from the field and current measurement assumes that the measured current is purely a conduction current.

4. INTERPRETATION OF THE MEASUREMENTS

We shall try to ascertain from the available registrations the cause of the sunrise effect. Since this is obviously a local and not a world wide effect, there can be no connection between these and field changes attributed to the thunderstorms generator. Theoretically, there remain two possibilities to explain an increase in the current associated with an increase in the field. First, a decrease of the columnar resistance to about one half its normal value and, second, an auxiliary generator effect. If it were a decrease of columnar resistance, that would imply an increase of about two fold in the conductivity. This increase must occur in the lower 3 km of the atmosphere, since it is in this region that the preponderant part of the resistance is concentrated. There is, however, no evidence that the ionization of the air near the surface is affected by sunlight nor it is a rule that the conductivity increases at sunrise. Most measurement of conductivity indicates a decrease at sunrise, which is attributable to an introduction of more nuclei by eddy diffusion that sets in about sunrise. That this decrease in conductivity is not confined to a shallow stratum in the earth's surface, but extends throughout the entire range of turbulence, was shown by Sagalyn and Faucher [3]. Such a decrease in conductivity is in harmony with our calculations according to Fig. 1, but the 25% increase in conductivity calculated for Fig. 2 would not be adequate to account for the 80 % increase in current. There are, therefore, several difficulties in the way of attributing the sunrise effect to a decreases of columnar resistance.

We shall now consider the second theoretical possibility, namely, that there exists an unknown generator. This generator is to be regarded as one, which moves charged particles against the direction of the electric field. Since the generator in question acts in such a way as to strengthen the normal field, it must be positive polar, i.e., it must have its positive pole in the air and its negative pole on the ground. In order to establish a positive pole in the air, it is necessary that a non-electric force moves charged particles from the earth's surface to a higher level

or moves negative charges from the air toward the earth. The latter case in realized in Fig. 3. Since this negative convection current ceases at about sunrise and occurs only in a smaller part of the cases with sunrise effect, this seems to be excluded as an explanation of the sunrise effect. The first case of positive charges carried upward can be brought about by eddy diffusion. In order that this effect may operate in the right way when low mobility ions are involved, there must be a second effect, which steadily supplies positive ions at the earth's surface. The more decimation of the positive space charge located at the earth's surface throughout a greater region would not give rise to increase of the field. This source of supply of positive ions is the electrode effect, which tends to maintain a positive space charge at the earth's surface. The continuous removal of large positive ions from this space charge permits a continuous replenishment by action of the electrode effect. The increased production of nuclei owing to eddy diffusion would tend to preserve the transformation of small ions into large ions by the electrode effect.

The local time variation of potential gradient on non-disturbed days was attributed by Brown [4] to the production and transfer of nuclei by means of eddy diffusion. One can ascribe all thee questions to a generator effect of eddy diffusion. Even the midday depression due to the diminution of nuclei can be explained as a result of a cessation of the generator effect and the consequent decrease in the supply of nuclei. Moreover, the eddy diffusion generator also explains the increase of current, while a simultaneous reduction of conductivity by eddy diffusion may be brought into harmony with the diurnal variation of potential gradient, even though this is directly contrary to a diurnal variation of the conduction current.

5. REFERENCES TO THE THEORY OF THE EDDY DIFFUSION GENERATOR IN PREVIOUS LITERATURE

Of all the publications known to the author, that of Holser [5] is the closest to views, which are presented in this paper. Holser, in an extensive program of measurements, recorded simultaneously the atmospheric electric elements on three mountain peaks and at sea, and,in an analysis of these results, he separated the local time effects from the universal time effects. In order to bring the diurnal variation of the current measured on the mountain peaks in harmony with the diurnal range of the conduction current at sea, he applied to the mountain peak measurements a correction formula, which was based partly on theory and was partly empirical. This correction involves both the following influences of eddy diffusion on the atmospheric electric current circuit:

1. The decrease of conductivity and consequently the increase of columnar resistance.
2. The generator effect of eddy diffusion by introduction of an auxiliary conduction current, which depends upon eddy diffusion.

These result in the following equation for the conduction current density on mountain peaks

$$i_{obs} = i [1 + A \sin(2\pi t/T)] \qquad (a)$$

Here i denotes the part of the vertical current density that comes from the thunderstorm generator and the term $i A \sin(2\pi t/T)$ is the part attributable to the eddy current generator. That the latter is proportional to the conduction current density of the thunderstorm generator comes from the circumstance that both generators are connected by virtue of the electrode effect, as it will be explained later more fully. The second term contains the coupling factor A and the daily variation of the eddy diffusion generator $\sin 2\pi t/T$, which is assumed to be sinusoidal over the interval T/2 from sunrise to sunset. The increase of the columnar resistance R_0 is taken into account by the additive term delta R, so that there occurs a relation between the conduction current density at sea i_{corr} and the conduction current density on mountains i

$$I_{corr}/ i = 1 + R/R_0 \qquad (b)$$

This, taken together with equation (a), gives the correction formula [5]

$$i_{corr} = i_{obs} (1 + R/R_0)/(1 + A \sin 2\pi t/T) \qquad (c)$$

A comparison of the current measurements corrected in this way for the mountain peaks with the simultaneously measured current density at sea were in remarkable agreement. This justifies the expectation that for stations lying in undisturbed positions in level country, it would also be possible to carry out a similar analysis.

When one goes back into the atmospheric electricity literature, he finds that the idea of the eddy diffusion generator in some form is older then the idea of the thunderstorm generator. Elster and Geitel [6] about fifty years ago sought to account for the negative charge of the earth as a result of the greater mobility and the greater diffusion coefficient for negative ions, which would give up their negative charge to the earth, whereas the positively charged ions would escape into the air. Following this, Elbert [7] took into account the carrying of the positive space charge, which has so gathered at the surface to higher levels by eddy diffusion and horizontal winds. The aim here was to find the source of the world wide atmospheric electric field. The criticism of Simpson [8] and Swann [9] led to the abandonment of Ebert's theory, especially because the horizontal distribution of charge would be too slow to account for world wide effect. Since research at that time was directed especially toward accounting for the world wide effects, the applicability of Elbert's theory to local phenomena was not considered.

Further clarification of the eddy diffusion generator is provided by Whipple [10], who attempted to consider eddy diffusion in calculating the electrode effect. His aim was to account for the decrease of the atmospheric electric field with altitude. Although Whipple was not concerned with the relation considered here, his paper does contain the generator effect of eddy diffusion. We will repeat this calculation,

however, with some slight modification. The modification consists of separating the thunderstorm generator from the eddy diffusion generator in order to investigate independently the properties of the latter. There exists between these two generators a coupling of the following sort. Before the eddy diffusion generator comes into play at sunrise, the positive space charge at the surface has already been established by the conduction current of the thunderstorm generator. It is this establishment of space charge, which makes it possible for the action of the eddy diffusion generator. When once in operation, the eddy diffusion generator can maintain itself by means of the electrode effect of its own conduction current. This feedback, as well as the dependance of the electrode effect of the thunderstorm generator, is taken into account in the calculation only in a very general way. This is introduced through space charge q_0 of the earth. We may regard this procedure as satisfactory for the stationary condition. This also admits the possibility of taking into consideration the space charge coming from soil respiration (as suggested by Elster, Geitel and Ebert). The transient conditions of the eddy diffusion generator, however, requires further provisions.

The symbols used in the following derivations have the following significance:

q - space charge density of the air,
c - convection current density,
λ - conductivity,
ϵ - dielectric constant,
x - altitude,
k - coefficient of eddy diffusion,
E - electric field,
i - conduction current density.

The subscript small "o" used with q, E and i designates the value of those elements at the earth surface. Since we need be concerned with only one space coordinate, div $E = dE/dx$, the symbol for designating the vector character of E and i is neglected here because we have to do only with the component in the x direction.

It is known that transfer of meteorological properties by eddy diffusion current c is proportional to the concentration gradient of the particular element considered. In our case this is the space charge that includes the constant k, which we have already designated as the turbulent mixing coefficient. Therefore, the eddy diffusion current is given an follows.

$$c = - k \, dq/dx \qquad (1)$$

Under stationary conditions, this convection current is exactly compensated by the conduction current

$$c + i = 0 \qquad (2)$$

It follows further from the well known relations $i = \lambda E$, div $E = dE/dx = q/\epsilon$ and from a consideration of equation (2)

$$\frac{di}{dx} = \frac{\lambda}{\epsilon} q = -\frac{dc}{dx} \qquad (3)$$

After differentiating equation (1) with respect to x, we obtain from equation (1) and (3) the differential equation for the space charge

$$\frac{d^2q}{dx^2} = \frac{\lambda}{\epsilon k} q \qquad (4)$$

The solution of this can be written in the form

$$q = q_o e^{-\sqrt{\frac{\lambda}{\epsilon k}} x} \qquad (5)$$

In the integration, ϵ, λ, and k are regarded as constant with respect to altitude. This is certainly not correct in the case of λ and k and, therefore, the solution can be only regarded as a first approximation. Furthermore, no upper boundaries of the eddy diffusion layer is assumed in this calculation. These restrictions are somewhat ameliorated by the circumstance that, according to equation (5), the space charge remains confined to the lower strait of air, even when the eddy diffusion extends to considerable altitudes. If, following Whipple, we assume, as a plausible value, $k = 1 \text{ m}^2 \text{ s}^{-1}$ and, as the time constant for the soil-air, $\epsilon/\lambda = 10$ min, then the space charge at an altitude of 25 km has already decreased to 1/3 of the value at the surface. Decrease to 1/10 of the initial value and 1/100 of the initial value would occur at 57 meters and 113 meters altitudes, respectively. From this it would appear that measurements on a tower 50 to 100 meters high should completely encompass the eddy diffusion generator.

From equation (5) we obtain, by simple integration, the equations for field strength and current density.

$$E = -\frac{q_0}{\epsilon} \sqrt{\frac{\epsilon k}{\lambda}} e^{-\sqrt{\frac{\lambda}{\epsilon k}} x}, \qquad (6)$$

$$i = -\frac{\lambda q_0}{\epsilon} \sqrt{\frac{\epsilon k}{\lambda}} e^{-\sqrt{\frac{\lambda}{\epsilon k}} x} \qquad (7)$$

Notice that the unconventional negative sign for the field and current in equations (6) and (7) comes from the practice in theoretical physics of writing div $E = q/\epsilon$ instead of usual for atmospheric electricity div $E = -q/\epsilon$.

If we put q_0 equal to 200 elementary charges or $3 \times 10^{-17} \text{ A s cm}^{-3}$, then we obtain the field strength $E_0 = -85 \text{ V m}^{-1}$ and the current density $i_0 = -1.25 \text{ A m}^{-2}$. These numerical results show that the part contributed by the eddy diffusion generator to the atmospheric electric field and the vertical current density is a considerable percentage of the measured values. Since the values for q_0 scatter widely, and since the range of variation in values for the turbulent mixing coefficient are large and difficult to assign, we will not draw any further quantitative conclusions from equations 5, 6, and 7. Instead, we will consider the characteristics of the eddy diffusion generator, which acts against the normal processes of the atmospheric electric current circuit.

According to equation (7), we should expect a decrease of the conduction current density with increased altitude, even though we have assume beforehand a stationary condition. A similar decrease in the field strength is indicated by equation (6), even though we have assumed constant conductivity. These remarkable results are an expression of the fact that we find ourselves, in a sense, inside of the eddy diffusion generator. In contrast to this, the current produced by the thunderstorm generator corresponds to a stationary condition of current density and, therefore, the field strength is inversely proportional to the conductivity. This provides us with a very simple means of testing whether the atmospheric electric measurements thus far made in the free atmosphere will reveal the presence of the eddy diffusion generator.

6. RESULTS WHICH POINT TO EDDY DIFFUSION GENERATOR

The first indication that the vertical current density decreases with height is found in an article by Wigand [11], who about 40 years ago had measured the atmospheric electric field and conductivity up to an altitude of 9 km. Wigand concluded from free balloon measurements that a decrease of the conduction current density with altitude existed and that this indicates that there must be a generator in the lower strata of the atmosphere[*].

Since this result was a contradiction of the world wide generator, the specialists disinclined to consider it[*] There was no investigation made during the ensuing 30 years to check these measurements.

In 1941 to 1943, Rossmann made a series of glider flights, during which the atmospheric electric field and the conductivity were measured simultaneously. Results of this investigation, since they are all reported together with the weather conditions, serve as excellent material for testing the view that there exists an eddy diffusion generator. It appears from these results that, during the prevalence of fine weather with well marked haze layers, the decrease of the electric field with altitude corresponds in general to the increase in conductivity. However, in the lower 500 to 1000 m, the field increase is roughly twice the conductivity increase. There are also cases where the field strength decreases from 100 or 150 V/m steadily to about 20 or 30 V/m at the upper boundary of the haze layer, although the conductivity is approximately constant within the haze layer. This indicates that the field strength and conductivity are no longer inversely proportional and, furthermore, that the current density decreases with increasing altitude. These are, precisely, the ear marks of the eddy diffusion generator. In addition to this general support, an auxiliary result is that eddy diffusion generator is strongly dependant upon the character of the prevailing weather and that much greater altitudes are involved than is to be expected of Whipple's calculation reproduced here. Both of these circumstances indicate that here is a promising field for investigation and that atmospheric electric measurements are especially suitable for penetrating this eddy diffusion problems, which can otherwise be approached only with difficulties. Israel [13] has repeatedly expressed the latter view primarily because he recognized the close correlation between the daily ranges of the atmospheric electric field and eddy diffusion phenomena.

For atmospheric electric investigations of eddy diffusion, it would be very important that, during the international geophysical year, the daily variation of the thunderstorm generator be determined for each day by making measurements of the electric field at sea and by also registering the atmospherics. In this way, the group of atmospheric electric research stations of the world would be in the important position that the effect of the thunderstorm generator could be eliminated from their registrations and that local effects could be investigated. In this circumstances, measurements at the surface alone would be adequate to determine the magnitude of the significant factors of the eddy diffusion generator. Since the method of measuring current requires some modifications, owing to the existence of the convection current, we shall in conclusion briefly discuss this modification. There is the possibility, on one hand, of calculating the conduction current from measurements of the field strength and conductivity. On the other hand, the total current, i.e., the sum of the conduction and the convection current, could be measured by the usual direct current method (horizontal collecting plate). To what extent the convection current is included in measurements made in this way depends upon the method of installation and the position of the collecting plate.

If the plate is installed parallel to the earth surface, then it will indeed include the descending convection current, which occurs in the after-noon, but will not include the ascending convection current that prevails in the forenoon. With such an arrangement, the forms of

vertical current at night and during the forenoon hours should agree with those, which are calculated. But in the afternoon, the current should exceed the calculated pure conduction current. Exactly such a result was found by Holser [5] in his atmospheric electric mea-surements on mountain peaks: "The computed and measured air-to-earth current densities were essentially the same at night. However, the currents exhibited systematic differences during the day, which are not satisfactorily explained. In particular, on Mt. Haleakala, the measured current exceeded the computed current in the late afternoon and early evening hours."

If the collecting plate in installed vertically, i.e.,that means perpe-ndicularly to the earth surface, then the calculated and the measured current should always agree, since in this case convection current reaches only the upper edge of the plate, whereas the conduction current flows into the entire surface of the plate. In this situation only a small part of the convection current enters the plate, but the conduction current is measured at its full value. With the perpendic-ular arrangement, however, the horizontal winds strike the plate with full effect so that we have a new kind of convection current to take into consideration. It is only necessary to study the special properties of this antenna in order to obtain a clear cut determination of the different currents.

For the determination of the conduction current, the antennas should collect as much conduction current as possible and as little convection current as possible. The simple way of realizing this would be a long horizontal wire antenna installed several meters above the earth's surface. This antenna collects the current lines from a large area in its surroundings, but it has very small collecting capacity for the convection current. It can be shown that the region, from which current in drawn by a horizontal antennas, is approximately equal to the product of its length and its height. Thus, for example, an antenna, which is 10 meters long and 5 meters of hight, collects all the cond-uction current, which would otherwise flow to an area of 5 times 10 m^2 of the earths surface. If this antenna wire is 2 millimeter in diameter, then its collecting surface for the convection current is only 0.02 m^2. The collected conduction current would therefore exceed the collected convection current by a figure of 2500. Since the magnifying factor of the antenna is the same for the field and for the current lines, there is no change in the matching conditions. One must only make sure that the time constant of the input impedance of the measuring apparatus shall be the same as the time constant of the soil-to-air. A farther advantage of the wire antenna is as follows: with relatively little demand on equipment, one can collect a current, which is 50 to 100 times that obtained with the usual collecting plate. This reduces the demand on high input resistance (good insulation).

When we wish to measure the vertical convection current, we have only to screen the collecting system with a Faraday cage, whose top is made of coarse mesh double wire screen, while the walls to protect from horizontal convection current are made of solid plate.

On the other hand, if one is to measure horizontal convection currents, the top of the Faraday cage should be made of solid metal and the walls out of coarse mesh screen. A cube or a cylinder installed inside this Faraday cage would be affected only by the horizontal convection current. Then, if the wind velocity were simultaneously measured, one could calculate the space charge.

These brief suggestions show that there are, even for surface stations a series of possibilities for making measurements, which will help to determine the eddy diffusion generator.

In conclusion, we will again briefly summarize the results, which lead to the assumption of the existence of a eddy diffusion generator:

1. The sunrise effect, that is the simultaneous increase of the atmospheric electric field and the vertical current density at sunrise.

2. The close correlation between the daily ranges of atmospheric electric field and eddy diffusion phenomena.

3. An explanation of the diurnal variation of the electric field in terms of the variation in conductivity alone leads to contradiction with the daily range in the vertical current density.

4. The fact that the field decreases with altitude more rapidly than is to be expected from the increase of conductivity, in other words, an indication that there is a decrease of the vertical current density with altitude.

5. Since eddy diffusion only brings about mechanical transportation of ions in order to have a generator effect, there must be some auxiliary source of ions.

6. One possible source of these ions is the electrode effect, but the Elster-Geitel-Ebert soil respiration or other ion forming effects may be active.

SECTION 2. GLOBAL CIRCUIT

1. At What Altitude Does the Global Atmospheric Electricity Circuit Connect? H. Israël and H.W. Kasemir, Extrait des Annales de Geophysique, Tome 5, fascicule 4, 1949.

2. The Current Efficiency of the Storm Generator with Regard to the Atmospheric Electrical Vertical Current of the Fair Weather Areas, H. W. Kasemir, Berichte des Deutschen Wetterdienstes in der UA-Zone, Nr. 38, Bad Kissingen, 1952.

3. The Storm as a Generator in the Atmospheric Electrical Circuit, H. W. Kasemir, Zeitschrift für GEOPHYSIK, Sonderdruck Jahrg. 25, Heft 2, 1959.

4. Theoretical Problems of the Global Atmospheric Electric Circuit, H. W. Kasemir,. (Proc.Conference on Electrical Processses in Atmospheres, Ed. H. Dolezalek and R. Reiter, Steinkopf Verlag, Darmstadt, 1977.

5. Current Budget of the Atmospheric Electric Global Circuit, H. W. Kasemir, J. Geophys. Res., Vol. 99, No. D5, pp.10,701-10,708. 1994.

6. The Atmospheric Electric Ring Current in the Higher Atmosphere, H. W. Kasemir, Pure and Applied Geophysics (PAGEOPH), Vol.84,1971/I, pp.76-88, Birkhäuser Verlag, Basel, 1971.

Extrait des ANNALES DE GÉOPHYSIQUE

Tome 5, fascicule 4. — Octobre-Novembre-Décembre 1949.

AT WHAT ALTITUDE DOES THE GLOBAL
ATMOSPHERIC ELECTRICITY CIRCUIT CONNECT?

BY H. ISRAËL AND H.W. KASEMIR

(Observatory Buchau a.F. of the National Weather Service Wuerttemberg-Hohenzollern)

Résumé. — *Les processus de l'électricité atmosphérique sont la manifestation d'un système mondial de courants dans un condensateur sphérique formé par la surface de la Terre et une haute atmosphère conductrice. Le peu de corrélation entre les phénomènes ionosphériques et électriques s'oppose à l'identification de la seconde armature avec l'ionosphère. On peut donc supposer que la conductibilité engendrée dans la strastosphère par le rayonnement cosmique suffit à établir l'équilibre électrique général.*

Pour le prouver, on détermine la croissance de la conductibilité avec l'altitude d'après les valeurs du rayonnement cosmique, de la pression et de la température. Si on suppose qu'à une certaine altitude, une couche d'épaisseur déterminée, possédant la conductibilité qui correspond à cette altitude, soit immergée entre des portions d'atmosphère non conductrices, on peut, par intégration d'une équation aux dérivées partielles, trouver le temps que mettra une charge introduite en un point de la couche pour se répartir uniformément sur toute cette couche.

On en déduit que la conductibilité est déjà assez grande vers 50 à 65 km d'altitude pour rendre possible l'égalisation électrique mondiale.

Summary. – *The atmospheric electrical processes are forms of a global power system in a spherical condenser, which is formed by the earth's surface and a strong conductive upper atmosphere. The nearly non-existent correlation between ionospheric and atmospheric electrical events speaks against an identification of this «secondary reinforcement» with the ionosphere. Therefore, it is necessary to presume that the conductivity of the stratosphere caused by cosmic ultra-radiation already suffices for the global atmospheric electrical circuit to close between thunderstorm and fair weather regions..*

To verify this, the increase of conductivity is determined with the altitude for the stratosphere according to cosmic-radiation values, pressure and temperature values. If one assumes that a layer of a certain thickness and conductivity relevant to the altitude in concern is embedded in this altitude while the remaining atmospheric particles are not expected to be conductive, the question regarding the time, within which a specific charge applied to this layer in one point distributes itself evenly throughout the entire layer, can be found through the integration of the relevant partial differential equation.

Accordingly, conductivity at an altitude of only 50-65 km is large enough to close the global atmospheric electrical circuit.

The current perception of the atmospheric electrical occurrence, as we know, is that of a large electrical power system, which – caused and controlled by the thunderstorm activity of the entire world – completely extends throughout the earth's atmosphere approximately up to the border of the ionosphere. The earth's surface and the highly-conductive upper atmosphere form a spherical condenser, in the center of which the storms act as generators. In the areas distant from storms, the produced charge separation in the form of the atmospheric electrical vertical flow counterbalances itself (1).

The obvious assumption that the ionosphere with its extraordinarily strong conductivity takes over the role of the «secondary reinforcement» is not satisfactory. If this was the case, then significantly clearer correlations between ionospheric events and atmospheric electrical phenomena than are actually present would have had to emerge.

Because the electrical conductivity resulting from the cosmic radiation rapidly increases with increased altitude, this enables the question – does the atmospheric electrical power system actually reach up to ionospheric altitude or does the closure of the circuit already occur at a lower altitude?

To answer this, we must first attempt to determine the condition for conductivity of the stratosphere.

H. Benndorf [2] did this during his time based on certain assumptions about a median mass absorption coefficient of cosmic radiation, about the decrease of pressure with altitude and about the presence of free electrons in the upper layers.

Since then, more precise values exist regarding the ionization value of the cosmic radiation at the various altitudes as well as regarding the buildup of pressure and temperature of the atmosphere, which have also been verified to a largely experimental level by the American V2 ascents. This allows a new calculation of the conductivity in the stratosphere.

If we identify the ionization through the cosmic radiation at the individual altitudes h relative to normal conditions (0°C; 760 mm) with $q_0(h)$, the ion density $n(h)$ can be calculated as a result and from this the conductivity $\Lambda(k)$ is calculated:

(1) $$\Lambda(h) = 2n(h)\,\varepsilon\,k(h)$$

$\varepsilon = 4.8 \cdot 10^{-10}$ ESE

k = ion mobility

$n(h)$ is given through the relationship

$$n(h) = \sqrt{q(h)/\alpha(h)}$$

$\alpha(h)$ = recombination coefficient

$q(h)$ may be determined according to

(2) $$q(h) = q_0(h)\,\frac{p}{p_0}\,\frac{T_0}{T}$$

p or p_0 = pressure at the altitude h or 0

T or T_0 = absolute temperature at the altitude h or 0

because it is analogously proportional to the density.

The mobility initially changes proportionally to the density with reduced pressure

(3) $$k(p,T) = k_0(p_0,T_0)\,\frac{p_0}{p}\,\frac{T}{T_0}$$

until free electrons emerge in increasing quantities with a low density.

The greatest uncertainty lays within the temperature and pressure dependence of the recombination coefficients α. J.J. Thomson [3] and E. Lentz [4] find direct proportionality to pressure. However, based upon the comparison with their conductivity measurements during the ascent of the "Explorer II", O.H. Gish and

K.L. Sherman [5] conclude with cosmic radiation values on the proportionality of α to the 0.28^{th} power of pressure. M.E. Gardner [6] finds proportionality to the third power from the pressure. – For the temperature dependency of the reunification coefficients α, O.H. Gish and K.L. Sherman assume reverse proportionality to the 7/3 power of the absolute temperature in proper correlation with the value of 2.5 with H.A. Erikson [7] and 2.2 with Phillips [8].

In light of this uncertainty, we would like to subsequently determine the relationship

(4)
$$\alpha(p, T) = \alpha_0(p_0, T_0)(p/p_0)^x (T_0/T)^y$$

and inspect both borderline cases

$$\text{a) } x = 1 \; ; y = 2.35$$
$$\text{b) } x = 0.3 \; ; y = 2.35$$

Then the conductivity $\Lambda(h)$ can be calculated for

(5)
$$\Lambda(h) = \frac{2\,\varepsilon\,k_0}{\sqrt{\alpha_0}} \sqrt{q_0(h)} \cdot (p_0/p)^{1+(x-1)/2} (T/T_0)^{1+(y-1)/2}$$
$$= \Lambda_0(h)(p_0/p)^r (T/T_0)^s$$

where, in the case of a or b, it is necessary to place:

$$\text{a: } r = 1 \; ; s = 1.675$$
$$\text{b: } r = 0.64; s = 1.675.$$

For $q_0(h)$, we will use the values recently published by A.V. Ganges, J.F. Jenkins and J.A. van Allen [9] from V2 ascents at a 40° northern latitude, which are represented in Figure 1. The values for pressure and temperature for the various altitudes are taken from the report from H.E. Newell Jr. [10].

The conductivity values calculated as such for both borderline cases α/p = const and $\alpha/p^{0.3}$ = const are contained in Table I and demonstrated in curves 1 and 2 of Figure 2.

As previously mentioned, equation 3 is only fulfilled for the mobility of the positive particles up to minimal pressures; while for the negative the product $p.k$ rapidly increases upwards from approx. 200 mm of pressure and beyond [11]. The reason is that with decreasing pressure, the medium life expectancy of free electrons in the air increases. The consequence is that the conductivity under otherwise equal conditions increases more rapidly with decreasing pressure than is to be expected according to equation 5. Compensation, for example, through a simultaneously, respectively increasing reunification, according to H. Benndorf [2], diverts in particular the required correction at the given location and finds as a correctional factor, with which the conductivity values calculated for minimal pressures must be multiplied, the expression (6).

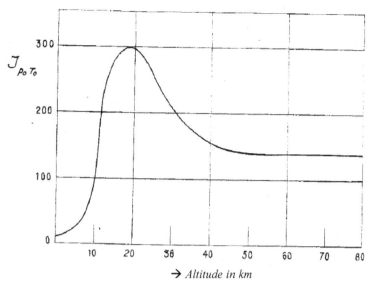

Fig. 1. – Ionization effect of ultra-radiation in various altitudes according to A.V. Gangnes, J.F. Jenkins and J.A. van Allen (9) at 41° northern latitude (Geiger counter measurement during V2 ascent). The values are based on normal conditions (0° C; 760 mm).

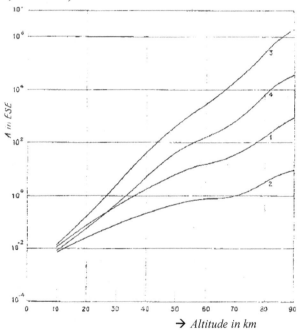

Fig. 2. – Conductivity in the stratosphere as a result of the ionization through ultra-radiation with consideration for the progression of pressure and temperature. The curves are explained as follows:

Curve 1: Conductivity without consideration for free electrons for α/p = const.

Curve 2: The same for $\alpha/p^{0.3}$ = const.

Curve 3: Conductivity with consideration for free electrons for α/p = const.

Curve 4: The same for $\alpha/p^{0.3}$ = const.

Ordinate amount is logarithmic.

The waviness of the curves is a result of the progression of temperature in the stratosphere.

(6) $$\xi(\rho/\rho_0) = 1 \div 5.7 \cdot 10^{-2} \cdot \rho_0/\rho \qquad \rho = \text{density}$$

This factor must be brought into the formulation until the density is so minimal that the negative carriers of electricity are practically only comprised of electrons. Then, ξ assumes the value of approx. 2,200.

The ξ values are also listed in Table I and in both of the last columns the conductivity is entered that is derived according to this correction, which must be added to the mobility of the negative carriers, for the individual altitudes (curve 2 and 4 from Fig. 2).

TABLE I

Calculation of the increase of conductivity in the stratosphere.

h = Altitude in km;

Λ_a = Conductivity for $x = 1$ (equation 4) and a $q_0(h)$ according to Fig. 1; information in ESU (Electrostatic Units);

Λ_b = The same for $x = 0.3$;

ξ = Correctional factor according to H. Benndorf (2) due to the presence of free electrons in air of a low density.

Λ'_a = Conductivity for $x = 1$ (equation 4) and a $q_0(h)$ according to Fig. 1; corrected due to the presence of free electrons; information in ESU;

Λ'_b = The same for $x = 0.3$;

Alt. in km	Λ_a	Λ_b	ξ	Λ'_a	Λ'_b
--	--	--	--	--	--
10	0.011	0.0065	1.2	0.0313	0.0078
20	0.075	0.0258	1.85	0.139	0.0476
30	0.347	0.0672	5.45	1.189	0.366
40	1.57	0.187	22.3	35.0	4.170
50	5.70	0.466	73.0	416	34.0
55	9.04	0.594	121	1090	71.6
60	14.2	0.748	201	$28.5 \cdot 10^2$	150
65	19.7	0.818	326	64.4	266
70	29.5	0.975	554	163.5	540
75	66.5	1.53	1290	$8.6 \cdot 10^4$	$0.198 \cdot 10^4$
80	180	3.02	2200	39.6	0.665
85	415	5.31	2200	91.2	1.17
90	936	9.63	2200	206	2.12
100	4340	25.7	2200	955	5.65

In order to deduce from this distribution of conductivity into the altitude, in which the atmospheric electrical balance occurs, we will make the following simplified formulation:

At time $t = 0$, a charge Q is supplied to a point laying in the atmosphere. P lies in a layer of given conductivity, which encompasses the earth concentrically. The transportation of charges from this layer to other atmospheric altitudes is excluded, therefore the rest is non-conductive.

VON H. ISRAEL UND H. W. KASEMIR   [ANNALES DE GÉOPHYSIQUE

For the mathematical treatment of the given problem, we will select the following mathematical formulation:

We examine a spherical condenser, the interior conductive surface of which is formed by the surface of the earth, which has a very high level of conductivity ($\lambda_e \to \infty$) for our problem. The outer condenser plate, of which the thickness and altitude above the earth we are still leaving available, has the conductivity λ. With the aid of the conducted calculation, we can then later decide at which altitude the conductivity of the atmosphere suffices in order to enable the interferences of the electrical balance to dissipate in an adequately rapid manner. A change to the potential of this layer would then become simultaneously noticeable throughout the earth in a respective change of the atmospheric electrical variable. The space between both spherical shells should be infused with a medium of a constant dielectric constant, the conductivity of which is small compared to that of the outer shell. Indeed, in reality we are dealing with conductivity that increases steadily with altitude and only the simplicity of the mathematical solution of the problem modified in this manner may justify the rough physical approximation.

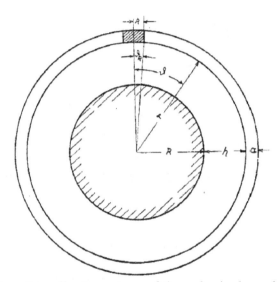

Fig.3. – Coordinate system of the spherical condenser.

We assume at time $t = 0$ that a storm feeds a charge quantity $+ Q$ to the upper conductive layer, which distributes itself evenly with time over the entire layer. The same large negative charge quantity $- Q$ creates its way to the earth. We are now requesting for the temporal law, with which this balancing process proceeds.

As a coordinate system, we logically select the spherical coordinates r, θ, φ, whereat we place the pole $\theta = 0$ in the location of the storm. The earth's radius is R, the distance between the earth and conductive layer h and its thickness a. Fig. 3.

Due to the fact that the present problem is rotationally symmetrical, we can limit ourselves to the examination in the section $\varphi = 0$. The self-induction of the system must be regarded as negligible with respect to resistance *or* conductivity. In other words, we would like to abstain from the development and removal of the magnetic field through the compensating current. With this, our balancing process is controlled by the differential equation of thermal conduction. With consideration that the derivation according to φ ($\varphi = $ const. = 0) as well as the derivation according to r ($r = $ const. $= R + h$) is omitted, this is simplified to

Imprimé par A. TAFFIN-LEFORT, à Lille (France). — *Published in France.*

(7)
$$T_e \frac{\partial U}{\partial t} = \frac{1}{(R-h)^2} \frac{1}{\sin \theta} \frac{\partial}{\partial \theta} \left(\sin \theta \frac{\partial U}{\partial \theta} \right).$$

This signifies the following

$U_{(\delta, t)}$ = voltage at the receptor point (R + h, θ) of the outer spherical shell for the time t

θ   = angle between the pole and receptor point

T_e = time constant / unit of area.

We have to determine the time constant T_e analogously to the physical foundations of the thermal conduction from the constants of the system (specific capacity and conductivity correlate to a specific thermal capacity and thermal conductivity). They are given through the capacity and conductivity of the volume element of our layer. The specific conductivity there is λ and that of our unit volume

(8)
$$\lambda_e = a\lambda.$$

The specific capacity C_K of our spherical condenser is the entire capacity divided by the surface. Therefore, according to the known formula for the spherical condenser capacity, the following applies

(9)
$$C_K = \frac{4\pi \varepsilon \frac{R}{h}(R-h)}{4\pi(R-h)^2} = \frac{\varepsilon R}{h(R-h)}.$$

in this case, ε signifies the dielectric constant of the intermediate layer, thus in our case, the air. The capacity of the volume element C_e also remains

(9 a)
$$C_e = C_K.$$

In this case, the difference between electrical and thermal current is expressed. While the heat quantity distributes itself spatially in the volume element in the case of the thermal current, the amount of electricity is collected on the underside of the upper condenser layer. For the operation of our balancing process, we want to disregard this current perpendicular to the surface of the condenser configuration due to the fact that we want to regard the thickness of the upper condenser plate as small compared to its extensive surface.

With 8.), 9.) and 9a.), we now achieve our time constant for

(10)
$$T_e = \frac{C_e}{\lambda_e} = \frac{\varepsilon R}{a \lambda h(R+h)}.$$

The solution of our differential equation 7.) is possible with the aid of spherical functions; and we achieve

(11)
$$U = \sum_0^\infty n \, e^{-n(n+1)t/T} A_n P_n.$$

With

$P_{n(\theta)}$ = Spherical functions

A_n = Constant coefficients,

$$T = T_e(R+h)^2 = \text{time constant} = \frac{\varepsilon R(R+h)}{a\lambda h}$$

in which the A_n is determined from the development of the initial state U_0 (t = 0) according to spherical functions.

(12)
$$U_0 = \sum_0^\infty n \, A_n P_n$$

with

(13)
$$A_n = \frac{2n+1}{2} \int_0^\pi U_0 P_n \sin \theta \, d\theta.$$

At time t = 0, U_0 is now 0 everywhere with the exception of a small area at the pole, in which the entire charge + Q is initially concentrated. The integrand from 13), therefore, has a value other than 0 only within this small area $0 < \theta < \theta_k$. Thus, we need to extend the integral in 13.) merely from 0 to θ_k.

We will choose this small area as a circular cylinder with the base πk^2 and the altitude a. (Fig. 3). Considering what has been said about the capacity of the unit of volume, the capacity of this cylinder is equal to

$$(14) \qquad c_k = \frac{\varepsilon R}{h(R - h)} \pi \, k^2.$$

And thus, the voltage that dominates there

$$(15) \qquad U_0 = \frac{Q}{c_k} = \frac{Q h(R + h)}{\varepsilon R \pi k^2}.$$

Or with the introduction of the spherical capacity $C = 4\pi\varepsilon (R + h)R/h$

$$(16) \qquad U_0 = \frac{Q}{C} \frac{4(R + h)^2}{k^2}.$$

In 13), therefore, we can pull U_0 in front the integral because this expression is independent from θ according to 16).

Moreover, the following applies for the small θ

$$P_n \rightarrow 1 \text{ for } \theta \rightarrow 0.$$

In this process, 7) obtains the form

$$(17) \quad A_n = \frac{2n + 1}{2} \frac{Q}{C} \frac{4(R - h)^2}{k^2} \int_0^{\theta_k} \sin \theta \, d\theta = (2n + 1) \frac{Q}{C} 2 \frac{(R + h)^2}{k^2} (1 - \cos \theta_k).$$

Now, if we develop $\cos \theta_k$ and consider that $\sin \theta_k = \dfrac{k}{R + h}$, we obtain

$$\cos \theta_k = \sqrt{1 - \sin^2 \theta_k} = \sqrt{1 - \left(\frac{k}{R - h}\right)^2} = 1 - \frac{1}{2}\left(\frac{k}{R + h}\right)^2 \doteq \cdots$$

and, therefore, according to 11.)

$$(18) \qquad A_n = (2n + 1)\frac{Q}{C}. \text{ For } k \rightarrow 0.$$

If we then introduce this value into the equation in 11), we finally achieve

$$(19) \qquad U = \frac{Q}{C} \sum_0^{\infty} {}_n (2n + 1) \, e^{-n(n+1)t/T} P_n.$$

This sequence is so strongly converged that we can cancel it for $T < t$ after both initial terms. The error made in this process is no more than 10%, which is quite tolerable, as the entire calculation itself is indeed only necessary for estimation.

We then obtain the simple approximation equation

$$(20) \qquad U = \frac{Q}{C}\left(1 + \frac{3 \cos \theta}{e^{2t/T}} + \cdots\right) \text{ for } T < t$$

For $t \rightarrow \infty$, the following remains

(21)
$$U_\infty = \frac{Q}{C}$$

independent from δ and t, i.e. the charge Q was distributed evenly throughout the upper part of the spherical condenser. We must resort to (19) for $t = 0$. Then it becomes

(22)
$$U_0 = \frac{Q}{C} \sum_0^\infty n \, (2\,n + 1)\, P_n.$$

The equation applies everywhere $U_0 = 0$ with the exception of the pole $\theta = 0$, in which U_0 becomes equal to ∞ because in this case we have concentrated the final charge Q to an infinitely small dot πk^2 with $k \to 0$. This result is not readily evident from 22). However, we do not need to concern ourselves with the evidence of that because we determined the coefficients of P_n in such a manner that this required initial status is represented by (22). Now, if we look at the voltage curve at both poles $\theta = 0$ and $\theta = \pi$, we can see that for $\theta = 0$ / $\cos \theta = 1$, the voltage of larger values seeks the value $U_\infty = Q/C$, while at the opposite pole $\theta = \pi$ / $\cos \theta = -1$, the voltage of the lower values increases to the value $U_\infty = Q/C$. (See equation 20) We can view the voltage difference between both poles with regard to the end voltage U_p/U_∞ as an indicator of the maximum percentage, at which the pole voltages equaled each other. From 20), we obtain the following for this

(23)
$$U_p/U_\infty = 6/e^{2t/T}.$$

If we determine the conductivity values Λ_a and Λ_b of Table I (comp. curves 3 and 4 in Fig. 2), then the time constants T_a and T_b for the individual altitudes listed in Table II can be calculated. One can easily derive from 23) that the time, within which U_p/U_∞ decreases to 1/10 or 1/100 of the initial value, is 2.05T or 3.2T.

TABLE II

The time constant $T = \varepsilon(R + h)R/ (a\Lambda h)$ for the various altitudes in the stratosphere, a conductivity pursuant to Table I or Fig. 2, curve 3 and 4 and a thickness a of the layer concerned of 1 km. The offset to 1/10 or 1/100 of the initial value occurs in t_{10} or t_{100} where

$$t_{10} = 2.05T \text{ and } t_{100} = 3.2T.$$

H in km	T_a sec	T_b sec
---	---	---
10	$25 \cdot 10^6$	$45.5 \cdot 10^6$
20	$1.21 \cdot 10^6$	$5.26 \cdot 10^6$
30	$57.4 \cdot 10^3$	$268 \cdot 10^3$
40	$2.3 \cdot 10^3$	$19.1 \cdot 10^3$
50	156	1950
55	54	825
60	19.2	364
65	7.9	193
70	2.85	85.5
75	0.51	22.1
80	0.105	6.2
85	0.042	3.26
90	0.017	1.68
100	0.004	0.58

174

If one omits the requirement for a layer thickness of $a = 1$ km, then the respective values for other layer thicknesses can be calculated in the simplest manner. In addition, it is possible to calculate the atmospheric altitudes into which must be entered in order to find a certain offset time with consideration for the actual increase in conductivity while ascending if the entire atmosphere from the bottom up to this altitude participates in the offsetting.

So what have we gained with this estimation?

Our thinking was that a layer was embedded into a non-conductive atmosphere at a certain altitude, which encompassed the earth concentrically and possessed a certain conductivity. If a charge is applied to this layer in one spot, it will distribute itself evenly throughout the entire layer in the given time. If this layer of the thickness a was now the secondary reinforcement of our atmospheric condenser, then this time would signify the phase delay at the antipodal point.

Now we know that the world time period of the atmospheric electrical potential gradient, as found on the oceans, runs synchronously to global thunderstorm activity – more precisely – that no noticeable phase difference can be observed between both. Due to the known «relaxation time» (time, within which the stationary state is established once again when changes to the atmospheric electrical conditions occur) (13) of approx. 10 to 15 minutes, we cannot determine a more precise synchronization than that which corresponds to this time from measurements taken near ground level.

This justifies our conclusion that a layer of the type observed above, the «time constant» of which is smaller than 10 to 15 minutes, would suffice in order to enable the global parallel of potential gradient and storms! As demonstrated in the result of the calculation, this is the case at the altitude between approx. 50 and 65 km.

Based on this, we can conclude that the ionosphere is *not* necessary for the global atmospheric electrical balance, but rather the conductivity present in the middle to the upper stratosphere is already sufficient.

An entirely different question is whether the charge balance actually [1] occurs in this manner and at this altitude. The observation made here in the form of a heat conduction problem cannot and should not reveal anything about this. A generalized treatment of the problem is necessary with consideration for the constant increase of conductivity upwards, which will be reported on later.

Manuscrit reçu le 25 juillet 1949.

---

[1] In this respect, this result is very evident as it can provide an explanation that solar or ionospheric effects can become electrically active in the atmosphere, though this is not a necessity (14, 15, 16). Without revealing anything about the mechanism of action, it can be expected that solar effects only become noticeable as atmospheric electricity if they penetrate deep into the atmosphere. However, generally the atmospheric electrical "sphere" is likely not touched.

BIBLIOGRAPHY

[1] H. ISRAËL und G. LAHMEYER, Studien ueber das atmosphaerische Potentialgefaelle I. Das Auswahlprinzip der luftelektrisch « ungestoerten » Tage. *Terr. Magn.* **53**, 573-586, 1948. H. ISRAËL, Die Tagesperiode des elektrischen Widerstandes der Atmosphaere. (Studien ueber das atmosphaerische Potentialgefaelle II). *Annales de Géophysique* (im Druck).

[2] H. BENNDORF, Ueber den durch die Hess-sche Hoehenstrahlung bedingten Ionisations- und Leitfaehigkeitszustand der hoeheren Atmosphaerenschichten. *Physikal. Zeitschr.*, **27**, 686-692, 1926.

[3] J. J. THOMSON, *Elektrizitaetsdurchgang in Gasen.* 1906.

[4] E. LENTZ, *Zeitschr. f. Phys.*, **76**, 660-678, 1932.

[5] O. H. GISH and K. L. SHERMAN, Information to be obtained from some atmospheric-electric measurements in the strastosphere. *Transactions of Edinburgh Meeting* (UGGJ), S. 382-395, Kopenhagen 1937.

[6] M. E. GARDNER, *Phys. Rev.*, **53**, 75 ff., 1938.

[7] H. A. ERIKSON, *Phys. Rev.*, **34**, 635 ff., 1925.

[8] P. PHILLIPS, cit. nach *Physics of the earth VIII*, New-York 1939.

[9] A. V. GANGNES, J. F. JENKINS and J. A. VAN ALLEN, The cosmic ray intensity of the atmosphere. *Phys. Rev.*, **75**, 57-69, 1949.

[10] H. E. NEWELL Jr., Upper atmosphere research with V 2 rockets. *Reprint of NRL Report R*, 3294, June 1948.

[11] A. F. KOVARIK, *Phys. Rev.*, **30**, 415 ff., 1910.

[12] J. J. THOMSON, *Phil. Mag.*, **47**, 337 ff., 1924.

[13] H. BENNDORF, Grundzuege einer Theorie des elektrischen Feldes der Erde. *Wiener Berichte* (IIa) **134**, 281-315, 1925 ; **136**, 175-194, 1927.

[14] J. SCHOLZ, Luftelektrische Messungen auf Franz-Josefs-Land waehrend des II.Internationalen Polarjahres 1932/33. *Transactions of the Arctic Institute of USSR*, **16**, 5-169, 1935. — Ders. *Gerl. Beitr. z. Geophys.*, **44**, 145-156, 1935.

[15] O. H. GISH, Evaluation and interpretation of the columnar resistance of the atmosphere. *Terr. Magn.*, **49**, 159-168, 1944.

[16] H. ISRAËL, Extraterrestrische Einfluesse auf das luftelektrische Feld. *Wiss. Arb.d. Dtsch. Met. Dienstes im franz. Bes. Geb.*, **1**, 62-67, 1947.

Special Print from

Reports of the German Weather Service in the US Zone

No. 38.

Bad Kissingen, 1952.

DK 551.594.21

The Current Efficiency of the Storm Generator with regard to the Atmospheric Electrical Vertical Current of the Fair Weather Areas

By H. W. Kasemir, Atmospheric Electrical Research Center Buchau a. F.

Summary: There are two methods of calculating the electrical fields in the atmospheric electricity, and there is a gap between them. For calculating the conditions of the fields and currents in connection with a thunderstorm it is common to use the image of a point charge above a conducting globe. For calculating the electrical situation in undisturbed areas it is successfull to use the homogenous field of a plate condenser. One coating is the earth surface, the other coating is represented by a layer in a greater altitude or by the inonosphere. It is the intention of this paper to close the gap between two methods.

Some corrections are to be made in these conceptions. The calculating of the fields in connection with thunderstorms must not be made in an electrostatic manner but according to the laws of the field of flow, and the increase of conductivity with the height has to be considered. This increase is essential for the fact, that the current delivered by the thunderstorm-generator does not flow to the ground in the neigbourhood of the thunderstorm (leakcurrent) but that a sufficient part of it is available for the air-earth-current in the areas far of the thunderstrom (profit-current, Nutzstrom).

Another correction is to be made in the image of the plate condenser. The upper coating, the so-called — equalizing layer (Ausgleichsschicht) — may not be defined as an equipotential layer. In the contrary, it has to be claimed that the equipotential layers are vertical to the equalizing layer in order to make the transverse flow possible, which distributes the profit-current of the thunderstorm-generator uniformly over the whole globe. The breaking down of potential-differences and by this the formation of a equipotential layer would be possible in the case, that the conductivity increases abruptly (i. e. at the surface of the earth). The ionosphere should not taken in consideration here — in a first approximation — because the main part oft the transverse flow is already in deeper strata of air.

While the derivation of the potential equation for the problem in discussion is carried out in closed form, the application is terminated on the proportion of profit-current to leak-current. Further discussions shall be given in following papers.

1. Introduction.

During the registration of the atmospheric electrical field, the striking difference between the relatively balanced diurnal variation of the field strength on the so-called undisturbed days and the high and strong fluctuations during precipitation, particularly during storms was noticed quite frequently. This led to a division of the atmospheric electricity into fair weather electricity and precipitation electricity for purely empirical reasons. As one began to recognize that the storm represents the atmospheric electrical generator and the fair weather areas with their compensating current represent the user, the division achieved its deep physical meaning.

This division into two separate parts was also expanded upon in the theoretical considerations. For the storm, an arrangement of spheres filled with space charges with their reflections on the conductive surface (1) were developed as a mathematical model image, while the undisturbed fair weather field was represented by the current in a spherical condenser, the upper boundary of which is provided by the ionosphere and the lower boundary by the earth (2). While each of these two model representations is distinguished by a great mathematical simplicity, it is difficult to find a transition between both, which ultimately must exist.

One attempted to close this gap through a physical consideration and assumed that the current produced by the storm generator flowed up to the ionosphere, distributes itself there evenly and then flows down vertically to the earth. This notion, however, is no longer a fitting presentation of the storm through spheres filled with space charges because it builds on the laws of electrostatics, while the current distribution of a storm depends on the conductivity of the immediate and remote environment, thus, of a material constant, which is not familiar in electrostatics.

The focus of this research is to create the mathematical connection between the two model images. In this process, it will be demonstrated that, in addition to the change that is experienced by the field and current curve of a storm cloud from the previously common view, the image of the current in the spherical condenser must also be modified. Of these, only the surface of the earth remains as an equipotential plane, while the question regarding the second "secondary reinforcement" of the condenser becomes invalid at greater altitudes. Namely, if one reaches altitudes above storm clouds (for example, 20 km), the vertical current and field lines begin to bow and are drawn into flatly expanded arcs to the individual storm centers. The equipotential planes are vertical to the surface of the earth in these areas. Thus, it is not possible to define this so-called compensating layer through an equipotential plane parallel to the surface of the earth. The ionosphere, however, is too high to be capable of being noticeably effective as a compensating layer in the former sense.

Furthermore, it inevitably arises that the earth can no longer be assigned the potential value of 0 as was previously customary, but rather the negative voltage of approx. 10^{10} volts. This potential value is defined against the equipotential plane, which lies on the medium voltage of the storm generator and has the value of 0. The entire electrical circuit would set itself to this value if the generator would cease to function and all of the present potential differences have balanced themselves. It is this potential that is given to the infinite remote in mathematics.

The following calculation that belongs in the framework of the research "Regarding the Current Theory of the Atmospheric Electrical Field" (3) (4) treats the storm as a current supply over the earth imagined as a conductive sphere, in which the conductivity increases exponentially with the altitude in the atmospheric space. The comparison of this calculation with that, which was previously common, shows that the increase of conductivity with altitude effectuates that areas distant from a storm can also be supplied by this current and that the entire current efficiency of the storm does not flow downward in direct proximity to the earth.

Because a mathematical solution is not known for the present problem, the potential equation is derived using the characteristic function of the associated differential equation. As a result, the mathematical calculation uses a somewhat larger space. Nonetheless, to avoid losing the overview, the simple example of electrostatics is emphasized in Section I and presented in detail in this physical train of thought. With the potential equation completely derived from the current theory, the key to the quantitative answer of all risen questions is provided. This will be presented in the example of the calculation of the leakage current and primary current of the storm generator.

II. The Calculation of the Current Efficiency of the Storm with a Constant Conductivity.

The potential function for this problem can be found in every textbook regarding electrostatics under the designation "the conductive sphere influenced by a point charge". However, the systematic approach using the development of the Laplace differential equation according to characteristic functions is almost never provided, but rather the much more elegant approach using the reflection on the sphere, in which the solution is anticipated and its accuracy of the fulfillment of the differential equation is verified with the prescribed marginal conditions. Because we are only dealing with the clarification of the physical content in this case, we would also like to follow this convention. However, in Section III, we will have to take the laborious path using the characteristic functions.

We will place a compilation of the designations and dimensions of the occurring variables at the top of our observations. The meanings are presented as follows:

$\emptyset$	=	Potential function [V]
E	=	$-\text{grad } \emptyset$ = field strength [V/m]
i	=	λE = current density [A/m^2]
λ	=	Conductivity [Siemens/m = 1/ohm·m]
ε	=	Dielectric constant [Farad/m]
Q	=	Point charge (representation of the storm charge) [C = Coulomb]
Q_s	=	Reflection of the point charge on the sphere [C]
Q_k	=	Spherical charge that supplements Q_k to the value of Q [C]
q	=	Charge density on the surface of the sphere [C/m^2]
I	=	Current production of the point source (Total current of the storm generator) [A]
I_V	=	Reflection of the point source on the sphere (leakage current of the storm generator) [A]
I_N	=	Primary current that supplements to the value of I [A]
$r,$	=	Polar coordinates [m], [✕], [✕]

ϑ, φ

$\mathbf{i}$	=	Distance of the point charge or point source from the coordinate neutral point [m]
$\mathbf{i_s}$	=	Distance of the reflection from the coordinate neutral point [m]
k	=	Spherical radius of the conductive sphere (earth) [m]
R_1	=	Distance of the grid point from the point charge or point source [m]
R_2	=	Distance of the grid point from the reflection [m]

When transitioning from electrostatics to the current, the following must be applied in the formulas [7]

I instead of Q

i instead of q

λ instead

all other variables are equal in both systems.

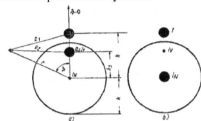

Figure 1

The point charge or point source with its reflection on a conductive sphere and its supplemental charge for

a. Constant conductivity of the medium

b. Exponentially increasing conductivity with altitude.

In Figure 1a, the arrangement of the charges and the meaning of the individual variables can be recognized without further explanation, and we can write the known potential equation for the point charge reflected on the sphere. The following applies:

$$\emptyset = \frac{Q}{4\pi\varepsilon}\frac{1}{R_1} + \frac{Q_s}{4\pi\varepsilon}\frac{1}{R_2} \qquad (2)$$

With

$$R_1 = \sqrt{r^2 + h^2 - 2rh\cos\vartheta}$$
$$R_2 = \sqrt{r^2 + h_s^2 - 2rh_s\cos\vartheta} \quad .$$

If the sphere should obtain the potential of 0 in the process, the following relationship must also exist:

$$Q_s = -Q\frac{k}{h} \quad \text{und} \quad h_s = \frac{k_2}{h} \quad , \qquad (3)$$

or in the case of the current:

$$\emptyset = \frac{I}{4\pi\lambda}\frac{1}{R_1} + \frac{I_V}{4\pi\lambda}\frac{1}{R_2} \qquad (4a)$$

with

$$I_V = -1\frac{k}{h} \quad . \qquad (4b)$$

The fact that the earth has the potential of 0, though generally accepted, it is however not stipulated by a compelling reason and we will see in a moment that it also not even the case. At this point, we must give precedence to another boundary condition. The storm does not produce a single-pole charge, but rather effects as a gravitational machine in a charge-divided manner. That means the positive charge surplus of the storm cloud cannot emerge by the fact that the equally sized amount of negative charge was transported to the earth by the precipitation. Thus, we must demand that the point charge Q and the earth charge are equally large. Naturally, the same also applies for the current production of the storm cloud and the earth.

We can achieve the charge or the current production of the earth by calculating field strength through differentiating of the equation (2) or (4a) according to r and as a result the surface charge density or the current density on the sphere is capable of the relationship $q = \varepsilon E$ or $I = \lambda E$ for $r = k$. Through integration throughout the entire surface of the sphere, we obtain the net charge or the current production of the sphere and see that this correlates to the reflection Q_s or I_V. According to (3) and (4b), Q_s and I_V are smaller than Q or I by the factor k/h. Therefore, we must still ascribe the difference amount between |Q| and |Q_s| to the sphere. Now, to avoid destroying the constancy of the potential on the sphere, we must evenly distribute this supplemental charge Q_N or I_N over the surface of the sphere. Thus, we can also represent them through a point charge or source in the center of the sphere. This additional charge or source naturally emits a negative potential on the sphere so that we must omit the adoption of an earth potential of 0 for the benefit of the claim that the positive and negative charges produced through the splitting of charges by the storm must be equal in size.

The charge density and the field strength proportional to it, which corresponds to the reflection charge Q_s, is inversely proportional to the third power of the radius E_1 or even R_2. Thus, the same law of the dependence on distance applies for the field strength just as with the reflection on the plane that is used most. Directly below the point charge (R_1 approx. 10km), the field strength is greater than on the antipodal point of the sphere by $2\cdot10^9$. That means that the surface charge respective to the reflection charge is virtually concentrated on the level of the surface of the sphere below the point charge. Naturally, the same applies when transitioning to the current source. The superposition of the charge or the current density that was distributed unevenly and evenly according to Q_N and Q_N, or I_V or I_N provide the field strength or the vertical current density of the earth.

Expanding on this, we can reach two definitions that form the core matter of the present paper.

1) The additional current source I_N at the center of the sphere with its even distribution of current density on the surface of the sphere represents the primary current of the storm generator because this is also available in the areas distant from storms as a contribution to the fair weather vertical current. It imparts a negative potential on the earth.

2) The reflection source I_V indicates the leakage current flowing down in the direct proximity of the storm because it virtually contributes nothing to the fair weather current. From a mathematical standpoint, the reflection source only serves to fulfill the stipulation of the constant potential of the earth despite the unsymmetrical influence by the storm.

These two current shares are calculated in such a manner that initially the potential equation is derived for the boundary condition of the earth's potential = 0. Then the value of the reflection source I_V is determined from this potential function. The supplementation of I_V to the value from I then results in I_N. The potential function of this I_N to be arranged at the center of the sphere must then be overlapped with the initially derived potential function to obtain the complete solution.

Furthermore, we will define the efficiency W of the storm generator by the ratio of primary current to total current.

$$W = \frac{|I_N|}{|I|} = 1 - \frac{|I_V|}{|I|} . \qquad (5)$$

From (4b), we will calculate the efficiency for

$$W = 1 - \frac{k}{h} = \frac{h-k}{h} .$$

h-k provides the altitude of the storm over the earth according to Figure 1, which we need to apply in this case with 10km, while the value of h is practically provided through the radius of the earth. Therefore, W = 0.16%, which directly shows that the calculation of the efficiency of the storm generator is insufficient without taking the conducting increasing with altitude. The previous estimate of the current balance of the storm generator (5) requires namely that this functions with 100% efficiency. The extent, to which this requirement is met, will be demonstrated in the next section.

III. Efficiency of the Storm Generator with Consideration for the Conductivity Increasing with Altitude.

To solve this problem, we must first establish the potential equation, which we obtain through the characteristic functions of the differential equation of the current theory. We can obtain the differential equation for the potential functions from the continuity equation of the stationary current

$$\text{div } i = 0 \qquad (6)$$

and the relationships

$$i = \lambda E \text{ und } E = -\text{grad } \phi. \qquad (7)$$

If we successively apply (7) to (6), we will obtain

$$\text{div}\lambda \cdot \text{grad}\phi = 0 \qquad (8)$$

or

$$\lambda \text{ divgrad } \phi + \text{grad}\lambda \quad \text{grad } \phi = 0. \qquad (9)$$

λ should increase with altitude according to an e-function, thus, it is given in the form

$$\lambda = \lambda_o \, e^{\,a(r-k)}$$

in which λ_0 has the value of the conductivity on the surface of the earth and a has the value $\ln 10/10$ km. In this manner, (9) is simplified to

$$\text{div grad } \phi + a \frac{\partial \phi}{\partial r} = 0. \qquad (11)$$

For a = 0, $\lambda = \lambda_0$ = const, and (11) transitions to the familiar Laplace differential equation of the potential theory

div grad $\emptyset = 0$.

We will engage this transition to the formulas of the potential theory with a = 0 in our calculation on a per case basis to review the correctness of the derived equations and to better recognize the meaning of the individual expressions.

The polar coordinates r, ϑ, φ are the coordinate system adapted to our problem. In addition, if we place the source point depicting the storm on the $\vartheta = 0$-axis, then rotational symmetry will dominate and all derivatives according to φ are equal to 0. If we now write the expression div grad $\emptyset$ in the polar coordinates, we will obtain the following from (11)

$$\frac{\partial^2 \phi}{\partial r^2} + \left(\frac{2}{r} + a\right) \frac{\partial \phi}{\partial r} + \frac{1}{r^2 \sin \vartheta} \frac{\partial}{\partial \vartheta} \left(\sin \vartheta \, \frac{\partial \phi}{\partial \vartheta}\right) = 0 . \qquad (12)$$

Now, a solution to this differential equation must be found with the marginal condition that, for r = k, the potential function $\emptyset$ assumes a constant value on the sphere and demonstrates a singularity at the location of the point source. To find this solution, we will first determine the characteristic function from (12) and strive to meet the marginal conditions through adequately overlapping these characteristic potentials.

To obtain the characteristic functions, we will develop the product approach for $\emptyset$

$$\phi \, (r, \vartheta) = R_{(r)} \cdot T_{(\vartheta)} , \qquad (13)$$

in which R should be function to be determined of r alone, while T should only depend on ϑ.

If we go into (12) with the approach (13), we will obtain

$$\frac{r^2}{R} \left(\frac{d^2 R}{dr^2} + \left(\frac{2}{r} + a\right) \frac{dR}{dr}\right) + \frac{1}{T \sin \vartheta} \left(\sin \vartheta \cdot \frac{dT}{d\vartheta}\right) = 0. \qquad (14)$$

As a result, T is given through the spherical functions P_a, while R has the form [6]

$$R_n = A_n \, (ar)^n \int_o^1 t^{n-1} \, (1-t)^{n+1} \, e^{-art} \, dt$$

$$+ B_n \, (ar)^n \int_1^\infty t^{n-1} \, (t-1)^{n+1} \, e^{-art} \, dt . \qquad (15)$$

In (15), An and Bn signalize constants capable of being randomly selected, which will serve us later in fulfilling the marginal conditions, while n is capable of penetrating all whole positive numbers from 0 to ∞.

$$n = 0; 1; 2; 3; \ldots\ldots \qquad (16)$$

In this process, we will also determine that, for n = 0, the first integral expression in (15) is a constant, which we can either achieve through a suitable selection from A_0 or take directly from (14), due to the fact that the second term of the equation is eliminated there for n = 0.

If we apply the following for shortening

$$f_n = (ar)^n \int_o^1 t^{n-1} \, (1-t)^{n+1} \, e^{-art} \, dt \quad \text{for } n = 1, 2, 3, \ldots.$$

$$f_o = 1 \qquad (17)$$

$$g_n = (ar)^n \int_1^\infty t^{n-1} \, (t-1)^{n+1} \, e^{-art} \, dt \quad \text{for } n = 0, 1, 2, \ldots. ,$$

every potential function that fulfills the differential equation (12) can be represented in the form

$$\phi = \sum_{n=o}^{\infty} (A_n f_n + B_n g_n) \, P_n . \qquad (18)$$

The integral of the functions f_n and g_n can be solved in a closed manner without difficulty. It is also possible to provide a differential expression for the functions g_n instead of an integral expression, which we can achieve either directly from (14) or with the aid of the integral from (17) used with the single-dimensional Laplace transformation. However, because we will only need the expressions for n = 0, thus f_0 and g_0 for the later estimate, we will need to refrain from providing the practical method of calculation for the evaluation of the integral in this case and turn to the physical meaning of these

180

...unctions or the equation (18). To be able to cross over to ...lectrostatics through the boundary $a \to 0$, we will write the functions ...n and g_n for $n = 0; 1; 2$ with the evaluated integrals. In this ...connection, the appropriate normalization constants are extracted ...rom A_n and B_n. We obtain

$$f_0 = 1$$

$$f_1 = \frac{3}{a}\left(1 - \frac{2}{ar} + \frac{2}{a^2r^2} - \frac{2}{a^2r^2}e^{-ar}\right) \tag{19}$$

$$f_2 = \frac{20}{a^2}\left(1 - \frac{6}{ar} + \frac{18}{a^2r^2} - \frac{24}{a^3r^3} + \left(\frac{6}{a^2r^2} + \frac{24}{a^3r^3}\right)e^{-ar}\right)$$

$$g_0 = a\ Ei\ (-ar) + \frac{e^{-ar}}{r}$$

$$g_1 = \frac{e^{-ar}}{r^2} \tag{20}$$

$$g_2 = \frac{e^{-ar}}{r^2}\left(\frac{1}{r} + \frac{a}{4}\right) .$$

Therefore, the sought potential function of our problem is found for the space 1. In the space 2, we can obtain them effortlessly by overlapping $\emptyset_2$ and $\emptyset_k$.

$$\emptyset_{G2} = \emptyset_2 + \emptyset_k = \sum_{n=0}^{\infty}(B_n + C_n)\ g_nP_n$$

$$\text{for } h \leq r \leq \infty . \tag{35}$$

We can see that the somewhat cumbersome development according to characteristic functions bears its fruit at this point. The question regarding the potential function of our problem amounts only to the simple provision of the constants from the boundary conditions. It would not be difficult as well to introduce other spherically symmetrical layers, in which the conductivity reacts differently, thus, e.g. the ionosphere or the mist layer (exchange layer). However, we will need to set this derivative aside and resolutely pursue the question regarding the current efficiency of the storm generator. According to the implementations made in Section II, the potential equation (33) must provide us with the leakage current of the storm generator. In this connection, it is necessary to calculate the current density on the sphere, which we obtain through the field strength, and then integrate it through the surface of the sphere. Through the differentiations from (33) according to r, we will obtain the following for $r = k$:

$$-\left(\frac{\partial\emptyset_{G1}}{\partial r}\right)_{r=k} = -\sum_{n=0}^{\infty}(A_nf_n^{'k} - C_ng_n^{'k})\ P_n = E_k , \tag{36}$$

and from this, we obtain the current density by means of the relationship

$$i_k = \lambda_0 E_k . \tag{37}$$

In the case of the following integration via the surface of the sphere, only the term of the sum provides a contribution with $n = 0$, while all remaining terms become 0.

Mathematically, this is expressed in the following manner. Due to the rotational symmetry, for the surface element d0, we can write

$$do = 2\pi k^2\ \sin\vartheta\ d\vartheta,$$

in which the integration must be extended from $\vartheta = 0$ to $\vartheta = \pi/2$. The volumetric head generation rate of the sphere I_V is provided through

$$I_V = \int_0^{\pi/2} i_k\ do ,$$

and with (36) and (37)

$$I_V = -\lambda_0 \sum_{n=0}^{\infty}(A_nf_n^{'k} - C_ng_n^{'k})$$

$$2\pi k^2 \int_0^{\pi/2} P_n\ \sin\vartheta\ d\vartheta . \tag{38}$$

In this regard, all the factors independent from ϑ have already been brought in front of the integral. According to (38), therefore, integrals Jn of the following form must be solved

$$J_n = \int_0^{\pi/2} P_n\ \sin\vartheta\ d\vartheta .$$

If we introduce the variable x through substitution instead of the angle ϑ

$$x = \cos\vartheta \quad , \qquad dx = -\sin\vartheta\ d\vartheta$$

then we obtain the following

$$J_n = \int_{-1}^{+1} P_n\ dx .$$

Due to $P0 = 1$, we can insert the spherical function of the zero order, from which we will obtain

$$J_n = \int_{-1}^{+1} P_0P_n\ dx .$$

A familiar phrase from the theory of the spherical functions (orthogonal condition) [8] states that

$$\int_{-1}^{+1} P_nP_m\ dx = 0 \quad \text{für } n \neq m \quad \text{and}$$

$$\int_{-1}^{+1} P_nP_n\ dx = \frac{2}{2n+1} .$$

As a result

$$J_0 = 2$$

$$J_n = 0 \quad \text{for } n \neq 0.$$

If we additionally taken into account that f0 is a constant and therefore $f'0 = 0$, (38) is simplified to

$$I_V = \lambda_0 4\pi k^2 C_0 g_0^{'k} . \tag{39}$$

We can see that only the spherical potential $\emptyset_k$ is provided by our potential function $\emptyset_{G1}$ and in turn from this only the term with $n = 0$ provides input to the current production of the sphere. This welcome mathematical simplification has the following physical meaning. It is understandable that the current source I cannot deliver a share to the current production outside of the sphere. As a result, the influence of the potential function $\emptyset 1$ of the point source I is removed. Furthermore, the reaction potential of the sphere dispersed into the sum of the potentials from a single pole ($n = 0$) is a resulting current production, while the multipoles are comprised of a number of equally-strong sources and depressions that mutually compensate each other with regard to a resulting current production. The single pole represents a current distributed evenly across the sphere that was converted into a distribution of current density adapted to the problem, in which the absolute value of the production of the single pole will not be changed.

From this, we need to retain that only the provision of the constants C_0 is necessary for determining the current production of the sphere. This also applies in the case, for which we will also take the ionosphere or other layers of varying conductivity into account. According to (34), we have the following

181

$$C_o = -A_o \frac{f_o^k}{g_o^k} \quad,$$

$$A_o = \frac{1}{4\pi h^2 \lambda_h} \frac{g_o^h}{f_o^h g_o'^h} \quad.$$

In this process, if we take into account that $f_o^k = f_o^h = 1$, then according to (39) we will have

$$I_V = -1 \frac{\lambda_o k^2 g_o^h g_o'^k}{\lambda_o h^2 g_o'^h g_o^k} \quad. \tag{40}$$

Furthermore, according to (17) we will have

$$g_o = \int_1^\infty \frac{t-1}{t} e^{-art} \, dt = a \, \text{Ei}(-ar) + \frac{e^{-ar}}{r}$$

and as a result

$$g_o' = \frac{dg_o}{dr} = -a \int_1^\infty (t-1) e^{-art} \, dt = -\frac{e^{-ar}}{ar^2} \quad.$$

Through application in (40), we obtain

$$\frac{-I_V}{1} = \frac{\lambda_o k}{\lambda_h h} \frac{1 - ahe^{ah} \, \text{Ei}(-ah)}{1 - ake^{ak} \, \text{Ei}(-ak)} \quad. \tag{41}$$

To check these results, we will transition to a constant conductivity with $a = 0$ and obtain the following from (41)

$$\frac{-I_V}{1} = \frac{k}{h} \text{ for } \lambda = \text{const.}$$

The conformity with equation (4b) Section II confirms the correctness of our calculation.

For the practical evaluation, we will develop integral logarithms in a semi-convergent series due to the fact that it is not tabulated for such large argument values as they appear in our case. By means of the relationship achieved through partial integration

$$\int \frac{e^{-z}}{z^n} \, dz = \frac{e^{-z}}{z^n} - n \int \frac{e^{-z}}{z^{n+1}} \, dz$$

we obtain

$$\text{Ei}(-z) = \int_1^\infty \frac{e^{-z}}{z} \, dz = \frac{e^{-z}}{z}$$

$$\left(1 - \frac{1}{z} + \frac{2!}{z^2} - \frac{3!}{z^3} \pm \ldots \ldots \right) \quad.$$

With $a = \frac{\ln 10}{10 \text{ km}}$ and h or k approx. $6\cdot10^3$, $z = ah$ takes on the order of magnitude of 10^3. Therefore, we can calculate the series expansion for Ei with the second term and obtain the following approximation formula from (41)

$$\frac{-I_V}{1} = \frac{\lambda_o k^2}{\lambda_o h^2} \quad.$$

Given, moreover, that $k = h$ aside from an error of a few 0/00, the simple and nonetheless sufficiently accurate approximation equation remains

$$\frac{-I_V}{1} = \frac{\lambda_o}{\lambda_h} \quad. \tag{42}$$

If we calculate the efficiency of the storm defined in Section II equation (5), we obtain

$$W = 1 - \frac{|I_V|}{|I|} = 1 - \frac{\lambda_o}{\lambda_h} \quad. \tag{43}$$

For a practical example, let us assume the head of the storm is at an altitude of 10km. At that level, the conductivity is approx. 10 times as great as on the surface of the earth. The following results from this

$W = 0.9 = 90\%$.

I.e. the primary current of the storm generator for the atmospheric electrical vertical current is about 90% of the total production, while the leakage current used in the tributary represents only 10%. We can see how radically this balance has changed upon taking the conductivity increasing with altitude into account when we compare it to our calculations according to Section II, and we recognize the effective medium in the conductivity, which utilizes the storm generator for the fair weather areas of the earth. This is expressed symbolically in Figures 1a and 1b by the fact that in the case of a local constant conductivity (Figure 1a) the reflection point I_V representing the leakage current through a heavy point is marked, while the primary current I_N is only reflected by a faint point in the center of the sphere. In contrast, for the case of the conductivity increasing with altitude (Figure 1b), the source point representing the leakage current I_V is faintly indicated, while the primary current I_N is marked heavily. The negative potential of the earth also increases with this increasing primary current. If we assume an average production of IA for a storm, the 1800 storms simultaneously occurring globally provide a total primary current of I_G of approx. 1600 A. With a surface conductivity of $\lambda_o = 2 \cdot 10^{-14} \frac{1}{\text{Ohm} \cdot \text{m}}$, according to the equation, the earth obtains

$$\varnothing_k = \frac{I_G}{4\pi\lambda_o} \frac{1}{k}$$

a negative potential of 10^{10} V. According to that, which was said in the introduction, this value is identical to the medium voltage of the storm generator.

LITERATURE

(1) Simpson, G., G. D. Robinson: The distribution of electricity in thunderclouds. II. Proc. Roy. Soc. A 1941, 281.

(2) Benndorf, H.: Grundzüge einer Theorie des elektrischen Feldes der Erde I und II. Sitz. Ber. Akad. Wien. IIa, 134, 281 (1925) und 136, 175 (1927).

(3) Kasemir, H.-W.: Zur Strömungstheorie des luftelektrischen Feldes I. Arch. Meteor. Geophys. Bioklim. A, 3, 84 (1951).

(4) Kasemir, H.-W.: Zur Strömungstheorie des luftelektrischen Feldes II. Arch. Meteor. Geophys. Bioklim. A, 5, 56 (1952).

(5) Israël, H.: Das Gewitter. Leipzig (1950).

(6) Kamke, E.: Differentialgleichungen, Lösungsmethoden und Lösungen, Bd. 1. Leipzig (1944).

(7) Ollendorff, F.: Potentialfelder der Elektrotechnik. Berlin (1932).

(8) Madelung, E.: Die mathematischen Hilfsmittel des Physikers. Berlin (1936).

JOURNAL

OF

GEOPHYSICS

Published on behalf of the

German Geophysical Association

by

B. Brockamp, Münster i. W.

with the collaboration of **A. Defant**, Innsbruck – **W. Dieminger**, Lindau b. Northeim
W. Hiller, Stuttgart – **K. Jung**, Kiel – **O. Meyer**, Hamburg
F. Möller, Mainz – **H. Reich**, Göttingen

Special Print

from the 1959 issue, no. 2

PHYSICA - VERLAG · WÜRZBURG

The Storm as a Generator in the Atmospheric Electrical Circuit I[1]

By H. W. Kasemir, Neptune, N.J.

Abstract: This paper treats the electric current flow inside and outside of a thunderstorm. The calculation starts with the current flow of a point source inbedded in a medium, where the conductivity increase with height as an e-function. The earth is introduced as a conducting layer by a suitable image source. The superposition of the current flow of a number of point sources with different polarity makes it possible to represent any given thundercloud. It is shown that the charge in the top of the cloud is about + 2400 C (Coulomb) if the strength of the current source is 1 Ampere. This is 100 times as much as the commonly accepted value of + 24 C, which is derived by the electrostatic theory. The much larger storage of charge makes it easier to understand that a thunderstorm is able to feed a number of lightning discharges, each consuming a charge of about 30 C, and in addition, to deliver a continuous current of about 1 A = 1 C/sec. to the ionosphere. 90% of the electric field of the large cloud charge is screened from the outside of the cloud by a negative surface charge of about − 2200 C. The surface charge is caused by the sudden change of conductivity at the boudary of the cloud and the surrounding air. An additional screening effect is excercised by a space charge, which is caused by the conductivity gradient of the air. The conductivity gradient is also responsible for the fact that a high percentage of the positive charge leaving the top of the cloud does not flow back to the ground in the immedeate neigborhood of the cloud but is lead up to the ionosphere. This means also that a high percentage of the negative charge coming down to the ground will spread over the whole surface of the earth as free charge. Only a small percentage is bound below the thunderstorm as an induced charge. This leads to the conclusion, different from the common concept, that the earth's surface and not the ionosphere is to be considered as the equalizing layer. The free charge would spread equally over the whole globe even if the ionosphere would not be present. This explains the fact that changes of the conductivity or the height of the ionosphere have practically no effect upon the atmospheric electric field at the ground.

The potential of the earth is not arbitrarily assumed to be zero, as it is usually done, but has the negative value which follows from the negative charge of the earth and its capacity. On the other hand, the potential of the ionosphere

---

[1] Dissertation Aachen 1954

will become zero with the logical assumption that the atmospheric electric generators produce an equal amount of positive and negative charges. These natural potential values of the earth and the ionosphere make the atmospheric electric circuit a closed system. Keeping the earth potential at zero would result in additional and unreasonable assumptions. These would be, either the potential of outer space depends on the world wide thunderstorm activity, or there must exist a continuous flow of negative charge from the ionosphere into outer space.

The representation of a thunderstorm by a fitting arrangement of point sources enables us to test the validity of a mathematically simpler model where the point sources are replaced by layer sources. It is shown that certain integrals, which express for example the profit current, the leakage current, the efficiency and so on lead to the same results by both calculus. The profit current of the thunderstormgenerator is defined as that part of the negative charge/sec delivered to the earth, which spreads equally over the whole earth surface and flows from there as the air earth current up to the ionosphere. It joins there the positive charge comming from the top of the cloud and completes the external closed circuit. On the other hand, the conduction current inside the thunderstorm is called the leakage current. The efficiency of the thunderstorm is defiend as the ratio of the profit current to the sum of the profit and the leakage current. The avarage thunderstorm has an efficiency of about 90%, if the low conductivity inside the cloud is taken into account.

The lines of current flow at greater distances from the cloud are the same for all types of thunderstorms and can be represented by the current flow of a dipole. The dipole moment can be calculated from the profit current and the conductivity gradient of the air. Assuming the conductivity increases steadly to infinity then the lines of current flow would reach far out into outer space. In reality the conductivity decreases behind the ionosphere whereby the lines of current flow are flattened. But charges penetrating the ionosphere follow different physical laws from those which are applied here. Therfore the calculation of their lines of current flow is beyond the scope of this paper. It is sufficient here to set the conductivity of outer space at zero and with that to limit the lines of current flow to the space between the ionosphere and earth.

185

The Meaning of the Characters used

e = Basis of the natural logarithm

j = $\sqrt{-1}$

x, R, ϕ = Cylinder coordinates

r, θ, ϕ = Spherical coordinates

ϕ = Potential function

Ψ = Current function

$\vec{E}$ = Field strength

$\vec{\imath}$ = Current density

i_0 = Current density on the surface of the earth

Q = Net charge

Q^u = Surface charge

Q' = effective charge

q' = Surface charge density

q = Space charge density

λ = $\lambda_0\, e^{2kx}$ = Conduction function

λ_0 = Conductivity on the surface of the earth

$2k$ = $\dfrac{\ln 10}{10\ \text{km}}$

ε = Dielectric constant

I = Total current of the storm generator

I_V = Leakage current of the storm generator

I_N = Primary current of the storm generator

U_K = Terminal voltage of the storm generator

w = Efficiency factor of the storm generator

W_0 = Resistance between +source layer and $+\infty$

W_i = Resistance between +source layer and earth
    or between + and - source layer

W_u = Resistance between -source layer and earth

W_a = Resistance of the entire atmosphere between the earth and $+\infty$

Vectors are marked with an arrow over the letter, e.g. $\vec{E}$

I. Introduction

The generator effect of the rain and storm clouds was chosen as the topic of the dissertation from the research projects executed by the Meteorological Observatory in Aachen due to the fact that the precipitation generator is the most important energy source of the atmospheric electrical circuit and, therefore, deserves special attention. Because the "atmospheric electrical generator" generally differs from the technical generator, i.e. the dynamo machine, it is necessary to define the term of the generator used in this case in the following manner:

Every charge carrier moved against the direction of an existing electrical field must be perceived as a generator or as a part of a generator. For through this movement against the electrical field, the charge carrier gains a potential energy, which is available to the system quasi as an electromotive force. As the simplest example for such a process, we will observe two point charges of opposite signs that are initially closely adjacent to each other and then pulled apart with a mechanical force. The potential energy gained in this process can be used to move the point charges back together even upon disengaging the mechanical force. In the case of conductivity of the intermediate medium, it is also possible – as well as the rule in atmospheric electricity – for other charge carriers to begin moving under the influence of the electrical field of both point charges and to dissipate the electric field by neutralizing the point charges. The potential energy of both point charges is used here as well.

The process described in this manner is the principle of each atmospheric electrical generator. The mechanical force moving the charge carriers can be from various sources. It can be provided by the wind, which e.g. moves the positive space charge stored above the surface of the earth, or by the exchange effect that this space charge conveys in the altitude. The force driving the cloud generator is gravitational force, which pulls the charged lightning bolt to the earth while the opposite charge attaches itself to the cloud elements and the remains in the cloud. Under the effect of the electrical field developing in this process, ions present in the atmosphere begin to move throughout the entire airspace surrounding the earth in order to again neutralize the formed cloud charges. In this manner, they form the atmospheric electrical vertical current. The current paths are defined in this process through the conductivity of the atmosphere and the earth itself increasing with altitude, which as a conductive sphere depicts an equipotential plane in this current diagram. It is the goal of this work to gather this current in a calculation, to affiliate it to the current ideas and measurement results, to promote the new knowledge in formulas and propositions that are as manageable as possible, and in this manner, to connect the variables characterizing the cloud generator.

One can properly categorize the observation of the precipitation generator according to 3 perspectives.

1. The charging mechanism of the precipitation particles.
2. The electrical conditions of the cloud.
3. The electrical conditions beyond the cloud in the entire atmosphere and on the earth.

The charging mechanism of the precipitation is not dealt with in this work. The division of the electrical current according to the points 2 and 3 occurs with regard to the term of the generator. Thus, all processes occurring within the generator fall under 2, while 3 depicts the current of the outer space, i.e. the user. This also corresponds then to the currently common bisection of the atmospheric electricity in disturbed and undisturbed conditions, through which all registering strips "disturbed" by lightning are separated from the "undisturbed" fair weather curves when evaluating the atmospheric electrical registrations. We will demonstrate that such a subdivision from mathematics itself will be offered as organic even in the case of the uniform mathematical approach to the entire electrical circuit.

The mathematical approach assumes the current diagram of a point source in a medium, the conductivity of which increases with altitude according to an e-function and ends with a representation of the storm generator through the arbitrary space charge or source areas of different polarity. As a result, it is possible to adjust the model diagram of the storm generator very precisely to the present measurements. The current of the equivalent dipole applicable for every model diagram is differentiated and discussed for the outer space.

The development of the equations for potential and current function, field and space charge distribution, etc. from the equations of Maxwell will be provided in a separate attachment and the final formulas from there will be included in the work upon request. This has proven to be practical in order to not disrupt the physical thought processes with lengthy mathematical derivations. On the other hand, the derivation of the formulas has not been provided to date in literature and is so important for the further expansion of theory that we should not renounce it.

II. Definition of the Problem and Overview of Literature

Today, we see that the problems of atmospheric electricity are treated according to two different disciplines of electrical science, namely according to electrostatics and the theory of electrical current.

At the beginning of the history of development, atmospheric electricity stood under the sign of electrostatics. Then as the conductivity of the atmosphere was discovered, the ideas and methods of calculation of the current theory achieve increasing space. In electrostatics, the electrical fields and potentials are reduced to space charges or surface charges and the field distribution is calculated in the case of a given charge arrangement, or the other way around, the charge arrangement is calculated in the case of a given field distribution. In the current theory, the primary causes of the electrical state are the current sources and the field and potential distribution are calculated from the arrangement of the current sources and the conduction function. Now, while the material constant of the medium – the dielectric constant – is constant in the entire atmosphere in electrostatics, the material constant of the current theory – the conductivity of the atmosphere – increases with altitude. If conductivity were constant, then electrostatics and current theory would lead to the same result for potential and field distribution. Due to the change of conductivity of the atmosphere, in the case of the current a space charge forms, which can also be calculated from the arrangement of the current sources and from the change of conductivity. These space charges would have to be given for the electrostatic calculation. However, they are most often overlooked because they must first be obtained in an indirect route using the current theory. For this reason, we will conduct the considerations according to the current theory and we will see from case to case how the electrostatic observations must be supplemented.

The Storm as a Generator in the Atmospheric Electrical Circuit I

The juxtaposition of the electrostatic and perception of the current theory may also be a reason for the fact that we find a mix of two measurement systems in the atmospheric electricity, namely the electrostatic CGS system and the practical System of Measurement. However, we do not intend to follow this custom, rather we will choose the practical system of measurement.

We would now like to gain a brief overview of the new literature based on some examples. Approximately 15 years ago, the issue regarding the polarity of the storm cloud was decided upon by the probe ascents of Simpson and Scrase [1] and Simpson and Robinson [2] to the extent that the storm cloud possesses a positive polarity, i.e. there is a positive space charge in the upper part of the cloud and a negative space charge in the lower part. …

Under the latter one still often finds a smaller positive space charge island, which cannot however reverse the positive polarity of the cloud. The size of the space charges was calculated from field measurements on the surface of the earth at approx. +24 C and -20 C for both main space charge centers and at +4 C for the small positive space charge island. The question about the polarity of the storm cloud was, therefore, of decisive importance because the charging mechanism for rain drops most known at that time, namely the Lenard effect, delivered a negative polar cloud, while the version of the storm cloud as a generator of the atmospheric electrical current for fair weather could only be maintained if positive charge carriers flowed from the head of the cloud, thus the cloud possesses positive polarity.

In addition to the proper polarity, the storm also had to display the ability to produce sufficient current to be able to deliver the vertical current discharging on the earth totaling approx. 1800 A. According to C.E.P. Brooks [3], because roughly 1800 storms occur simultaneously across the earth, every individual storm produces for an average current of approx. 1 A. This production of current was able to be proven by Wait [4] according to the scale by determining the current density from field and conductivity measurements above the storm cloud. In connection with the space charge determination, the question arises, what is the relation between the space charge of the storm cloud and the production of current and whether the space charges calculated based on electrostatic principles are correct. A single lightning bolt discharges namely an average charge of 30 C, although there is also lightning with 160 C. With regard to the fact that lightning sequences of 30 seconds are not uncommon in strong storms, the cloud charge of ± 25 C seems really low. With a precipitation down velocity of 8 m/second, particles would have fallen 240 m in 30 seconds. Thus, it would have been impossible for it to recharge the space charge spheres separated by a distance of 3 km. Even more important is the objection that the charge of the storm cloud described above is clearly not sufficient for explaining the atmospheric electrical field of fair weather.

The information from 1 A 'Current Production of the Storm Cloud' states that the charge of 30 C necessary for an average lightning bolt can be made available every 30 seconds by the storm generator. This absolutely corresponds with the observed lightning sequence. Now, if the lightning discharge is located between the poles of the generator, then it represents a short circuit and the charge quantity destroyed by the lightning will be deprived of the fair weather vertical current. In this process, it would be necessary to examine whether such a storm is capable of distributing the main share of its current production over the entire earth so that it can provide a respective share to the vertical current of fair weather areas that are somewhat distant. An initial contribution to this question was already provided by the author in a paper regarding the current efficiency of the storm generator [5].

The Storm as a Generator in the Atmospheric Electrical Circuit I

The expansion of the current provided by the storm over the entire earth touches upon another issue, which can be outlined with the keyword "equalization layer". The general idea is that the positive charge delivered by the storm flows toward the ionosphere, distributes itself there evenly due to the high ionospheric conductivity around the entire earth and then flows down to the earth. However, because particularly the lower layers of the ionosphere are subject to diurnal variations, one could expect that these variations would have to become noticeable on the earth's surface in the fair weather field as well. The absence of such influences provided a reason to analyze the question, whether the atmospheric electrical cicuit cpnnection does not occur in lower laying layers. Following a rough approximation calculation by H. Israël and H. W. Kasemir [6], one could presume that this compensating layer is located at an altitude of approx. 50 to 60 km, thus significantly below the ionosphere. H. Wichmann [7], however, uses the variations of the conductivity in the upper layers to explain some of the peculiarities in the diurnal variation of the atmospheric electrical field that he introduced. The calculation conducted in this case reveals that the current paths by all means do extend up to the ionosphere, that the conductivity there, however, is so great compared to the surface-level layers that a change of conductivity cannot become effective in the upper atmospheric layers for the current density and the field strength on the earth's surface. This fact is emphasized even more clearly through the interpretation conveyed by the calculation. The earth is charged namely negatively through the storm generator, in which the main portion of this charge is distributed evenly over the surface of the earth. The diversion of the negative charge from the earth is then influenced by the high ohmic surface-level atmospheric layers. Changes to the resistance of the very low ohmic atmospheric layers of the upper atmosphere naturally remain without an effect on the total resistance and only the total trsistance is responsible for the current density.

The problems touched upon here briefly are not the points of origin of the following calculation, although we will provide more detail throughout the course of the mathematical approach. A whole other series of additional questions were also able to be tied in, which can automatically be revealed and answered however during the systematic analysis, such that we would like to review this calculation at this point.

III. The Current of a Point Source in the Medium with a Variable

Conductivity

The basic building block for the subsequent storm representations through a random number of point sources is the current of an individual point source in a medium, the conductivity of which increases according to an e-function in the direction of the positive x-axis. To create a connection from the onset between the abstract mathematical terms and the practical application, we can envision the point source as an image for a charged cloud element or precipitation particles. We can also represent an entire cloud segment, e.g. the positively charged head of the cloud, through a point source. In the process, the point source would no longer be allowed to shrink to a small point, but rather would have to be envisioned as a sphere with a radius of 2 km. The application of the calculation must naturally be limited then to the outer space.

The calculation is conducted for a constant current production of the point source and we must prove to what extent our application examples satisfy this requirement. For the head of the cloud, this notion is readily plausible for the duration of a storm, which is expressed with particular clarity in a formulation selected by J. Küttner [8]. According to this formulation, the precipitation acts on the cloud elements in such a manner that it guides a charge of an unspecified sign to the earth, while the opposing charge attaches to the cloud element and remains behind in the cloud. In this process, a state of equilibrium is introduced, in which the loss of charge of the cloud is kept in balance by the ensuing conductive current of the atmosphere and the charge increase is kept in balance by the acting precipitation.

In the case of the individual cloud element itself, the notion of a constant current production is certainly not applicable. Namely, if a cloud element is charged by a falling precipitation particle, it will be completely or partially discharged by the ensuing conductive current until it is recharged by another precipitation particle. However, because this process occurs continuously on countless cloud elements, we can safely postulate the constant current production for a cloud element in the statistic average.

In the case of the precipitation particle itself, the loss of charge caused by the conductive current can be offset by a continual recharge if the charge device functions appropriately. At this point, it is necessary for us to take the rapid local variation into account. This can lead to a quasi stationary image if we just pursue a single precipitation particle for a brief distance on its path of descent, and then return up this path to catch the next precipitation particle on the same path, and then the next, etc.

In this manner, we remain practically at the same location – merely that the precipitation particles supplant each other. Thus, we can adhere to the image of the stationary point source with a constant current production in this case as well.

We can conduct the calculation of the present problem in two different coordinate systems, namely the cylinder or spherical coordinate system – depending on if we want to later depict the earth through a plane or through a sphere. The first approach has the benefit of simplicity because it leads to a closed solution of the differential equation. However, it is limited to the approximate surrounding of the storm, in which the curvature of the earth may be ignored. The second approach comprises the current around the entire earth, although it leads to a cumbersome solution displayed as a sum. We would therefore prefer to calculate with cylinder coordinates x, R, ϕ and use the calculation already conducted in [5] in spherical coordinates r, θ, ϕ only for the flow of the fair weather vertical current encompassing the earth.

We can achieve the best idea of the current of our point source through a discussion or graphic depiction of the potential and current function. The laborious approach to derive these functions from Maxwell's differential equation is listed in Section I of the appendix. The current production I of the point source is given for this calculation, which is location in the center of the cylinder coordinate system x, R, ϕ, and the conductivity $\lambda = \lambda_0 e^{2kx}$ as a function of the space. λ_0 is the conductivity of the plane $x = 0$. Without going into the calculation at this point, we will take the following from the Appendix AI (28): The equation of the potential function for a current source in a medium, the conductivity of which increases with altitude.

(1)
$$\Phi = \frac{I}{4\pi\lambda_0} \frac{e^{-k(\sqrt{x^2+R^2}+x)}}{\sqrt{x^2+R^2}}$$

or even in spherical coordinates r, θ, ϕ:

(2)
$$\Phi = \frac{I}{4\pi\lambda_0} \frac{e^{-k(1+\cos\theta)r}}{r}$$

In addition to the familiar dependency of the potential function on the distance r, the effect of the conductivity emerges in the factor $e^{-kr(1+\cos(\cdot))}$.

If $k = 0$, thus $\lambda = \lambda_0$ = const., then the e-function becomes 1, and we obtain the following familiar equation from (2):

(3)
$$\Phi = \frac{I}{4\pi\lambda_0} \frac{1}{r}.$$

for the potential function of a current source in a medium with a constant conductivity. This equation immediately reminds us of the potential equation of a point source Q, which we obtain if we replace the factor $I / 4\pi\lambda_0$ with $Q / 4\pi\varepsilon$ in (3). Now we know [9] that we can achieve such current from potential fields in electrostatics if we replace Q and ε with I and λ. However, a much deeper connection hides behind this analogy, which will lead us to the relationship between the space charge and the current production of the storm cloud.

In this context, we will observe a raindrop with the radius a and the current production I, which should be located at the altitude h with the conductivity λ. Near the raindrop, we can perceive the conductivity as a constant. According to (3), the following potential equation applies:

(4)
$$\Phi = \frac{I}{4\pi\lambda} \frac{1}{r}.$$

From this potential equation, we can calculate the field strength on the surface of the drop and from that the surface charge density q' and obtain

$$q' = \frac{\epsilon I}{4\pi\lambda} \frac{1}{a^2}$$

Through integration over the entire surface, we obtain the charge Q of the drop for

(5)
$$Q = \frac{\epsilon}{\lambda} I,$$

and with that, we have achieved the correlation between the current production and charge.

H.W. Kasemir

Charge = (time constant) x (current production).

The factor ε/λ namely is the time constant of the atmosphere at the location of concern. At higher altitudes, where this time constant with an increasing conductivity becomes increasingly small, a raindrop has a small charge with the same current production. Furthermore, it is noteworthy that the ratio of the charge to the current production is not dependent on the size of the sphere, through which this relationship applies for a most minute cloud element as well as for larger cloud areas.

With (5), therefore, we can also introduce the current production I, the charge Q in the potential equations (1), (2), and (3) and with (3), we come to the familiar formula of electrostatics

(6)
$$\Phi = \frac{Q}{4\pi\epsilon}\frac{1}{r}.$$

The comparison of (4) and (6) thus demonstrates that not only a mere analogy exists between the potential equations of electrostatics and the current theory, but rather it implies that in the case of the current of a point source with the current production I, on which a charge Q is located, which stands with the current production I in the correlation given through (5). We can perceive the potential function ϕ as a voltage drop of a current stream over the resistance of a certain spatial element, although we could just as well trace the potential function to the effect of the charge Q in the point source. This close correlation is substantiated, from a mathematical standpoint, in the fact that Poisson's equation also applies in the current theory, that the space charge in this case is therefore also derived from the spatial divergence and the surface charge from the divergence of the area of the field strength. As a result, we are able to compare the calculations conducted according to electrostatics with those of the current theory.

For the individual raindrops-current source, we can create the following image. A very high conductivity is present in the inside of the drop compared to the outer space. Due to the fact that a sudden increase of conductivity exists on the surface according to the outer space, at this point an accumulation of the charge carriers occurs, which are in route from the inside of the drop to the outer space. This accumulation then perpetuates the formation of the surface charge, namely for a positive source to a positive surface charge and vice versa. The image becomes even clearer if we start at the surface charge. In this case, we can say that, under the field influence of the charge Q in the conductive outer space the charge carriers of the same sign migrate away from the charge Q, whereby the oppositely charged migrate to the charge Q. As long as the value of the charge is maintained by some process, a stationary current I is applied in the outer space, the total value of which we will call the current production I of the source. Now, if the precipitation cloud is comprised of several individual charged particles, we must add up the current strengths and the charges of the individual drops to the total current production or the net charge. In this process, therefore the relationship (5) applies for the entire cloud area as well.

For the upper positive space charge sphere of the Simpson storm model, we can calculate the space charge for a current production of 1A in the following manner.

H.W. Kasemir

The conductivity of the cloud is approx. one third of the surface conductivity [10]. For the time constant, we therefore obtain a value of $\varepsilon/\lambda = 2400$ seconds and according to (5) the value of 2400 C for the space charge. This is approx. 100 times that of the charge calculated by Simpson and Robinson from the field strengths on the surface of the earth. Now the effect of this space is largely shielded by a surface charge that is formed on the cloud surface due to the jump in conductivity present there. We need to calculate this surface charge for the following simplified conditions. A current source of the current production I and the charge Q lies in the center of a sphere K, the conductivity of which should be λ_2. The outer space will have the constant conductivity λ_1. According to A V (71), we obtain the following equation for the surface charge Q^0 of the cloud sphere:

(7)
$$\dot{Q}^0 = -\left(1 - \frac{\lambda_2}{\lambda_1}\right) Q \, .$$

We can comprehend the emergence of the surface charge in the following manner for example. Because the conductivity of the outer space is greater than that of the inner space of the sphere, an accumulation of those charge carriers, which migrate into the sphere, emerges on the surface of the sphere. Because these charge carriers have the opposite sign of the source point charge, thus the surface charge also has the opposite sign. It shields the central source point charge to the degree that only the difference of these two charges Q and Q^0 is effective in the outer space. We can identify this effective charge difference with Q' and we obtain the following simple correlation from (7):

(8)
$$Q' = Q + Q^0 = \frac{\lambda_2}{\lambda_1} Q \, ,$$

which states:

The effective charge is the cloud charge reduced in the correlation of the cloud conductivity to the atmospheric conductivity.

For our practical example, we will assume the value of the atmosphere at an altitude of 6 km as the outer conductivity. The conductivity there is approx. 4 times as great as on the surface of the earth. Because only approx. one third of the surface conductivity is present on the inside of the cloud, therefore, the 2400 C of the positive space charge sphere are reduced to a twelfth of their effectiveness. For the outer space, it seems as though only a charge $Q' = 2400/12$ C $= 200$ C is present. The shielding surface charge is the main reason for the fact that the electrostatic calculation from the surface field strengths provides values that are much too small for the cloud charge.

A second reason is the omission of the space charge in the atmosphere, which is formed due to the conductivity increasing with the altitude. We will return to this in more detail at the end of the section.

We will now follow a thought process from Holzer and Saxon [11], who only calculate with the effective charge Q' for their calculation of the storm generator. If we select the conductivity of the inner and outer space of the sphere equal in strength ($\lambda_1 = \lambda_2$), then naturally the surface charge will become Q^0 $= 0$ according to (7). A jump of conductivity on the surface of the sphere indeed no longer exists. The value of Q' is revealed for the charge of the current source according to (5). Therefore, this means that:

H.W. Kasemir

The effective charge Q' corresponds to a charge of the current source, which would be set, if the conductivity of the outer space is also assumed in the cloud instead of the low cloud conductivity.

This procedure has the great benefit that the conductivity of the storm cloud, which was previously only rarely measured, does not need to be known in order to calculate the fields of the outer space. The applicability of the calculation, however, is limited to just the outer space. The field and potential distribution on the insider of the cloud, just as the true cloud charge, cannot be determined through this calculation. In the case of a comparison of lightning and cloud charge, however, one must revert to the space charge actually present in the cloud.

Previously, for the calculation of the surface charge of the cloud sphere, we made a simplification that the conductivity of the outer space is constant and corresponds to the value as it is found at the altitude of the spherical central point in a cloudless sky. We will now omit these limitations and take the conductivity of the outer space increasing with altitude into account for the calculation of the effective charge. We will take the following equation from Appendix A VI (74) for this:

$$(9) \qquad Q' = \frac{\epsilon}{\lambda_m} I \frac{3}{2 (2ak)^3} \left[(1 + 2ak) e^{-2ak} - (1 - 2ak) e^{2ak} \right] .$$

In this connection, λ_m refers to the value of the conductivity in the center of the sphere and a is its radius. In the case of our practical example, the positive space charge sphere of the Simpson storm model has the effective charge Q' of 300 C. Thus, just as with the approximation calculation ($Q' = 200$ C), this also reveals an approx. 10 times greater value than that of the electrostatic calculation of 24 C. The last task of this section should be to uncover the reason for this discrepancy.

For this purpose, we will observe the current line diagram of a point source reflected in Figure 1a and 1b for a conductivity that is constant and one that increases with the altitude.

The Storm as a Generator in the Atmospheric Electrical Circuit I

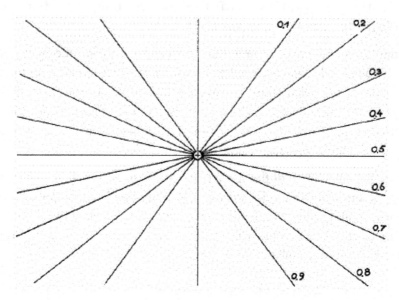

Figure 1a: Current line diagram of a point source in a
medium with a constant conductivity

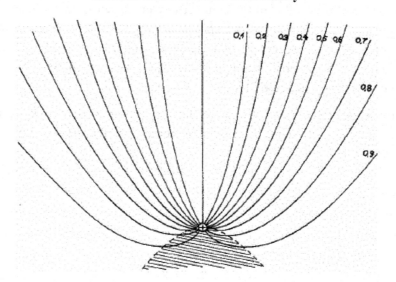

Figure 1b: Current line diagram of a point source in a medium with a conductive increasing
in the direction of the x-axis

The equation of the current function Ψ associated to Figure 1b reads as follows according to
Appendix AI (35) in cylinder coordinates x, R, ϕ:

$$(9a) \qquad \Psi = I\left[1 - \frac{1}{2}\left(1 + \frac{x}{\sqrt{x^2 + R^2}}\right) e^{-k\left(\sqrt{x^2 + R^2} + x\right)}\right]$$

H.W. Kasemir

and in spherical coordinates r, θ, ϕ

(10)
$$\Psi = I \left[1 - \frac{1 + \cos \theta}{2} e^{-kr(1+\cos \theta)} \right] .$$

For $k = 0$, (10) transitions to the equation of the current function of a point source of a constant conductivity (Figure 1a).

(11)
$$\Psi = I \frac{1 - \cos \theta}{2} .$$

For a better understanding of the figures, we would like to remind that the same fraction of the total current always flows in the rotational objects limited by two adjacent current lines, which is always $1/10 \, I$ in the presented figures. Simultaneously, the current lines also provide the field direction.

While the current lines extend out radially and in a straight line from each other in all directions in the case of a constant conductivity in Figure 1a, we recognize in Figure 1b how the current lines that initially flowed downward bend under the influence of the conductivity increasing with the altitude until finally all current lines have assumed the direction of the positive x-axis. This is an extremely import result because it contains the reason for the fact that the charges emitting from the head of the storm do not flow in direct proximity to the earth, but rather preferably flow to the ionosphere to expand there and only then to flow back to the earth as a equally dispersed vertical current. In this manner, it becomes possible that a storm can also provide a significant contribution to the vertical current in its antipodal point on the earth. The proportionally low field effect below the storm will also become understandable. Because the current lines preferably flow toward the positive x-direction, the current line density and thus the field strength next to the storm is not as great as with a constant conductivity. (An electrostatic calculation is equivalent to a calculation of the current theory with a constant conductivity because $\lambda = 0$ is also a constant conductive value.) Thus, if the charge of the storm cloud is calculated in an electrostatic-manner from the surface field strength like in [1] and [2], out of necessity it must be too small.

This emerges even more clearly if we take a closer look at the space charge conditioned by the change of conductivity. According to Appendix AI (32), the space charge distribution q is provided in cylinder coordinates through the following equation:

(12)
$$q = - \frac{k^2 \epsilon I}{2\pi \lambda_0} \frac{e^{-k(\sqrt{x^2+R^2}+x)}}{\sqrt{x^2+R^2}} \left[1 + \frac{x}{\sqrt{x^2+R^2}} + \frac{x}{k \cdot (x^2+R^2)} \right]$$

and in spherical coordinates:

$$(13) \qquad q = -\frac{k^2 \epsilon I}{2\pi \lambda_0} \frac{e^{-kr(1+\cos\theta)}}{r} \left[1 + \cos\theta + \frac{\cos\theta}{kr} \right]$$

The space charge is consistently negative in the area of the positive x-half-plane. In contrast, in the negative x-half-plane there is cone-shaped, positively-charged space below the current source, which has the negative x-axis as a rotational axis. The space charge density moves toward 0 on the border of this positive space charge cone, to then assume a negative sign outside of the cone. The border line of the cone is marked with a hashed line in Figure 1b and the total positively-charged cone space is lightly shaded. From equation (13), we will immediately take the equation of conditions for the borderline. If the space charge q should be 0, the bracket term must be removed, for which the following is revealed:

$$(14) \qquad \cos\theta + 1 + \frac{\cos\theta}{kr} = 0$$

We can explain the establishment of this positive and negative space charge in the same way we explained the surface charge above. If charge carriers are forced to penetrate in areas with a lower conductivity, an accumulation of these charge carriers will occur, which leads to a space charge formation of the same sign. In Figure 1b, the positive charge carriers of the source now penetrating into the negative x-area migrate against the conductive gradient in areas with a worsening conductivity. Therefore, the positive space charge with the same name as the source forms there. As soon as the current paths, however, have a positive x-component instead of a negative one, they migrate to areas with an increasing conductivity. The positive charge carriers can therefore migrate away easier than they can toward an object. There is no longer any reason for an accumulation. It is exactly the opposite in the case of negative charge carriers, for they migrate in the opposite direction as that of the positive charge carriers. They form a negative space charge where the positive charge carriers migrate away without obstruction and vice versa. This is nothing other than a descriptive explanation of the general formula derived in Appendix AI (4)

$$(15) \qquad q = \epsilon \vec{E} \frac{\text{grad } \lambda}{\lambda} \ .$$

If we take into account that $\vec{E}$ grad λ is the scalar product of both vectors $\vec{E}$ and grad λ, then the immediate result is that the space charge is positive if $\vec{E}$ and grad λ have the same direction and vice versa. If the field strength $\vec{E}$ stands vertically on the conductive gradient grad λ, the space charge is equal to 0.

If we are still aware of the fact that the current lines simultaneously provide the direction of the field strength, we will directly recognize the applicability of (15) in Figure 1b. The conductive gradient is always in the positive x-direction. The field strength vector shows the current lines to the positive source. In the positive space charge cone, field strength and conductive gradient are directed oppositely in the entire remaining space. The field strength stands vertically on the conductive gradient on the borderline of the positive space charge cone. Therefore, the space charge in this case is equal to 0.

The following physical interpretation is given for the flow of current. Charge carriers, which are emitted from the positive source downward into the negative x-area, are restrained by the positive space charge of the cone in its course downward and attempt to swerve to the side. This effect is supported as well by the negative space charge present on the sides and in the upper space, such that the current lines of these charge carriers that were originally flowing downward finally merge into the positive *x*-direction. For a quantitative evaluation of this knowledge with regard to the storm generator, we must first take the effect of the earth's surface into account in our formulas, which will occur in the next section.

We can retain the following as the most important results of this section:

1. A current source of the current production I has the charge $Q = \frac{\epsilon}{\lambda} I$.

Through this relationship, it is possible to calculate the charge of a storm cloud from the current production and vice versa and to apply the relate results of the calculation of electrostatics and current theory to each other.

2. A surface charge Q^0, which shields the effect of the cloud space charge Q outward for the most part (90%), forms on the surface of the space charge cones of the Simpson storm model due to the jump in conductivity at the border of the cloud atmosphere and the cloud-free atmosphere. The sum of space charge Q and the counter-polar surface charge Q^0 provides the effective charge Q', which can be appropriately used for the calculation of the potential and field distribution in the outer space. It is namely then that the conductivity of the cloud is eliminated from the formula. It does not even need to be known for the calculation of the outer fields.

3. An atmospheric space charge forms outside of the cloud due to the conductivity increasing with the altitude through the current in such a way that finally all current lines exiting form the current source bend in the direction of the positive x-axis, i.e. in areas with a high conductivity.

The reason for this can be seen by the fact that a high percentage of the cloud charge flows toward the ionosphere and expands there around the entire earth. As a result, areas that are further away from the storm can be supplied with current.

IV. The Precipitation Cloud as a Point Source above the Conductive Plane

After we have personally observed a cloud pole in the previous section, we would like to come to the complete electrical circuit of the cloud generator in its simplest form in this section. We would like to present it mathematically through a positive point source above a conductive plane. In reality, the storm generator has multiple space charge or source areas of a different polarity that are arranged one on top of the other. However, if the calculation for the current of a source point exists, the current is simply revealed for a random arrangement of source points through a superposition. Later, we will see that, with the exception of the cloud itself and its approximate environment, even the complicated storm structure only features the current diagram of a single point source. We will return to this in greater detail in Section VIII with regard to the current of the equivalent dipole.

We must envision the conventional cloud generator in such a way that the precipitation permanently carries a negative charge from the cloud to the earth. Therefore, the cloud obtains a constant positive charge influx of +*I* and the earth obtains a constant charge influx of -*I*, which then represents the current production of the positive and negative source. However, the precipitation itself, i.e. the convection current, will not be gathered by the mathematical image. Because it represents a negative space charge, however, provided it is located in the area between the cloud and the earth, we must attempt to estimate its effect. Based on measurements of the field direction and the precipitation polarity at Zugspitze, J. Küttner concludes that the precipitation itself does not produce any significant field of its own because precipitation and field have a different polarity in the overwhelming majority of the cases. Accordingly, the field cannot originate from the precipitation charge, but rather only from the counter-polar cloud charge. This experimental finding becomes understandable by the fact that due to its high velocity of descent the precipitation represents a greatly extended space charge, which cannot compete with the compact cloud charge due to its low charge density. This phenomenon known as a mirror-image effect [12] is a frequently observed fact, such that we would also like to ignore the field effect of the convection current in the calculation.

For a constant conductivity $\lambda_0 = \lambda_h$, -I_N becomes 0 and $-I_V = -I$, i.e. in this case, the storm would provide no contribution to the fair weather field. The total current production of the earth would be depleted in order to absorb the current flowing from the cloud generator to the earth.

We can also form an analogous concept on the basis of electrostatics. The positive charge of the storm cloud +Q would bind the entire negative ground charge –Q without the negative space charge of the atmosphere in the storm surroundings and preferably concentrate on the part of the earth's surface lying below it. However, a negative space charge forms through the increase of conductivity of the atmosphere with the altitude, aside from the positive space charge cone below the source, which shields the effect of the positive cloud charge +Q on the earth. As a result, a part of the negative ground charge –Q is bound through the cloud charge, while the other part is distributed evenly on the surface of the earth and provides the contribution of the cloud generator to the atmospheric electrical fair weather field.

This negative space charge of the atmosphere, which is preferably arranged on the side and above the cloud pole, explains the already frequently observed minimal remote effect of the storm cloud with regard to the surface field strength through its shielding or compensating effect. However, the questions, which are related to the compensating layer or to the influence on the surface field strength through the uneven distribution of the storm on the earth, are answered at this point.

In the electrostatic interpretation as well as in that associated with current theory, at the moment, the ionosphere as a compensating layer does not even occur. New light will be shed on this result later through Figure 7 of Section VIII, in which the current lines of the entire electrical circuit are presented with or without the presence of the ionosphere. The current path in the area between the earth and the ionosphere is the same in both images. The ionosphere is not even necessary for the even distribution of the non-bound ground charge $-Q_N$ or of the non-bound primary current $-I_N$. The earth itself is the effective compensating layer. The process is the same as for a charged sphere, on which the charge distributes itself evenly above the surface of the sphere. With this line of vision, it is easy to understand that, first of all, the changes of the conductivity in the ionosphere do not have the expected influence on the electrical field on the earth's surface, and secondly, that even an irregular distribution of the storm on the earth cannot influence the fair weather field. We could concentrate all storms at a single location of the earth without hesitation and not disturb the even distribution of the non-bound primary current $-I_N$ or the charge $-Q_N$.

H.W. Kasemir

The calculation itself has been conducted in the appendix, such that we can limit ourselves at this point to the indication of the mathematical approach. The potential function for a point charge above a conductive plane can be obtain through the known procedure of the reflection of the charge on the plane. This method also leads to the goal in the current theory, even if the conductivity – as in our case – increases with a positive x-direction. However, the current production of the image source according to the figure is then smaller than that of the original source. In the case of a constant conductivity or with electrostatic problems, the original charge or source and the reflection are inversely equal in size. If we identify the current production of the image source with $-I_V$ and that of the original source as usual with I, the correlation between the two exists (Appendix AII (39))

(16)
$$-I_V = -\frac{\lambda_0}{\lambda_h} I ,$$

i.e.

The volumetric heat generation rate of the image source reacts to the original source like the conductivity on the plane to the conductivity on the location of the original source.

This simple mathematical relationship is of great significance in its effect on the atmospheric electrical circuit. Only through this is it possible that the outer electrical circuit is formed, which distributes the charges delivered by the generator evenly on the earth and as a result also forms a fair weather field in areas far from the storm. We will now analyze in detail how this electrical circuit materializes.

According to the definition of the cloud generator provided in the introduction and that, which was mentioned above regarding the charge influx to the cloud and to the earth, the current production of the cloud pole and the earth is inversely equal in size, namely $+I$ or $-I$. Equation (16) reveals that of the total current production $-I$ of the earth, only a fraction $-I_V$ of this is used to absorb the charges flowing from the cloud pole to the earth. We will call this fraction $-I_V$ "negative leakage current". The other part of $-I$, which we will identify with "negative primary current" $-I_N$, is distributed evenly over the surface of the earth and represents the contribution of the cloud generator to the atmospheric electrical fair weather vertical current. The following relationship applies for it according to (15):

(17)
$$-I_N = -I - (-I_V) = -\left(1 - \frac{\lambda_0}{\lambda_h}\right) I .$$

Through the effect of the conductivity increasing with the altitude, a significant fraction of the cloud current $+I_N$ is detracted to the positive x-direction toward $+\infty$, and only the remainder flows down to the earth, namely $+I_V$, where it then also binds a respectively small portion $-I_V$ of the total current production $-I$ to the earth. The free portion $-I_N$ is then the negative primary current of the generator.

H.W. Kasemir

This result applies even for a complicated storm structure from multiple source areas and inevitably results from the division of the electrical circuit in the currents $+I_V$, $-I_V$ and the currents $+I_N$, $-I_N$ according to equations (16) and (17). Therefore, a division according to the term of the generator is suggested through the mathematical approach to the problem, and we would now like to clearly emphasize both electrical circuits in the current line diagram of point source Figure 2a as well as in the equivalent circuit diagram Figure 2 associated with electrical engineering.

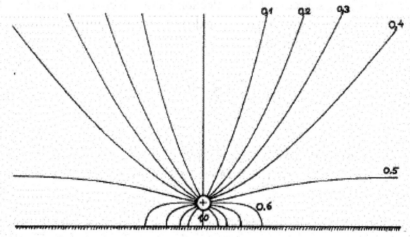

Figure 2a: Current line diagram of a point source above a conductive plane

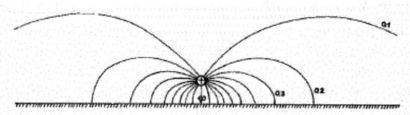

Figure 2b: Field line diagram of a point source above a conductive plane

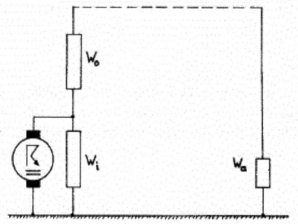

Figure 2c: Equivalent circuit diagram of the simplified storm generator

Compared to Figure 2a, Figure 2b illustrates the field line diagram of a point charge above a conductive plane. A division of the field line cannot be noticed here. They all pull towards the earth in greater or small arcs.

The current line diagram is limited to the approximate environment of the point source. In this process, the altitude of the point source above the earth's surface is selected in Figure 2a in such a way that $\lambda_h = 2\lambda_0$. According to (16) and (17), for that reason half of the current production of the point source is omitted for the primary current as well as for the leakage current. This is expressed in the current line diagram by the fact that half of the current lines with the figures of 0 to 0.5 flow into the positive x-direction toward $+\infty$ according to the primary current $+I_N$, while the other half of the current lines 0.6 to 1 flow down to the earth according to the leakage current $+I_V$. Seen from the earth, the current lines of the negative leakage current $-I_V$, which are identical to those of the positive leakage current $+I_V$, in the approximate environment of the cloud pole ascent to it. The negative primary current $-I_N$, in contrast, was distributed over the entire surface of the earth and its current line density became so small in the process that its presentation in Figure 2a is no longer possible. The current lines of the negative primary current extend vertically form the surface of the earth to positive infinity and unite there with the current lines of the positive primary current.

To simplify the concept for the construction of the equivalent circuit diagram pursuant to electrical engineering, we need to avoid connecting the current over infinity by the fact including the term of the ionosphere as that of the upper connecting layer (hashed line in Figure 2c). Then the positive and negative primary current will not flow to infinity, but rather only up to the ionosphere in order to neutralize there. With this introduction of the ionosphere, the old concept of the current in the spherical condenser ionosphere – earth for the fair weather vertical current regains its meaning. This spherical condenser current occurs in our case as a partial element in the current course of the primary current.

In the circuit pursuant to electrical engineering, Figure 2c, we have marked the resistance symbol W_i for the resistance of the atmospheric space between the cloud pole and earth, through which the leakage current flows from the positive cloud pole to the earth. Parallel to W_i is the resistance of the primary electrical circuit, which consists of the partial elements W_0 between the cloud pole and ionosphere and resistance switched behind that W_a of the entire remaining atmosphere. The circuit diagram serves in the main to be able to arrange the respective terms from electrical engineering to the three-dimensional current line diagram. Its fruitfulness will be shown later if the calculate the resistance values from the atmospheric conductivity and therefore are able to dominate the current of the storm generator without solutions to differential equations.

Before concluding this section, we must still deal with the earth potential and the ground charge. In electrical science, the potential if defined as the work, which must be provided upon charging the unit, in order to bring them from infinity to the starting point in the given electrical field. In this connection, the potential value of the infinite is set to 0^2.

---

[2] This determination expresses that the effect of the field producing charges is decreased to 0 at an infinite distance. The potential function decreases with $1/r$ and $1/r^2$. This determination loses its sense if, as in the case of an infinitely large plate condenser, the charges themselves extend to infinity and their sums are infinitely large. Then one must define the 0-value of the potential function differently. Because in our case, however, the charge quantities are limited and all lie in the finite range, we must keep the potential value of 0 by definition at infinity.

H.W. Kasemir

The calculation in cylinder coordinates reveals the following term (A II (40)) for the potential function of the source point above the conductive plane:

$$(18) \quad \Phi_1 = \frac{I}{4\pi\lambda_0}\left[\frac{e^{-k(\sqrt{(x-h)^2+R^2}+x+h)}}{\sqrt{(x-h)^2+R^2}} - \frac{e^{-k(\sqrt{(x+h)^2+R^2}+x+h)}}{\sqrt{(x+h)^2+R^2}}\right]$$

However, we must take note here that while this potential function reveals the entire current production of the positive source $+I$, although from the negative current production $-I$ it only takes the leakage current $-I_V$ into account. The potential function Φ_2 of the primary current $-I_V$ will therefore still (18) have to be overlapped. Because, however, the earth may no longer be depicted as a plane, but rather as a sphere, we would like to discuss the two potential functions separately.

In (18), for $x = 0$ we also obtain $\Phi_1 = 0$. Our conductive plane, therefore, obtains the potential value of 0 and because it reaches into infinity, this is compatible with the potential value of 0 in the infinite. However, we must ask ourselves - will the potential value remain intact if we introduce a large conductive sphere instead of the conductive plane for the representation of the earth? Will our relationship (16) and (17) also remain applicable, upon which we have built the conclusions of this section? The cumbersome calculation in spherical coordinates for this case was conducted in [5], and it was demonstrated that the equations (16) and (17) are exceptionally beneficial approximation equations, which we can continue to use without hesitation. As well, [5] shows us that in the case of the current of a point source above a conductive sphere, it obtains the potential value of 0 as well if we only take the portion $-I_V$ of the negative current production of the earth into account.

If we supplement the current production of the earth by the primary current $-I_N$ to $-I$, however, then the earth obtains a negative potential through this. According to (5), we can calculate the negative ground charge for the current production $-I_N$ and obtain the potential of the earth Φ_E with the capacity of the earth $C = 4\pi\varepsilon R_0$ (R_0 = the radius of the earth) for

$$\Phi_E = -\frac{-Q_E}{C} = -\frac{I_N}{4\pi\lambda_0 R_0}$$

In this connection, I_N however means only the portion of the fair weather vertical current originating from a storm. To obtain the actual potential of the earth, we must add up the primary currents of all simultaneously occurring storms on the earth in I_N. If we wanted to follow the previous custom and randomly set the earth's potential to 0, I_N would also become 0 and constant contradictions would result in the calculation with basis ideas and experiences. For example, from the storm we would have 'more + as − charge'. The fair weather vertical current as well as the normal atmospheric electrical field would become 0. A field would extend through the ionosphere, which would have a permanent migration of positive charge carriers as a result, etc. However, if we assume the negative potential for the earth provided by the calculation as real and trace the potential curve with the altitude, we will see that the potential over the fair weather areas merely has an infinitesimally small negative value at the altitude of the ionosphere, while that, which is above the storm areas, is positively weak. Later, we will revisit this in more detail based on Figure 5. This low potential difference would balance out to the value of 0 through the presence of the highly-conductive ionosphere. Therefore, the ionosphere as well as space lies on the potential of 0. There is no potential difference between both, i.e. a migration of the electrical charge carriers to or from will not occur. (We will naturally exclude the processes that do not belong in the atmospheric electrical circuit, e.g. corpuscular radiation from the sun, etc.) All atmospheric electrical processes take place within the ionosphere virtually like in a Faraday Cage. An effect in space is shielded by the ionosphere. Because the ionosphere has the potential value of 0, it can also be selected as a reference potential just like that of the distant infinitely. These concepts are substantiated by the present mathematical calculation of the atmospheric electrical circuit.

The most important results of this Section IV are:

1. Under the influence of the conductivity increasing with altitude, the current of the cloud generator dissolves into two circuits, the leakage current and the primary current circuit.

2. The leakage current I_V represents the conductive current in the storm cloud and its vicinity. It must be calculated from the total current production I of the cloud through the simple formula $I_V = \lambda_0/ \lambda_h$ I. (λ_0 = conductivity on the surface of the earth, λ_h = conductivity at the altitude of the cloud pole).

3. The negative primary current $-I_N$ is distributed evenly over the surface of the earth and flows as a contributor of the storm for the fair weather vertical current to the ionosphere. There it unites with the positive primary current $+I_N$, which flows from the cloud pole to the ionosphere.

4. The earth, and not the ionosphere or upper atmosphere, must be perceived as the equalizing layer for the negative primary current. The even distribution of the primary current on the earth would occur even without the presence of the ionosphere. An influence of changes of conductivity of the ionosphere on this distribution and therefore on the atmospheric electrical surface field is not there. An irregular distribution of the storms over the earth to the atmospheric electrical fair weather field is likewise without influence.

5. The old concept regarding the flow of the fair weather vertical current in the spherical condenser ionosphere-earth appropriately exists despite the dot-shaped current source of the storm generator. The current in the spherical condenser ionosphere-earth occurs as the last part of the primary electrical circuit.

6. Like space, the ionosphere also has the potential value of 0 and shields the atmospheric electrical circuit against space, such that no loss of charge carriers occurs or current paths do not extent to space. The earth has a negative potential.

V. The Storm as a Dipole Source

In this section, we will look at the representation of the storm cloud, as it is provided according to the test ascents by Simpson, Scrase and Robinson [1] [2]. In this process, we will omit the small positive space charge island at the base of the storm cloud and reproduce only the positive and negative main space charge center through a positive and negative point source. The charge mechanism of this bipolar cloud generator has not been fully clarified up to this point. In general, it is indeed assumed that the precipitation falling from the head of the cloud, which remains positive, forms the negative space charge at the base of the storm cloud and that from there the current flows to the earth primarily as a conductive current. The continually falling precipitation is no longer substantial in addition to the conductive current. J. Küttner [8] believes that two charge producing generators are effective, namely the snow dipole, which produces the positively charged storm head, and the rain dipole, which form the negative charge at the base.

We do not wish to engage in a discussion regarding these possibilities at this point, but rather for our purpose, to accept the fact of a positive charge influx in the head and of a negative at the base of the storm head as a given. According to the detailed explanations of the previous section, the calculation of this storm model is possible without difficulty. We only have to overlap the current of a positive and of a negative source point. The space charge effect of the charged precipitation particles during the descent is omitted for the same reasons as in the previous section. The earth itself does not emerge here as the source, but rather it lies in the electrical circuit of both of the sources located in the atmosphere. The electrical inflow and outflow of current to the earth surface must be equal to each other. In this regard, however, a penetration of the convective precipitation current on the earth must be taken into account. The arrangement of an equally positive and negative source in the atmosphere, thus, reflects the initial state of the storm cloud, in which the precipitation has not yet reached the earth. As soon as this is the case, we would have to include an additional generator in the manner of the previous section.

The current line diagram of our bipolar generator is reflected in Figure 3a. We can see that the current line dissolves into three groups at this point. The first group goes from the positive source upward to infinity and provides the current path of the + primary current. The second group goes from the positive to the negative source. Through this flow of current, the leakage current of the storm generator is presented. It flows preferably in the area of the storm cloud between the positive and negative space charge. The third group of current lines gravitates from the negative source to the earth. In contrast to the correlations reflected in Figure 2a for the single-pole generator, this current line group belongs to the primary electrical circuit, which clearly emerges from the equivalent circuit diagram pursuant to electrostatics in Figure 3b. The fourth group, which adds the outer current path to a closed circuit, is the one that depicts the homogenous current of the fair weather areas between the earth and ionosphere or infinity. These current lines cannot be reflected as in Figure 2a due to their very low density. They are compiled in the current portion, which flows through the resistance W_a in Figure 3b.

We will now go over how to derive the individual current portions of the bipolar generator from the primary and leakage current defined in the previous section. According to the subdivision of the current lines into 3 groups as made above, we will divide the entire space into 3 parts, the upper space, the middle space and the lower space. The upper space (index 1) represents the space that extends from the positive source to the ionosphere or to infinity. The middle space lies between the positive and negative source (index 2). In this space, there is mainly the storm cloud. The lower space (index 3) is the area between the negative source and the surface of the earth. The current production of the positive and negative source is $+I$ or $-I$. The altitude of the positive source above the surface of the earth is h_1 and the locally dominating conductivity λ_1. Accordingly, h_2 and λ_2 relate to the negative source. λ_0 refers to the conductivity on the surface of the earth. Now we must add the currents of both sources flowing in the individual spaces and obtain in space 1 the primary current of the bipolar generator I_{BN} for:

$$(19) \qquad I_{BN} = \left(\frac{\lambda_0}{\lambda_2} - \frac{\lambda_0}{\lambda_1} \right) I \, ,$$

and in space 2 the leakage current I_{BV} for:

H. W. Kasemir

$$(20) \qquad I_{BV} = -\left[1 - \left(\frac{\lambda_0}{\lambda_2} - \frac{\lambda_0}{\lambda_1}\right)\right] I \, ,$$

In space 3, we obtain once again the primary current I_{BN} though only with the opposite sing, because the current density reversed compared to space 1.

Let us now turn to a brief discussion of the results based on a practical example. The positive source is located at an altitude of 7 km. There, the conductivity is approx. 5 times as great as on the earth's surface. The negative source should lie at an altitude of 3 km, where the conductivity has increased to approx. twice the value compared to the surface of the earth. The primary current is calculated for this according to (19) for:

$$I_{BN} = \left(\frac{1}{2} - \frac{1}{5}\right) I = 0{,}3 \, I$$

And the leakage current according to (20) for:

$$I_{BN} = -0.7 \, I$$

From this, we can see that the leakage current, which is used on the inside of the storm cloud, is roughly twice that of the primary current. The negative sign indicates that the leakage current has the opposite direction of the primary current. If the negative precipitation were to fall to the earth, we would have to select the model image of the single-pole generator handled in the previous section. The leakage results in this case in 0.2 I and the primary current in 0.8 I. The efficiency is therefore worse with the bipolar generator. According to J. Kuttner, if we envision the storm from both single generators combined with the snow and rain dipole, due to the different polarity, the rain generator is switched opposite to the snow generator. The efficiency worsened in this case. If we set the charge production in the storm at approx. 1.5 C/second = 1.5 A, then the portion flowing to the ionosphere is approx. 0.5 A at the onset of the storm (bipolar illustration) and for a complete development or toward the end of the storm 1.2 A (single-pole illustration). Whether these numerical values correspond well with the dispersion of the measurements by Wait [4], we would like to place emphasis on the general conclusions. From our model image, we can draw a conclusion for the efficiency of a storm cloud, namely that it becomes better, the higher the positive source and the lower the negative source lies. The further the distance between both space charges, the greater the resistance is between them and therefore the leakage current sinks and the efficiency increases.

The most important results of Section V are:

1. By simple additions of the potential fields and currents of the sources of a multiple-source cloud generator one can easily obtain the potential function and currents even for storm models that have a complicated structure. In the case of the bipolar generator, which consists of an equally strong + and − source in the head and at the base of the storm cloud, the leakage current is calculated for:

$$I_{BV} = -[1 - (\lambda_0/\lambda_2 - \lambda_0/\lambda_1)] \, I$$

And the primary current for

$$I_{BN} = (\lambda_0/\lambda_2 - \lambda_0/\lambda_1) \, I \, .$$

(λ_1 is the conductivity at the altitude of the positive cloud pole,

H. W. Kasemir

λ_1 is the conductivity at the altitude of the negative cloud pole).

2. The efficiency of the bipolar generator becomes even better the further the two point sources are from each other. However, it is still worse than the single-pole generator, which only has one positive pole in the cloud, while the other is pulled down to the earth.

H. W. Kasemir

The Storm Generator in the Atmospheric Electrical Circuit II

By H. W. Kasemir, Neptune

Résumé: La publication traite du courant électrique au dedans et hors d'un nuage orageux. La calculation commence avec le courant d'une source punctuelle située dans un médium dont la conductibilité augmente avec l'hauteur selon une fonction exponentielle. La terre est considerée comme une plaine conductive par un procédé de réflection appropié. L'addition des courants de plusieurs sources punctuelles de différente polarité permet de représenter les conditions de charge et de courant dans quelconque nuage orageux. Donné un pouvoir de courant de 1 ampère il résulte que la charge dans la tête d'un nuage orageux sera environ 2400 coulombs. C'est cent fois la valeur de 24 coulombs supposée jusqu'à présent comme elle était calculée à la base de la théorie électrostatique. Ce contenu de charge considérablement plus grand explique, qu'un nuage orageux puisse créer une suite de décharges d'éclairs, dont chaque éclair consomme environ 30 coulombs, et, en surplus, puisse continuellement produire un courant de 1 ampère = 1 coulomb/seconde jusqu'à la ionosphère. Mais le champ électrique de cette charge de nuage est abrité à l'extérieur à 90% par une charge de surface du signe opposé d'une grandeur d'environ − 2200 coulombs, à cause du chute de conductibilité à la surface du nuage. Un autre effet abritant est aussi représenté par la charge spaciale qui se forme à la proximité de l'orage par raison de la conductibilité augmentant avec l'hauteur. Cette augmentation avec l'hauteur est aussi la raison qu'un grand pourcent de la charge positive émanant de la tête d'orage ne retourne pas à la terre tout près de l'orage mais monte à la ionosphère. Appliqué à la charge négative descendant à la terre cela signifie qu'une grande partie de la charge négative sera distribuée sur toute la surface du globe et que seulement une mince partie reste ficée au dessous de l'orage. Cela résulte dans l'image que − déviant de l'usage commun jusqu'à présent − la surface de la terre et non pas la ionosphère doit être considérée comme "couche d'égalisation (= Ausgleichsschicht)": Aussi si la ionosphère n'existait pas la charge libre serait distribuée également par la surface du globe. Par cette raison on peut comprendre que des variations de l'hauteur ou de la conductibilité de la ionosphère n'auront à peu près aucune influence sur le champs èletrique près de la surface de la terre.

Dans cette thèse le potentiel du globe a non plus la valeur zéro comme supposée jusqu'à alors mais plutôt une valeur négative calculée de la charge négative et la capacité du globe. D'autre part, le potential de la ionosphère résulte dans zéro, supposé que les générateurs électriques atmosphériques produisent des charges positives et négatives des mêmes quantités. Donné ces valeurs du potentiel au globe et à la ionosphère le circuit électrique atmosphérique devient un système complet. Si l'on insistait à partir de l'image que le globe a le potentiel zéro, on arriverait aux hypothèses additionelles et sans fondement concernant le potentiel du space mondial ou concernant un courant perpétuel et unipolaire de la ionosphère au space mondial. La représentation de l'orage par un ordre approprié de sources punctuelles permet d'examiner la validité de la représentation plus simple au sens mathématique par des sources-couches, (Quellschichten) la coupe finite des sources-couches est indentique avec la négligence des distributions marginales. Il est évident que, pour obtenir certains notions intégrales comme "courant productif (Nutzstrom)", "courant de perte (Verluststrom)", "effet (Wirkungsgrad)" etc., les deux modes de calculation produiront les mêmes résultats. Le courant productif du générateur orageux est défini comme le part de la charge orageuse négative qui se répand sur la terre et se distribue également au surface mondial et, enfin, retourne de la terre à la ionosphère comme contribution de l'orage au courant vertical. Ici, à la ionosphère, il se joint avec le courant des charges positives coulant de la tête orageuse à la ionosphère, tellement formant le complet circuit extérieur. Au contraire, le courant conducteur au dedans de l'orage est défini comme courant de perte parce qu'il ne contribue pas au champs électrique de beautemps. L'effet de l'orage est défini comme le ratio du courant productif dividé par le courant total. Considérant la conductibilité diminuée au dedans du nuage l'orage moyen devrait avoir un effet d'environ 90%.

VI. The Storm as a Source Layer

The depiction of the storm generator through source points has produced an array of knowledge; however, it remains limited to the currents flowing in the storm. We cannot answer the question regarding the voltage of the cloud pole because the potential of an infinitely small point source therein will itself become infinitely large. Both the question regarding the resistances as well as that regarding the performance of the generator remain unanswered. In this case, the only resort is to randomly determine an appropriately selected equipotential area as a marginal area of the cloud pole and to define its potential value as the voltage of the generator.

H. Israël [13] takes a different approach for the calculation of the electrical portion of energy from the total thermodynamic energy of the storm. He presents the electrically-active areas of the storm cloud as source layers rather than as point sources. For the requirement of a purely homogenous current, i.e. when ignoring the edge dispersion, an extraordinarily simple model image emerges for the calculation, for which all previously inaccessible variables can be acquired. The potential of a source layer, namely, will not become infinite because the finite source strength and therefore the finite charge are not required to concentrate in an infinitely small point, but rather can expand on a finite area. The resistances can also be easily calculated because the individual current tubes have a constant cross-section in the case of the homogenous current such that the current only depends on a single coordinate, namely the x-coordinate.

It does, however, raise the question whether the requirement of the homogenous current, which presents a compulsory intervention in the natural current, does not fundamentally falsify the results. Through the precise calculation of the source point current, we are able to review the reliability of the depiction of the swelling layer with the homogenous current. In this process, we will see surprisingly that both types of calculations lead to exactly the same result.

To derive the formulas, we do not need this time to take the cumbersome approach using the characteristic functions of the differential equation, but rather we can derive all necessary equations from the equivalent circuit diagram in electrical engineering. We will begin again with the depiction of the single-pole generator. At the altitude h_1, there is the positive source layer with the total current production I, which is evenly distributed on the area F. The surface of the earth presents the negative source layer with the total current production $-I$.

The space charge effect of the falling precipitation, which carries the charge to the source layers, will be omitted from this diagram.

The current line diagram of the swelling layer generator is depicted in Figure 4a.

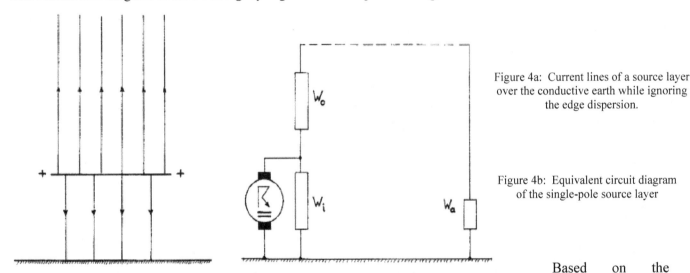

Figure 4a: Current lines of a source layer over the conductive earth while ignoring the edge dispersion.

Figure 4b: Equivalent circuit diagram of the single-pole source layer

Based on the previous sections, we know that the current I flowing from the positive source separates into a portion I_N, which flows out to the ionosphere or into + infinity, and into a portion I_V, which flows from the source layer to the earth. The current production of the earth also separates into a current $-I_V$, which flows from the part of the earth's surface located under the positive source layer and a current $-I_N$, which - while dispersing evenly over the entire surface of the earth – flows out to the ionosphere or infinity. The portion of current $-I_V$ serves to neutralize the portion $+I_V$ of the positive source layer, while the portion $-I_N$ represents the contribution of the generator to the atmospheric electrical fair weather vertical current and serves to neutralize $+I_N$.

Based on the ratios for the point source discussed in detail, we can design a completely analogous equivalent circuit diagram according to Figure 4b with the resistances W_i between the source layer and the earth, W_0 between the source layer and ionosphere and W_a as the resistance of the remaining atmospheric space earth-ionosphere. If we initially only analyze the cylindrical element of the atmospheric space with the cross-section F of the source layer, which extends below and above this from the earth to the ionosphere or to infinity, with the resistances W_i and W_0. The separation of the current production $+I$ into the portions $+I_V$ and $+I_N$ must correspond to the variables of the resistances W_i and W_0. These resistances can be calculated from the conduction function of the atmosphere through an easy integration.

I/λ is the specific resistance of a mass volume of atmosphere at the altitude x. Therefore, we obtain the resistance W of a cylinder form the cross-section F and the length h_n to h_m for

$$(21) \qquad W_{nm} = \frac{1}{\lambda_0 F} \int_{h_n}^{h_m} e^{-2kx} dx = \frac{1}{\lambda_0 2kF} \left(\frac{\lambda_0}{\lambda_n} - \frac{\lambda_0}{\lambda_m} \right).$$

This results in the following:

$$(22) \qquad W_0 = \frac{1}{\lambda_0 2kF} \frac{\lambda_0}{\lambda_1} \quad {}_3$$

and

$$(23) \qquad W_i = \frac{1}{\lambda_0 2kF} \left(1 - \frac{\lambda_0}{\lambda_1} \right).$$

Because the current I_V and I_N have to behave inversely like the resistances W_i and W_0, (22) and (23) produce the following result

$$(24) \qquad \frac{I_N}{I_V} = \frac{W_u}{W_0} = \frac{\lambda_1}{\lambda_0} - 1.$$

If we also take into account that $I_N = I - I_V$, we obtain

$$(25) \qquad I_V = \frac{\lambda_0}{\lambda_1} I \qquad \text{und} \qquad I_N = \left(1 - \frac{\lambda_0}{\lambda_1} \right) I,$$

thus, the same relationship as with the point source.

At first glance, it may seem surprising that the same result emerges for the division of the total current strength in the leakage and primary current in the case of the point source and the swelling layer. However, we can understand easily if we consider that we can imagine the swelling layer is formed from nothing but individual point sources. In this connection, even the edge dispersion would also be taken into account. Therefore, we can see that the omission of the edge dispersion, i.e. the compulsory introduction of a homogenous current does not change anything in the ratio of primary and leakage current.

---

[3] Because it makes practically no difference whether we extend W_0 from the source layer to infinity or only to the ionosphere, the integration was extend to infinity with consideration for the simpler formula.

H. W. Kasemir

Due to the fact that, in our cylindrical element, we are aware of the current strength as well as the resistances in the individual segments, it is also possible to calculate the voltage drop over a given element of resistance and thus the potential function of the current. The following applies for the area between the earth and the source layer:

(26)
$$\Phi_u = \frac{I}{\lambda_1 F 2k}\left(1 - \frac{\lambda_0}{\lambda}\right) \quad \text{for} \quad 0 \leq x \leq h_1 ,$$

and the following applies for the area between the swelling layer and the ionosphere:

(27)
$$\Phi_0 = \frac{I}{\lambda F 2k}\left(1 - \frac{\lambda_0}{\lambda_1}\right) \quad \text{for} \quad h_1 \leq x \leq \infty .$$

If we apply $\lambda = \lambda_1$, we obtain the terminal voltage U_k of the storm generator from both equations for

(28)
$$U_k = \frac{I}{\lambda_1 F 2k}\left(1 - \frac{\lambda_0}{\lambda_1}\right) .$$

For the numerical values $I_N = 1\,A,\ h_1 = 7\,km,\ \lambda_1 = 5\lambda_0,\ \lambda_0 = 2 \cdot 10^{-14}\,\frac{1}{m\,Ohm}$,

$2k = \frac{\ln 10}{10\,km},\ F = 50\,km^2$ yields the terminal voltage of value $U_k = 8.7 \cdot 10^8$ volts.

Until now, we have only observed the current of the original source with the current production $+I$ and its underlying element of the earth's surface with the current production $-I_V$. Thus, we also need to supplement the leakage current $-I_V$ flowing from the earth through the primary current $-I_N$ to the total current production. As already mentioned, however, we can no longer calculate the potential function in cylinder coordinates in this case, but rather we must revert to the presentation in spherical coordinates. We will obtain the potential ϕ_N of the fair weather vertical current for

(29)
$$\Phi_N = \frac{I_{NG}\, e^{2kR_0}}{4\pi\lambda_0}\left[2k\, Ei_{(-2kr')} + \frac{e^{-2kr'}}{r'}\right]_\infty^r$$

or in sufficiently precise approximation

(30)
$$\Phi_N = \frac{I_{NG}\, e^{-2k(r-R_0)}}{4\pi\lambda_0}\cdot\frac{1}{r} .$$

In the process, I_{NG} indicates the sum of all primary currents of the storms occurring simultaneously on the earth, R_0 the radius of the earth and r the spherical coordinates. Upon transitioning to our cylinder coordinates, we have the relationship $r = R_0 + x$.

In Figure 5, the course of ϕ_N is marked with a hashed line; the course of ϕ_u and ϕ_0 is marked with a hash-dotted line. Finally, the overlapping of both potential functions dominating in the storm is reflected through the solid line.

H. W. Kasemir

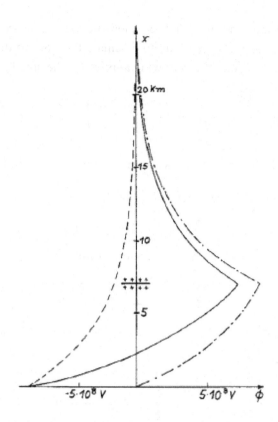

Figure 5: Potential curve with the altitude in the fair weather area and in the storm (single-pole generator)

We can see that the potential in the storm area begins on the surface of the earth with the value $-7.5 \cdot 10^8$ V, increases rapidly with altitude, then penetrates through 0 at an altitude of approx. 2.3 km and achieves the maximum value of $+7.2 \cdot 10^8$ V in the source layer. The potential decreases with an increasing altitude over the source layer and approximates the value 0 asymptotically. At an altitude of 100 km, the potential has achieved the value of 0 with the exception of one vanishing remainder. The potential curve ϕ_N of the fair weather area marked with a hashed line also decreases heavily from the surface value $-7.5 \cdot 10^8$ V with the altitude and strives toward the potential value of 0. This fact already discussed in Section IV can be seen clearly in the potential diagram of Figure 5.

The most important results from Section VI:

1. Because the presentation of the storm generator essentially only reveals the current course in both current spheres through point sources, though it does not permit the calculation of the terminal voltage of the generator, the resistances and the performance, instead of a point source, a source layer with a compulsory homogenous current is used as a model image.

2. The calculation of the potential and current function can be arranged just as extraordinarily easy as that of the terminal voltage, resistances, etc. Moreover, it is revealed that both storm representations common in literature lead to the same result with regard to the division of the total current production into primary and leakage current.

3. The potential curve presented graphically in the storm model and in the fair weather area illustrates the idea of the atmospheric electrical current in the Faraday cage of the ionosphere explained in Section IV.

VII. The Storm as a Dipole Source Layer with Consideration of the Decreased Cloud Conductivity and the Efficiency of the Various Storm Models

According to the statements in Section V regarding the source layer and point source, the transfer of the results from the bipolar point source to the bipolar source layer does not pose any difficulties. With regard to the calculation of primary and leakage current, the point source and the source layer lead to the same result. In addition, the resistances and the terminal voltage of the generator can be calculated with the depiction of the source layer. We will continue to define the efficiency w of the generator as the ratio of primary current to leakage current $w = \dfrac{I_N}{I}$. Taking the previous sections into account, we come to the following formula arrangement. s

For the source point and the source layer, the following applies:

(31)
$$I_V = \frac{\lambda_0}{\lambda_1} I \; ; \quad I_N = \left(1 - \frac{\lambda_0}{\lambda_1} \right) I \; ; \quad w = 1 - \frac{\lambda_0}{\lambda_1} \; .$$

The following applies for the source layer alone:

(32)
$$U_K = \frac{I}{\lambda_1 \, 2 k F} \left(1 - \frac{\lambda_0}{\lambda_1} \right) \; ;$$

$$W_i = \frac{I}{\lambda_0 \, 2 k F} \left(1 - \frac{\lambda_0}{\lambda_1} \right) \; ; \quad W_0 = \frac{I}{\lambda_0 \, 2 k F} \, \frac{\lambda_0}{\lambda_1} \; .$$

The following applies for the bipolar source from point sources or source layers:

(33)
$$I_{BV} = \left(1 - \frac{\lambda_0}{\lambda_2} + \frac{\lambda_0}{\lambda_1} \right) I \; ; \quad I_{BN} = \left(\frac{\lambda_0}{\lambda_2} - \frac{\lambda_0}{\lambda_1} \right) I \; ;$$

$$w = \frac{\lambda_0}{\lambda_2} - \frac{\lambda_0}{\lambda_1} \; .$$

The following applies alone for the bipolar source layer:

$$U_K = \frac{I}{2kF} \left(\frac{1}{\lambda_0} + \frac{1}{\lambda_1} - \frac{1}{\lambda_2} \right) \left(\frac{\lambda_0}{\lambda_2} - \frac{\lambda_0}{\lambda_1} \right) ;$$

$$W_u = \frac{1}{\lambda_0 \, 2kF} \left(1 - \frac{\lambda_0}{\lambda_2} \right) ;$$

(34)

$$W_i = \frac{1}{\lambda_0 \, 2kF} \left(\frac{\lambda_0}{\lambda_2} - \frac{\lambda_0}{\lambda_1} \right) ;$$

$$W_0 = \frac{1}{\lambda_0 \, 2kF} \frac{\lambda_0}{\lambda_2} .$$

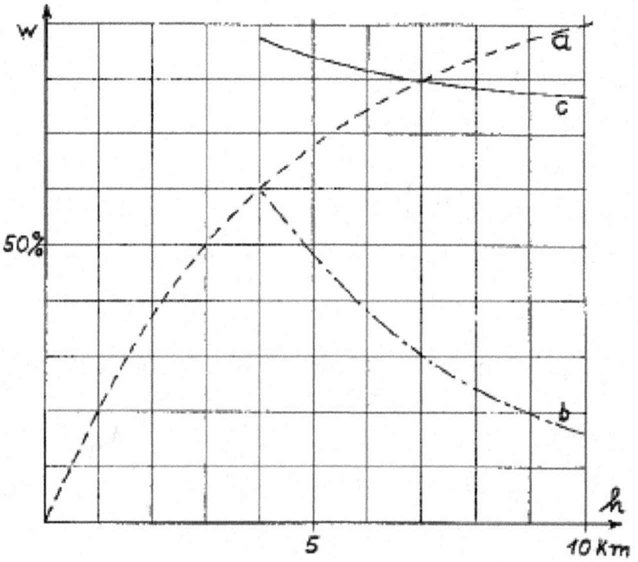

Figure 6: The efficiency of the storm generator dependent upon the altitude of the source areas over the surface of the earth
a) The single-pole generator
b) The bipolar generator
c) The bipolar generator with consideration for the lower cloud conductivity

In Figure 6, the efficiency for the single-pole generator (hashed line) and the bipolar generator (hashed-dotted line) is presented graphically in its dependence on the altitude over the surface of the earth. In the process, a constant distance of both source layers (or source points) of 4 km is assumed for the bipolar generator. Figure 6 demonstrates that the single-pole generator has an even better efficiency the higher the positive pole is above the surface of the earth. That is quite understandable because the inner final resistance W_i becomes increasingly greater with increasing altitude and the outer resistance lying above the source becomes increasingly smaller such that the primary current increases at the expense of the leakage current. In the case of the bipolar generator, the curve first begins with the altitude of the positive source layer of 4 km because the negative layer just comes into contact with the earth with a distance between the positive and negative source layer of 4 km. The efficiency at this point is 60% - precisely that of the single-pole generator. It becomes continually worse with an increasing altitude. If the positive pole has reached the altitude of 10 km and the negative pole is accordingly at an altitude of 6 km, the efficiency has decreased to 16%.

This result is somewhat contradictory to the experimental findings that the tropical storms, which extend up to an altitude of 10 and more kilometers, constitute a primary part of the atmospheric electrical generator. This can be recognized in the universal time diurnal variation of the atmospheric electrical potential gradient, in which the application of the tropical storm centers rises significantly. Therefore, it would be incomprehensible if the tropical storm were to have a poor efficiency. According to our previous calculation, however, this only applies for the formation of a storm with two space charge centers, while the efficiency of the single-pole generator increases with altitude. Thus, we must take another critical look at the requirements for the calculation of the bipolar generator. In doing so, one can immediately object to the fact that the decreased conductivity of the cloud air was not taken into account.

All previous results imply that the efficiency increases with a growing inner final resistance. This is the case if the oppositely charged sources are a far from each other as possible and if the conductivity of the intermediate medium is low. Thus, we must also quantitatively analyze how strong this decreased conductivity affects the efficiency of the bipolar generator.

The calculation for this is not any more difficult, although it is somewhat more cumbersome and therefore it is conducted in the appendix. From that point, we will assume the equations A IV (58) (59) (60) (67) for the resistances W_0, W_u, W_i and the efficiency w. In this context, h_1 and h_2 stand for the altitude of the positive or negative source layer. λ_0, λ_1, λ_2, λ_3 is the conductivity of the earth's surface, at the altitude of the upper source layer, at the altitude of the lower source layer respectively in the cloud.

This reveals the following result:

$$(35) \qquad W_0 = \frac{1}{\lambda_0 \, 2kF} \frac{\lambda_0}{\lambda_1}$$

$$(36) \qquad W_u = \frac{1}{\lambda_0 \, 2kF} \left(1 - \frac{\lambda_0}{\lambda_2} \right)$$

$$(37) \qquad W_i = \frac{h_1 - h_2}{\lambda_3 \, F}.$$

The primary current of the bipolar generator according to A IV (65) is still:

$$(38) \qquad I_{BN} = \frac{W_i}{W_i + W_0 + W_u} \, I$$

and the leakage according to A IV (66) is:

$$(39) \qquad I_{BV} = - \frac{W_0 + W_u}{W_i + W_0 + W_u} \, I$$

Thus, the efficiency for the following is revealed:

H. W. Kasemir

$$(40) \qquad w = \frac{h_1 - h_2}{h_1 - h_2 + \dfrac{\lambda_3}{2k\lambda_0}\left(1 + \dfrac{\lambda_0}{\lambda_1} - \dfrac{\lambda_0}{\lambda_2}\right)}$$

The equation (38) and (40) were achieved in the calculation conducted in the appendix according to the method of the overlapping of the current of two single-pole generators, as has already been described in detail in Section V for the bipolar point source generator. According to the equivalent circuit diagram of the bipolar generator provided in Figure 3, we can take them directly from the circuit diagram. In this case, we are dealing with the branching of current in both resistors switched in parallel W_i and $W_0 + W_u + W_a$. In this process, it is necessary to ignore the resistance W_a against $W_u + W_0$ because, in reality, it is approx. 6 decimal powers small than $W_u + W_0$. We can obtain the resistance W of the parallel switch from W_i and $W_u + W_0$ for:

$$(41) \qquad W = \frac{W_i (W_0 + W_u)}{W_i + W_0 + W_u}$$

Due to the fact that the total current I as well as both partial currents IBV and IBN always produce the generator voltage on the resistances they flowed through, the immediate result is:

$$I_{BN}(W_u + W_0) = I \frac{W_i (W_0 + W_u)}{W_i + W_0 + W_u}$$

or the following as well:

$$I_{BN} = \frac{W_i}{W_i + W_0 + W_u} I$$

In this simple and clear calculation, the value of the previous considerations is displayed, which enable the return of the three-dimensional current of the atmospheric electrical circuit to a simple circuit diagram associated with electrical engineering. The problem that seems to be nearly unsolvable with the current – taking the cloud conductivity in the calculation into account as well – can be writing in a few lines based on the equivalent circuit diagram.

According to equation (40), the efficiency of our bipolar generator is entered as a thick, solid line in Figure 6 with the above-selected constant distance 4 km between the swelling layers (taking the decreased cloud conductivity $\lambda_3 = \frac{1}{3}\lambda_0$ into account). Our expectations have been fully met. Through the increased inner final resistance, the efficiency has increased substantially and in the process, it has also become relatively independent from the altitude. In the case of a change in altitude of 4-10 km, the efficiency only drops from 87% to 78%, while it decreases from 60 to 16% without consideration for the decreased cloud conductivity. We can see just how fundamentally important it is to take the decreased conductivity in the storm cloud into consideration.

Even if the efficiency calculated at the end represents a maximum value and will turn out somewhat lower in reality due to the omission of the edge dispersion, which indeed is located in the area of normal atmospheric conductivity, we will still be able to maintain that on average we will have to take an efficiency of 60-70% into account even for the bipolar storm generator.

H. W. Kasemir

The most important results from Section VII:

1. The compilation of the formulas for the calculation of the storm generator with one or two source points or swelling layers on p. 72.

2. The efficiency becomes even greater the stronger the inner final resistance becomes. This increases with an increasing distance of the poles or with the decrease of the conductivity between the poles.

3. The efficiency of the storm generator is calculated under consideration for the decreased cloud conductivity. Based on a graphic representation of the efficiency for the various ratios in Figure 6, one can conclude with an average efficiency of 60 to 70%.

4. Particularly in the case of a calculation of the tropical storm, which extends up to altitudes of 10 km and higher, taking the low cloud conductivity into account is necessary. Without this, the tropical storms would only have the poor efficiency of 16%. Due to the low cloud conductivity, the inner resistance of the storm generator increases and, with that, its efficiency increases to approx. 80%.

5. The return of the atmospheric electrical three-dimensional current to an equivalent circuit diagram associated with electrical engineering enables an extraordinarily simple calculation of even a complicated storm model, in which even the decreased conductivity of the cloud can be taken into account.

VIII. The Long-Distance Effect of a Randomly Structured Storm

Previously, we have observed the current in the storm generator itself or the fair weather vertical current from a very distant storm. Thus, we must now concentrate on the intermediate area, in which the current lines expand form the center of the storm in order to finally be able to merge into the current lines of the negative primary current running vertical to the earth's surface. In the current line diagrams in Figure 7a and 7b, we will later see that this area lies high above the altitude of the ionosphere if we do not take the ionosphere into account in the calculation. If we apply the high conductivity of the ionosphere to our calculation as well, current lines will no longer penetrate beyond it and we will arrive at the diagram of the current in a very narrow spherical condenser. We can then bend this up without noticeable error in a plate condenser equal to the surface of the earth's sphere with a radius twice as large as the radius of the earth.

Observed from such a distance, our source point or source layer would virtually shrink with its reflection to a dipole at an altitude of 7 km above the surface of the earth. Therefore, we want to try to derive the potential and current line function for a dipole current with the given conductivity distribution. This can be done, as demonstrated in the appendix, through a transition with the development of the potential function (3) in a Taylor series, and we obtain the equation according to A III (48, 49) for the potential function ϕ_D of our dipole

$$(42) \qquad \Phi_D = \frac{I_N}{4\pi\lambda_0 k} \; \frac{x(1 + k\sqrt{x^2 + R^2})\, e^{-k(\sqrt{x^2+R^2}+x)}}{(x^2 + R^2)^{3/2}}$$

219

or in polar coordinates:

$$(43) \qquad \Phi_D = \frac{I_N}{4\pi\lambda_0 k} \frac{\cos\theta\,(1+kr)\,e^{-k(1+\cos\theta)r}}{r^2}.$$

The expression I_N/k is the dipole moment M of the dipole. It is the primary current I_N of the source point or the swelling layer related to k. When transferring the source point current to the dipole current, therefore, we must apply the following:

$$(44) \qquad M = \left(1 - \frac{\lambda_0}{\lambda_1}\right)\frac{I}{k} - \frac{I_N}{k}$$

Upon transitioning $k = 0$, we achieve the familiar formula for the dipole current in a constant conductivity.

If we analyze the x = 0-plane, i.e. the surface of the earth, we can see from the equation (42) that the potential function assumes the constant value 0. As a result, we can replace this equipotential area with the conductive surface of the earth without changing the potential function. If we form the orthogonal trajectories (Appendix A III (52)) from equation (30), we will obtain the equation of the current function Ψ_D.

$$(45) \qquad \Psi_D = I_N\left\{1 - \frac{1}{2}\left[\cos\theta\,(1+\cos\theta) - \frac{\sin^2\theta}{2kr}\right]e^{-2kr(1-\cos\theta)}\right\}.$$

This reveals the following interesting fact. If we set $\theta = 180°$, we will obtain $\Psi_D = I_N$. With $\theta = 180°$, we have gathered all current lines emitting from and merging into the dipole. If the positive and negative source, of which the dipole is comprised, were equally great as, e.g. in the case of the dipole current, it is in a constant conductivity; then we would have to obtain the value of 0 for Ψ_D. However, in this case the value $\Psi_D = I_N$ demonstrates that the is unsymmetrical, i.e. that the current production of the positive source outweighs that of the negatives by the amount of I_N. In this respect, that is not surprising because we formed the dipole from 2 unequally sized sources prior to our transition. Therefore, this characteristic was preserved despite the transition. To obtain a current balance, we must allow the additional current - I_N to flow from the surface of the earth, over which it is evenly dispersed, just as with the single-pole source. Thus, all considerations of the previous sections remain intact.

If we intend to calculate the equivalent current dipole for the storm generator formed by 2 source points or swelling layers, we must simply overlap the equivalent current dipoles of the positive and negative source. That formally implies that we must use the primary current I_{BN} (33) for this combination to calculate the equivalent dipole-moment. This becomes

$$(46) \qquad M = \frac{I_{BN}}{2k} = \frac{I}{2k}\left(\frac{\lambda_0}{\lambda_2} - \frac{\lambda_0}{\lambda_1}\right).$$

H. W. Kasemir

With this calculation scheme, it is easily possible to provide the equivalent dipole-moment for storm models, which are formed by more than 2 source points or swelling layers. Only the primary current of the storm model is decisive for the current at a somewhat greater distance, such that all types of storm illustrations flow together into one uniform current diagram.

Figure 7b: Current lines of the equivalent dipole with consideration for the ionosphere

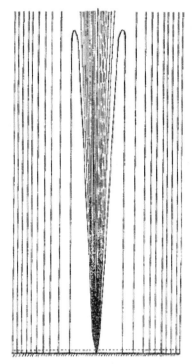

Figure 7a: Current lines of the equivalent dipole current without consideration for the ionosphere.

In Figure 7a, the closed electrical circuit of a storm generator is presented without consideration for the ionosphere. The ionosphere is marked with a hashed line. We can immediately see that the current paths extend far beyond the ionospheric altitude only to bend sharply and thrust vertically toward the earth's surface. Therefore, with the exception of the storm area itself, a homogenous current like that in a plate condenser is present in the space between the ionosphere and the earth. The fact that the distance of the vertical current lines from each other consistently decreases with increasing distance from the storm is justified as follows.

The same fraction of the total current merges into every circular ring that is cut from the surface of the earth by two adjacent current lines during rotation. To ensure that the current density remains constant in the individual circular rings, these areas must be equal. This can only be achieved through radii that become steadily smaller, through which the current lines move closer together with an increasing distance from the dipole source.

The current line diagram of the dipole current is depicted in Figure 7b if one sets the conductivity outside of the ionosphere equal to 0. In this process, the current lines in the ionosphere, however, are not calculated, but rather their probable path is developed based on the following considerations. Due to the dramatically increased conductivity upon penetrating the ionosphere, the diagonally merging current lines would be refracted to normal. However, because the current lines exiting the head of the storm and to a higher degree from the earth ascent vertically and therefore have nearly the normal direction, the penetration into the ionosphere would not significantly change the course of the current lines. Upon exiting the ionosphere in space, however, the conductivity increases heavily and as a result, the current lines refract

221

from the normal at this point. The current lines exiting the head of the storm and the earth would close in flatly expanded arcs through space. Under the assumption that space has the conductivity of 0, these arcs would even condense toward the space in the ionosphere. The conductivity of space of 0 would have to be interpreted in such a manner in this case that only an ionization of the medium would cease due to the lack of atoms or molecules, but rather also that ions or electrons could not penetrate the border between the ionosphere and space. The justification for these assumptions, however, has not been given, such that in any case, an expulsion of the current lines into space must be taken into account. However, because we are dealing with freely movable charge carriers in this case, the magnetic field of the earth would have to be considered for the calculation of their current paths. Such analyses, however, go beyond the framework of this paper, so that the current lines in Figure 7b are marked under the (physically incorrect) assumption of the conductivity of 0 in space.

The representation of the earth through a plate with the size of the surface of the earth, as in Figure 7a and 7b, is actually no longer reliable. In this case, we would have to represent the earth through a sphere. The mathematical effort for this, however, is so incomparably greater that we have foregone this in our work. The point of this current line diagram, namely the equal distribution of the current density on the surface of the earth, would also remain for the depiction of the earth as a sphere. With the presence of multiple storms, the individual current line diagrams would overlap and the current densities on the earth would add up to the resulting vertical current density.

The most important results from Section VIII.

1. In distances from the storm, which are large compared to the altitude of the head of the storm over the earth's surface, the potential and current lines of the single and multiple-pole storm model come together in a uniform diagram of the dipole current.

2. The dipole-moment M is calculated from the resulting primary current I_N of the storm model through the simple formula $M = I_N / k$. (k is the constant of the conduction function $\lambda = \lambda_0 e^{2kx}$).

3. Without the presence of the ionosphere, the current paths would ascend up to areas far above the ionosphere. When considering the ionosphere, the current paths fold due to the abrupt jump in conductivity and protract in flat arcs within the ionosphere up to the point that they leave the ionosphere to descend vertically toward the earth.

4. The homogenous current of the vertical current between the earth and ionospheric altitude depends on whether the jump in conductivity in the ionosphere is taken into account or not.

This research was conducted at the Meteorological Observatory in Aachen under the guidance of Prof. Dr. H. Israël, and the author would like to express his sincere gratitude to Prof. Israël at this point for the constant support and countless thought-provoking discussions.

H. W. Kasemir

Mathematical Appendix

A I. The Current of a Point Source in a Medium with the Conductivity $\lambda = \lambda_0 \, e^{2kx}$

Given:

The current production of the point source I.

The conductivity of the space $\lambda = \lambda_0 \, e^{2kx}$.

The conductivity of the $x = 0$ plane λ_0.

The current source lies in the middle point of the cylinder coordinate system x, R, ϕ.

Unknown:

The potential function ϕ.

The current function Ψ.

The space charge function q.

1. The Characteristic Function [Proper Function] of the Differential Equation of the Electrical Current in Cylinder Coordinates (Rotational Symmetry)

To obtain the differential equation of the current, we will assume the continuity equation of the stationary current

$$\text{(1)} \qquad \text{div } \vec{\imath} = 0$$

and the correlation

$$\text{(2)} \qquad \vec{\imath} = \lambda \vec{E} .$$

The result of (1) and (2)

$$\text{(3)} \qquad \text{div } \lambda \vec{E} = \lambda \text{ div } \vec{E} + \vec{E} \text{ grad } \lambda = 0 .$$

Due to the fact that $\vec{E} = q / \epsilon$, we obtain the general equation for the space charge from (3)

$$\text{(4)} \qquad q = -\frac{\epsilon}{\lambda} \vec{E} \text{ grad } \lambda .$$

With $\vec{E} = -\text{grad } \Phi$ we also obtain the following from (3)

$$\text{(5)} \qquad \text{div grad } \Phi + \frac{1}{\lambda} \text{ grad } \lambda \text{ grad } \Phi = 0 .$$

Due to the fact that λ only depends on the x-coordinate, only the x-coordinate of 0 differs from the vector grad λ. Thus, only the product of both x-coordinates remains from the scalar vector product grad λ grad ϕ. With $\dfrac{\text{grad } \lambda}{\lambda} = 2k$ we obtain the following from (5)

H. W. Kasemir

(6)
$$\text{div grad } \Phi + 2k \frac{\partial \Phi}{\partial x} = 0 .$$

To obtain the characteristic function of the differential equation, we will make the product approach for ϕ

(7)
$$\Phi = f_{(R)} g_{(x)} ,$$

where f alone should be a function of R and g alone a function of x. Because rotational symmetry dominates for the problems in question here, we can set the generality ϕ and the derivation ϕ to 0 without limitation. If will still write out div grad ϕ in cylinder coordinates, then we obtain the following from (6) and (7)

(8)
$$\frac{1}{f}\left(\frac{d^2 f}{d R^2} + \frac{1}{R}\frac{df}{dR} \right) + \frac{1}{g}\left(\frac{d^2 g}{d x^2} + 2k \frac{dg}{dx} \right) = 0 .$$

Thus, this equation must apply if we place the arbitrarily selectable constant value $(jy)^2$ for the first expression in (8).

(9)
$$\frac{1}{f}\left(\frac{d^2 f}{d R^2} + \frac{1}{R}\frac{df}{dR} \right) = (jy)^2 .$$

If equation (9) is applied to (8), this results in a linear homogenous differential equation with constant coefficients for g.

(10)
$$\frac{d^2 g}{d x^2} + 2k \frac{dg}{dx} - y^2 g = 0 .$$

Its solution is

(11)
$$g = A\, e^{(\sqrt{k^2+y^2}-k)x} + B\, e^{-(\sqrt{k^2+y^2}+k)x} .$$

The letters A and B are randomly selectable constants that we will use to adapt the solution of the predefined marginal conditions.

The equation (9) is the differential equation of a cylinder function of the 0 order Z_0 with the argument $y R$.

(12)
$$f = Z_{0(yR)} .$$

With equation (7), we obtain the following as the solution for our potential function

(13)
$$\Phi = \left[A\, e^{(\sqrt{k^2+y^2}-k)x} + B\, e^{-(\sqrt{k^2+y^2}+k)x} \right] Z_0 .$$

For the constant R, the e-function standing near A increases with a growing x and decreases near B, whereby the first expression reveals itself as the outer pole function ϕ_a and the second as the inner pole

224

function ϕ_i. We need to understand that the outer pole function is a potential function, the sources of which lie over the observed area in a positive x-direction. In the case of the inner pole function, the sources lie below in a negative x-direction. With regard to the latter calculation, we will separate (13) into the equation of the outer pole functions and into that of the inner pole functions. If we still take into account that A and B are freely available, i.e. that they can still be perceived as functions of the parameter y, then every potential function can be expressed in an integral representation according to y in the characteristic functions of the differential equation. We obtain the integral representation for outer pole functions

$$(14) \qquad \Phi_a = \int_0^\infty A_{(y)} e^{(\sqrt{k^2+y^2}-k)x} Z_{0(yR)} \, dy \, ,$$

and for the inner pole functions

$$(15) \qquad \Phi_i = \int_0^\infty B_{(y)} e^{-(\sqrt{k^2+y^2}+k)x} Z_{0(yR)} \, dy \, .$$

2. The Potential Function of a Point Source

We will place the point sources into the coordinate neutral point of the cylinder coordinate system x, R, ϕ. As such, the potential function of the point source is an inner pole function of the form (15) for positive x and an outer pole function of the form (14) for negative x. For the $x = 0$ plane, both potential functions must merge into each another. Consequentially, we obtain the first conditional equation for the functions $A_{(y)}$ and $B_{(y)}$. From $\phi_a = \phi_i$ for $x = 0$, we obtain the following according to (14) and (15)

$$(16) \qquad A_{(y)} = B_{(y)} \, .$$

We will achieve the second conditional equation if we can find an integral representation in cylinder functions for the source strength I in the $x = 0$ plane. This result must correspond with the calculation of the distribution of the source strength on the $x = 0$ plane from the potential equations (14) and (15). If we designate the current density on the $x = 0$ plane with i_0, then the following applies

$$(17) \qquad i_0 = -\lambda_0 \left[\left(\frac{\partial \Phi_i}{\partial x} \right)_{x=0} - \left(\frac{\partial \Phi_a}{\partial x} \right)_{x=0} \right] \, .$$

Consequentially, with (14) and (15) we obtain

$$(18) \qquad i_0 = 2\lambda_0 \int_0^\infty A_{(y)} \sqrt{k^2+y^2} \, Z_{0(yR)} \, dy \, .$$

According to [9], we obtain an integral representation for a random distribution of the source strength in the form of

(19)
$$i_0 = \int_0^\infty J_{0(yR)} \, y \int_0^\infty i_{(z)} J_{0(yz)} \, z \, dz \, dy \, .$$

As a result, J_0 is the cylinder function of the first type of the order 0. If we divide the source strength I evenly on a very small circle (radius a), which we will allow to shrink into one point later, then the following equations apply for the weighting function $i_{(z)}$

(20)
$$i_z = \frac{I}{\pi a^2} \qquad 0 \leq z \leq a \, ,$$

$$i_z = 0 \qquad a < z \leq \infty \, .$$

As a result, the inner integral from (19), which we will indicate with G, becomes

(21)
$$G = \frac{I}{\pi a^2} \int_0^a J_{0(yz)} \, z \, dz = \frac{I}{\pi a^2} \left. \frac{y \, z \, J_{1(yz)}}{y^2} \right]_0^a \, .$$

With the series expansion of the cylinder function J_1

$$J_1 = \frac{\frac{1}{2} y z}{1} - \frac{\left(\frac{1}{2} y z\right)^3}{1! \, 2!} \pm \cdots$$

we obtain the following for $z = a \to 0$

(22)
$$G = \frac{I}{\pi a^2} \frac{y \, a \, y \, a}{2 \, y^2} = \frac{I}{2 \pi} \, .$$

If we place this value in (19), we achieve the integral representation of the source density in the form of

(23)
$$i_0 = \frac{I}{2 \pi} \int_0^\infty y \, J_{0(yR)} \, dy \, .$$

Through the comparison with (18), we achieve

(24)
$$A_{(y)} = \frac{I}{4 \pi \lambda_0} \frac{y}{\sqrt{k^2 + y^2}}$$

and

H. W. Kasemir

(25)
$$Z_{0(yR)} = J_{0(yR)} \ .$$

Thus, the functions $A_{(y)} = B_{(y)}$ are determined and we obtain the representation of the source point potential as an outer pole function from (14)

(26)
$$\Phi_a = \frac{I}{4\pi\lambda_0} \int_0^\infty \frac{y}{\sqrt{k^2+y^2}} \, J_{0(yR)} \, e^{(\sqrt{k^2+y^2}-k)x} \, dy,$$

and the representation as the inner pole function from (15)

(27)
$$\Phi_i = \frac{I}{4\pi\lambda_0} \int_0^\infty \frac{y}{\sqrt{k^2+y^2}} \, J_{0(yR)} \, e^{-(\sqrt{k^2+y^2}+k)x} \, dy \ .$$

Both potential representations can be brought into a universal form by solving the integral

(28)
$$\Phi = \frac{I}{4\pi\lambda_0} \frac{e^{-k(\sqrt{x^2+R^2}+x)}}{\sqrt{x^2+R^2}} \ .$$

The potential equation is as follows in polar coordinates r, θ, ϕ with the correlation $r = \sqrt{x^2+R^2}$ and $\cos\theta = x/\sqrt{x^2+R^2}$

(29)
$$\Phi = \frac{I}{4\pi\lambda_0} \frac{e^{-kr(1+\cos\theta)}}{r} \ .$$

Thus, we have found the sought potential equation for a point source, which is located in a medium with conductivity increasing in the positive x-direction.

3. The Space Charge Equation

Through the differences from (28) according to x, we obtain the field strength in the x-direction.

(30)
$$-\frac{\partial\Phi}{\partial x} = E_{(x)} = \frac{Ik}{4\pi\lambda_0} \frac{e^{-k(\sqrt{x^2+R^2}+x)}}{\sqrt{x^2+R^2}} \left[1 + \frac{x}{\sqrt{x^2+R^2}} + \frac{x}{k(x^2+R^2)} \right],$$

and therefore, the equation for the space charge according to (4)

(31)
$$q = -\frac{k^2\epsilon I}{2\pi\lambda_0} \frac{e^{-k(\sqrt{x^2+R^2}+x)}}{\sqrt{x^2+R^2}} \left[1 + \frac{x}{\sqrt{x^2+R^2}} + \frac{x}{k(x^2+R^2)} \right],$$

or in the polar coordinates r, θ, ϕ

H. W. Kasemir

(32)
$$q = -\frac{k^2 \epsilon I}{2\pi\lambda_0} \frac{e^{-kr(1+\cos\theta)}}{r} \left[1 + \cos\theta + \frac{\cos\theta}{kr} \right].$$

4. The Current Function

We can obtain the current flow through a plane if we integrate the components of the current density situated on the plane vertically across this plane. To calculate the current function, we will choose a circular disc with the radius R', which lies parallel to the $x = 0$ plane. From (30), we obtain the current density in the x-direction for

(33)
$$i_{(x)} = \lambda E_{(x)} \quad,$$

and the current function through integration over the disc

(34)
$$\Psi = \frac{Ik}{2} \int_0^{R'} \left[1 + \frac{x}{(x^2+R^2)^{1/2}} + \frac{x}{k(x^2+R^2)} \right] \frac{e^{-k(\sqrt{x^2+R^2}-x)}}{(x^2+R^2)^{1/2}} R\,dR.$$

The integral can be solved in a closed form and we obtain the current function for the equation

(35)
$$\Psi = I \left[1 - \frac{1}{2}\left(1 + \frac{x}{\sqrt{x^2+R^2}} \right) e^{-k(\sqrt{x^2+R^2}-x)} \right].$$

or in the polar coordinates r, θ, ϕ

(36)
$$\Psi = I \left[1 - \frac{1+\cos\theta}{2} e^{-k(1-\cos\theta)} \right]. \quad .$$

(In (35), R is used again instead of R'.)

AII. The Potential Function of a Point Source over a Conducting Plane

Given:
The point source with the current production I at the altitude h over the plane $x = 0$.
The conduction function in the space $\lambda = \lambda_0 e^{2kx}$.

Unknown:
The potential function ϕ.

By moving the point source to the x-axis at the altitude h, the potential equation (28) changes to

H. W. Kasemir

$$(37) \qquad \Phi_1 = \frac{I}{4\pi\lambda_h} \; \frac{e^{-k\left(\sqrt{(x-h)^2+R^2}+x-h\right)}}{\sqrt{(x-h)^2+R^2}} .$$

We are attempting to achieve the solution to our problem through reflection on the plane and obtain the potential function Φ_2 of the reflection source with the current production $-I_V$ by moving the source point to the depth $-h$. According to (28), the following applies

$$(38) \qquad \Phi_2 = \frac{-I_V}{4\pi\lambda_{-h}} \; \frac{e^{-k\left(\sqrt{(x+h)^2+R^2}+x+h\right)}}{\sqrt{(x+h)^2+R^2}} .$$

By overlapping Φ_1 and Φ_2, we obtain the potential function for the point source over the conducting plane.

$$\Phi = \Phi_1 + \Phi_2 .$$

Φ must be constant on the $x = 0$ plane. This condition can be fulfilled if we give Φ the special value of 0.

Thus, we obtain an equation of conditions for the currently still freely available $-I_V$. For $\Phi = 0$, (37) and (38) result in

$$(39) \qquad -I_V = -\frac{\lambda_0}{\lambda_h} I .$$

Thus, the potential equation of the source point over a conducting plane becomes

$$(40) \qquad \Phi = \frac{I}{4\pi\lambda_0} \left[\frac{e^{-k\left(\sqrt{(x-h)^2+R^2}+x+h\right)}}{\sqrt{(x-h)^2+R^2}} - \frac{e^{-k\left(\sqrt{(x+h)^2+R^2}+x+h\right)}}{\sqrt{(x+h)^2+R^2}} \right] .$$

AIII. The Dipole Current

I. The Potential Function of a Dipole Current

To obtain the potential equation of a dipole current from (40), we will let that h move towards 0. Then we can develop the terms in brackets into a Taylor series. If we designate the first term with A and the second with B, then we obtain

$$A = \frac{e^{-k\sqrt{x^2+R^2}}}{\sqrt{x^2+R^2}} - \frac{h}{1!}\frac{\partial}{\partial x}\frac{e^{-k\sqrt{x^2+R^2}}}{\sqrt{x^2+R^2}} + \cdots$$

$$B = \frac{e^{-k\sqrt{x^2+R^2}}}{\sqrt{x^2+R^2}} + \frac{h}{1!}\frac{\partial}{\partial x}\frac{e^{-k\sqrt{x^2+R^2}}}{\sqrt{x^2+R^2}} + \cdots$$

If we still taken into that

(42)
$$\frac{\partial}{\partial x}\frac{e^{-k\sqrt{x^2+R^2}}}{\sqrt{x^2+R^2}} = \frac{x(1+k\sqrt{x^2+R^2})e^{-k\sqrt{x^2+R^2}}}{(x^2+R^2)^{3/2}}$$

then

(43)
$$\Phi = \frac{2hI}{4\pi\lambda_0}\frac{x(1+k\sqrt{x^2+R^2})e^{-k(\sqrt{x^2+R^2}+x)}}{(x^2+R^2)^{3/2}}.$$

12 Geophysics Magazine. 25

Apart from one constant factor k, the dipole moment $2hI$ represents the primary current I_N of the storm generator. We can recognize that immediately if we develop the e-function into a series, which we can cancel with the second term due to $h \to 0$, in the equation

(44)
$$I_N = \left(1 - \frac{\lambda_0}{\lambda_h}\right)I = (1 - e^{-2kh})I$$

(45)
$$I_N = \left(1 - 1 + \frac{2kh}{1!} \mp \cdots\right)I,$$

or even

(47)
$$I_N = 2khI.$$

Thus, according to (43), the potential ϕ of the dipole, which should be formed from the original and reflection source,

(48)
$$\Phi = \frac{I_N}{4\pi\lambda_0 k}\frac{e^{-k(\sqrt{x^2+R^2}+x)}x(1+k\sqrt{x^2+R^2})}{(x^2+R^2)^{3/2}}.$$

or in the polar coordinates $r,\ \theta,\ \phi$

H. W. Kasemir

(49)
$$\Phi = \frac{I_N}{4\pi\lambda_0 k} \frac{\cos\theta\,(1+kr)\,e^{-kr(1+\cos\theta)}}{r^2} \quad .$$

2. The Current Function of a Dipole Source

The calculation of the current function Ψ of the dipole source can be calculated easier in polar coordinates according to the equation (49). The integration of the current density i occurs via a spherical cap with a constant r up to the angle θ_1. In general, the following applies

(50)
$$\Psi = \int_0^r i\,d0 = -\int_0^{\theta_1} \lambda\frac{\partial\Phi}{\partial r} 2\pi r^2 \sin\theta\,d\theta \quad .$$

Thus, with (49) we obtain

(51)
$$\Psi = \frac{I e^{-kr}}{2r}\int_0^{\theta_1} \left[k^2 r^2 + 2(1+kr) + (1+kr)kr\cos\theta\right]\cos\theta\,e^{-kr(1-\cos\theta)}\sin\theta\,d\theta \quad .$$

The integral can be solved in a closed form. If we place only θ again instead of θ_1 after setting the parameters, we obtain the current function of the dipole current

(52)
$$\Psi = I\left\{1 - \frac{1}{2}\left[\cos\theta\,(1+\cos\theta) - \frac{\sin^2\theta}{kr}\right]e^{-kr(1-\cos\theta)}\right\}$$

A IV. Calculation of the Source Layer Generator

1. The Single Pole Generator

Given:
The source layer with the current production I and the plane F at the altitude h above the surface of the earth. The conductivity at the altitude h is λ_I.
The conduction function of the space $\lambda = \lambda_0\,e^{2kx}$.

Unknown:
The resistors between the source layer and $+\infty$ W_0
        and between the source layer and earth W_1.
The leakage current between the source layer and earth I_V.
The primary current between the source layer and I_N.

231

The resistance W of a cylindrical piece of the air space between the altitude h_n and h_m is calculated for

(53)
$$W = \frac{1}{F} \int_{h_n}^{h_m} \frac{dx}{\lambda} = \frac{1}{\lambda_0 \, 2 \, k \, F} \left(\frac{\lambda_0}{\lambda_n} - \frac{\lambda_0}{\lambda_m} \right).$$

This results in

(54)
$$W_u = \frac{1}{\lambda_0 \, 2 \, k \, F} \left(1 - \frac{\lambda_0}{\lambda_1} \right).$$

and

(55)
$$W_0 = \frac{1}{\lambda_0 \, 2 \, k \, F} \frac{\lambda_0}{\lambda_1}.$$

The currents I_V and I_N are inversely proportionate to the resistances W_u and W_0 they passed through.

(56)
$$\frac{I_N}{I_V} = \frac{W_u}{W_0}.$$

The following results with $I_N + I_V = I$

(57)
$$I_N = \frac{W_u}{W_u + W_0} I, \qquad \text{und} \qquad I_V = \frac{W_0}{W_u + W_0} I.$$

2. The Source Layer Bipolar Generator with Consideration for the Lowered Cloud Conductivity

Given:
The positive source layer at the altitude h_1 with the plane F and the current production I. Conductivity has the value λ_1 at the altitude h_1.
The negative source layer at the altitude h_2 with the plane F and the current production $-I$. Conductivity has the value λ_2 at the altitude h_2.
The conductivity in the space between the altitudes h_1 and h_2 has the value λ_3. In the remaining air space, the conductivity is given through the function $\lambda = \lambda_0 \, e^{2kx}$.

Unknown:
The primary and leakage current I_N and I_V, and the efficiency factor w.
The resistors
W_u in space 3 between the earth and the negative source layer,

H. W. Kasemir

W_i in space 2 between the positive and negative source layer,
W_0 in space 1 between the positive source layer and $+\infty$.

According to (53), W_u and W_0 are given through

(58)
$$W_0 = \frac{1}{\lambda_0 \, 2 \, k \, F} \frac{\lambda_0}{\lambda_1} \,,$$

and

(59)
$$W_u = \frac{1}{\lambda_0 \, 2 \, k \, F} \left(1 - \frac{\lambda_0}{\lambda_2} \right).$$

The resistance W_i is calculated for

(60)
$$W_i = \frac{h_1 - h_2}{\lambda_3 \, F} \,.$$

From equation (57), we obtain the leakage current and the primary current of the positive source I_V^+ and I_N^+ if we take into account that the bottom resistance is given in this case by connecting W_u and W_i in a series.

(61)
$$I_V^+ = \frac{W_0}{W_u + W_i + W_0} \, I \,,$$

and

(62)
$$I_N^+ = \frac{W_u + W_i}{W_u + W_i + W_0} \, I \,.$$

The leakage and primary current of the negative source I_V^- and I_N^- can be calculated in a similar manner, only that in this case, the upper resistance is given by connecting W_0 and W_i in a series.

(63)
$$I_V^- = - \frac{W_0 + W_i}{W_u + W_i + W_0} \, I \,.$$

(64)
$$I_N^- = - \frac{W_u}{W_u + W_i + W_0} \, I \,.$$

We obtain the primary current of the bipolar generator I_{BN} through the overlapping of the primary currents of the positive and negative source layer in space 1.

(65)
$$I_{BN} = \frac{W_i}{W_u + W_i + W_0} I ,$$

and the leakage current I_{BV} through the overlapping of the leakage current of the positive source and the primary current of the negative source layer. In this process, it is necessary to take into account that in this case the leakage current of the positive source layer must be applied with a negative sign because the direction of the current and the positive coordinate direction oppose each other.

(66)
$$I_{BV} = - \frac{W_u + W_0}{W_u + W_i + W_0} I .$$

The efficiency factor is given by the correlation of the primary current to total current production and is calculated according to (65) and (58) to (60) for

(67)
$$w = \frac{h_1 - h_2}{h_1 - h_2 + \frac{\lambda_3}{\lambda_0 2k} \left(1 + \frac{\lambda_0}{\lambda_1} - \frac{\lambda_0}{\lambda_2} \right)} .$$

A V. The Surface Charge of a Sphere with the Radius K and the Conductivity λ_2 located in a Medium with the Conductivity λ_1. A Current Source with the Current Production I is situated in the Middle of the Sphere.

Given:
A sphere with the radius K and the inner conductivity λ_2. A current source with the current production I and the charge Q is located in the middle of the sphere. The outer space has the conductivity λ_1.

Unknown:
The surface charge Q^0 of the sphere.

From the continuity of the current to the boundary layer of the sphere K follows

(68)
$$i = \frac{I}{4 \pi K^2} = \lambda_1 E_1 = \lambda_2 E_2 .$$

The surface charge density q' of the sphere is calculated from the plane divergence of the field strength

(69)
$$q' = \epsilon (E_1 - E_2) = \frac{I \epsilon}{4 \pi K^2} \left(\frac{1}{\lambda_1} - \frac{1}{\lambda_2} \right) .$$

The integration via the surface results in the net charge ma for

H. W. Kasemir

$$(70) \qquad Q^0 = I \epsilon \left(\frac{1}{\lambda_1} - \frac{1}{\lambda_2} \right) .$$

Because the charge Q of the current source and its current production are connected through the correlation $I = \lambda_2 / \varepsilon Q$, we obtain the following from (70)

$$(71) \qquad Q^0 = - \left(1 - \frac{\lambda_2}{\lambda_1} \right) Q .$$

A VI. The Net Charge of a Sphere with the Space Charge Density q

Given:

The middle of the sphere with the radius a lies in the middle of the spherical coordinate system r, θ, ϕ. The volumetric heat generation rate of the sphere is I.

The space charge density of the sphere is given through
The conduction function $\lambda = \lambda_m e^{2kr \cos \theta}$
The conductivity in the middle of the sphere λ_m.

$$q = \frac{I \epsilon}{\frac{1}{3} \pi a^3 \lambda} .$$

Unknown:
The net charge of the sphere Q.

The charge q' is given by the following on a circular disc vertical to the $\theta = 0$ axis with the radius $a \sin \theta$ and the thickness $a \sin \theta \, d\theta$

$$(72) \qquad q' = \frac{I \epsilon}{\frac{1}{3} \pi a^3 \lambda_m} e^{2ka \cos \theta} \pi a^2 \sin^2 \theta \, a \sin \theta \, d\theta .$$

The net charge Q of the sphere emerges through the integration of q' via the sphere.

$$(73) \qquad Q = \frac{\epsilon}{\lambda_m} \frac{I \pi a^3}{\frac{4}{3} \pi a^3} \int_0^\pi \sin^3 \theta \, e^{-2ka \cos \theta} \, d\theta .$$

The integral can be solved in a closed form and we obtain

$$(74) \qquad Q = \frac{\epsilon}{\lambda_m} I \frac{3}{2(2ka)^3} \left[(1 + 2ka) e^{-2ka} - (1 - 2ka) e^{2ka} \right] .$$

235

H.W. Kasemir

Literature

[1] Simpson, G.C., Scrase, F.J.: The distribution of electricity in thunderclouds I. Proc. Roy. Soc. 309-352, 1937.

[2] Simpson, G.C., Robinson, G.D.: The distribution of electricity in thunderclouds. II. Proc. Roy. Soc. 281-329, 1941.

[3] Brooks, C.E.P.: The distribution of thunderstorms over the globe. Geoph. Mem. London 3 Nr. 4, 1925.

[4] Wait, G.R.: Aircraft measurements of electric charge carried to ground through thunderstorms.
Thunderstorm Electricity. The University of Chicago Press. Edited by H.R. Byers.

[5] Kasemir, H.-W.: Die Stromausbeute des Gewittergenerators in Bezug auf den luftelektrischen Vertikalstrom der Schönwettergebiete. Ber. D. Wetterd. US-Zone, 428-434, 1952.

[6] Israël, H., Kasemir H.-W.: In welcher Höhe geht der weltweite luftelektrische Ausgleich vor sich. Ann. Geoph. 5, 313-324, 1949.

[7] Wichmann, H.: Die Weltgewittertätigkeit und das luftelektrische Feld der Erde. Arch. Met. Geoph. Biokl. 290-302, 1951.

[8] Küttner, J.: Die elektrischen und meteorologischen Vorgänge in der Basis von Gewitterwolken.
Das Gewitter. H. Israël. Akad. Verlagsges. Leipzig, 1950.

[9] Ollendorff, F.: Potentialfelder der Elektrotechnik. Verlag Julius Springer, Berlin 1932.

[10] Israël, H., Kasemir, H.-W.: Studien über das atmosphärische Potentialgefälle VI. Beispiele für das Verhalten luftelektrischer Elemente bei Nebel. Arch. Met. Geoph. Biokl. 71-85, 1952.

[11] Holzer, R.E., Saxon, D.S.: Distribution of electrical conduction currents in the vicinity of thunderstorms. J. Geoph. Res. 207-216, 1951.

[12] Simpson, G.C.: Atmospheric electricity during disturbed weather. Geoph. Mem. 84, 1-51, 1949.

[13] Israël, H.: Bemerkungen zum Energie-Umsatz im Gewitter. Geof. Pura e Appl. 3-11, 1953.

Session 6

GLOBAL CIRCUIT AND TEN-YEAR PROGRAM

Chairman: *B. Vonnegut*

Remark by Editors: for additional information to some of the problems discussed in this session, see session 8 b.

Theoretical Problems of the Global Atmospheric Electric Circuit *)

H. W. Kasemir

With 6 figures and 2 tables

Abstract

A current-flow model of the global atmospheric electric circuit is proposed which deviates from the commonly used spherical capacitor model in the following points. (a) It does not stop at the ionospheric level but extends to infinity. (b) The zero potential is placed at infinity and the earth carries a driving potential of about − 300 kV, the ionosphere has a potential of about − 30 μV. The conductivity is assumed to increase exponentially to infinity. This model has the advantage that the fair and foul weather part of the atmospheric electric circuit can be treated separately and later superimposed to the complete picture. This allows us to calculate the fair-weather condition with the simple assumption that the earth is the negative current source and infinity is the only current sink. We don't have to specify that only thunderheads are the generators for the positive current, but may include later any generator such as the austausch generator or *Frenkel*'s general cloud generator. Furthermore, because the ionosphere is not a priori assumed to be an equipotential layer of infinite conductivity as it is in the spherical capacitor model, it will be possible to extend this model to the inclusion of the geomagnetic field. Mathematical solutions for several atmospheric-electric fair-weather problems are given and discussed. The conductivity function includes the austausch layer with day and night variations on a global scale. The fundamental significance of the power density $W = Ei$ (the product of field and current density) at the ground as a global parameter is pointed out, a parameter which has been used by *R. Reiter* with great success in his correlation of terrestrial atmospheric electric and solar events. The independence of the power density at the mountain top from diurnal variations of the austausch layer is established. The problem of mapping up and down of electric disturbances caused either by conductivity variations or convection currents is discussed. It seems possible that potential differences found at a 10 km altitude level are caused by convection currents in or below the 10 km region but are not necessarily caused by potential difference at the ionospheric level. Procedures are given to determine if a potential difference at a certain level is caused by a disturbance above or below this level.

Introduction

A simplified circuit diagram of the global atmospheric electric circuit is shown in Fig. 1. The two current sources of equal strength but opposite polarity are the earth and the thunderheads. All thunderstorms active at any one time on the earth are combined in the circuit diagram in the positive terminal, marked *Th*, which is located in the horizontal resistor branch. Other generators, not shown in Fig. 1, as for instance, the austausch generator may be added if their importance on a global scale is established. The vertical resistor branch represents the current flow of the fair-weather areas. The circuit is closed at infinity which has the potential value zero. The potential of the thunderstorm terminal is chosen to be 100 MV and that of the earth − 300 kV. With the assumption that the conductivity in-

---

*) Invited Paper.

237

creases exponentially by a factor 10 for every 10 km altitude gain the equipotential lines are drawn in Fig. 1 for $\phi = -3$ kV at 20 km altitude, for $\phi = -30\,\mu$V at ionospheric altitude and for $\phi = 0V$ at infinity. Note that the zero potential dips down from infinity in the region of the thunderhead like a funnel from a tornado cloud and encloses the positive terminal. It crosses below the thunderhead, the resistor connecting the thunderhead with ground. Every fair-weather equipotential line by being lower than zero has to dip down in the thunderstorm region and intersect the resistor below the crossing point of the zero potential. As illustrated by the arrows in Fig. 1 we have two paths for the current to flow from the positive terminal "Thunderhead" to the negative terminal "Earth". The direct path is through the resistor between Th and Earth. This resistor represents the volume of air between the thunderstorm and ground. The second path is through the resistor between the thunderstorm and infinity and then back to the earth by the vertical resistor branch R_2, R_1 which represents the fair-weather areas.

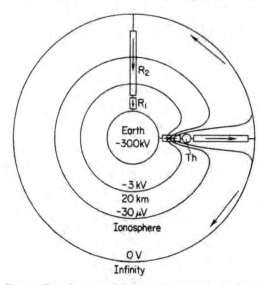

Fig. 1. Current flow diagram of the global atmospheric electric circuit

This circuit deviates from the widely used spherical capacitor model in the following points: first it extends to infinity, second the zero potential is at infinity not at the earth surface, and third, the earth has a -300 kV potential. The advantage of the circuit used here is that for a theoretical treatment the calculation of the current flow can be split into two parts. First the fair-weather circuit. Here the earth is the current generator with the driving voltage $\phi_a = -300$ kV and infinity is the sink which absorbs the current. From a physical point of view it would be senseless to impose in this model on the generator earth the driving voltage zero and keep it zero regardless of the current output. It would be equally meaningless to attach the driving voltage to the ionosphere which does not act as a sink or source and has in the current-flow circuit no more significance than any other altitude level. It is in most cases possible to switch from the capacitor model to the current-flow circuit by subtracting the ionospheric potential from the potential function.

The second part would be the foul-weather circuit with the thunderheads acting as the current source and again infinity acting as the sink. In this case the earth has to be introduced as an equipotential layer with a potential value which follows from the condition that in and out flowing current should balance. A superposition of these two current flow patterns would give the complete picture.

In this paper fair-weather problems are discussed. Quasistationary conditions are assumed. which means that time changes of the current source are always slow compared to the largest time constant in the circuit. All problems are governed by the continuity equation of the conduction current div $\bar{i} = 0$. The influence of the earth's magnetic field on the conductivity in the ionosphere and space is neglected. This is done to keep the mathematical effort on a moderate level. However, it should be kept in mind that an application of the equations and results of this paper to the ionosphere has a purely ficticious value. This is true not only for the current-flow circuit presented here, but applies just as well to the spherical capacitor model or any other calculation which treats the conductivity in the ionosphere as a scalar quantity. Conditions in the ionosphere, magnetosphere, and exosphere which will also apply to atmospheric electric problems in these regions have been described by *Obayashi* and *Maeda* (1965). A first attempt to incorporate the geomagnetic field in the fair-weather current was made by *Kasemir* (1971).

The current flow in the thunderstorm region has been discussed by *Holzer* and *Saxon* (1951), with the ionosphere treated as an infinite conductive equipotential layer, and by *Kasemir* (1952 and 1959), with an exponential conductivity extended to infinity. In these papers the influence of the geomagnetic field is neglected. A very interesting calculation of the field and current flow from the thunderhead through the ionosphere and into the magnetosphere has been worked out by *Park* and *Dejnakarintra* (1973). Here the influence of the geomagnetic field on the current flow in the ionosphere has been taken into account. In the magnetosphere magnetic field lines are assumed to be equipotential lines.

In the next section of this paper, the fair-weather current in a two-layer conductivity is discussed. In the austausch layer the conductivity is constant and above the austausch layer it is exponential. This section has the purpose to acquaint the reader with the somewhat different terminology, illustrate Fig. 1, and serve as a reference for the following sections. In the third section, the conductivity undergoes a day and night pattern, the electrical power density is introduced as global parameter and the mapping-up of the austausch layer is discussed. The fourth section treats the power density at a mountain top and the last section the problem of up and down mapping of potential waves. The significance of potential differences of the integrated field obtained by radiosonde balloon flights is discussed.

All potential functions Φ are calculated by a method given by *Kasemir* (1963). Φ is composed of two functions M and N. N is given by the conductivity function λ. With λ_0 being the conductivity at the earth surface $N = (\lambda/\lambda_0)^{1/2}$. M enforces the boundary condition and has to fulfill the differential equation

$$\frac{\Delta M}{M} = \frac{\Delta N}{N}.$$

The potential function Φ which fulfills the boundary condition and the current continuity equation $\text{div } i = 0$ in an environment with the conductivity λ is then given by

$$\Phi = \frac{M}{N}.$$

If, for instance, as in the fourth section, λ is given by a quadratic equation $\lambda = \lambda_0(mz + 1)^2$, $\Delta N = 0$, and any electrostatic solution U which fulfills the boundary condition and $\Delta U = 0$ could serve as M. In the case of an exponential increase of the conductivity as in the second and third section from now on, M has to fulfill the differential equation $\Delta M = kM$. The eigen functions of this equations, for instance in a spherical coordinate system, are the *Hankel* functions $H_{n+\frac{1}{2}(r)}$ multiplied by *Legendre*'s polynominals $P_{n(\cos\theta)}$. Therefore, in general, M is given in the form

$$M = \sum_0^{\prime} A_n H_{n+\frac{1}{2}(r)} P_{n(\cos\theta)}.$$

List of Symbols:

Coordinate systems:

x, y, z = cartesian coordinates $[m, m, m]$;
z, R, ϕ = cylindrical coordinates $[m, m, \sphericalangle]$;
r, θ, ϕ = spherical coordinates $[m, \sphericalangle, \sphericalangle]$;

   a = earth radius $[m]$;
   b = radius of the top of the austausch layer or exchange layer $[m]$ in the next section;
     = radius of the spherical shell $[m]$ in the last section;
   $h = b - a$ = thickness of the austausch layer $[m]$ in the next section;
     = height of the spherical shell above earth $[m]$ in the last section.

Electric parameters (some additional definitions in the text):

$\Phi_{(x,y,z)}$ = potential function [V];
Φ_a = earth potential [V];
Φ_i = ionospheric potential [V];
$\bar{E}$ = electric field vector [V/m];
F_0 = electric field at the ground [V/m];
$\bar{\imath}$ = current density vector [A/m^2];
λ = conductivity [1/Ωm];
$s = 1/2k = 10^4$ m/ln 10 = scale height of exponential conductivity [m];
$W = \bar{E} \cdot \bar{\imath}$ = power density [W/m^3];
I = current output of the earth [A];
R = global resistance [Ω];
R_c = columnar resistance [Ω m^2];
j = convection current density [A/m^2];
J = net convection current [A].

Definitions:

$\bar{E} = -\,\mathrm{grad}\,\Phi$;
$\bar{\imath} = -\lambda\,\mathrm{grad}\,\Phi$;
$\Phi = \int_{\infty}^{x,y,z} \bar{E}\,d\bar{s}$.

The Fair-Weather Current with a Spherical Symmetric Conductivity in the Austausch Layer and Above

The conductivity of the ground layer (austausch layer) reaching from $r = a$ to $r = b$ is λ_a, being constant in this region, $b - a = h$ is the thickness of the austausch layer. This layer is called space 1 and the atmospheric electric parameters in it are identified by the index 1. Above this layer in space 2, extending from $r = b$ to $r = \infty$, the conductivity is given by the equation

$$\lambda_2 = \lambda_b \frac{b^2}{r^2} \exp\left[(r - b)/s\right]. \tag{1}$$

The index 2 identifies the parameters in space 2. The problem has spherical symmetry. Therefore only the r component of the field and current vector are different from zero. The factor b^2/r^2 in [1] is 1 for the first 100 km altitude above the earth surface with a maximum error of about 0.03% so that [1] gives the desired exponential increase of the conductivity with altitude.

In space 1 and 2 we have the following equations for the different atmospheric electric parameters:

Columnar resistance:

$$R_{c_1} = \frac{a}{\lambda_a}\left(1 - \frac{a}{b}\right); \qquad R_{c_2} = \frac{s}{\lambda_b}. \tag{2}$$

Global resistance:

$$R_1 = \frac{1}{4\pi\lambda_a}\left(\frac{1}{a} - \frac{1}{b}\right); \qquad R_2 = \frac{s}{4\pi b^2 \lambda_b}. \tag{3}$$

Potential function:

$$\Phi_1 = I\left[R_1 + R_2 - \frac{1}{4\pi\lambda_a}\left(\frac{1}{a} - \frac{1}{r}\right)\right]; \qquad \Phi_a = I\left[R_1 + R_2\right] = IR. \tag{4}$$

Field; Current density:

$$E_1 = \frac{I}{4\pi\lambda_a}\frac{1}{r^2}; \qquad i_1 = \frac{I}{4\pi}\frac{1}{r^2}. \tag{5}$$

Potential function:

$$\Phi_2 = I R_2 \exp\left[-(r-b)/s\right]. \tag{6}$$

Field; Current density:

$$E_2 = I \frac{R_2}{s} \exp\left[-(r-b)/s\right]; \quad i_2 = \frac{I}{4\pi} \frac{1}{r^2}. \tag{7}$$

If the exponential conductivity would extend all the way to the ground (no austausch layer) eqs. [1], [3], [6] and [7] may be used by simply exchanging the letter b by a. The index 2 can be omitted. In this case it is:

Global and columnar resistance:

$$R = \frac{s}{4\pi a^2 \lambda_a}; \quad R_c = \frac{s}{\lambda_a}. \tag{8}$$

Potential function:

$$\Phi = I R \exp\left[-(r-a)/s\right]; \quad \Phi_a = I R. \tag{9}$$

Field; Current density:

$$E_r = I \frac{R}{s} \exp\left[-(r-a)/s\right]; \quad i_r = \frac{I}{4\pi r^2}. \tag{10}$$

Conductivity:

$$\lambda = \lambda_a \frac{a^2}{r^2} \exp\left[(r-a)/s\right]. \tag{11}$$

To obtain an idea how well these equations lead to reasonable values for the different atmospheric electric parameters we have to choose certain key values and calculate the other parameters according to the eqs. [2] to [10]. It is not claimed that the numerical values chosen here are in exact accordance with the latest experimental results. They are rounded off to the first significant digit. They may be adjusted to fit the individual preference or recent measurements. We set for the chosen parameters:

$a = 6.3 \times 10^6$ m = earth radius;
$4\pi a^2 = 500 \times 10^{12}$ m^2 = earth surface;
$i = -2 \times 10^{-12}$ A/m^2;
$I = 4\pi a^2 i = -1000$ A = earth current output (thunderstorms);
$\lambda_a = 2 \times 10^{-14}$ 1/Ωm = ground conductivity with austausch layer (continental);
$\lambda_e = 4 \times 10^{-14}$ 1/Ωm = ground conductivity without austausch layer (oceanic);
$s = 10^4$ m/ln 10 = scale height;
$h = 0, 2.5, 5$ km thickness of austausch layer.

The fair-weather field at the ground would follow from *Ohm*'s law $E = i/\lambda_a$, $E_a = -100$ V/m or $E_a = -50$ V/m for the continental or oceanic value respectively. (Note the negative sign of the physical sign convention $E = -\text{grad } \Phi$; i.e. vector E pointing downwards is called negative).

Table 1

h [km]	λ_a [1/Ωm]	λ_b [1/Ωm]	R_1 [Ω]	R_2 [Ω]	R [Ω]	Φ_a [kV]	Φ_b [kV]	Φ_{20} [kV]
0	4×10^{-14}				218	-218		-2.18
2.5	2×10^{-14}	7.1×10^{-14}	254	123	377	-377	-123	-2.18
5	2×10^{-14}	12.6×10^{-14}	508	69	577	-577	-69	-2.18

Table 1 gives the values for λ_a (conductivity at the earth surface), λ_b (conductivity at the top of the austausch layer), R_1 (global resistor of the austausch layer), R_2 (global resistor above the austausch layer), $R = R_1 + R_2$ complete global resistor, Φ_a (earth potential), Φ_b (potential at the top of the austausch layer), Φ_{20} (potential at 20 km altitude) for three different thicknesses of the austausch layer $h = 0, 2.5, 5$ km. For clarification of the different resistors see also Fig. 1.

It is not surprising that in all three cases the 20 km potential is the same, because the current density and the conductivity above the austausch layer are the same and the integration of the field starts from infinity to the point of observation. The different values in the earth potential are caused by the increased resistor values in the austausch layer. In other words the earth potential adjusts to the increased resistor in the austausch layer in such a way that the same current can be driven through the austausch resistor. That the driving voltage adjusts to the load resistor is a general feature of any current generator. It shows quite clearly that the earth potential (ionospheric potential) depends not only on the world thunderstorm activity but also on the existing global resistor. It may very well be that the global diurnal variation of the field at the oceans is not caused predominantly by a variation of the world thunderstorm activity, which may even be constant, but by the diurnal variation of the global resistor (*Kasemir*, 1950). We have to keep this possibility in mind, if we, for instance, try to compare the world thunderstorm activity determined by a spherics network with the diurnal variation of the field at the oceans or the earth potential (ionospheric potential). *Mühleisen* and coworkers determined the ionospheric potential Φ_i with numerous radiosonde flights integrating the measured field up to balloon peak altitude and extrapolating it to ionospheric altitude. They found an average value of $\Phi_i = 280$ kV, most of the individual flight values were between $\Phi_i = 220$ kV and 350 kV with extremes down to 160 kV and up to 1580 kV (*Mühleisen*, 1971). If we attach the negative sign to these values and call it the earth potential Φ_o we see that the values given in Table 1 are in good agreement with the experimental ones.

If we compare the potential value Φ_{20} km with the earth potential we see that 99% or more of the potential drop from the earth surface to infinity occurs already in a layer of 20 km thickness above the earth surface.

If we calculate the global resistance R_{20} for the region from 20 km to infinity by replacing λ_b by λ_{20} km in [3] we obtain $R_{20} = 2.18\,\Omega$ which is 1% or less from the total global resistance R in Table 1. This again means the earth potential is practically controlled by the global resistance of a layer between ground and 20 km altitude. Therefore, it appears that fair-weather electricity in the classical sense is confined to a layer between the earth surface and about 20 km altitude.

The Fair-Weather Current with a Day-Night Pattern of the Conductivity in the Austausch Layer and Spherical Symmetry Above

We will extend the model discussed in the previous section to include the day and night variation of the conductivity in the austausch layer. At the dayside of the globe with the austausch in full swing the conductivity in the austausch layer is low and consequently a part of the air-earth current is deflected to the nightside where the conductivity is higher. The field at the dayside will be increased because of the low conductivity but this increase is somewhat counterbalanced by the decrease in the current density. At the nightside the opposite effect will take place, an increase of the current density and with it also an increase in the field. The field increase however, is reduced by the high conductivity. Above the austausch layer where the conductivity is not affected by the austausch and may be assumed to be the same at the day- and nightsides, the surplus current of the nightside will tend to flow to the dayside until in higher altitude an equal current density distribution over the whole globe is established. It is the purpose of this paragraph to calculate the field and current distribution of this problem.

It may be said right at the beginning that this model is still incomplete, for two reasons. First: the austausch over the ocean surface is very much reduced or absent. This has the result that even on the dayside some of the illuminated earth surface maintains its nightside conductivity and this reduces the overall day-night difference in the field and current flow. It would require a more complex equation for the conductivity than the one we will use later to model the continental areas on the earth surface. However, the simple day-night sub-division is sufficient here to clarify the principle effect on the diurnal variation of the current flow. Second: In addition to producing a conductivity change the austausch acts as a generator, *Kasemir* (1956). The mathematical treatment of the austausch generator effect is not given in this paper.

To obtain a still relatively simple closed solution for the potential function of our problem we express the conductivity as the square of a sum of three eigen functions of the differential equation

$$\Delta N - k^2 N = 0, \qquad [12]$$

$$N = (f - \alpha g + \beta H \cos \phi), \qquad [13]$$

with

$$f = \frac{a}{r} \exp\left[k(r-a)\right]; \quad g = \frac{a}{r} \exp\left[-k(r-a)\right]; \quad H = \frac{a^2}{r^2} \frac{kr+1}{ka+1} \exp\left[-k(r-a)\right]. \quad [14]$$

The conductivity is then given by

$$\lambda = \lambda_0 N^2 = \lambda_0 (f - \alpha g + \beta H \cos \phi)^2. \qquad [15]$$

The function f in [15] squared is the same function we have used in eq. [1] of the previous section to represent the pure exponential increase of the conductivity, λ_0 is the conductivity at the ground in this case and has the numerical value $\lambda_0 = 4 \times 10^{-14} \, 1/\Omega m$. The functions g and h are inpole functions which vanish with increasing r. The purpose of g is to reduce the conductivity in the austausch layer and the purpose of $H \cos \phi$ is to introduce the day-night variation. The two constants α and β allow us to adjust the strength of each effect individually. Note that f, g and H are normalized to a (radius of the earth) and are 1 at the earth surface. Therefore the conductivity at the ground λ_a is given by

$$\lambda_a = \lambda_0 (1 - \alpha + \beta \cos \phi)^2. \qquad [16]$$

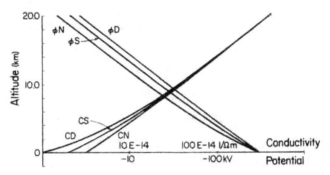

Fig. 2. Day and night pattern of the conductivity and the potential function versus altitude

CD = Day conductivity ΦD = Day potential
CS = Sunset conductivity ΦS = Sunset potential
CN = Night conductivity ΦN = Night potential
$(10 E - 14 = 10 \times 10^{-14})$

In the following discussion and in Figs. 2 and 3 we indicate the dayside noon, the sunset and the midnight values by the letters *D, S, N* respectively. In our numerical examples we choose $\alpha = 0.3$ and $\beta = 0.2$. This results in a dayside conductivity at noon at the ground of $\lambda D = 1 \times 10^{-14} \, 1/\Omega m$ and a night conductivity at midnight of $\lambda N = 3.2 \times 10^{-14} \, 1/\Omega m$. In Fig. 2 the three conductivities are plotted against altitude and marked CD = Day-Conductivity, CS = Sunrise-Conductivity and CN = Night-Conductivity. We see in Fig. 2 that the difference between the three conductivities is already small at 10 km altitude (about 3%) and has practically vanished at 20 km altitude.

The potential function Φ for such a conductivity pattern with earth as the current generator with the driving voltage Φ_a is given by

$$\Phi = \Phi_a \frac{(1 - \alpha) g + \beta H \cos \phi}{f - \alpha g + \beta H \cos \phi}. \qquad [17]$$

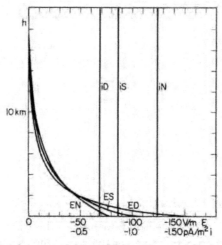

Fig. 3. Day and night pattern of the field E and current density i versus altitude h

ED = Day field   iD = Day current density
ES = Sunset field   iS = Sunset current density
EN = Night field   iN = Night current density

We may take advantage here of the fact that the thickness of the layer we are interested in — let's say from the earth surface up to the ionosphere — is very small compared to the earth's radius. Therefore, with an error of less than 2% we may set $r = a$ and $r - a = z$. z is our new vertical coordinate counted from the earth's surface. The functions f, g, and H reduce to the exponential functions $\exp kz$ and $\exp(-kz)$, and eq. [17] simplifies to

$$\Phi = \Phi_a e^{-2kz} \frac{1 - \alpha + \beta \cos \phi}{1 + (\beta \cos \phi - \alpha)e^{-2kz}}. \tag{18}$$

In a similar good approximation we obtain for the vertical (radial) component of the electric field E and the air-earth current density i the equations

$$E = 2k\Phi_a e^{-2kz} \frac{1 - \alpha + \beta \cos \phi}{[1 + (\beta \cos \phi - \alpha)e^{-2kz}]^2}, \tag{19}$$

$$i = \lambda_0 2k\Phi_a(1 - \alpha + \beta \cos \phi) \tag{20}$$

and for the global resistor R and the columnar resistor R_c

$$R = \frac{1}{4\pi a^2 2k\lambda_0(1 - \alpha)}; \quad R_c = \frac{1}{2k\lambda_0(1 - \alpha + \beta \cos \phi)}. \tag{21}$$

With our chosen numerical values $\Phi_a = -300$ kV, $2k = \ln 10/10^4$ m; $\lambda_0 = 4 \times 10^{-14}$ 1/Ωm we obtain for the field and the current density at the ground $ED = -138$ V/m; $ES = -98.5$ V/m; $EN = -76.7$ V/m; $iD = -1.38$ pA/m^2; $iS = -1.93$ pA/m^2; $iN = -2.48$ pA/m^2.

The columnar resistance is

$$R_cD = 2.18 \times 10^{17} \, \Omega\text{m}^2; \quad R_cS = 1.55 \times 10^{17} \, \Omega\text{m}^2; \quad R_cN = 1.21 \times 10^{17} \, \Omega\text{m}^2$$

and the global resistor

$$R = 310 \, \Omega.$$

These values fit reasonably well the commonly accepted values in the literature. (Note the negative sign of the field and current density due to the sign convention). The one exception is that the night current is stronger than the daytime current which is contrary to experimental results. It shows that

244

our model is not complete. The missing part is probably the austausch generator effect which will be treated in another publication.

The potential functions ΦD, ΦS, and ΦN are plotted as the second triple of curves in Fig. 2. We see that they all start from the −300 kV ground value and approach zero in higher altitudes. However, contrary to the conductivity the day and night values diverge from each other and from the sunrise value in the lower 10 km layer and remain so up to the ionosphere. The constant separation in the logarithmic scale of Fig. 2 means that the relative difference between the day and night value remains the same whereas the absolute difference becomes smaller by a factor 10 with each 10 km gain in altitude. For instance, at 10 km altitude we have an absolute difference between the day and night time potential of 12 kV, at 20 km altitude this difference is reduced to 1.2 kV, and at ionospheric altitude the difference will be reduced to 12 μV.

Fig. 3 shows the field and current triple against altitude. The most striking feature is that the current stays constant with altitude, even though the day and night values differ from each other by almost a factor 2. One would expect that in some higher altitudes horizontal components would occur which result in an equalization of the current density around the globe. However, this equalization does not occur in the first 100 km altitude as is also evidenced by eq. [20]. That these large differences in the current density do occur is shown in simultaneous current radiosonde flights reproduced in Figs. 4 and 5. The average current between 10 and 20 km altitude at Eastern Test Range, Boulder, Hachijojima, and Kayoshima, are 1, 1.5, 1.75, and 2.6 pA/m² (Fig. 4). The first two of these stations are located in North America and the last two in Japan. Even on the same continent large differences are found as is shown in Fig. 5. The average current between 15 and 25 km altitude at Duluth, White Sand, and Boulder (all in the USA) are 0.9, 1.85, and 3.6 pA/m². The most likely explanation for the different current densities is the existence of widely different columnar resistances at the different stations.

Another striking feature in the few examples given in Figs. 4 and 5 is that the current density even in altitudes up to 25 km is by no means constant. Variations in the current density up to 30% are not unusual. These current variations with altitude in the individual flight indicate the presence of convection currents in the respective altitude range. The electrical convection current in the 20 km altitude layer is not yet investigated and deserves a concentrated research effort. Another possible cause of the current variation would be the sideways drift of the radiosonde during its ascent through localized regions with different columnar resistances. However, the two simultaneous radiosonde records shown in Fig. 6 show the same variation even though the stations are 1000 km apart. This does not point to a local variation of the columnar resistance.

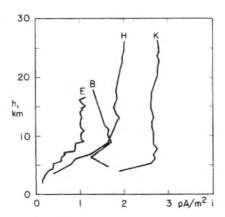

Fig. 4. Current density i versus altitude h. 14 February 1972
E = Eastern Test Range, Florida USA, 17:15−19:36 world time, 12:15−14:36 local time;
B = Boulder, Colorado USA, 19:42−21:22 world time, 12:42−14:22 local time;
H = Hachijojima, Japan, 18:00−   :   world time, 03:00−   :   local time;
K = Kagoshima, Japan, 18:00−   :   world time, 03:00−   :   local time

245

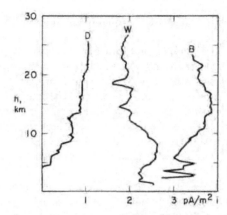

Fig. 5. Current density i versus altitude h, 16 February 1972.
D = Duluth, Minnesota, USA, 14:00 – 16:30 world time, 08:00 – 10:30 local time
W = White Sands, New Mexico, USA, 13:51 – 15:29 world time, 06:51 – 08:29 local time;
B = Boulder, Colorado, USA, 13:45 – 15:03 world time, 06:45 – 08:03 local time

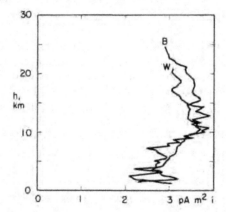

Fig. 6. Current density i versus altitude h, 17 February 1972
B = Boulder, Colorado, USA, 14:07 – 15:17 world time, 07:07 – 08:17 local time;
W = White Sands, New Mexico, USA, 13:45 – 15:04 world time, 06:45 – 08:04 local time

A very interesting result is obtained if we calculate the field E_a and current density i_a at the ground from [19] and [20] and multiply these two parameters. This product gives the electrical power density W_a.

$$W_a = E_a i_a = \lambda_0 (2k\Phi_a)^2 .$$ [22]

We see that the power density in our model is independent of the local conductivity variation. It is the same everywhere at the earth surface and proportional to the square of the driving voltage Φ_a. By rearranging we obtain from [22]

$$\Phi_a = \frac{1}{2k} \left(\frac{E_a i_a}{\lambda_0} \right)^{\frac{1}{2}} .$$ [23]

[23] offers a way to check the earth potential at the ground as determined by a field radiosonde ascent. However, more important is the fact that $E_a i_a = W_a$ and $(E_a i_a)^{\frac{1}{2}}$ are global parameters which would be better suited to detect global relations than the field or current density by themselves. *Reiter* and co-workers deserve the credit for introducing the product Ei in the atmospheric electric measuring technique. They have used the product $E_a i_a$ with good success already for some time at the mountain top Zugspitze to correlate terrestrial electric activity with solar events (*Reiter*, 1969, 1971, 1972). As this promises to be a powerful method, we will discuss the conditions on a mountain top in the next section. Other

246

variations of [23] may be obtained if we introduce from eqs. [9] and [10] the field E_{a0} and the current density i_{a0} in regions without an austausch layer. It follows from [9] and [10],

$$E_{a0} = 2k\Phi_a, \quad i_{a0} = \lambda_0 2k\Phi_a$$

and with [23]

$$E_{a0} = \left(\frac{E_a i_a}{\lambda_0}\right)^{\frac{1}{2}}; \quad i_{a0} = (\lambda_0 E_a i_a)^{\frac{1}{2}}. \tag{24}$$

[24] links the field and the current E_{a0} and i_{a0} at places without austausch, let us say at the oceans or arctic regions, with the field and the current on the continents. The condition for this relation is that eq. [22] is at least approximately true if large areas of the earth like the oceans and arctic regions do not participate in the day-night pattern; and second that the austausch causes only a conductivity variation but does not act as a generator. The second condition especially may not be fulfilled. Therefore, the results in this section should be interpreted only as a first step in the analysis of the interrelation of continental and oceanic field and current patterns. However, we have gained valuable suggestions in which direction the theoretical analysis and data collection should be extended.

The Electrical Power Density at a Mountain Top

The mountain is represented mathematically by a hemispherical or a cylindrical boss on an infinite horizontal plane. It is possible to extend the calculation with a slightly greater mathematical effort to spheroidal shapes, i.e., a tall half spheroid or a flat disc of circular or cylindrical form, however, the results will be similar to the simpler forms treated here. The plane and the boss are a source of a negative current. In a large (horizontal) distance away from the boss the current density i_0 is constant. The conductivity at the ground is λ_0 and consequently the field at the ground F_0 is given by

$$F_0 = i_0/\lambda_0. \tag{25}$$

The conductivity λ has a parabolic increase with altitude z and is given by

$$\lambda = \lambda_0 (mz + 1)^2 \tag{26}$$

The parabolic increase instead of the exponential one is chosen here because it gives a better representation of the austausch layer and also leads to simpler mathematical expressions. The factor m in [26] may contain a time factor which reflects the conductivity variation of the austausch layer during the day and night. We may set m to be given by

$$m = m_0(1 + \alpha \cos \omega t). \tag{27}$$

It will be shown that the electrical power density W given by the product of the field and current density Ei at the top of the mountain is independent of the day and night pattern of the conductivity. Cartesian coordinates x, y, z will be used with the z axis (altitude) in the vertical direction. The axis of the cylindrical boss is in the y direction so that the cylindrical coordinate R is given by

$$R = (x^2 + z^2)^{\frac{1}{2}}.$$

In spherical coordinates the equivalent relation is $r = (x^2 + y^2 + z^2)^{\frac{1}{2}}$. "$c$" is the radius of the cylindrical or hemispherical boss. Field or current components in x and z direction are indicated by the index x or z.

For the cylindrical boss we have the following equations:

$$\Phi = \frac{F_0}{m}\left[1 - \frac{mz}{mz + 1}(1 - c^2/R^2)\right], \tag{28}$$

$$E_z = F_0\left[\frac{1 - c^2/R^2}{(mz - 1)^2} + \frac{2c^2 z^2}{(mz + 1)R^4}\right], \tag{29}$$

$$i_z = i_0\left[1 - c^2/R^2 + 2c^2 z^2(mz + 1)/R^4\right], \tag{30}$$

$$E_x = F_0 \frac{z}{mz + 1} \frac{2c^2 x}{R^4}. \tag{31}$$

$$i_x = i_0 2c^2 xz(mz + 1)/R^4. \tag{32}$$

At the ground and at the boss surface for $z = 0$ or $R = c$ we obtain from [28]

$$\Phi = \Phi_0 = \frac{F_0}{m}$$

Φ_0 corresponds to Φ_a, the driving voltage of our previous section. In higher altitudes $z \to \infty$, the potential approaches zero. At the top of the boss where $R = z = c$, $x = 0$ we obtain from [29] to [32]

$$E_z = 2F_0/(mc + 1); \quad i_z = 2i_0(mc + 1); \quad E_x = i_x = 0. \tag{33}$$

Note that for constant conductivity $m = 0$ the field and current concentration factor $E_z/F_0 = i_z/i_0 = 2$. The electrical power density follows from [33]

$$W = E_z i_z = 4F_0 i_0. \tag{34}$$

$F_0 i_0$ is the electrical power density on the earth surface away from the cylindrical boss where conditions are undisturbed by the presence of the boss. Eq. [34] means that the power density at the top of the boss is four times as much as on the undisturbed part of the plane and that this relation is independent of the conductivity variation with altitude i.e. independent of m. Therefore, if the power density on the earth surface is proportional to a global pattern under conditions discussed in the previous section the power density at the mountain top will follow the same pattern with an additional constant concentration factor. This factor is given by the square of the field concentration factor of the same problem with altitude-constant conductivity. We may expect, therefore, that in case of the spherical boss, where the electrostatic field concentration factor is 3, that the energy concentration factor is 9. That this is true can easily be verified. The potential function for the spherical boss is

$$\Phi = \frac{F_0}{m} \left[1 - \frac{mz}{mz + 1} \left(1 - \frac{c^3}{r^3} \right) \right]. \tag{35}$$

The field and current density at the top of the boss follows from [35]. It is

$$E_z = 3F_0/(mz + 1); \quad i_z = 3i_0(mz + 1) \tag{36}$$

and the power density

$$W = E_z i_z = 9F_0 i_0. \tag{37}$$

The energy concentration factor is 9 as expected and we may conclude that in general for other shapes of the boss the energy concentration factor is the square of the field concentration factor of the equivalent electrostatic problem.

The field distortion of a mountain in an environment with an altitude dependent conductivity has for some time been a problem in atmospheric electricity which defied a simple analytic solution. It may be mentioned here that the horizontal and vertical field distortion above the mountain can be easily obtained from [28] and [35] as shown for the cylindrical case in [29] to [32]. Such orographic distortions become important for electric field measurements from an airplane or balloon over mountainous terrain.

Up and Down Mapping

At a spherical shell concentric with earth, $r = b$, shall exist a surface current source which produces a potential Φ_b at this level given by

$$\Phi_b = A \cos \theta. \tag{38}$$

248

At the earth surface $r = a$ with $a < b$ the potential shall be constant. The conductivity λ shall increase exponentially from the earth surface to infinity.

$$\lambda = \lambda_a \frac{a^2}{r^2} e^{2k(r-a)} = \lambda_a N^2 . \tag{39}$$

We would like to know the solution of the following problem: In which manner does the potential of the shell map up and down. For instance, if the maximum potential difference at antipode points at 10 km altitude above the earth is 20 kV, how much is the difference attenuated if we reach ionospheric level. On the other hand if at the ionospheric level there is a potential difference of 20 kV at antipode points, how much of a potential difference would we measure at the 10 km level. The fair-weather current produced by the earth generator may be superimposed to the current flow of the spherical shell to obtain a more complete picture. This calculation deals with potential, field, and current distribution of the shell only. It may be considered as the first step to the solution of the austausch generator problem.

The shell subdivides space into two parts. The first one, indicated by the index 1 on the electrical parameters, is the region between the earth surface and the shell, $a \leq r \leq b$. The second one, indicated by the index 2, is the space outside the shell, $b \leq r \leq \infty$. We define four *Hankel*-type functions F, G, K, L.

$$F = \frac{b^2}{r^2} \frac{kr-1}{kb-1} e^{k(r-b)} ; \quad G = \frac{b^2}{r^2} \frac{kr+1}{kb+1} e^{-k(r-b)} ; \quad K = \frac{a^2}{r^2} \frac{kr+1}{ka+1} e^{-k(r-a)} ; \quad L = \frac{a}{r} e^{k(r-a)} . \tag{40}$$

If the argument r of the function is $r = a$ or $r = b$ this is indicated by the index (a) or (b) in brackets. Note that

$$F_{(b)} = G_{(b)} = K_{(a)} = L_{(a)} = 1 .$$

The potential function Φ_1 and Φ_2 in space 1 and 2 are given by

$$\Phi_1 = \frac{L_{(b)}}{L} \frac{F - F_{(a)}K}{1 - F_{(a)}K_{(b)}} A \cos \theta , \tag{41}$$

$$\Phi_2 = \frac{L_{(b)}}{L} \frac{G - F_{(a)}K}{1 - F_{(a)}K_{(b)}} A \cos \theta . \tag{42}$$

$A \cos \theta$ in [41] and [42] is the driving potential Φ_b at the shell $r = b$ as given by [38] and the two fractions in front of it give the attenuation in space 1 and 2 as a function of r. If we now introduce the approximation used before $r/a = r/b \to 1$, the new coordinate $z = r - a$, which gives the altitude above the earth surface, and $h = b - a$ which is the altitude of the driving shell above the earth surface, the eqs. [41] and [42] simplify to

$$\Phi_1 = \frac{1 - \exp(-2kz)}{1 - \exp(-2kh)} A \cos \theta , \tag{43}$$

$$\Phi_2 = \exp[-2k(z-h)] A \cos \theta . \tag{44}$$

Boström et al. obtained an equation equivalent to [43] for the two-dimensional case (*Boström* et al., 1973). They treat the more general case of time varying potential waves in the ionosphere including the time independent solution as a special case for the general problem. The eq. [4.7] on page 14 of their report,

$$V = \frac{\exp(r_1 z) - \exp(-r_2 z)}{\exp(r_1 a) - \exp(-r_2 a)} \cos kx , \tag{43a}$$

converts to [43] if we consider that their scale factor k of the x coordinate is a very small number if $\cos kx$ represents a wave around the globe. In good approximation it is $r_1 \to 0$; $r_2 \to 1/H$ and with "$1/H$" equivalent to our "$2k$" and "a" equivalent to our "h" the two eqs. [43a] and [43] agree with each other.

Prior to *Boström* et al., *Volland* (1972) calculated the mapping down of the potential wave generated by the Sq current in the ionosphere. He used a spherical coordinate system and a three-term expression of the potential function in spherical harmonics. His wave is sunrise-sunset oriented being zero at noon and midnight and at the poles. The altitude dependence of the potential wave is given by a term $1 - \exp(-2kz)$ which again is in agreement with [43].

The maximum and minimum of the potential in each level is obtained for $\theta = 0$, $\cos\theta = 1$ and $\theta = \pi$, $\cos\theta = -1$. Therefore the maximum difference D_1 and D_2 in space 1 and 2 is given by

$$D_1 = 2A \frac{1 - \exp(-2kz)}{1 - \exp(-2kh)}, \qquad [45]$$

$$D_2 = 2A \exp[-2k(z - h)]. \qquad [46]$$

The factor $2A$ is the maximum potential difference in the shell itself. Let's assume first that the shell is at $h = 10\,km$ and $2A = 20\,kV$. The decay of this difference with increasing altitude — mapping up — is given by [46]. We see that the difference D_2 is attenuated by a factor 10 with each altitude gain of 10 km. At ionospheric level $z = 100\,km$ the potential difference has attenuated from the 20 kV at the 10 km level to 20 μV. This is in agreement with the result of the third section showing that the absolute potential difference decays in the same fashion at the potential itself.

In we now turn to the problem of mapping down we have to use eq. [45]. Let's assume again that the potential difference at two antipode points in the ionosphere is 20 kV. h is now 100 km and $z = 10\,km$. From [45] follows that at the 10 km level we have still a potential difference of 18 kV. Therefore in this case we have an almost unattenuated mapping down of the absolute difference. Only close to the ground level, i.e. in the last few kilometers altitude the potential difference vanishes. However, with this perfect mapping down we will not be able to determine if the disturbance is only 10 or 90 km above our measuring level. If the disturbance would be at 20 km altitude with the same potential difference of 20 kV it would map down to the 10 km level with a potential difference of 18.2 kV.

We will calculate now the strength of the current source and the charge, which will produce the postulated potential wave with an amplitude of $A = 10\,kV$, and the field and current distribution at the ground and with altitude. From [41] and [42] or in good approximation from [43] and [44] we calculate the components E_{1z}, E_{2z}, i_{1z} and i_{2z} of the field and current vector in the r or z direction. It is

$$E_{1z} = -\frac{2k\exp(-2kz)}{1 - \exp(-2kh)} A\cos\theta\,; \qquad E_{2z} = 2kA\cos\theta\exp[-2k(z - h)], \qquad [47]$$

$$i_{1z} = -2k\lambda_0 \frac{A\cos\theta}{1 - \exp(-2kh)}\,; \qquad i_{2z} = 2k\lambda_0 A\cos\theta\exp(2kh). \qquad [48]$$

The surface current source density j and the surface charge density q at the shell for $z = h$ follows from [47] and [48]

$$q = \varepsilon(E_{2z} - E_{1z}) = 2k\varepsilon A\cos\theta\left(1 + \frac{1}{\exp(2kh) - 1}\right), \qquad [49]$$

$$j = i_{2z} - i_{1z} = 2k\lambda_0 A\cos\theta\left[1 + \frac{1}{\exp(2kh) - 1}\right]\exp(2kh). \qquad [50]$$

By integrating q or j over half of the spherical shell we obtain the net charge Q or the net current J:

$$Q = 2\pi b^2 k\varepsilon A\left[1 + \frac{1}{\exp(2kh) - 1}\right], \qquad [51]$$

$$J = 2\pi b^2 k\lambda_0 A\left[1 + \frac{1}{\exp(2kh) - 1}\right]\exp(2kh). \qquad [52]$$

Table 2 lists the maximum surface charge and current source density q and j, and the net charge and current Q und J for shell altitudes of 10, 20 and 100 km.

Table 2

h [km]	q [pC/m²]	j [pA/m²]	Q [C]	J [A]
10	18.4	0.83	2300	1.04×10^2
20	21	9.1	2520	1.14×10^3
100	22	9.2×10^8	2540	1.15×10^{11}

The current source density j would be proportional to a convection current density which delivers the current. We see from Table 2 that for a shell altitude of 10 km the required convection current density is moderate, $j = 0.8$ pA/m². The presence of a convection current is indicated in many current radiosonde ascents and can be recognized by the fact that the conduction current is not constant with altitude. The deviation of the conduction current from a constant value is proportional to the convection current. Fig. 6 shows two simultaneous current radiosonde flights at Boulder and White Sands which are about 1000 km apart. Both flight records show a conduction current increase in the 5 to 15 km layer of about 1 pA/m² with a slow decrease to 25 km altitude. Such increases and decreases of the conduction current with altitude are by no means unusual, but occurred in a more or less pronounced way in a large number of flights. It is interesting to note that a similar but smaller bulge can be seen in a graph of the air-earth current density versus altitude obtained by *Hake, Pierce,* and *Viezee* (1973), by calculating the air-earth current density from the field and conductivity and averaging these values over a large number of radiosonde flights.

A current source shell in 20 km altitude would require already 9 pA/m² convection current to produce a potential difference of 20 kV in this level. Also the net current of about 1000 A is already quite substantial, whereas the required net current of 100 A in the 10 km level is certainly in the realm of continental convection currents. Almost unrealistically high are the values of the current source density $j = 0.1$ mA/m² and the net current $J = 115 \times 10^9$ A at ionospheric altitude, which are required to produce a potential wave of 10 kV amplitude.

Introduced by *Fischer* (1962), it is today common practice to extrapolate the potential at the peak altitude of a field-measuring radiosonde to the ionospheric level. If the balloon is flown to 20 km altitude or higher this extrapolation is not necessary because the correction gained by extrapolation would be below the measuring accuracy of the radiosonde. For the determination of the earth potential a 20 km peak altitude should be quite sufficient. However, conclusions drawn from extrapolated potential differences with regard to conditions in the ionosphere, *Mühleisen* (1971, 1972), *Fischer* and *Mühleisen* (1972), can be accepted only with great reservation. Without additional safeguards there is no way to show that potential differences measured at, let's say, a 10 km altitude level are not caused by disturbances just above or even below this level. Therefore it may be advisable to discontinue the pratice of extrapolating potential values to the ionosphere. Instead, the field should be integrated to balloon or airplane peak altitude and the potential value reported as the 10 km potential or to whatever altitude the measurement was actually carried out.

Fortunately, the difference between up and down mapping provides a simple means to check if potential differences are caused from above or below the peak altitude. If the potential difference is larger at a level below peak altitude than at peak altitude, the disturbance is below. If the differences are the same or the peak altitude difference is the greater one, the disturbance is above. An even better safeguard is to fly a current radiosonde simultaneously. As long as the current density is not constant with altitude − and this is usually the case, not an exception − the supposition for extrapolating the peak potential does not exist.

The field at the earth surface $z = 0$ can be calculated from the first equation in [47]. If the disturbance is higher than 20 km the exponential term in the denominator can be neglected. For $\cos \theta = 1$ we obtain the amplitude of the potential wave:

$$A = E_{1z}/2k .\qquad\qquad [53]$$

Lobodin and *Paramonov* (1972) analyzed a large number of field measurements from 8 stations at different latitudes during aurora. The field deflection at the most northern station (80° N) was negative

251

(atmospheric electric sign convention) and of the order of $-30\,\text{V/m}$. The deflection decreased with decreasing latitude and showed a positive maximum at the most southern station (66° S). If we interpret these field deflections as caused by a negative charge deposited in northern and a positive charge deposited in the southern hemisphere we may use [53] to calculate the amplitude of the corresponding potential wave. With $2k = 1n\,10/10\,\text{km}$ it would be $A = -130\,\text{kV}$. The net charge to produce such a wave follows from [51] and would be in the order of $\mp\,30000$ to $\mp\,33000\,\text{C}$. Note that the amount of charge does not depend very much on the altitude level in which the charge has been deposited. It would be interesting to compare the calculated potential difference of such a potential wave with the data obtained by *Mühleisen* from simultaneous flights at the equator and at Weissenau, Germany (estimated latitude 48 N). The calculated difference ΔV would be $-62\,\text{kV}$. Disregarding the sign this would agree with the maximum difference of 60 kV reported by *Mühleisen* for the potential difference extrapolated to ionospheric altitude. For a better comparison the ionospheric potential values would have to be restored to the balloon peak altitude values. (Note that the calculated wave maps down almost unattenuated, but the extrapolation to the ionosphere may reduce the measured potential difference at the 10 km level by a factor 2 or more).

Acknowledgment

The author wishes to express his thanks to Professor *D. Olson*, University of Minnesota, Duluth; Chief Meteorologist *I. Kühnast*, Eastern Test Range, Florida; Dr. *W. Webb*, White Sands, New Mexico; and Dr. *I. Shimizu*, Meteorological Agency, Tokyo, Japan, for providing the data of the air-earth current radiosonde flights used in Figs. 4, 5 and 6.

References

1. *Boström, R., U. Fahleson, L. Olansson,* and *G. Hallendal*, Theory of time varying atmospheric electric fields and some applications to fields of ionospheric origin. Dept. Plasma Physics, Roy. Inst. Tech. (see page 14, eq. [4.7]) (Stockholm, 1973). – 2. *Fischer, H. J.*, Die elektrische Spannung zwischen Ionosphäre und Erde. Diss. T. H. (Stuttgart, 1962). – 3. *Fischer, H. J.* and *R. Mühleisen*, Met. Rundsch. **25**, 1, 6 (1972). – 4. *Hake, R. D., E. T. Pierce,* and *W. Viezee*, Stratospheric Electricity. Stanford Research Institute Report, Project 1724 (see page 57, Fig. 16) (1973). – 5. *Holzer, R. E.* and *D. S. Saxon*, J. Geophys. Res. 207 (1951). – 6. *Kasemir, H. W.*, Arch. Met. Geophys. Biokl. **III**, 84 (1950). – 7. *Kasemir, H. W.*, Die Stromausbeute des Gewittergenerators in bezug auf den luftelektrischen Vertikalstrom der Schönwettergebiete. Ber. d. Wetterd. d. U.S. Zone 6 (38): 428–434, 1952. – 8. *Kasemir, H. W.*, Arch. Met. Geophys. Biokl. **9**, 357 (1956). – 9. *Kasemir, H. W.*, Z. Geophys. **25**, 33 (1959). – 10. *Kasemir, H. W.*, USA ELRDL Tech. Report 2394 (1963). – 11. *Kasemir, H. W.*, Pure and Applied Geophysics **84**, 76 (1971). – 12. *Lobodin, T. W.* and *N. A. Paramonov*, PAGEOPH **100**, 167 (1972). – 13. *Mühleisen, R.*, Z. Geophys. **37**, 759 (1971). – 14. *Mühleisen, R.*, Elektrische Felder in der Ionosphäre, abgeleitet aus luftelektrischen Messungen (Klein-Heubacher Berichte **15**, 361 (1972)). – 15. *Obayashi, T.* and *H. Maeda*, Electrical state of the upper atmosphere. Problems of Atmospheric and Space Electricity (Amsterdam, 1965). – 16. *Park, C. G.* and *M. Dejnakarintra*, J. Geophys. Res. **78**, 28, 6623 (1973). – 17. *Reiter, R.*, PAGEOPH **72**, 259 (1969). – 18. *Reiter, R.*, PAGEOPH **86**, 142 (1971). – 19. *Reiter, R.*, PAGEOPH **94**, 218 (1972). – 20. *Volland, H.*, J. Geophys. Res. **77**, 10, 1961 (1972).

Discussion

Ogawa, Kyoto, Japan:

What have been the altitude profiles of the air-earth current density? The air-earth current density is strongly influenced by the underneath orography. In other words, in a mountain area, the electric current increases by the reduction of the columnar resistance. Did you check the underneath orography when you compared the air-earth current density profiles at the different places? Did you check the time of the flight of the day when you compared the air-earth current at different places? Did you check the weather on the day of those flights?

Kasemir, Boulder, Colorado, USA:

The first point was whether I took into account the orographic shape of the land over which the balloon flies. In Boulder all the weather comes from the west; the balloon rises and is blown to the east, and in the east there is what we call the "high plains", that is high-altitude flat country. We very seldom have wind coming from the east or bad weather. All the balloon flights I showed in Boulder are in fair weather and over flat country. The same is approximately true for White Sands. Now I do not know the conditions of the Japanese flights. At the South Pole we have a high plain situation and the flights were made during fair weather. I was not there, but probably *Cobb* can answer this question. There was a second point, concerning time. The flights are made at the same universal time. Some of them, as the Boulder flights, were early in the morning, one to three o'clock. The Japanese flights would be about 7 hours earlier, i.e. 18:00 to 20:00 on the day before. There was a third question about weather conditions: The flights were fair-weather flights.

Author's address:

H. W. Kasemir
NOAA, ERL, APCL
Boulder, Colorado 80302
USA

JOURNAL OF GEOPHYSICAL RESEARCH, VOL. 99, NO. D5, PAGES 10,701–10,708, MAY 20, 1994

Current budget of the atmospheric electric global circuit

Heinz W. Kasemir

Colorado Scientific Research Corporation, Longmont

Abstract. The paper provides a mathematical calculation based on theoretical physics to estimate the current contribution for different types of charge generators existing in the atmospheric electric global circuit. The Earth is represented by a sphere and has a constant potential due to its high conductivity. The generators are composed of one or more point current sources or point charges placed above the sphere. The resulting potential, field, and current flow distributions are obtained by a superposition of the corresponding values of the individual point sources. A new physical model of the global circuits is presented. The calculation of the current budget of the global circuit based on this model shows the generator outputs balanced by an additional current source which provides the Earth with an equally distributed charge density on it surface and gives it a negative potential value. This negative potential enables the Earth to drive, in fair weather areas, a current up to the ionosphere, which then represents the contribution of the generators to the global circuit.

1. Introduction

The main purpose of this paper is to devise a mathematical scheme based on theoretical physics that enables us to calculate the current contribution to the electric global circuit for different types of generators. The generators always have two terminals which produce an equal amount of positive and negative charge or current. The terminals are usually given as point sources and sinks or an arrangement of those. In some cases, for instance in a positively-charged rain cloud, where the rain brings down negative charge to the ground, the Earth surface represents the negative terminal, while the cloud represents the positive terminal.

The assumptions in our calculation are that the conductivity increases exponentially with altitude and that the Earth is represented by a conductive sphere. The field E, current density i, and potential functions Φ are given as solutions of Laplace equation div grad $\Phi = 0$ for the electrostatic case (including the current flow case with constant conductivity), or as solutions of div $i = 0$ in case of stationary flow problems, or as solutions of div $c = 0$ where c (x,y,z, t) is the Maxwell current in case of time-dependent flow problems. The electrostatic and stationary current flow solutions will be necessary boundary conditions for the more general time-dependent flow problems of the Maxwell current.

The solutions given and discussed in the next sections throw a different light on the current budget of the global circuit. A fundamental change in the generally accepted model of the global circuit (the conventional model) as well as a reassessment of the contribution of the different generators to the global circuit will be necessary. The changes which have to be made are based on two rather obvious requirements: (1) whatever generator we assume, we have to make sure that it produces the same amount of positive and negative charges or current and (2) that in calculations of the global current flow we have to represent the Earth by a sphere and not by a plane surface. The conventional model does not satisfy these conditions and has, in addition, some other physical errors or unwarranted assumptions in the global current flow which will be pointed out later. The new model of the global circuit proposed in this paper, called the physical model, avoids these errors.

In the next two sections we first treat the electrostatic problem of a point charge above the conductive sphere and then the electrodynamic problem of a point source above the conductive sphere. In the following sections we use these solutions to discuss the contribution of different types of generators to the global circuit.

2. Point Charge Above a Conductive Sphere

If we represent the Earth by a conductive sphere our first problem would be to calculate the potential function Φ of a point charge Q above a conductive sphere of radius k with the condition that at the sphere surface the potential function Φ assumes a constant potential value Φ_k.

A well-known solution of this problem, using the mirror image method, is given in most text books on electrostatics and electrodynamics [e.g., *Smythe*, 1950; *Ollendorff*, 1932]. In Figure 1 the sphere is shown as a large circle with the radius k and centered at the origin of the spherical coordinate system r, ϑ. The charge above the sphere marked Q is located at $r = h_1$, $\vartheta = 0$, and its mirror image marked Q_b at $r = h_2$, $\vartheta = 0$. The radial distances R_1 and R_2 from the charges Q and Q_b to the point of observation at r, ϑ are functions of the spherical coordinates, given by the equations

$$R_1 = \sqrt{r^2 + h_1^2 - 2rh_1 \cos\vartheta} \tag{1}$$

$$R_2 = \sqrt{r^2 + h_2^2 - 2rh_2 \cos\vartheta} \tag{2}$$

The potential functions Φ_1 and Φ_2 of Q and Q_b can now be written in the concise form

$$\Phi_1 = \frac{Q}{4\pi\epsilon}\frac{1}{R_1}; \quad \Phi_2 = \frac{Q_b}{4\pi\epsilon}\frac{1}{R_2} \tag{3}$$

Paper number 93JD02616.
0148-0227/94/93JD-02616$05.00

254

KASEMIR: CURRENT BUDGET OF THE ATMOSPHERIC ELECTRIC GLOBAL CIRCUIT

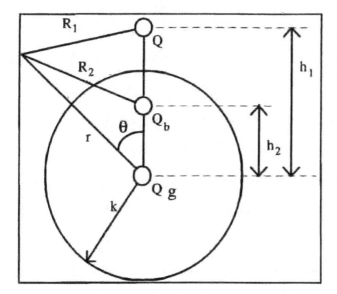

Figure 1. Sketch of a point charge above a conductive sphere: r, ϑ, spherical coordinates; k, radius of the sphere; Q, point charge above the sphere; Q_b, mirror charge of Q on the sphere; Q_g, global charge; contribution to the global circuit.

and the sum of Φ_1 and Φ_2 gives us the potential function Φ_{12} as

$$\Phi_{12} = \frac{Q}{4\pi\varepsilon}\frac{1}{R_1} + \frac{Q_b}{4\pi\varepsilon}\frac{1}{R_2} \qquad (4)$$

The requirement that the conductive sphere has to have an equipotential surface can be satisfied under the following conditions:

$$\text{for } r = k : \Phi_{12} = \Phi_1 + \Phi_2 = 0 \qquad (5)$$

$$h_1 h_2 = k^2 \qquad (6)$$

$$Q_b = -\frac{k}{h_1}Q \qquad (7)$$

The image charge Q_b (in electrostatic called the bound charge indicated by the subscript b) represents the surface charge density integrated over the whole surface of the sphere. In technical terms Q_b would be the net charge on the negative terminal of the generator. Since h_1 is always larger than k, the magnitude of Q_b is always smaller than that of Q. Therefore we have to add another charge Q_g which is shown in Figure 1 as a little circle in the center of the sphere, to satisfy our requirement that the positive charge and negative charge at the terminals of the generator should be of the same magnitude. This requirement leads to the condition

$$Q_g + Q_b = -Q \qquad (8)$$

and replacing Q_b by (7)

$$Q_g = -(1 - \frac{k}{h_1})Q \qquad (9)$$

The other requirement to be fulfilled is that the sphere should retain a constant potential value. It is not necessary that the value has to be zero. This leads us immediately to the well-known example of the charged sphere with a constant charge density q_s at the sphere surface, and the potential function

$$\Phi_3 = \frac{Q_g}{4\pi\varepsilon}\frac{1}{r} \qquad (10)$$

The potential function Φ of a point charge above a conductive sphere is now given by

$$\Phi = \Phi_1 + \Phi_2 + \Phi_3 = \qquad (11a)$$

$$\frac{Q}{4\pi\varepsilon}\frac{1}{R_1} + \frac{Q_b}{4\pi\varepsilon}\frac{1}{R_2} + \frac{Q_g}{4\pi\varepsilon}\frac{1}{r}$$

and the potential value Φ_k of the sphere , where for $r = k$: $\Phi_{1k} + \Phi_{2k} = 0$

$$\Phi_k = \frac{Q_g}{4\pi\varepsilon}\frac{1}{k} = -(1 - \frac{k}{h_1})\frac{Q}{4\pi\varepsilon}\frac{1}{k} \qquad (11b)$$

In difference to equations (8), (11a), and (11b) the conventional model does not have the global contribution Q_g. Therefore positive and negative charges are not balanced, and the Earth potential is zero.

The advantage of the electrostatic solution is the simplicity of the equations and the lucidity of the physical concept. Since the same mathematical structure is used in the electrodynamic case, where the equations are more complex, the electrostatic calculation can be used as a valuable guideline. For altitudes up to 500 m above ground (corona blanket), where the conductivity σ can be considered to be approximately constant, we have only to substitute for the parameters Q, Q_b, Q_g, ε, in the electrostatic equations the electrodynamic parameters I, I_b, I_g, σ, to obtain the solution for the current flow problem.

One example of an electrostatic problem is the cloud-to-ground lightning discharge. Its lifetime of about 1 s is much too short for the conductivity of the air to have any influence of the charge distribution at the Earth surface or on the lightning channel. At completion of the ground discharge, the channel charge Q and the surface charges Q_b and Q_g can be calculated by the electrostatic equations. This would be the starting charge distribution for the following time-dependent recovery curve of the lightning discharge.

To adapt the lightning problem to our point charge model, we assume that the positive charge remaining on the lightning channel can be approximately represented by a point charge Q located at the channel at an altitude H_1 above ground. The negative charge discharged to ground is then $-Q$, which splits up into the local charge Q_b and the global charge Q_g. With $h_1 = k + H_1$, we obtain from equations (7) and (9)

$$Q_b = -\frac{k}{k + H_1}Q; \quad Q_g = -\frac{H_1}{k + H_1}Q \qquad (12)$$

and the ratio Q_g/Q_b is then given by

$$\frac{Q_g}{Q_b} = \frac{H_1}{k} \qquad (13)$$

Equation (13) is a very important equation in global circuit analysis. We will use it here only to show that the contribution of a cloud-to-ground lightning discharge to the global circuit is insignificant. The Earth radius is $k = 6367$ km. If we choose for H_1 the value 6.367 km , we obtain $Q_g = 0.001\ Q_b$. This means that from the negative charge $-Q$ brought down to the ground by a cloud-to-ground discharge, the contribution to the global circuit Q_g is only 0.1% of $-Q$, also practically negligible. The bound charge Q_b is 99.9% of $-Q$ and remains underneath the storm. If we double the value of H_1 to 12.7 km, we obtain $Q_g = 0.002\ Q_b$. The altitude of 12.7 km is already too high for a cloud-to-ground discharge, bringing negative charge to the Earth, and the contribution Q_g to the global circuit would still be negligible. It is interesting to note here that *Williams* and *Heckman* [1993] came to the same conclusion through an analysis of worldwide lightning data in reference to the diurnal variation of electric parameters of the global circuit.

2.1 Conversion to Cylindrical Coordinates With Earth Represented by a Plane Surface

In calculations of the global circuit the Earth is frequently represented by a plane surface. This conversion shows that such an approximation excludes contributions to the global circuit. We keep the charge Q and its height H_1 above the sphere fixed. Moving the center point of the sphere down to infinity converts the sphere into the tangential plane, which is the new presentation of the Earth. The contact point of the plane and the sphere is the zero point of the cylindrical coordinate system R, z. From Figure 2 we obtain the following relations:

$$R_1 = \sqrt{R^2 + (z - H_1)^2} \qquad (14)$$

$$R_2 = \sqrt{R^2 + (z + H_2)^2} \qquad (15)$$

$$h_1 = k + H_1 \quad ; \quad h_2 = k - H_2 \qquad (16)$$

Replacing in equations (6), (7), and (9) h_1 and h_2 by the expression given in equation (16), we obtain after slight rearrangement

$$H_1 - H_2 = \frac{H_1 H_2}{k} \qquad (17)$$

$$Q_b = -\frac{1}{1 + \dfrac{H_1}{k}} Q \qquad (18)$$

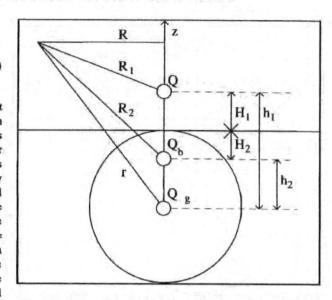

Figure 2. Conversion of spherical coordinates r, ϑ to cylindrical coordinates z, R.

$$Q_g = -\frac{H_1}{k + H_1} Q \qquad (19)$$

If we let k go to infinity, we obtain

$$H_1 = H_2 ; \quad Q_b = -Q ; \quad Q_g = 0 \qquad (20)$$

The most important result is $Q_g = 0$, which means that the Earth potential and all contributions to the global circuit are zero. This is also consistent with $Q_b = -Q$. If the local charge Q_b is already equal in amount to Q, there is no need or no place for Q_g to balance the charges. This emphasizes the fact that representing the Earth by a plane gives us only the local charge on the Earth surface and explains the fact that the conventional model doesn't contain an input to the global circuit. However, we can also take advantage of this result in the next section 3. by using cylindrical coordinates for the calculation of the potential functions $\Phi_1 + \Phi_2$ of the point source I and its mirror image I_b. Spherical coordinates have to be used for the calculation of the potential function Φ_3 of global contribution I_g.

3. Point Current Source Above a Conductive Sphere

This calculation follows, in essence, the same scheme as used in the electrostatic case in section 2. However, the conductivity σ, which is the counterpart of ε, is not constant but increases exponentially with altitude.

$$\sigma = \sigma_0\, e^{a(r-k)} = \text{conductivity} ; \qquad (21)$$

$$\sigma_0 = \text{conductivity at the ground}$$

KASEMIR: CURRENT BUDGET OF THE ATMOSPHERIC ELECTRIC GLOBAL CIRCUIT

A list of symbols and their counterparts (in parentheses) in electrostatics is given below.

a = 1/scale height;
I = current source above sphere (Q);
I_b = local current (Q_b);
I_g = global current (Q_g);
i = current density (field E).

Other symbols such as potential function Φ, spherical coordinates, etc., are the same as in section 2. For the geometrical placement of the current sources (charges) see Figure 1.

The potential function Φ of the current flow problem is of the form analog to (11a)

$$\Phi = \Phi_1 + \Phi_2 + \Phi_3 \qquad (22)$$

Φ_1 is the potential function generated by the point source I above the sphere, Φ_2 the potential function of the local current I_b, and Φ_3 that of the global current Ig. Because Φ_1 and Φ_2 are only effective in the local environment of the source, the earth may be represented by a plane surface. The solution to this problem is given by the equations (30) to (40). However, for the calculation of the global potential function Φ_3 we have to represent the Earth by a sphere. The solution of this problem is given in the equations (23) to (29).

$$\Phi_3 = \frac{I_g}{4\pi\sigma r}\left[1 - are^{\,ar}Ei(ar)\right] \qquad (23)$$

$$E_g = \frac{I_g}{4\pi\sigma r^2} \; ; \quad i_g = \sigma E_g = \frac{I_g}{4\pi r^2} \; ; \qquad (24)$$

Furthermore, we obtain from the calculation given by Kasemir [1952] the equation

$$I_b = -I\frac{\sigma_o k}{\sigma_{h_1} h_1}\frac{1 - ah_1 e^{\,ah_1}Ei(ah_1)}{1 - ake^{\,ak}Ei(ak)} \qquad (25)$$

The integral logarithm Ei(x) [*Abramowitz and Stegun*, 1964], is for x = ar defined by

$$E_i(ar) = \int_1^\infty \frac{1}{t}e^{-art}dt \qquad (26)$$

Differentiation to r interchanging differential and integral operator leads to

$$\frac{dEi(ar)}{dr} = \frac{e^{-ar}}{r} \qquad (27)$$

We expand Ei by continuous partial integration into a power series.

$$Ei(ar) = \frac{e^{-ar}}{ar}[1 - \frac{1!}{ar} + \frac{2!}{(ar)^2} - \frac{3!}{(ar)^3} + \cdots] \qquad (28)$$

and obtain from (28) the relation

$$1 - are^{\,ar}Ei(ar) = \frac{1!}{ar} - \frac{2!}{(ar)^2} + \frac{3!}{(ar)^3} + \cdots =$$
$$\sum_{n=1}^{m}(-1)^{n-1}\frac{n!}{(ar)^n} \qquad (29)$$

This semiconverging series converges for large "ar" values very rapidly. If we assume that the conductivity increases by a factor 10 for every altitude gain of 10 km above ground then a=ln10/10km. Our lowest r value is the radius of the Earth k=6378 km, then ar=1469. Using only the first term of the Σ in (29) would cause an error of about 0.1%.

In this case, the upper boundary would be m = 1. Equations (23) and (25) simplify to

$$\Phi_3 = \frac{I_g}{4\pi\sigma}\frac{1}{ar^2} \qquad (23')$$

$$I_b = -I\frac{\sigma_o}{\sigma_{h_1}}\frac{k^2}{h_1^2} \qquad (25')$$

The potential functions Φ_1 and Φ_2 of the original source and its mirror image are given in expansions in harmonic functions. Since they decay very rapidly with distance from the source where the Earth can still be considered as a plane surface, we apply here the same transformation from spherical to cylindrical coordinates as was used at the end of section 2. We obtain then the following equations and definitions (see Figure 3):

$$\sigma = \sigma_o e^{az} \qquad (30)$$

$$R_1 = \sqrt{R^2 + (z-H)^2} \; ; \quad R_2 = \sqrt{R^2 + (z+H)^2} \qquad (31)$$

$$\Phi_1 = \frac{I}{4\pi\sigma_o}\frac{e^{-\frac{a}{2}R_1}}{R_1}e^{-\frac{a}{2}(z+H)} \qquad (32)$$

$$\Phi_2 = \frac{I_b}{4\pi\sigma_o}\frac{e^{-\frac{a}{2}R_2}}{R_2}e^{-\frac{a}{2}(z-H)} \qquad (33)$$

$$I_b = -I\frac{\sigma_o}{\sigma_H} \qquad (34)$$

$$\Phi_2 = -\frac{I}{4\pi\sigma_o}\frac{e^{-\frac{a}{2}R_2}}{R_2}e^{-\frac{a}{2}(z+H)} \qquad (33')$$

257

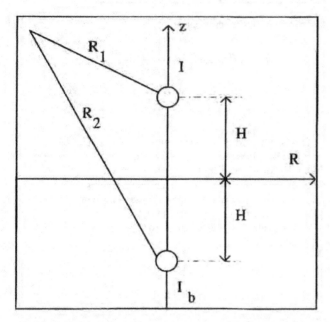

Figure 3. Sketch of a point source above a conductive sphere: z, R, cylindrical coordinates; I, point source above a conductive plane, I_b, mirror source of I on the conductive plane.

Using (32) and (33'), we define the local potential function Φ_{12}

$$\Phi_{12} = \Phi_1 + \Phi_2 = \frac{I}{4\pi\sigma_0} e^{-\frac{a}{2}(z+H)} \left(\frac{e^{-\frac{a}{2}R_1}}{R_1} - \frac{e^{-\frac{a}{2}R_2}}{R_2} \right) \quad (35)$$

It is easy to recognize that at the plane where $R_1 = R_2$, the potential function $\Phi_{12} = 0$ since the two expressions of R_1 and R_2 cancel each other. Furthermore, (35) agrees with equation (41) in the mathematical appendix of *Kasemir*, [1959] where the same problem has been solved in a different way.

The field components E_z and E_R are obtained by a straightforward differentiation of Φ_{12} to z and R. To avoid an unwieldy equation, we define

$$f_1 = \frac{a}{2} + \frac{z-H}{R_1^2}\left(\frac{aR_1}{2} - 1\right) \quad ; \quad f_2 = \frac{a}{2} + \frac{z+H}{R_2^2}\left(\frac{aR_2}{2} - 1\right) \quad (36)$$

Then E_z is given by

$$E_z = \frac{I}{4\pi\sigma_0} e^{-\frac{a}{2}(z+H)} \left(f_1\frac{e^{-\frac{a}{2}R_1}}{R_1} - f_2\frac{e^{-\frac{a}{2}R_2}}{R_2} \right) \quad (37)$$

and at the plane where $z = 0$ and $R_1 = R_2$

$$E_{zo} = -\frac{I}{4\pi\sigma_0} \frac{e^{-\frac{a}{2}(R_1+H)}}{R_1^3}(aR_1 - 2)H \quad (38)$$

$$E_R = -\frac{I}{4\pi\sigma_0} e^{-\frac{a}{2}(z+H)}$$

$$\left[\left(\frac{a}{2}R_1 - 1\right)\frac{Re^{-\frac{a}{2}R_1}}{R_1^3} - \left(\frac{a}{2}R_2 - 1\right)\frac{Re^{-\frac{a}{2}R_2}}{R_2^3} \right] \quad (39)$$

and at the plane with $R_1 = R_2$,

$$E_{Ro} = 0 \quad (40)$$

The complete solution of our problem is obtained by the superposition of the local potential function Φ_{12} given by equation (35) and the global potential function Φ_3 given by (23) or in good approximation by (23'). However, note that Φ_{12} is given in cylindrical coordinates R,z and is limited to an area close to the generator where the Earth can be approximately represented by a plane surface.

4. Numerical Evaluation and Conclusions

It is not the purpose of this paper to evaluate a large number of experimental data or even claim that the presently assumed values for height, ground field or current, scale height of conductivity, etc., agree or do not agree with the data measured or used by other scientists. The numerical evaluations should be considered only as few examples to show how the theoretical equations can be applied to experimental data or research.

4.1. The Rain Cloud

Scale height of conductivity $1/a$; $a = \ln 10 / 10$ km. The height of point source over the Earth surface is H = 5km. The electric field underneath the cloud at the ground is $E = -5$ kV/m. From equation (31) for $R = 0$ and $R_1 = H$,

$$E_{zoo} = -\frac{I}{4\pi\sigma_0}\frac{aH-2}{H^2}e^{-aH} \quad ; \quad (41)$$

$$I = -\frac{4\pi\sigma_0 H^2 e^{aH}}{aH-2}E_{zoo} = 0.031A.$$

The equivalent charge

$$Q = \frac{\varepsilon}{\sigma_H}I = 4.34 \ Cb. \quad (42)$$

The local current

$$I_b = -\frac{\sigma_0}{\sigma_H}I = -0.010 \ A. \quad (43)$$

The global current

$$I_g = -I - I_b = -0.021 \ A. \quad (44)$$

The global current density

$$i_g = 4.1 * 10^{-5} \ pA/m^2. \quad (45)$$

KASEMIR: CURRENT BUDGET OF THE ATMOSPHERIC ELECTRIC GLOBAL CIRCUIT

The global electric field

$$E_g = 2\,mV/m. \tag{46}$$

It would require 50,000 rain clouds of this type going on continuously on the Earth to produce the fair weather current density $i = 2\,p\,A/m^2$.

The equations given here are new and quite powerful. They show the key parameters which should be measured or checked, for instance, conductivity with altitude up to 7 km or more (aircraft, balloon), altitude, and location of cloud (radar, theodolite 3 stations at the ground). If aircraft is available, this could be equipped with conductivity and electric field recorders. Electric field measurement at the ground would then provide the very desirable redundant measurements.

These are quite moderate requirements for a research project. It is really surprising how little attention is given to electrified rain clouds as compared to lightning and thunderstorms.

4.2 Thunderstorm

The generally accepted generator for the global circuit is the worldwide thunderstorm activity with about 1800 thunderstorms being active at any one time. The simple model is the bipolar thunderstorm with the positive source in the upper part and the negative source in the lower part of the cloud. Figure 4a shows the current flow lines of such a bipolar arrangement and Figure 4b the equivalent technical circuit diagram [Kasemir, 1959, p. 61].

Obviously, most of the current lines go from the positive to the negative pole (inside the cloud). This current was called in the previous section the local current. Here we may call it the intracloud current I_c instead of I_b. In Figure 4b it is the current through the resistor Wi connecting directly the positive and negative terminal of the generator. The current from the negative pole to the ground (previously called I_g) is already part of the global circuit. If the Earth is represented by a sphere, this current would spread out equally over the Earth surface and represent the contribution of the storm to the fair weather

current. It would also raise the Earth to a negative potential. If the Earth is represented by an infinite plain, the current will spread out to zero. The remaining flow lines from the positive pole upward will go into the ionosphere. How and where they connect with the fair weather current is still an open question. This is indicated in Figure 4b by the horizontal dashed line which represents the ionosphere.

The calculation of the bipolar thunderstorm model is not difficult. If we are interested in a local problem which is represented in the left vertical branch in Figure 4b, we add the potential function of the positive source Φ_{12+} to the potential function Φ_{12-} of the negative to obtain the potential function Φ of our problem. For this we use equation (35) with appropriate height and strength of the individual source. For the calculation of the contribution to the global circuit we use equation (23) or in good approximation (23').

The parameters used for the following numerical evaluation are based on thunderstorm data obtained during the 2-year lightning suppression project of the Atmospheric Physics and Chemistry Laboratory (APCL) of the National Oceanic and Atmospheric Agency (NOOA) at Boulder, Colorado. The thunderstorms investigated were of small or moderate size with a top height of about 9 km above ground. The negative charge center was at about 5 km and the positive at about 7 km above ground. The maximum field between the center was at about 6 km and of a strength of 300 kV/m. From these data we calculate first the current output of the two current sources $\pm\,I$, then the net current between the poles I_c and the contribution to the global circuit I_g, and finally, the maximum field at the ground and the charge Q_{cg} on a cloud-to-ground lightning channel. The charge per unit length q on the channel is given by $q = k\Phi$ with $k = 25\,pF/m$ and Φ the thunderstorm potential value at the height z. The length of the lightning is, in our case 6 km, given by the zero crossing of the potential function Φ, at this altitude. The net charge Q_{cg} on the channel is obtained by integrating q from z = 0 to z = 6 km: given $H_1 = 7\,k$, $H_2 = 5km$, $E_z\,max = 300\,kv/m$; calculated strength of current sources $= \pm\,I = \pm\,0.152A$; current between poles $= I_c = -0.133\,A$; global current contribution $= I_g = -0.019\,A$; maximum field at

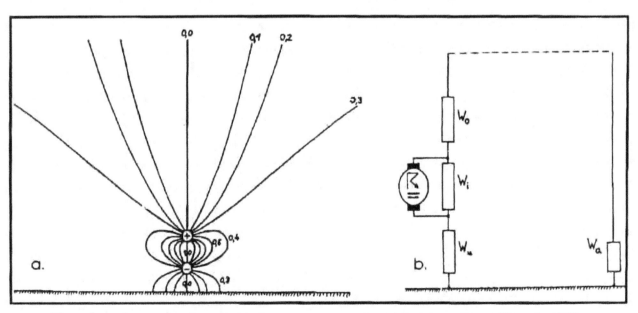

Figure 4. (a) Current flow lines of a bipolar thunderstorm and (b) technical circuit diagram [Kasemir, 1959].

ground $= E_{zm} = 15$ kv/m; net charge of ground lightning $Q_{cg} = -9.7$ C. Since the lightning is in essence an electrostatic phenomenon, we have to use equation (9) to calculate the contribution to the global circuit, which is 0.006 C.

The most surprising result is the low value of the strength of the current source $I = 0.152$ A and the extremely low value of the contribution to the global circuit of $I_g = -0.019$ A. If we accept the average of the fair weather current density of $i = -2p$ A/m², multiply this with the surface of the Earth $S = 509 * 10^{12}$ m² , and divide this by the number of thunderstorms $N = 1800$, we obtain the required contribution of the single storm $I_g = -0.57$ A. The discrepancy here is a factor 30. If this can be resolved by a greater number of storms, or stronger storms, is open to doubt. Intracloud lightning and corona on precipitation particles seem to act like safety valves to put an upper limit on fields inside the storm. Hail and maybe graupel may bring negative charge to the ground, but the author is not aware of any published data on this topic. Another possibility is a completely different generator like, for instance, the Austausch (eddy diffusion) generator proposed by the author *Kasemir* [1956] to supply the missing contribution to the global current. We have to leave this here as a problem open for discussion and to future research.

5. The Global Circuit

From the equation (23') we calculate the ground and ionospheric potential first for the required fair weather current density $i = -2$ pA/m². Multiplying by the Earth surface, we obtain the global current $I_g = -1022$ A, and with $r = k =$ Earth radius and (23') we obtain the ground potential $\Phi = -43.6$ kV and the ionospheric potential at 100 km above the ground $\Phi_i = -4.23$ µV. If we compare these values with the values used in the conventional global circuit $\Phi = 0$, $\Phi_i = 300$ kV/m, it is obvious that there is a fundamental difference in the concept of the physical and conventional global circuit.

Only a few of the most striking errors of the conventional global circuit shall be mentioned: (1) A negatively charged Earth cannot have zero potential and still drive a fair weather current up to the ionosphere. (2) The assumed ionospheric potential of 300 kV does not provide a potential difference with Earth. An equally distributed charge in the ionosphere lifts the whole interior including Earth to the ionospheric potential, and with the Earth potential zero added to the ionospheric potential, the Earth assumes also the ionospheric potential of 300 kV. There are no field lines or current flow lines between the Earth and the ionosphere (principle of the Faraday cage). (3) A positive charge from the thundercloud or a negative charge from the fair weather areas flowing up to the ionosphere does not bend around and distribute itself over the ionosphere. It would not need a highly conductive but a good insulating ionosphere to produce this spreading out effect. In reality, the conductivity in the ionosphere increases steadily with altitude even as it assumes tensor character under the influence of the Earth magnetic field [see *Volland*, 1984, Figure 5]. This author also gives a concise description of the conventional model of the global circuit.

The ionospheric potential above the bipolar thunderstorm model of the previous section is about +8.6 mV and over fair weather areas about -4.3µ V. This means the ionosphere is

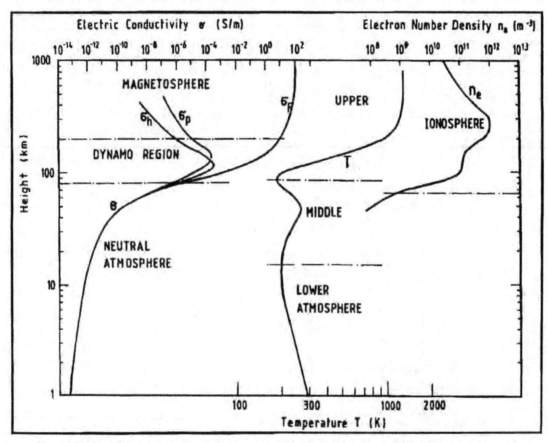

Figure 5. Electric conductivity and electron number density as functions of altitude [*Volland*, 1984].

neither an equipotential nor an equalizing layer. The only equipotential and equalizing layer is the Earth. It has also the large negative potential to drive the fair weather current. The proposed physical model of the global circuit is shown in Figure 4b. The dashed line between generator branch and fair weather branch indicates that this is an open circuit. The connection between the two branches occurs in the ionosphere or magnetosphere. The role of the equipotential and equalizing layer is transferred from the ionosphere down to the Earth surface. This frees us from the necessity to impose unrealistic conditions such as very low conductivity and high potential values on the ionosphere, which will never be accepted by ionospheric or magnetospheric scientists.

References

Abramowitz, M., and I.A. Stegun (Eds.), Handbook of mathematical functions, in *National Bureau of Standards*, 227 pp., 1964.

Kasemir, H.W., Die Stromausbeute des Gewittergeneratons in Bezug auf den luftelektrischen Vertikalstrom der Schönwettergebiete, *Ber. Dtsch. Wetterdienstes, 38*, 428-434, 1952.

Kasemir, H.W., Der Austauschgenerator, *Arch. Meteorol., Geophys. Bioklimatol., 9*(3), 357-370, 1956.

Kasemir, H.W., Das Gewitter als Generator im luftelektrischen Stromkreis, *Z. Geophys.,25* (2), 33-96, 1959.

Ollendorff, F., *Potentialfelder der Elektrotechnik*, Springer-Verlag, New York, 1932.

Smyth, W.R., *Static and Dynamic Electricity*, McGraw-Hill, New York, 1950.

Volland, H., *Atmospheric Electrodynamics*, Springer-Verlag, New York, 1984.

Williams E., and S. Heckman, *J. Geophys. Res., 98*(D3), 5221-5234, 1993.

H. W. Kasemir, Colorado Scientific Research Corporation, 8093 Anchor Drive, Longmont, CO 80501-7721.

(Received May 18, 1993; revised September 13, 1993; accepted September 13, 1993.)

Reprint from the Review
PURE AND APPLIED GEOPHYSICS (PAGEOPH)
Formerly 'Geofisica pura e applicata'

Vol. 84, 1971/I   BIRKHÄUSER VERLAG BASEL

The Atmospheric Electric Ring Current in the Higher Atmosphere

By Heinz W. Kasemir[1])

Summary – The atmospheric electric current flow in the ionosphere was discussed in a qualitative way at the UGGI General Assembly at Berkeley, California in 1963. The following picture emerged: The atmospheric electric fair weather current leaves the earth in a radially outward direction. As it enters the higher regions of the atmosphere and the ionospheric it is increasingly influenced by the earth's magnetic field. Because the main part of the current is crowded into the polar regions, the current density over the equatorial belt is small. A circular movement around the earth's axis results in an overall flow pattern tentatively termed, 'the atmospheric electric ring current'. An attempt to calculate this current flow soon made it clear that the generally used simplification of the one-dimensional case with slanted magnetic field lines is not adequate – not even as a first approximation. The same is true for the assumption usually made in magnetohydrodynamics that the current follows approximately the magnetic field lines. An essential feature of the atmospheric electric ring current is that in equatorial regions the flow is forced across the magnetic field lines, the component along the lines being zero. A calculation is discussed that treats the magnetic field lines as those of a true dipole field with the corresponding tensor character of conductivity. The results of the calculation are presented as graphs of the density distribution of the ring current, the space charge distribution, the current flow, and equipotential lines.

Zusammenfassung – Der luftelektrische Stromfluss in der Ionosphäre ist in qualitativer Weise während der UGGI Tagung in Berkeley California, 1963 diskutiert worden. Hierbei hat sich das folgende Bild ergeben: Der luftelektrische Schönwetterstrom fliesst von der Erdoberfläche nach ausswärts in radialer Richtung. Sobald er in die höheren Atmosphärenschichten und dann in die Ionosphäre kommt wird er in zunehmendem Masse vom erdmagnetischen Feld beeinflusst. Der Hauptteil des Stromes wird in die Polarzonen abgedrängt, wodurch die Stromdichte über dem Äquatorgürtel verhältnismässig klein wird. Zu gleicher Zeit wird eine kreisförmige Bewegung um die Erdachse ausgelöst, was ein Strombild ergibt, das versuchsweise der luftelektrische Ringstrom genannt wird. – Bei der Berechnung dieses Stromflusses ergab sich bald, dass die allgemein üblichen Vereinfachungen des eindimensionalen Falles mit homogenem, schräg einfallendem Magnetfeld nicht brauchbar sind, nicht einmal in erster Näherung. Dasselbe gilt für die Annahme, die gewöhnlich in der Magnetohydrodynamik gemacht wird, nämlich dass der Stromfluss angenähert dem magnetischen Felde folgt. Eine wichtige Eigenschaft des luftelektrischen Ringstromes ist es, dass der Strom über dem Äquatorgürtel gezwungen ist quer über die magnetischen Feldlinien zu fliessen, wobei die Stromkomponente in Richtung der Feldlinien gleich 0 ist. In der hier durchgeführten Rechnung wird das magnetische Feld als wahres Dipolfeld behandelt mit dem einer solchen Feldverteilung entsprechenden Tensorcharakter der Leitfähigkeit. Die Ergebnisse der Rechnung werden an Hand von graphischen Darstellungen der Ringstrom- und Raumladungsdichte und der Strom- und Äquipotentiallinien diskutiert.

[1]) Atmospheric Physics and Chemistry Laboratory, NOAA Environmental Research Laboratories, Boulder, Colorado, 80302, USA.

The Atmospheric Electric Ring Current in the Higher Atmosphere

1. Critical review of the atmospheric electric spherical capacitor model

For the last half century, the classical model of the global atmospheric electric current flow has been that of the leaky spherical capacitor, where the ionosphere is assumed to be an equipotential layer of infinite conductivity. If one intends to calculate the atmospheric electric current flow in the ionosphere, the validity of this assumption has to be carefully re-examined, and if it turns out that such an assumption is invalid, a new model of the global atmospheric electric current flow has to be developed. In this section, we shall examine the old model in detail and make the necessary correction.

The two electrodes of the spherical capacitor – the ionosphere and the earth – are supplied with positive and negative charges respectively by the worldwide thunderstorm activity. The leakage current between the earth and the ionosphere constitutes the air-earth current of the fair weather areas. Its voltage drop over the resistance of the unit volume of air is known as the fair weather atmospheric electric field. It is tacitly assumed that the poles of the thunderstorm generator are connected directly to the ionosphere and the earth. This implies that the flow of the positive supply current from the top of the storm to the ionosphere is confined to a small column and does not influence the radial outward flow of the negative fair weather current. That this is true to a very good approximation has been shown by the author [1][2] in a detailed calculation. The reason for the concentration of the positive supply current in a small cylindrical volume above the thunderhead is the rapid increase in conductivity with altitude. Another fortuitous effect of the conductivity increase is that the negative charge brought down to earth is not bound to the area below the storm but released to spread out equally over the earth's surface [1]. This is true for both the charge brought down by precipitation as convection current and the charge brought to earth by conduction current. The second reason for the equal distribution of charge over the surface of the earth is the small resistivity of the earth's crust compared with the very high resistivity of the enveloping air. A negative charge delivered from the base of the storm to the ground would find less resistance to flow all the way around the earth to the antipode point, then to penetrate only a few centimeters into the air envelope. We therefore, are perfectly justified, in treating the earth as a conductor in the global circuit or as an electrode in the capacitor model. But this is not true for the ionosphere. There is no sharp steplike decrease of about 10 powers of ten between the conductivity of the stratosphere and ionosphere as there is, for instance, between the earth's surface and the adjacent air. On the contrary, the conductivity increases steadily in the direction of the current flow. This tends to lead the current in the direction of the conductivity gradient and impedes a sideways spread. The radially outward flowing fair weather current entering the ionosphere will therefore follow the increasing conductivity and penetrate the ionosphere instead of

---

[2]) Numbers in brackets refer to References, pages 87/88.

bending sideways to return to the thunderstorm head, as has been clearly demonstrated by the author [1]. Hence the ionosphere does not constitute an equipotential layer in the global current flow and cannot be replaced by the outer electrode of the spherical capacitor model.

Somewhat surprisingly, it is not even necessary to introduce the ionosphere as an outer electrode. When, for instance, in the spherical capacitor model the ionosphere as an infinite conducting layer is removed, the current flow in the lower regions of the atmosphere and stratosphere does not change at all. The introduction of the ionosphere as an outer electrode did not come from any physical necessity but was caused by the desire to establish a closed circuit in a limited space.

Using only the physical facts of which we are sure, we can construct the model of the global current flow as follows. To represent the flow of the fair weather current, the earth is assumed to be the negative current source and the conductivity of the air is given as a function increasing with altitude. The current flow then is radially outward; it would not stop at the ionosphere but flow outward to infinity. To represent one thunderstorm a positive current source is placed at the location of the top of the storm. The potential and field distribution is calculated with the same conductivity function as above and the boundary condition that the earth is an equipotential layer. Each thunderstorm is then represented by such a positive current source. Superposition of the potential functions of all the positive point sources and the negative earth source results in the complete picture of the global current flow.

The additional condition that the sum of the positive point sources and the negative earth source are of equal strength, follows from the fact that both current sources are fed by the thunderstorms, which can produce only positive and negative charges of the same amount.

The picture that emerges is about as follows. The earth emits from the whole surface a weak but constant stream of negative charge flowing radially outward. Imbedded in this eternal drift are short-lived point sources of positive charge at the locations of the thunderstorm heads, which spring into existence and die as does the light of fireflies on a moonlit night. The current flow of the positive point source is funneled into a tube of very small diameter, which stretches upward from the thunderhead to the ionosphere. How and where the positive and negative streams of charge meet to form a closed circuit depends on the conductivity in and beyond the ionosphere. Under the assumption that the conductivity increases indefinitely with distance from the earth and that the influence of the earth's magnetic field can be neglected, the author [1] has shown that the negative fair weather current goes far out into interplanetary space before being neutralized by the positive current flow emerging from the thunderstorm heads.

From about 90 km altitude upward, however, the influence of the earth's magnetic field on the atmospheric electric current flow becomes increasingly dominant. The deviation from the radial path caused by the earth's magnetic field is discussed in the two sections that follow.

2. Calculation of the influence of the earth's magnetic field on the atmospheric electric fair weather current

From a physical point of view, the production and the dissipation of positive and negative charges in the atmospheric electric circuit are tied together. The generator effect of the thunderstorm can be pictured as one of charge generation and separation. The first will provide the precipitation particle in the cloud with surplus negative charge, whereas the remaining positive charge is attached to the cloud elements. Gravity forces, which cause the precipitation to fall to the ground, will separate the positive and negative charges, producing a constant supply of positive charge, (i.e., a positive current source) in the top of the thundercloud and a constant supply of negative charge (i.e., a negative current source) at the ground. This provides simultaneously positive and negative current sources of equal strength.

In a mathematical treatment of the problem, we are not bound to calculate the current flow of a positive and a negative current source simultaneously but we can split the problem into two parts, first calculating the current flow of the negative current source, with infinity acting as a sink, and then calculating the current flow of the positive current source, again with infinity acting as a sink. Superposition of the two solutions will furnish the picture of the two current sources producing positive and negative charges simultaneously. We will make use of this superposition principle in this paper to calculate the flow of the fair weather current under the influence of the earth's magnetic field.

At altitudes of about 100 km and higher above the earth's surface, the magnetic field exerts a steadily increasing influence on the movement of charged particles. The gross influence on the current flow can be mathematically expressed by treating the conductivity as a tensor. When the electric field vector is pointing in the same direction as the magnetic field vector, the current flow is not hampered and the conductivity (longitudinal conductivity) is the same as defined in atmospheric electricity. When the electric field vector is at right angles to the magnetic field vector, the current flow in the direction of the electric field is impeded. This effect can be viewed as though the conductivity across the magnetic field lines (Pedersen conductivity) were smaller than the longitudinal conductivity.

If the electric and magnetic fields are not in the same direction, a deflection of the current occurs at right angles to the plane determined by the magnetic and electric field vectors. The components of the current and the conductivity in this direction are called the Hall current and the Hall conductivity. We will see later that a circular Hall current forms around the earth, which we may call the atmospheric electric ring current.

It may be mentioned here that since, due to the rotational symmetry, the circular Hall current does not produce polarization charges, the consequence of the polarization charges – the impeded Hall current and the Cowling conductivity – are absent in our problem.

In trying to find a mathematical solution to a rather complex and unwieldy problem, one usually has the choice between a complicated solution, which represents the physical object rather well, or a more manageable and elucidating solution, which requires some severe simplifications in the physical framework. In our case the two-dimensional, instead of the three-dimensional, problem is solved. This means that the earth is not represented by a sphere but by an infinite cylinder with its axis perpendicular to the earth's axis, and that the earth's magnetic fiield is not that of a dipole but that of a dipole line. This simplification is acceptable because of the rotational symmetry of the three-dimensional problem. The equipotential and current flow lines of a cross section of the cylinder will be very similar to those in the three-dimensional rotationally symmetric case.

Less justified is the assumption that the longitudinal and Pedersen conductivities are constant with respect to the distance from the cylinder's surface. The ratio of the two conductivities and the conductivity values can be adjusted so that they fit the conditions in the ionosphere between the altitudes of 150 and 300 km. The calculation will then show the tendency of the current flow in this region. In nature, however, below the ionosphere there are the stratosphere and the troposphere through which the current must pass and where the conductivity is so low that the effect of the magnetic field can be neglected. This means that the longitudinal and the Pedersen conductivities are equal and that a buffer region exists which will somewhat influence the current flow in the ionosphere. In interplanetary space outside the ionosphere the conductivity changes again. This would require another modification of the present calculation. We may regard the treatment here, therefore, as a first but necessary step to a more complete solution.

With the simplifications outlined above we can formulate our problem as follows:

Consider an indefinite cylinder of radius R_0 and charge per unit length q imbedded in a medium with longitudinal and Pedersen conductivities λ and μ. The cylinder is an equipotential surface. The magnetic field is given by the potential function of a magnetic dipole line. A steady state condition is imposed by the equation

$$\operatorname{div} \bar{\imath} = 0, \tag{1}$$

where $\bar{\imath}$ is the current density. The potential and streamline function of our problem are given by the real and imaginary part of a function W, which is the solution of the differential equation (1) with the boundary condition that the potential function ϕ – the real part of W – is zero at the cylindrical surface.

We use the potential and streamline function of the magnetic dipole line to define the coordinates l, t, and z, which are related to the cylindrical coordinates R, φ, and z by

$$l = R_0^2 \sin \varphi / R; \quad t = R_0^2 \cos \varphi / R; \quad z = z. \tag{2}$$

This coordinate system is depicted in figure 1a. The coordinates l, t, and z comprise a curvilinear orthogonal system with the metric tensor components g_{ll}, g_{tt}, and g_{zz}

given by

$$g_{ll} = g_{tt} = R_0^2(l^2 + t^2)^{-2}; \quad g_{zz} = 1. \tag{3}$$

For calculating the divergence and gradient operator, the h parameters are used, which are derived from

$$h_l = (g_{ll})^{1/2}; \quad h_t = (g_{tt})^{1/2}; \quad h_z = (g_{zz})^{1/2}. \tag{4}$$

From (3) and (4) we obtain

$$h_l = h_t = R^2/R_0 = R_0^2/(l^2 + t^2); \quad h_z = 1. \tag{5}$$

The physical components E_l, E_t, and E_z of the electric field vector are given by

$$E_l = -\frac{1}{h_l}\frac{\partial \Phi}{\partial l}; \quad E_t = -\frac{1}{h_t}\frac{\partial \Phi}{\partial t}; \quad E_z = -\frac{1}{h_z}\frac{\partial \Phi}{\partial z}. \tag{6}$$

Because the potential function ϕ is constant in the z direction, E_z is zero. With the unit vectors in the direction of the three coordinates denoted by $\vec{l}$, $\vec{t}$, and $\vec{z}$, the current density vector is given by

$$\vec{i} = \vec{l}\,\lambda\,E_1 + \vec{t}\,\mu\,E_t + [\vec{H}/|\vec{H}| \times \vec{t}]\,\omega\,E_t. \tag{7}$$

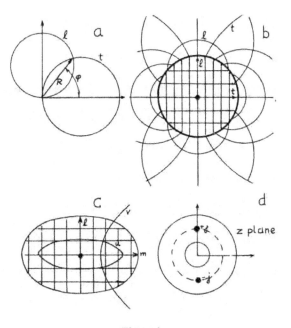

Figure 1

a. Cylinder coordinates R, φ and magnetic dipole coordinates l, t. b. Inversion of the magnetic dipole coordinates at the cylinder R_0. c. Expansion of the inverted l, t coordinates into the l, m coordinates and the elliptical u, v coordinates. d. Two line charges at $\pm j$ in the cylindrical condenser

The last term in (7) is the Hall current i_H, ω being the Hall conductivity and $\vec{H}$ the magnetic field vector. Since ω as well as E_t do not depend on z, we have

$$\operatorname{div} i_H = \frac{\partial}{\partial z}(\omega\, E_t) = 0.$$

(8)

From (1), (7), and (8) it follows that

$$\operatorname{div}(\vec{l}\, \lambda\, E_l + \vec{t}\, \mu\, E_t) = 0.$$

(9)

Using the h parameters, we can write the divergence operator in the l, t, z coordinate system as

$$\frac{\partial}{\partial l}(\lambda\, h_t\, h_z\, E_l) + \frac{\partial}{\partial t}(\mu\, h_l\, h_z\, E_t) = 0,$$

(10)

and, with (5) and (6),

$$\lambda \frac{\partial^2 \phi}{\partial l^2} + \mu \frac{\partial^2 \phi}{\partial t^2} = 0.$$

(11)

With the substitution

$$m = (\lambda/\mu)^{1/2}\, t = \beta\, t,$$

(12)

(11) transforms to

$$\frac{\partial^2 \phi}{\partial l^2} + \frac{\partial^2 \phi}{\partial m^2} = 0.$$

(13)

Equation (13) is the well known Laplace differential equation for which a solution that fits our boundary condition must be found.

Following step by step the different substitutions, we see that by the coordinate transform (2) the space outside the cylinder with radius R_0 is inverted into the space inside the cylinder. The coordinates l, and t, which are two sets of cylinders, are transformed into two sets of l, t planes inside the cylinder R_0 that intersect at right angles (figure 1b). Because of the different conductivities along and across the magnetic field lines, the spacing of the planes in the l and t direction is not the same. This is altered by substitution (12), which expands the spacing of the t coordinate to that of the m coordinate (figure 1c). The m and l coordinates are now evenly spaced, and we are in Laplacian space. However, the price we have to pay for the simpler differential equation is that our circular cylinder is deformed by the t expansion into an elliptical cylinder (figure 1c), resulting in a more complicated boundary condition.

Inverting the space outside the cylinder into the inside transforms the sink at infinity into a line source at the center of the cylinder. If, for instance, our cylinder is a source of negative current, the line source in the center has to absorb the negative current, i.e., it has to be a positive line source. Expanding the coordinate system from the t to the m spacing does not change the magnitude or the position of the line source. Our problem then is to find the potential function of a line charge inside an elliptical cylinder.

A solution to this problem can be obtained by conformal transformation from the complex potential function of two positive line charges at points $\pm j = (-1)^{\frac{1}{2}}$ in the complex plane inside a concentric cylindrical capacitor (figure 1d). The complex potential function of one line charge inside a grounded cylindrical capacitor has been calculated by Courant and Hilbert [2]. A simple superposition yields the potential-function W of two line charges:

$$W = -\frac{q}{2\pi\varepsilon}\ln\frac{-\vartheta_{1(w-k_1)}\,\vartheta_{1(w-k_2)}}{\vartheta_{4(w+k_1)}\,\vartheta_{4(w+k_2)}},\tag{14}$$

where q = line charge density per unit, length ϑ_1, ϑ_4 = theta functions, w = complex argument of the theta functions, and k_1, k_2 = constant complex numbers indicating the position of the line charges.

If the cross section through the cylindrical capacitor is called the complex z plane and c' is the complex number representing the location of the line charge, the following relation exist between z and w, and between c' and k:

$$z = \exp(2\pi j w); \quad c' = \exp(2\pi j k).\tag{15}$$

The location of the two line charges are $c' = \pm j$. Therefore, from (15),

$$k_1 = +\tfrac{1}{4}; \quad k_2 = -\tfrac{1}{4}.\tag{16}$$

The conformal transformation

$$z = \exp(u + j v),\tag{17}$$

or, with (15),

$$w = (u + j v)/2\pi j\tag{18}$$

transforms the cylindrical capacitor into the elliptical cylinder shown in figure 1c. The letters u and v in (17) and (18) designate the elliptical coordinates in this figure. They have the following relation to the l and m coordinates:

$$m = c\cosh u\cos v; \quad l = c\sinh u\sin v,\tag{19}$$

where c is the eccentricity of the elliptical cylinder and is defined by

$$c = R_0(\beta^2 - 1)^{1/2}.\tag{20}$$

By the transformation, the two line charges in figure 1d move into the center of figure 1c and combine into one line charge. Everything inside the dashed circle in figure 1d is removed from the inside of the elliptical cylinder. Only the space between the outer solid and the dashed circle in figure 1d is transferred to the inside of the ellipse in figure 1c.

From (2), (12), and (19), we obtain

$$\left.\begin{aligned}
\cosh u\cos v &= \beta R_0\cos\varphi/R(\beta^2 - 1)^{1/2},\\
\sinh u\sin v &= R_0\sin\varphi/R(\beta^2 - 1)^{1/2}.
\end{aligned}\right\}\tag{21}$$

Using the summation formula of complex hyperbolic functions and (18), we can obtain a direct relation between the argument w of the theta function and the cylindrical coordinates R and φ of our problem, namely,

$$2\pi j\, w = \cosh^{-1} R_0(\beta \cos \varphi + j \sin \varphi)/R(\beta^2 - 1)^{1/2}. \tag{22}$$

3. Discussion of the atmospheric electric current flow in and beyond the ionosphere

The real part of (14) is the potential function and the imaginary part of (14) is the streamline function of the current flow from a charged circular cylinder under the influence of a magnetic dipole line. A graph of the equipotential and streamlines is given in figure 2. The ratio of the longitudinal to the Pedersen conductivity is chosen to be 9:1. The choice is arbitrary and the ratio is somewhat large for ionospheric conditions, but the advantage is that the effect of the magnetic field on the current flow can be seen very clearly. The solid lines in figure 2 represent equipotential lines, the dashed lines streamlines; the dashed dotted line is one magnetic field line. The equipotential lines are not concentric circles around the cylinder, as they would be in the absence of the magnetic field, but elongated lines that are widely spaced over equatorial and narrowly spaced over polar regions. Close to the cylinder they will become more and more circular, to conform finally with the circular shape of the cylinder's surface. We may expect, therefore, in the troposphere, and may be partly

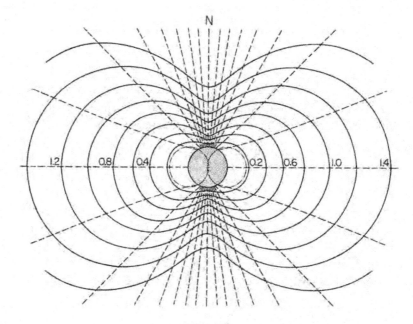

Figure 2
Equipotential lines (solid lines) stream lines (dashed lines) and magnetic field line (dashed dotted line) of a cylindrical negative current source under the influence of a magnetic dipole line

in the stratosphere, the equipotential surfaces of the fair weather field to be concentric spheres around the earth, but in the higher stratosphere, in the ionosphere, and beyond they will change to flat disclike bodies with the small diameter at the pole and the large diameter at the equator.

Since it appears that the old concept of the ionosphere being an equipotential larger will turn out to be incorrect, it is not advisable to make the ionospheric potential one of the key parameters in the global circuit. Also, the practice of extrapolating the ionospheric potential from field measurements with a radiosonde is based on the assumption that the earth's magnetic field has no effect on the atmospheric electric current flow at higher altitudes. It would be much better if the field were integrated only up to the maximum balloon altitude, because that is really the true experimental result. Extrapolation to the ionosphere has only fictitious value.

The next interesting result is the concentration of the streamlines at the poles and their dispersion at the equator. This effect will be damped by the low conductivity in the tropospheric and lower stratospheric layers, where the influence of the earth's magnetic field is negligible. Surprisingly, however, this effect has been found at the earth's surface [3]. It is known as the atmospheric electric latitude effect. The explanation – with some reservations by the authors – is the lower columnar resistance and therefore the higher current density in polar regions. We may call this the conductive latitude effect to distinguish it from the magnetic latitude effect. It would be interesting to carry out experiments to determine to what extent each of these two effects are responsible for the current concentration at the poles. Such experiments would consist of high altitude balloon sounding at the equator and in the polar regions to determine simultaneously the potential difference between the ground and the maximum balloon altitude and the current density at both places. If the current density were higher at the pole, but the integrated field to a common altitude level were the same at both places, then the conductive latitude effect would be at work. If, however, both the current density and the integrated field were higher at the pole than at the equator, then the magnetic latitude effect would be responsible.

It seems fairly certain that the low conductivity in the troposphere and the lower stratosphere prevents the magnetic latitude effect from penetrating to the surface in full magnitude. In the ionosphere and farther out, however, the magnetic field increasingly dominates the atmospheric electric current flow, and the magnetic latitude effect is present in full force. This means that at some altitude below the ionosphere there is a transition region in which a horizontal poleward current and field component will appear. This component would be opposite to the current flow one would expect in the absence of the magnetic field. Indeed, the current stream in the fair weather polar region would tend to flow in the direction of the main thunderstorm regions at the equator and not away from them, a phenomenon that could be the basis for another experimental check on the validity of the electro-magnetic current flow theory of atmospheric electricity.

There is one more rather unusual feature in figure 2 that is contrary to atmospheric electric concepts: the current flow lines are not at right angles to the equipotential lines; or in other words, the electric field and the current flow are not in the same direction. This is a well-known fact in ionospheric physics and magnetohydrodynamics. However, the rule that the current flow tends to follow the magnetic field lines is not evident in figure 2. In the equatorial regions the flow goes right across the magnetic field lines. Nonetheless, the small current density in these regions indicates the additional resistance imposed by the magnetic field, so that most of the current is deflected to the polar regions. If the circular cylinder emits negative current, one would expect that a negative space charge would be deposited over the equatorial region, causing weaker fields and current densities; over the polar regions, one would expect positive space charge densities, causing the opposite effects.

Figure 3 shows the lines of equal space charge density and confirms this expectation. This figure also illustrates in much more detail the space charge distribution. Note that this is not a space charge generated by the Hall current but is built up to minimize the energy consumption of the overall current flow.

The highest positive space charge concentration occurs, surprisingly, at about $\frac{1}{2}$ earth radius above the north and south poles. The strongest negative space charge

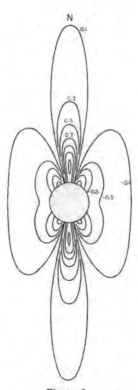

Figure 3
Lines of constant space charge density of the current flow of a negative charged cylinder under the influence of a magnetic dipole line

H. W. Kasemir

[2] R. COURANT and D. HILBERT: *Methods of Mathematical Physics*, Vol. 1, (Interscience Publishers, New York 1953). (Reference to page 386–388).

[3] O. H. GISH and K. L. SHERMAN: *Latitude-effect in electrical resistance of a column of atmosphere*, Trans. Amer. Geophys. Union. 19th Ann. Meetg. (1938), 193–199.

[4] H. ISRAËL: *Atmosphärische Elektrizität*, Teil II (Akademische Verlagsgesellschaft, Leipzig 1961). (Reference to page 32–34.)

SECTION 3. THUNDERSTORM ELECTRICITY

1. The Thundercloud, H.W. Kasemir, Problems in Atmospheric and Space Electricity, Ed. S. Coroniti, Elsevier Publishing Co., Amsterdam, London, New York, 1965.

2. Charge Distribution in the Lower Part of the Thunderstorm, H. W. Kasemir, Proc. International Conference on Cloud Physics, 1968, Toronto, Canada.

3. Charge Distribution in Thunderstorms, H. W. Kasemir, Proceedings in Atmospheric Electricity, Lothar H. Ruhuke and John Latham, Eds., A. Deepack Publishing, 1983.

4. Modification of the Electric field of Thunderstorms, H.W. Kasemir and H.K. Weickmann, Proceedings of the International Conference on Cloud Physics, 1965.

5. Lightning Suppression by Chaff Seeding, H. W. Kasemir, Final Report on Project USAFCOM 67-95906, June 1968.

6. The Behavior of Chaff Fibers in a Thunderstorm and their Effect on the Electric Field, H. W. Kasemir, Atmospheric and Chemistry Laboratory, Environmental Service Administration, Boulder, Colorado, NOAA, no date (n.d.).

7. The Effect of Chaff Seeding on Lightning and Electric Fields of Thunderstorms, H. W. Kasemir, Proc. International Conference on Weather Modification, Tashkent, USSR, 1973.

8. Accelerated Decay of Thunderstorm Electric Fields by Chaff Seeding, F. J. Holiza and H. W. Kasemir, J.Geophys. Res. Vol. 79, No.3, January 1974.

9. Corona Discharge and Thunderstorm Fields, H. W. Kasemir, Proc. Conference on Cloud Physics and Atmospheric Electricity in Issaquah, Wash. AMS, Boston, MA, 1978.

Session III.4

THE THUNDERCLOUD

H. W. KASEMIR

*U.S. Army Electronics Research and Development Laboratory,
Fort Monmouth, N.J. (U.S.A.)*

PRELIMINARY REMARKS

Even a casual survey of the literature concerned with the electric phenomena of a thundercloud yields an enormous amount of detailed data. Less effort has been spent on integrating these data into an overall picture. There are several reasons for this state of affairs: The complexity of the problem itself; our lack of knowledge as to which of the large number of charge-generating effects in the thundercloud is the predominant one; the difficulty of measurements in the thundercloud itself, and therefore the meager and incomplete data on charge and field distribution inside the storm; the ambiguity of conclusions drawn from measurements at the ground; and the lack of confidence that laboratory experiments represent a fair duplication of the cloud conditions and that insignificant effects at the laboratory scale do not become important in cloud dimensions.

A survey article about our present knowledge of thunderstorm electricity may on one hand give a most complete list of all the pertinent data in an objective manner, or it may on the other hand value and select these items which seem to be the most significant and try to integrate these into an interrelated structure. The second way reflects by necessity the subjective view of the author, but leads more naturally to a future plan of research in this area. Since at this conference ample opportunity is offered during the discussion and rebuttal period to comment, supplement, or criticize this paper, the danger of a too one-sided representation is minimized, and therefore the second way is chosen.

REQUIREMENTS OF A THUNDERSTORM THEORY
AFTER ISRAËL, MASON AND CHALMERS

We will start with a survey of the facts which a thunderstorm theory should fulfil. ISRAËL (1961) put forward the following conditions[1]. (*1*) The electrification process shall be of such a kind that it leads by the segregating effect of gravity and (vertical) airstream to the well-known charge distribution of a positive polar thundercloud. (*2*) Considered as a current source, the electrification process shall be able to produce a current density

---

[1] Translation of the author.

of about 10^{-12} A/cm² (10^{-8} A/m²), or a net current of 0.5–1 A/thunderstorm. (3) The direction of the current has to be so that negative charge is carried to the ground.

MASON (1957) came to the following requirements: (1) The average duration of precipitation and electrical activity from a single thunderstorm cell is about 30 min. (2) The average electric moment destroyed in a lightning flash is about 110 Ckm, the corresponding charge being 20–30 C. (3) The magnitude of the charge which is being separated immediately after a flash, by virtue of the fall speed v of the precipitation elements, is of the order of 8000/v C, where v is the fall velocity of the particles relative to the air, in m/sec. (4) In a large, extensive cumulus, this charge is generated and separated in a volume bounded by the -5 °C and the -40 °C levels and having an average radius of perhaps 2 km. (5) The negative charge is centered near the -5 °C isotherm, while the main positive charge is situated some kilometers higher up, a subsidiary positive charge may also exist near the cloud base, being centered at or below the 0 °C level. (6) The charge generation and separation processes are closely associated with the development of precipitation, particularly in the form of graupel. The precipitation particles must be capable of falling through updrafts of several meters per second. (7) Sufficient charge must be generated and separated to supply the first lightning flash within 12–20 min of the appearance of precipitation particles of radar-detectable size.

CHALMERS (1957) requires that every theory satisfy the following conditions: (1) The process must give a positive upper charge and a negative lower charge. (2) The process must give a rate of separation of charge of up to several amperes. (3) The process must operate at temperatures below the freezing point. (4) The process must be connected with precipitation in the solid form. (5) The process must operate in the nimbostratus clouds much less effectively, if it operates at all, than in cumulonimbus clouds.

These statements, with the implementation of the experimental results and theoretical deductions leading to them, represent a fair cross-section of our present knowledge of the electrical processes in the thundercloud. We will now discuss them in more detail and try to correlate them.

THE ELECTRIC MOMENT OF THE THUNDERCLOUD AND THAT OF A LIGHTNING DISCHARGE

The electric moment M of the thundercloud is found by SIMPSON and ROBINSON (1940) from field measurements at the ground to be:

$$M = 72 \text{ Ckm (24 C at 6 km, } -20 \text{ C at 3 km altitude)}$$

by GISH and WAIT (1950) from field measurement above the thunderstorm:

$$M = 234 \text{ Ckm (39 C at 9 km, } -39 \text{ C at 3 km altitude)}$$

and by MALAN (1952) again from ground measurement:

$$M = 200 \text{ Ckm (40 C at 10 km, } -40 \text{ C at 5 km altitude)}$$

THE THUNDERCLOUD

Comparison with the average electric moment $M = 110$ Ckm destroyed in a cloud stroke (Mason's point 2) shows that the Simpson–Robinson value of the electric moment of the cloud $M = 72$ Ckm is somewhat low, since the lightning flash cannot discharge more charge than the cloud originally obtains. The values determined by Gish and Wait, and Malan, fit very well if one assumes that the lightning flash is able to discharge about half of the cloud charge.

BROOK et al.(1962) determined the negative charge discharged by ground strokes and found an average of 19 C for discrete flashes (lightnings without continuing junction streamers between main strokes), and 34 C for hybrid flashes (lightnings with long continuing junction streamers between main strokes). Also, this indicates that the main negative cloud charge should be at least higher than 34 C. If we assume again that about half of the negative cloud charge is discharged to the ground by the lightning stroke, we obtain 38 and 68 C, respectively, for the lower negative cloud charge. Therefore the value of 40 C for the lower negative cloud charge is rather small. This seems to be even more so if one considers that the negative space charge has to feed both cloud and ground flashes. Also, the assumption that half of the cloud charge is neutralized by a lightning flash may not be correct. The fraction may be considerably less. If so, it would be easier to understand that a thunderstorm will produce lightnings every 30 sec or sometimes even faster without exhausting its charge supply. Therefore, it becomes an important task to determine the ratio of the charges neutralized by a lightning flash to the cloud charges. This may be achieved either by calculation, by laboratory experiments, or by field measurements, which determine the electric moment of the lightning flash and the cloud simultaneously. Also, we will see in the next chapters that the lower negative charge is more likely to be -340 C instead of -40 C.

THE DISCREPANCY BETWEEN THE CLOUD CHARGE AND THE CONDUCTION CURRENT

We will now connect the cloud charges Q with the conduction current I produced by the storm. We have here the simple relationship (KASEMIR, 1959):

$$Q = TI \tag{1}$$

$T = \varepsilon/\lambda$ is the time constant of air in the neighborhood of the charge. It should be pointed out that this is a basic relationship which is independent of the geometrical shape of the charged body. It can be applied as well to the charged raindrop, graupel, or snow crystal as to the positive or negative net space charge in the cloud, as long as the conductivity in the space occupied by the charge is constant.

Conductivity measurements in thunderstorms, to the knowledge of the author, have never been carried out. Therefore we will use the values obtained by PLUVINAGE (1946) and ISRAËL and KASEMIR (1952) in non-precipitating clouds and fog. On the average the conductivity is lowered by a factor 1/3 inside the cloud from the value of clear air in the corresponding altitude. If we put the center of the positive cloud charge at 9 km altitude and the center of the negative space charge at 3 km altitude, the corresponding values for the conductivity in clear air are $\lambda_9 = 45 \cdot 10^{-14}$ $1/\Omega m$, and $\lambda_3 = 7.8 \cdot 10^{-14}$ $1/\Omega m$. (The indices 9 and 3 refer to the altitudes of 9 and 3 km.) With the factor 1/3 for the conductivity inside the cloud and the dielectric constant $\varepsilon = 8.86 \cdot 10^{-12}$ A sec/V m, we obtain $\lambda_9 = 15 \cdot 10^{-13}$ $1/\Omega m$ and $\lambda_3 = 2.6 \cdot 10^{-14}$ $1/\Omega m$, and with the relation $T = \varepsilon/\lambda$ the time constant $T_9 = 60$ sec, and $T_3 = 340$ sec. The current output of the storm as quoted from Israël (point 2) is 0.5–1A. These values originate from Gish and Wait's measurements above the storm (GISH and WAIT, 1950).

If we consider that the upper pole of the thunderstorm generator feeds not only the current flow up to the ionosphere, but also the conduction current inside the cloud to the negative pole, we may be justified in using for our calculation the higher value of 1A. With the above-estimated time constant and equation *1*, we obtain for the upper positive cloud charge $Q_9 = 60$ C and for the lower negative cloud charge $Q_3 = -340$ C. These values, especially the second one, are in striking disagreement with the values of SIMPSON and ROBINSON (1940), GISH and WAIT (1950) and MALAN (1952). The upper charge is higher by 50%; the lower charge by 750%.

Such a discrepancy can hardly be caused by the fact that the data used are from different storms, so it is obvious that we missed some essential link. Therefore, let us reconsider the facts.

(*1*) The value of 0.5–1 A output current is based on 15 flights (GISH and WAIT, 1950) and is tied in with the overall supply current to the fair weather regions. A reduction of this value by a factor 1/2 –1/10 would upset the precarious balance between supply and consumption in the world-wide atmospheric current flow. So let us assume that the figures 0.5–1 A are correct. Still it would be very desirable to have this important figure based on more extensive data.

(*2*) Conductivity measurements in thunderstorms are not made at all, and one may question the validity of transferring the values of non-precipitating clouds to thunderstorms. There is a most urgent need for direct measurements. However, GUNN (1948) reports from his field measurements in thunderstorms a sudden field increase by entering the cloud, which certainly indicates a decrease in conductivity inside the cloud. Also, it is hard to imagine that the cloud particles of a thundercloud, contrary to these from non-precipitating clouds, would not catch air ions and so decrease the conductivity, or that a yet unknown ionization effect is present in the thundercloud. Therefore, the chosen conductivity values seem to be at least probable.

(*3*) Finally, there remains the check on the calculation of the cloud charges from field records by Simpson and Robinson, Malan, and Gish and Wait. These cloud charges were calculated according to electrostatic theory. It is at this point that, in the opinion of the author, the solution of the disagreement lies.

THE CLOUD CHARGE CALCULATED ACCORDING TO THE THEORY OF CURRENT FLOW

As outlined in detail by KASEMIR (1959), the theory of current flow has to be applied to

THE THUNDERCLOUD

calculate the cloud charges, i.e., space and surface charges due to the change of the conductivity have to be taken into account. We will make this clear in a simple example. A current source of the strength $I = -1A$ is placed in the center of a sphere of radius R and conductivity $\lambda_i = 2.6 \cdot 10^{-14}$ $1/\Omega m$. Outside the sphere the conductivity should be $\lambda_0 = 7.8 \cdot 10^{-14}$ $1/\Omega m$. The sphere shall represent the lower part of the thundercloud. The charge at the current source is given by equation 1, $Q = IT_1 = -340$ C. If the conductivity λ_0 outside the sphere is larger than λ_i inside, at the boundary of the sphere a charge Q_1 will form, which is given by:

$$Q_1 = -\left(1 - \frac{\lambda_i}{\lambda_0}\right) Q \quad \text{(KASEMIR, 1959),} \tag{2}$$

or $Q_1 = -\frac{2}{3}Q = +230$ C. The surface charge Q_1 will reduce the field of the original charge Q outside the sphere to $1/3$ of its value, but leaves the field inside the sphere unchanged. This means that someone, who would calculate the charge inside the sphere from field measurement outside, would obtain only a fraction, in our example one third, of the correct value, if he does not take into account the screening effect of the surface charge. The outside field looks as if it were generated by a charge Q_2 given by:

$$Q_2 = Q + Q_1 = \frac{\lambda_i}{\lambda_0} Q; \quad Q = \frac{\lambda_0}{\lambda_i} Q_2 \tag{3}$$

This relationship has been derived and used by HOLZER and SAXON (1951) in their calculation of the current flow in the vicinity of a thunderstorm. It provides an easy means to correct as a first approximation the values for the cloud charges derived from electrostatic theory. Q_2 represents the charge calculated in the electrostatic fashion; for instance, the values of Gish and Wait, and Malan, of -40 C. If we choose the ratio of free air to cloud conductivity $\lambda_0/\lambda_i = 3$, as in our example, we obtain for the corrected cloud charge by equation 3 $Q = -120$ C. This value is still too small against -340 C.

We have taken into account the difference between the conductivity values inside and

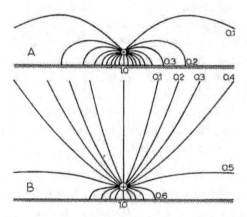

Fig. 1. Field and current lines of a point source above a conducting plane in a medium with (A) constant conductivity and (B) conductivity increasing with altitude.

outside the cloud, that means the influence of the surface charge. But we have not considered the effect exerted by the space charge due to the increase of conductivity with height. We will do this in a qualitative way only, with the help of Fig. 1A and 1B (KASEMIR, 1959). Fig. 1A shows the current or field lines of a current source above a conducting plate in a medium of constant conductivity. Fig. 1B shows the current or field lines for the same current source, but this time in a medium with a conductivity increasing with altitude according to an e function. The current source is at such an altitude above ground that the conductivity λ_h has increased to twice its ground value λ_0. We see that in Fig. 1B only half of the field lines reach the ground, whereas in Fig. 1A all field lines do. This effect can be expressed as the influence of the space charge, which will be formed because of the increase of the conductivity with height. This space charge tends to bend the current and field lines away from the ground to regions with increasing conductivity, and therefore weakens the field at the ground.

There is a simple relationship between the conductivity λ_0 at the ground and the conductivity λ_h at the height h, the current I_0 which reaches the ground, and the current I of the current source, given by:

$$I_0 = \frac{\lambda_0}{\lambda_h} I \qquad (4)$$

In our example, $I_0 = \frac{1}{2}I$. This means, if we determine the current source or its equivalent charge from field measurement at the ground by electrostatic theorems, i.e., without taking into account the screening effect of the space charge, the result will be necessarily too low.

More detailed calculations according to the theory of current flow are worked out by HOLZER and SAXON (1951) and by the author (KASEMIR, 1959). However, it is sufficient for our purpose here to show that the discrepancy between the cloud charge values calculated from field records and those calculated from the current output may be fully or at least partly explained by the erroneous application of electrostatic theorems to problems of current flow. The correct values will be those resulting from the current flow theory, namely, $+60$ C for the upper, and -340 C for the lower charge.

THE RECOVERY CURVE

One of the difficulties in thunderstorm theories based on the low values of cloud charges of, for instance, ± 40 C has been the fact that the average lightning flash discharges 20–30 C and that this charge is replaced by the storm in about 10–20 sec, as indicated by the recovery curve. Precipitation falls with a velocity of about 8 m/sec (rain) to 16 m/sec (graupel). In 20 sec it will advance 160–320 m to the ground. This is quite inadequate to bridge the gap of 4000–5000 m between the positive and negative space charge in the cloud and to restore the destroyed electric moment. To avoid the necessity of the charged precipitation falling through the whole distance of 5000 m in 10 or 20 sec, WILSON (1921) envisaged the following mechanisms (Fig. 2).

THE THUNDERCLOUD

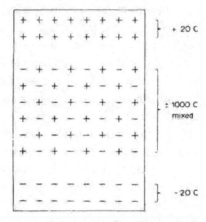

Fig. 2. Charge distribution in a thundercloud after WILSON (1921).

In the middle of the cloud there is a large volume where positive charge (on cloud elements) and negative charge (on precipitation particles) are of the same amount of about $\pm 1000\,C$. As a whole, this volume is electrically neutral; no space charge is formed and no field is generated. If the downward movement of the precipitation segregates the two charges, there will be a build-up of a positive space-charge area at the top of the volume and a negative space-charge area at the bottom. The middle part still remains neutral.

With the charge accumulation, a field is generated that grows so strong that the downward motion of the precipitation is retarded or stopped until a lightning flash occurs. Then the field is reduced and the downward motion begins again. This cycle will be repeated after each lightning flash.

From this picture evolved Mason's point 3, that a charge $Q = 8000/v$ Cm/sec or with $v = 8$ m/sec, $Q = 1000$ C should be separated immediately after a lightning flash. There are quite a number of questionable points in this mechanism:

(1) With an average charge of $7.5 \cdot 10^{-2}$ e.s.u. (GUNN, 1947) on a precipitation particle of 2-mm radius, a field of $3.2 \cdot 10^6$ V/m would be required to balance the gravity force, i.e., to keep the precipitation from falling. Such a field strength can never be reached in the cloud because, by much lower values, point discharge will start on the precipitation particles and inhibit a further field increase. Even if the generator effect in the cloud is strong enough to over-ride the limiting effect of point discharge, lightning discharge will set in and put an effective stop to a further field increase. The maximum field ever measured in a thundercloud just before a lightning flash occurs was $0.34 \cdot 10^6$ V/m (GUNN, 1948). This is about a factor 10 lower than the field strength of $3.2 \cdot 10^6$ V/m required for balancing the gravity force.

(2) If the electric field is able to retard or stop precipitation, a rain gush after each lightning stroke should be a regular feature. Nevertheless, such a sudden increase in the rain intensity is observed occasionally, but the interpretation offered by VONNEGUT (1962) explains this phenomenon by an increase of the coagulation coefficient in the strong field of the lightning discharge.

H. W. KASEMIR

(*3*) TAMURA (1955) discussed in his paper some further objection against Wilson's mechanism. His strongest point, however, is that he could show in an analytical calculation based on the theory of current flow that the fast recovery curve results, even by a continuous charge-separating mechanism, from the natural decay of the lightning charges and the screening effect of the space charge which forms because of the increase of conductivity with altitude. The recovery curve has to be considered as a transient phenomenon, in which the electric field pattern changes from the electrostatic distribution to that given by the theory of current flow. This means, as can be easily recognized from a comparison of Fig. 1A and 1B, a fast weakening of the field strength at the ground with time. As Tamura's calculation could also explain finer details in the recovery curve, for instance, the difference in the shape for near and distant lightnings, which were reconfirmed by measurements of SMITH (1958), there is not much doubt that his explanation is the right one. Therefore, Mason's point *3* based on Wilson's interpretation of the recovery curve would better be replaced by a statement concerning the probable convection current, maybe in the formulation of Chalmers', "The electrification process should be able to produce a convection current in the order of several amperes."

THE MAXIMUM FIELD STRENGTH IN THE THUNDERCLOUD

Tamura used the following values for his numerical evaluation of his thunderstorm model:

Altitude of the positive pole, $a = 7$ km.
Conductivity at a, $\lambda_a = 1.8 \cdot 10^{-3}$ e.s.u. $= 1.62 \cdot 10^{-13}$ Ω^{-1}m^{-1}.
Altitude of the negative pole, $b = 5$ km.
Conductivity at b, $\lambda_b = 1.2 \cdot 10^{-3}$ e.s.u. $= 1.08 \cdot 10^{-13}$ Ω^{-1}m^{-1}.
Conductivity at the ground, $\lambda_0 = 4 \cdot 10^{-4}$ e.s.u.
Conductivity as a function of the altitude z, $\lambda_0 e^{2Kz}$.
$K = 0.11$ 1/km.
Strength of current source, $I = 1.8$ A.
Positive cloud charge, $Q_a = 80$ C.
Negative cloud charge, $Q_b = -120$ C.

If we calculate the field strength E_c for the midpoint between the positive and negative cloud charge according to the theory of current flow, we obtain the formula:

$$E_c = -\frac{I}{\pi(a - b)^2}\left[\frac{1}{\lambda_a} + \frac{1 + K(a - b)}{\lambda_c}\right] \tag{5}$$

λ_c is the conductivity of the midpoint between the two poles. This point is given by:

$$c = \frac{a + b}{2}$$

To keep formula 5 simple, the influence of the ground has been neglected, which will introduce a small error. If we insert Tamura's values for a, b, I, K, λ_a, and λ_c ($\lambda_c =$

THE THUNDERCLOUD

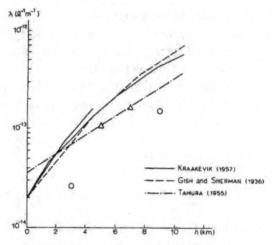

Fig. 3. Conductivity λ versus altitude h.

$1.32 \cdot 10^{-13}\ \Omega^{-1}m^{-1}$ is calculated from the formula $\lambda = \lambda_0 e^{2Kz}$), we obtain for the field strength in the middle of the two poles $E_c = 2.2 \cdot 10^6$ V/m. If we remember that in this calculation the conductivity inside the cloud is assumed to be equal to the conductivity in free air at the same altitude, and correct this by taking only $^1/_3$ of the free air value, the field will increase by a factor 3, i.e., $E_c = 6.6 \cdot 10^6$ V/m. Both of these values for E_c will never be reached in the thunderstorm, as was argued before.

The too strong midpoint field in Tamura's thunderstorm model will not affect the validity of his conclusions about the recovery curve, but it will lead us to the importance of another parameter, namely, the distance of the generator poles in the thunderstorm.

Fig. 3 shows the conductivity of clear air as a function of altitude for the first 10 km as measured by the Explorer II (GISH and SHERMAN, 1936) and by KRAAKEVIK (1957). The values chosen in our thunderstorm model are marked as circles, and the values of Tamura as triangles. It is immediately obvious from this figure that it is not so much the lower clear air values for the conductivity, but mainly the small distance of only 2 km between the positive and negative cloud charge that is responsible for the high field value. In our thunderstorm model this distance is 6 km and the midpoint field turns out to be $3.2 \cdot 10^5$ V/m. As can be seen from formula 5, the field is inversely proportional to the square of this distance, which would result in a difference of the field values of almost a factor 10 between these two thunderstorm models. This shows how important it is to choose the right distance between the poles, and brings up the problem of how the distance could be measured or deduced from other data. A first approximation could be obtained from meteorological radiosonde data with reference to Mason's points *4* and *5*, which place the negative charge center at the $-5\ °C$ level and the positive charge not higher than the $-40\ °C$ level. We will content ourselves here with this short remark because this topic is in the realm of the charge-generation-and-separation process and its connection with the meteorological aspects of the thunderstorm, which belong to papers dealing with the thunderstorm theory proper.

H. W. KASEMIR

It should be mentioned that the author (KASEMIR, 1959) assumed the conductivity in the cloud to be $^1/_3$ of the ground value, which results in positive and negative cloud charges of ± 2400 C and in a midpoint field strength of about $E_c = 20 \cdot 10^6$ V/m. This shows immediately that the assumed values for the conductivity inside the cloud were chosen far too low. All the numerical evaluations based on the wrong conductivity value would have to be corrected accordingly.

What can we learn from these mistakes?

(*1*) The conductivity inside the thunderstorm is of paramount importance to link the current and the cloud charges. Conductivity measurement inside the thunderstorm would be a necessary task of future research.

(*2*) The maximum field strength in the cloud should be determined theoretically and experimentally. It should be checked in every theoretical thunderstorm model.

(*3*) The altitude of the charge centers in the cloud should be determined tentatively from the temperature records of a nearby radiosonde.

(*4*) There seems to be a very small margin between the minimum field strength given by the current output of the thunderstorm and the maximum field strength measured and estimated by the breakdown field. This suggests a field-limiting mechanism, which may very well turn out to be the lightning discharge.

THE FIELD INCREASE AT THE BEGINNING OF THE STORM

Our next remarks will pertain to Mason's point 7, which requires that within 12–20 min from the appearance of precipitation particles the charge accumulation has reached such a magnitude that the first lightning occurs. We represent the thunderstorm by the circuit diagram of Fig. 4 as used by ISRAËL (1961) and KASEMIR (1959), but supplement it by the capacitor C, which simulates the cloud charges; and the spark gap L, which provides the possibility of a cloud discharge. R_0, R_i, and R_u are the resistances between the ionosphere and the cloud top, the cloud top and base, and the cloud base and ground,

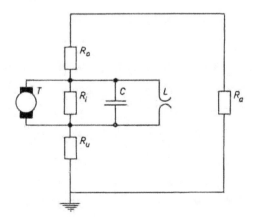

Fig. 4. Circuit diagram of a thunderstorm.

THE THUNDERCLOUD

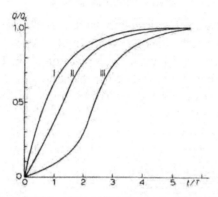

Fig. 5. Normalized cloud charge Q/Q_s versus normalized time t/T for (*I*) convection current is constant, (*II*) convection current increases linearly until $t/T = 1$, then is constant, and (*III*) convection current increases linearly until $t/T = 2$, then is constant.

respectively. R_a is the resistance of the integrated fair-weather area of the whole world between the ionosphere and the ground. For a 50 km² thunderstorm area, Israël calculates the following numerical values: $R_0 = 4.8 \cdot 10^8 \ \Omega$, $R_i = 2 \cdot 10^8 \ \Omega$, $R_u = 8 \cdot 10^8 \ \Omega$. The parallel connection of R_i and $R_0 + R_u + R_a$ results in a resistor $R = 1.74 \cdot 10^8 \ \Omega$. This means that the time constant of the circuit is slightly less (10%), as $T = CR_i$. If we calculate the resistances for our thunderstorm model, we obtain $R_0 = 2 \cdot 10^7 \ \Omega$, $R_i = 3 \cdot 10^9 \ \Omega$, $R_u = 3 \cdot 10^9 \ \Omega$, and $R_a = 145 \ \Omega$. The difference is caused by the different altitudes for the cloud charge centers and different conductivity values. For the resulting resistor R of the parallel connection, we obtain $R = R_i/2$; i.e., the time constant of the circuit is $T = CR_i/2$. To be consistent with the here-chosen thunderstorm model, we will use the second value for our discussion.

If I is the current output of the generator, the differential equation connecting I, Q, R, and T is easily obtained:

$$I = \frac{Q}{T} + \frac{dQ}{dt} \tag{6}$$

The solution is:

$$Q = e^{-t/T} \int_0^t I e^{x/T} dx \tag{7}$$

with the boundary condition $Q = 0$ for $t = 0$, and x the time variable of the integral.

If we let $I = I_0 = $ constant, the integral can be solved; and we obtain:

$$Q = I_0 T (1 - e^{-t/T}) \tag{8}$$

The equation shows how fast the charges in the cloud are built up if the precipitation current sets in at $t = 0$ with full force. This is certainly not the case in nature, and we will discuss later the effect of a gradually increasing current.

For t large compared to T, we reach the steady-state condition given by $Q_s = TI_0$. This is the formula 1 used before and throughout in the literature, and equation 8 gives us an estimate under what conditions we are justified to do so. Fig. 5, curve *I*, shows a

286

H. W. KASEMIR

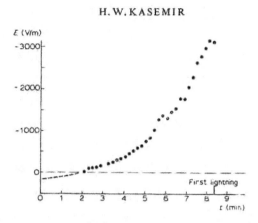

Fig. 6. Electric field E versus time t after initial radar return ($t = 0$). (After REYNOLDS and BROOK, 1956.)

graph of Q/Q_s against t/T. We see that Q reaches 86%, 95%, and 99% of its final value in a time 2, 3, or 4 times the time constant. In our model the time constant T_1 for the upper charge was 60 sec = 1 min. This means that in 3 min the charge has already reached 95% of its final value. This is too short a time to fit Mason's 12–20 min. For the lower charge with the time constant $T_2 = 340$ sec, the 95% value will be reached in 17 min, which agrees rather well with Mason's requirement.

But we have made two very crude approximations: (1) The circuit does not reflect the different conductivities at the positive and negative pole of the generator. It could be extended by subdividing R_i and C. But as Mason's 12–20 min are deduced from measurement at the ground, where the field of the lower charge is predominant, we will restrict our results to the lower charge and keep the simple circuit of Fig. 4. (2) By setting the time constant $T = T_2$, which is the value inside the cloud, we have neglected the influence of the resistor $R_0 + R_a + R_u$, i.e., the conduction current of the thunderstorm to the ground. If we assume that the current output of the thunderstorm is split equally between the current inside and outside the thunderstorm, it follows that $R_i = R_0 + R_a + R_u$, which is also the result of the numerical calculation above. Then the time constant T would be equal to $T_2/2$. This means that the 95% value of Q would be reached in 8.5 min. This is exactly a time recorded by REYNOLDS and BROOK (1956). The field–time graph is given in Fig. 6.

We will now investigate the case that the current generator does not start with the maximum current I_0, but that the current I increases linearly with time until the value I_0 is reached and then remains constant at this value. At first, then, the current is given by:

$$I = I_0 \frac{t}{aT} \tag{9}$$

where a is an arbitrary factor. For $a = 1$, the current will reach its final value for $t = T$ and for $a = 2$ for $t = 2T$, i.e., in about 2.8 or 5.4 min in our example. The solution of equation 7 is then:

$$Q = \frac{I_0 T}{a} \left(\frac{t}{T} - 1 + e^{-t/T} \right) \tag{10}$$

287

THE THUNDERCLOUD

In Fig. 5, curves II and III represent the charge accumulation for $T = 2.8$ min and $a = 1$ or 2, respectively. The 95% value will be reached in 10 and 13 min, respectively. This is about the lower time limit of Mason.

The significance of this calculation is not so much the agreement or disagreement between measured and calculated numerical data. One may even be surprised that such a crude simulation of a thunderstorm as that given by the circuit diagram of Fig. 4 yields numerical results which fit relatively well with the measured data. The significance of this brief calculation is the result that the time required for an accumulation of charge necessary to ignite the first lightning flash does not depend only on the convection current, i.e., the charge generated and separated; but also on the conduction current, i.e., the conductivity inside and outside the cloud, which tends to reduce the effect of the convection current.

Furthermore, a comparison between Fig. 6 and curves I and III of Fig. 5 shows that there is quite a similarity between the measured curve of Fig. 6 and the calculated curve III of Fig. 5. As curve III represents the field increase for a convection current generator which does not start with full force, but gradually, the result seems first to be rather trivial. However, the interesting point is that in the measured curve of Fig. 6 the field increase does not taper off and assume a final value, as would be the case if steady-state conditions are reached. Judging from Fig. 6, as compared to curve III from Fig. 5, about half of the charge accumulation of the steady-state condition is reached at the time of the first lightning.

A similar point of view is taken by TAMURA (1955) in his paper of the analysis of the recovery curve. He states, "It is quite likely that the condition for a lightning discharge, in general, will be attained when accumulation of charges is going on and not at the state when accumulation of charges has ceased. Conditions of initiation of lightning discharges may be quite delicate; that is, heterogeneous distribution of charge may strengthen a local electric field sufficiently large to initiate a discharge, even when the general tendency of the field is still unfavorable for a discharge. However, it is very difficult to expect that a discharge occurs some time after the electric field intensity, or the accumulation of charge has reached the stationary state."

If we now consider that an average thunderstorm produces a lightning flash every 20 seconds, which destroys 20 coulomb charge, this would result in an average current loss of 1 A by lightning discharges. With an additional current loss of 1 A by the conduction current, the balance would be held by a convection current of 2 A. Under these conditions at least 2 A convection current of Chalmers' point 2 could be accommodated.

GENERAL REMARKS ON CHARGE GENERATION AND SEPARATION

Mason made a quantitative study of a number of proposed charge-generating-and-separating effects and came to the conclusion that all of them ". . . are open to objection on quantitative grounds and/or because they do not fit the known facts about the meteorological and electrical behavior of thunderstorms. The only other likely mechanism suggested by laboratory experiments and which appears to fit, at least qualitatively, the observed facts, is the charge generation associated with the formation of rime and which might be associated with the growth of graupel pellets in thunderstorms." Graupel certainly is a form of precipitation, which is typical of thunderstorms, but not of a nimbus-stratus or a snow cloud. Any electrification process connected with it would fit perfectly Chalmers' points 3, 4, and 5.

Furthermore, KUETTNER (1950) found in his thunderstorm research at the Zugspitze that the strongest electric fields were always associated with the appearance of graupel. This relationship was so pronounced that it led Kuettner to the postulation of the graupel dipole.

In 1948, WICHMANN (1948) published a monograph on the basic problems of the physics of thunderstorms. This book is remarkable in several ways. One of them is the close interrelation between the meteorological and the electrical phenomena in the thunderstorm; another one is that he used the riming of graupel as the main electrification process. He developed a thunderstorm model in which all three charge centers in the cloud are accounted for. This is a point that is quite often neglected.

Many theories of charge generation and segregation in thunderstorms content themselves with showing that the precipitation particle acquires a negative charge, and by falling down to lower regions in the cloud leaves the positive charge in the upper part of the cloud. This implements more or less the reasoning that the negative charge center is formed by the precipitation itself. If the precipitation would evaporate at the base of the cloud or, as in Wilson's theory, would stop there by virtue of the electric field, such a mechanism would be acceptable. But as the precipitation reaches the ground and disappears from the picture, an effect has to take place that enables the precipitation to unload its negative charge in the lower part of the cloud. In other words, the charging mechanism, which places the negative charge on the precipitation in the cloud top, should reverse its sign in the lower part of the cloud. We may even go further and request another reversal at the base of the cloud to account for the lower positive charge center.

It should be mentioned here that the discharge of the precipitation particle by conduction current would not place the precipitation charge in the surrounding air, as sometimes is assumed, but is such a process that leaves no space charge behind.

If we assume only one sign reversal and represent the lower positive charge center by the out-falling precipitation, we would obtain not a positive charge center concentrated at the base of the cloud, but a column of positive charge stretched out from the base of the cloud all the way down to the ground, with a comparatively low charge density. The precipitation which reaches the ground will be discharged, while the negative charge remaining in the cloud accumulates continuously. So, very soon the negative cloud charge out-weighs the positive precipitation charge. The field at the ground will be that of a negative charge overhead. (In the commonly used atmospheric electric nomenclature, it is a negative field.) But, in general, the positive ground field in the center of the storm is thought to be caused by the lower positive space charge center. Such a relation will not be possible if we identify the lower positive space charge with the precipitation charge.

THE THUNDERCLOUD

This leads to some doubt that this positive charge center can be interpreted as precipitation charge. The same reasoning can be applied to each of the three charge centers in the thunderstorm, with the result that none of them can be explained by precipitation charge because it would lack the feature of concentration and accumulation. These charge centers are more likely the accumulated charge left behind, with the opposite charge carried away by precipitation. The consequence of such a mechanism is that we have to assume two sign reversals of the charge-generation process to explain three charge centers of alternating sign with one precipitation carrier. Alternately, each sign reversal can be substituted by a different carrier with a new charge-generating effect or by a completely different charge-generating-and-separating effect not using precipitation.

CONCLUSIONS

The following picture of the electric structure of an average thundercloud emerges from a survey of the literature.

The charge distribution of a tripolar storm, positive charge at the top, negative charge in the middle, and positive charge at the base, still remains. The amount of charge has to be altered drastically if it is to be consistent with the 1 A conduction current from the top of the thunderstorm to the ionosphere. The new numerical values would be 60 C for the upper positive cloud charge, -340 C for the negative charge, and about 50 C for the lower positive charge. The value for the lower positive charge is thereby only a guess, but it has to be considerably higher than the old value of 5 C.

The charge accumulation in the cloud depends on the conduction current, the conductivity, and the breakdown field; the breakdown field depends also on the spacing of two opposite charge centers.

The conductivity inside a thundercloud has never been measured, and the breakdown field has been measured just once. The acquiring of these data should be one of the most pressing tasks of future research.

The thunderstorm model, from which the above numerical values for the cloud charges have been calculated, is based on data taken from lists of thunderstorm requirements of Israël, Mason, and Chalmers. As these lists in turn represent the evaluation of data throughout the literature, the combination of them should give a solid base for this thunderstorm model. The parameters chosen are as follows: convection current, 3 A; conduction current to the ionosphere and the ground, 1 A; conduction current between the positive and negative space-charge centers, 1 A; average current required for cloud discharges, which destroy 20 C every 20 sec, 1 A; and location of the upper positive cloud charge at 9 km altitude and of the negative charge at 3 km altitude. The conductivity inside the cloud was estimated to be one-third of its clear air value at the same altitude level; the breakdown field assumed to be of the order of $5 \cdot 10^5$ V/m.

The calculated cloud charges of this model do not agree with numerical values of ± 40 C or $+25$ C, -20 C, $+5$ C, as given by Gish and Wait, Malan, and Simpson and

Robinson. The difference comes from the application of the electrostatic theory to the calculation of the cloud charges from field measurement at the ground and above the thundercloud, while in the here-proposed thunderstorm model the cloud charges are calculated from the conduction current and the conductivity inside the cloud according to the theory of current flow.

Furthermore, this model does not support point 3 of Mason's list, which requires that $8000/v$ C should be separated immediately after each lightning flash (v is the fall velocity of precipitation in m/sec). This requirement is based on Wilson's explanation of the recovery curve. Instead, Tamura's explanation of the recovery curve, based on the theory of current flow, is adopted.

The conclusion is that the theory of current flow has to be used to correlate the different electrical parameters of the thunderstorm, not the electrostatic theory.

From the field increase at the beginning of the storm as well as from the presence of lightning strokes it follows that steady-state conditions are not reached in the "stationary" field or current flow of the storm. The field-limiting parameter will be the breakdown field. The breakdown field may be a common parameter (within limits) to all kinds of thunderstorms, which again urges research in this direction.

A lightning flash, which neutralizes 20 C, would destroy about half of the cloud charge (± 40 C) in the electrostatic model of the thunderstorm, but only one-tenth of the negative charge of the current-flow model. It would be a worthwhile task to determine the ratio of the electric moment of a lightning flash to that of the cloud charges.

Field records of a lightning flash at the ground will reflect the charge distribution in the thundercloud in much more detail, as we are able to evaluate with our present measuring technique and knowledge. The main handicap here is the lack of information of the geometrical form of the lightning flash, especially inside the cloud. This information may be gained by infrared photography or a fast-scanning radar device. It should be available for the evaluation of the lightning field records. Also, more theoretical and laboratory research to link the lightning flash to the charge distribution in the thundercloud would be highly desirable.

An even worse situation exists between the correlation of the meteorological and electrical aspects of the thunderstorm. This is due to the fact that there is no agreement as to which of the many charge-generating-and-separating effects are the dominating ones. One of the most promising effects is that caused by the electrification (riming) of graupel. The quantitative evaluation of a number of proposed electrification mechanisms by Mason points in this direction. Electrification by graupel precipitation would also fit several points of Chalmers' list, and it is most strongly supported by observations and measurements of Kuettner on a large number of thunderstorms of the Zugspitze in the German Alps. It is amazing how much information with extremely simple instruments Kuettner could obtain at the mountain observatory, working either inside the cloud or close to the base. This recommends a more extensive thunderstorm research at mountain observatories, where the much-needed measurements of the conductivity inside the cloud or the breakdown field strength could be carried out without too much difficulty.

DISCUSSION

SUMMARY

A survey is given over the requirements which a thunderstorm theory should fulfil to conform with the experimental data, gained by researchers all over the world, and with the conclusions drawn from these experiments. The different characteristic parameter of thunderstorms and lightning discharges are connected with each other and correlations and contradictions between them are pointed out. It is shown where gaps in the theory or in fundamental measurements exist and in which way present ideas have to be changed or extended. Some suggestions for future research are offered.

REFERENCES

BROOK, M., KITAGAWA, N. and WORKMAN, E. J., 1962. *J. Geophys. Res.*, 67 : 649.

CHALMERS, J. A., 1957. *Atmospheric Electricity*. Pergamon, Oxford, 327 pp.

GISH, O. H. and SHERMAN, K. L., 1936. *Natl. Geograph. Soc., Tech. Papers, Stratosphere Series*, 2 : 94.

GISH, O. H. and WAIT, G. R., 1950. *J. Geophys. Res.*, 55 : 473.

GUNN, R., 1947. *Phys. Rev.*, 2 : 71, 3 : 181.

GUNN, R., 1948. *J. Appl. Phys.*, 19 : 481.

HOLZER, R. E. and SAXON, D. S., 1951. *J. Geophys. Res.*, 56 : 207.

ISRAËL, H., 1961. *Atmosphärische Elektrizität, II*. Akademische Verlagsgesellschaft, Leipzig, 503 pp.

ISRAËL, H. und KASEMIR, H. W., 1952. *Arch. Meteorol. Geophys. Bioklimatol., Ser. A*, 5(1) : 71.

KASEMIR, H. W., 1959. *Z. Geophys.*, 25(2) : 33.

KRAAKEVIK, H., 1957. *The Electrical Conductivity and Current Density in the Troposphere*. Thesis, Univ. of Maryland, 195 pp.

KUETTNER, J., 1950. *J. Meteorol.*, 7 : 322.

MALAN, D. J., 1952. *Ann. Géophys.*, 8 : 385.

MASON, B. J., 1957. *The Physics of Clouds*. Oxford Univ. Press, London, 502 pp.

PLUVINAGE, P., 1946. *Ann. Géophys.*, 2 : 13, 160.

REYNOLDS, S. E. and BROOK, M., 1956. *J. Meteorol.*, 13 : 376.

SIMPSON, G. C. and ROBINSON, G. D., 1940. *Proc. Roy. Soc. (London), Ser. A*, 177 : 281.

SMITH, L. G., 1958 (Editor). *Recent Advances in Atmospheric Electricity*. Pergamon, Oxford, 631 pp.

TAMURA, Y., 1955. *Geophys. Res. Papers (U.S.)*, 42 : 190.

WICHMANN, H., 1948. *Grundprobleme der Physik des Gewitters*. Wolfenbüttler Verlagsanstalt, Wolfenbüttel, 118 pp.

WILSON, C. T. R., 1921. *Phil. Trans. Roy. Soc. (London), Ser. A*, 221 : 73.

Discussion

Dr. Kasemir, in his closely reasoned paper on the structure of a thunderstorm, computed cloud charges from conduction currents and the conductivity inside the cloud according to the theory of current flow. In doing so, he obtained a structure which he felt was more consistent with the common body of accepted fact about thunderstorms, even though it was inconsistent with some reported values of cloud charge. Dr. Kasemir's application of the conduction theory produced a very full discussion of the pros and cons of using the conduction theory as compared to the electrostatic theory for explaining charge distribution and the slope of the charge regeneration curve after a lightning flash.

B. VONNEGUT: A very real question is whether we can use the term "conductivity" within clouds. The term conductivity implies that the flow of current in the cloud is proportional to the electric field; in other words, that it is ohmic. The conductivity of a cloud presumably will depend on the population within the cloud of charged particles which can move under the influence of an electric field. Fast ions, which are primarily responsible for conduction, will become attached to cloud particles

DISCUSSION

and will precipitate at a rate that is proportional to the applied field. Therefore, the conductivity will depend upon the applied electric field. In the case of the more intense fields, there will be further departures from an ohmic behavior because of the formation of ions by point discharge and by other ionization processes. We should be cautious in assuming that a simple concept of "conductivity" within a cloud will enable us to compute the charges that are moving within it.

H. W. KASEMIR: Dr. Vonnegut has said the following: "If we have cloud particles in the air and we have a strong electric field, these cloud particles represent a kind of recombination area." This would mean that the conductivity is even lower than it is normally. We still should call it "conductivity" because, even though the conductivity may not be the ohmic, it may be a function of the field. Professor Freier treated this question mathematically. The question is whether Professor Freier's calculation is electrostatic or not. I think the electrostatic aspect comes in at the point where he integrates over boundaries representing field lines and equipotential lines of an electrostatic problem or the equivalent current flow problem with constant conductivity. The equations he uses certainly are current flow equations, but the boundaries he integrates over are, in reality, not constant with time; this fact, in my opinion, introduces the big difference.

G. FREIER: I would like to make some comments with regards to the inference that my analysis of the conductivity was an electrostatic one. I used all of Maxwell's equations except the fact that curl $E = 1/c \frac{dB}{dt}$. I assumed that curl E was equal to zero; in other words, there is no appreciable electric field due to electromagnetic induction. As soon as curl $E = 0$, E may be derived from electrostatic potentials. Curl H has a value, and the analysis was to simply take div curl $H = 0$, so that in my first equation, the divergence of all terms in the brackets was equal to zero: div curl $H = \operatorname{div} \frac{1}{c} \left[\frac{\delta E}{\delta t} + 4\pi\lambda E + 4\pi\rho V \right] = 0$. As soon as I had the divergence of the quantity in brackets, I could integrate it over a region of space and convert it to the surface integral surrounding that space. I then selected a surface which had undirectional fields through the entire surface and made measurements on one of those surfaces. The conductivity or the relaxation time enters into the term λ times E. By this trick, I could eliminate the space variables. Maxwell's equations contain both space and time, and I had to find a way to get rid of the space variables in order to study the time varying phenomena. I then simply measured the relaxation times of the fields, and found these were much shorter time constants. I interpreted this by saying that the convection currents ρV, whatever it may be, generates the charge and persists as a constant value; then, a relaxation time is observed.

Another thing with respect to Vonnegut's remarks about the ohmic behavior, from the analysis I would also conclude that the system was quite ohmic, for the simple reason that when we plot these relaxation curves, we find that they are very close to exponential. This indicates that the currents have to be proportional to E, not rather than to any higher power of E. I can only come to one conclusion about this analysis which is that the conductivity is quite high in thunderstorms.

B. J. MASON: I would like to agree with practically everything that Professor Freier has just said. Of course, his analysis is not electrostatic; he has in fact done the proper current treatment. Quite independently of him, I have used very much the same treatment in my paper. There is no real difference between him and Dr. Vonnegut. I agree entirely with Dr. Vonnegut that λ will, in fact, effectively contain the ionization currents produced by point discharge and other things, and indeed they are the main drain on the charging system. So in the absence of any measurements of λ, which is the conductivity inside a cloud, the very sensible thing to do is to study the recovery of the field, as Professor Freier has done, and to show by careful plotting that it is in fact ohmic in perhaps most cases. Of course, from these relaxation times you do get an effective value of λ, which turns out to be something like $2 \cdot 10^{-3}$ e.s.u., giving a recovery period of about 40 sec. From a theoretical point of view, that is probably the best estimate of λ we have at the moment. In other words, the effective conductivity in a thunderstorm will be at least an order of magnitude larger than that which we think we measure at the same level outside the cloud.

R. E. HOLZER: In his paper, Dr. Kasemir indicated that the thunderstorm must be looked at as a current pattern rather than as an electrostatic pattern. In fact, one cannot infer a charge distribution in a thunderstorm unless he has a rather detailed knowledge of the distribution of conductivity in the medium. However, it is worth noting that there are circumstances in which it is possible to use an electrostatic theory of thunderstorms of the type that you find in the standard text books. In the case

DISCUSSION

of lightning discharges, the duration of the discharge is short in comparison with the relaxation time in any part of the medium. The difference in field before and after the flash represents a field change which is representative of the electrostatic structure discharged by the storm, and I think it is quite appropriate to use an electrostatic approach in this case.

H. W. KASEMIR: In reply to Professor Holzer's remark, I agree that right after the lightning flash we have the electrostatic field distribution which changes during the recovery curve over to that of the current flow.

J. A. CHALMERS: This question of whether the particles in the cloud are being held up by the field, or whether it is a "conductivity", relies on the same thing. You can put them both into the $4\pi\lambda E$ term. The force on the charged particles in the field amounts to giving them a contribution to the conductivity.

B. J. MASON: I would agree with the last remark of Dr. Chalmers that mathematically these two are similar. Physically, they imply something very different. If you are going to have these large drops of particles, hailstones, or whatever they are, held in the electrostatic field of a few thousand volts per cm, the charge on these particles has to be enormous. This is a very difficult physical requirement, as Dr. Kasemir pointed out.

T. W. WORMELL: Would Professor Holzer agree, that one can apply electrostatics not only to the field changes due to the lightning discharges, but also to the initial slope or the recovery curve, measured in the first half second after the flash? It is this which gives you the rate at which the cloud regenerates electric moment when there is no opposing field. This was interpreted by Wilson and by Mason and has been criticized by Kasemir in terms of a particular picture.

R. E. HOLZER: I did not mean to make my remarks quite so restricted. The basic point was that, in any situation in which the interval involved in the field changes is small in comparison to the relaxation time of the medium, the electrostatic approximation is a reasonable one.

H. W. KASEMIR: I do not agree in this case. Right after the lightning flash is finished, the electrostatic field distribution changes over that of the current flow and influences the initial slope of the recovery curve. Even if the charge-generating effect of the cloud and the decay of the lightning charges by conduction current could be compensated by a suitable supply current, you would measure a field decay at the ground beginning right after the flash, which is caused by the change from the electrostatic to the current-flow field distribution. Therefore, one can apply the electrostatic theory to the calculation of the electric field distribution right after the flash, but not to the time variation of this field, i.e., the slope of the recovery curve.

B. VONNEGUT: I wish to express my agreement with Dr. Kasemir. When we observe the electric field after a lightning stroke, whether it be from the ground or an aeroplane, we see that the electric field is changing; we will all agree that the reason for this is that there must be charge moving in the cloud and in the storm above us. The questions are where and what is the charge that is moving? There are clearly several different possible motions of charge that may be causing the observed change of field. The first of these is the movement of charged particles within the cloud, which may be causing the development of charge responsible for electrification and lightning. However, this is certainly not the only current that is flowing. For example, the measurements that were made by Gish and Wait from an airplane above the storm show that there is a large electric field following the lightning stroke above the cloud. The field above the cloud does not return to zero after the lightning, but, on the other hand, it may go to values quite as high in the opposite direction. Therefore, it is clear that just after the lightning stroke there are large currents flowing above the cloud as well as within it. Any interpretation in which we assume that it is purely a conduction current or that it is purely a charging current which is flowing may very well be in error. There must be a combination of a number of complicated currents, some of which are dissipative and some which contribute to the charging of the storm.

G. FREIER: I disagree with Dr. Vonnegut's statement. In the thundercloud, the fields are a linear superposition of lots of effects. Nevertheless, in the type of analysis which I have described here, one can treat the problem as a superposition of the fields; whether the field goes positive or negative has no bearing on the problem. One particular cell can give this change, while other cells may be doing something else, or may be in a more or less equilibrium or quiescent state.

J. A. CHALMERS: Dr. Kasemir pointed out, many years ago, the question of whether you consider it as a current system or as an electrostatic system amounts to the same thing. One is more convenient than the other in some cases. If you are considering it as an electrostatic system, you must take in all the charges. You cannot just consider the simple charges in the cloud, you have to consider other charges in other places; and when you have a variation of conductivity with height, you have space

DISCUSSION

charges to ensure continuity of current. Thus, whether it is "electrostatic" or "current" is not a fundamental difference, but a difference of point of view.

B. J. MASON: Dr. Kasemir misrepresents C. T. R. Wilson; the Wilson picture of the charge production and separation is an over-simplification, but on the whole it is perfectly valid. It takes into account the production of charge on the particles given by the right hand term in Professor Freier's equation. Against this is that much of that charge at any one time will be masked by a counter conduction current containing point discharge and all the other things, which is the middle term of the equation; thus, at any one time, most of the charge on the precipitation will be masked by this counter current. Wilson pictures that, over a considerable period, both the terms build up until we get a field strong enough for the first lightning flash. Then, in the interval between that lightning flash and the next one, there is a sedimentation of the precipitation particles; most of the total charge still remains masked and we get 20 C or so at the bottom and 20 C at the top. There is no reason to believe that the particles have to be held up by the electric field, as though the whole thing comes to a stop and everything is in a state of suspended animation. This is not at all necessary; the whole thing can continue, with the particles falling all the time. After all, the lightning flash takes a very short time compared to the recovery. There is no need to assume that the charges on the particles must be so large, or that they are held suspended in the field.

T. W. WORMELL: The Wilson picture was that the separation occurs until the reduced rate of separation just balances the steady dissipation currents (he was non-committal, about whether the electric forces on the particles are comparable with the weight or not). Anyone who has recorded fields of thunderstorms at a distance will know that in only moderately active storms, the field is frequently very nearly steady for some time before the next lightning flash occurs. The next point is that immediately after an internal discharge in the cloud, the field E is presumably very small; therefore, there is no appreciable force stopping the separation of the particles and there is no conduction current dissipating them. In other words, the rate of change of the field, which corresponds on the average to a time constant of seven seconds, is just the rate at which a thunderstorm can regenerate electric moment in the absence of any field trying to stop the separation, and in the absence of any internal field trying to dissipate the charges which have already been separated.

H. W. KASEMIR: I am glad to hear that I misinterpreted Wilson's picture of the recovery curve. What really bothered me was the stopping of the rainfall 10–20 sec after each lightning flash. However, if the precipitation particles are not stopped or retarded by the electric field, the question of what effect reduces the rate of charge generation or separation during the recovery curve remains. This problem should be considered in detail by every charge generation and separation theory that does not accept Tamura's explanation of the recovery curve in lieu of Wilson's.

T. W. WORMELL: I want to summarize briefly the main problems discussed in this session (Chapter III). In this session, the Conference turned for the first time to the consideration of atmospheric electrical phenomena which arise in bad weather conditions. The papers and discussion thus covered a wide range, although the basic problem of the main mechanism of generation of thunderstorm charges was specifically excluded and postponed to a later session devoted solely to this topic.

The session was opened by four invited papers. The first, under the general title "Generation of electric charges outside thunderclouds", dealt with a variety of topics; in particular, with point-discharge current in the atmosphere, with electric fields due to steady rain and to clouds not producing precipitation, with precipitation electricity and, finally, with the effects occurring near the lower boundary, the earth, due to space charge and convection, and complicated by the increase in conductivity very close to the ground.

The second paper surveyed work on electrical charges associated with precipitation and paid particular attention to the influence of electrostatic forces on the relative motion of two drops when close together and, hence, on the growth of precipitation by coagulation.

A third paper surveyed techniques for investigating the electrical properties of clouds with aircraft. The difficulties, in some cases unsolved, were well brought out and some examples of observational results were presented.

The fourth paper discussed the general model of a thundercloud and attempted to specify the requirements which any theory concerning the basic electrical mechanism or mechanisms must fulfil in order to be in agreement with observations. It emphasized the necessity of picturing the thunderstorm and its surroundings as a current system and of considering the effects in the electric field of the space charges thus produced.

DISCUSSION

The ensuing discussion was lively and demonstrated that the general field covered by this session is one in which many workers are actively engaged in observations and interpretation. The discussion on point discharge, considered both as a laboratory experiment, and as a natural phenomenon, emphasized the complexities of this field. We are still, after 40 years or so of effort, uncertain of the total contribution of the point-discharge currents from a forest, or from a tree-covered countryside, to the total vertical electric current under a storm. Possibilities for a new observational approach were discussed.

Several speakers emphasized the violent electrical effects which can accompany dust storms and the necessity to include a possible contribution from such phenomena to the total exchange of electric charge between earth and atmosphere.

The electrification of so-called steady rain, whose cause is still somewhat of a mystery, was discussed at some length as was the significance of electrical forces in the coalescence process leading to the growth of precipitation. This discussion perhaps served primarily to emphasize the complexity of the phenomenon and the need for extreme caution if gross errors are to be avoided in computations.

A discussion of the technical problems which arise when endeavouring to measure the fundamental electrical parameters in and near clouds emphasized the extreme difficulty of avoiding grossly spurious effects if the aircraft enters clouds containing precipitation. It was even suggested that we need to reconsider the question of whether such a concept as conductivity has a meaning inside a cloud.

Turning to the thundercloud, discussion developed as to the significance of the form of the recovery of the field after each lightning discharge; and the necessity to think of the storm as a current system and not just as a problem in electrostatics was recalled. In other words, the field we observe is the result not only of the primary charges in the thundercloud but also of the whole system of space charges produced in an atmosphere whose conductivity varies with height from the ground.

Finally, we were reminded of electrical problems in the atmospheres of other planets and that the time is approaching when they will become susceptible to observations, which will doubtlessly lead to many reactions on our understanding of both basic physics and of our own atmosphere.

The whole discussion in this session was lively and wide-ranging. It became clear that divergences of view exist, that there are gaps in our observational knowledge and that our understanding of what has been observed is obscure. This general field is clearly one where investigation is still active and fruitful.

From Proceedings of the International Conference on Cloud Physics, August 26-30, 1968, Toronto, Canada, pp.668-672.

CHARGE DISTRIBUTION IN THE LOWER PART OF THE THUNDERSTORM

Heinz W. Kasemir
Atmospheric Physics and Chemistry Laboratory
Environmental Science Services Administration
Boulder, Colorado

1.   Introduction: It is rather difficult to obtain reliable and sufficient data of atmospheric electric parameters inside of a thunderstorm. On the other hand, there is an urgent need for such data because the correct choice between the large number of proposed thunderstorm theories can be based only on experimental results obtained inside or around a thunderstorm. The airplane is by far the superior platform for thunderstorm research because it has enough altitude range and furnishes the necessary lift and power to operate adequate instrumentation. Most important of all, the airplane has the speed to reach .an individual storm in time and make a number of penetrations so that the time history of the storm can be recorded in a consecutive series of measurements. This point will be well demonstrated in the later discussion.

Of the electric parameters, the electric field is the most suitable to measure. However the distribution of electric charges in space and time and not the field is the key parameter of thunderstorm electrification. It is the main objective of this paper to show how the charge distribution can be calculated from the recording of the three components of the electric field. The limitations of the process and the experimental results of actual recordings will be discussed.

2.   Instrumentation and data collection: For the electric field measurement two cylindrical field mills have been used. One field mill mounted on the nose of :he airplane measures the vertical component Fz and the horizontal component Fx in the direction right to left wing tip. The second field mill is mounted between the wings on the fuselage and measures the other horizontal field component Fy in the direction nose to tail as well as the component Fx. A detailed description of the instrument and its calibration procedure is given by Kasemir (1964, 1967). The field components together with other flight data such as altitude, temperature, turbulence and so on are recorded on an 8 channel galvanometer recorder as well as on a magnetic tape recorder. After the flight the electric field data are digitized for computer evaluation.

In general the following simple flight plan was adopted. The airplane would start at the beginning of a strong development of cumulus-towers and select the one which either showed the first stages of icing at the top or the beginning of light precipitation at the base. A flight path was selected close to the base through the center of the rain curtain. This flight pass was followed back and forth throughout the whole lifetime of the storm. In general one path below the storm lasted about 2 minutes. The tear drop turn at the end of each run requiired also about 2 minutes so that consecutive runs are separated by a 4 minutes time interval. This is a close enough sequence to follow the charge development in the storm as will be seen later on. With a lifetime of about 40 minutes for the single cell thunderstorm about 10 consecutive passes could be flown if the airplane reached the storm at its beginning.

3. <u>Evaluation of measurements and discussion of results</u>: The evaluation of the field data is based on the field pattern of a single point charge as discussed by Kasemir (1964). The main field maxima of the combined Fx and Fz vectors was treated as if caused by a substitute point charge representing the main bulk thunderstorm charge. The amount of the first substitute charge and its space coordinates are calculated from a best fit of the recorded data. The exact pattern of this first substitute charge is then calculated and subtracted from respective field component records. The remaining field curves are then treated in the same way to determine charge centers of second, third and following order until all of the field records are reduced to values less than 1% of the original curves. This set of charges is only a first approximation because each substitute charge is calculated disregarding the influence of the following as yet not determined charges. Using this first approximation the data are re-evaluated now taking into account the combined influence of all the charges to obtain an improved second approximation, This iterative method is repeated until the solution has been found. Theoretically this process will lead to an exact replica of any given charge distribution. In practice the analysis is limited by the accuracy of the measurement. This means that because field strength diminishes with the square of the distance from the charge, the nearby charges dominate the field records and determine the sensitivity setting of the field mills.

The consequence is that the positive charges in the top of the storm are not revealed in the analysis of the examples given later on. However, this restriction is not inherent either in the analytical method or in the measuring capabilities. It is rather a consequence of the specific flight path which was chosen to detect the strongest fields for the purpose of chaff seeding in a research program of lightning suppression. This situation is similar to that of sitting in a concert close to the trumpets of an orchestra and trying to detect the weaker instruments when the trumpets are blown with full force. For thunderstorm research a flight pass half way up and to one side of the cloud would probably be best to detect the whole composition of the thunderstorm charges.

The computer program has been tested by an artificially composed field pattern calculated from a known charge distribution. Already the first approximation did not deviate more than 20% from the exact values and the second approximation was accurate to 1%. A detailed discussion of this analysis and the problems is given in a separate paper.

From the calculated results models have been constructed which illustrate the charges and their position with regard to the individual flight pass. Fig. 1 is a perspective drawing of one flight pass model. The heavy horizontal wire is the flight path of a length of 5 km. The circle represents the calculated negative charge. The area of the circle is proportional to the amount of charge, in this case -0.41 Cb. The rectangular bent wire, which holds the circle up, represents the x and z distances from the flight path. Positive charges are represented by triangles. .In this case there is only one positive charge to the left of the negative charge and down from the flight path. The area of the triangle is proportioned to the amount of positive charge, here +0. 05 Cb.

Fig. 2 shows a photograph of three models of passes underneath a storm on the 25th of July 1967. The storm was already diminishing and the main negative charge of 7. 5 Cb in the base of the cloud is well established at the first pass. There appears also a second negative charge center further along the flight pass of -0. 87 Cb. Furthermore, a positive charge of +0. 5 Cb and a

298

minute negative charge of -0. 06 Cb are located below the pass. Minute charges in the order of 1 to 2 % of the maximum charge are of a fictitious nature due to the limited accuracy of the measurements and should be disregarded. The second pass shows besides the minute charges only one negative main charge 15% smaller than at the first pass and 14% lower, whereas in the third pass the main negative charge has increased to 14. 3 Cb, i. e. 100% compared to the first pass with an increase of altitude of the main charge, and the question arises can this wobbling of the charge be due to a computational or measuring error. In theory, the computer calculation converges to any desired accuracy, however there are certain critical charge distributions which require a denser field of data points for evaluation than one measurement each 50 m which has been used here. The third flight pass with two negative charge centers one above the other may be such a case.

Two rather large positive charge centers appear in the third pass which have no precedents in the first and second pass. A more detailed study of the third pass is now being carried out to determine how reliable the analysis is. In the meantime the results are reported without corrective surgery.

A very clear cut case of the charge development in a rain shower on the 28th of July 1968 is shown in Fig. 3.

This storm was one of the rare occasions, where the very beginning of the charge build-up was recorded. No rain was coming out of any one of the large cumuli clouds populating the sky. Therefore the biggest and most promising tower of a large bank of growing cumuli was selected and a flight path chosen, which lead directly under the center of this tower in an altitude of 2700 m about 100 m below the base of the cloud. The first flight passes were uneventful, i. e. only the fair weather field was recorded. On the sixth pass very light rain was falling out of the cloud base and with it the field increased. The evaluation shows a charge distribution of a small negative and positive charge center above and small negative charge center below the flight pass.

This is the only flight pass which shows a negative charge below the airplane. At the return flight No. 7, the lower negative charge has disappeared and the upper positive and negative charge increased in strength. From the seventh pass on, i. e. 8 minutes after the appearance of the first raindrops the charge development becomes more organized. A rapidly growing negative charge center appears always at the same spot about 1300 mm above the flight pass. It grows from 0. 2 Cb to 1. 2 Cb in about 20 minutes.

It should be mentioned that due to the screening surface charge at the cloud-air boundary, the amount of the calculated substitute point charge may be too low by a factor approximately proportioned to the ratio of the conductivity inside to outside of the cloud. A correction factor 3 to 10 depending on the conductivity values has to be applied.

Summary:
It seems to be possible to calculate the charge distribution in thunderstorms from the records of the three field components measured from an airplane which flies a known path through or below the storm. Consecutive passes will give the charge development with time. The choice of the path determines if the calculation results in a selective or composite picture of the charge

distribution. The calculation procedure and the obtainable results are demonstrated on two examples of actual flight records. More flights will be analyzed in the near future. The investigation reported in this paper is a part of the results obtained during lightning suppression research project, which is a joint enterprise of the Atmospheric Physics and Chemistry Laboratory, ESSA, Boulder, Colorado.

References:

Kasemir, H. W. The Cylindrical Field Mill" Tech. Rpt ECOM 2526, 1964.
Kasemir, H. W. 1967: ''Measurement of Atmospheric Electric Parameters", NCAR TN-29, pp 107-115.

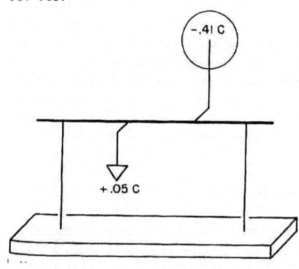

Fig. 1. Model of the charge distribution in a rain shower calculated from electric field components recorded from an airplane flying underneath the storm.

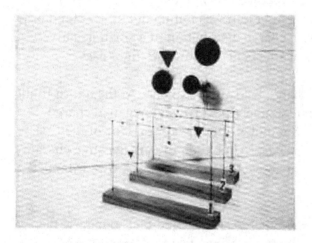

Fig. 2. Model of charge distribution in a thundershower obtained from three consecutive flight passes underneath the storm.

Fig. 3. Model of charge distribution in a rain shower obtained from 17 consecutive flight passes underneath the storm.

THUNDERSTORMS

CHARGE DISTRIBUTION IN THUNDERSTORMS

Heinz W. Kasemir

Colorado Scientific Research Corporation, Berthoud, Colorado, USA

1. INTRODUCTION

The classical model of the electric charge distribution in thunderstorm is that developed by Simpson and Scrase [1] and by Simpson and Robinson [2]. It is based on measurements of the vertical component of the electric field made from balloons ascending into and penetrating thunderstorms. The analysis of these measurements led to a charge distribution in the thundercloud where the positive space charge of about +24 Coulomb is located in the upper part of the cloud, the negative charge of about -20 Coulomb in the lower part and a smaller positive - so called pocket - charge of about +5 Coulomb at the base of the cloud. The charge centers are at 6, 3, and 1.5 km respectively. This tripolar thunderstorm model with the charge centers arranged vertically one above the other has been the most commonly used scheme for the charge distribution in a thundercloud.

Different values of the net charge and the altitude of the center of the positive and negative main charges have been suggested for instance by Gish and Wait [3] +39 C at 9.5 km and -39 C at 3 km, by Malan [4] +40 C at 10 km and -40 C at 5 km, and by Kasemir [5] +60 C at 6 km and -340 C at 3 km. A slightly different charge arrangement has been proposed by Küttner [6] based on measurements at the mountain peak Zugspitze (Germany). He identified the "graupel-dipole" as a dipole with its centerpoint at about 4 km and the negative charge above the positive. The "snow-dipole" is a dipole with its center at about 5 km but horizontally displaced from the graupel dipole with the positive charge above the negative one. The two negative charges being at about the same level could be combined into one large volume of negative space charge with one positive charge below but offset to one side and another positive charge above but offset to the other side. The whole charge arrangement would resemble that of a tilted

301

quadrupole. From the analysis of the five storms reported in this paper all five have a more or less pronounced offset of the upper positive against the lower negative charge.

The slanted dipole or quadrupole pattern of the thunderstorm charges is a necessary assumption to explain the occurrence of slanted lightning discharges. Therefore, this result fits very well with the emphasis placed in the recent literature on the slanted or even horizontal lightning discharge. [7],[8],[9]. Küttner's model shows how a slanted charge distribution may result from two charging mechanisms working side by side but each one using vertically falling precipitation particles for charge separation. Another way to obtain a slanted charge arrangement is windshear. The positive upper charge ascending with the growth of the cloud may be blown sideways when entering a windshear region. A good example of this case is storm 2 discussed in this paper later on.

The results of the data analysis given below deviate from the classical model in two respects. First, the arrangement of the charge centers is slanted with an angle of 20 to 80 degrees from the vertical and second the amounts of charge or dipole moments is greater than in the Simpson-Scrase model by a factor 50 to 100. This would mean, for instance, that charge generation should be 50 to 100 times more effective than assumed now and that the current contribution of the average storm to the global circuit may be much larger than the 1/2 to 1 ampere quoted in the literature.

II. MEASUREMENTS.

A network of 25 field mills have been installed at Kennedy Space Center scattered over an area of 7 x 10 km as shown in Fig. 1. The locations of the field mills are indicated by the numbers 1 to 25. The shore line of the Atlantic Ocean is shown on the right side (east) and the outline of the two lakes, Banana River halfway to the middle, and Indian River on the left side (west) of Fig. 1. Furthermore, the two main roads, Kennedy Parkway running north-south and NASA Parkway running east-west are drawn for easier orientation in the Kennedy Space Center area.

The field mills monitor continuously the potential gradient. The measured values are digitized and transmitted to a central station where the data are scanned and stored at a rate of one scan per second. About every minute

THUNDERSTORMS

one set of data is evaluated by a computer that draws on a surface map of the KSC area gradient contourlines of ± 1, ± 2, ... kV/m. The sign is usually given in the approximate center of a set of these gradient rings. A reproduction of such a contour map with one contourline of -1 kV/m is given in Fig. 2. At the right side of the figure the gradient values of all 25 field mills are printed out. This is an example of the data used in the multipole analysis. It may be mentioned that the multipole analysis is completely different from the method used to calculate the contour lines. This will explain a contradiction between the contour maps and the multipole analysis discussed later on.

III. ELECTRICAL TRACKING OF THUNDERSTORMS AND SHOWER CLOUDS.

The potential gradient measured by the 25 field mills has been evaluated with respect to the pole center and the amount and polarity of net charge and dipole moment of 5 storms occurring on 15 July 1977 and on 24 and 25 July 1978 over or in the neighborhood of the Kennedy Space Center area. The projection of the pole center on the earth surface during the lifetime of these storms is shown in Fig. 1 by the three tortuous lines marked "Track 1" (25.07.78), "Track 2" (15.07.77), and "Track 3" (24.07.78). Each minute the charge center has been recalculated and projected to the ground. Connection of these projection points by straight lines forms the tracks. The beginning of each track is indicated by a small circle and the end by an arrow. Track 1 starts about 0.7 km north of the crossing of Kennedy Parkway and NASA Parkway and ends about 5 km south of it. The storm lasted 20 minutes from 15:50 to 16:10 GMT. Track 2 starts about 0.5 km N-E of field mill site 14 close to the east coast. After three unsuccessful attempts to move further east out over the ocean the storm changed direction by 180 degrees and moved across the KSC area in a south-west direction. It died out after 28 minutes about 1 km west of field mill site 22. The average speed of the first storm was about 18 km/h and that of the second storm about 30 km/h. Both storms did not produce lightning discharges. They were electrified shower clouds.

The third storm (17.10 - 17.39 GMT) was of a different nature. Track 3 starts south of Titusville-Cocoa airport on the west side of the Indian River, that is about 10 km outside the KSC area. After 8 minutes the track jumps over the Indian River to the west side of the KSC area where it stays for 16 minutes

303

close to the coastline of the Indian River. Thereafter, it jumps back to the other side of the river to the area south of Titusville and remains there. Contrary to this track, the contour maps during this time showed only one, occasionally two, negative gradient rings -1 to -2 kV/m centered around a point x_0, y_0 well inside the KSC area (Fig. 2) that give the impression of an electrified shower-cloud with its center between field mill 18 and 22. Re-examination of the method used to calculate the contour maps showed that this method cannot and never claimed to be capable of tracking thunderstorms outside the field mill network.

Several features indicate that we are dealing here with 3 different storms that we will identify as storm 3a, 3b, and 3c respectively. First, the river was crossed between storms 3a and 3b in one minute and between storms 3b and 3c in three minutes indicating a storm speed of 600 or 200 km/h respectively. These values are by about a factor of ten too high to be part of the track of one storm. Second, storms 3a and 3c produced a fair number of lightning discharges whereas storm 3b produced none. Combined into one storm this would give a very unusual lightning pattern. Third, the net charge and the dipole-moment of the storm 3a was larger by a factor 10 and those of storm 3c by about a factor 5 than those of storm 3b. Finally, a very detailed description of the meteorological conditions, radar and visual observations, occurrence of lightning discharges and so on, is given in the "Summary of Weather: Monday 24 July 1978" of the KSC Weather Office. The observations confirm the occurrence of the storms 3a and 3c on the west side of the Indian River as determined by the multipole analysis.

IV. NET CHARGE AND DIPOLE-MOMENTS.

A cartesian coordinate system is used with x and y being the surface co-ordinates and z the altitude above ground. The values x_0, y_0, and z_0 are the co-ordinates of the centerpoint of the multipoles. The analysis is limited here to the representation of the charge distribution in a thunderstorm by a monopole Q and a dipole D with its three components: horizontal dipole in x direction Dx, horizontal dipole in y direction Dy, and vertical dipole in z direction Dz. A composite picture of such a charge arrangement is shown in Fig. 3. Vector addition of the three dipole components would result in the dipole vector

THUNDERSTORMS

$\bar{D}$ with the absolute value $D = (Dx^2 + Dy^2 + Dz^2)^{1/2}$. The tilt angle θ of $\bar{D}$ against the vertical z axis is given by $\cos\theta = \dfrac{D_z}{D}$.

Table 1 presents a summary of the evaluated data of all storms. In the columns 1 to 9 is listed storm identification, date, time interval over which the data are averaged, average net charge, dipole moment, tilt angle against the vertical, pole center altitude, temperature at pole center, altitude of 0°C level and comments.

TABLE 1

1	2	3	4	5	6	7	8	9
	Date	Time Interval [Min.]	Charge [C]	Dipole [C km]	Tilt [Degree]	Pole Cent. Alt. [km]	Cent. Temp [°C]	Alt. °C [km]	Comments
Shower Cld.2	15.07.77	1-27	-4	79	79	8.4	-30	4.5	Strong Wind Shear
" " 3b	24.07.78	8-21	0	71	32	8.3	-29	4.6	
" " 1	25.07.78	1-20	-2	9	51	4.4	-1	4.3	Weak Shower
Thunderst. 3a	24.07.78	1-4	-401	1934	17	4.8	-1	4.6	End of Storm 3a
Thunderst. 3c	24.07.78	25-29	-30	704	22	10	-40	4.6	Beginning of Storm 3c

The parameters listed in Table 1 show some remarkable features. The net charges of the showerclouds as well as of the thunderstorms are all negative. The net charges of the thunderstorms are by a factor 10 to 100 greater than those of the showerclouds. This is also true for the dipole moments. If we assume for the distance between the dipole charges a minimum 2 or a maximum of 5 km this results in ±1000 C or ±400 C dipole charge for a dipole moment of 2000 C km. This is approximately the dipole moment at the end of storm 3a. If for a separation of 2 km we add half of the net charge $Q = -400$ C to the negative dipole charge and subtract the other half from the positive dipole charge this storm could be represented by an asymmetric tilted dipole with +800 C in the upper and -1200 C in the lower part of the storm. If the distance between the dipole charges is 5 km the corresponding values would be +200 C in the upper part and -600 C in the lower part. These values are considerably higher than usually quoted in the literature. This large reservoir of charge makes it easier to understand that a thunderstorm can supply the 30 C for an average lightning discharge every 10 to 20 seconds without seriously depleting the thunderstorm charge.

ATMOSPHERIC ELECTRICITY

All storms - showerclouds as well as thunderstorms - show a more or less pronounced tilt angle of the dipole axis from the vertical (see column 6 in Table 1). In showercloud 2 this angle is 79°, i.e., the positive and negative charges are arranged almost side by side and not on top of one another. This is probably due to a strong wind shear. A radiosonde launched two hours later showed that the wind changed from 20° N-W to 75° N-E in the altitude from 7 to 9 km. The average pole center was at 8.4 km. The two thunderstorms 3a and 3c show smaller tilt angles than the showerclouds. Nevertheless the tilted dipole charge may lead to an explanation of slanted lightning discharges. The strongest field, and therefore at least one location for the origin of lightning, will be found between the upper positive and lower negative charge of the dipole. If these charges are not vertically arranged but have an offset in the horizontal direction the electric field vector will also have a tilt angle. Since the lightning discharge will grow best in the direction of the field vector a slanted lightning channel could be expected. Indeed the argument could be reversed. If slanted lightning discharges are observed, as they are according to reports in recent literature, it would follow that the charge distribution in the thunderstorm should contain regions where positive and negative charge centers are arranged with a sideways slant.

The pole centers of showerclouds 2 and 3b were found at 8.4 and 8.3 km altitude, at temperatures of -30 and -29°C, respectively. The weak shower cloud 1 had its pole center at 4.4 km altitude at the -1°C temperature level. It is most interesting that in storm 3a representing a storm at the end of its lifetime the pole center has dropped to 4.8 km or -1°C level, whereas storm 3c that shows the beginning of the second thunderstorm has its pole center at 10 km altitude or the -40°C level.

There are a few tentative conclusions that we can draw from this limited set of data. The electrification process of a thunderstorm that produces the main charge centers starts at 8 to 10 km altitude at temperature levels between -30 and -40°C. However, the pole center does not remain at this height altitude but sinks down during the lifetime of the storm to about 4.5 km and the -1°C level. The upper positive charge is of an order of +200 to +800 C and the lower negative charge of -600 to -1200 C. These charges are not vertically arranged

THUNDERSTORMS

one on top of the other but skewed at a tilt angle of 20° or more. Wind shear seems to contribute to the offset.

Consequently, computer models of electrified clouds should be capable to incorporate a tilt angle. In other words, they should be three dimensional. The charge generating and separating mechanisms should be capable of producing 50 to 100 times as much charge as hitherto assumed, and should not be bound to a certain temperature level.

ACKNOWLEDGEMENT

This research was supported by NASA Kennedy Space Center, Contract CC 80190A

REFERENCES

1. Simpson, G.C., Scrase, F.J.: "The distribution of electricity in thunderclouds I." Proc. Roy. Soc. A, 161, 309-352, 1937.

2. Simpson, G.C., Robinson, G.D.: "The distribution of electricity in thunderclouds II." Proc. Roy. Soc. A, 177, 281-329, 1941

3. Gish, O.H., Wait, G.R.: "Thunderstorms and the earth's general electrification." J.G.R. 55, 473-484, 1950

4. Malan, D.J.: "Les décharges dans l'air et la charge inférieur positive d'un nuage orageux." Ann. Géophys. 8, 385-401, 1952

5. Kasemir, H.W.: "The thundercloud." Problems of atmospheric and space electricity, ed. C. Coroniti, Elsevier Publ. Co., New York, 1965.

6. Küttner, J.: "The electrical and meteorological conditions inside thunderclouds." J. Met. 7, 322-332, 1950

7. Krehbiel, P., M. Brook, R. McCrory: "Lightning ground stroke charge location from multistation electrostatic field change measurements." Electrical Processes in Atmospheres, ed. Dolezalek, Reiter, Steinkopff Verlag Darmstadt 1977.

8. Few, A.A., T.L.Teer, D.R.MacGorman: "Advances in a decade of thunder research." Electrical processes in Atmospheres. ed. Dolezalek, Reiter, Steinkopff Verlag, Darmstadt 1977.

9. Taylor, W.L.: "A VHF technic for space-time mapping of lightning discharge processes." J.G.R., Vol. 83, No. C7, 3575-3583, 1978

10. Bevington, P.R.; "Data reduction and error analysis for the physical sciences." McGraw Hill Book Co. New York, 1969.

MODIFICATION OF THE ELECTRIC FIELD OF THUNDERSTORMS

HEINZ W. KASEMIR AND HELMUT K. WEICKMANN
U. S. Army Electronics Laboratories, Fort Monmouth, New Jersey

1. INTRODUCTION

It is well known that corona discharge appears at sharp points of conductors even if the electric field is far below the breakdown value. Based on this fact, a method was proposed by H. Weickmann (1963) to limit the growth of the electric field of a thunderstorm by dispersing 50 to 100 pounds of chaff inside the cloud. This may prove to be a feasible method of keeping the electric field below the value necessary to ignite lightning discharges. This paper describes the first results with chaff seeding.

2. THEORETICAL CONSIDERATIONS

The thunderstorm is a current generator. It produces a certain amount of charge per second, which is equivalent to a current output I. The current I is split up in three parts, which represent the contribution to the fair weather current I_1, the leakage current inside the storm I_2, and the charge dissipated by lightning discharges I_3. The current I of an average thunderstorm is about three amperes, with one ampere for each of the three partial currents.

Figure 1 shows a simplified circuit diagram of the thunderstorm generator. The two generators G_1 and G_2 supply the positive and the negative charge. The leakage current I_2 inside the storm flows through the shunt resistor R_s. The fair weather current I_1 follows the path through R_1 up to the ionosphere, through R_f--the resistance of the fair weather areas--from the ionosphere to the ground and through R_g from the ground to the negative pole of the thunderstorm generator. To represent lightning discharges, two spark gaps--CD for cloud discharges and GD for ground discharges--are provided in the circuit diagram through which the lightning current I_3 is flowing. Parallel to the spark gaps, two glow lamps G,G are connected, which show the influence of the corona discharge induced by chaff seeding. The effect of chaff seeding on lightning discharges can be deduced immediately from the circuit diagram. If the ignition voltage of the glow lamp is smaller than the breakdown voltage of the spark gap, and the continuous current I_4 through the glow lamp equal to or greater than the integrated intermittent current I_3, then a spark-over will never occur. The glow lamp acts as a voltage limiter so that the breakdown voltage of the spark gap is not reached.

In application to the thunderstorm it appears feasible that lightning discharges can be prevented if 1) the corona discharge on the chaff needles starts at a lower electric field than is necessary to ignite lightning discharges, 2) the corona discharge current is of the same magnitude or preferably larger than the current dissipated by lightnings, and 3) the chaff can be dispersed at the right place and at the right time in the thundercloud.

To check the first requirement, the field augmentation factor at the tips of the chaff needle has been calculated. The cylindrical body of the chaff fiber can be approximated by a slim spheroid. If d is the length of the spheroid, r the radius of curvature, and E the electric field at the tips of the spheroid, and F the electric field into which the spheroid is placed, then the following relation holds:

$$E = \frac{1}{10} \cdot \frac{dF}{r} \tag{1}$$

with the field augmentation factor $f = \frac{1}{10} \cdot \frac{d}{r}$.

For a chaff fiber of d = 30 mm and r = .0125 mm, the augmentation factor f becomes 240. This means if the breakdown field is assumed to be 3×10^6 V/m, then according to Eq. (1) a thunderstorm field of only 12,500 V/m is necessary to start corona discharge at the ends of the chaff fiber. As the fieldstrength capable of igniting a lightning discharge is assumed to be of the order of 350,000 to 1,000,000 V/m, the corona fieldstrength is at least by a factor 30 lower. Therefore the first requirement is certainly fulfilled.

For the calculation of the corona discharge current, a formula derived by S. Chapmann (1958) is used:

$$i = adkF^2/2, \tag{2}$$

where

i = cornoa discharge current,
a = 1.4×10^{-11} F/m,
d = 3×10^{-2} m (length of chaff fiber)
k = 2×10^{-4} m²/V sec (mobility of ions)
F = 2×10^4 V/m.

With the given values, a corona discharge current of 1.7×10^{-8} ampere per chaff fiber will result. Each pound of chaff contains about 2×10^6 fibers, so that 50 pounds of chaff would produce about 1.7 ampere discharge current. This would be an adequate amount of corona current to dissipate the amount of charge otherwise destroyed by lightning flashes. It may be mentioned that the corona discharge current increases with the square of the field. If we assume a field F = 100,000 V/m instead of 20,000 V/m, which is still sufficiently below the breakdown field, then the corona current would increase by a factor 25 to 42.5 amperes. There is no indication in the literature that even a very severe storm will produce as a whole such a large current. We may conclude that the second requirement can also be fulfilled by chaff seeding.

The third requirement--to disperse the chaff at the right time and at the right place in the storm--is a formidable problem. There are two areas in a storm where chaff seeding would be most effective. One area is between the upper positive and the lower negative charge, where a maximum field exists and the intra-cloud discharges are generated; and the other area is below the negative charge and the lower positive charge pocket, where the ground discharges originate (Kasemir, 1960). It should be mentioned, however, that the last statement is based on inadequate experimental data of the field and charge distribution in a thunderstorm, and that the first step in an effective chaff-seeding program should be a thorough investigation of the development, movement, and decay of the charge centers during the life history of a thunderstorm.

3. INSTRUMENTATION

The aircraft used in the chaff-seeding tests was a two-engine propeller-driven airplane C-47. It was equipped with a standard chaff dispenser ALE 24, two cylindrical field mills, and communication equipment to the ground station. At the ground stations the electric field of the storm and of the lightning flashes was recorded and the position of the lightnings plotted on a map. In addition, an M-33 tracking radar was installed at ground station No. 1 to record the flight path of the airplane. The chaff dispenser was able to eject one package containing about two million chaff fibers every one-half second, i.e., every 50 m of the flight path. Usually 50 to 60 packages were ejected during one run.

The cylindrical field mill is described in detail by Kasemir (1964, 1965). The outstanding features of the instrument are that it records two components of the electric field simultaneously and automatically cancels out the influence of the electric charge of the airplane on the field measurement. The noise level is extremely low, equivalent to a field of about 50 mV/m. A multiple range switch covered five powers of ten in ten steps. The largest field that could be measured was 500,000 V/m.

4. TEST FLIGHTS

The following procedure for chaff-seeding flights was worked out: The airplane would fly just below the base of a well developed growing cumulus cloud, and the electric field would be recorded. From the field records the plane could be guided during the flight to the areas with maximum fields. If the field-strength surpassed the value of 20,000 V/m, the plane would return on the same path, and the area with the strong field would be seeded. After seeding, the plane would continue to fly back and forth on the same path so that the influence of the corona discharge on the electric field could be recorded.

Five missions of this kind were carried out. Three showed a definite decrease of the electric field after seeding; even so, the storm was in its first stage and continued to develop. On the fourth flight a lightning strike occurred after seeding. The field of the lightning was recorded, but its location was not observed. It is therefore possible that the field discharging effect was due to the lightning and not to the chaff seeding. On the fifth flight a large area below an anvil but outside the main trunk of the storm was seeded, but no field decreasing effect could be detected.

In Fig. 2 the field records of the three successful chaff seeding flights are reproduced. The ordinate displays the vertical field component and the abscissa the time. It took about seven minutes to cross the base of the cloud and another three minutes to make a 180-degree turn to retrace the previous path. The time when chaff has been dropped is marked by the words, "chaff drop," and the penetration of the seeded area by the words, "seeded area." It can be seen by the record of the flight on 22 July in Fig. 2a that after the chaff has been dispersed the electric field has dropped in the seeded area to a much smaller value. The same effect took place on the flight of 31 July (Fig. 2b). Here the airplane kept flying back and forth on the same path several times after seeding to check if the electric field would recover with time. A small increase can be seen after the fifth traverse, but the field never recovered to its preseeded value. On the third test flight on 1 August (Fig. 2a), the chaff was dropped in a field of about 40,000 V/m. A strong corona discharge started right after the chaff had been dropped. This can be seen by the oscillations in the field record. Also, the M-33 tracking radar was saturated with high frequency noise every time it followed the airplane into the seeded area. Unfortunately the field mill amplifier broke down during the second return flight, so the field decrease could not be recorded.

5. CONCLUSIONS

A few critical remarks are offered in conclusion: There is not much doubt that the electric field in a thunderstorm can be influenced by chaff seeding. However, a large number of flights are needed to establish a more reliable pattern of the field generation and its spatial and time variation during the lifetime of a thunderstorm. Furthermore, the seeding method needs some improvement. The dispersion of chaff should be continuous and should also cover a broader area. A strip of about 50 m width and 3 km length constitutes only a small fraction of the base of a thunderstorm.

6. REFERENCES

Chapmann, S., 1958: Corona-point-discharge in wind and application to thunderclouds, Recent Advances in Atm. Electricity, Perg. Press, 277-288.

Kasemir, H. W., 1960: A contribution to the electrostatic theory of a lightning discharge, J. Geophys. Res., 65, No. 7, 1873-1878.

Kasemir, H. W., 1964: The Cylindrical Field Mill, ECOM Technical Report 2526, Fort Monmouth, New Jersey, 18 pp.

Weickmann, H. K., 1963: A realistic appraisal of weather control, J. Appl. Math. Phys., 14, Fasc. 5, 528-543.

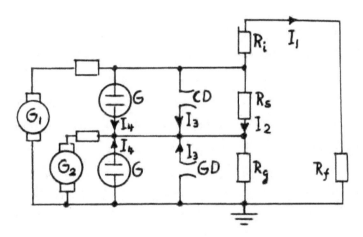

Fig. 1. Simplified Circuit Diagram of Thunderstorm Generator

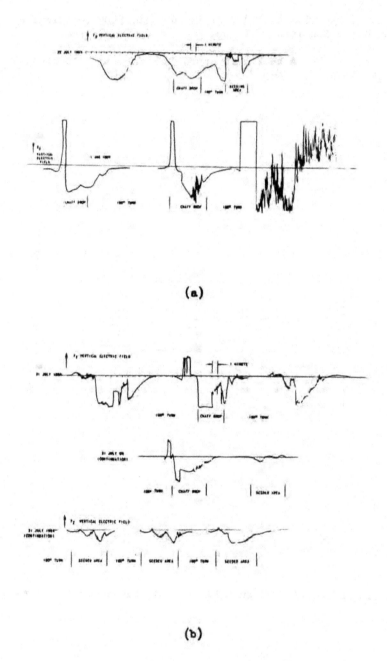

(a)

(b)

Fig. 2. Effect of Seeding on Thunderstorm Electric Field

LIGHTNING SUPPRESSION BY CHAFF SEEDING

Heinz W. Kasemir
Atmospheric Physics and Chemistry Laboratory
ESSA Research Laboratories
Environmental Science Services Administration
Boulder, Colorado

FINAL REPORT

June 1968

USAECOM Project Order: 67-95906
Amendment: 68-95920

CONTENTS

INTRODUCTION

The principle idea on which lightning suppression by chaff seeding is based can easily be demonstrated by a laboratory experiment. If a grounded plate or sphere is placed in the neighborhood of the charged sphere of a Vande Graff machine, sparks of 10 to 20 cm length can be drawn. However if a sharp point for instance a chaff needle, is attached to the charged sphere, corona discharge will occur before the voltage is high enough to cause a spark over to the grounded sphere. If the corona current is strong enough to balance the supply current of the machine the voltage necessary to ignite a spark will never be reached. The same effect is used in a thunderstorm to suppress lightning discharges by chaff seeding.

1.   The equivalent circuit diagram of a thunderstorm:

Fig. 1 shows the equivalent circuit diagram of a thunderstorm. On the left side the generator symbols G_1 and G_2 represent the positive and negative charge generation in the top and the base of a storm. The unmarked resistor symbols (rectangular small boxes) at the hot terminals of the generators indicate that we are dealing with current generators. This means that charge is produced at a constant rate whether it can dissipate or not. Three different means are provided in the circuit diagram for the dissipation of charge. These are, from left to right in Fig. 1, the two glow lamps G the two spark gaps CD and GD and the ohmic resistors R_s, R_g in parallel to R_i, R_f. The ohmic resistors represent the conduction currents of a thunderstorm. R_s is the shunt resistor between the hot terminals of the two generators; it is determined by the conductivity of the column of air between the upper positive and the lower negative charge inside the storm. R_g is the resistance of the air column between the base of the cloud and the ground, and R_i the resistance of the air column between the top of the storm and the ionosphere. This branch of the circuit is connected to ground by R_f which represents the resistance between the ionosphere and the earth in

the fair weather areas. With the exception of R_f, the resistors
have comparatively high values (in the order of hundreds of
Meg ohms) so that the charge produced by the generators can
not leak away very fast. As a consequence, the voltage on the
terminals builds up to high values until a spark is ignited through
the spark gaps CD and or GD. These spark gaps represent the
lightning discharges in the real thunderstorm, CD representing cloud discharge and GD ground discharge. The glow lamps
of the circuit diagram have no natural equivalent in the thunderstorm. They represent the artificially introduced corona discharge of the chaff fibers. If the voltage across the glow lamps
reaches their breakdown value the lamp will ignite and keep the
voltage very effectively at this value. Any further increase in
the current output of the generators will be shunted by the glow
lamps.

Two important conclusions can be drawn from the circuit
diagram. If sparks (lightning discharges) shall be prevented,
the ignition voltage of the glow lamps (threshold field for corona
discharge on the chaff fibers) should be lower than the breakdown
voltage of the spark gaps (breakdown field necessary to start
lightning discharges). Furthermore the glow lamps should be

able to dissipate a current of the same order as the current output of the generators. Applied to the thunderstorm this means that the corona current of the chaff fibers should be equal in amount to the charge dissipated by lightning strikes.

2. Laboratory experiments:

Chaff fibers of different lengths and thicknesses were placed in a plate condensor and measurements carried out of the threshold field necessary to start corona discharge. Fig. 2. The corona current as a function of fiber length for different fields is shown in Fig. 3. The onset field for corona discharge on a chaff fiber of 10 cm length is between 25 and 30 kV/m, depending slightly on the thickness (radius of curvature on the tip) of the fiber, Fig. 2. The corona current of the onset field is less than 0. 1μ A but increases rapidly with an increase of field strength. For a field of 70 kV/m the corona current increased to 1. 0μ A.

What do these values mean in regard to lightning suppression by chaff seeding? The breakdown field strength at the ground is 3000 kV/m. As this value depends linearly on pressure it may be reduced by about a factor 2 at cloud altitude. We may also account for the field concentration on a precipitation particle by decreasing the required outside field strength by another factor of 3. This brings the 3000 kV/m down to 500 kV/m as the necessary field value to start a lightning discharge. As will be seen later in Fig. 5 fields as strong as 300 kV/m have been recorded below a cloud without lightning occurring in this area for a period of about half an hour. This shows that stronger fields are required to ignite a lightning discharge and lends some support to the theoretical value of 500 kV/m. If we compare this value with the onset field of corona discharge of 30 kV/m or even the 1μ A corona current field of 70 kV/m, we see that the corona fields are lower by a factor of 17 or respectively of 7 as the threshold field of the lightning discharge. This shows that the first require- ment - corona field smaller than lightning field - is certainly fulfilled.

The second requirement would be that the corona current produced by the chaff fibers is of the same magnitude as the current consumed by lightning discharges. A moderate thunder- storm will produce a lightning discharge every 30 seconds. The average lightning strike discharges about 30 coulombs. This represents an average continuous supply current of about 1 ampere.

To match this value we would need 1,000,000 chaff fibers each producing 1μ A corona current. 2,000,000 fibers is the content of a 1 pound package of ordinary x band chaff. This chaff however proved to be unsatisfactory because of the short length and the bird nesting effect (clumping together of the fibers). However as will be discussed later, it is well within the state of the art to disperse such amounts of chaff of the necessary length from a specially designed chaff dispenser. The average lightning supply current of 6 ampere of a severe storm with about one lightning every 5 seconds can still be matched by corona current if the number of chaff fibers is increased to 6,000,000. Even with only 1,000,000 fibers a 6 ampere corona current will be produced if the electric field increases to about 200 kV/m. This is still only half of the field strength necessary to start lightning discharges. Therefore we may also conclude that the second requirement - corona current equal to lightning supply current - can be fulfilled.

3.   Field tests:

For the field test the C-47 airplane was equipped with the following instruments:

a.   Two field mills measuring the three components of the electric field.

b. Two chaff dispensers.

c. One corona discharge indicator.

d. Sensors for different meteorological and airplane parameters including a strip chart and tape recorders.

Each field mill records two different components of the atmospheric electric field and automatically eliminates the influence of the airplane charge on the measurement. A detailed description of this instrument is given in ECOM Technical Report 2526, 1964, by H. W. Kasemir.

The new chaff dispenser contains the chaff not as needles cut to a certain length and pressed a million a piece in little packages, but the chaff is a long strand of conductive fibers wound up on a reel. 10 reels constitute one dispenser unit housed in one wing tank. During operation the ten strands are forced out through ten guide holes at great speed, and before leaving the tank completely are chopped by a helical chopper into needles of a preset length. This design has several features which are crucial for lightning suppression. (1) The chaff is emitted continuously. Bird nesting, i. e. bunching together in clumps of several 100 or 1,000 needles, is completely eliminated. (2) The chaff is distributed more evenly behind the airplane because a continuous stream of needles and not individual packages emerge from the airplane. (3) It is possible to experiment with different lengths of needles, the needle length depends on the speed of the chopper, which is easily adjustable.

The corona discharge indicator measures the electromagnetic emission of the chaff fibers as soon as they emerge from the chaff dispenser. The range should be limited to 50 to 100 meters and the indication selective to corona discharge on the chaff only, i. e. corona discharge on the airplane itself should not be indicated. The last point is difficult to establish and needs a more detailed study. Otherwise the instrument seems to be working properly. One essential point has already been confirmed, namely that corona discharge occurs on the chaff needles if the electric field in the atmosphere surpasses a threshold value of about 25 kV/m, which is in agreement with the laboratory tests.

The following flight procedure has been worked out. The airplane would fly below developing thunderstorms or shower clouds and would hunt for areas with electric fields about 30 kV/m. If such an area was found and the field pattern had been established by several passes through this area, chaff would be ejected on two to four runs and corona discharge and the electric field were recorded by continuous passes back and forth through the seeded area until either corona discharge or the strong electric field disappeard. Fig. 4 and 5 are typical examples of such flight records.

Fig. 4 shows the corona discharge on the upper trace and the vertical field component on the lower trace. Seeding and the seeded area are marked as such. Eleven passes have been made below the storm. On the second, third, and fifth pass chaff was dispersed. On the second and fourth pass corona discharge is small and irregular. At the third pass the plane missed the previous seeded area completely and no corona discharge was recorded. But it seems that the chaff needles spread out very rapidly and after five to ten minutes the whole area is solidly filled with corona discharge until the fields drop below 20 kV/m.

Fig. 5 shows the decay of a strong electric field after chaff seeding. It may be pointed out that between the first and the second pass more than 20 minutes elapsed due to the fact that the area was lost and could not be re-located earlier. This proved to be fortunate, because it shows that during this time the field remained at its high value of about 300 kV/m.

After the area was found again chaff seeding began at the third and following passes. The decay of the field can be recognized three minutes after chaff seeding started. Ten minutes thereafter the field has completely collapsed.

The Flagstaff experiments have established that corona discharge is generated if chaff needles of 10 cm length are dispersed in the electric field of thunderstorms exceeding values of 30 kV/m.

It seems highly probable that the decay of strong electric fields is caused or accelerated by corona current produced by the chaff needles.

To study the effect of lightning suppression by chaff seeding an airplane is required, which is capable of penetrating the storm and locating the birthplaces of lightnings. The airplane used in the reported tests was limited to areas below the storm and outside heavier precipitation and turbulence.

4. Field configuration of an infinite seeded layer in a constant electric field.

A theoretical calculation has been carried out to answer the question: is the electric field in a thunderstorm augmented at the boundary of the chaff seeded area? The principal factors and limitations of these calculations shall be discussed on the very simple model consisting of an infinite sheet of chaff seeded area exposed to a homogeneous electric thunderstorm field. Later on the field augmentation factor for other models will be calculated. Fig. 6 gives a cross-sectional view of our problem. Inside the two horizontal lines is the chaff seeded area A_2 with the field strength F_2, the current density i_2, and the conductivity λ_2. If we assume steady state conditions the problem is governed by the condition of the continuous current flow.

$$\text{div } \overline{i} = 0 \tag{1}$$

Equation (1) reduces in the one dimensional case to

$$i_1 = i_2 = i = \text{constant} \tag{2}$$

using

Ohm's law $\qquad i_1 = \lambda_1 F_1$ and

$$i_2 = \lambda_2 F_2 \tag{3}$$

it follows from equation (2) and (3)

$$F_1 = i/\lambda_1 = \text{constant}; \quad F_2 = i/\lambda_2 = \text{constant} \tag{4}$$

or $\qquad F_2/F_1 = \lambda_1/\lambda_2$

Equation (4) tells us that in the not seeded area A_1 the electric field F_1 is constant throughout the area and up to the boundary between the areas A_1 and A_2, i.e. in this case there is no field augmentation at the boundary in spite of the negative and positive surface charge layer at the upper and lower boundary of area A_2. From an electrostatic point of view this may be at first sight a somewhat surprising result, because the negative surface charge at the upper boundary may be expected to increase the field in area A_1 at least in the immediate neighborhood of the charge. However if we remember that the field generated by an infinite layer of charge does not decrease with distance we see that for instance in area A_1 the field of the negative surface charge is completely cancelled by the opposite field of the positive surface charge of the lower boundary and only the original field F_1 remains. Inside area A_2 the fields of the two surface charges augment each other and because they are of opposite direction to the original field F_1, they will weaken F_1 in area A_2.

If we are not dealing with infinite but with finite layers of positive and negative surface charge, then the fields of these charges decay with distance and will not cancel each other completely in the outside area A_1. In this case we may expect a field augmentation at the boundary. This effect will be less pronounced

326

if the seeded area resembles a horizontal layer i. e. its horizontal dimensions are large compared to the vertical one. The field augmentation will be more pronounced if the opposite is true. The first case will be encountered if the chaff is dispersed from an airplane and the second case, if the chaff is dispersed from a drop sonde.

However, before we calculate the field augmentation for different shapes of the seeded area, a few remarks shall be made on how well corona discharge on the chaff fibers can be represented by an increased conductivity λ_2 in the seeded area.

5.   The relationship between the corona discharge on chaff fibers and the conductivity of the seeded area:

The conductivity of the air stems from the ionization of air molecules by cosmic rays and radioactive emanation of the ground. The ion production is given by the ionization constant q, which is in the order of 5 to 20 x 10^6 ion pairs per second per cubic meter. One single chaff fiber with a corona current of 1 micro ampere will produce 6 x 10^{12} ion pairs per second, i. e. it would be equivalent to the natural ion production of a volume of 3 x 10^5 m^3 or a cube of 68 m length, width and height. If we require that the conductivity in the cube should be increased three times, then the volume ionized by one chaff fiber would shrink to $10^5 m^3$ and the cube length to about 46 m. A cloud volume of $4 \times 4 \times 4 \ km^3 = 64 \times 10^9 m^3$ contains 64×10^4 cubes of $10^5 m^3$ volume. This means that about a half million chaff fibers would be enough to increase the conductivity in the cloud by a factor three or decrease the field by a factor three.

In estimating the increase of conductivity by an increase of ion production we have not taken into account the increased loss by recombination. As the negative ion stream is moving upwards from the chaff fiber and positive ion stream moving downwards ions of opposite polarity do not meet. Consequently there is no ion loss by recombination of small ions. Only if the upward moving negative ion stream meets the downwards moving positive ion stream produced by another chaff fiber float- ing at a higher level there will be recombination of small ions. However, this recombination or inter-mixing will be beneficial for a continuous current flow, as will be discussed later.

A heavy loss of small ions will occur in the first few seconds by attachment to the cloud droplets until the droplets are charged to capacity and reject further attachment of ions of the same sign. An area of negative space charge will form

to more than two layers of chaff fiber is evident and leads immediately to Fig. 6. The charge distribution of Fig. 7 resembles very strongly that of the polarization of a dielectric material exposed to an external field. The local fields of the internal dipoles cancel each other by the sheer number of dipoles and the reduction of the primary field inside the dielectric can be interpreted as the effect of the surface charge at the boundary of the dielectric material.

In the case of chaff seeding for the purpose of lightning suppression we require in addition to the field reducing effect that an enhanced current flow is carried through the seeded area. This would necessitate that at the contact area - marked by a dashed line in Fig. 7 - between the positive space charge of the upper chaff fiber and the negative space charge of the lower chaff fiber a strong recombination between the opposite charged cloud droplets takes place or that this contact area is penetrated by a sufficient number of positive small ions coming down from the upper chaff fiber and of negative small ions coming up from the lower chaff fiber. Even if the small ions recombine rapidly with cloud droplets or small ions of opposite polarity they reduce the space charge pockets and stimulate a continuous corona discharge. For a continuous current flow through the seeded area it is only necessary that each chaff fiber carries the current through its own region of influence.

A very rough estimate of the size of this region of influence leads to a volume of 70m height and 35m square cross-section. This estimate is based on the reduced average lifetime and the velocity in a field of 70 kV/m of the small ion. One other way is to assume a spherical shape for the space charge pocket with the space charge density given by maximum droplet charge and the number of cloud droplets per m^3. The size of the sphere should be such as to produce a field of 40 kV/m at the position of the chaff fiber so that the thunderstorm field of 70 kV/m is reduced to 30 kV/m, which is about the onset field for corona discharge. Both of these estimates of the region of influence agree surprisingly well with each other. If we compare the height of the cube assigned to each fiber of 46m as calculated before, we see that a certain overlapping of the region of influence of the different chaff fibers will occur.

A numerical analysis to determine the region of influence more accurately is outside the scope of this paper. However three effects shall be mentioned, which tend to reduce the little pockets of space charge, thereby helping to prevent the quenching of the corona discharge, and support the current flow. These effects are: turbulent mixing, recombination of positive and negative charged cloud droplets and wash-out by precipitation. The enhanced coagulation of oppositely charged cloud droplets may even lead to increasing precipitation. This would apply also to rain drops falling through the seeded area. The rain drop will pick up a number of highly charged cloud droplets passing through - lets say - the upper negative space charge region. It becomes itself negatively charged. Entering the positive space charge pocket below, the rain drop will now coagulate with the positive charged cloud droplets, lose its negative charge and becomes positively charged. This process will continue alternately until the rain drop leaves the seeded area. Beside reducing the space charge pockets this will also increase the growth rate of the rain drop.

With regard to the purpose of the following calculation we will assume that an enhanced current flow can be carried through

331

the seeded area. Given a sufficient overlapping of the region of influence of the single chaff fibers, the large number of small ions liberated by corona discharge, the recombination of charged cloud droplets, the effect of turbulent mixing and the washout of the space charge pockets by precipitation is lumped together in one material parameter, namely, the conductivity λ_2 inside the seeded area, with the resulting effect, that the conductivity inside the seeded area is increased by corona discharge.

Before we return to the calculation of the field augmentation at the boundary of the seeded area a short remark shall be made on the boundary surface charge. In theory the surface charge is confined to an infinitesimal thin layer, in reality the surface charge will be spread out into a layer of several tens of meters inside the cloud and several hundreds of meters outside the cloud. A good example for the spreading of surface charge layer is the space charge layer which forms above the earth surface if the thunderstorm field is strong enough to generate corona discharge at the ground. However it should be emphasized that the thickness of the surface charge layer does not determine the thickness of the seeded area. At the ground the corona points are limited to the earth's surface whereas chaff seeding requires that the chaff fibers should be dispersed through a large volume of the thundercloud. Therefore conclusions drawn from the corona discharge at the ground below thunderclouds may be applied with some caution to the surface but not at all to the volume of the seeded area. In the following calculation the surface charge layer is treated as one of negligible thickness i. e. the enlargement of the seeded area by the spread out of the surface charge is dis-regarded.

6.   Field concentration at the boundary of different shapes of the seeded area in a constant electric field.

The calculation is carried out for the following problem. A body of a given shape with the constant conductivity λ_2 is imbedded into an environment of the constant conductivity λ_1. The whole system is exposed to a homogeneous electric field F_1 in the direction of the z axis of a cartesian coordinate system x, y, z. Find the field F_2 inside the body and the maximum field F_m at the boundary of the body. The body shall be represented by an infinitely long elliptical cylinder with its axis in the x direction. The solution shall be given for

333

ac) the small axis of the elliptical cross-section is in the z direction

bc) both axis are equal, in which case the elliptical cylinder becomes a circular cylinder.

cc) the large axis of the elliptical cross-section is in the z direction.

The same problem shall be solved when the body is represented by a spheroid, which changes from

as) a flat disc to

bs) a sphere to

cs) a prolonged spheroid

The cases ac to bc may be encountered if the chaff is dispersed continuously from an airplane and the cases as to cs if the chaff is dispersed discontinuously in little packages by an airplane, dropsonde or rocket.

The derivation of the potential function is not given here because the solution to the problems ac to cc can be obtained from the text book "Static and Dynamic Electricity" by Smythe (1) if the dielectric problem is transfored into a current flow problem by substituting the condictivities λ_1 and λ_2 for the two dielectric constants ϵ_1 and ϵ_2. The solution to problem as to cs is given by Kasemir (2).

A very simple equation has been derived for the maximum field concentration F_m at the boundary of the seeded area as well as the attentuated field F_2 inside the seeded area. These equations are valid for all the above state problems and are as follows:

$$\frac{F_2}{F_1} = \frac{\lambda_1/\lambda_2}{\lambda_1/\lambda_2 + (1-\lambda_1/\lambda_2)\rho} \tag{5}$$

and

$$\frac{F_m}{F_1} = \frac{1}{\lambda_1/\lambda_2 + (1-\lambda_1/\lambda_2)\rho} \tag{6}$$

From (5) and (6) also follows

$$F_2 = \frac{\lambda_1 F_m}{\lambda_2} \tag{7}$$

According to the earlier notation it is

F_1 = original field outside the seeded area

λ_1 = conductivity outside the seeded area

λ_2 = conductivity inside the seeded area

ρ = form factor, which depends only on the geometrical shape of the seeded area. (Fig. 8)

With the horizontal half axis a and the vertical half axis b of either the elliptical cylinder or the spheroid, ρ is given for the elliptical cylinder by

$$\rho_c = \frac{a}{a + b} \tag{8}$$

and for the spheroid by

$$\rho_s = \frac{a^2}{a^2 - b^2} \left(1 - \frac{b}{\sqrt{a^2 - b^2}} \tan^{-1} \frac{\sqrt{a^2 - b^2}}{b}\right) \tag{9}$$

For a $\to \infty$ the elliptical cylinder as well as the spheroid degenerate into a horizontal infinite layer. This case has been briefly discussed before. Note that ρ_c and ρ_s approach 1. From (5).

$$\frac{F_2}{F_1} = \frac{\lambda_1}{\lambda_2} \tag{10}$$

is obtained for $\rho = 1$. The field F_2 inside the seeded area is reduced in proportion to $1/\lambda_2$ the increase of the conductivity λ_2. From (6) follows for $\rho = 1$

$$F_m = F_1 \tag{11}$$

The maximum field F_m is equal to the original field F_1 outside of the seeded area, i. e. there is no field augmentation at the boundary.

If we go to the other extreme and let $b \to \infty$ then the elliptical cylinder deteriorates into a vertical infinite layer and the spheroid into infinitely long circular cylinder. In this case ρ_s and ρ_c approach zero. From (5) and (6) we obtain

$$F_2 - F_1 \qquad\qquad (12)$$

and

$$\frac{F_m}{F_1} = \frac{\lambda_2}{\lambda_1} \qquad\qquad (13)$$

These are somewhat unexpected results. Equation (12) says that the field inside the seeded area is the same as outside the seeded area no matter how much we increase the conductivity inside the seeded area. This means that continuous seeding from a dropsonde, where we may generate a body of seeded area like a prolonged spheroid, would have almost no field reducing effect. Furthermore we learn from equation (13) that the field concentration on the ends of the prolonged spheroid is proportional to the inside outside conductivity ratio of the seeded area. Both effects are favorable for generating lightnings and adverse to lightning suppression. Therefore the manner in which a cloud is seeded will have some effect on the outcome i. e. if lightnings are suppressed or prematurely triggered.

The field concentration at the ends of an infinitely long cylinder is not a physical reality. Therefore the field augmentation given by equation (13) has to be considered as a maximum value, which will never be obtained. To give an idea how close we may come to the maximum value with reasonable shapes of the seeded area, Fig. 9 shows the field augmentation F_m/F_1 as a function of the ratio b/a which represents the ratio of the vertical to the horizontal dimension of the seeded area. The curves of Fig. 9 are calculated with the assumption that $\lambda_2 = 2\lambda_1$, i. e. the conductivity inside the seeded area is twice as much as the conductivity outside. The maximum field augmentation is then 2. For an axial ratio of $b/a = 10/1$ the field augmentation is 1.96 and for $b/a = 5/1$ it is 1.92. This means that even for moderately slim spheroids the field augmentation at the ends of the spheroid is close to its maximum value. However it should be remembered that the surface charge of the mathematical model is confined to a very thin layer whereas in reality the surface charge is spread out over an area of several 10's to several 100's of meters thickness. This would smooth out the field concentration and, because of the rapid drop of the field with increasing distance from the tip of the spheroid, result in an average field of lower intensity.

If we reduce the axial ratio further to $b/a = 1$ the spheroid takes on the shape of a sphere. Here we have the maximum field concentration factor 1. 5 and the weakening factor of 0. 75 for the field inside the sphere. If we assume that a field F_1 of 70 kV/m is necessary to produce a corona current strong enough to consume that part of the thunderstorm current output which would otherwise dissipate in lightning discharges, we have a maximum field at the boundary of 105 kV/m and an inside field of 52 kV/m. The maximum field is still by a factor 5 lower than the threshold field necessary to initiate lightning discharges and the inside field about 80% above the threshold field to start corona discharge. Even if we double the fields to the values, inside field $F_2 = 104$ kV/m, thunderstorm field $F_1 = 140$ kV/m and maximum field at the boundary of the seeded area $F_m = 210$ kV/m, we are still more than a factor 2 below the lightning threshold field. The current flow through the seeded area has also doubled and should load down the thunderstorm generator enough to prevent further increase of the field. It may be expected that an equilibrium state of charge production and dissipation will be reached inside the above mentioned field values.

The maximum field given by Equation (6) is an upper limit. This value will be reduced by the fact that the surface charge at the boundary is not restricted to a very thin layer but will spread out over a volume of 100 or more meter thickness. Furthermore it will be reduced since the primary thunderstorm field F_1 is not constant but drops off at the poles of the generator as a field generated by space charges will do. This will be shown in the next portion of this text for a seeded area in the form of a sphere. It would be more tedious than difficult to carry out the same calculation for all the different shapes of the seeded areas used here. However, the results will be principally the same. Therefore the example of the sphere is sufficient to demonstrate the reducing effect, which the inhomogeneous field of space charge layers has on the field concentration at the boundary of the seeded area.

7. Field concentration at the boundary of a seeded area (sphere) in the inhomogeneous field of a dipol space charge:

A calculation has been carried out of the field concentration at the top of a sphere which is exposed to an inhomogeneous field distribution. The conductivity inside the sphere is λ_2 and outside the sphere λ_1.

As emphasized before the thunderstorm has to be considered as a current generator, i. e. its primary characteristic is the current source density w not the space charge. The relation between the current source density ω and the space

charge density q is

$$\omega = \frac{\epsilon}{\lambda} \, q$$

i. e. the space charge density q depends on λ if the current

source density is given. This will be the first of the con-

ditions to impose on the potential function which solves our

problem, that with a given current source density the space

charge density inside the seeded area shall be reduced in

proportion to the increase of the conductivity. The second

and third condition is that the potential function Φ_1 outside

and Φ_2 inside the sphere are identical at the sphere surface,

and that the radial current flow through the sphere surface

is continuous.

We divide the thunderstorm into four layers of equal

thickness numbered I, II, III, and IV from top to bottom,

and place the zero point of the cartesian coordinate system

x, y, z as well as that of the later used spherical coordinate

system r, θ, φ in the middle of the storm. In the highest layer

I the current source density ω increases from zero at the top

of the layer to its maximum value ω_0 at the bottom of this

layer, Fig. 10.,

$$\omega = \omega_0 \frac{2a - z}{a} ; \ a \leq z \leq 2a ; \text{Layer I} \tag{15}$$

In layer II and III the current source density decreases linearly with z, goes through zero in the middle of the storm and reaches its maximum negative value at the bottom of layer III

$$\omega = \omega_0 \frac{z}{a} ; \ -a \leq z \leq + a ; \text{Layer II and III} \tag{16}$$

In layer IV the current source density drops from its negative maximum value at the top to zero at the bottom of the layer

$$\omega = \omega_0 \frac{2a + z}{a} ; \ -2a \leq z \leq -a ; \text{Layer IV} \tag{17}$$

Fig. 10 shows the primary charge, field and potential configuration of the given current-source distribution. The parameters are normalized, field, current source and charge density to its maximum value, and the distance z and later the radius R of the seeded sphere to the thickness of the layers a. The positive and negative pole of the thunderstorm generator i. e. the highest positive and negative charge concentration are then at $z = \pm 1$. The maximum field e -1 occurs in the middle between the positive and negative charge for $z = 0$, and tapers off with increasing positive or negative z.

The equations of the normalized field e and potential φ for our thunderstorm model can be obtained by simple integration

from current source distribution (15), (16), and (17). They
are as follows:

$$e = -\frac{(z-2)^2}{2}$$

$$\varphi = \frac{(z-2)^3}{6} + 1$$

Layer I; $1 \leq z \leq 2$ (18)

$$e = \frac{z^2}{2} - 1$$

$$\varphi = -\frac{z^3}{6} + z$$

Layer II and III; $-1 \leq z \leq +1$ (19)

$$e = -\frac{(z+2)^3}{2}$$

$$\varphi = \frac{(z+2)^3}{6} - 1$$

Layer IV; $-2 \leq z \leq -1$ (20)

If we now introduce in this thunderstorm model a seeded area in
the shape of a sphere with radius R and conductivity λ_2, we obtain
the potential function φ_1 outside the sphere

$$\varphi_1 = \left[r - \frac{r^3}{10} + \frac{1-\alpha}{2+\alpha}\left(1 - \frac{R^2}{10}\right)\right] P_1 - \left(\frac{1-\alpha}{3\alpha+4}\ \frac{3R^7}{r^4} + r^3\right)\frac{P_3}{15} \tag{21}$$

and the potential function φ_2 inside the sphere

$$\varphi_2 = \frac{1}{\alpha} \left[r - \frac{r^3}{10} + \frac{2(\alpha-1)}{2+\alpha}(1 - \frac{R^2}{10})r \right] P_1$$

$$- \left[\frac{3(1-\alpha)}{3\alpha+4} + 1 \right] \frac{r^3 P_3}{15}$$

(22)

In equation (21) and (22) it is

$$\alpha = \frac{\lambda_2}{\lambda_1} = \text{ratio of conductivities inside to outside of the sphere}$$

$r; \theta$ = polar coordinates

$P_1; P_3$ = Legendre's polinomials, which are functions of $\cos \theta$

The potential functions φ_1 and φ_2 fulfill all the boundary conditions outlined above and are therefore the unique solutions to our problem. Differentiation in respect to r yields the r component e_{1r} and e_{2r} outside and inside the sphere.

$$e_{1r} = - \left[1 - \frac{3r^2}{10} + \frac{2(\alpha-1)}{2+\alpha} \frac{R^3}{r^3}(1 - \frac{R^2}{10}) \right] P_1$$
$$- (\frac{1-\alpha}{3\alpha+4} \frac{4R^7}{r^7} - 1)\frac{r^2 P_3}{5}$$

(23)

and

$$e_{2r} = - \frac{1}{\alpha} \left[1 - \frac{3r^2}{10} + \frac{2(\alpha-1)}{2+\alpha}(1 - \frac{R^2}{10}) \right] P_1$$
$$+ \frac{7}{3\alpha+4} \frac{r^2 P_3}{5}$$

(24)

At the surface of the sphere $r = R$, we obtain from (23) and (24) the r component of the field e_{1R} and e_{2R}.

$$e_{1R} = - \left(\frac{3\alpha}{2+\alpha} - \frac{R^2}{10} \; \frac{5\alpha+4}{2+\alpha} \right) P_1 + \frac{7\alpha}{3\alpha+4} \; \frac{R^2 P_3}{5} \qquad (25)$$

and

$$e_{2R} = - \frac{1}{\alpha} \left(\frac{3\alpha}{2+\alpha} - \frac{R^2}{10} \; \frac{5\alpha+4}{2+\alpha} \right) P_1 + \frac{7}{3\alpha+4} \; \frac{R^2 P_3}{5} \qquad (26)$$

From (25) and (26) it is easy to see that

$$e_{1R} = \alpha e_{2R} \text{ or } \lambda_1 e_{1R} = \lambda_2 e_{2R}$$

which is the condition for the continuous current flow through the surface of the sphere.

If we differentiate e_{1R} with respect to θ and set this equation equal zero we obtain from (25) the point of the sphere surface where the maximum field occurs. With $P_1^{\; 1}$; $P_3^{\; 1}$ being the differentials of the Legendre's polinomials we obtain from (25)

$$0 = \left[\left(\frac{3\alpha}{2+\alpha} - \frac{R^2}{10} \; \frac{5\alpha+4}{2+\alpha} \right) P_1^{\; 1} - \frac{7\alpha}{3\alpha+4} \; \frac{R^2 P_3^{\; 1}}{5} \right] \sin \theta$$

As the expression in the brackets is not in general zero it follows that $\sin \theta = 0$

$$\theta = 0 \text{ or } 180^{\circ}$$

i. e. the field maxima occur at the upper and lower point of the sphere. At these points, P_1 and P_3 are ± 1 and we obtain the maximum field e_{1R} max from (25)

$$\pm e_{1R} \text{ max} = - \frac{3\alpha}{2+\alpha} + \frac{R^2}{10} \left(\frac{5\alpha+4}{2+\alpha} + \frac{14\alpha}{3\alpha+4} \right) \qquad (27)$$

To compare this result with that obtained above for the homogeneous field we set $\lambda_2 = 2\lambda_1$; $\alpha = 2$, which reduces (27) to $\pm e_{1R}$ max $= -1.5 + 0.58 \, R^2$ (28)

For $R \ll 1$ the field concentration at the top of the sphere is 1.5 which was also the case for the homogeneous field. This result is to be expected because if the radius of the sphere is small compared to a, the sphere is surrounded by a practically homogeneous field. If the radius of the sphere grows to a, the field concentration at the top decreases to about 1. This is the result which was predicted above.

The dashed line in Fig. 10 shows the field curve e_s as a function of z if a sphere of radius $R = 1$ and $\lambda_2 = 2\lambda_1$ is brought into the thunderstorm model. The external field rises with the approach of the upper and lower point of the sphere to about 0.9, which is not quite twice the value 0.5 of the thunderstorm field at this point. Nevertheless even a field concentration of 0.9 is still less then the maximum field 1.0 of the storm center for $z = 0$.

Inside the sphere the field is reduced to 0.44 at the top and bottom and increases to about 0.73 in the center of the sphere.

With respect to a real thunderstorm we have about the following situation. If the maximum field of the storm has grown to 100 kV/m and the storm is seeded we have a maximum field concentration of 150 kV/m at the boundary of the seeded area and 75 kV/m inside, if the sphere shaped area is small compared to thunderstorm dimensions. As the area growth, the maximum field at the boundary drops to about 90 kV/m and the inside field varies between 44 and 73 kV/m. The strong field gradient at the boundary has a tendency to suck the chaff fibers by electric force into the unseeded area and therefore to enlarge the seeded area

The maximum fields are below lightning igniting values by a factor 2 to 3 and the minimum fields are still above the corona on set field. Therefore there is not much danger that lightnings will be generated instead of suppressed by chaff seeding. At the same time the fields are not reduced to values too small to maintain corona discharge. The key factor here is the small corona onset field of about 30 kV/m, which for instance can never be achieved by ordinary silver iodide seeding.

The real thunderstorm does not consist of infinite layers of space charge, as used in our model, but is rather limited in the horizontal as compared to the vertical extension. However if we cut away the sides and limit the space charge layers to the inside of a cylindrical box the thunderstorm field will drop off more rapidly from its maximum value as shown in Fig. 10. In consequence the field concentration at the upper and lower point of the seeded area will drop off faster with the growth of the area so that the overall field levelling effect will set in earlier and will be more pronounced.

8. Charge distribution in the lower part of the thunderstorm:

a. Instrumentation and data collection: For the electric field measurement two cylindrical field mills have been used. One field mill mounted on the nose of the airplane measures the vertical component F_z and the horizontal component F_x in the direction right to left wing tip. The second field mill is mounted between the wings on the fuselage and measures the other horizontal field component F_y in the direction nose to tail as well as the component F_x. A detailed description of the instrument and its calibration procedure is given by Kasemir (1964), 1967). The field components together with other flight data such as altitude, temperature, turbulence and so on are recorded on an 8 channel galvanometer recorder as well as on a magnetic tape recorder. After the flight the electric field data are digitized for computer evaluation.

In general the following simple flight plan was adopted. The airplane would start at the beginning of a strong development of cumulus-towers and select the one which either showed the first stages of icing at the top or the beginning of light precipitation at the base. A flight path was selected close to the base through the center of the rain curtain. This flight pass was followed back and forth throughout the whole lifetime of the storm. In general one path below the storm lasted about 2 minutes. The tear drop turn at the end of each run required also about 2 minutes so that consecutive runs are separated by 2 minute time interval. This is a close enough sequence to follow the charge development in the storm as will be seen later on. With a lifetime of about 40 minutes for the single cell thunderstorm about 10 consecutive passes could be flown if the airplane reached the storm at its beginning.

b. Evaluation of measurements and discussion of results: The evaluation of the field data is based on the field pattern of a single point charge as discussed by Kasemir (1964). The main field maxima of the combined F_x and F_z vectors was treated as if caused by a substitute point charge representing the main bulk of the thunderstorm charge. The amount of the first substitute charge and its space coordinates are determined from a best fit of the recorded data. The exact field pattern of this first substitute charge is calculated and subtracted from the respective field component records. The remaining field curves are then treated in the same way to determine charge centers of second, third and following order until all of the field records are reduced to values less than 1% of the original curves. This set of charges is only a first approximation because each substitute charge is calculated disregarding the influence of the following as yet not determined charges. Using this first approximation the data are re-evaluated now taking into account the combined influence of all the charges to obtain an improved second approximation. This iterative method is repeated until the final solution has been found. Theoretically this process will lead to an exact replica of any given charge distrituion. In practice the analysis is limted by the accuracy of the measurement. This means that because field strength diminishes with the square of the distance from the charge, the nearby charges dominate the field records and determine the sensitivity setting of the field mills.

The consequence is that the positive charges in the top of the storm are not revealed in the analysis of the examples given later on. However this restriction is not inherent either in the analytical method or in the measuring capabilities. It is rather a consequence of the specific flight path which was chosen to detect the strongest fields for the purpose of chaff seeding. This situation is similar to that of sitting in a concert close to the trumpets of an orchestra and trying to detect the weaker instruments when the trumpets are blown with full force. For thunderstorm research a flight pass half way up and to one side of the cloud would probably be best to detect the whole composition of the thunderstorm charges.

The computer program has been tested by an artificially composed field pattern calculated from a known charge distribution. Already the first approximation did not deviate more than 20% from the exact values and the second approximation was accurate to 1%. A detailed discussion of this analysis and the problems involved will be given in a separate paper.

From the calculated results models have been constructed which illustrate the charges and their position with regard to the individual flight pass. Fig. 11 is a perspective drawing of one flight pass model. The heavy horizontal wire is the flight path of a length of 5 km. The circle represents the calculated negative charge. The area of the circle is proportional to the amount of charge, in this case -0.41 Cb. The rectangular bent wire, which holds the circle up, represents the x and z distances from the flight path. Positive charges are represented by triangles. In this case there is only one positive charge to the left of the negative charge and down from the flight path. The area of the triangle is proportioned to the amount of positive charge, here + 0.05 Cb.

Fig. 12 shows a photograph of three models of passes underneath a storm on the 25th of July 1967. The storm was already diminishing and the main negative charge of 7.5 Cb in the base of the cloud is well established at the first pass. There appears also a second negative charge center further along the flight pass of -0.87 Cb. Furthermore a positive charge of + 0.5 Cb and a minute negative charge of -0.06 Cb are located below the pass. Minute charges in the order of 1 to 2% of the maximum charge are of a fictitious nature due to the limited accuracy of the measurements and should be disregarded. The second pass shows besides the minute charges only one negative main charge 15% smaller than at the first pass and 14% lower, whereas in the third pass the main negative charge has increased to 14.3 Cb, i.e. 100% compared to the first pass with an increase of altitude, and the question arises can this wobbling of the charge be due to a computational or measuring error. In theory the computer calculation converges to any desired accuracy, however there are certain critical charge distributions which require a denser field of data points for evaluation than one measurement each 50 m which has been used here. The third flight pass with two negative charge centers one above the other may be such a case.

Two rather large positive charge centers appear in the third pass which have no precedents in the first and second pass. A more detailed study of the third pass is now being carried out to determine how reliable the analysis is. In the meantime the results are reported without corrective surgery.

A very clear cut case of the charge development in a rain shower on the 28th of July 1968 is shown in Fig. 13.

This storm was one of the rare occasions, where the very beginning of the charge build-up was recorded. No rain was coming out of any one of the large cumuli clouds populating the sky. Therefore the biggest and most promising tower of a large bank of growing cumuli was selected and a flight path chosen, which lead directly under the center of this tower in an altitude of 2700 m about 100 m below the base of the cloud. The first flight passes were uneventful, i. e. only the fair weather field was recorded. On the sixth pass very light rain was falling out of the cloud base and with it the field increased. The evaluation shows a charge distribution of a small negative and positive charge center above and a small negative charge center below the flight pass.

This is the only flight pass which shows a negative charge below the airplane. At the return flight No. 7, the lower negative charge has disappeared and the upper positive and negative charge increased in strength. From the seventh pass on i. e. 8 minutes after the appearance of the first raindrops the charge development becomes more organized. A rapidly growing negative charge center appears always at the same spot about 1300 mm above the flight pass. It grows from 0.2 Cb to 1.2 Cb in about 20 minutes.

It should be mentioned that due to the screening surface charge at the cloud-air boundary, the amount of the calculated substitute point charge may be too low by a factor approximately proportional to the ratio of the conductivity inside to outside of the cloud. A correction factor 3 to 10 depending on the conductivity values has to be applied.

Summary: It seems to be possible to calculate the charge distribution in thunderstorms from the records of the three field components measured from an airplane which flies a known path through or below the storm. Consecutive passes will give the charge development with time. The choice of the path determines if the calculation results in a selective or composite picture of the charge distribution. The calculation procedure and the obtainable results are demonstrated on two examples of actual flight records. More flights will be analyzed in the near future.

9. Status Quo and Future Investigations of Lightning Suppression by Chaff Seeding: Lightning suppression by chaff seeding is based on the idea of discharging the thunderstorm by corona discharge generated on the chaff fibers before the electric field in the storm can reach the necessary strength to ignite lightning discharges. The following essential facts have been established:

a. The onset field of corona discharge on a 10 cm long chaff fiber is about 30 kV/m, which is about a factor of 17 lower than the threshold field necessary to ignite lightning discharges, which is assumed to be 500,000 V/m.

b. About five pounds of chaff, which contain 10,000,000 chaff fibers, will produce in a field of 70 kV/m a 10 ampere corona current which should be adequate to counterbalance the current output of an average thunderstorm (estimated to be 3 ampere). Here it is assumed that the chaff can be dispersed in a large volume of the cloud and not in a thin sheet.

c. The corona current increases with the square of the field, so that doubling the field to 140 kV/m will produce at least four times as much corona current. Even a field of 140 kV/m is still by a factor of 3 below the ignition field of lightning discharges. This square law dependence will be an effective limitation to a further field increase.

d. A chaff fiber will float horizontally. However the electric force of a 25 kV/m field is already strong enough to align the chaff fiber with the electric field against aerodynamic forces. The maximum corona current will be produced if the chaff fiber is directed along the field lines.

e. The following instruments have been developed for field experiments:

(1) The cylindrical field mill to measure the three components of the electric field from an airplane flying underneath or through a thunderstorm.

(2) A chaff dispenser to emit chaff fibers of an arbitrary length without birdnesting.

(3) A corona discharge indicator, effective in a range of about 100 m.

f. During actual flight underneath a thunderstorm the following facts could be established:

(1) Corona discharge is generated if chaff fibers are dispersed in a field greater than 30 kV/m and continues to be present until this field drops below the 30 kV/m level.

(2) Stronger fields in the order of 100 to 300 kV/m decay much more rapidly with seeding than without chaff seeding.

All of these preliminary tests and theoretical considerations, which are discussed in more detail above, proved that the basic idea of lightning suppression by chaff seeding is sound, well founded, and inspires confidence in future work.

Future Plans: Two basic problem areas require thorough investigation before actual lightning suppression by chaff seeding should be attempted. They are (1) the movement of small ions liberated by corona discharge in the environment of the cloud, and (2) the distribution and buildup of charge in a thundercloud. The first investigation can be carried out in a large cloud chamber equipped with a plate condensor and high voltage supply capable of generating fields in the order of 100 kV/m over a distance of about 10 m. This will necessitate a voltage generator of 1,000,000 Volt with a current output in the order of 10 micro amperes.

The second investigation has to be carried out with at least one and preferably two airplanes equipped with the field

recording system already used in the preliminary tests.
One airplane has to be able to fly above the thunderstorm
and the second must be capable of cloud penetration. How
the charge development in the storm can be computed from
the field records is discussed in section 8 . The computation
here is carried out with a digital computer. However, it is
anticipated that the field records of thunderstorms with a
large number of lightning discharges can be successfully
analyzed only with a specifically designed analog computer.

Our knowledge of the charge and field distribution in a
thunderstorm is very scanty, but this knowledge is a necessity
for the performance of effective chaff seeding. The actual
chaff seeding should be carried out with three airplanes of which
two shall be capable of cloud penetrations and be equipped with
chaff dispensers. It is expected that at least two regions in the
cloud contain lightning igniting fields. The higher region will
be the birthplace of the cloud discharges and the lower region
that of the ground discharges. Both areas should be seeded
simultaneously. The third airplane will monitor the field and
charge generation in the cloud and direct the operation. As
pointed out in reference 4, it is not necessary to penetrate the
cloud to calculate the charge distribution of the storm. It would

even be preferable if the monitoring airplane remains outside the storm. However, it has to have a ceiling of about 30,000 feet (9 km) and be capable of carrying 500 pounds of equipment.

The airplane operation has to be complimented by a lightning monitoring ground network of at least 3 ground stations. These stations will be equipped with short range lightning direction finders so that each lightning can be identified with its individual storm. This equipment is still to be designed because existing lightning locating systems seem inadequate at short ranges of about 10 to 50 miles.

References

1. Smyth, William R. "Static and Dynamic Electricity". McGraw-Hill Book Company, New York, Toronto, London, 1950.

2. Kasemir, H. W. "Zur Strömungstheorie des luftelektrischen Feldes II." Archiv fur Meteorologie, Geophysik Bioklimatologie. V, Heft 1, 56-70 1952.

3. Kasemir, H. W. "The Cylindrical Field Mill". Tech. Rpt ECOM 2526, 1964.

4. Kasemir, H. W. "Measurement of Atmospheric Electric Parameters". NCAR TN-29, pp 107-115. 1967.

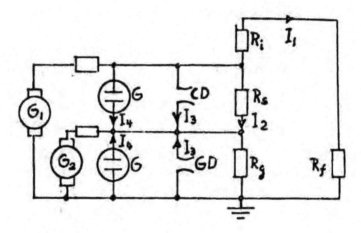

Fig. 1. Simplified Circuit Diagram of Thunderstorm Generator

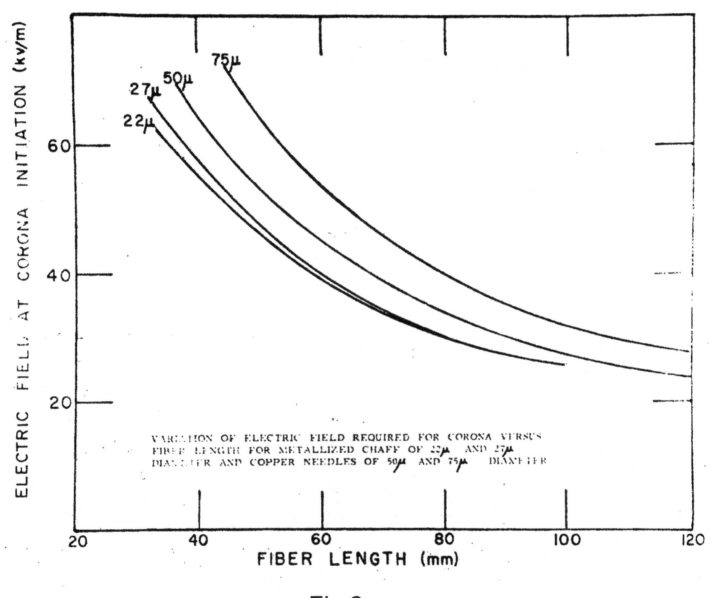

Fig 2

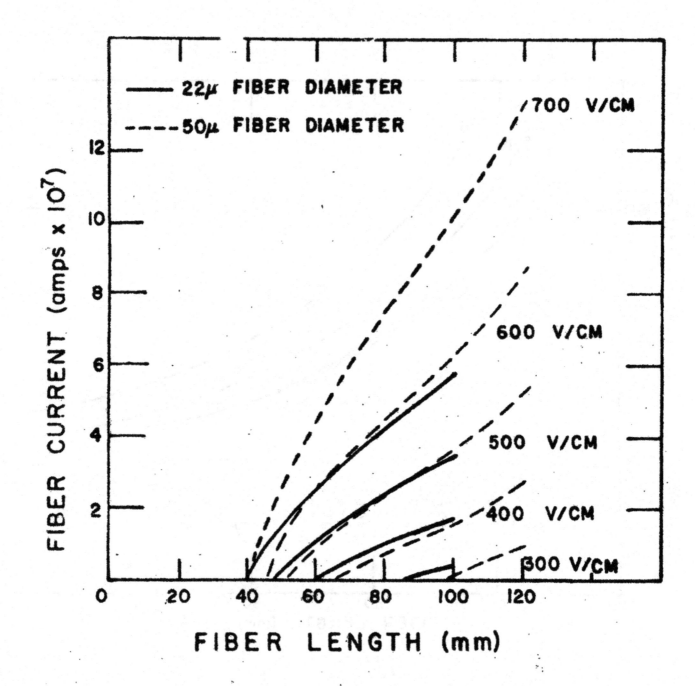

VARIATION OF CORONA CURRENT WITH FIBER LENGTH AT GIVEN
ELECTRIC FIELD INTENSITIES FOR 22μ DIAMETER METALLIZED
CHAFF NEEDLE AND FOR 50μ DIAMETER COPPER NEEDLE.

Fig.3

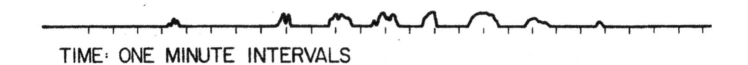

TIME: ONE MINUTE INTERVALS

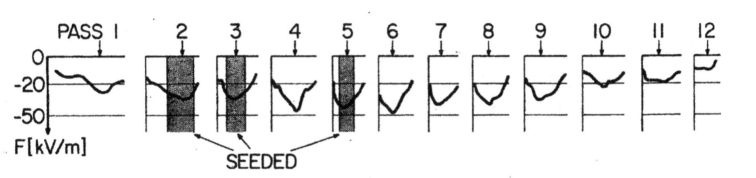

SEEDED

CORONA DISCHARGE GENERATED BY CHAFF SEEDING
2 August 1966

Fig. 4

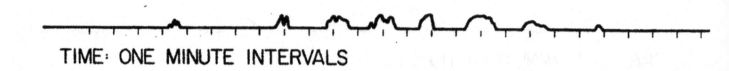

TIME: ONE MINUTE INTERVALS

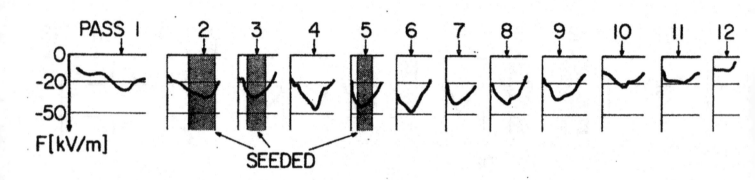

CORONA DISCHARGE GENERATED BY CHAFF SEEDING
2 August 1966

Fig. 5

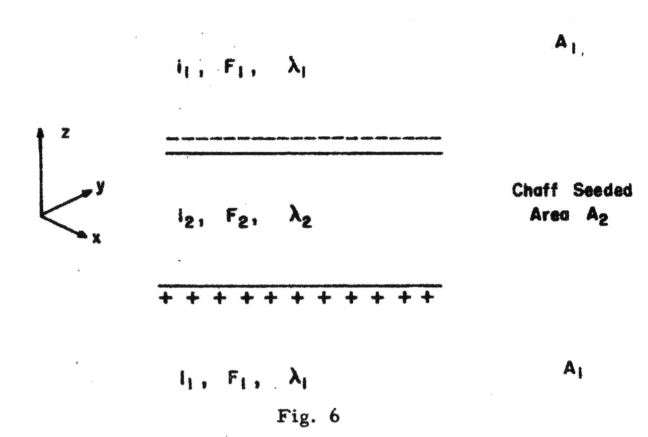

Fig. 6

Sketch of seeded area A_2 inbedded in not seeded area A_1

in not seeded area A_1
i_1 = current density
F_1 = electric field
λ_1 = conductivity

in seeded area A_2
i_2 = current density
F_2 = electric field
λ_2 = conductivity

- - - and $\overline{+ \; + \; +}$ negative and positive surface charges

at the boundaries of the seeded area.

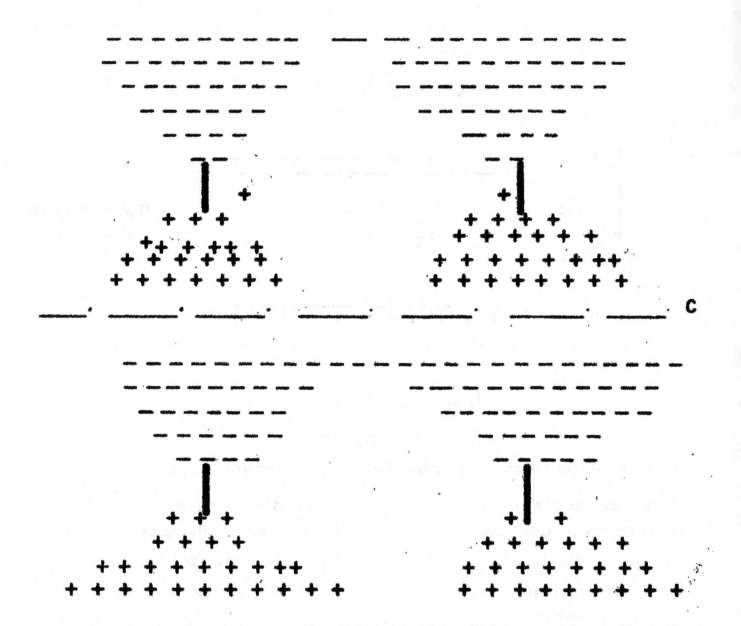

Fig. 7

Positive (+) and negative (-) space charges forming at
the upper and lower end of a chaff fiber.

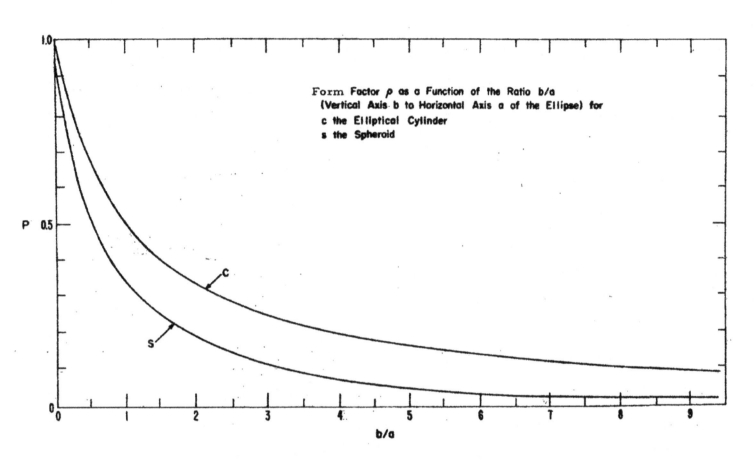

Fig.8

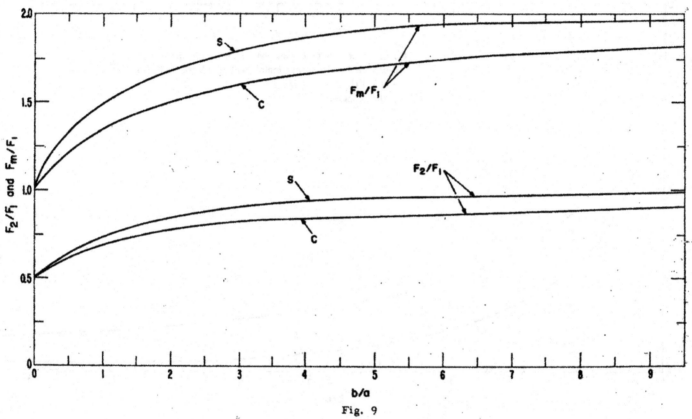

Fig. 9

Field concentration F_m/F_1 at the boundary field attenuation. F_2/F_1 inside the seeded area.

S = spheroid   C = cylinder

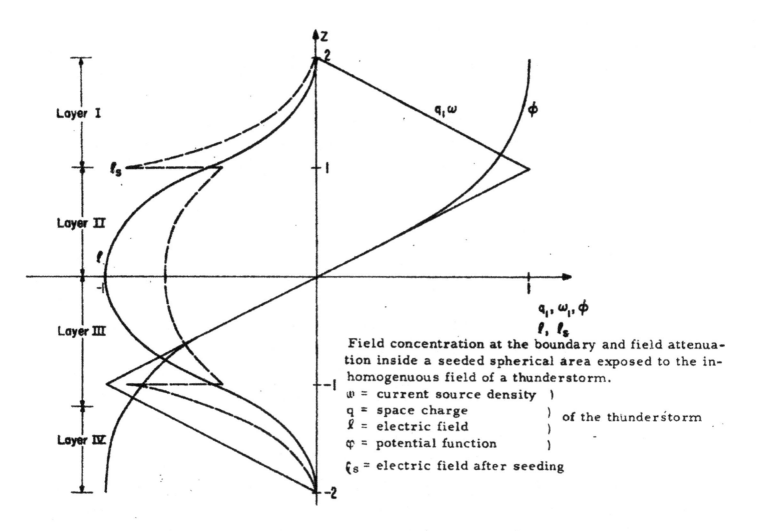

Field concentration at the boundary and field attenuation inside a seeded spherical area exposed to the inhomogenuous field of a thunderstorm.

ω = current source density)
q = space charge) of the thunderstorm
ℓ = electric field)
φ = potential function)

ℓs = electric field after seeding

Fig.10

Figure Legends

Fig. 1.   Simplified Curcuit Diagram of Thunderstorm Generator

Fig. 2.   Electric Field at Corona Initiation (kV/m)

Fig. 3.   Fiber Current (amps x 10^7)

Fig. 4.   Corona Discharge Generated by Chaff Seeding
                 2 August 1966

Fig. 5.   Field Decay After Chaff Seeding 1 August 1966

Fig. 6.   Sketch of seeded area A_2 imbedded in not seeded
                 area A_1

in not seeded area A_1	in seeded area A_2
i_1 = current density	i_2 = current density
F_1 = electric field	F_2 = electric field
λ_1 = conductivity	λ_2 = conductivity

- - - and $\overline{+\ +\ +}$ negative and positive surface charges
at the boundaries of the seeded area.

Fig. 7.   Positive (+) and negative (-) space charges forming at
                 the upper and lower end of a chaff fiber.

Fig. 8.   Form factor ρ as a function of the ratio $b/_a$ (Vertical
                 axis b to horizontal axis a of the ellipse)
                 for   c = the elliptical cylinder
                        s = the spheroid

Fig. 9.   Field concentration $F_m/_{F_1}$ at the boundary field
                 attenuation $F_2/_{F_1}$ inside the seeded area.

                 s = spheroid   c = cylinder

Fig. 10.   Field concentration at the boundary and field attenuation
                 inside a seeded spherical area exposed to the inhomogenuous
                 field of a thunderstorm.
                 ω = current source density)
                 q = space charge   ) of the thunderstorm
                   = electric field   )
                 φ = potential function   )
                 ζ_s = electric field after seeding

372

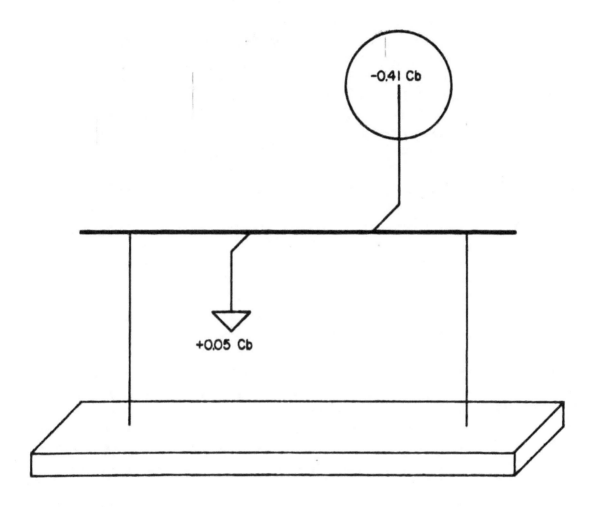

Fig. 11

Model of the charge distribution in a rain shower calculated from the electric field components recorded from an airplane flying underneath the storm.

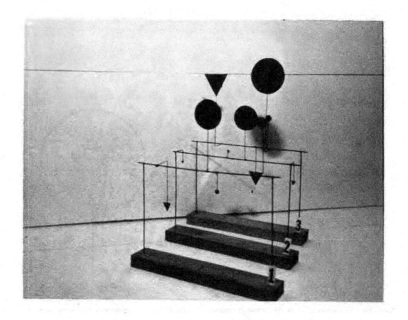

Fig. 12

Fig. 12. Model of charge distribution in a thundershower obtained from three consecutive flight passes underneath the storm.

Fig. 13

Fig. 13. Model of charge distribution in a rain shower obtained from 17 consecutive flight passes underneath the shower.

374

Fig. 11. Model of the Charge Distribution in a rain shower calculated from electric field components recorded from an airplane flying underneath the storm.

Fig. 12. Model of charge distribution in a thundershower obtained from three consecutive flight passes underneath the storm.

Fig. 13. Model of charge distribution in a rain shower obtained from 17 consecutive flight passes underneath the shower.

USCOMM - ERL

THE BEHAVIOR OF CHAFF FIBERS IN A THUNDERSTORM AND THEIR EFFECT ON THE ELECTRIC FIELD

Heinz W. Kasemir
Atmospheric Physics and Chemistry Laboratory
Environmental Science Services Administration
Boulder, Colorado

In the following calculation it is assumed that the corona discharge generated at the tips of the chaff fiber by the electric field of the cloud will produce an increase of conductivity inside the seeded area. This seemingly straightforward assumption still contains several ramifications and unsolved problems. For instance, corona discharge will only be produced at the tips of the chaff needle if the needle is lined up in the direction of the electric field. This is not "a priori" the case because the aerodynamically stable floating position of the needle is horizontal whereas the electric field in the cloud is usually vertical. The electric field however exerts a force on the fiber to turn it into the direction of the field. This can be easily demonstrated by hanging a chaff needle on a thin nylon thread inside a plate condenser so that the needle is in the horizontal position parallel to the plates. If a voltage difference is applied to the plates generating a field of about 30 kV/m the needle will turn to a vertical position even against the tensile strength of the nylon fiber, which has to bend into a hook to accommodate the vertical position of the needle. A brief sketch of the calculation of the torque exerted on the needle by the electric field is given. The needle is represented by a slim spheroid with long axis a and short axis b. The eccentricity is $c = (a^2-b^2)^{1/2}$. If this needle is placed in an electric field K with the long axis a in the field direction, and if z is the coordinate along the long axis with the zero point in the center of the spheroid, the differential charge dQ on a ring shaped surface element is given by

$$dQ = -2 \pi \epsilon K c^2 z \, dz/a^2 Q_{1a} \tag{1}$$

where ϵ = dielectric constant - 8.86 x 10^{-12} A sec/Vm, and $Q_{1a} = a/2c \ln(u + c)/(u-c) - 1$. If the chaff needle is floating at an angle α with respect to the direction of the external field F, the field component K in the direction of the long axis is given by $K = F \cos \alpha$. The differential torque dM exerted on the needle is $dM = F \, dQ \, y$. The distance y is given by $y = \sin \alpha$. Integration with respect to z from - a to +a yields the torque M.

$$M = 4 \pi \epsilon c^2 Q F^2 \sin \alpha \cos \alpha /3Q_{1a} \tag{2}$$

For slim spheroids one may set c = a and $Q_{1a} = \ln(2a/b) - 1$. From (2)

$$M = 4 \pi \epsilon a^3 F^2 \sin \alpha \cos \alpha /3 (\ln(2a/b) - 1) \tag{3}$$

Note that the torque increases with the cube of the needle length and with the square of the field. For the angles $\alpha = 0$ and $\alpha = 90^o$ the torque is zero. In the first case the lever length y is zero; in the second case the field component K is zero. The maximum torque is for $\alpha = 45^o$. For a chaff needle of 10 cm length in a field of 30 kV/m the maximum torque is M = 1.74 x 10^{-6} (A sec V = Newton meter). This is about the torque necessary to overcome the aerodynamically stable horizontal floating position. Since the torque increases with the square of the field the position in the direction of the electric field becomes very firm for fields larger than 30 kV/m so that even strong air turbulence can deflect the chaff needle only for brief moments.

During the experiments with the plate condenser it was noted that for fields of about 65 kV/m and more the needle would float up to negatively charged plate, even carrying the thin nylon thread. The following explanation is offered for this phenomenon. Negative corona discharge will start at a lower field or potential than positive corona discharge. The ratio between positive and negative corona onset potential or onset field is about 1.4. Negative corona discharge will start as soon as the breakdown field of about 3 x 10^6 V/m is surpassed at the lower point of the chaff needle. This will happen when the plate condenser field is about 30 kV/m. The chaff needle will lose negative charge by corona discharge and will attain a positive net charge. The positive charge will reduce the field at the negative tip and increase the field at the positive tip of the needle until the required difference in the corona onset fields is established so that equal amounts of positive and negative corona current flow from the needle. If the external field is strong enough the coulomb force acting on the charged needle may overcome the gravity force and lift the needle to the upper plate. To calculate the positive net charge on the needle we replace it again by a slim spheroid. The field E_a at the tip is given by

$$E_a = Q/4 \pi a \rho \tag{4}$$

In this equation ρ is the radius of curvature at the tip and a the half length of the needle. If E_b is the breakdown field of about 3×10^6 V/m the difference between the positive and negative corona onset field is $0.4\,E_b$. The needle has to be charged to such an extent that half of this field is produced at each tip. $E_a = 0.2\,E_b$ or with

$$Q = 0.8\,\pi\,\epsilon\,a\,\rho\,E_b \tag{5}$$

If the external field is F, the coulomb force C on the needle is

$$C = F \cdot Q = 0.8\,\pi\,\epsilon\,\rho\,a\;F\,E_b \tag{6}$$

Within the mass of one needle and g the gravitational constant the gravitational force G and is given by

$$G = mg \tag{7}$$

Let (6) be equal to (7) and we obtain the electric field which will just cancel the gravitational pull.

$$F = mg/0.8\,\pi\,\epsilon\,a\;\rho\;E_b \tag{8}$$

For a numerical evaluation we may set: $m = 10^{-6}$kg; $g = 9.81$ m sec^{-2}, $\epsilon = 8.86 \times 10^{-12}$ A sec/Vm; $a = 5 \times 10^{-2}$ m; $\rho = 5 \times 10^{-5}$m; $E_b = 3 \times 10^6$ V/m. For these values we obtain from (8) $F = 59$ kV/m. The experimental result was $F = 65$ kV/m, which seems to be in good agreement with the calculation.

It is difficult to estimate the effect of the electric force on a mass of chaff needles in a thunderstorm. In a very crude approximation one would expect that chaff dispersed below the storm would be pulled towards the negative main charge if the field is strong enough. Chaff dispersed above the storm would be kept floating by the positive main charge underneath, until the field is weakened by the discharging process, so that the gravity force overcomes the coulomb force. However the author is well aware of the fact that up and down drafts, i.e., aerodynamic forces will play an important role in the distribution of the chaff needles.

For the next calculation it is assumed that a large number of chaff fibers is dispersed over a certain volume inside the cloud and that the overall effect of the corona discharge is such that the conductivity inside the seeded area is increased. To what extent this assumption is correct is still an open question. Small ions, which emerge from the corona point, will attach themselves to cloud elements and become immobile. A space charge pocket will form in front of the corona point which will hinder and finally block the passage of the ion stream from the corona point, if the space charge pocket can grow unchecked. There are several mechanisms which tend to neutralize or destroy the accumulation of the space charge pocket. If two chaff needles are lined up along a field line one above the other, the two points of the two needles facing each other will produce space charge pockets of opposite sign, which will mutually cancel to some extent the field reducing effect at the chaff needle points.

The opposite charged cloud elements in the overlapping space charge regions of the two chaff fibers will coalesce rather readily and neutralize the charges. Furthermore the space charge pockets will be swept out by precipitation or mixed by turbulent air motion. All these effects help to keep the corona discharge going and produce an enhanced current flow through the seeded area. In the mathematical representation they are lumped together in one material constant, namely the conductivity, with the result that the conductivity inside the seeded area is increased.

At the boundary of the seeded area there is no production of neutralizing space charge pockets after the last row of needles. Therefore here at the surface of the seeded area the boundary needles produce a surface charge layer of similar amount and shape as a body which has a higher conductivity than the cloud. In the mathematical treatment the surface charge is confined to an infinitesimally thin layer. In reality the surface charge will be spread out over a layer of several tens of meters thickness inside the cloud and several hundreds of meters outside the cloud. A good example for the spreading of the surface charge layer is the space charge layer which forms above the earth surface if the thunderstorm field is strong enough to generate corona discharge at the ground. It should be emphasized however that the thickness of the surface charge layer does not determine the thickness of the seeded area. At the ground the corona points are confined to the earth's surface whereas chaff seeding requires that the chaff fibers should be dispersed through a large volume of the thundercloud. Therefore conclusions drawn from the corona discharge at the ground below thunderclouds may be applied with some caution to the surface but not to the volume of the seeded area. In the following calculation the surface charge layer is treated as one of negligible thickness i.e., the enlargement of the seeded area by the spreading of the surface charge is disregarded.

The calculation is carried out for the following problem. A body of a given shape with the constant conductivity λ_2 is imbedded in an environment of the constant conductivity λ_1. The whole system is exposed to a homogeneous electric field F_1 in the direction of the z axis of a cartesian co-ordinate system x, y, z. Find the field F_2 inside the body and the maximum field F_m at the boundary of the body. The body shall be represented by an infinitely long elliptical cylinder with its axis in the x direction. The solution shall be given for

    ac) the small axis of the elliptical cross-section is in the z direction.
    bc) both axes are equal, in which case the elliptical cylinder becomes a circular cylinder.
    cc) the large axis of the elliptical cross-section is in the z direction.

The same problem shall be solved when the body is represented by a spheroid, which changes from

    as) a flat disc to
    bs) a sphere to
    cs) a prolonged spheroid.

The cases ac to bc may be encountered if the chaff is dispersed continuously from an airplane and the cases as to cs if the chaff is dispersed ciscontinuously in little packages by an airplane, dropsonde, rocket, or gun shell.

The derivation of the potential function is not given here because the solution to the problems ac to cc can be obtained from the textbook "Static and Dynamic Electricity" (1) if the dielectric problem is transformed into a current flow problem by substituting the conductivities λ_1 and λ_2 for the two dielectric constants ϵ_1 and ϵ_2. The solution to problems as to cs is given (2).

A very simple equation has been derived for the maximum field concentration F_m at the boundary of the seeded area as well as the diminished field F_2 inside the seeded area. These equations are valid for all the above stated problems and are as follows:

$$F_2/F_1 = \frac{\lambda_1/\lambda_2}{\lambda_1/\lambda_2 + (1-\lambda_1/\lambda_2)} \qquad F_m/F_1 = \frac{1}{\lambda_1/\lambda_2 + (1-\lambda_1/\lambda_2)\,\rho} \tag{9}$$

From (9) also follows

$$F_2 = F_m \lambda_1/\lambda_2 \tag{10}$$

where

    F_1 = original field outside the seeded area
    λ_1 = conductivity outside the seeded area
    λ_2 = conductivity inside the seeded area
    ρ = form factor, which depends only on the shape of the seeded area
    a = horizontal axis of elliptical cylinder

The equations for p_c of the elliptical cylinder and p_s for the spheroid are

$$p_c = a/(a+b) \; ; \; p_s = \frac{a^2}{a^2 + b^2}\left(1 - \frac{b}{(a^2 - b^2)^{1/2}}\tan^{-1}\frac{(a^2 - b^2)^{1/2}}{b}\right) \tag{11}$$

For $a \to \infty$ the form factors p_c and p_s approach 1 and the elliptical cylinder as well as the spheroid degenerate into a horizontal infinite layer. From (9) for p = 1.

$$F_2/F_1 = \lambda_1/\lambda_2 \tag{12}$$

The field F_2 inside the seeded area is reduced in proportion to $1/\lambda_2$. Furthermore from (9) follows, for p = 1,

$$F_m = F_1 \tag{13}$$

The maximum field F_m is equal to the original field F_1 outside of the seeded area, i.e., there is no field augmentation at the boundary.

If we go to the other extreme and let $b \to \infty$ then the elliptical cylinder deteriorates into a vertical infinite layer and the spheroid into an infinitely long circular cylinder. In this case p_c and p_s approach zero. From (9) we obtain

$$F_2 = F_1 \; ; \text{ and } F_m/F_1 = \lambda_2/\lambda_1 \tag{14}$$

These are somewhat unexpected results. Equation (14) says that the field inside the seeded area is the same as outside no matter how much we increase the inside conductivity. This means that continuous seeding from a dropsonde, where we may generate a body of seeded area like a prolonged spheroid, would have almost no field reducing effect. Furthermore we learn from equation (14) that the field concentration on the ends of the prolonged spheroid is proportional to the ratio of the conductivities inside and outside of the seeded area. Both effects are favorable for generating lightnings and adverse to lightning suppression. Therefore the manner in which a cloud is seeded will have some effect on the outcome i.e., if lightnings are suppressed or prematurely triggered.

If we set b/a = 1 the spheroid takes on the shape of a sphere. Here we have the maximum field concentration factor 1.5 and the weakening factor of 0.75 for the field inside the sphere. If we assume that a field F_1 of 70 kV/m is necessary to produce a corona current strong enough to consume that part of the thunderstorm current output which would otherwise dissipate in lightning discharges, we have a maximum field at the boundary of 105 kV/m and an inside field of 52 kV/m. The maximum field is still by a factor 3 to 5 lower than the threshold field necessary to initiate lightning discharges and the inside field about 80 percent above the threshold field to start corona discharge. Even if we double the fields (inside field F_2 = 104 kV/m, thunderstorm field F_1 = 140 kV/m and maximum field at the boundary of the seeded area F_m = 210 kV/m) we are still about 100 kV/m, below the lightning threshold field, even if this is assumed to be only 300 kV/m. The current flow through the seeded area has also doubled and should load down the thunderstorm generator enough to prevent further increase of the field. It may be expected that an equilibrium state of charge production and dissipation will be reached inside the above mentioned field values.

The maximum field F_m given by equation (9) is an upper limit. This value will be reduced by the fact that the surface charge at the boundary is not restricted to a very thin layer but will spread out over a volume of 100 or more meter thickness. Furthermore the primary thunderstorm field F_1 is not constant but has its maximum between two space charge regions and drops off towards the center points of the charge. If the seeded area approaches these center points the maximu field at the boundary will drop off in a similar manner. These two effects will reduce the danger of inadvertently igniting a lightning discharge.

References

1. Smythe, William R, 1950: Static and Dynamic Electricity. McGraw-Hill Book Company, New York, Toronto, London.

2. Kasemir, H. W., 1952: Zur Strömungstheorie des luftelektrischen feldes II. Archiv für Meteorologie, Geophysik Bioklimatologie, V, Heft 1, 56-70.

The Effect of Chaff Seeding on Lightning and Electric Fields of Thunderstorms

H. W. Kasemir
Atmospheric Physics and Chemistry Laboratory
National Oceanic and Atmospheric Administration
Boulder, Colorado 80302

Abstract:

Corona discharge appears at the ends of a chaff fiber in an electric field of 30 kV/m or more. If a pound of chaff consisting of about 2 million fibers is released from an airplane in a thunderstorm field of about 60 kV/m, a corona current of about one microampere is generated at the ends of each fiber. The net current of the two million fibers adds up to about 2 ampere and the charge released into the cloud in a time period of 100 seconds would be of the order of 200 Coulomb. The net current as well as the net charge are comparable to the values of an average thunderstorm and will tend to mask or neutralize the thunderstorm field. The application of this concept to modifying lightning or thunderstorm fields encounters two main difficulties. One is the distribution of chaff fibers in the high field areas of the cloud fast enough to be effective in a large enough cloud volume, and the second one is the possibility that ions liberated by corona discharge are trapped by the cloud droplets in the immediate neighborhood of the individual chaff fiber. This could generate a concentrated space charge around the fiber, which may quench the corona discharge before a large amount of charge is released.

To investigate these problems a number of field experiments have been planned to study the effect of chaff seeding on the electric fields and the lightning discharges of thunderstorms. After some preliminary tests in Flagstaff, Arizona, the first of these experiments was carried out in Boulder, Colorado, in 1972. The results seem to indicate that the decay of the electric field underneath a thunderstorm is accelerated by about a factor five by chaff seeding as compared to the natural decay of the field of a not seeded storm. Instrument development, test procedures, and data analysis of this field decay experiment as well as its significance to lightning modification will be discussed.

1. Introduction

The field decay experiment Boulder 1972 may be considered as the first experimental step in lightning suppression by chaff seeding. The physical concept of this method is based on the idea that corona discharge on a large number of chaff fibers releases so many ions into the cloud that the conductivity of the air is increased and consequently the field is decreased. The high field value of about 500 kV/m necessary to ignite lightning discharges may never be reached /1/, /2/. This picture is largely oversimplified and many expected ramifications are discussed in /1/ and more unexpected difficulties are anticipated.

One of the favorable aspects of chaff seeding is the fact that corona discharge on the chaff fiber will start in an electric field of about 30 kV/m which is more than a factor 10 below the lightning igniting field — and that the corona current (ion production) increases with the square of the field. /3/ Therefore the discharging effect of the chaff fibers intensifies with the growth of the thunderstorm field. With respect to the decay-speed of the field after seeding we may expect that high fields on the order of 120 kV/m or more — which is four times the corona onset field — decay fast, that lower fields in the order of 60 kV/m — i.e., twice the corona onset field — decay slow, and that for fields below 30 kV/m there should be no effect from chaff seeding on the normal variation of the thunderstorm field. The objective of the 1972 chaff seeding experiment was to confirm these patterns.

2. Experiments

The chaff is dispersed from an airplane making repeated passes under a thunderstorm. The airplane is equipped with a chaff dispenser, field mills, /4/, a corona indicator, and the necessary accessory equipment such as a strip chart and magnetic tape recorder, altitude recorder and so on. The test procedure was to select one growing cumulus cloud which already showed the emergence of light precipitation at the base. The electric field was then monitored during continuous passes back and forth underneath the storm. If the field reached values of the order of about 50 kV/m two consecutive passes were seeded (or not seeded) and the field recording continued until the storm dissipated. The decision to seed or not to seed was based on the requirement that data should be collected of an equal amount of seeded and not seeded storms. With the clear cut physical objective and the necessary monitoring equipment no elaborate randomization process is needed.

The graphs presented in Figures 1 to 3 show the field underneath the storm versus time. For each pass the maximum value of the negative vertical field component was listed with the corresponding time and these peak values are connected in the graph by straight lines. The time interval between consecutive passes ranges from 3 to 9 minutes (with an average of 4 minutes). In Figure 1 the average field of four seeded storms and in Figure 2 of four not seeded storms, is shown in the upper half of the figure. In the lower half of the figure the field of each individual storm is given with a reduced vertical scale. To calculate the average field values the seeded storms have been synchronized with respect to the first seeding run, and the not seeded storms with respect to their maximum field value. Seeding events are marked on the curves by the letter S.

It is evident from a comparison of Figures 1 and 2 that the field decay of the seeded storms after the second seeding run (Fig. 1) is much faster than the field decay of the not seeded storms (Fig. 2). The accelerated field decay of the seeded storm is 6 kV/m/min as compared to the decay of the not seeded storm of 1.2 kV/m/min. If the field in the seeded storm is reduced to 34 kV/m the accelerated decay stops because the corona discharge on the chaff fibers stops, and further decay continues at a rate of 1.5 kV/m/min which is about the value of the not seeded storm.

Storm 4 merits special attention because the field did increase after the first seeding run and started to decrease only after the second seeding run. Furthermore, the field increase before seeding is much faster than in all of the other examples. This means we are dealing here with a rapidly growing charge which was still far before an equilibrium state. Therefore, the first seeding run was not effective enough to reverse the sharp field increase to a field decrease. However, the most significant feature of this test was that seeding was done inside the cloud instead of below the cloud as in all the other cases. One has to expect that a large number of ions liberated by the corona discharge are captured by the cloud droplets, made immobile, and therefore removed from their role of increasing the conductivity of the cloud. From purely theoretical consideration it is not possible to decide if this effect will completely counteract the purpose of chaff seeding or if it is only an insignificant side effect. This one test seems to indicate that chaff seeding causes an increase of conductivity even inside the cloud and consequently a decrease in the field but there may be a delay in the overall process.

Figure 3 shows the field decay caused by chaff seeding of a storm with an extremely high field value of 300 kV/m. The field maintained its high value for about 18 minutes when chaff seeding began. After 4 seeding runs, which required about 5 minutes, a dramatic field decay occurred at the high rate of 58 kV/m/min. The field dropped in about 5 minutes to the low value of 10 kV/m and remained below this value for the rest of the flight. This is about 10 times the decay rate of the previously discussed storms.

The field decay experiment Boulder 1972 has been repeated in 1973 with similar results. The meterological conditions in 1973 have been somewhat unusual with respect to the thunderstorm activity in Colorado. The storms appeared much later in the season, i.e., in the second half of the summer. They had generally a shorter life time and produced higher fields, which changed more rapidly than during the storms in 1972. The last factor may reduce the average field decay rate of the not seeded storms somewhat. However, even with the greater field variability in 1973 the decay rate of the seeded storms was always faster than that of the not seeded storms. Ten storms produced fields underneath the base in the range of 100 to about 300 kV/m. This will furnish some high field control storms to the one seeded case discussed above, and also more seeded cases. The exact figures of the decay rate of the seeded and not seeded high field storms have to await the proper evaluation of the data. However the faster decay rate of the high field as compared to the low field seeded storms was already apparent by a preliminary review of the records.

4. The significance of the field decay experiments with regard to lightning suppression

The link between the field decay experiment and lightning suppression is the strong electric field necessary to initiate a lightning discharge. If the generation of high fields of about 500 kV/m in the cloud can be prevented or their time of existence shortened it is a reasonable assumption that lightnings will not occur or that at least the number of lightnings per storm will be very much reduced. The most effective way to achieve lightning suppression would be to seed the regions in the storm where the highest fields occur. These maximum fields will occur between two charge centers of opposite polarity. Ground discharges should be initiated between the main negative charge center in the base of the cloud and the positive pocket charge in or at the base of the cloud.

The numerous field records obtained during the field decay experiments revealed the fact that the old tripolar thunderstorm model of Simpson, Scarse, and Robinson /5/,/6/ should be corrected. This well known model has the three charge center arranged vertically one above the other with the smallest — the positive pocket charge of about 4C — at the base of the cloud, the main negative spherical space charge of about -20C with its center at about 3 km above ground and the spherical positive space charge of about 24C at about 6 km above ground.

From our flight records it follows that the positive base charge is first of a much larger magnitude — in the average about half of but sometimes even larger than the negative space charge — that its lateral extension is much larger, in the order of several kilometers, and that the positive base charge center is located about 1 km above cloud base. The negative space charge center is 1.5 to 2 km above base and its lateral distance from the positive base charge in the order of 4 km. Therefore it seems that these two opposite base charges are arranged more side by side than on top of each other. This picture conforms much better with the charge arrangement given by Reiter /7/. He also states that the horizontal dimension of the positive base charge is several kilometers and that the maximum lightning activity occurs in the region of the positive base charge. (According to his fig. 117, page 276, at the boundary between the positive and negative charge.)

Furthermore our flight records show usually a certain time delay in the growth of the positive base charge. Positive fields — indicating positive charge overhead — appear when the negative charge is already well established, and with time the positive field grows stronger whereas the negative field decays. It happens sometimes that at the end of the storm the positive charge completely dominates in the base of the cloud so that only positive fields are recorded.

In conclusion we may state the following:

(1) The field decay experiments Boulder 1972 and 1973 show that electric fields underneath a thunderstorm decay more rapidly with chaff seeding than without seeding.

(2) The higher the seeded fields the more rapid is the decay.

(3) The seeding took place underneath the cloud in clear air sometimes in rain.

(4) There is some question if seeding inside the cloud has the same or appreciably the same effect on the electric field as it has in clear air. The one in cloud seeding shows also an accelerated field decay after seeding. However this one flight is not enough by far to establish this result as a general fact. More in-cloud seeding flights are needed.

(5) For lightning suppression location and timing of the seeding runs are important. From the analysis of the field records obtained underneath the cloud it seems that the best place to seed would be about 1 km above cloud base between the positive and negative base charges. This area can be of determined in time from field records of flights underneath the cloud.

382

References

/1/ Kasemir, H. W., 1973: Lightning suppression by chaff seeding. NOAA ERL Technical Report.

/2/ Holitza, F. J., Kasemir, H. W., 1973: Accelerated decay of thunderstorm electric fields by chaff seeding. J. G. R. (in press).

/3/ Chapman, S., 1958: Corona-point-discharge in wind and application to thunderclouds. Recent Advances in Atmospheric Electricity, Pergaman Press (pp. 277-287).

/4/ Kasemir, H. W., 1972: The Cylindrical Field Mill. Meteor. Rundschau, 25, 2, pp. 33-38.

/5/ Simpson, G. C., Scarse, F. J., 1937: The distribution of electricity in thunderclouds. Proc. Met. Roy. Soc., A, 161, pp. 309-352.

/6/ Simpson, G. L., Robinson, G. D., 1940: The distribution of electricity in thunderclouds. Proc. Met. Roy. Soc., A, 177, pp. 281-329.

/7/ Reiter, R., 1964: Felder, Ströme und Aerosole in der unteren Troposphäre. Dietrich Steinkopt Verlag. Darmstadt (page 276, fig. 117).

Figure Legends

Figure 1. Field F versus time T. Seeded cases.

Figure 2. Field F versus time T. Control cases.

Figure 3. Field F versus time T. Seeded cases.

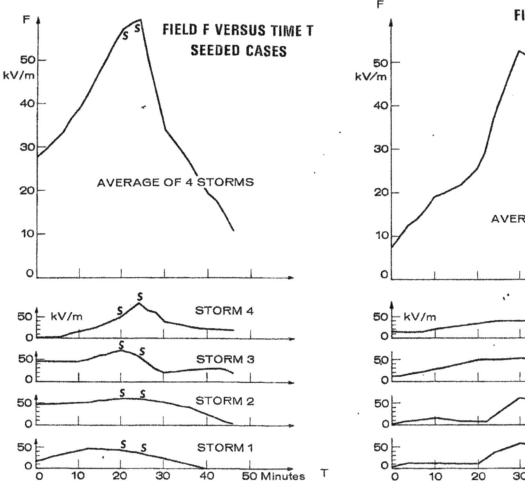

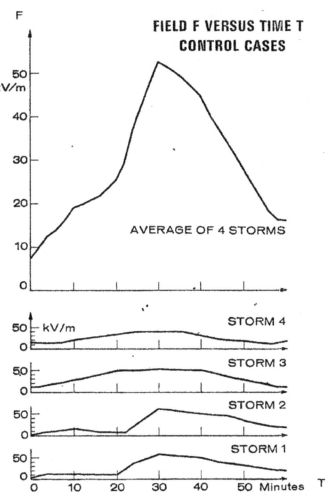

Figure 1. Field F versus time T.
Seeded cases.

Figure 2. Field F versus time T.
Control cases.

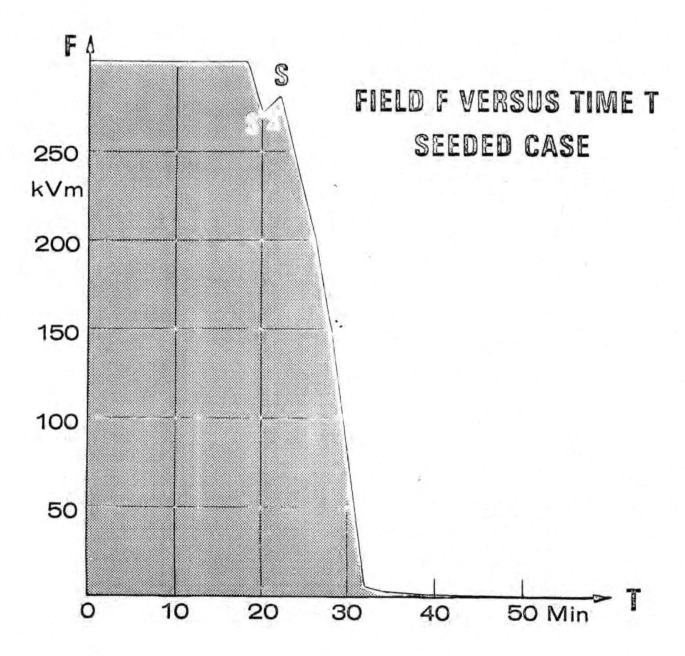

Figure 3. Field F versus time T.
Seeded cases.

Accelerated Decay of Thunderstorm Electric Fields by Chaff Seeding

F. James Holitza and Heinz W. Kasemir

Atmospheric Physics and Chemistry Laboratory
National Oceanographic and Atmospheric Administration
Boulder, Colorado 80302

The field decay experiment conducted during the summer of 1972 in Colorado by the Atmospheric Physics and Chemistry Laboratory of the National Oceanic and Atmospheric Administration may be considered as the first experimental step to lightning suppression by chaff seeding. An airplane equipped with an electric field measuring system and a chaff dispenser flew underneath a developing thunderstorm, continuously monitoring the electric field. If the field exceeded 50 kV m^{-1}, chaff was dispersed during the next two passes underneath the cloud, and the field decay was recorded until the storm dissipated. Each seeded storm was matched by a control (unseeded) storm to obtain the field decay rate under normal conditions. The field tests showed that chaff seeding in moderate fields of 50 kV m^{-1} accelerated the field decay by a factor of 5. In one case a stronger field of 300 kV m^{-1} was seeded, and the field decay rate was faster by a factor of 10 than that of the seeded storms having moderate fields. The physical background of this process, its limitations, and its significance to the problem of lightning suppression are discussed.

Experiments have been carried out to reduce the electric field growth in Colorado thunderstorms. They may be considered as the first step toward lightning suppression by chaff seeding. The physical concept of this method is based on the idea that corona discharge on a large number of chaff fibers releases so many ions into the cloud that the conductivity of the air is increased and consequently the field is decreased to such an extent that the high field value necessary to initiate lightning discharges is never reached. The physical principle of chaff seeding may be demonstrated by a simple laboratory experiment. If the spherical terminal of a high-voltage generator is placed above a grounded sphere (Figure 1a), spark discharge will occur between the two spheres if the breakdown field is reached. In our model the charged sphere represents the thunderstorm, the grounded sphere represents the earth, and the sparks represent cloud to ground lightning discharges. If a chaff fiber is put into the air gap between the spheres, the sparks will stop immediately but will start again as soon as the chaff fiber is removed. (The chaff fiber is attached to a teflon rod, well insulated from the ground and short enough so that it does not touch either one of the spheres.) This spark-suppressing effect can readily be explained. If the generator terminal is charged negatively, a large number of positive ions are liberated by corona discharge at the upper end of the chaff fiber and flow upward to the negatively charged terminal, and an equal number of negative ions are liberated at the lower end of the chaff fiber and flow down to the grounded sphere (Figure 1a). This ion flow provides a semiconductive path between the two spheres and loads down the generator so much that the high voltage necessary for a spark discharge cannot be maintained.

There are two important conditions that have to be fulfilled in this experiment. First, the corona discharge from the chaff fiber must begin at a field that is lower than the breakdown field. Second, the corona current produced by the chaff fiber should be of the same order of magnitude as or higher than the current delivered by the generator.

The field necessary for spark discharge is of the order of 3000 kV m^{-1}. The field necessary to start corona discharge on the chaff fiber is about 30 kV m^{-1}, which is about a factor of

100 lower than the breakdown field. Therefore in the laboratory experiment the first condition is certainly fulfilled. Transferring this requirement to thunderstorm conditions, we have to compare the 30 kV m^{-1} corona onset field to the lightning igniting field of about 500 kV m^{-1}. Here the corona onset field is lower than the lightning igniting field by about a factor of 17. This is less by a factor of 6 than the conditions of the laboratory experiment but still good enough for the chaff seeding method to work efficiently.

The second condition requires that the corona current be of the same order of magnitude as that produced by the generator. If in the laboratory experiment a Van de Graff generator is used, the generator current is of the order of 1 μA. A corona current of 1 μA is produced by the chaff fiber in a field of about 60 kV m^{-1}. Therefore long before the flashover field of 3000 kV m^{-1} is reached, the corona current will load down the generator. Even if the generator current is doubled, the corona current will increase with the square of the field and will match the doubled generator current in a field higher by a factor of $2^{1/2}$ than 60 kV m^{-1}, i.e., in a field of 85 kV m^{-1}. Therefore in the laboratory experiment the second requirement is also fulfilled. The application of the second requirement to thunderstorm conditions encounters a number of presently unsolved problems. The generator current of the average thunderstorm is estimated to be of the order of 3 A, 2 A being dissipated by conduction current and 1 A in lightning discharges. If we assume that 30 C are destroyed by the average lightning stroke, the 1-A lightning current would provide the charge for one lightning stroke every 30s. If one of the more severe storms produces a lightning stroke every 5 s, the required lightning supply current would be 6 A. The thunderstorm generator current and the average lightning charge are only rough estimates and may vary within wide limits.

The next uncertainty is connected with the conductivity inside the cloud. It is a fair assumption that the conductivity inside the cloud is less than the conductivity outside. This is caused by the fact that a certain percentage of the small ions are captured by the cloud elements and are very much reduced in their mobility. The same fate will meet the ions liberated by corona discharge; however, the percent captured by cloud elements and the distance of cloud penetration of the

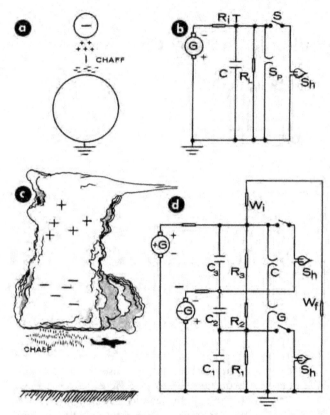

Fig. 1. (a) Corona discharge on a chaff fiber between charged spheres. (b) Circuit diagram to a. (c) Charge distribution in a thunderstorm. (d) Circuit diagram of a thunderstorm.

remaining ones are open questions. The ideal situation occurs when enough ions liberated from the chaff fibers can travel to the cloud charge centers and actively neutralize them. If this is not the case, chaff has to be dispersed throughout the cloud with such density that the region of influence of the individual chaff fibers overlaps, so that a conductive path between two opposite charge centers is established.

One may expect that a certain field concentration occurs at the boundary of the chaff seeded region in a manner similar to the field concentration at the surface of a conductive body in an electric field. This problem has been treated analytically by *Kasemir* [1973], who showed that when the vertical dimension of the seeded volume is not much greater than the horizontal dimension, the field concentration will be moderate and take a value of the order of 2. However, if the dimension of the seeded volume in the direction of the field is much larger than the dimension perpendicular to the field, then the field concentration may be high enough to trigger lightning discharges.

If it is assumed that the main direction of the fields in a thunderstorm is vertical, a horizontal line of chaff dispersed from an airplane would be the best way to avoid dangerous field concentration, whereas seeding from a dropsonde may cause elongated seeded volume with higher field concentration at the ends.

These problems, mentioned here briefly, have been discussed in more detail by *Kasemir* [1973]. However, this discussion is of necessity somewhat speculative, and for a decisive answer one has to wait until the necessary laboratory and field experiments can be performed. We may obtain a minimum estimate of the amount of chaff necessary to balance the average lightning supply current of 1 A by dividing it by the corona current of one chaff fiber. In a field

of 60 kV m⁻¹ the corona current of one chaff fiber is about 1 µA. Therefore 1 million fibers could generate 1 A corona current. This amount is equivalent to about 100 g of chaff and can be dispersed in a few minutes. Inside the storm the fields are probably higher by a factor of 3 or more, and so the necessary chaff would be reduced. However, if volume seeding should be required, the amount would increase again, fast dispersion becoming an additional problem.

The value of the corona current of 1 µA produced by a 10-cm-long chaff fiber in a field of 60 kV m⁻¹ has been obtained by extensive laboratory tests by B. Phillips and one of us. This value agrees well with the one we may calculate using the semiempirical formula published by *Chapman* [1958, equation 11]:

$$i = alkE^2/2 \qquad (1)$$

where i is the corona current in amperes; a is the empirical constant, equal to 2×10^{-11} m⁻¹; l is the length of chaff fiber, equal to 0.1 m; k is the mobility of ions, equal to 2×10^{-4} m² V⁻¹ s⁻¹ at sea level and 3.8×10^{-4} m² V⁻¹ s⁻¹ at thunderstorm base altitude; and E is the thunderstorm field, equal to 60 kV m⁻¹. Inserting the numerical values in (1), we obtain $i = 0.72$ µA at sea level and $i = 1.37$ µA at thunderstorm base altitude.

As a first step to lightning suppression the influence of chaff seeding on the electric field below a thunderstorm has been tested in the 1972 field decay experiment. Before we turn to a description of the test procedure and the discussion of the results, we will clarify the effect of chaff seeding on lightning and on the electric field below the storm by a circuit diagram that shows the important electrical properties of the thunderstorm (Figure 1d). As an introduction to the thunderstorm circuit let us first consider the circuit diagram of the laboratory experiment in Figure 1b. On the left side is the generator symbol G. The negative pole of the generator is connected through the internal resistor R_i, indicating a current source, to the capacitor C. In our laboratory experiment the short-circuit current was of the order of 1 µA. The capacitor C represents the capacitance between the two spheres. If there were no other load at the terminal point T connected to the generator, the voltage at T would rise with the time constant R_iC to the open circuit voltage and remain at this value. However, there are three more paths in the circuit diagram through which charges delivered by the generator can bleed to ground. These are the load resistor R_L, the spark gap S_p, and, connected through switch S, the glow lamp Sh. All these circuit elements are connected in parallel to the capacitor C. The load resistor R_L represents the resistance of the air volume between the two spheres and any insulation resistance of the charged sphere to ground. In the laboratory experiment, R_L is assumed to be high in comparison with R_i, so that its voltage-reducing effect at the terminal T is negligible. In the thunderstorm this resistor is low enough to bleed away about ⅔ of the charge delivered by the thunderstorm generator. This is the resistor that should be drastically reduced by the corona discharge of the chaff fiber. This effect is represented in the circuit diagram by the glow lamp Sh. Finally, there is the spark gap S_p, which accounts for the spark in the laboratory experiment and for the lightning in the thunderstorm.

It is easy to understand the operation of this circuit. If the generator is switched on, the capacitor C is charged up until the voltage reaches the breakdown voltage of the spark gap S_p, if we assume that the switch S is open. The ensuing spark discharges the condenser C and stops if the charge of the condenser is exhausted. The condenser is then charged up again by the generator until the next spark occurs. This sequence of events will repeat itself as long as the generator is operating. If we now close the switch S, the glow lamp is ignited as soon as its breakdown voltage is reached and will continue to bleed away the charges delivered by the generator. The voltage at the terminal T will never reach the value for sparkover if the ignition voltage of the glow lamp is lower than the breakdown voltage of the spark gap. This is equivalent to the condition stated above that the corona onset field is lower than the lightning igniting field. The second condition, that the corona current should be of the same order of magnitude as the thunderstorm generator current, means that the glow lamp must be capable of absorbing as much current as the generator is able to produce. The closing and opening of switch S is equivalent to bringing the chaff fiber between the two spheres or taking it away.

The thunderstorm circuit diagram in Figure 1d is composed of two interconnected units of the circuit type shown in Figure 1b. Additional features are the two resistors W_i and W_f that connect the top of the thunderstorm to the ionosphere and the ionosphere to the earth. The current flow through these resistors represents the contribution of the individual storm to the maintenance of the fair weather field and current. Furthermore, the load resistor of the negative charge generator $-G$ and the capacitor to ground is subdivided into the two parts R_1, R_2 and C_1, C_2 to represent the cloud volume from the negative charge center in the cloud to cloud base and the air volume from cloud base to ground. The purpose of this subdivision is to clarify the effect of chaff seeding below the cloud base, which will lead up to the field decay experiment discussed later on.

To facilitate relating the circuit diagram to the charge distribution in thunderstorms, Figure 1c shows a bipolar thunderstorm with positive charge in the top and negative charge in the base of the storm. The heights of the charge centers in Figure 1c are the same as the heights of the terminals of the positive or negative charge generator in the circuit diagram in Figure 1d. In Figure 1c an airplane is drawn below the cloud, and the released chaff fibers are shown as short vertical lines below the cloud base to demonstrate one of the seeding runs in the field decay experiment.

The spark gap marked C in Figure 1d between the positive and negative generator terminals symbolizes cloud discharges, and the spark gap G between the negative terminal and ground symbolizes ground discharges. The latter representation is incomplete insofar as the positive pocket charge in the base of the cloud and its corresponding generator are omitted from Figures 1c and 1d. This positive charge center in the base of the cloud is essential for the initiation of ground discharges and would have to be included if lightning suppression were discussed. However, the amount of the lower positive charge and its location in the cloud are not even approximately known. Furthermore, they are not essential for the discussion of the field decay experiment and thus are omitted here.

If it is assumed that the simple distribution of the positive and negative main charge centers is correct, chaff should be dispersed between the two charges to suppress cloud discharges. For ground discharge suppression the space between the negative and the elusive positive pocket charges in the base of the cloud should be seeded. In both cases the further assumption has to be made that either line or volume seeding will be effective in the cloud environment.

In the field decay experiment, seeding was done in the field of the negative space charge underneath the cloud. Therefore the glow lamp in Figure 1d is connected at the junction of resistors R_2 and R_1, and its influence on the spark gap G is reduced by the resistor R_2. A strong lightning suppression effect on ground discharges will not result from this type of seeding for the following reason. A necessary requirement for a lightning discharge is the high threshold field of about 500 kV m^{-1} inside the cloud that produces the original breakdown. If this field is maintained over a short distance of the order of 100 m, the field concentration at the tips of the lightning channel is already so high that the lightning will continue to grow in a comparatively weak external field. This behavior is evidenced by the fact that a ground discharge is capable of penetrating toward the ground in spite of the weak fields closer to the ground, which are only of the order of 1–10 kV m^{-1}. Chaff seeding underneath the cloud would merely increase the region of weaker fields between ground and cloud base, but it is doubtful that it can also significantly reduce the field inside the cloud. In other words, more of the potential energy required for the growth of a lightning discharge is furnished by the electric field inside the cloud than by the field underneath the cloud. However, a reduction of the voltage drop over the resistor R_1 or its equivalent, a reduction of the field underneath the cloud base, can be expected. It was the purpose of the field decay experiment to investigate if such an accelerated field decay can be created by chaff seeding.

One of the favorable aspects of chaff seeding is the fact that corona discharge on the chaff fiber will start in a comparatively low electric field of about 30 kV m^{-1} and that the corona current (ion production) increases with the square of the field. Therefore we may expect that high fields of the order of 120 kV m^{-1} or more, which is 4 times the corona onset field, decay faster than lower fields of the order of 60 kV m^{-1}, which is twice the corona onset field, and that for fields below 30 kV m^{-1} there should be no effect from chaff seeding on the normal variation of the thunderstorm field.

EXPERIMENTS

The chaff is dispersed from an airplane making repeated passes under a thunderstorm. The airplane is equipped with a chaff dispenser, field mills [Kasemir, 1972], a corona indicator, and the necessary accessory equipment, such as a strip chart and magnetic tape recorder, altitude recorder, and so on. The test procedure was to select one growing cumulus cloud that already showed the emergence of light precipitation at the base. The electric field was then monitored during continuous passes underneath the storm. If the field reached values of the order of about 50 kV m^{-1}, two consecutive passes were seeded (or not seeded), and the field recording continued until the storm dissipated. The decision to seed or not to seed was based on the requirement that data should be collected for an equal amount of seeded and unseeded storms. With the clear-cut physical objective and the necessary monitoring equipment, no elaborate randomization process was required.

We define two categories of storms investigated, namely, category 1 for fields above 100 kV m^{-1} and category 2 for fields between 30 and 100 kV m^{-1}, and split each category into the seeded storms and the control storms, which were not seeded.

The following data have been collected and will be discussed: category 1, one seeded storm and no control storms; category 2, four seeded storms and four control storms. Fields of 100 kV m⁻¹ or more are very seldom encountered below the storm. The data of the one seeded storm has been obtained in an earlier mission by *Kasemir* [1973] and is included here in the discussion. The graphs presented in Figures 2 and 3 show the field underneath the storm versus time. For each pass the maximum value of the negative vertical field component was listed with the corresponding time, and these peak values are connected in the graph by straight lines. The time interval between consecutive passes ranges from 3 to 9 min, the average being 4 min. In Figures 2a and 2b the average field of four category 2 storms is shown in the upper half of the figure, whereas the field of each individual storm is given with a reduced vertical scale in the lower half of the figure. To calculate the average field values, the seeded storms have been synchronized with respect to the first seeding run, and the control storms with respect to their maximum field value. Seeding events are marked on curves by the letter *S*.

It is evident from a comparison of Figures 2a and 2b that the field decay of the seeded storms after the second seeding run (Figure 2a) is much faster than the field decay of the control storms (Figure 2b). The accelerated field decay of the seeded storm is 6 kV m⁻¹ min⁻¹, as compared with the decay of the control storm of 1.2 kV m⁻¹ min⁻¹. If the field in the seeded storm is reduced to 34 kV m⁻¹, the accelerated decay stops because the corona discharge on the chaff fibers stops, and further decay continues at a rate of 1.5 kV m⁻¹ min⁻¹, which is about the value of the control storm. For the following discussion we will keep in mind that in category 2 storms the rate of field decay after seeding is about 6 kV m⁻¹ min⁻¹, which is an increase by a factor of 5 over the average rate of field decay of the four control storms.

In the lower part of Figure 2a the field curves of the individual storms are arranged according to their maximum field values of 56, 62, 72, and 80 kV m⁻¹ in ascending order; i.e., the field curve with the lowest maximum field of 56 kV

m⁻¹ is at the bottom, and the curve with the highest maximum field of 80 kV m⁻¹ is at the top of the four curves. Even though caution was taken not to overstress the significance of the data, there seems to be an indication that the field decay is faster for the higher maximum fields.

Storm 4 merits special attention because the field did increase after the first seeding run and started to decrease only after the second seeding run. Furthermore, the field increase before seeding is much faster than that in all other examples. Thus we are dealing here with a rapidly growing charge buildup that has not reached an equilibrium state. Therefore two seeding runs were necessary to reverse the sharp field increase to a field decrease. However, the most significant feature of this test was that seeding was done inside the cloud instead of below the cloud as it was in all the other cases. One has to expect that a large number of ions liberated by the corona discharge are captured by the cloud droplets, made immobile, and therefore removed from their role of increasing the conductivity of the cloud. From a purely theoretical consideration it is not possible to decide if this effect will completely counteract the purpose of chaff seeding or if it is only an insignificant side effect. This one test seems to indicate that chaff seeding causes an increase in conductivity inside the cloud in spite of the ion capture and consequently causes a decrease of the field.

Figure 3 shows the field decay caused by chaff seeding of a category 1 storm [*Kasemir*, 1973]. The field in this case had an extremely high value of 300 kV m⁻¹ and had maintained its high value for about 18 min when chaff seeding began. After 3 seeding runs, which required about 5 min, a dramatic field decay occurred at the high rate of 58 kV m⁻¹ min⁻¹. The field dropped in about 8 min to a value below 30 kV m⁻¹ and remained below this value for the rest of the flight. This is about 10 times the decay rate of category 2 storms. Because the corona current increases with the square of the field, such

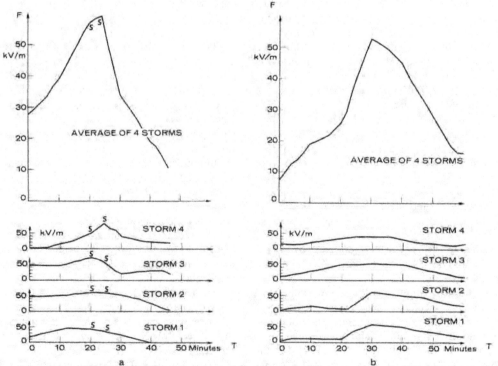

Fig. 2.   Field *F* versus time *T* for (*a*) seeded cases and (*b*) control cases. Upper part shows average of four storms: lower part shows field pattern of the individual storms 1 through 4.

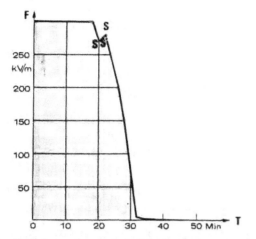

Fig. 3.   Field decay after chaff seeding (S) of a category 1 storm.

a result could be expected from a theoretical point of view. However, it is very satisfying to see the fast decay happen in an actual experiment.

DISCUSSION

The data presented in the previous discussion indicate that chaff seeding has a discharging effect on the electric field of a thunderstorm. Since lightning is generated by an electric field, the same method also seems to be a good approach to lightning suppression. The pilot experiments in lightning suppression would follow a procedure similar to the field decay experiment with more extensive seeding and monitoring requirements. In the field decay investigation the difficult task of recording location and magnitude of each individual lightning stroke is not necessary, and seeding can be confined to the part of the storm where the field decay is investigated. The whole experiment can be carried out with a single airplane.

But even though our experiment is restricted to the part below the storm, the thunderstorm charge that generates the field at the monitoring airplane is spread out over several cubic miles, and still the data indicate that the field of such a charge can be reduced in the seeded area in a matter of minutes with two seeding runs. If we visualize that the chaff fibers emerge from the airplane in a very thin line compared to thunderstorm dimensions, this is truly a remarkable result and indicates a fast dispersion of the liberated ions and also the generation of an amount of charge comparable to the thunderstorm charge in a comparatively short time. The following discussion is devoted to these two points, dispersion and generated net charge.

There are two effects that may account for the fast dispersion, the mobility of ions and the movement of the chaff fibers themselves. Small ions have a mobility k of about $k = 4 \times 10^{-4}$ m^2 V^{-1} s^{-1} in the altitude where chaff seeding occurred. In an electric field E they travel with a velocity $v = kE$. If the field is for instance 50 kV m^{-1}, as it is in the chaff seeding run of category 2 storms, positive ions would speed to the negative part of the cloud with a velocity of $v = 20$ m s^{-1}. This means that ions released at the beginning of the first run would have traveled in clear air a distance of 4.8 km in the time period of 4 min, when the second seeding run begins. Since the corona discharge is continuous, the stream of ions is also continuous, the result being that the volume of air

between the airplane path and the cloud is filled with positive ions. As soon as the ions reach the base of the cloud at least a certain percent are captured by the cloud droplets, and a surface layer of positively charged cloud droplets is formed. Other positive ions coming up will be deflected sideways, the result being a wider spread of the positive space charge. The attenuation of the field underneath the cloud will not be very different whether the ions accumulate at the cloud base or penetrate further into the cloud.

The second effect that may help to ionize the air of a larger air or cloud volume is the movement of the chaff fibers themselves [Kasemir, 1973]. Because negative corona discharge begins at a lower field than the positive corona discharge, the chaff fiber will initially lose negative charge and become positively charged until a balance between the positive and negative corona currents is reached. This leaves the chaff fiber with a positive net charge and subject to Coulomb force. Laboratory tests showed that gravity force pulling the chaff fiber down will be overcome by Coulomb force pulling the chaff fiber up if the field is greater than −60 kV m^{-1}. This force will cause the chaff fibers to move toward the negative charge centers in the cloud, especially inside the cloud where the fields are frequently larger than 60 kV m^{-1}. In higher fields of 100 kV m^{-1} or more the traveling speed of the chaff fiber may become several meters per second. The positively charged chaff fibers are repelled from positive charge centers. Therefore negative charge centers should be seeded from underneath and positive from overhead.

Coulomb forces may overcome gravitational forces, but they are not strong enough to overcome updrafts or downdrafts. Since the positive charge center is assumed to be in the upper part of the cloud, it may be possible to seed the positive charge center by releasing the chaff in a strong updraft.

The above discussion implies that the charge released by a thin line of chaff fibers may spread out in about 4 min in clear air over a volume of several cubic kilometers. We must next determine whether the amount of charge released in this time period is already an appreciable fraction of the average thunderstorm charge. The corona current on a single chaff fiber in a field of 60 kV m^{-1} is about 1×10^{-6} A. During one chaff seeding run about 3 million chaff fibers are dispersed. The net current of all of these fibers is about 3 A. This results in a net charge of 720 C released in a 4-min period. During the second run again 3 million chaff fibers are dispersed, so that 4 min after the second run 2160 C of charge will be generated, if all the chaff fibers produce the 1-μA corona current undiminished during this time period. This charge is considerably greater than the one in the negative charge centers, which is assumed to be of the order of −24 to −320 C. Therefore we may conclude, according to the estimates given above the net generated charge and the dispersion velocity, that the results of the field decay experiment are not in contradiction with physical principles.

REFERENCES

Chapman, S., Corona-point-discharge in wind and application to thunderclouds, in *Recent Advances in Atmospheric Electricity*, pp. 277–287, Pergamon, New York, 1958.

Kasemir, H. W., The cylindrical field mill, *Meteorol. Rundsch.*, 25, 33–38, 1972.

Kasemir, H. W., Lightning suppression by chaff seeding, *ERL Tech. Rep. ERL 284-APCL 30* Nat. Oceanic and Atmos. Admin., 1973.

(Received December 28, 1972;
revised October 2, 1973.)

CORONA DISCHARGE AND THUNDERSTORM FIELDS

Heinz W. Kasemir

National Oceanic and Atmospheric Administration
Atmospheric Physics and Chemistry Laboratory
Boulder, Colorado 80303

1.   INTRODUCTION

The electric charge released from the ground by point or corona discharge underneath a thunderstorm plays a dominant role in the charge transfer between thunderstorms and earth. In a review of the global current balance in the two leading textbooks of atmospheric electricity by Chalmers (1957) and Israel (1962) the charge carried by corona current per km^2 and year is about -100 to -300C as compared to +30C transported by precipitation and -20C by lightning. Another important aspect of the corona current is its limiting effect on the maximum field value at the ground to about ±10 kV/m. Without corona discharge maximum fields may be as high as ±50 to ±100 kV/m, Toland and Vonnegut, (1957).

The importance of corona discharge in the electric budget of a thunderstorm was recognized by Weber 1866-1892 and investigation and discussion of the problem was continued by Wilson (1920), Wormel (1927, 1930), Schonland (1928), to name a few. A very recent research of the thunderstorm field modified by corona discharge was carried out by Standler and Winn (1978). Two major difficulties are encountered in all these investigations. First, the application of laboratory experiments on point discharge of a single point to the gross effect of corona discharge of a large variety of corona points furnished by vegetation and artificial constructions on the earth surface underneath a storm. Second, the lack of a mathematical theory. This paper will give a simple mathematical model of the problem.

2.   CORONA CURRENT AT THE GROUND

The basic relation between corona current I_b and electric field E_b at the ground is usually expressed by a power law.

$$I_b = c(E_b - M)^n \qquad (1)$$

c is a constant, M the onset field of corona and the exponent is a positive number. A best fit to experimental data could be achieved by Kirkman and Chalmers (1951) with N = 1.1. Such a choice would give the constant c the odd dimension $A(V/m)^{-1.1}$. Other variations of the power law are the quadratic one used by Whipple and Scrase (1936) and many others

$$I_b = c(E_b^2 - M^2), \qquad (2)$$

the relation found by Warburg (1899)

$$I_b = c E_b (E_b - M) , \qquad (3)$$

and the cubic relation found by Ihawar and Chalmers (1967) for corona discharge from trees and used by Standler and Winn (1978)

$$I_b = c E_b (E_b - M)^2 \qquad (4)$$

It should be mentioned that in the above equations the current $I_b [A]$ sometimes refers to the current flowing from an individual point and at other times I_b refers to a corona current density with the dimension A/m^2 and represents the corona current per unit area underneath a thunderstorm. Before applying laboratory experiments on a single corona point to the conditions underneath a thunderstorm, we would have to determine the number of corona points and their activity per square meter of the earth surface underneath the storm. Research in this direction has been carried out by Wilson (1925), Whipple and Scrase (1936), Chalmers and Mapleson (1955), Ihawar and Chalmers (1965), and Standler and Winn (1978). It is outside the space and time limit of this paper to give an adequate list of references not to mention a detailed discussion of work already done on this problem.

3.   SOLUTION OF THE TIME DEPENDENT CORONA CURRENT.

Here we will assume that a relation between corona current density I_b and the field E_b at the ground under steady state conditions can be given in the form of equation (1) to (4). These equations will enter our calculation as boundary conditions. The calculation itself is concerned with the solution of the differential equation of the space and time dependent uni-polar current flow (corona current) with a driving force constant in time. In other words, the current output of the thunderstorm generator is assumed to be constant for the considered time period. The space dependance is restricted to the vertical coordinate z, i.e., the altitude of the considered space should be small with regard to the horizontal extension. The goal of the calculation is to obtain equations for the driving current, the corona constant and the onset field as functions of field measurements at the ground and aloft. The determination of the parameters c and M is then based on measurement of the thunderstorm field and would provide an excellent check on the application of the laboratory results to thunderstorm problems. The corona current density combined with the field contour lines of the field maps of Kennedy Space Center would provide the net current and the net charge that can be attributed to corona discharge. The time dependent solution will provide information when steady state conditions are approximately obtained and how strong the recovery curve after a lightning discharge is influenced by corona discharge at the ground.

It is realized that a restriction of the calculation to a time constant generator current and to one dimensional space is rather severe. However, analytic solutions of non-linear differential equations are normally difficult to obtain.

If the scheme for such a solution under simple assumptions is worked out the restrictive simplification can be relaxed at the cost of more complex solutions.

Symbols and units:

$E_{(z,t)}$ = electric field [V/m]

$I_{(z,t)}$ = corona current density [A/m²]

J = thunderstorm generator current density [A/m²]

ρ = ε div E = space charge density [As/m³]

ε = permittivity [As/Vm]

k = mobility of ions [m²/Vs]

M = corona onset field [V/m]

c = corona constant [A/V²]

z = space coordinate giving altitude above ground [m]

t = time [s]

Boundary values of E, I, and t are indicated by the subscript b. Since our boundary conditions equation (1) to (4) are given at the ground for z = 0, E_b, I_b, and t_b are ground values.

In general, E, I, and J are vectors. In one dimensional space only the z component is different from zero. Therefore, the vector symbol -- usually a bar or arrow above the letter -- is omitted.

Basic equations.

Current continuity equation

$$\text{div } I = -\partial\rho/\partial t \qquad (5)$$

Since no sinks or sources for the current I are specified in this equation, this means that a change of the corona current with altitude can be accomplished only by a deposit to or withdrawal from the existing space charge density, or the current has to be supplied by the boundary, in our case at the ground. If no current is supplied at the boundary, (5) describes the decay of an existing space charge. In steady state conditions with $\partial\rho/\partial t = 0$ only the boundary supplied current is present. In one dimensional space from div I = $\partial I/\partial z = 0$ follows I = I_b = constant in the considered space. The case of space charge being picked up by precipitation and carried to ground would represent a generator effect (source or sink) in the considered space and is therefore not included in this calculation.

$$\rho = \varepsilon \text{ div } E = \varepsilon \, \partial E/\partial z \qquad (6)$$

is one of the basic electrostatic equations and links the space charge density with the divergence of the electric field.

$$I = k E \rho = k E \varepsilon \, \partial E/\partial z \qquad (7)$$

is the definition of the uni-polar current flow (corona current). The corona current density in space is given by the product of the space charge density ρ, the driving field E and the mobility of the ions k. In the right hand side of (7), ρ is expressed by ε div E or in our case by $\varepsilon \, \partial E/\partial z$.

Substitution of (6) and (7) in (5) gives in a more general form the differential vector equation

$$\text{div } (\varepsilon\frac{\partial E}{\partial t} + k \varepsilon E \text{ div } E) = 0 \qquad (8)$$

with the solution

$$\varepsilon\frac{\partial E}{\partial t} = k \varepsilon E \text{ div } E = J \qquad (9)$$

The "integration constant" J is a forcing vector function of space and time coordinates with the restriction that div J = 0. This indicates that J is given by boundary conditions, and is a solution of an electrostatic problem. In our one-dimensional space, J is at best a function of time since div J = $\partial J/\partial z = 0$. For our further discussion, we will assume that J is also independent of time. It represents then a constant generator current density. Replacing in (9) the divergence operator by the partial differentiation to z we obtain the differential equation

$$\varepsilon\frac{\partial E}{\partial t} + \varepsilon k E \frac{\partial E}{\partial z} = J \qquad (10)$$

To solve this equation, we apply the characteristic method described in detail for instance in the textbook Methoden der Mathematischen Physik II by Courant-Hilbert (1937). With the auxiliary parameter "s" we have the three characteristic equations

$$\frac{dt}{ds} = \varepsilon; \quad t - t_b = \varepsilon s \qquad (11)$$

$$\frac{dE}{ds} = J; \quad E - E_b = Js \qquad (12)$$

$$\frac{dz}{ds} = \varepsilon k E = \varepsilon k (E_b + Js) ;$$

$$z - z_b = \varepsilon k (E_b s + J\frac{s^2}{2}) \qquad (13)$$

The differential equation (10) has to be valid at the boundary, too. We replace E by E_b, $\partial E/\partial t$ by dE_b/dt_b, and the corona current $\varepsilon k E \partial E/\partial z$ by I_b. Then our boundary differential equation is

$$\varepsilon\frac{dE_b}{dt_b} + I_b = J \qquad (14)$$

or integrating

$$\varepsilon \int_{E_{bo}}^{E_b} \frac{dE_b}{J-I_b} = \int_0^{t_b} dt_b = t_b \qquad (15)$$

The integral on the left side of (15) can be solved analytically for boundary conditions (2) and (3) where the quadratic relation between

393

I_b and E_b is used. The cubic relation of (4) would lead to an elliptical integral. For the general power relation given in (1) $I_b = c(E_b-M)^n$ the integral has no analytic solution. It is certainly possible to evaluate the integral by numerical methods, series expansion, or other means. However, the necessity to use such methods is usually a good indication that we are trying to force a boundary condition on the mathematical model that does not fit into the physical representation of the model. In our case, the one dimensional model is not capable to represent a three dimensional point discharge at the ground. There will be a boundary layer above the earth surface where the one dimensional model loses its meaning and has to be replaced by a three dimensional model. From each corona point on trees, bushes or other vegetation an individual stream of ions flows upwards into the atmosphere. At a certain altitude these individual streams merge into one compact ion blanket that has approximately a constant ion density in the horizontal direction. From this altitude on upwards, the one dimensional model can be applied. The consequence of this picture is that really none of the boundary relations given in (1) to (4) is applicable to a one dimensional model. There is, however, a certain affinity of boundary condition (2) to our model, as will appear later, and we will continue the calculation with this boundary condition.

To solve the integral on the left side of (15) we have first to calculate the roots of the quadratic equation in the denominator of the argument and split the argument by partial fractioning into two terms

$$\frac{1}{J-I_b} = \frac{1}{2Ac}\left(\frac{1}{E_b-A} - \frac{1}{E_b+A}\right); \quad A = \left(M^2+\frac{J}{c}\right)^{\frac{1}{2}} \quad (15a)$$

The integral of each term is the natural logarithm of the denominator. The solution of (15) is

$$-\frac{\epsilon}{2Ac} \ln \frac{E_b-A}{E_{bo}-A} \frac{E_{bo}+A}{E_b+A} = t_b \quad (16)$$

We may invert (16) and express E_b as a function of t_b. With the abbreviation

$$a = \frac{2Ac}{\epsilon} \quad (17)$$

$$E_b = A \frac{E_{bo}+A+(E_{bo}-A)\exp(-at_b)}{E_{bo}+A-(E_{bo}-A)\exp(-at_b)} \quad (18)$$

If in (18) $t_b = 0$ then $E_b = E_{bo}$ and if $t_b = \infty$ then $E_b = A$. Therefore E_{bo} is the ground field at the time $t_b = 0$. This value has to be given as part of the boundary condition. For $t_b \to \infty$ we approach steady state conditions, and the final value of E_b is

$$E_{bs} = (M^2+J/c)^{\frac{1}{2}} = A \quad (19)$$

Another important parameter is the time constant $T = 1/a$ of the exponential function in (18). With (17) and (19) it is given by the expression

$$T = \epsilon/2c \, E_{bs} \quad (20)$$

It is obvious from equation (18) that the approach to steady state condition is not given by a simple decay of an exponential function, however, an exponential decay is involved in a more complex manner.

With the calculation of the boundary condition accomplished, we can now return to the characteristic equation (11), (12) and (13) and obtain the complete solution for E as a function of z and t. The procedure followed here is first to calculate s as a function of E and z from (12) and (13) eliminating E_b. With $z_b = 0$ this leads to the equation

$$s = \frac{E}{J} - \sqrt{\left(\frac{E}{J}\right)^2 - \frac{2z}{k\epsilon J}} \quad (21)$$

The negative sign of the root has to be taken since for z = 0 at the boundary s = 0. Substituting s in (11) and (12) we obtain for E_b and t_b the equations

$$E_b = \sqrt{E^2 - \frac{2zJ}{\epsilon k}} \quad (22)$$

and

$$t_b = t - \frac{\epsilon}{J}\left(E - \sqrt{E^2 - \frac{2zJ}{\epsilon k}}\right) \quad (23)$$

We may now substitute (22) and (23) either in (16) or (18) to obtain the final solution. The beauty of the scheme is that it will work for any given boundary equation. The drawback is that the resulting equation is rather complex and cannot be put in a form that E is given explicitly as a function of t and z. The best we can do is use (16) and obtain t as a function of E and z.

$$t = \frac{\epsilon}{J}\left(E - \sqrt{E^2 - \frac{2zJ}{k\epsilon}}\right)$$

$$-\frac{\epsilon}{2Ac} \ln \frac{\sqrt{E^2-\frac{2zJ}{k\epsilon}} - E_{bs}}{E_{bo} - E_{bs}} \frac{E_{bo}+E_{bs}}{\sqrt{E^2-\frac{2zJ}{k\epsilon}} + E_{bs}} \quad (24)$$

Steady state condition with $t \to \infty$ is given when the logarithm on the right side goes to infinity due to the nominator of the first fraction going to zero. This gives for the field E_s in steady state the equation

$$E_s^2 = E_{bs}^2 + \frac{2zJ}{k\epsilon} \quad (25)$$

This equation has already been given by Wilson (1925) and again derived by Chalmers (1967). If we set $\partial E/\partial t = 0$ in (10) for steady state, (25) follows as an easy solution of the truncated equation (10).

4.   DISCUSSION

Our main interest will be focussed on the possibility to use field measurements at different altitudes to obtain unknown parameters such as the thunderstorm current density J or to determine corona parameters such as c or M for which values are reported in the literature that differ often by one or several powers of ten. Furthermore, it would be very valuable to determine the same parameter in different ways to check the consistency of the theoretical model or in different sense, the validity or limitation of the calculation.

4.a.   Thunderstorm current density J.

The easiest way to determine the thunderstorm current density is by measuring the electric field in steady state at different altitudes. If we measure E_1 at z_1 and E_2 at z_2, it follows from (25)

$$J = \frac{k\varepsilon}{2(z_1-z_2)} (E_{1S}^2 - E_{2S}^2) \qquad (26)$$

Since J is constant with altitude, this equation provides an opportunity to check first the validity of the model by carrying out measurements in different altitudes and second by extending the measurement all the way to the ground to determine the layer close to the ground where it can be expected that the data will not fit the equation.

There is a striking similarity between the empirical boundary relation $I_b = c(E_b^2 - M^2)$ given in (2) and equation (26) based solely on theoretical calculation. Since in steady state the current density J has to stay constant all the way to the ground, it is equal to I_b which links both equations (26) and (2). It is tempting to use dimensional analysis to speculate about the construction of c using physical parameters. The constant c and the expression $k\varepsilon/2(z_1-z_2)$ in (26) have the same dimension. The numerical value of c is in the range of 10^{-16} to 10^{-15} A/V^2. The product of $k\varepsilon/2$ has a value of about 4.5 to 9×10^{-16} A m/V^2. The first value is calculated with an assumed mobility k of 1 and the second of 2×10^{-4} m^2/Vs. The numerical values of $\varepsilon k/2$ and c are of comparable magnitude. Therefore, we may set tentatively

$$c = \frac{\varepsilon k}{2d} \qquad (27)$$

where d represents an as yet undetermined length of about 1 m. Speculation may carry us further to interpret d as the ratio of the average effective area A to the average height h of an individual corona point. So we would have

$$c = \frac{\varepsilon k h}{2A} \qquad (27a)$$

However, there are other possibilities to explain the unit length d. We will conclude this discussion with the remark that ε connecting field and space charge and k the mobility of ions are natural building blocks for the corona constant c.

4.b   Corona onset field M.

It is rather difficult to give a general definition of the corona onset field. It depends on the sharpness and the height above the ground of the corona point and quite often on the sensitivity of the current meter. Values given

in the literature range from .8 kV/m by Whipple and Scrase (1936) to 3 to 5 kV/m by Standler and Winn (1978). In a descriptive way we may say that M is the field strength at which enough corona points go into corona discharge and produce an average corona current density that is stronger than the conduction current density. We may use (26) to determine two different J_1 and J_2, and with the corresponding ground fields E_{b_1} and E_{b_2} obtain from (2) an equation for the determination of M or M^2.

$$M^2 = \frac{J_1 E_{b_2}^2 - J_2 E_{b_1}^2}{J_1 - J_2} \qquad (28)$$

4.c   Corona constant c.

Since J in steady state is equal to I_b at the ground and J and M are determined in (26) and (28), the corona constant c is given by rewriting (2).

$$c = \frac{J}{E_b^2 - M^2} \qquad (29)$$

From the different relations (1) to (4) and the widely scattering values reported in the literature with regard to the corona constant c one may conclude that c is not a constant at all. From the discussion in 3.a it is also obvious that severe averaging is involved. What is the average height of the corona points on a specified earth surface; what is the average effective area of the individual point; how many corona points has a pine tree; etc.? Measurements of the corona current on a single point at different heights, on live trees, on tree arrays and so on, have been made by many investigators. However, there are too many open questions how well these measurements represent larger surface areas to supply a solid base for the determination of c. The definition as well as the experimental determination of c and M is still a problem to be solved.

4.d   Numerical evaluation.

The very valuable paper of Standler and Winn (1978) contains a number of graphs showing the field at the ground and the field measured from a captive balloon raised to about 150 to 200 m above ground and lowered again. There are balloon sounding during steady state conditions at KSC on 8 July, 1976 and at Langmuir mountain observatory on 15 August 1975 and 27 July, 23 and 26 August 1976, that allow evaluation of the data. The evaluation is necessarily crude due to the error - about 10% - introduced by reading the graphs. Nine soundings at KSC and 12 at Langmuir observatory have been evaluated using the equations given in this paper. From the data of Langmuir observatory, it was possible to determine in one case the corona onset field to M = 5.2 kV/m   using equation (28). Standler and Winn quoted a value of M = 5 kV/m. The close agreement may be due to chance. For the KSC data, the corona onset-field was assumed to be 2.5 kV/m. Table I lists for KSC and Langmuir the average thunderstorm current density J, its minimum and maximum value J min., J max., obtained by equation (26), the average corona constant c, and its minimum and maximum value, c min and c max, obtained by equation (29), and the time constant T calculated by (20) using average values for c and J.

TABLE I

	J pA/m²	J min pA/m²	J max pA/m²	C 10⁻¹⁶ A/V²	C min 10⁻¹⁶A/V²	C max 10⁻¹⁶A/V²	T s
KSC	438	183	577	2.09	.5	3.8	7.3
Langmuir	419	133	842	.3	.12	1.25	24

REFERENCES

Chalmers, J. A. 1957, sec. ed. 1967, Atmospheric Electricity, Pergamon Press, London.

Chalmers, J. A., Mapleson, W. W. 1955, Point Discharge currents from a captive balloon, J. Atm. Terr. Phys. 6, 149-159.

Courant, R., D. Hilbert, 1937, Methoden der Mathematischen Physik II, Julius Springer Verlag, Berlin

Ihawar, D.S., Chalmers, J. A., 1965, Point discharge from multiple points, J. Atm. Terr. Phys. 27, 367-371.

Israël, H. 1961, Atmosphärische Elektrizität II, Geest and Portig Verlag, Leipzig

Kirkman, J.R., Chalmers, J.A., 1957, Point discharge from an isolated point, J. Atm. Terr. Phys. 10, 258-265.

Schonland, B.F.J., 1928, The interchange of electricity between thunderclouds and the earth, Proc. Roy. Soc. A. 118, 252-262.

Toland, R. B., B. Vonnegut, 1977, Measurement of Maximum electric field intensities over water during thunderstorms, J.G.R., 82, No. 3, 438-440.

Standler, R.B., W.P.Winn, 1978, Effects of coronas on electric fields beneath thunderstorms, (submitted to A J R M S)

Warburg, E., 1899, Spitzenentladung, Wied. Ann. 67, 69-83.

Whipple, F.J.W., F.J.Scrase, 1936, Point discharge in the electric field of the earth, Geophys. Mem. Lond. 68, 1-20.

Wilson, C.T.R., 1925, The electric field of a thundercloud and some of its effects, Proc. Phy. Soc. Lond. 37, 320-370

Wormell, T.W., 1927, Currents carried by point discharge beneath thunderclouds and showers, Proc. Roy. Soc. A, 115, 443-455.

Wormell, T.W., 1930, Vertical electric currents below thunderstorms and showers, Proc. Roy. Soc. A, 127, 567-590.

SECTION 4. LIGHTNING PHYSICS

1. Qualitative Overview of Potential, Field and Charge Conditions in the Case of a Lightning Discharge in the Storm Cloud, H. Kasemir, Das Gewitter: Ergbnisse und Probleme der modernen Gewitterforschung, ed. H. Israel, Akademische Verlagsgesellschaft Geest & Portig,. Leipzig, 1950.

2. A Contribution to the Electrostatic Theory of a Lightning Discharge, H.W. Kasemir, J. Geophys. Res., Vol.65, No. 7, July 1960.

3. Static Discharge and Triggered Lightning. H.W. Kasemir, Proc. 8th International Aerospace and Ground Conference on Lightning and Static Electricity, Fort Worth, Texas. June 1983,

4. The Field Equations of the Radiating Dipole, H.W. Kasemir, Atmosphärische Elektrizität, Teil II, Felder, Ladungen und Ströme p.409, H.Israel ed., Akademische Verlagsgesellschaft Geest & Portig, Leipzig, 1961.

5. The Electric Field of Lightning Discharges as an Observational Tool in Thunderstorm Research, H. W. Kasemir, U.S. Army Electronics and Development Laboratories, Forth Monmouth, New Jersey, NOAA, no date (n.d.).

6. Lightning Hazards to Rockets during Launch I, H. W. Kasemir, Technical Report ERL 143-APCL 11, US Dept. of Commerce Publication, 1969.

7. Lightning Hazards to Rockets during Launch II, H. W. Kasemir, Technical Report ERL 144-APCL 12, US Dept. of Commerce Publication. 1970.

8. Basic Theory and Pilot Experiments to the Problem of Triggering Lightning Discharges by Rockets, H. W. Kasemir, Technical Memorandum ERL APCL 12, US Dept. of Commerce Publication, 1971.

9. Theory and Experiments to the Problem of Rocket Triggered Lightning Discharges, H. W. Kasemir, Proc. Conference on Weather Modification of AMC, June 1972, Rapid City, SD, pp. 237-238, 1972.

10. Lightning Suppression by Chaff Seeding and Triggered Lightning, H. W. Kasemir, Weather and Climate Modification, pp. 612-629, Ed. Wilmot Hess, John Wiley & Sons, Inc. 1974.

11. Lightning Suppression by Chaff Seeding at the Base of Thunderstorm, H.W. Kasemir, F. J. Holitza, W. E. Cobb and W. D. Rust, J. Geophys. Res., Vol 81, No. 12, 1976.

12. Triggered Lightning, D. W. Clifford and H. W.Kasemir, IEEE Trans. on Electromagnetic Compatibility, Vol. EMC-24, No. 2, 1982.

13. Electrostatic Model of Lightning Flashes Triggered from the Ground, H. W. Kasemir, Proc. International Conference on Lightning and Static Electricity. Dayton, OH, 1986.

14. Analysis of Cloud, Ground, and Triggered Lightning Flashes from Gradient Measured at the Ground, H. W. Kasemir, Addendum, International Aerospace and Ground Conference on Lightning and Static Electricity, NOAA Special Report, U.S. Department of Commerce, Environmental Research Laboratories, 1988.

15. Energy Problems of the Lightning Discharge, H. W. Kasemir, Proc. VII. Int. Conf. on Atm. Electricity, June 1984, Albany N.Y., AMS.

16. Breakdown Waves in Return Stroke and Leader-Step Channels, L. W. Parker, H. W. Kasemir, Proc. VII. Int. Conf. on Atm. Electricity, June 1984, Albany N.Y., AMS.

17. Electrostatic fields of Ground-Triggered Lightning, L. H. Ruhnke, H. W. Kasemir, Proc. Int. Conf. on Atm. Electricity, Uppsala, Sweden, 1988.

From the book Das Gewitter (Results and Problems of Modern Thunderstorm Research), ed. H. Israel, Akademische Verlagsgesellschaft, Leipzig, 1950.

Qualitative Overview of Potential, Field and Charge Conditions in the Case of a Lightning Discharge in the Storm Cloud

By H.-W. Kasemir, Atmospheric Electrical Research Center in Buchau a. F.

Summary: It is possible to get the characteristics of the field and potential of lightning in a thundercloud by developing a pattern demonstrating the electric equipotential lines similar to those used in the representation of mountains on maps. Supposing a distribution of charges as given by the Wilson-Simpson thunderstorm model, one can derive the configuration of lightning as it is found during the different phases of development of a thunderstorm. A thunderstorm starts with air discharges before the beginning of heavy rain. This stage is followed by lightning to the earth with one or several partial discharges in such a rate as the negative space charge in the lower part of the thundercloud is diminished by precipitation. At the end of precipitation, i.e. during the decay of the thunderstorm, air discharges become again predominant. This scheme of sequence of lightning forms will be modified by the positive space charge center at the basis and the different concentration of the space charges in the diverse parts of the thundercloud. The positive space charge at the basis is also chiefly responsible for the deflection of the primary vertical lightning channel into a horizontal direction. Therefore, this lightning gliding along the bottom of the cloud is not to be interpreted as discharges between two different thunderstorm centers, but are caused by the positive space charge at the bottom of the cloud.

I. Introduction

When we observe the various works of the last 30 years regarding lightning and its consequences, we are left with an image that demonstrates a basic scheme in addition to numerous individual occurrences after a lightning discharge takes place in general. We owe the essential developments to the images with the slow and rapid rotating camera and the measurements of the electrical and magnetic field that lightning produces on the surface of the earth close to and distant from its location of origin. In particular, the electromagnetic wave field – known as atmospheric radio interference – has been examined in extensive series of measurements. However, in order to properly evaluate the electrical measurements now, a theoretical bridge is needed, which extends from the change of the atmospheric electrical field at the location of measurement to the electrical events within the lightning itself. This electrical configuration of a lightning discharge is determined through the arrangement and size of the space charge in the storm cloud. These are, so to speak, the base from which the lightning grows out.

If the theoretical correlation between the fields jump at the location of the measurement and the field structure on the lightning itself was only provided up to this point in roughest approximation, this could have two reasons. On one hand, a theoretical calculation of the lightning field requires a most-precise awareness of the charge arrangement in the storm cloud. On the other hand, the execution of the calculation requires a significant mathematical effort. The first requirement was provided in approximate terms by Simpson, Scrase and Robinson through the ascents of probes into storm clouds. Based on these measurements, an attempt shall be made in the present work to provide a purely qualitative overview of the field and charge structure of a lightning discharge based on a demonstrative model.

The knowledge gained in the process can serve as preparation for an exact mathematical calculation. However, throughout the course of this work, we will see that some information regarding the conditions for the various types of lightning and their correlation to certain space charge distributions in the storm cloud can be achieved from such a purely qualitative observation.

II. Graphic Diagram of a Potential Distribution with the Aid of a Relief using Contour Lines

We obtain a very vivid image of a potential function through the familiar depiction in a relief using contour lines[1]). In this case, the equipotential lines are expressed through the contour lines of a mountain range, while the field strength is reflected correspondingly in the steepness of the incline; a fact that is already expressed through the word 'potential gradient'. We would like to clarify the essential points with a simple example.

In Figure 1, the potential relief of an elongated conductor in a homogenous electrical field is depicted. To associate the mathematical formula equally with our physical problem, we need to envision the homogenous field as the atmospheric electrical field between the earth and ionosphere and the elongated conductor as lightning; even if the field of a storm cloud – as we will later see – differs significantly from the simplest case of the homogenous field selected here.

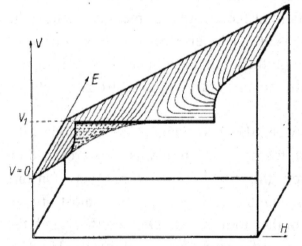

Figure 1. Potential relief of an uncharged conductor in a homogenous field.

---

[1] A depiction as it is used for the known functions in the function tables from Jahnke – Emde in a very clear manner.

Due to the fact that in Figure 1, the one coordinate is used for the potential value V and the second is used for the altitude H above the surface of the earth, only one coordinate remains in a three-dimensional illustration for the surface of the earth E. This illustration is therefore limited to two-dimensional or rotationally symmetrical potential fields.

The potential function of the uninterrupted homogenous field is formed by an inclined plane in Figure 1, the tendency of which is a gauge for the field strength. Now, if we bring an elongated conductor into this field in the direction of the field strength, we will achieve the deformation of the potential relief demonstrated in Figure 1. Experimentally, we can produce this figure with a transversely-tensioned rubber membrane, through the middle of which we will poke a knitting needle by no more than half of its length. If we then press this needle in a horizontal position such that its upper end stretches the elastic membrane downward, while the lower half pulls the membrane upward, in a very nice way we will obtain a visual model of Figure 1. In this context, we must still imagine that the rubber membrane is cut open at the needle level, whereat only the rear half is then depicted in Figure 1. Now, we can qualitatively extract the potential, field and charge distribution from the figure.

The constant potential V_1 dominates on the wire as it is required in the case of a conductor in electrostatics. This condition can also be adapted according to the finite conductivity of the lightning path. If the lightning channel still has a noticeable resistance – even if only a small one – then a voltage difference dominates in it between its ends according to this resistance and the current flow. Therefore, we may not push the needle down parallel to the baseline H, but rather only to a slight incline against it. Indeed, even the various resistances of individual lightning sections can be reproduced by bending the needle accordingly. However, we should not go further into these details at this point so as not to obscure the clarity of the figure.

The steepness of the potential plane is, as mentioned above, proportional to the field strength. We see that the greatest gradients occur at the ends of the conductors and accordingly in this case the greatest field strengths must be expected.

If we were to allow the conductor to grow in length, the tensioned elastic membrane would continue to steeply ascend from the ends of the wire and accordingly increasingly larger field strengths will develop at these points. And because an electrical discharge continues to develop on its own when a certain field value is exceeded, namely the so-called breakdown field strength, the lightning would continue to grow outwards in the homogenous field until this process is limited by the earth or ionosphere. This phenomenon was first presented in detail by Toepler and the formula was discovered that the lightning pushes forward the field necessary for its continual expansion.

Another significant point can be easily recognized in the depiction, namely the charge distribution on the lightning. Because the charge density is proportional to the field strength on the surface of a conductor and the field strength is provided in this case by the gradient of the potential mountain range, from Figure 1 we can see that the charge density is greatest at the end of the lightning, then decreases to 0 after the middle and changes its sign to achieve the maximum value again at the other end, though with the opposite sign.

A further significant observation particularly adapted to the lightning discharge is associated to this. Due to the great speeds with which the lightning expands, one can only take the electrons as the carrier of the charge transport, which are released by the collisional ionization within the lightning channel. In the short periods of the lightning discharge, the space charge elements of the cloud practically remain seated in their position due to their low mobility. As a result, the lightning cannot absorb a charge from anywhere nor discharge itself barring contact with the earth. Only a separation of the charge emerges in it according to the influence of the field. Perceived as a whole body, the lightning remains uncharged. The potential value of the lightning is determined by the condition that cumulative charge $= 0$.

If we attempt in Figure 1 to integrate the inclination values of the elastic membrane on the wire, which are indeed proportional to the charge density, across the entire length of the wire whereby these values must be rated as negative with the ascension of the elastic membrane and positive with its descent, then we will achieve the entire charge value 0.

Due to the symmetry of the configuration, the one half of the wire provides exactly the same integral value as the other, though merely with the opposite sign.

In contrast, a case is depicted in Figure 2, for which only inclination values of a sign occur. The integral provides a finite value, i.e. perceived as a whole, the wire is not uncharged, but rather carries a negative charge. Compared to Figure 1, it is on the lower potential V_2. In order to obtain the entire charge of 0 in this case as well, we would have to lift the wire and then finally arrive at the position

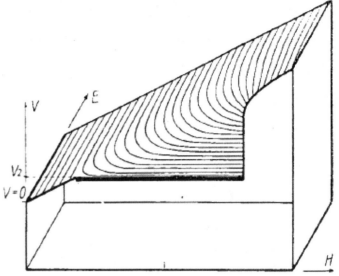

Figure 2. Potential relief of a charged conductor in a homogenous electrical field.

presented in Figure 1. With some degree of perception, it is possible to estimate the position – cumulative charge = 0 – even in the case of inhomogeneous fields.

The results discussed here, which are thoroughly familiar in potential theory, will be expanded to the discharge of lightning in the field of a storm cloud in the following chapters.

III. The Potential Relief of the Wilson-Simpson Thunderstorm Model

To now study the field and potential conditions for the lightning discharge in the storm cloud, it is necessary to first present the potential relief of the Wilson-Simpson thunderstorm model, which is broadly recognized today. We are no longer dealing with the homogenous field of a plate condenser in this case, but rather with an inhomogeneous field of alternating direction, which is established by the positive and negative center of the main space charge in the upper and lower portion of the storm cloud. Simpson's measurements revealed that an additional smaller and concentrated positive space charge island forms in the base of the storm cloud under the large negative space charge in the case of several storms.

The assumed charge distribution in the storm cloud is reflected according to the Simpson space charge spheres in the upper part of Figure 3 and the potential relief associated therewith is reflected below that. To enable a better allocation of the space charges to their potential values, the storm cloud is rotated by 90° against the normal position so that the head of the cloud is on the right side of Figure 3 and the base is on the left side. The normal atmospheric electrical field, which is overlaid by the field of the storm cloud, is so weak compared to the latter that it is no longer even significant. The potential scale is minimized in Figure 3 to the extent that the infinitesimally small grade of the rear edge of the potential plane, which is no longer disturbed by the storm, can no longer by depicted.

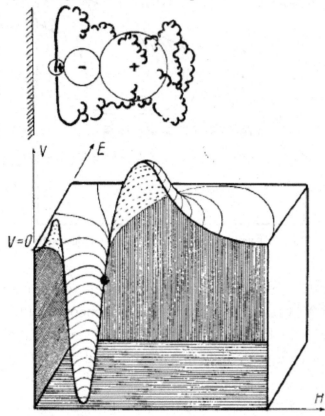

Figure 3. Potential relief of the Wilson-Simpson thunderstorm model.

Changes in the charge distribution with the development of the storm cloud can be easily traced in Figure 3. The buildup of the field begins with the shedding of precipitation particles in the upper portion of the storm cloud. Through the ensuing separation of charges, two space charge centers of identical size form, in which the negative charges accumulate on the descending precipitation particles while the positive remain in the head of the storm cloud. Thus, the upper positive potential mountain and the underlying negative potential cavity would establish themselves identically in the potential relief, whereby both increase in amplitude and mutual distance with time and the lower descent of the precipitation particles. If the storm is only poorly developed, in the borderline case, after reaching a certain field strength the precipitation will be held suspended or the charge buildup finds a limitation through the depletion of ions.

With the then potentially instating first lightning bolt, a reduction of the field and a strong reformation of ions occurs, which leads to a reinstating separation of the charge. However, in the case of stronger storms, if the condensation is so severe that larger precipitation particles (hail) forms, which falls against the effect of the electrical field - and potentially the updraft to the surface of the earth, a part of the negative charge will discharged to the ground with the descending precipitation. In this case, we will disregard the fact that the precipitation can be recharged near the earth by the approaching positive ion current. The positive and negative space charge remaining in the cloud is, therefore, no longer equally large. In our case, the positive potential mountain would surpass the negative potential cavity. A variety of additional factors will certainly cause a displacement of the proportions between positive and negative space charge.

Additionally, the concentration of space charges has a great impact on the shape of the potential mountain. According to the Simpson model image of storm charges, a charge of 24 Cb of positive polarity is enclosed in an upper large sphere (Figure 3) with a radius of 2 km. 20 Cb of negative polarity charge is concentrated in the middle sphere with a radius of 1 km, while the small lowest sphere with a radius of 0.5 km carries a positive charge of 4 Cb. From Figure 3, we can see that the negative main charge of 20 Cb impresses a much deeper hole into the potential mountain range with the maximum value of approx. $180 \cdot 10^6$ V than the positive main charge of 24 Cb is capable of rendering in altitude to its associated potential mountain. This is because the maximum value is only $110 \cdot 10^6$ V. (The equipotential lines of the main space charges possess a potential distance of $10 \cdot 10^6$ V.)

A lightning discharge between both main space charges would then store small space charge islands of the opposite sign, which spread over time in the main space charge and are neutralized by it. Because the lightning penetrates the space charge island somewhat far, it is conceivable that the small positive space charge island stems from the positive heads of the cloud lightning, which penetrated the negative main space charge, at the base of the storm cloud.

An additional unilateral effect of the charge balance of the storm cloud occurs through the conduction current to the ionosphere, which sets in to a larger degree due to the much greater mobility of the ions in the thin atmospheric layers and attempts to discharge the positive space charge present in the head of the storm. As a result, the building potential mountain would partially disintegrate.

This paper does not set out to determine the space charge distribution in the storm cloud and its temporal variation with the progress of the storm, and for the following thought processes, it should merely be noted that the positive and negative space charge centers, which are initially identical in size due to their origin, are changed in size and position through various influences.

IV. The Potential Relief of Lightning in a Storm Cloud

We would now like to proceed to the observation of the deformation of the potential relief through the developing lightning. For this purpose, we will first take a storm cloud as the simplest case, which comprises only an upper positive and lower negative space charge. In this case, the lightning would originate at the point of the greatest field strength (marked with a small black full circle in Figure 3) and push horizontally into the positive potential mountain with its negative end while its positive end seeks to bypass the potential cavity. Such a lightning discharge is reflected in Figure 4 and Figure 5 in two different phases of development.

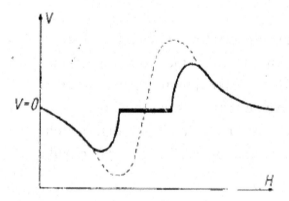

Figure 4. Initial stage of lightning in a bipolar space charge.

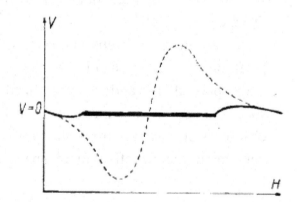

Figure 5. Final state of lightning in a bipolar space charge.

For further observations, it should suffice to present only the sections in the figures. Figure 4 depicts lightning at the moment, in which it has reached approximately half of its final length. The potential curve undisturbed by the lightning is shown with a dotted line prior to its development. We recognize that the field strength (steepness) at the ends of the lightning still increased substantially compared to the field strength present at its point of origin. Thus, the lightning is transgressed with increased force until it has reached its final state (Figure 5). In this case, the field strength at the ends of the lightning has become so weak that it no longer suffices for the ionization of the air molecules. The lightning dies before reaching the surface of the earth. This is the cloud lightning - described in English literature as air discharges - that protrude slightly from the cloud but are not capable of penetrating through to the surface of the earth. In the case of equal positive and negative space charge, lightning could never reach the earth due to its simple polar cloud.

We can make this fact physically plausible in Figure 3 if we trace the field strength curve from the small black circle. If we go down from that point to the earth, we will initially have a strong potential downward descent, thus positive field strength. This field strength, however, will become 0 at the center of the negative space charge (lowest point of the potential cavity) and assumes strong negative values by further approaching the earth. The fact that the field strength becomes positive again after penetrating the lower positive space charge island no longer plays a roll, insofar as the lightning is no longer even capable of penetrating this positive space charge island. Therefore, lightning would find a favorable field for its earth-facing head from the direction of the black circle as long as it travels down the potential mountain range. However, if it has penetrated down to the middle of the negative space charge, the field strength will reverse its sign and it must start towards an opposing field, which slows it down before finally stopping it.

The same considerations can be made for the transgression of the upper end of the lightning in the positive space charge in the head of the storm cloud except that the opposite sign must logically be applied in this case for the field and the space charge.

Now, it would be conceivable that the velocity of protrusion of the positive end of the lightning is greater than that of the negative and that an earthbound lightning bolt materializes in this manner if one assumes its point of origin is higher. Such a case is reflected in Figure 6.

However, the condition discussed at the end of Chapter II will be applied at this point, namely that the cumulative charge of lightning prior to making contact with the earth is 0; although that is no longer the case with the position of the lightning reflected by the solid thick line in Figure 6. Such a potential distribution is physically not conceivable. According to that which was previously discussed, the lightning will sink into the position characterized by the

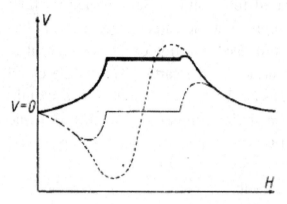

Figure 6. Potential image of lightning from the head of the storm cloud.

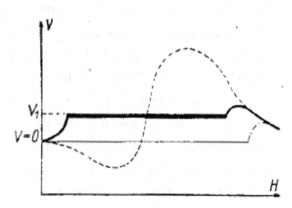

Figure 7. Earth-bound lightning from two unequally sized space charges.

thin solid line and therefore, concludes at a higher point of origin and different velocity of protrusion of its ends in the position identified in Figure 5. In this example, we can see the limitation of a lightning discharge very clearly. Lightning is never capable of penetrating a potential mountain or boring itself into the other side of a potential cavity as the field strength at its ends would pass through 0 in this process, though the lightning already stops after dropping below the penetration field strength of some 10^6 V/m. Therefore, the lightning can never eject additional paths from the area of two equally sized space charges of a differing polarity.

The situation would be different in the case unequally sized space charges or the type of cloud comprising three space charge centers. We should clarify the first case in a figure. If we assume that a large part of the negative space charge was transported to earth through rainfall, then the potential cavity became flatter (dotted line in Figure 7), thus enabling lightning to break through beyond this towards the earth.

The solid line in Figure 7 shows the lightning shortly before reaching the surface of the earth. Prior to making contact with the surface, it has the positive voltage V_1 versus the ground and discharges so much positive charge on the earth upon contact until it is on the earth potential 0. This lightning phase unfortunately identified with return stroke has a charge transport in the same direction as that of the head of the lightning bolt facing the earth prior to coming into contact with the earth. Therefore, it is recommended to exclusively use the word "main stroke" for this lightning phase.

As a result, we can generally conclude that lightning, which brings a positive charge to the earth, originated from a positively polar cloud, in the head of which the positive charge is located, from which however a part of the negative charge located in the base of the cloud is diverted to the earth through rainfall.

Through the sudden discharge of the lightning channel toward the earth, the field strength increased sharply at the upper end of the lightning, as we see from the dotted position of the lightning in Figure 7. The growth into the positive space charge portion thus receives a strong push – a condition that plays a leading role for the explanation of the partial discharges of earthbound lightning. With this, it becomes obvious why the discharge form of the partial discharges was observed very frequently in earthbound lightning and never with cloud lightning. The decisive factor is the contact with a conductor – in this case the earth – and the associated increase of field strength on the other end of the lightning. Earthbound lightning with a partial discharge are already almost on the earth potential from the onset, thus preventing a significant change to field strength at the upper head of the lightning bolt.

Throughout the development of our theoretical storm, we will have to expect, therefore, the following discharge forms of lightning. The storm begins before the rainfall with cloud lightning, transforms into earthbound lightning with one or less partial discharges and has earthbound lightning with several partial discharges following strong rainfall and then, after the strong rain has stopped, concludes in reverse order once again with cloud lightning. This regression to cloud lightning is contingent upon the fact that the upper positive cloud charge is reduced with the strong earthbound lightning and the discharge toward the upper ionosphere through the conduction current so that the positive space charge reduced in this manner equals the size of the negative charge reduced by the rainfall.

The extent of development of this regression to cloud lightning depends on the duration of the charge separation of the storm cloud upon dying out.

Indeed, it is clear that image of the development of the lightning throughout the course of a storm revealed here is heavily schematized such that we may not expect 100% compliance of the labeled sequence in nature.

A strong variation of the depicted lightning formation is included through above-mentioned lower positive space charge island, which is formed under the negative charge in strong storms. The potential relief of a lightning bolt from this area is shown in Figure 8.

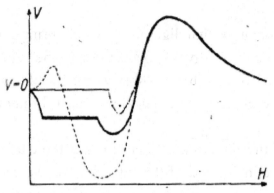

Figure 5. Earthbound lightning from the base of the storm cloud.

The position of the lightning prior to reaching the earth is depicted as the solid line, the position after making contact is the dotted line. The potential line prior to the development of the lightning is presented as hashed (see Figure 3). These lightning bolts are most likely all earthbound lightning and bring a negative charge to the earth. The former case is a result of the fact that the upper end of the lightning has sufficient room in the negative potential cavity to advance itself, and that the lower end of the lightning already reached the earth before the upper end is restricted by the ascending potential mountain. The unevenness of both space charges becomes effective at this point. The cumulative length and the temporal duration of these negative earthbound lightning bolts are shorter than those of the positive.

Another effect of this positive space charge island on the positive earthbound lightning should be briefly mentioned. This small but steep potential mountain at the base of the cloud forms a barrier for the lightning descending from the upper positive space charge (first example from section IV), which pushes toward the earth with its positive end and is thus ejected from the space charge island of the same name. This lightning will attempt to swerve sideways and run parallel to the lower edge of the cloud until it finally does strike toward the earth with sufficient strength of the upper positive space charge and partially rained-out negative space charge.

We see that this lightning running parallel to the base of the cloud does not need to discharge between two adjacent storm centers, but rather that a sideways diversion of originally vertical cloud lightning is possible through the effect of the lower positive space charge island.

Before concluding our observation, we must address an additional variation of the potential mountain range in the storm cloud, which clearly shows the numerous possibilities for a different field structure of a storm cloud available, even then if we think of them as comprised of 3 space charge centers. This concerns the different spatial concentration of otherwise equally great space charges. If we assume according to the Wilson-Simpson thunderstorm model that the upper positive space charge fills a sphere with a radius of 2 km while the lower equally great negative space charge must be concentrated in a sphere with a radius of only 1 km, we obtain a potential mountain range, in which – as mentioned above – the depth of the negative potential cavity far surpasses the altitude of the positive potential mountain. In contrast, the lightning has many more possibilities of working its way into it due to the significantly larger spatial expansion of the potential mountain and this end of the lightning could have only partially penetrated the potential mountain while the opposite lightning head has already bypassed the potential cavity and made contact with the earth. In this way, it is possible that earthbound lightning can result from two equally large, though differently concentrated space charges contrary to our previous statement.

Therefore, if we have already discussed a theoretical storm with the change of the cloud lightning to earthbound lightning and once again to cloud lightning without considering the impact of a different charge concentration, this occurred with the intention of presenting the necessary dependence of the development of lightning from the volume of the individual space charges and their temporal variation. Thus, it is the procedure for processes frequently applied in physics, which are subject to the influence of many variable elements, of arbitrarily keeping the one or the other element constant in order to better understand the impact of the element being observed. Therefore, it is necessary to be clear that we have to deal with the interaction of several factors in reality, of which in this case we have certainly only dealt with a fraction. Thus, in conclusion it must be stated that this work is not intended to provide a valid scheme for the development of lightning in the course of a storm, but rather presents an attempt of being able to see the connection and the linking of the formation of lightning with the development and breakdown of the space charges in the storm cloud based on a graphic model image.

JOURNAL OF GEOPHYSICAL RESEARCH   VOLUME 65, No. 7   JULY 1960

A Contribution to the Electrostatic Theory of a Lightning Discharge

HEINZ W. KASEMIR

U. S. Army Signal Research and Development Laboratory
Fort Monmouth, New Jersey

Abstract. The electrostatic treatment of the field and charge distribution of a lightning discharge leads to the result that the charge distribution of a lightning stroke is composed of (1) the influence charges induced by the electric field of the thundercloud with the net charge zero, and (2) the net charge, which results from the potential difference of the lightning stroke and the ground before the lightning hits the ground. The first kind of charge distribution is that of a cloud discharge and that of the first leader of a ground discharge. The second kind of charge distribution is a feature of the main or return stroke of a ground discharge only. It is the charge distribution of a charged body and independent of the field distribution in the thundercloud. We can therefore apply to the return stroke the well-known electrodynamic theory of transients on a transmission line.

INTRODUCTION

The lightning discharge is an electrodeless spark, which grows in the electric field of the thundercloud. This electrostatic field of the cloud furnishes the energy and determines, in accordance with the laws of the electrodynamic theory and gaseous discharges, the development of the lightning flash. The relationships among the numerous physical elements involved are usually given by differential equations. In a mathematical sense, then, the theory of the lightning discharge would be the solution of a set of differential equations by certain electrostatic initial or boundary conditions. These boundary conditions would be given by the potential function ϕ_1 of the thundercloud for the time $t = 0$ just before the lightning starts, and the potential function ϕ for the time $t = T$ when the lightning stroke is completed.

The potential function ϕ can be obtained by the superposition of the original potential function ϕ_1 of the charge distribution in the thundercloud and a secondary potential function ϕ_2 of the charge distribution at the lightning channel. ϕ_2 can be calculated from ϕ_1 and the geometric form of the lightning channel by treating the lightning as a conductor in an electrostatic field.

We know, from photographic investigations with the Boys camera, that especially the lightning to the ground consists of different phases such as the stepped leader and the main or return stroke. In a multiple ground stroke this sequence may be repeated several times. As these phases, even the single steps of the stepped leader, are separated from one another by a rest period, we may be justified (at least as an approximation) in framing each phase by electrostatic boundary conditions.

It is the purpose of this paper to discuss these electrostatic boundary conditions in greater detail and to see how we can improve the existing conception of the lightning discharge. We will find that the application of the electrodynamic theory of traveling waves on transmission lines to the main stroke of the lightning discharge is justified.

Bewley [1951] has presented a fine treatment of this electrodynamic part of the problem. The present author's concept of the events in the cloud and of the leader stroke is different from Bewley's, but it is, with some modification, the same for the main stroke. Therefore the electrodynamic treatment of the main stroke will be referred to Bewley's book.

Another very interesting electrodynamic theory of the ground stroke is given by *Hill* [1957]. Even though this theory reflects very much the present concept of the lightning stroke, it deviates considerably in the treatment of the cloud as a conductor and in the charge distribution along the lightning channel from the treatment in the present paper. In opposition to Bewley's assumption of reflected waves, Hill considers the main stroke an aperiodic discharge. Finally, the calculation of the electrostatic boundary condition presented here is based on a paper by *Kasemir* [1950], read at a conference on thunderstorm electricity in 1948 at Buchan, Germany; an extract was published in Israel's *Das Gewitter*, to which the reader is referred for a more detailed discussion of the electrostatic aspect of the lightning discharge.

DISCUSSION

The charge distribution on a conducting spheroid in an arbitary, given electric field. The development of any theory usually involves considerable mathematical calculation, which is the most laborious part of the theory. Since it is a hindrance to have the physical discussion of the results interrupted by lengthy mathematical calculations, they are not included here.

The lightning is represented by a prolonged spheroid, which is placed in the electric field of the thundercloud. The problem is assumed to be of rotational symmetry, whereby the spheroid lies in the rotational axis. The electric field of the thunderstorm is given by its potential function ϕ_1.

For electrostatic considerations, the lightning stroke is a conductor. On its surface an influence charge distribution will form which generates a secondary potential function ϕ_2 of such a kind that the superposition of ϕ_1 and ϕ_2 results in a constant potential ϕ_L at the surface of the lightning stroke. The value of ϕ_2 is determined for a cloud stroke by the condition that the net charge of the stroke is zero. This condition is usually neglected in present theories, but it follows from the fact that the mobility of cloud-charge particles is far too low for them to take an active part in the lightning discharge. The charge movement in the lightning channel is supported only by free electrons produced by the ionization of collision in the channel itself. They move from one end of the lightning stroke, leaving behind a surplus positive charge, to the other end, where they form a surplus negative charge. In this way they generate the charge distribution on the channel which causes the potential function ϕ_2. For the ground stroke we do not have the condition that the net charge of the lightning channel has to be zero. In the electrostatic sense, the earth is a conductor of great capacity, and the lightning can draw any necessary charge from it. After contact with the earth the lightning will assume earth potential.

Rather than discuss the potential function ϕ_2, which is generated by the charge distribution on the lightning channel, we will deal with the charge distribution itself. With the help of several graphs we will start from simple cases and advance to the complete picture. In the graphs the lightning is always represented by a heavy vertical line, and the charge distribution on it is shown by the shaded areas. The potential function ϕ_1 of the given electric field is presented by a solid line ϕ_1. We will see that the charge distribution is proportional to ϕ_1, but of opposite polarity. Therefore, with a convenient scale factor, the envelope of the shaded areas appears to be the mirror image of the potential function ϕ_1 with respect to the lightning channel. To obtain this simple relationship, the potential ϕ_L of the lightning stroke has to be chosen as the reference potential, which means that it is convenient to set $\phi_L = 0$. From $\phi_L = \phi_1 + \phi_2 = 0$ it follows that $\phi_1 = -\phi_2$. The potential function ϕ_2 at the surface of the lightning, which is generated by the influence charge, is of the same amplitude as, but of opposite polarity from, the potential function ϕ_1 generated by the cloud charges.

In (*a*) of Figure 1 we see the charge distribution of a charged spheroid. The net charge is equally distributed along the channel. The charge per unit length is constant. This is also the charge distribution on a charged transmission line and, as we will see later, on the main stroke of a lightning flash to ground. The potential function ϕ_2 at the surface of the spheroid is constant, and therefore $\phi_1 = \phi_2$ is also a constant; see Figure 1(*a*). Figure 1(*b*) shows the charge distribution on the spheroid in a constant field. The charge at the mid point of the spheroid is zero and increases linearly by progressing to the upper end of the spheroid. It decreases linearly with the progress to the lower end of the spheroid. In this way we have the positive influence charge on the upper half and the negative influence charge on the lower half of the spheroid. The net charge is zero.

If the stroke starts from the base of the cloud and advances with its lower end to the ground

413

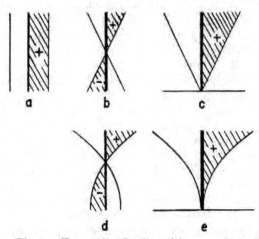

Fig. 1. Charge distribution: (a) on a charged spheroid; (b) on an uncharged spheroid in a homogeneous electric field; (c) on a charged spheroid in a homogeneous electric field; (d) on an uncharged spheroid in an inhomogeneous electric field; (e) on a charged spheroid in an inhomogeneous electric field.

and its upper end into the cloud, (b) represents the charge distribution on the leader stroke just before it makes contact with the ground. After the contact with the ground, the lightning has assumed ground potential, but the charge distribution along its channel is again proportional to ϕ_1. Because ϕ_1 at the ground is zero, the charge density at ground point of the channel also has to be zero. The new charge distribution is shown in (c). We can construct this charge distribution of (c) by superimposing that of (a) to that of (b). Hereby, (a) would give the charge distribution deposited on the channel by the main stroke. Before advancing to the lightning stroke in an inhomogeneous field, we should like to point out the difference between this concept and the present model of a lightning stroke.

In Figure 2, (a) shows qualitatively, in the first four successive pictures, how the leader stroke advances to the ground with the deposit of negative charge along its channel [Schonland, 1938]. This charge is delivered by the negative space charge center of the cloud. As soon as the leader reaches the ground the negative charge is discharged into it.

In the mathematical treatment [Hill, 1957], the cloud base and the earth are represented by parallel conducting planes and the leader stroke

by the lower half of a conducting spheroid with its midpoint at the cloud base. In Figure 2, (b) and (c) show the charge distribution before and after contact with the ground, according to Hill. The crucial point here is the assumption of the cloud base as a conductor. We know, from conductivity measurements in clouds, that its conductivity is even less than that of clear air, i.e., the cloud is indeed a very good insulator. To overcome this discrepancy, a widespread hypothetical streamer process inside the cloud is assumed to enable the cloud charge to feed the leader strokes of the successive partial discharges of a ground flash. It may be of interest to cite here a comment from the Russian literature. The following is a somewhat free translation from one of the publications of Trevskoi [1954].

We cannot regard a cloud as a conductor. The conductivity of air inside the cloud is close to zero because the ions are attached to the cloud particles. For this reason, the concept that by multiple strokes the charges required for the following stroke flow to the exit area from other cloud areas must be admitted as not corresponding to reality. Of course it could be assumed that the lightning inside the cloud is so highly branched that its ramification permits a conducting connection among the charged particles, but such an assumption is highly improbable.

From a mathematical point of view, there is not much difference between the representation of the leader stroke [Figs. 1(b) and 2(c)]. We can always replace the conducting cloud base of Figure 2(b) by the mirror image on it and arrive at Figure 1(b). In this way we can avoid the introduction of the cloud as a conductor. But the transition from the leader to the main stroke is completely different in the two concepts, as can be seen from Figures 2(c) and 1(c). In Figure 1(c) there is a strong influence charge deposited along the lightning channel. In Figure 2(c) the charge of the channel is zero. It is remarkable that Bruce and Golde [1941] completed the qualitative pictures of Schonland in such a way as to give the main stroke a positive charge. This is shown in the last four pictures of Figure 2(a). This concept would come closer to the theory presented here. We will now return to the charge distribution of a spheroid in an inhomogenous field.

In (d) of Figure 1 is shown the charge dis-

HEINZ W. KASEMIR

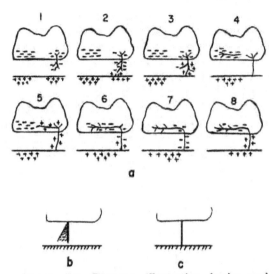

Fig. 2. (a) Diagram illustrating leader and main strokes to the first and subsequent strokes in a sequence of eight pictures after *Schonland* [1938] and *Bruce and Golde* [1941]. (b) Charge distribution of the leader stroke according to the theory of E. L. Hill. (c) Charge zero on the completed main stroke according to the theory of E. L. Hill.

in a thundercloud. In the preceding paragraph we discussed the charge distribution for an arbitrary potential function ϕ_1. We will now apply the results to the potential function ϕ_1 of a thundercloud, basing our discussion on the potential function of the old Simpson-Wilson thunderstorm model. This consists of three spheres vertically arranged, one at the top of the other and filled with homogeneously distributed space charge [Fig. 3(a)] [*Kasemir*, 1950]. The top sphere, with positive charge, represents the main positive space charge in the top of the thundercloud; the middle sphere, with negative charge, represents the main negative space charge center in the lower part of the cloud; and the lowest sphere represents the positive charge pocket at the base of the cloud. Figure 3(b) shows the potential function ϕ_1 at the central axis z resulting from such an arrangement. The heavy vertical line represents a cloud stroke. The shaded areas again give the charged distribution on the lightning channel. With the right scale factor, the envelope of these areas is again the mirror image of the potential curve ϕ_1 with

tribution on a spheroid for an inhomogeneous field of the potential function ϕ_1. Again we see that the charge distribution is given by the mirror image of the potential curve, but here this is true only as an approximation. The error becomes small for very slim spheroids, and, because the lightning stroke is very long indeed compared with its diameter, the approximate mirror-image relationship between the potential function ϕ_1 and the charge distribution is more justified than the representation of a real lightning stroke with its many bends and branches by a prolonged spheroid. The net charge is again zero. In Figure 1, (d), analogous to (b), would represent the charge distribution of a cloud stroke in an inhomogeneous field. To obtain the charge distribution of a ground stroke we have to superimpose again a uniform charge along the channel as shown in (a), so that the charge density and the potential of the contact point with ground is zero. This will result in a charge distribution as shown in (e). The laws of charging a long transmission line to a constant potential, therefore, hold also for the main stroke of lightning in an inhomogeneous field.

The charge distribution on a lightning stroke

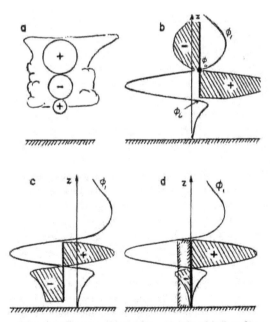

Fig. 3. Charge distribution: (a) in the thunderstorm model of Simpson Wilson; (b) on an intracloud stroke; (c) on a leader stroke of a cloud to ground discharge; (d) on the ground discharge after completion of the main stroke.

SYMPOSIUM ON SFERICS AND THUNDERSTORM ELECTRICITY

respect to the lightning channel. The lightning itself has the potential ϕ_L, which is given by the crosspoint of the vertical line and the potential curve ϕ_1. The cloud stroke may start from this point marked by a black circle in Figure 3(b), and grow to both sides until it crosses the potential curve ϕ_1 again. At this point it will stop, because there is no potential difference between the tips of the stroke and the surrounding air. Consequently, the electric field at the ends of the lightning stroke will drop to zero. The lightning will stop growing even before these points are reached, namely, when the field strength at the tips drops below the breakdown field strength necessary for ionization. This and many other features are pointed out by Kasemir [1950] and will not be repeated here, but we may mention that, at the upper end of a cloud stroke which reaches into the positive space charge center of the cloud, negative influence charge is accumulated, and positive influence charge is accumulated at the lower end of the cloud. This gives the cloud stroke a negative polarity. The opposite is true for the ground stroke, as we can see from (c) and (d) in Figure 3. Positive influence charge at the upper end and negative at the lower end gives the ground stroke positive polarity. This polarity rule holds for the vast majority of lightning strokes, according to the experimental data from several workers in this field: for example, Pierce [1955], Workman and Brook [1958], and Kasemir [1956].

Figure 3(c) shows the charge distribution of the leader stroke just before the contact with the ground is made, and (d) shows that of the main stroke. The difference between these charge distributions, indicated by the hatched rectangle in (d), is the contribution of the main stroke. The distribution of the influence charge deposited by the leader remains unchanged by the main stroke. This leads us to two conclusions that may be very valuable for the evaluation of field records of lightning strokes: (1) The charge distribution on the cloud stroke and on the leader of a ground stroke reflects very closely the potential function, and with this the charged distribution of the thundercloud. (2) The charge distribution of the main stroke of a ground discharge can be separated from the influence charge of the leader and is uniform

along the lightning channel. It is independent of the inhomogeneous field of the thundercloud, and the net amount of this charge distribution is given only by the capacity, i.e., the length of the lightning channel and the potential difference of the leader stroke and the ground. As a result of this uniform charge distribution we may apply the calculus of transients on transmission lines to the main stroke of a ground discharge. As this theory is well known and already applied to the lightning flash by Bewley [1951] we can confine our remarks to the following. The mechanism of charging an open transmission line is that of reflected waves. If the conductivity of the lines is very high, the damping is very low and the wave travels almost with the speed of light. With decreasing conductivity, the traveling speed decreases and the damping increases until the process changes to an aperiodic form. The conductivity of the lightning channel is certainly much lower than that of a metallic wire. It is also a function of the current that has flowed through the channel, and therefore is not a constant but a function of time. Furthermore, we have to consider a loss of energy by corona discharge along the lightning channel, which makes the mathematical solution of this problem very complex and difficult. Therefore we will mention here only some experimental evidence that may point to the concept of reflected waves. Some photo-

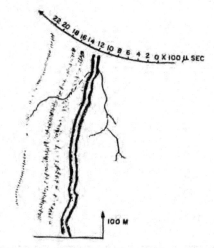

Fig. 4. Luminosity of the main stroke after Schonland, Malan, and Collens [1935]; dark strips indicate reflected waves.

416

HEINZ W. KASEMIR

graphs with the Boys camera show dark, regularly spaced strips imbedded in the bright band of the main stroke [*Schonland, Malan, and Collens*, 1935, Fig. 4]. These strips could indicate the alternating weakening and strengthening of the current flow by the successive reflected waves.

Another check would be the frequency of the radiated electromagnetic wave. One cycle would be completed by the time necessary for the electric surge to travel once up and down the channel. The traveling speed of the main stroke is in the average 6×10^9 cm/sec. If we assume the length of the lightning channel to be 3 km, we would arrive at a frequency of 10 kc/s, in good agreement with the experimental data.

REFERENCES

Bewley, L. V., *Traveling Waves on Transmission Systems*, John Wiley & Sons, New York, 1951.

Bruce, C. E. R., and R. H. Golde, The lightning discharge, *J. Inst. Elec. Engrs., 88*, 487–505, 1941.

Hill, E. L., Electromagnetic radiation from lightning strokes, *J. Franklin Inst., 263*, 107–119, 1957.

Kasemir, H. W., Qualitative Übersicht über Potential-, Feld- und Ladungsverhältnisse bei einer Blitzentladung in der Genritterwolke, in *Das Gewitter*, by H. Israel, Akademische Verlagsgesellschaft Geest und Portig, Leipzig, pp. 112–125, 1950.

Kasemir, H. W., Electrical thunderstorm types derived from lightning discharges, paper given at January meeting of the American Meteorological Society, New York, 1956.

Pierce, E. T., Electrostatic field changes due to lightning discharges, *Quart. J. Roy. Meteorol. Soc., 81*, 211–228, 1955.

Schonland, B. F. J., *Proc. Roy. Soc. London, 164*, 132, 1938.

Schonland, B. F. J., D. J. Malan, and H. Collens, Progressive lightning, II, *Proc. Roy. Soc. London, A, 152*, 595, 1935.

Tverskoi, P. N., The current status of the theory of thunderstorm electricity, *Priroda, 43* (2), 30–40, 1954.

Workman, E. J., and M. Brook, Thunderstorm electricity, *Signal Corps Research Contr. Nr DA36-039 SC-71145*, 1958.

417

STATIC DISCHARGE AND TRIGGERED LIGHTNING

Heinz W. Kasemir

Colorado Scientific Research Corporation, Berthoud, Colorado

ABSTRACT

The mechanism of a lightning discharge as introduced by Schonland[1]* and coworkers 1938 is analogous to a spark discharge between two metallic electrodes. Charges are collected by the leader stroke from the negatively charged cloud volume (negative electrode) and carried either to ground (positive electrode) producing a ground discharge or to the positively charged cloud volume producing a cloud discharge. We may call this the unidirectional or charged leader theory. A different mechanism has been suggested by the author, Kasemir[2] 1950, based on a uncharged leader (net charge zero) that works only with induced charge separation in the conductive lightning channel itself. This leader growth to both sides from a nucleus in the high field regions between oppositely charge cloud volumes. One end penetrating into the positive space charge carries negative induction charges and the other end penetrating into the negative space charge region carries positive induction charge. The net charge of the leader is zero. We will call this bi-directional or uncharged leader theory. This leader does not collect cloud charges but uses only the electric field energy provided by the charged cloud for its mechanism. An energy calculation shows that the unidirectional leader requires energy input above that provided by the cloud. Therefore, even beside the unexplained charge collecting mechanism the commonly used unidirectional leader is not a workable physical process. On the other hand, the bi-directional leader does not require charge collection and obtains enough energy from the electric field of the thunderstorm to feed the energy consuming processes of the lightning discharge, such as ionizing and heating the channel production of electromagnetic fields and radiation, etc. Applied to the problem of triggered lightning the bi-directional leader theory furnishes unexpected but far reaching results. For instance, it is shown that all so-called static discharges encountered on an aircraft are triggered lightning. Since triggered lightning outnumbers by far the rather rare hit by natural lightning and since warning devices, danger zones, etc., are quite different for the two types of accidents, a more intense study of the rather neglected triggered lightning is recommended. An intriguing controversy between pilots and scientists on charge and energy problems of static discharges will be discussed and resolved in this paper. These problems always center on the questi n: "Where does the lightning discharge obtain its charges to produce the measured currents or fields, if it is triggered far away from a thundercloud?" These problems are resolved by the uncharged leader theory, which explains that lightning does not have to collect charges from the environment be it a thunder cloud or a volume of clear air. The charges are generated in the lightning channel itself by ionization and then separated in the channel by the induction mechanism.

418

INTRODUCTION

It is the purpose of this paper to correct a mistake made in the theory of the lightning discharge which may be labelled the cloud charge collecting leader usually known as stepped leader, cloud to cloud, intra cloud, or cloud to air discharge. All these discharges are essentially of the same type. The correction includes the following points. The channel does not carry any net charge collected from the cloud. It works only with induced charges generated inside the channel, separated by the external cloud field and energy for growth and other energy consuming phenomena is obtained solely from the electric field. From this new point of view we can obtain the answer to many enigmatic problems. All static discharges are lightning triggered by the airplane. Critical objections that the aircraft cannot provide charge nor energy to produce a lightning discharge become irrelevant, since the lightning discharge obtains the energy from the field and manufactures the charges inside its channel by ionization. Charge separation is then accomplished by the field. However the net charge of the channel remains zero. Since the static discharges outnumber by far the rare hit of an airplane by natural lightning a new look at the lightning hazard to airplanes appears to be necessary. The difference between the hit by a natural discharge and a triggered discharge includes occurrence, danger zones and in consequence development of warning devices, direction of research, etcetera. This will be different for the two types of accidents caused by a triggered lightning or by a naturally occurring lightning discharge.

THE LEADER STROKE

There are two basically different physical concepts of the leader stroke of a lightning discharge, one introduced by Schonland (1) 1938 and the other one by Kasemir (2) 1950. To follow and comprehend the discussion in this paper it is essential to understand thoroughly the differences between these two concepts.

THE LEADER BASED ON CLOUD CHARGE COLLECTION - Schonland's concept of the stepped leader of cloud to ground discharge has been developed and diversified over the last 48 years. It includes in essence also the different types of cloud discharges, for instance, intra cloud, cloud to cloud, and cloud-air discharge. Detailed descriptions are given in the literature and text books. Uman (3) 1969, Golde (4) 1977, Volland (5) 1982. Figure 1a shows a brief sketch of a stepped leader advancing from the cloud to ground. The leader has collected negative charge from the cloud and stored along its channel with a charge per unit length of about q=1C/km. If contact with ground is made the negative charge flows into the ground in the return stroke. This process may be repeated several times by a dart leader-return stroke sequence bringing more negative charge to ground, each time discharging new areas of the negatively charged cloud volume by the upward or sideways growing branched lightning channel. After the lightning is terminated a portion of the cloud charges and the charges on the lightning channel have flown into the ground. Figure 1b.

In Figure 1c the same discharging mechanism is shown for an intracloud discharge. In this case it is an upward moving streamer which collects negative charge from the lower part of the cloud and moves it to the upper positive part of the cloud. This direction negative to positive is favored by Smith (6) 1957. Ogawa and Brook (7) deduce from field records at the groundin unidirectional descending streamer carrying positive charge downwards Figure 1d. However, in a footnote they mention the possibility of a bidirectional streamer moving upwards and downwards as shown in Figure 2c and 2d.

THE LEADER BASED ON INDUCED CHARGES - The leader based on induced charges has been introduced by the Kasemir (2), for the following reasons: 1. The thundercloud has to be considered as an excellent insulator. The cloud charges are immobile and fixed to there place in the time period of a lightning discharge of about one second. They cannot take an active part in the discharge. 2. Even if they are made mobile - Uman (3) suggests for this purpose a break down field extending over a larger region between positive and negative charged cloud volumes - the charges would follow the electric field lines, i.e., they would spread as far apart as possible, especially if we are dealing with unipolar charge. There is no physical law which makes even mobile negative (or positive) cloud charges concentrate in a lightning channel. This would require energy that cannot be drawn from the cloud. The energy calculation of this problem is discussed in the next section. 3. The hypothetical charge collecting mechanism by a widespread filament system of the lightning has never been explained.

In the new theory the inoization process inside the channel generates a large number of positive ions and electrons. A slight shift of electrons along the channel following the external field will create a surplus of electrons at one end of the channel, forming the required negative induced charge, and at the same time leaving at the other end of the channel a surplus of positive ions, generating in this way the required positve induced charge. The net charge of the channel still remains zero. The cloud charge collecting mechanism is completely avoided.

Figure 2a shows the charge distribution on the uncharged stepped leader corresponding to Figure 1a of the charged leader. Figure 2b shows the charge distribution on the lightning channel after the return stroke - or the dart leader- return stroke sequence is finished. Note here the marked difference between the corresponding Figure 1b, where the lightning channel is completely discharged, and Figure 2b, where the lightning channel carries a maximum of positive charge. When the ground discharge has brought -30 coulomb to ground then +30 coulomb of charge remains on the lightning channel. During the whole discharge the cloud charges haven't moved or participated in the discharge in any way. We may mention that Figure 2b shows also the charge distribution of a rocket triggered lightning for a rocket fired from the ground but grounded by a trailing wire. Figure 2c and 2d show the charge distribution at the beginning and at the end of a cloud discharge. Here too the lightning has the maximum charge density on the channel at the end of the discharge. The cloud charges remain at their place. In the exact meaning of the word "discharge", neither cloud nor ground discharges "discharge" the cloud. What they do in fact is to transport the opposite charge into the cloud volume. During the life time of the lightning strike this charge is still confined within the lightning channel. The actual neutralization between cloud and lightning charge occurs after the lightning is over. This neutralization process may take a minute or more rather than a second or less. After the lightning loses its conductivity and free electrons are attached to air molecules the air ions will disperse in the neighborhood of the channel,

neutralize oppositely charged cloud elements or attach themselves to neutral cloud elements or precipitation particles. We will end up with a cloud volume containing positive and negative charged cloud elements or precipitation particles. This will cause increased coagulation and result in the rain gush often observed after a lightning discharge. The idea that the lightning causes the rain gush has been promoted for some time by C. Moore. An excellent description of the whole process is given by Moore and Vonnegut (8). It will be difficult to explain this effect for a ground discharge by the charged leader, which just removes negative cloud charge from the lower part of the thundercloud. (see Figure 1b).

ENERGY PROBLEMS OF THE LEADER STROKE

Energy calculations are not often applied to lightning or thunderstorm processes. They are, however, quite useful to check if a suggested mechanism is energy consumming or producing. Lightning should consume thunderstorm energy since in turn the lightning has to provide energy for all the different processes necessary to or connected with its growth. These are mainly ionizing and heating the channel, producing electromagnetic radiation, building a magnetic field, etcetera. In general we may consider the lightning discharge as an energy converter.

ENERGY BALANCE OF THE CHARGED LEADER - To calculate the energy balance of the charged leader we need a mathematical model and a scheme of the calculation. With reference to Figure 1a we represent the negatively charged volume of the thundercloud by a sphere of radius R filled with negative charge Q of constant charge density ρ. The leader is represented by a spheroid filled with negative charge Q_1 with constant charge per unit length q. The centerpoint of the leader is positioned at the lowest point of the sphere surface and the long axis is radially directed. We calculate now the electric energy U of the system before the leader has formed and the energy V after the leader has formed to the length L. The difference W of U and V

$$W = U - V \qquad (1)$$

will tell us if the formation of the leader has consumed energy, U>V, W>0, or if the formation has increased the energy of the system U<V,W<0. The second result indicates a non workable mechanism and the first result a workable one. U is the energy of the charged sphere. If ϕ is the potential function inside the sphere we integrate the product $\rho\phi$ over the volume of the sphere. This gives with the factor 1/2 the energy U.

$$U = 4\pi R^5 \rho^2/15\varepsilon = 3Q^2/4\pi\varepsilon R. \qquad (2)$$

The energy V consists of two terms. First we have after removing the charge Q_1 from the rim of the sphere the main portion of Q_γ in the now slightly smaller sphere of radius R_1. For the calculation of the energy U_1 of the smaller sphere we use also Equation (2) replacing R by R_1.

$$U_1 = 4\pi\varepsilon R_1^5 \rho^2/15\varepsilon. \qquad (3)$$

Furthermore, we have to concentrate Q_1 into the spheroid representing the leader. If C_1 is the capacity of the spheroid and K the concentration energy, we have

$$K = Q_1^2/2C_1. \qquad (4)$$

From Equation (1), (2), (3), and (4) we obtain the final result

$$W = U-U_1-Q_1^2/2C_1. \qquad (5)$$

$U-U_1$ represents now the energy of a very thin spherical shell charged with the charge Q_1. We can either make the transition $R_1 \rightarrow R$ using Equations (2) and (3) or use right away the capacity formula (4) replacing C_1 by $C=4\pi\varepsilon R$. C is the capacity of the spherical shell. The result is then

$$W = -Q_1^2(1/C_1-1/C)/2. \qquad (6)$$

Since C is always greater than C_1, the energy difference is negative and consequently the charged leader mechanism is not a workable concept. From a physical point of view the result is self evident, since - as mentioned above - the concentration of charge requires always an energy input.

ENERGY BALANCE OF THE INDUCTION LEADER - J. A. Stratton (9) gave in his textbook, "Electromagnetic Theory," under section number 2.13 a theorem on the energy of uncharged conductors in the following verbal and mathematical formulations: "The introduction of an uncharged conductor into the field of a fixed set of charges diminishes the total energy of the field", or expressed in a mathematical formula

$$W = \{\int_{V_0} \varepsilon E^2 dV + \int_{V_1} \varepsilon(E-E')^2 dV\}/2. \qquad (7)$$

W is again the difference of the energy U before and the energy V after the introduction of the conductive spheroid representing the leader of a lightning discharge. E and E' designate the field vector before and after the body is brought into the field. It shall be understood that Stratton's expression "uncharged conductor's" means that the net charge on the conductor is zero. Induced charges are permitted and present since they are the cause of the diminishing of the field energy. This theorem, mathematically expressed in Equation (7), proves the point we wanted to make which is:

The leader mechanism introduced by Kasemir 1950, which is based on an uncharged bidirectional growing channel carrying only induced charges can draw energy for its growth and connected energy consuming phenomena from the energy stored in the electric field of a thunderstorm.

We apply now Stratton's Equation (7) to a slim spheroid presenting the leader for the case shown in Figure 2a. The leader is restricted in length so that the cloud field can be assumed constant in the neighborhood of the leader. The equation of the potential function of the uncharged conductive spheroid exposed to a constant electric field is generally known or can be found in any textbook of potential theory or electricity, for instance, Smythe (10) "Static and Dynamic Electricity." To obtain the field E' and solve the integrals of Equation (7) requires only straightforward calculus. Use partial integration to solve the second integral.

The spheroid has the long axis a, the short axis b, and the eccentricity c. The solution of the first integral I_1 in (7), which has to be taken over the inside of spheroid V_0 is

$$I_1 = 4\pi\varepsilon E^2 ab^2/6. \qquad (8)$$

The solution of the second integral I_2 taken over the outside of the spheroid is

$$I_2 = 4\pi\varepsilon E^2 a^3(1-b^2 Q_0/ac)/6Q_1 \qquad (9)$$

$$Q_0 = -\frac{1}{2}\ln\frac{a+c}{a-c} \; ; \quad Q_1 = \frac{a}{2c}\ln\frac{a+c}{a-c} - 1 \qquad (9a)$$

Q_0 and Q_1 are Legendre's Polinomials of

second kind and order 0 and 1. Inserting and combining Equation (8) and (9) in Equation (7) we obtain

$$W = 4\pi\varepsilon E^2 a^3 (1-b^2(Q_0-Q_1)/a^2)/6Q_1. \quad (10)$$

or in good approximation for slim spheroids,

$$W = 4\pi\varepsilon E^2 a^3/6Q_1 \; ; \quad b \ll a. \quad (10a)$$

STATIC DISCHARGE AND TRIGGERED LIGHTNING

We are now in a position to take a new look on lightning strikes to aircraft, static discharges and triggered lightning, and the problems involved in energy and charge supply. Harrison (11) reported on a controversy between pilots and scientists representing observation on one side and theory on the other side. Since the confusion existing about static discharges and triggered lightning cannot be better illustrated than by this discussion, we quote here from a paper of Clifford (12) who gives the following description of the controversy:

"The pilots almost unanimously argeed that there are two distinct types of lightning observed in flight. The most common variety usually occurs while flying in precipitation at temperatures near freezing. This type is preceded by a build up of static noise in the communication gear and the presence of corona (St. Elmo's fire) can be observed if the flight is at night. The build up may continue for several seconds before the discharge terminates the static and corona.

The second variety occurs abruptly without warning. It is most likely to be encountered in or near thunderstorms, in contrast to the former variety which is more likely to be experienced in precipitation that has no connection with thunderstorms. Pilots tend to believe that the slow build up type of discharge is not a true lightning strike but rather a discharge of excess charge build up on the aircraft by flying through precipitation. The non-thunderstorm type greatly outnumbers the other. Both kinds can create a brilliant flash and a boom which can be heard throughout the airplane.

The response of the scientists to the pilots' static discharge theory has been universally negative. They insist that insufficient charge can be stored on an aircraft to produce a discharge which looks and sounds like lightning. Scientists are even more emphatic that insufficient energy could be contained in such a static charge buildup to produce any visible evidence such as burn marks, pitting or other damage on the aircraft. Yet, the pilots continue to insist that the aircraft is discharging and that the discharges do manifest themselves by bright noisy arcs and (not all pilots are sure about this) visible damage. The controversy has been characterized as a difference in view between scientists of long standing and pilots of long sitting."

This is a remarkable and illustrative description of the problem of the triggered lightning. Scientists who still adhere to the charged leader theory of the natural lightning discharge will have extreme difficulties in solving the charge supply or energy problems of the static discharge. What they obviously don't realize is that the same problem exists with the natural as with the triggered lightning. Using the uncharged leader theory there is no problem in any of these cases. What the pilots label static discharge is nothing else than a triggered lightning. The energy is furnished from the electric field. It would not matter if the field is produced by a thundercloud, by electrified shower clouds or the debris clouds after a thunderstorm which are known to be highly electrified. Even snowstorms which produce very few natural lightning discharges or non at all generate high electric fields capable of providing the energy of a triggered lightning. The starting nucleus is the airplane itself. Since it is much larger than a precipitation particle it can trigger lightning discharges in clouds where the field is not strong enough for a precipitation particle to do so. The pilots emphasize the point that flying through precipitation close to the zero degree level, or more precisely, through a cloud region containing a mixture of ice and water enhances the possibility of a static discharge. This too, is a valid observation since flying through sleet or wet snow produces very effective triboelectric charging of the aircraft. This charging of the aircraft will not be able to furnish enough charge or energy for a lightning discharge but it will help in the critical stage of converting corona into a lightning discharge. As soon as a sufficient long filament is established the external electric field can take over and provide the necessary energy for further growth. This help in starting the discharge furnished by a large nucleus and friction charge will enable the aircraft to trigger lightning in relative weak fields where the ensuing lightning is also weak and it may not be easy to detect marks on the aircraft. However, this may lead to the wrong conclusion that triggered lightning -or static discharges - are generally harmless. This will not be true if the relatively weak leader reaches ground and the aircraft is exposed to the return stroke. Furthermore, not knowing how to differentiate a triggered lightning from a hit by a natural lightning it is a fair conclusion that many lightning discharges triggered in a thunderstorm are mistaken for lightning hits. If the trigger mechanism works in the weaker fields of non-thunderstorm clouds there is not reason to assume that it wouldn't work even better in the stronger fields of thunderstorms. The question here is how to distinguish between a triggered and the hit of a natural lightning. The step leader of a natural lightning has still a velocity of about 10^5 m/s. A leader coming towards the airplane would bridge 100 meter in 1 millisecond or 1000 meter in 10 milliseconds. This is too short a time for a human eye to focus on or resolve any detail. Therefore the statement of the pilots that, "The second variety occurs abruptly without warning," is again correct. The streamers emerging from an airplane to meet or intercept a lightning, shown in most drawings of such an event, are certainly not based on observation.

With the exception of the case that the aircraft was already in corona before the lightning hit we have here a simple rule to differentiate the triggered lightning from the hit by a natural lightning.

Triggered lightning (static discharge) is preceded by corona discharge, which lasts long enough to be perceptible by human senses, and which is stopped at the occurance of the lightning discharge. The hit by a natural lightning occurs abruptly without a preceding perceptible corona discharge.

The distinction between the triggered and not triggered lightning variety is important since in accordance with pilots observations the first greatly outnumbers the second type. (Estimates of the author about 100 to 1.) Lightning hits can occur almost anywhere inside, underneath or in very close proximity of a thunderstorm. Triggered lightning is bound to occur in zones of high electric fields inside or in close proximity to the cloud but these zones are not limited to thunder clouds, they occur also in other non thundering but highly electrified clouds. The two bolts of lightning triggered by the Apollo 12 rocket are the best documented examples of such a situation. The probability of the occurrence of triggered lightning is enhanced by certain types of precipitation. Much effort in laboratory tests, lightning analysis, occurrence, location, and design of warning devices is spent on natural lightning, but surprisingly little effort is directed to the study of triggered lightning, which accounts for the majority of lightning accidents.

LABORATORY EXPERIMENTS OF TRIGGERED LIGHTNING

It has been made clear in the discussion of the preceding sections that triggered lightning is what we may term an electrodeless discharge. That means the discharge channel is not conected or originates at one metallic electrode of a high voltage generator and proceeds to the other electrode. However the discharge may start on a floating metallic or non-metallic nucleus, such as an airplane or a precipitation particle, and use the electric field energy provided by the generator for its growth. The distinctive difference between a spark starting from one electrode and an electrodeless discharge is that the spark has ready access to draw current or charge from the generator, restricted only by the internal resistance of the generator, whereas the electrodeless discharge can draw no current or charge from the generator and is limited to the use of the energy confined in the field. This makes the two types of discharges quite different in mechanism and discharge characteristics. Therefore the spark discharge between two electrodes is not a good model of a leader of a natural lightning nor of a triggered lightning.

It is easy to see that the spark between a pointed electrode and a plate inspired the model of the charged leader. The spark emerges from the pointed electrode, which may have been a piece of wire connected to one terminal of the generator. The other terminal is connected to the plate. The spark easily draws charge from the generator and deposits it along the channel during its growth to plate. Making contact with the plate the spark will discharge the charge in the channel and the charge of the capacitor of the generator. Schonland's drawing of the first leader to ground and the return stroke shown in Figure 1a is an exact transference of the events taking place during a spark discharge between two electrodes energized by a charged capacitor. This picture has set the mold for the model of the lightning discharge up to the present time.

The spark between the two electrodes of a high voltage generator has in most cases been the experimental tool to investigate the properties and effects of a lightning discharge. Since the model was in error right at the beginning, the results of these laboratory experiments should be interpreted with caution. Especially all events which require charge or current supply in short time (microseconds) or have a high frequency content (MHz) will be influenced or even dominated by the internal impedance of the generator. The technical resistor, for instance, is a linear circuit element and technical circuit problems are based on the solution of linear differential equations. The resistance or conductivity of a lightning channel is very definitely not a linear circuit element and the treatment of dynamic lightning problems require the solution of non-linear differential equations. Indiscriminate applications of the telegraph equation to lightning problems, for instance, may lead to erroneous conclusions. To the knowledge of the author no laboratory experiment has been carried out which reflects the situation of a lightning discharge. However there is an experimental determination of the field necessary to trigger a lightning discharge by the orbiter (space shuttle) Kasemir, Perkins (15) which shows at least the initial stage of a triggered lightning as seen from the point of view of this section - an electrodeless discharge. A model of the orbiter was supported on three teflon posts in a large plate condensor capable of producing static electric fields of maximum 330 kV/m. Figure (3). The distance of the belly of the orbiter to the lower plate was 165 mm and that of the tip of the tail fin to the upper plate 280 mm. The dimensions of the upper plate were 5x5 m, and for the lower plate the ground was used, so for all practical purposes, it was an infinite conductive plate. The distance between upper plate and ground was 750 mm so that in the area of the model the field was uniform (with the orbiter absent.) The

main point here is that the orbiter was well insulated from either electrode, i.e., it could not obtain charge or current from the generator. The following measurements were carried out. The voltage at the upper plate was raised until the first small corona discharge appears. In this case it was in a field of 153 kV/m. Then the voltage was raised until the corona was strong enough to be photographed. The drawing in Figure 3 was made when the field was 280 kV/m. In Figure (4) flash over occured in a field of about 304 kV/m.

Figure (4) is the reproduction of a photograph taken on color film. The reproduction of the color photo of Figure (3) doesn't show the corona points at the fuselage of the orbiter and is here replaced by a technical drawing.

There are several interestng comments one can make to this experiment.

1. If the field is kept constant at a value between 153 and 304 kV/m corona is maintained as long as the field lasts. Since corona cannot be sustained out of a charge the airplane has accidentally aquired or out of any charge collected from the air, or from the field generator we have to have at least two corona points on opposite sides of the airplane. One point releases positive charge into the air - we call this the positive corona point - and the other one releasses negative charge, so that the airplane remains essentially uncharged. There is no demand that the airplane has to deliver charge to the corona points.

2. It is well known that negative corona starts at a slightly lower field than positive corona. In this case the negative corona point will charge the airplane with positive charge. The result is that the field at the positive point is increased until positive corona starts too. At the same time the field at the negative corona point is decreased so that the negative charge release at this point is diminished. This is an automatic balancing effect which keeps the two corona currents equal. A similar balancing mechanism is also effective on a leader stroke which keeps it essentially uncharged.

3. More critical than the different corona starting fields is the difference in the exposure of the corona points. It is obvious from Figures (3) and (4) that the corona point at the tail fin of the orbiter is

more exposed, i.e., the field concentration factor there is higher than on any point on the fuselage where the field concentration is weak. This generates many corona points, even at the nose of the orbiter, to counter balance the strong tail fin corona.

4. We may mention here the effect of precipitation or triboelectric charging. This continuous charge supply to the aircraft will cause a certain imbalance in the corona currents. If for instance negative charge is supplied, the current of the negative corona point will get stronger. This will support filament formation (discussed in 5) and may turn corona into lightning.

5. Flash over is preceded by formation of distinct filaments in the otherwise diffuse corona glow. They are short in length at the beginning, about 5 to 10 cm, and of short duration, about 1/2 seconds. We see here the first sign of the changeover from a cold to a hot discharge, or from diffuse corona to a slim lightning channel. As soon as one of these jumping filaments reaches the upper plate, flash over occurs. Corona glow may reach the upper plate but that causes no flash over and has no visible effect on the form or intensity of the discharge.

6. As soon as flash over occured the discharge ceases to represent lightning since the discharge is now governed by the electric circuit of the generator. The filament may have reached several meter length if not limited by the upper plate.

7. It will be of interest here to report on an observation made by the author on his many thunderstorm penetration flights during NOAA's (National Oceanic and Atmospheric Agency) lightning suppression program. During cloud penetrations at 6 to 10 km altitude in fields of more than 100 V/m, numerous slim and weak discharges of about 10 to 30 meter length could be seen in the cloud in a distance up to about 100 or 200 meters from the airplane. These discharges were called flicker discharges or baby lightnings. They were obviously natures attempt to start lightning discharges. They produced no visible strikes on the field records. However the field mills were running on the 300 kV/m range and fields of a few kV/m would not have been visible on the records.

8. The transition from corona to channel formation is one of the most crucial but practically unexplored aspects in triggered lightning

research. It may be difficult to generate electric fields strong enough and over a long enough distance. It may however be possible to use thunderstorm fields at suitable locations either at mountain tops or using high towers. Another approach is to trigger lightning from the ground by a rocket with a trailing wire connected to ground. However the best and most appropriate measurements could be obtained from a suitably equipped aircraft. Since research along these avenues is still in progress and will be reported on this conference the reader can obtain the newest information from the relevant papers.

CONCLUSIONS

1. Two different mechanism for the leader stroke of a cloud to ground discharge have been suggested. One by Schonland 1938 (1), which will be called "The charged leader," and another one by Kasemir 1950 (2), which will be called, "The uncharged leader."

2. The charged leader collects charges from the cloud, stores them in the channel, and carries these charges to the ground in case of a cloud to ground discharge, or to the oppositely charged cloud volume in case of an intra-cloud discharge.

3. The uncharged leader does not collect charges from the cloud. It produces positive (positive ions) and negative (electrons) charges inside its channel by ionization. These charges are separated in the channel by the external cloud field to form the induced charges according to the minimum energy theorem of physics (Thompson's theorem). Consequently one half of the channel carries positive and the other half negative induced charges. The net charge of the channel is and remains zero. Therefore the label, "The uncharged leader."

4. The charged leader actually discharges the cloud during its lifetime. The uncharged leader carried only negative induced charges into the positive charged cloud volume and positive induced charges into the negative charged cloud volume. During the lifetime of the leader the induced charges remain in or very close to the channel. The actual discharging or neutralization of the cloud charges occurs after the lightning discharge is over.

5. The charged leader requires energy to concentrate the cloud charges in its channel. This energy cannot be obtained from the cloud since the charge flow does not follow the field

lines produced by the cloud. Consequently there is an energy deficit in the charged leader concept. Furthermore the introduction of the cloud charge collecting mechanism has never been explained or justified. In conclusion:

The charged leader is not a workable physical concept.

6. The formation of the uncharged leader diminishes the energy of the cloud (Stratton's theorem). In other words it liberates energy from the electric field of the cloud and converts it into the energy consuming processes of the lightning channel such as ionization, channel heating, electromagnetic radiation, etcetera.

7. The uncharged leader is based on the well-known physical laws or concepts, ionization for producing positive and negative charges inside the channel, and induction for separating these charges into a positive and a negative induced charge on the upper and lower half of the channel. This concept of the uncharged leader can, applied to the problem of static discharges, resolve existing controversies, guide research, and predict or explain the outcome of future experimental results or measurement.

8. One conclusion of the uncharged leader concept is that all static discharges observed on airplanes are triggered lightning. With the knowledge of the mechanism of a triggered lightning and its characteristics, we have a simple rule to decide if we are dealing with triggered lightning or an accidental hit by natural lightning. The triggered lightning is preceded by observable corona discharge and the hit by natural lightning is not. The triggered lightning is restricted to the zones of high fields in the cloud and the probability of occurrence is enhanced by certain types of precipitation. The triggered lightning is not restricted to thunderstorms but can happen also in non-thunderous electrified clouds including snowstorms. One of the most important future research objectives is the determination of the conditions that convert corona discharge into a lightning channel.

REFERENCES

1. B.F.J.Schonland, "Progressive Lightning IV," Proc. Roy. Soc. (A) 164, 132 - 150, 1938

2. H.W.Kasemir, "Qualitative Uebersicht ueber Potential -, Feld -, und Ladungsverhaltnisse bei einer Blitzentladung in der Gewitterwolke," in "Das Gewitter" by Hans Israel, Akad. Verlags. Ges. Geest and Portig K.-G., Leipzig, 1950

3. M.A.Uman, "Lightning," McGraw-Hill Book Company, New York, 1969

4. R.H.Golde, ed. "Lightning I, II," Academic Press, New York, 1977

5. H.Volland, ed. "Handbook of Atmospherics I, II," LRC Press Inc., Boca Raton, Florida, 1982

6. L.G.Smith, "Intracloud Lightning Discharges," Quart. J. Roy. Meterol. Soc., 83, 103-111, 1957

7. T.Ogawa, M.Brook, "The Mechanism of the Intracloud Lightning Discharge," J.G.R., Vol. 68, 24, 5141-5154, 1964

8. C.B.Moore, B.Vonnegut, "The Thundercloud," in "Lightning I," ed. R.H.Golde, Academic Press, New York, 1977

9. J.A.Stratton, "Electomagnetic Theory," McGraw-Hill Book Co., New York, 1941

10. W.R.Smythe, "Static and Dynamic Electricity," McGraw-Hill Book Co., New York, 1950

11. H.T.Harrison, "UAL Turbojet Experience with Electrical Discharges," UAL Meteorological Circular No. 57, United Airlines, Chicago, Ill., January 1965

12. D.W.Clifford, "Aircraft Mishap Experience From Atmospheric Electricity Hazards," AGARD Lecture Series No., 110, North Atlantic Treaty Organization, 1980

13. D.W.Clifford, H.W.Kasemir, "Triggered Lightning," IEEE, Vol. EMC-24, No. 2, 112-122, 1982

14. L.W.Parker, H.W.Kasemir, "Airborne Warning Systems for Natural and Aircraft-Initiated Lightning," IEEE, Vol. EMC-24, No. 2, 137-158, 1982

15. H.W.Kasemir, F.Perkins, "Lightning Trigger Field of the Orbiter," NASA, KSC Contract CC 6964A, Final Report, 1978

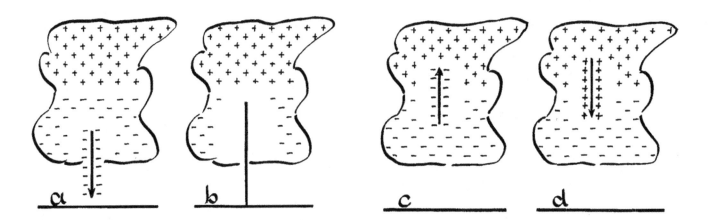

Fig. 1. Charged Distribution on Charged Leader

    a. Stepped leader
    b. After return stroke
    c. Cloud discharge advancing upwards
    d. Cloud discharge advancing downwards

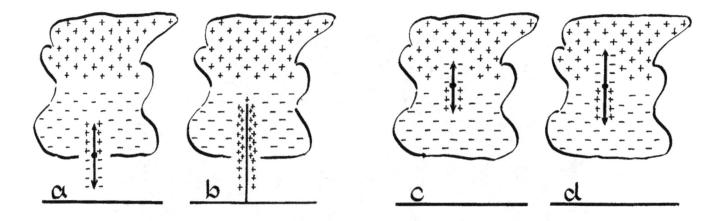

Fig. 2. Charged Distribution on Uncharged Leader

    a. Stepped leader
    b. After return stroke
    c. Cloud discharge beginning stage
    d. Cloud discharge end stage

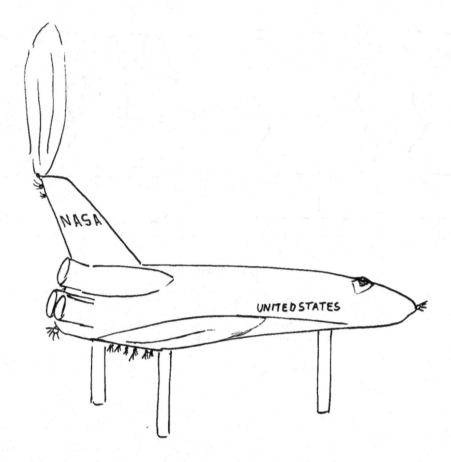

Fig. 3. Corona Discharge on Orbiter

Fig. 4. Flashover on Orbiter

The Field Equations of the Radiating Dipole

H.W. Kasemir

Atmosphärische Elektrizität, Teil II, Felder, Ladungen und Ströme p.409, H.Israel ed.,
Akademische Verlagsgesellschaft Geest & Portig, Leipzig, 1961.

FIELD EQUATION OF THE RADIATING DIPOLE*

Meaning of symbols used

$\vec{E}$ — electric field strength
$\vec{H}$ — magnetic field strength
q — space charge density
$\vec{i}$ — current density
Φ — retarded scalar-potential function
$\vec{A}$ — retarded vector-potential function
$\pm Q$ — dipole charges
$2h$ — separation between dipole charges
$2hQ$ — dipole moment
c — speed of light
t — time
x, y, z — rectangular coordinate system
r, φ, ϑ — polar coordinates
$\varepsilon = \mu = 1$ (electrostatic c. g. s. measuring system)

Vectors are indicated by an arrow above the relevant symbol (e. g., $\vec{E}$).

Maxwell's equations will form the starting point of our discussion. We shall make the following simplifying assumptions: let $\varepsilon = \mu = 1$ in the entire medium. Assume the conductivity of the medium is so small that outside the dipole there are no conduction currents. Further, let us separate the space charges (stationary thunderstorm field) occurring in the medium. We thus have

$$\operatorname{div} \vec{E} = 4\pi q = 0. \tag{515}$$

With these restrictions Maxwell's equations take on the simple form

$$\operatorname{curl} \vec{H} - \frac{1}{c}\frac{\partial \vec{E}}{\partial t} = 0, \tag{516}$$

$$\operatorname{curl} \vec{E} + \frac{1}{c}\frac{\partial \vec{H}}{\partial t} = 0. \tag{517}$$

---

* The following derivation of the widely used field equation of the radiating dipole was kindly performed by my former co-worker Dr. H. W. Kasemir, now in Boulder, Colorado, USA. At this stage I would once more like to express my gratitude for his kind cooperation.

From the fact that no free magnetic charges appear, it follows

$$\operatorname{div} \vec{H} = 0. \tag{518}$$

Thus, the magnetic field strength can be represented as the curl of a vector $\vec{A}$:

$$\vec{H} = \operatorname{curl} \vec{A}. \tag{519}$$

This vector $\vec{A}$ plays the same role in a magnetic field as the potential Φ in an electrostatic field. It is therefore also called the vector potential of the magnetic field. Substituting (519) into (517), we obtain

$$\operatorname{curl}\left(\vec{E} + \frac{1}{c}\frac{\partial \vec{A}}{\partial t}\right) = 0. \tag{520}$$

This equation states that the parenthetical expression represents an irrotational vector, which can be derived from a potential function by taking the gradient Φ

$$\vec{E} + \frac{1}{c}\frac{\partial \vec{A}}{\partial t} = -\operatorname{grad}\Phi, \quad \text{or} \quad \vec{E} = -\frac{1}{c}\frac{\partial \vec{A}}{\partial t} - \operatorname{grad}\Phi. \tag{521}$$

To recognize the significance of the potential function Φ, we shall first treat briefly the static case. For this we set t and all derivatives with respect to t equal to zero. Then (521) simplifies to

$$\vec{E} = -\operatorname{grad}\Phi, \tag{521a}$$

i. e., Φ becomes the electrostatic potential. From (521) we see that in the general electrodynamic case the field strength $\vec{E}$ is made up of two parts, one of which is given in a well-known fashion as the gradient of a potential function Φ, while the other is computed from the vector potential $\vec{A}$. Now, the relationship between the vector potential $\vec{A}$ and the current density i in a radiating dipole is the same as that between the scalar potential Φ and the charge density q. However, i and q are not independent of one another, but rather must satisfy the continuity equation

$$\operatorname{div} \vec{i} = -\frac{dq}{dt}.$$

This expression holds for the vector and scalar potential, so that we here introduce the following equation of condition

$$\operatorname{div} \vec{A} = -\frac{1}{c}\frac{\partial \Phi}{\partial t}. \tag{522}$$

This equation is certainly plausible, but not compulsory. It will be justified when later the solution satisfies the specified boundary conditions. Ordinarily its use is based on the fact that grad Φ as the constant of integration of equation (520) as well as Φ are readily at one's disposal, and thus also, for example, via equation (522).

From (516), (519), and (521) we obtain

$$\text{curl curl } \vec{A} + \frac{1}{c^2}\frac{\partial^2 \vec{A}}{\partial t^2} + \frac{1}{c}\frac{\partial}{\partial t}\text{grad } \Phi = 0 \,,$$

and with (522)

$$\text{curl curl } \vec{A} + \frac{1}{c^2}\frac{\partial^2 \vec{A}}{\partial t^2} - \text{grad div } \vec{A} = 0 \,.$$

Now, according to the laws of vector algebra

$$\text{curl curl } \vec{A} = \text{grad div } \vec{A} - \Delta \vec{A} \,,$$

so that the differential equation for $\vec{A}$ simplifies to

$$\Delta \vec{A} - \frac{1}{c^2}\frac{\partial^2 \vec{A}}{\partial t^2} = 0 \,. \tag{523}$$

In a similar way we get from (515) and (521)

$$-\frac{1}{c}\text{div}\frac{\partial \vec{A}}{\partial t} - \text{div grad } \Phi = 0$$

and allowing for (522)

$$\Delta \Phi - \frac{1}{c^2}\frac{\partial^2 \Phi}{\partial t^2} = 0 \,. \tag{524}$$

The Δ-operator is a short way for writing div grad. In the vector equation (523) it means that this equation consists of three equations, one for each of the components A_x, A_y, and A_z.
$\vec{A}$ has the same direction as the current density $\vec{i}$ of our dipole. Taking the dipole moment axis along the z-axis, the current will flow in the z-direction. The x- and y-components of the current as well as the vector $\vec{A}$ vanish. Accordingly, equation (523) reduces to

$$\text{div grad } A_z - \frac{1}{c^2}\frac{\partial^2 A_z}{\partial t^2} = 0 \,. \tag{523a}$$

For a spherical wave in a polar coordinate system r, ϑ, φ, the derivatives with respect to ϑ and φ vanish. We thus have

$$\text{div grad } A_z = \frac{1}{r}\frac{\partial^2 (r A_z)}{\partial r^2} \,,$$

and (523a) becomes

$$\frac{1}{r}\frac{\partial^2 (r A_z)}{\partial r^2} - \frac{1}{c^2}\frac{\partial^2 A_z}{\partial t^2} = 0 \,. \tag{525}$$

Substituting $r A_z = B$ and $r/c = T$, we can reduce (525) to the simpler form

$$\frac{\partial^2 B}{\partial t^2} - \frac{\partial^2 B}{\partial T^2} = 0 \,. \tag{526}$$

431

As can be readily shown by substitution, this differential equation is solved for any arbitrary function f of argument $(t-T)$. Also the derivatives of f with respect to t or T satisfy the differential equation since they, too, are functions of argument $(t-T)$. To simplify our later computation of Φ from (522), we shall choose as a solution for B the first derivative of f with respect to t, adding the constant factor $1/c$. Thus,

$$B = \frac{1}{c}\frac{\partial}{\partial t}f_{(t-T)}$$

or

$$A_z = \frac{1}{cr}\frac{\partial}{\partial t}f_{(t-r/c)} . \qquad (527)$$

At a distance r the argument $(t-r/c)$ gives rise to a time lag of the wave by the propagation time $T = r/c$, i. e., at this distance the prevailing state of the wave $(r=0)$ at time t will arrive later by the propagation time of the wave. Because of this retardation $\vec{A}$ and Φ are also called the retarded potentials. In what follows we shall no longer write the argument of the function f, but stipulate that every time derivative of the function f can be replaced by a derivative with respect to r under addition of the factor $-c$. We thus obtain the expression

$$\frac{\partial f}{\partial t} = -c\frac{\partial f}{\partial r} . \qquad (528)$$

To compute the potential function Φ we return to equation (522). Since only the z-component of the vector $\vec{A}$ is different from zero, we get

$$\operatorname{div}\vec{A} = \frac{\partial A_z}{\partial z} = \frac{\partial A_z}{\partial r}\frac{\partial r}{\partial z} = \frac{z}{r}\frac{\partial A_z}{\partial r} . \qquad (529)$$

According to equations (522), (527), and (529)

$$-\frac{1}{c}\frac{\partial \Phi}{\partial t} = \frac{z}{r}\frac{\partial}{\partial r}\left(\frac{1}{cr}\frac{\partial f}{\partial t}\right) = \frac{z}{r}\left(-\frac{1}{cr^2}\frac{\partial f}{\partial t} + \frac{1}{cr}\frac{\partial}{\partial r}\frac{\partial f}{\partial t}\right) \qquad (530)$$

and allowing for (528)

$$\frac{\partial \Phi}{\partial t} = \frac{z}{r}\left(\frac{1}{r^2}\frac{\partial f}{\partial t} + \frac{1}{cr}\frac{\partial^2 f}{\partial t^2}\right) . \qquad (531)$$

The solution of the above equation is straightforward. It is

$$\Phi = \frac{z}{r}\left(\frac{f}{r^2} + \frac{1}{cr}\frac{\partial f}{\partial t}\right) . \qquad (532)$$

Remembering $z/r = \cos\vartheta$, it can be shown relatively easily that the above solution for Φ also satisfies (524).

To recognize the physical significance of the function f, we let the time variation in f become so small, temporarily, that the second term in the parenthesis of (532) can be disregarded. Then,

$$\Phi = f\frac{z}{r^3} = f\frac{\cos\vartheta}{r^2} . \qquad (533)$$

In electrostatics, however, the potential function Φ_e of a dipole with dipole moment $2hQ$ is given by

$$\Phi_e = 2hQ\frac{\cos\vartheta}{r^2}. \tag{534}$$

Comparison with (533) shows that the function f represents the dipole moment $2hQ$, and in the case of radiating dipole also comprises the time function according to which the dipole oscillates.

If the static dipole moment is $2hQ_0$ and the time function is given, for example, as a sine oscillation with angular frequency ω, we get

$$f_{(t-r/c)} = 2hQ_0 \sin \omega\,(t - r/c)\,.$$

It should be emphasized, however, that the solution for $\vec{A}$ and Φ holds for any arbitrary time function and is not only restricted to sine oscillations. Therefore, the solution is also valid for a jumplike change in the dipole moment or the current during lightning.

Expressing f in terms of $2hQ$, we have from (527) and (532)

$$A_z = \frac{2h}{cr}\frac{\partial Q}{\partial t}, \tag{535}$$

$$\Phi = \frac{z}{r}\left(\frac{2hQ}{r^2} + \frac{1}{cr}\frac{\partial Q}{\partial t}\right). \tag{536}$$

The field strength can be readily calculated from the retarded potentials $\vec{A}$ and Φ in accordance with (521). We shall carry this out for the field components E_r and E_ϑ in a spherical system of coordinates r, ϑ, φ. We have

$$\left\{\begin{aligned}
A_r &= \frac{2h}{cr}\cos\vartheta\,\frac{\partial Q}{\partial t} \\[2mm]
A_\vartheta &= -\frac{2h}{cr}\sin\vartheta\,\frac{\partial Q}{\partial t} \\[2mm]
\mathrm{grad}_r\,\Phi &= \frac{\partial\Phi}{\partial r} \\[2mm]
\mathrm{grad}_\vartheta\,\Phi &= \frac{1}{r}\frac{\partial\Phi}{\partial\vartheta} \\[2mm]
\frac{z}{r} &= \cos\vartheta\,.
\end{aligned}\right. \tag{537}$$

The vector equation (521) yields the following two equations for the two field components

$$E_r = -\frac{1}{c}\frac{\partial A_r}{\partial t} - \mathrm{grad}_r\,\Phi\,, \tag{538}$$

$$E_\vartheta = -\frac{1}{c}\frac{\partial A_\vartheta}{\partial t} - \mathrm{grad}_\vartheta\,\Phi\,. \tag{539}$$

Remembering that (528) also holds for the "retarded charges" $Q_{(t-r/c)}$, we therefore obtain

$$\frac{\partial Q}{\partial t} = -c\frac{\partial Q}{\partial r},$$

and substituting (537) into (538) and (539), we finally have

$$E_r = 2\cos\vartheta\left(\frac{2hQ}{r^3} + \frac{2h}{cr^2}\frac{\partial^2 Q}{\partial t^2}\right), \tag{540}$$

$$E_\vartheta = \sin\vartheta\left(\frac{2hQ}{r^3} + \frac{2h}{cr^2}\frac{\partial Q}{\partial t} + \frac{2h}{c^2 r}\frac{\partial^2 Q}{\partial t^2}\right). \tag{541}$$

In the plane perpendicular to the dipole axis ($\vartheta = \pi/2$) we have $\cos\vartheta = 0$ and $\sin\vartheta = 1$. Here, the r-component E_r of the field strength vanishes, while E_ϑ simplifies to the well-known field equation

$$E_\vartheta = \frac{2hQ}{r^3} + \frac{2h}{cr^2}\frac{\partial Q}{\partial t} + \frac{2h}{c^2 r}\frac{\partial^2 Q}{\partial t^2}. \tag{542}$$

Supplement: For completeness, the above expression derived by Kasemir for the electrical, magnetic, and electromagnetic "disturbances" produced by lightning is also given below in other coordinate systems and different forms in accordance with the choice of the fundamental constants.
Electrostatic CGS-system

$$E = 1\cdot\frac{2h}{r^3}\cdot Q + \frac{1}{c}\cdot\frac{2h}{r^2}\cdot\frac{dQ}{dt} + \frac{1}{c^2}\cdot\frac{2h}{r}\cdot\frac{d^2 Q}{dt^2}. \tag{543}$$

Electromagnetic CGS-system

$$E = c^2\cdot\frac{2h}{r^3}\cdot Q + c\cdot\frac{2h}{r^2}\cdot\frac{dQ}{dt} + 1\cdot\frac{2h}{r}\cdot\frac{d^2 Q}{dt^2}. \tag{544}$$

International System
a) when only μ_0 occurs:

$$E = \frac{\mu_0 c^2}{4\pi}\cdot\frac{2h}{r^3}\cdot Q + \frac{\mu_0 c}{4\pi}\cdot\frac{2h}{r^2}\cdot\frac{dQ}{dt} + \frac{\mu_0}{4\pi}\cdot\frac{2h}{r}\cdot\frac{d^2 Q}{dt^2}, \tag{545}$$

b) when only ε_0 occurs:

$$E = \frac{1}{4\pi\varepsilon_0}\cdot\frac{2h}{r^3}\cdot Q + \frac{1}{4\pi\varepsilon_0 c}\cdot\frac{2h}{r^2}\cdot\frac{dQ}{dt} + \frac{1}{4\pi\varepsilon_0 c^2}\cdot\frac{2h}{r}\cdot\frac{d^2 Q}{dt^2}, \tag{546}$$

c) when besides ε_0 and μ_0 there is the characteristic impedance of the vacuum:

$$E = \frac{1}{4\pi\varepsilon_0}\cdot\frac{2h}{r^3}\cdot Q + \frac{\mu_0 c}{4\pi}\cdot\frac{2h}{r^2}\cdot\frac{dQ}{dt} + \frac{1}{4\pi c}\sqrt{\frac{\mu_0}{\varepsilon_0}}\cdot\frac{2h}{r}\cdot\frac{d^2 Q}{dt^2}, \tag{547}$$

where c — light speed; ε_0 — dielectric constant of vacuum or capacitivity; μ_0 — permeability of vacuum; $\sqrt{\mu_0/\varepsilon_0}$ — characteristic impedance of vacuum (approx. 377 ohm).
The meaning of the remaining symbols is given in Section 68a.

The electric field vector E defined here is normal to the propagation direction and parallel to the dipole axis. The magnetic field vector H is normal to E as well as to the propagation direction, and its magnitude in

434

the different measurement systems is given, in CGS, by:

$$H = \frac{2h}{r^2} \cdot \frac{dQ}{dt} + \frac{1}{c} \cdot \frac{2h}{r} \cdot \frac{d^2Q}{dt^2},$$

(548)

or in the International System

$$H = \frac{1}{4\pi} \cdot \frac{2h}{r^2} \cdot \frac{dQ}{dt} + \frac{1}{4\pi c} \cdot \frac{2h}{r} \cdot \frac{d^2Q}{dt^2}.$$

(549)

For large r, we have in the International System:

$$\frac{E}{H} = \sqrt{\mu_0/\varepsilon_0} = {\sim}377 \text{ ohm}$$

(550)

The Electric Field of Lightning Discharges as an
Observational Tool in Thunderstorm Research

Heinz W. Kasemir

U. S. Army Electronics Research and Development Laboratories
Fort Monmouth, New Jersey

ABSTRACT

The stationary electric field inside a thunderstorm, which is generated
by the movement of charged precipitation particles, causes the lightning discharges. The distribution of the stationary field in the cloud is reflected
in the electric field of the lightning stroke. Different types of lightning
strokes (cloud or ground discharges) are connected with different kinds of
field distributions in the thundercloud. This leads to certain conclusions
of the precipitation pattern and the development stage of the thunderstorm.

Electric field and spherics records of different types of thunderstorms
are discussed, and useful parameters in the field records are pointed out to
determine characteristic electric features of the thunderstorm.

- - -

1. Charge Distribution in the Thundercloud

Investigations of the electrification of precipitation started as early
as 1885.[1] Since then, a steady stream of theories[2-9] on the charge generation in thunderstorms has been proposed and discussed. None of them explains
all of the different observations on thunderstorm electricity; and objections
have been raised from different points of view such as polarity, charge-
generating power, efficiency by liquid as well as solid precipitation, laboratory experiments, and many others. Therefore the theory of a lightning discharge has to be based on probable assumptions of the electric field and
charge distribution in the thunderstorm and has to exhibit a large flexibility
to facilitate adaptation to the various thunderstorm models.

There is general agreement that the thunderstorm has a positive polarity,
meaning that the main positive charge is located in the top, and the main
negative charge in the bottom of the thundercloud. Below the negative charge
exists a smaller positive-charge pocket. This tripolar charge distribution
is clearly depicted in the thunderstorm model of Simpson and Robinson[10] (Fig.
1a). Other models have been proposed by Schonland,[11] Wickmann,[12] and Vonnegut,[9]
as shown in Figs. 1b, c, and d, respectively. The first three models (a, b,
and c) are based on charge separation by precipitation, and the last model, d,
on charge separation by convective air currents. The positive polarity of the
storm is common to all of them, but they differ from each other in the shape,
size, and arrangement of the individual charge centers.

The amount of charge for the positive and negative main charge center is
given by Simpson and Robinson as +24 coulombs and -20 coulombs. Schonland
assumes the values of ±40 coulombs. The author[13] calculated a positive charge

of +60 coulombs and a negative charge of -340 coulombs for the main charge centers. These values resulted from the requirement that the average thunderstorms should deliver about one ampere conduction current to the ionosphere. This brief summary shows the wide variety in the arrangement and in the amount of thunderstorm charges postulated for the different models. We will see in the next chapter what influence this will have on the lightning-discharge theory.

2. The Cloud Discharge

If the charge accumulation in the thundercloud reaches such a value that the generated electric field exceeds the breakdown value, a lightning discharge will occur. It is not necessary that the breakdown value of the field be reached over the entire distance of the flash. It is sufficient when this value is surpassed in a small area, which we may call the starting point of the flash. We can obtain more detailed information from a graph of the potential function ϕ of the cloud charges as shown in Fig. 2. Here we place two equally large spheres in the thunderstorm, one above the other, which are filled respectively with positive or negative charge of constant charge density. The potential function ϕ (abscissa) through the center line of the storm as a function of altitude h (ordinate) is shown in Fig. 2b. The potential function ϕ is calculated according to the electrostatic theory. As pointed out by the author,[13] the theory of current flow has to be used for a quantitative calculation of the potential and field distribution of the thunderstorm. But for our qualitative discussion here, the electrostatically derived potential function may serve as well.

The slope of the curve $d\phi/dh$ in Fig. 2b, with respect to the h coordinate, is proportional to the field strength. The maximum field strength F_{max} is at the midpoint between the spheres. This will be the starting point of the flash. Near the center points of the spheres, the field strength F passes through the value zero (marked $F = 0$ in Fig. 2b) and reverses its sign on the way further up or down. A lightning flash, which starts at F_{max} and grows upwards and downwards, will find an adverse field for further growth after passing through the centers of the space charges. The lightning will stop growing when the field strength at its tips no longer exceeds the breakdown value F_b. The corresponding potential value of ϕ_b (Fig. 2b), defined as the difference between the lightning potential and that of the potential function[14] ϕ at the location of the lightning tips, is given by the approximate formula

$$\phi_b = 2.5 \text{ m} \cdot F_b. \tag{1}$$

We may draw an important conclusion from Figs. 2a and b. The lightning flash, represented by the heavy vertical line in Fig. 2b, will never reach the ground. The field strength at the tip will be less than the breakdown value when the tip reaches the base of the cloud. The lightning will remain a cloud discharge. A ground discharge cannot be generated by a charge distribution given in Fig. 2a. To produce ground discharges, the lower positive space charge has to be present, as we will see later. Therefore if we record thunderstorms with cloud discharges, only the power positive space charge is missing.

From the electrostatic point of view, the lightning channel can be considered as a conductor which is exposed to the electric field of the cloud charges. Therefore we have negative influence charges at its upper half and positive influence charges on its lower half. The polarity of a cloud discharge is negative and the net charge is zero. The charge per unit length q on the lightning channel is shown in Fig. 2c (dashed line). It is the mirror image of the potential function ϕ with regard to a vertical line representing the lightning channel. An approximate formula[14] connecting q and ϕ is given by

$$q = -c\,\phi. \qquad (2)$$

The proportionality factor c can be interpreted as the capacity per unit length of the lightning channel. It has the approximate value of 25×10^{-12} F/m.[14]

It may be worthwhile to point out some differences between the commonly used concept of the lightning discharge and that developed here. In general, it is assumed that the lightning flash starts in the center of, say, the negative space charge. By the development of a large network of fine streamers, the lightning collects the cloud discharges and transports them along its channel, either to the ground (ground discharge) or to another space-charge center of opposite polarity (cloud discharge). The lightning stops if the supply of cloud charge is exhausted. After the completion of the flash, that part of the cloud penetrated by it, as well as the lightning channel itself, is discharged.

In the concept proposed here, the lightning starts in between two charge centers, where the electric field has a maximum value. (In the center of the space charge, the field would be approximately zero.) The direction of channel growth is up as well as down under the influence of the field. The lightning does not collect cloud charges, and therefore it is not necessary to introduce a hypothetical network of fine streamers. There is only a charge separation in the highly ionized lightning channel itself to build up the necessary influence charges. The concept is that of an uncharged conducting rod exposed to an electric field. The lightning flash stops, not because the cloud charges are exhausted, but because the field strength at its tips does not exceed the breakdown value. After the lightning stops, the lightning channel is filled up with influence charges. Each half of the channel carries influence charges of opposite polarity to the surrounding cloud charges.

The net charge of the whole channel is zero. After the cloud flash is completed, a slow neutralization process begins between the influence charges of the lightning channel and the surrounding cloud charge of opposite polarity. Only this process should rightly be called the discharge of the cloud by the lightning.

3. The Ground Discharge

If we add the lower positive charge pocket to the positive and negative main charge centers, we obtain the charge distribution of the Simpson-Robinson thunderstorm model. This is shown in Fig. 3a, side by side with the corresponding potential function ϕ in Fig. 3b. We see that two areas with strong electric fields exist, marked sc and sg in Fig. 2b. The upper one, sc, is the starting point for cloud discharges, and corresponds to the point

F_{max} in Fig. 2b. The lower one, sg, is due to the presence of the positive pocket charge and is the starting point for ground discharges. In a similar manner as the cloud flash, the ground flash grows to both sides until its lower end reaches the ground. This will happen only if the field potential F_b strength at the lower tip remains greater than or equal to the breakdown, or $\phi_1 - \phi_0$ is always greater than ϕ_B (see Fig. 3b), where ϕ_1 is the potential of the lightning channel, and ϕ_0 the potential of the ground.

If the breakdown field could not be maintained on the downward travel of the lower tip, the lightning would stop growing and end up as an air discharge. The equation

$$\phi_1 - \phi_0 = \phi_b = 2.5 \text{ m } F_b = -7.5 \times 10^6 \text{ V/m} \tag{3}$$

gives the minimum potential difference against ground, which the leader of a ground flash must have to reach the ground. In Eq. 3 is set $\phi_1 = 0$: $F_b = -3 \times 10^6$ V/m. It is easy now to imagine the conditions for the charge distribution of a thunderstorm which produces predominantly ground discharges and few or no cloud discharges. They are as follows: 1) The upper positive charge center should be weak; 2) the lower negative charge center should be strong; 3) the lower positive pocket charge should exist. Figure 3c shows the potential function of such a charge distribution.

If the lower end of the lightning channel touches the ground, the lightning will assume ground potential ϕ_0. This is accomplished during the phase of the main or return stroke. With respect to the h - ϕ graph in Fig. 3b, it means that the lightning channel (heavy vertical line) shifts to the right from the ϕ_1 to the ϕ_0 value. The new position and the new charge distribution (broken line) are shown in Fig. 3d. We see that after the completion of the main stroke the lightning channel carries, on its upper part, a large amount of positive influence charge, and on its lower part a smaller amount of negative influence charge. The net charge on the channel is no longer zero, but positive. This means that during the main stroke the channel has acquired positive charge from the ground or that negative charge has been conducted into the ground. The amount of this charge Q_m is represented in Fig. 3d by the rectangular hatched area. It can easily be calculated from the capacity c of the channel and the potential difference $\phi_1 - \phi_0$ before and after the main stroke.

$$Q_m = c \ (\phi_1 - \phi_0). \tag{4}$$

If we assume that the length of the channel is 4 km, that the capacity per unit length has the previous value of 25×10^{-12} F/m, and $\phi_1 - \phi_0$ has the minimum value of -7.5×10^6 V, we obtain for the minimum charge brought to earth by the first main stroke $\phi_{min} = -0.75$ coulomb. This value may have to be doubled or tripled according to the number of branches of the considered lightning flash, as each branch would increase the capacity. But even so, the value of Q_{min} has the right order of magnitude. This is remarkable insofar as Q_{min} is based only on the value of the breakdown field strength at the ground and the length of the lightning channel, but does not imply any data from the thunderstorm. For successive main strokes in a multiple flash, the value of Q_{min} would be lower because the breakdown field in the pre-ionized channel is lower, and only the length added to the channel by the considered multiple stroke to the channel has to be taken into account.

Each successive leader main stroke sequence will extend the channel higher up or sideways into the cloud until the whole negative charged area of the cloud is penetrated by the flash. Therefore, ground discharges with a large number of multiple strokes indicate a large area of the negative cloud charge. Flashes with one main stroke indicate a small region of the negative charge center. It is interesting to review the different thunderstorm models in Fig. 1 from this point of view. Schonland's and Vonnegut's models, Fig. 1b and d, is apt to give high-order multiple strokes. Simpson-Robinson's and Wichmann's model, Fig. 1a and c, will generate ground flashes with only one or a low-order of multiple strokes.

4. Relation between Properties of the Electric Field of a Lightning Discharge and Thunderstorm Parameters

The large number of charge-generating theories and the failure to explain all the electrical features of different types of thunderstorms by one of them suggest the idea that several kinds of charging mechanism are at work in different parts of the thunderstorm. Kuttner[15] observed on the mountain observatory at the Zugspitze that strong electric fields are always associated with graupel precipitation. This correlation was so pronounced that it led him to the introduction of the graupel dipole, with a negative polarity, which produces part of the negative cloud charge and the positive space charge pocket at the base of the cloud. Kuttner attributed the upper positive charge and the other part of the negative charge in the higher level of the cloud to the snow dipole.

The author[13] pointed out that it is doubtful that any of the three cloud charges can be identified with the charge riding on the precipitation particles. The precipitation charge lacks the ability to accumulate and concentrate in a certain cloud volume. The cloud charges can be better interpreted as the charges left behind, while the precipitation carries away the charge of opposite sign. This leads to the conclusion that either two sign reversals in the precipitation-charging mechanism are required to account for a tripolar thundercloud or that at least two, if not three, different charge mechanisms are present which may work on different precipitation particles. The second alternative would then agree with Kuttner's snow and graupel dipole. With respect to the lightning discharge, this would mean that cloud discharges reflect the intensity of the charging mechanism in the upper part, and ground discharges the intensity of the charging mechanism in the lower part of the cloud. A record of the polarity of the flashes at the ground (negative for cloud and positive for ground discharges) would give an easy means to follow the development of the respective charge centers and their corresponding precipitation particles during the life history of the storm.

With the strong tie between the lower positive space charge pocket and ground discharges, the following relation may be stated as a working hypothesis. Thunderstorms with ground discharges contain precipitation in the form of graupel; those without ground discharges do not. It is not necessary that the graupel pellet reaches the ground still as graupel; it may be melted to a raindrop on the way down. But its existence at the base of the cloud is required. These deductions could easily be checked by investigation of the warm thunderclouds (storms where the top does not reach the zero-degree altitude level) of the tropical regions. Because these storms do not contain graupel as precipitation particles, they should not be able to produce ground discharges.

As shown earlier, the positive polarity of the ground discharge and the negative polarity of the cloud discharge follow from the positive polarity of the thundercloud charges. Falconer and Schaefer[16] and Vonnegut and Moore[17] report measurement of strong negative electric fields, indicating a negative charge overhead, from the cirrus deck accompanying the Worcester tornado. If this implies that a tornado is topped with a negative charged anvil above the upper positive charge center, cloud flashes with positive as well as negative polarity would be produced. This would make a tornado easily identifiable from the polarity records of lightning flashes.

A characteristic feature for the activity of a thunderstorm would be the number of lightning discharges produced in a given time interval. Brook and Kitagawa[18] defined as activity index A the frequency of lightning occurrence in a period of five minutes. The time interval of lightning discharges in weak, intermediate, and heavy storms is about 5 minutes, 30 seconds, and 3 seconds, respectively. An activity index of 1, 10, and 100 would then result. One could develop this method further by differentiating between the number of cloud and ground flashes. The cloud flash index A_c would then indicate the activity of the upper positive charge center, and the ground flash index A_g that of the lower positive charge center. Jones[19] used the number of spherics per second emitted from a tornado for his tornado identification and warning system. From his data we may infer that a tornado produces about six lightnings per second. This would result in a tornado activity index of 1800.

References

1. Elster, J., Geitel, H.: "Uber die Elektrizitats-entwicklung bei der Regenbildung," Ann. Phys. n. Chem. 25, 121-131, 1885.

2. Sohnke, W.: "Der Urosprung der Gewitter-elektrizitat und der gewohnlichen Elektrizitat der Atmosphare," Ann. d. Phys. 28, 1886.

3. Lenard, P.: "Uber die Elektrizitat der Wasserfalle," Wied. Ann. d. Phys. 46, 584-636, 1892.

4. Simpson, G. C.: "On the Electricity of Rain and Its Origin in Thunderstorms," Phil. Trans. Roy. Soc., A 209, 397-413, 1909.

5. Wilson, C. T. R.: "Some Thundercloud Problems," F. Franklin Inst. 208, 1-12, 1929.

6. Findeisen, W.: "Uber die Entstehung der Gewitterelektrizitat," Meteor. Z., 210-215, 1940.

7. Frenkel, J.: "Influence of Water Drops on the Ionization and Electrification of Air," J. Phys., Moscow 10, 151-158, 1946.

8. Workman, E. J., Reynolds, S. E.: "A Suggested Mechanism for the Generation of Thunderstorm Electricity," Phys. Rev., 74, 709, 1948.

9. Vonnegut, B.: "Possible Mechanism for the Formation of Thunderstorm Electricity," Proc. Conf. Atmos. Elec., 1954, Geophys. Res. Paper No. 42, 169-181, 1955.

10. Simpson, G. C., Robinson, G. D.: "The Distribution of Electricity in Thunderclouds," Proc. Roy. Soc., A 177, 281-329, 1940.

11. Schonland, B.: "Progressive Lightning IV. The Discharge Mechanism," Proc. Roy. Soc., A, 168, 455-469, 1938.

12. Wichmann, H.: "Grundprobleme der Physik des Gewitters," Wolfenbuetteler Verlagsanstalt, 1948.

13. Kasemir, H. W.: "The Thundercloud." Paper given at the Third International Conference on Atmospheric Electricity (in print).

14. Kasemir, H. W.: "The Lightning Discharge," USAELRDL Technical Report 2401, USAELRDL, Fort Monmouth, New Jersey, Dec 63.

15. Kuttner, J.: "The Electrical and Meteorological Conditions Inside Thunderclouds," J. of Met., 7, 322-332, 1950.

16. Falconer, R., Schaefer, V.: "Cloud and Atmospheric Electrical Observations of the Formative Stages of the Worcester, Massachusetts, Tornado," Bull. Amer. Met. Soc., 35, 9, 1954.

17. Vonnegut, B., Moore, C.: "Giant Electrical Storms," Naval Research Contract NONR-1684(00), 1958.

18. Brook, M., Kitagawa, N.: "Some Aspects of Lightning Activity and Related Meteorological Conditions," J. Geophys. Res., 65, 4, 1203-1210, 1960.

19. Jones, H.: "Research on Tornado Identification," Signal Corps Research Contract DA 36-039 SC-64493, 1956.

U. S. DEPARTMENT OF COMMERCE
Maurice H. Stans, Secretary

ENVIRONMENTAL SCIENCE SERVICES ADMINISTRATION
Robert M. White, Administrator
RESEARCH LABORATORIES
Wilmot N. Hess, Director

ESSA TECHNICAL REPORT ERL 143-APCL 11

Lightning Hazard to Rockets During Launch I

HEINZ W. KASEMIR

NASA Contract No. 906 - 61 - 34 - 00 - 98 - 0 - 00 - 4 - 25 J - 0 - XA00 - 000

ATMOSPHERIC PHYSICS AND CHEMISTRY LABORATORY
BOULDER, COLORADO
December 1969

LIGHTNING HAZARD TO ROCKETS DURING LAUNCH I.

Heinz W. Kasemir

The launch of a rocket through thunder, or shower, clouds carries with it a definite lightning hazard - not so much because the rocket will be hit accidentally by natural lightning but because the rocket will trigger off its own lightning as soon as it enters strong electric fields inside the cloud. The larger the rocket the higher the field concentration factor is at the ends of the rocket. The chance that lightning be ignited by the intruding rocket, even in clouds that do not contain fields strong enough to produce natural lightning discharges, increases in the same proportion. It is the purpose of this report to discuss some of the fundamental problems involved in such a situation.

The nature of charge generation and its distribution in different types of clouds is still an open question. For lack of a more accurate and detailed picture, the model of the tripolar thunderstorm that is generally accepted and presented in the literature has to be considered our working hypothesis. Nevertheless, the conclusions drawn from this model can be readily adjusted to any other model, if actual measurement in the cloud suggests a modified version.

The charge distribution and field lines of a tripolar thunder-storm model are given in a qualitative way in figure 1a. The positive and negative main space charges are located in the upper and lower part of the storm. Below the main negative space charge at the base of the storm is another positive space charge pocket. The field at the ground of this arrangement of charges is shown in figure 1b. If the storm would move over the recording station at the ground, the trace of the P. G. (potential gradient) or field recorder would appear similar to that of figure 1b. Only a small fraction of the field lines reach the ground; most of the field lines terminate inside the storm. Note also that at the points marked zero in figure 1b the field changes polarity and goes through zero. This is by no means an indication that the field in the cloud is also zero. To deduce the field in the cloud from ground measurements is an extremely difficult task because the fraction of field lines reaching the ground depends on a number of factors usually not known; their existance is often not realized. However the electric field inside the cloud is the determining agent for the generation of natural lightning discharges, as well as arti-ficially triggered off lightning discharges by rocket penetration, and should be the key parameter of a lightning warning system.

The field at the ground will be weak if (a) positive and negative space charges in the cloud are of equal amount, (b) their mutual distance is small, (c) the distance to ground is large,

445

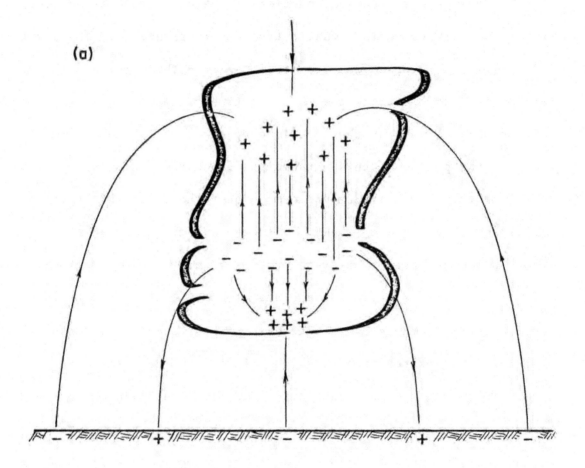

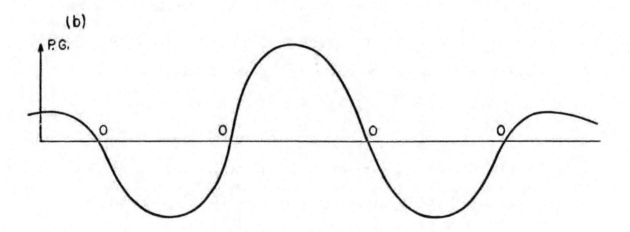

Figure 1.　(a) Tripolar thunderstorm model

　　　　　　(b) Potential gradient at the ground below thunderstorm

(d) the space charges are spread more horizontally than vertically,
(e) the conductivity in the cloud is much smaller than in free air,
which results in a screening surface charge at the cloud-air
boundary, (f) the conductivity of the free air increases with
altitude, which results in a screening space charge in free air,
and (g) a screening space charge blanket exists, resulting from
corona discharge at the ground.

At present the only safe means to determine the field inside
the cloud is to measure it from an airplane equipped with instruments
recording all three components of the electric field vector and
capable of cloud penetration. Even this means requires expert
knowledge because the airplane has to penetrate the cloud at the
right altitude to measure the maximum fields. In the middle of the
positive or negative space charge regions, for instance, the field
will be close to zero. This can be seen from figure 2a and b, which
represent the potential function ϕ and electric field E versus
altitude h through the center of the storm. There are two critical
altitudes with maximum field values marked by the letters g and
c in figure 2a and b; c is the birth place of cloud discharges
and g that of ground discharges. These maximum fields always
occur between two space charges of opposite polarity. We have
now to determine the threshold value of the electric field necessary
to ignite a lightning discharge and how much this threshold value
is lowered by the intruding rocket.

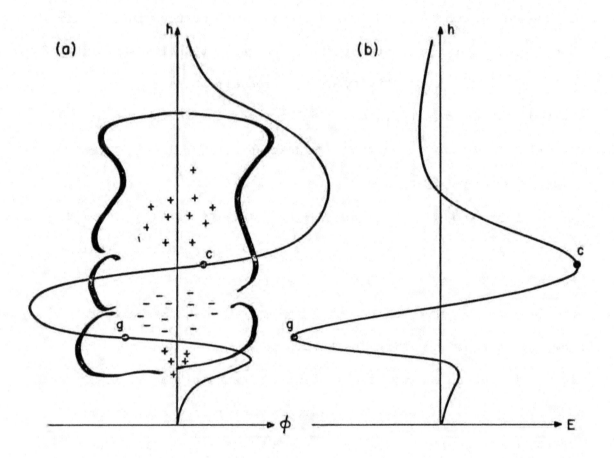

Figure 2. (a) Potential function versus altitude h through the
              middle of the storm

          (b) Electric field E versus altitude h

The break-down field in air under normal pressure is about 3×10^6 V/m. At the altitude of lightning origin the air pressure is only about half of the value at sea level, and the breakdown field may drop to 1.5×10^6 V/m. Furthermore, we have to consider the field concentration caused by the introduction of a conductive body. A precipitation particle, for instance a raindrop, has the field concentration factor of 3 at its highest and lowest points, if we assume that we can treat the raindrop as a conductive sphere. This means that the field around a raindrop falling through an external field of 500 kV/m would just reach the breakdown value of 1.5×10^6 V/m at its top and bottom. Small filaments of corona discharge will emerge from these two points, adding conductive protuberances to the precipitation particle. At the tips of these protuberances still higher field concentration factors will exist, causing further growth accompanied by increasing fields. Once started, this process will rapidly convert the corona filaments into a lightning discharge. There are other factors involved in igniting a lightning discharge that are not outlined in the above oversimplified description. For instance, raindrops are known to deform in strong electric fields into a spheroidal shape that would result in a higher field concentration factor, or two precipitation particles can be close enough together so that they are linked by their corona filaments forming an elongated body. For igniting a lightning discharge, it will also be necessary that the breakdown field value is surpassed over a certain distance ahead of the growing corona filament. Otherwise the corona discharge may remain just that or even be quenched by its own space charge production. This effect can be offset by a sudden field increase caused by turbulent motion of the air, which brings pockets of opposite space charge close together. These are finer details, mostly of a speculative nature, and are not within the scope of this report.

We are interested in a rough estimate for the electric field intensity necessary to produce lightning discharges and will accept for the time being the value 500 kV/m as a maximum. It may be mentioned in this respect that the author recorded from an airplane a field strength of 290 kV/m right under the base of a cumulonimbus for about 30 min and that this cloud did not produce a single lightning discharge. This means that the lightning igniting field value has to be higher than 290 kV/m and the right value will lie somewhere between 300 and 500 kV/m.

The upper end of the cloud discharge, which penetrates into the positive space charge region of the cloud, accumulates negative induction charge. Also, the lower end of the cloud discharge penetrates into the negative space charge region of the cloud and accumulates positive induction charge. These induction charges are formed by electrons flowing in the highly ionized channel from its lower into its upper part under the influence of the external electric field. This flow of electrons results in a surplus of positive ions in the lower part and builds up a negative surplus charge in the upper part of the channel. It is not necessary or even physically plausible that the lightning has to collect cloud charges by a wide spread streamer process, funnel these charges into the channel, and transport them from one charge center of the cloud to another or to ground. We are dealing here only with a charge separation inside the lightning channel. The cloud charges remain practically stationary during the short lifetime of the lightning discharge. They generate only the electric field that furnishes energy for the lightning. There is much confusion and misinterpretation on this point in the literature.

The amount and distribution of these influence charges are determined by the external field of the cloud and the length and shape of the lightning channel. The current that flows in the channel of a cloud discharge to build up these induction charges is in the order of 100 ampere. This would also be the current that flows through the rocket, if it, instead of a precipitation particle, triggers the cloud discharge. We will see later that a large rocket with a much larger field concentration factor can ignite a cloud discharge in much weaker fields. Correspondingly the induction charges, as well as the current producing them, would be less.

We turn now to the discussion of a ground discharge that would start at point g in figure 2a or b. The leader stroke, which is the first part of the ground discharge and occurs before it makes contact with the ground, will form in a manner similar to that of the cloud discharge. The difference is that the upper part of the

451

leader stroke will grow into the negative cloud charge region accumulating positive induction charge, and the earth bound part growing through the lower positive space charge pocket will carry negative induction charge. Currents in the leader stroke of a ground discharge will be of the same order as those in a cloud discharge but will flow in the reverse direction. The situation however changes drastically as soon as the leader stroke touches ground. The lightning channel has the large negative potential of its origin in the cloud, which is in the order of many millions of volts against ground, and if ground contact is made the lightning is recharged to ground potential during the return stroke. In this phase lies the greatest danger to the rocket if it happens to trigger off a lightning discharge at point g, that grows into a ground discharge. Being in the middle of the lightning channel the rocket will experience the strong current surge of 10000 A or more of the return stroke. The rocket, however, with its high field concentration factor may trigger a comparatively weak ground discharge, and the current may be smaller than that of the average natural return stroke.

For the calculation of the field concentration factor, we may substitute for the rocket a conducting spheroid of equivalent length and thickness. The formula for the field concentration at the ends of an uncharged spheroid in an electric field is not as well known as that for a sphere.

It is

$$A = \frac{c^2}{b^2\left[\dfrac{a}{2c} \ \ln \ \dfrac{a + c}{a - c} \ -1\right]} \quad , \quad\quad (1)$$

where

a = long axis of the spheroid,

b = short axis of the spheroid,

$c = (a^2 - b^2)^{1/2}$ = excentricity,

E = electric field at the end points,

F = external field of the cloud,

$A = \dfrac{E}{F}$ = field concentration factor.

The only uncertain parameter in equation (1) is the length of the rocket, because it is not known to what length the exhaust flames can be considered a conductor. If we assume for numerical evaluation that the thickness of the rocket is 2b = 10 m = 33 feet, and the length is 2a = 100m = 330 feet, the field concentration factor is about 50. If we add another 100 m to the length of the rocket to include the exhaust flames it becomes 150. This would mean that cloud fields as low as 10 to 30 kV/m have to be considered potentially dangerous. Compared with the lightning igniting fields of about 300 to 500 kV/m, one has to expect that practically every non-thundering rain shower is capable of producing fields strong enough for rocket ignited lightning.

U. S. DEPARTMENT OF COMMERCE
Maurice H. Stans, Secretary

ENVIRONMENTAL SCIENCE SERVICES ADMINISTRATION
Robert M. White, Administrator
RESEARCH LABORATORIES
Wilmot N. Hess, Director

ESSA TECHNICAL REPORT ERL 144-APCL 12

Lightning Hazard to Rockets During Launch II

HEINZ W. KASEMIR

NASA Contract No. 906-61-34-00-98-0-00-4-25 J-O-XA00-000

ATMOSPHERIC PHYSICS AND CHEMISTRY LABORATORY
BOULDER, COLORADO
January 1970

For sale by the Superintendent of Documents, U. S. Government Printing Office, Washington, D. C. 20402
Price 25 cents

LIGHTNING HAZARD TO ROCKETS DURING LAUNCH II.

Heinz W. Kasemir

1. INTRODUCTION

In designing a lightning warning system for rocket launch
areas, one has to take into account that the greatest danger to
a rocket is not an accidental strike by natural lightning but
the high probability that a rocket will trigger its own lightning
discharge, if it enters areas inside a cloud with strong electric
fields (Kasemir, 1969). Fields in the order of 20 to 50 kV/m may
be enough to produce rocket triggered lightning discharges, and
fields of this magnitude can be found in a small cumulonimbus,
or rain shower, which does not contain natural lightning discharges.
This situation is aggrevated by the fact that the electric fields
are sustained inside the cloud for some time (one-half hour) after
the rain or lightning has stopped. On the other hand, there are
precipitating clouds, namely of the nimbostratus type, that do
not generate electric fields in excess of a few 100 V/m. It would
be perfectly safe to launch a rocket through such a cloud.

The consequence of these facts is that meteorological observation,
including radar, is not sufficient for a lightning warning system
at rocket launch areas - the key parameter of such a warning system

455

has to be the electric field inside the cloud. If this has been measured, one has to know in addition the threshold value of the field for the different types of rockets that will cause (a) corona discharge or (b) lightning discharge by rocket penetration.

The field distribution and its time history inside different types of clouds, the corona threshold field, and the lightning threshold field are the three particular research areas that have to be explored before an effective warning system can be designed or even before an investigation of the possibility of discharging the cloud can be successfully carried out.

It is the purpose of this report to discuss a program of investigation of the above mentioned research areas. Section 2 deals with the determination of electric fields inside a cloud and section 3 with artificially triggered lightnings by test rockets.

2. DETERMINATION OF ELECTRIC FIELDS INSIDE THE CLOUD

There are three ways to determine an electric field inside a cloud:

(a) By direct measurement from an airplane penetrating the cloud (Fitzgerald, 1967; Cobb and Holitza, 1968).

(b) By analysis of field measurement from an airplane outside the cloud (Kasemir, 1968; Schuman, 1969).

456

(c) By analysis of field measurement taken at ground level.

The direct measurement of the field inside the cloud from an airplane seems to be the best method; however there are two problems.

First, it is relatively easy to measure the two components of the electric field at right angles to the flight path (i.e., the up-down and the left-right components). It is quite difficult to measure the field component in direction of the flight (i.e., the nose-tail component) during precipitation because precipitation particles impinging on the sensitive segment of the field mill facing into the rain generate a strong electrostatic signal that saturates the nose-tail field component amplifier . It is very important that all three components of the electric field vector be measured with the same accuracy; therefore, one of the first items of a research program would be to improve existing equipment so that all three field components can be recorded, even in precipitation, with equal accuracy.

The second problem is that the maximum fields inside the cloud, which are the main interest, can only be located by a trial and error method. This problem can be solved by a second airplane, which stays outside the cloud and determines from its field measurement the charge distribution and the altitudes of maximum fields inside the cloud. The absolute value of the maximum field cannot be calculated from field measurements outside the cloud because the conductivity inside the cloud is not known, but the second plane could guide

the first plane to the areas of maximum field inside the cloud, and

from comparison of the measured and calculated field, the conductivity

inside the cloud can be obtained. If from a number of such test

flights it follows that the conductivity inside the cloud is fairly

constant, or at least has a constant ratio to the conductivity outside

the cloud at the same altitude, the field distribution inside the

whole cloud can be calculated. The obvious advantage of an analytical

calculation is that the complete charge and field distribution

inside the cloud can be obtained from one pass outside the cloud,

whereas the field measurement of one cloud penetration will give only

the field distribution along that one pass. There is no assurance

that such a pass would lead through the area of maximum field or

that there are not several areas with high fields in the cloud. The

detection of these areas with high fields is the main purpose of the

investigation.

Theoretically an analysis of the field measurement along

the pass inside the cloud should result in a complete charge and

field distribution throughout the whole cloud; however this requires

that all three field components be measured with the same accuracy

also in rain (which with the present state of the art cannot be done)

and that the airplane pass the different charge centers in distances

of similar magnitude. Otherwise the field of the closest charge center

becomes overwhelmingly large and blanks out the field of the other charge centers. A plane flying outside the cloud at midaltitude can fulfill this condition much easier than a plane penetrating the cloud. My conclusion, based on the discussion above, is that two airplanes are necessary to obtain the complete charge and field distribution in a cloud. The first airplane has to be capable of cloud penetration and should be equipped with instruments for recording the vertical and at least one horizontal component of the electric field. The second airplane remains outside the cloud at about a 20,000 ft. altitude and should be equipped to measure all three components of the electric field with about the same accuracy.

The third method mentioned earlier, field measurements taken at several ground stations at its best has several severe handicaps. The requirement that the stations should be at similar distances from the different charge centers can obviously not be fulfilled for the charges at the top and the base of the cloud. Consequently, a high demand of accuracy and dynamic range is put on ground instrumentation. Furthermore one charge center in the cloud has four parameters: namely, the three coordinates and the charge magnitude. Field measurements at four different ground stations are required to calculate these four parameters. If the cloud contains three charge centers, twelve ground stations would be required, and if two tripolar clouds are located in an area of the ground network, twenty-four stations would be necessary for an analysis. It is assumed that the conductivity in the cloud is known, that local disturbances like spacecharge pockets from exhaust fumes, from corona discharge of power lines, or a nearby surf are absent and that a sufficiently accurate solution of the twenty-four simultaneous equations is possible. All these assumptions are probably not fulfilled; however, if a flight program is carried out to determine the maximum fields in the different types of clouds, a comparison of the ground and flight results can be used to determine the benefit and the limitations of a ground network.

3. ARTIFICIALLY TRIGGERED LIGHTNINGS BY TEST ROCKETS

The necessary equipment to artificially trigger lightning by test rockets is:

(a) Two airplanes equipped with field mills to determine the maximum electric fields in the cloud.

(b) Test rockets equipped with corona discharge point and transmitter to ground.

(c) Lightning plotting system.

The preliminary tests should be carried out with small rockets driven by compressed air and equipped with parachutes, so that they can be launched from a mobile station and be recovered after descent, if they are not destroyed by the triggered lightning.

A corona discharge point would be mounted on the tip of the rocket and the onset of corona discharge transmitted to the ground station. Using Kasemir's (1969) equation 1, the external field to produce corona discharge can be calculated. The two airplanes will check the clouds for their maximum fields and if these surpass the corona onset field, the rocket is launched into the cloud. The main purpose of this test is to determine how much the corona breakdown field has to be surpassed in order to turn corona discharge into lightning discharge. Another objective of this task is to determine how long a cloud remains discharged by a triggered lightning.

It is well known that after a lightning discharge the electric field of a thunderstorm at the ground recovers in about 10 sec to its predischarge value. This fast recovery curve has been - and to some extent still is - a big mystery in the thunderstorm charging mechanism. A medium thunderstorm may produce one lightning discharge every 20 sec, which would indicate that the charging mechanism after a lightning stroke is accelerated to restore the destroyed charge in about 10 sec.

The questions that arise here are: will the charge in the shower cloud also be restored in about 10 sec after an artificially triggered lightning, or what is the recovery time in this case; how long a time interval has to elapse before a second lightning will be triggered by a second test rocket. Such an investigation will be useful in establishing the feasibility of discharging shower clouds by rockets.

Another variation of the test program will be to use test rockets of different sizes and to equip some of them with trailing wires of different lengths. This will prove the validity of equation 1 of Kasemir (1969). It will also explore the possibility of replacing larger rockets by smaller rockets trailing a wire.

Large rockets with the larger field concentration factor will trigger lightning discharges in much weaker fields than small rockets.

If a cloud has to be tested for a launch of a large rocket the test rocket should be of the same size. If this is economically not feasible, the test rocket may be of smaller size with a trailing wire making up for the length deficiency. It is to be expected that such a test rocket will go into corona discharge at the same field as the large rocket but will need much larger fields to trigger lightning discharges.

To test whether lightning is triggered by the test rocket, a lightning plotting system has to be developed which will consist of three ground stations located at the corners of an equilateral triangle with a side length of about 15 miles. The direction of the lightning stroke will be obtained with crossed loop antenna at each station and the position by triangulation.

The tests to trigger lightning by penetrating rockets have to be carried out during disturbed weather conditions where the presence of other thunderstorms in the test area is quite likely. The lightning plotting system provides the means to differentiate between triggered lightning and that produced by surrounding thunderstorms.

4. REFERENCES

Cobb, W. E., F. J. Holitza. (1968), A note on lightning strikes to aircraft, Monthly Weather Rev. 9, No 11, 807-808.

Fitzgerald, D. R. (1967), Probable aircraft "triggering" of lightning in certain thunderstorms, Monthly Weather Rev. 95, No. 12, 835-842.

Kasemir, H. W. (1968), Charge distribution in the lower part of the thunderstorm, Proc. of Intern. Conf. on Cloud Phys., Toronto, Canada, 668-672.

Kasemir, H. W. (1969), Lightning hazard to rockets during launch I, ESSA Tech. Rept. ERL 143-APCL 11 (U.S. Government Printing Office, Washington, D. C.)

Schuman, E. (1969), The inference of the charge distribution in thunderstorms from airborne measurements of the electric field, J. Appl. Met. 8, No. 5, 820-824.

U.S. DEPARTMENT OF COMMERCE
National Oceanic and Atmospheric Administration
Environmental Research Laboratories

NOAA Technical Memorandum ERL APCL-12

BASIC THEORY AND PILOT EXPERIMENTS
TO THE PROBLEM OF TRIGGERING
LIGHTNING DISCHARGES BY ROCKETS

H. W. Kasemir

Atmospheric Physics and Chemistry Laboratory
Boulder, Colorado
April 1971

TABLE OF CONTENTS

ABSTRACT

The objective of this contract was to determine the feasibility of discharging electrified clouds with triggered lightning. As a first step towards this goal, the conditions have been calculated under which lightning may be triggered by different types of rockets. For instance the breakdown field value is reached at the tip of the Apollo-Saturn V rocket in a cloud field of about 10 kV/m, whereas a cloud field of about 125 kV/m is necessary to produce breakdown field at the folding fin aircraft rocket Mighty Mouse tipped with a trigger sphere of about 3.5 cm radius. The field concentration factor η at the tip of the rocket which augments the external cloud field to the field at the rocket tip is a function of the ratio of the length of the rocket a to the radius of curvature at the tip ρ. In the case of the Apollo rocket, $a/\rho = 1000$ and in the case of the Mighty Mouse rocket, $a/\rho = 34$, with $a = 120$ cm and $\rho = 3.5$ cm. To simulate the Apollo by the Mighty Mouse rocket, it is necessary either to increase the length a or decrease the radius of curvature ρ of the Mighty Mouse rocket in such a way that $a/\rho = 1000$. This leads to $a = 35$ m or $\rho = 1.2$ mm. It is shown in section 1.4 that the length a may be further reduced to about 26 m. The radius of curvature ρ determines the type of discharge. For $\rho > 2$ cm the discharge will be in the form of a spark and for $\rho < 2$ cm in the form of corona discharge. The corona current is a function of the electric field strength. Using the corona current on a sharp needle as a field indicator, an instrument has been designed (measuring payboard) which can be attached to the Mighty Mouse rocket [L. H. Ruhnke, 1971]. This instrument measures and transmits to ground the electric field values along the rocket path.

Abbreviations and Symbols

a [m] = long axis of spheroid

b [m] = short axis of spheroid

c [m] = eccentricity of spheroid

ρ [m] = radius of curvature at the tip of the spheroid

u, v, φ [m, m, $\}$] = spheroidal coordinates

R, z, φ [m, m, $\}$] = cylindrical coordinates

h_i (i = u, v, φ) = reciprocals of the root of the metric
                     coefficients g_{ii}.

Φ [V] = potential function

E [V/m] = electric field in general

E_n [V/m] = electric field component in the direction of the
           spheroidal coordinate n

E_a [V/m] = electric field at the surface of the spheroid n = a

E_T [V/m] = electric field at the tip of the spheroid

E_b [V/m] = breakdown field value

F [V/m] = external field (cloud field)

$n = E_T/F$ = field concentration factor at the tip of the
           spheroid

Q [A sec = C] = charge of the spheroid

p [C/m^2] = surface charge density

q [C/m] = surface charge per unit length

ε [C/Vm] = dielectric constant

BASIC THEORY AND PILOT EXPERIMENTS TO THE PROBLEM OF TRIGGERING LIGHTNING DISCHARGES BY ROCKETS

H. W. Kasemir

1. ELECTROSTATIC THEORY OF THE CHARGED ROCKET PENETRATING A THUNDERCLOUD

In the following theoretical calculation, the rocket is represented by a slim spheroid with the long axis a, the small axis b, and the eccentricity $c = (a^2-b^2)^{\frac{1}{2}}$. The problem based on electrostatic theory is solved in three steps. First, the charged rocket with the external field being zero; second, the uncharged rocket in the external field F; and third, the charged rocket in the external field F. The purpose of the calculation is to determine the field at the tip of the rocket. The highest field concentration which will cause either corona-, spark-, or lightning-discharge exists at this point.

1.1 The Charged Rocket

The calculation is performed in spheroidal coordinates u, v, ϕ, which have the following relations with the cylindrical coordinates R, z, ϕ.

$$\frac{z^2}{u^2} + \frac{R^2}{u^2-c^2} = 1; \quad \frac{z^2}{v^2} - \frac{R^2}{c^2-v^2} = 1 \qquad (1)$$

$$z = u \cdot v/c; \quad R = (u^2-c^2)^{\frac{1}{2}}(c^2-v^2)^{\frac{1}{2}}/c \ . \qquad (2)$$

The differential lengths ds_u, ds_v, ds_ϕ in the u, v, ϕ direction are given by

$$ds_u = h_u du; \quad ds_v = h_v dv; \quad ds_\phi = h_\phi d\phi. \qquad (3)$$

with

$$h_u = (u^2-v^2)^{\frac{1}{2}}(u^2-c^2)^{-\frac{1}{2}}; \quad h_v = (u^2-v^2)^{\frac{1}{2}}(c^2-v^2)^{-\frac{1}{2}};$$

$$h_\phi = (u^2-c^2)^{\frac{1}{2}}(c^2-v^2)^{\frac{1}{2}}/c. \tag{4}$$

The h_i parameters ($i = u,v,\phi$), defined in (4), are the reciprocals of the root of the metric coefficients g_{ii}, $h_i = 1/\sqrt{g_{ii}}$. E_u, E_v, and E_ϕ as well as in the calculation of the surface element $dO = ds_v \, ds_\phi$ on the surface of a given spheroid $u = a$. If $\phi_{(u,v,\phi)}$ is a given potential function, the field component in the u direction E_u is obtained by

$$E_u = -\frac{1}{h_u}\frac{\partial\phi}{\partial u} \; . \tag{5}$$

(Note that in the atmospheric electric sign convention, the right side of (5) is positive instead of negative.)

If Q is the net charge of a charged spheroid and ε, the dielectric constant, the potential function of a charged spheroid is given by

$$\phi = \frac{Q}{4\pi\varepsilon}\frac{1}{2c}\ln\frac{u+c}{u-c} \; . \tag{6}$$

From (4), (5), and (6), follows the field in the u direction

$$E_u = -\frac{1}{h_u}\frac{\partial\phi}{\partial u} = +\frac{Q}{4\pi\varepsilon}\frac{1}{(u^2-v^2)^{\frac{1}{2}}(u^2-c^2)^{\frac{1}{2}}} \; . \tag{7}$$

If we set in (7), $u = a$, $v = c$, $a^2-c^2 = b^2$, we obtain the field E_a on the tip of a charged spheroid,

$$E_a = \frac{Q}{4\pi\varepsilon}\frac{1}{b^2} \; . \tag{8}$$

Equation (8) has an outstanding peculiarity. The field
at the tip does not depend on the length of the spheroid
(long axis a) but only on its thickness (short axis b). One
would expect that the longer the spheroid and, as a consequence, the sharper the tip, the greater is the field at the
tip. Equation (8) implies that the increasing effect of the
smaller radius of curvature for longer spheroids on the field
is equally offset by the fact that the net charge Q is distributed over the larger surface area of the longer spheroid.
This reveals a certain weakness in (8) if it is applied, for
example, to the Apollo rocket. There is no doubt that the
radius of curvature, ρ, on the rocket tip has more influence
on the field than has the width of the rocket base. Thus,
if the field at the rocket tip is to be calculated, the
radius of curvature at the tip of the substitute spheroid
should be matched first to that of the rocket. The long axis
is of secondary importance and the short axis is the least
significant parameter. Using the well known relation $b^2 = a\rho$,
we replace b in (8) giving

$$E_a = \frac{Q}{4\pi\epsilon} \frac{1}{a\rho} \quad . \tag{9}$$

Solving the above equation for Q, gives

$$Q = 4\pi\epsilon \ a\rho E_a \quad . \tag{10}$$

The radius of curvature ρ and the length of the rocket
proper is known. However, it is not well known how much
additional effective length is added by the flame tail. The
electric effect of the flame is two-fold. First, the flame
is a highly ionized gas which tends to discharge any charged
body with which it is in contact. From the atmospheric
electric point of view, the flame can be pictured as a strong

470

radioactive probe which bleeds away electric charge until the
charge density and the field at the probe or the flame is
zero. Indeed, the flame collector was the forerunner of the
radioactive collector as a technique for the measurement of
the atmospheric electric potential gradiant.

The second effect is the charging mechanism of the flame
similar to that of the exhaust fumes of piston and jet engines.
The cause of this charging effect is not known in detail. Probably there are a number of different mechanisms working simultaneously, such as difference in diffusion between electrons
and ions, friction charge of carbon particles in the exhaust
pipe, etc. The overall result is that charge of one sign is
carried away by the exhaust, whereas the charge of the opposite sign accumulates at the engine. However, due to the
previously discussed discharging effect of the ionized gases,
a part of the flame charge is shunted back to the engine and
this part increases in proportion to voltage produced by the
accumulated engine charge. An equilibrium is reached in a
few seconds between charging and discharging current. After
a brief transient this exchange leaves the rocket with a
final net charge and the flame tail uncharged. In the
electrostatic picture, the charge stops at the end of the
rocket and the flame tail does not add to the length of the
charged rocket. If the discharging current cannot counterbalance the charging current before the breakdown field
strength at the tip of the rocket is reached, corona or
spark discharge will start at the tip supplementing the
discharging current so that any further accumulation of
charge on the rocket is stopped.

Therefore, the maximum charge which the rocket can hold
is given by (10) if we set $E_a = 3.10^6 V/m = E_b$, where E_b is
the breakdown field at the rocket tip. The radius of curvature at the tip of the Apollo rocket is $\rho = 0.1m$ and the
rocket length is $2a = 100m$.

471

With these values, the maximum charge of the Apollo rocket given by (10) is Q_{max} = 1.9 mC. This value is only 1/3 to 1/10 of the maximum charge, usually assumed (see *"Analysis of Apollo 12 Lightning Incident"* MSC-01540 NASA Feb. 1970) and it is most probable that the actual value is only a fraction of Q_{max}. This makes it immediately clear that such a small charge cannot be determined from field measurements at the ground during lift-off since the field generated by the charges of the steam clouds with charges of approximately 100 mC is about 50 times as strong as that generated by the rocket charge. Furthermore, as long as the flame tail of the rocket touches ground after lift-off and the rocket is still under the influence of the strong field of the steam clouds, the charge distribution and the net charge of the rocket is vastly different from that in free flight. Therefore, the free-flight charge can only be determined if the rocket is at least several hundred meters above ground.

The best method to obtain the rocket charge in free flight would be to measure the surface charge density on two selected points on the rocket itself. However, as measurement on the Apollo rocket is not permitted, the only remaining method is to measure the electric field of the rocket charge from an airplane flying as close as possible to the rocket. Two conditions must be met before successful measurements can be made. First, the atmospheric electric fair-weather field has to be low and stable. This is usually the case during fair weather above the exchange layer, i.e., above about 3-km altitude. Second, the field-measuring instruments must be extremely sensitive and carefully balanced so that they do not respond to the electric field of the airplane charge. We may obtain an estimate of the required sensitivity by calculating the field at a distance of about 5 km and an altitude of about 2.5 km using the maximum rocket charge Q = 1.9 mC. At the nearest approach between rocket and airplane, i.e., when both vehicles are at the same altitude, the maximum horizontal field is E = 0.5 V/m. Therefore, the field-measuring instrument should have a sensitivity of about 0.1 V/m with a range of 5 V/m in full deflection.

During the launch of Apollo 13, an attempt was made to determine the charge of the rocket by field measurements from the NASA-6 Beachcraft. This airplane was equipped with field mills having a sensitivity of about 1 V/m (full deflection on the most sensitive range, 50 V/m). The flight path was timed so that the closest approach to the Apollo 13 occurred as the vehicle passed the flight altitude of the airplane. It was fortunate that the fair-weather field was low (10 V/m), constant and had only a vertical component. The horizontal fair-weather field component was zero and a field of the rocket of about 1 V/m would have been just detectable. However, no field due to the passing of the rocket was detected which indicates that the rocket charge was lower than 3.8 mC. This is not a surprising result when compared with the maximum charge of 1.9 mC deduced from theory.

It would be much more satisfying if even a small response on the airplane field record due to the passing of the Apollo rocket could be obtained. This would require either that the sensitivity of the field mills increased by at least a factor 10. The second method seems to be possible when considered solely from the standpoint of the instrument capability. In the laboratory, fields of 0.1 V/m or even less have been measured with the cylindrical field mill. The airplane, however, produces additional electrical noise. The main contribution comes from the electric charge generated by the exhaust fumes. How much this noise can be reduced is still an open question.

The electric charge on the rocket will not, by itself, interfere with the operation of electronic equipment inside the rocket. However, if the rocket charge is large enough to generate an appreciable fraction of the breakdown field at the tip of the rocket, only a moderate cloud field is necessary to trigger off lightning discharges which will interfere with the proper operation of electronic equipment inside the rocket. It is for this reason that it is so important to determine the free-flight charge of the rocket.

Whether the rocket charge will enhance or reduce the effect of an external cloud field depends on the amount and polarity of the rocket charge and the direction of the external field.

1.2 The Uncharged Rocket in the External Field F

Using the spheroidal coordinate system defined in the previous section, the potential function ϕ of the uncharged spheroid in a homogeneous external field F is given by

$$\phi = Fu \frac{v}{c} \left(1 - \frac{\tanh^{-1}\frac{c}{u} - \frac{c}{u}}{\tanh^{-1}\frac{c}{a} - \frac{c}{a}} \right), \tag{11}$$

From (4) and (5) it follows that the field in the u direction is given

$$E_u = -\frac{1}{h_u}\frac{\partial\phi}{\partial u} - F \frac{v\sqrt{u^2-c^2}}{c\sqrt{u^2-v^2}} \left[1 - \frac{\tanh^{-1}\frac{u}{c} - \frac{u}{c}}{\tanh^{-1}\frac{u}{a} - \frac{u}{a}} \right.$$

$$\left. + \frac{c^3}{u(u^2-c^2)(\tanh^{-1}\frac{c}{a} - \frac{c}{a})} \right]. \tag{12}$$

The field at the spheroid surface E_a is obtained by setting $u = a$. With $a^2 - c^2 = b^2$,

$$E_a = -F \frac{vb}{c\sqrt{a^2-v^2}} \frac{c^3}{ab^2(\tanh^{-1}\frac{c}{a} - \frac{c}{a})} \cdot \qquad (13)$$

Let us discuss first the field at the tip of the spheroid E_T, which we obtain from (13) if we set $v = c$,

$$E_T = -F \frac{c^3}{ab^2(\tanh^{-}\frac{c}{a} - \frac{c}{a})} \cdot \qquad (14)$$

The same argument concerning the importance of the radius of curvature ρ at the tip for the field E_T given in the previous section applies here. Therefore, the small axis b, and the eccentricity c are substituted for by the long axis a and the radius of curvature ρ. Using the relation $b^2 = a\rho$, we obtain from (14)

$$E_T = -F \frac{\frac{a}{\rho} - \sqrt{\frac{\rho}{a}}}{\tanh^{-1}\sqrt{1 - \frac{\rho}{a}} - \sqrt{1 - \frac{\rho}{a}}} \cdot \qquad (15)$$

Figure 1 shows the field concentration factor at the tip of the spheroid $\eta = E_T/-F$ for a large range of values $2a/\rho$. For the application of (15) to the triggered lightnning problem either by an Apollo rocket or by the folding-fin aircraft rocket, we are faced with the same problem as in the previous section 1.1, namely, do we add the flame tail to the length of the rocket and if so, with what length? It will be shown in the next section 1.3 that regardless of the conductive length of the flame tail, the length of the rocket determines the axis a of the substitute spheroid.

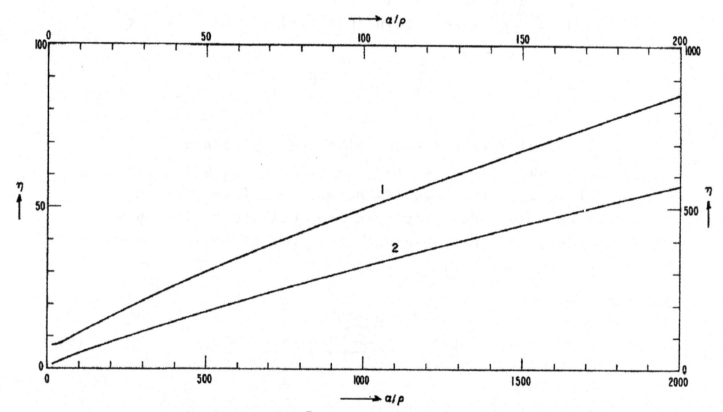

Figure 1. Normalized field $\eta = {}^{E}T/F$ at the top of an uncharged spheroid in a homogeneous field as a function of a/p with "a", the long axis of the spheroid, and "ρ", the radius of curvature at the top. The left and upper scale belongs to curve 1. The right and lower scale to curve 2.

The same also applies approximately to the folding-fin aircraft rocket where the corona discharge at the fins produces a similar effect as the flames at the exhaust pipes of the Apollo rocket.

For the Apollo rocket la/ρ = 1000, and from figure 1 we obtain a field concentration factor η = 317. This means an external cloud field of 9.5 kV/m would produce the breakdown field of 3000 kV/m at the tip of the rocket. A similar value for dangerous cloud fields of 7.5 kV/m is given by Brook, Holmes and Moore in the *"Analysis of Apollo 12 Lightning Incident"* NASA MSC-01540 (Brook, et al.; 1970) and a value of 10-30 kV/m is given by the author in *"Lightning Hazard to Rockets during Launch I"*, ESSA Tech. Rept. ERL 143-APCL 11, Kasemir, 1969).

To place this value of about 10 kV/m within the framework of meteorological events, some rough average field values for different weather conditions are listed in table 1.

Table 1. Average Fields under Different Weather Conditions.

Weather Condition	Field at Ground	Field underneath Cloud	Field inside Cloud
Fair	100-200 V/m	Diminishing with altitude	
Moderate Rain	500 V/m	5 kV/m	15 kV/m
Rain Shower	1-2 kV/m	10-20 kV/m	30-60 kV/m
Thunderstorm	2-10 kV/m	20-100 kV/m	60-300 kV/m

It should be emphasized that the above list is only a crude guideline and even where limits are given in the different columns, 100 percent higher values of the upper limit and 50 percent lower values of the lower limit may be encountered.

The field values below the cloud are based on measurement by the author of about 60 shower cloud and thunderstorm flights.

During these flights, a field of 290 kV/m was recorded underneath a rain cloud, which did not produce lightning discharges and, therefore, could not be classified as thunderstorms. This value is more than 10 times as much as the listed maximum average value of 20 kV/m, which shows very clearly that large deviations from the average values may occur.

The values inside the cloud are based on the assumption that the conductivity inside the cloud is only one third of the value outside the cloud and that the field in first approximation increases proportionally to the decrease in conductivity. There are scientists, for instance, Freier, (1968) and Evans (1969), who claim that the conductivity inside the thundercloud is even greater than outside. This would imply that the fields inside the cloud are weaker than those outside. However, with such an assumption, it would be hard to explain why lightning discharges are usually generated inside the cloud and not outside since they will start at points with the highest fields. Without going into further discussion of the many controversial issues of thunderstorm theories involved, we postulate the field values inside the storm as given in table 1. Frequently the fields in thunderstorms should be higher than the maximum average values of 300 kV/m. They should reach up to 500 kV/m in limited areas of the cloud because this is generally assumed to be the field necessary to start a lightning discharge.

We see from table 1 that the 10 kV/m cloud field necessary to produce breakdown at the tip of the Apollo rocket may even be produced by a moderate rain cloud. It will certainly occur inside a shower cloud. It is interesting to note that ground fields of about ± 3 kV/m were recorded in the KSC area during the Apollo 12 launch (Kasemir, 1971), indicating that this was one of the more electrified shower clouds on the borderline between shower and thundershower.

It may be mentioned here that the breakdown field at the tip of a rocket does not necessarily imply a triggered lightning discharge. Under what conditions the discharge takes the form of a corona-, spark-, or lightning-discharge depends, furthermore, on the electric energy provided by the cloud to the growing streamer and probably several other unknown parameters. We will come back to this problem area in section 2.

The last topic of this section is to determine the cloud field capable of producing the breakdown value at the tip of the folding fin aircraft rocket. These rockets were topped by small spheres of different sizes with radii of 2.5, 3.5, and 5 cm to inhibit corona discharge. With the length of the rocket of 1.20 m or 1.50 m, the following values of a/ρ were obtained: a/ρ = 48, 34, and 28. The last value belongs to the 5-cm sphere which had to be mounted on an extension rod of 20 cm to clear the launching tube. This increased the length of the rocket from 1.20 to 1.40 m. We see from figure 1 that the field concentration factor is 29, 24, and 20 respectively. This would necessitate cloud fields at the tip of the rockets.

Experiments were conducted when the field underneath the cloud was 100 kV/m or more even though the field inside the cloud would have been 300 kV/m if the factor 3 is taken into account. In these cases, lightnings were always triggered at least by the first rocket. At the end of the experiments, the threshold value of 100 kV/m was relaxed to 60 kV/m underneath the cloud. Even in this case, lightnings were triggered which shows that the fields inside the cloud were at least by a factor of 2 stronger than that underneath the cloud. These few remarks about the experimental results shall suffice here to indicate the application of the theory. A detailed discussion of the experiments will be given in section 3.

1.3 The Charged Rocket in the External Field F

This section is the most important one in our theoretical discussion because: (1) it will give the proof that the length of the flame tail of the Apollo rocket has only a minor influence on the field at the tip of the rocket; (2) it will furnish the missing link in the determination of the effective length of the lightning triggering folding-fin aircraft rocket, and (3) will supply the basis for the simulation of a large rocket by a small rocket with a trailing wire. All three problems have the common property that on some given point on the conductive body a surface charge density of zero is enforced by either the conductive flame at the exhaust of the Apollo rocket, or by corona discharge at the fins of the FFAR or at the end of the trailing wire. In the last two cases, the initial field necessary to start corona discharge has been neglected.

From a mathematical point of view, all three problems fall in the category of the charged spheroid in a homogeneous external field. The solution is obtained by the superposition of the potential function of the charged spheroid, as given in section 1.1, and of the potential function of the uncharged spheroid in the homogeneous field F as given in section 1.2. Using (6) and (11), the potential function ϕ for this problem is given by

$$\phi = \frac{Q}{4\pi\varepsilon} \frac{1}{2c} \ln \frac{u+c}{u-c} + Fu\frac{v}{c}\left(1 - \frac{\tanh^{-1}\frac{c}{u} - \frac{c}{u}}{\tanh^{-1}\frac{c}{a} - \frac{c}{a}}\right). \tag{16}$$

The superposition principle is not only valid for the potential functions but also for fields, surface charge densities, and other parameters. Our first task will be to calculate the surface charge q per unit length on the spheroid. We will do this separately for the case of the charged spheroid with the surface charge per unit length q_1, and for the case of the uncharged spheroid in the homogeneous field with the surface charge per unit length q_2. We will then superimpose the two surface charges q_1, and q_2 and determine the point where the two densities cancel each other.

The surface charge density p_1 in the first case is given by

$$p_1 = \varepsilon \, E_a \, , \qquad (17)$$

where E_a is the field at the surface given by (7) with u = a

$$p_1 = \frac{Q}{4\pi b} \frac{1}{(a^2 - v^2)^{\frac{1}{2}}} \, . \qquad (18)$$

For the surface element $dO = ds_v \, ds_\phi$, it follows from (3) and (4), again with u = a, that

$$dO = \frac{(a^2 - v^2)^{\frac{1}{2}}}{(c^2 - v^2)^{\frac{1}{2}}} \frac{b(c^2 - v^2)^{\frac{1}{2}}}{c} dv d\phi = \frac{b}{c}(a^2 - v^2)^{\frac{1}{2}} dv d\phi \, . \qquad (19)$$

The charge of Q_{12} of a ring segment on the spheroid surface between v_1 and v_2 is obtained by integration,

$$Q_{12} = \int_0^{2\pi} \int_{v_1}^{v_2} p_1 dO = \int_0^{2\pi} \int_{v_1}^{v_2} \frac{Q}{4\pi b} \cdot \frac{b}{c} dv d\phi \, . \qquad (20)$$

The solution of (20) is quite elementary and we obtain

481

$$Q_{12} = \frac{Q}{4\pi c} 2\pi \, (v_2 - v_1) = \frac{Q}{2c}(v_2 - v_1) \, . \qquad (21)$$

We introduce the cylindrical coordinate z defined by (2) and set u = a. With v = cz/a, we obtain from (21)

$$Q_{12} = \frac{Q}{2a}(z_2 - z_1) \, . \qquad (22)$$

If we keep $z_2 - z_1$ at a unit length, and divide (22) by it, we obtain the charge per unit length,

$$q_1 = \frac{Q}{z_2 - z_1} = \frac{Q}{2a} \, . \qquad (23)$$

From (23), it follows that the charge per unit length of a charged spheroid is constant regardless of the thickness of the spheroid. The shrinking of the surface of the ring segment towards the ends of the spheroid is compensated by an increase in the surface charge density.

The calculation of the charge per unit length q_2 of the uncharged spheroid in a homogeneous field follows similar steps. With (13) and (19), we obtain for the charge Q_{12} of a ring segment,

$$Q_{12} = - \frac{\varepsilon F c}{a \left(\tanh^{-1} \frac{c}{a} - \frac{c}{a} \right)} \int_0^{2\pi} \int_{v_1}^{v_2} v \, dv \, d\phi \, . \qquad (24)$$

The integration of (24) is again elementary and by solving the double integral, we obtain

$$Q_{12} = - \frac{2\pi \varepsilon F c}{a \left(\tanh^{-1} \frac{c}{a} - \frac{c}{a} \right)} \frac{v_1^2 - v_1^2}{2} \, . \qquad (25)$$

Transforming to cylindrical coordinates by the use of (2) yields

$$Q_{12} = - \frac{2\pi\epsilon Fc}{a\left(\tanh^{-1}\frac{c}{a} - \frac{c}{a}\right)} \frac{z_2 + z_1}{2} (z_2 - z_1) \ . \qquad (26)$$

We again divide by $z_2 - z_1$, keeping this distance at unit length to obtain the charge q_2 per unit length. At the same time, we introduce the height of the middle of the ring segment $z = z_2 + z_1/2$ as a new coordinate. Using this transformation, we obtain

$$q_2 = - \frac{2\pi\epsilon Fcz}{a\left(\tanh^{-1}\frac{c}{a} - \frac{c}{a}\right)} \ . \qquad (27)$$

The charge per unit length q_2 is zero at the midpoint of the spheroid and increases linearly to its negative maximum at the upper end and to its positive maximum at the lower end of the spheroid. Figure 2 shows the charge per unit length q_1 of a charged spheroid (fig. 2a), q_2 of an uncharged spheroid in a homogeneous field (fig. 2b). The charge per unit length, q_3 in figure 2c, is obtained by the superposition of q_1 and q_2. We see that the point of zero charge density is shifted upwards. The amount and the direction of the shift depends on the amount and polarity of the charge q_1. If, for instance, the polarity of q_1 is negative, the point of zero charge density is shifted downwards as shown in the sequence of the figures 2 e,f,g. There the amount of q_1 is so adjusted that the point of zero charge density is at the lower tip of the spheroid. This would be the condition for the lightning triggering FFAR or, also, for the rocket with a trailing wire. Figure 2c illustrates the charge distribution on the Apollo rocket with a conductive flame tail. The rocket itself would be represented by the upper quarter of the spheroid (length a^1) whereas, the flame tail would be represented by the lower three quarters of the spheroid (length a + d).

It is valid to question whether the flame tail can be represented by the lower part of a spheroidal conductor because an ionized volume of hot gas differs in many respects from a conductor, for instance, by rapid changes in conductivity, diffused boundaries, etc. However, it will be shown that the upper part of the spheroid, which represents the rocket, determines almost exclusively the field at the tip of the rocket. Therefore, the length and the shape of the flame tail are of minor importance, and by the same reason, the difference between a hot gas and a conductor becomes immaterial.

To show this, we will calculate the field at the tip of the spheroid with a charge distribution as given in figure 2c and compare it with the field at the tip of a spheroid with a charge distribution as shown in figure 2d.

In the first case, the flame tail is three times as long as the rocket and in the second case, the rocket and the flame tail are of the same length. The second case has the charge distribution of the uncharged spheroid in a homogeneous field as can be realized by comparing figures 2b and 2c. If the parameters in this case are marked by a prine ', the field at the tip, E_T' follows from (14),

$$E_T' = - F \frac{c'^3}{a'b'^2 \left(\tanh^{-1} \frac{c'}{a'} - \frac{c'}{a'} \right)} \quad . \tag{28}$$

Because we have to keep the radius of curvature ρ constant, we transform c' and b' into a' and ρ by the relations

$$b'^2 = a'/ \; ; \; \frac{b'^2}{a'} = \rho, \; \frac{c'}{a'} = \sqrt{1 - \frac{\rho}{a'}} \quad . \tag{29}$$

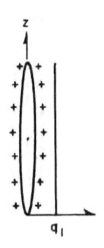

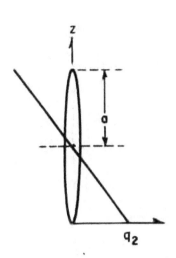

a)   a positively charged spheroid

b)   an uncharged spheroid
     in a homogeneous field

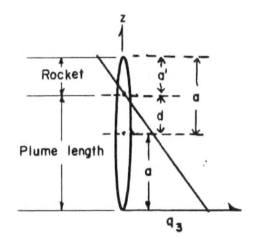

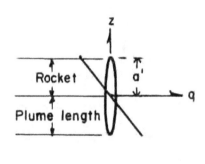

c)   a positively charged spheroid
     in a homogeneous field
     (Apollo Rocket)

d)   an uncharged spheroid
     half the length as in
     figure 2b in a homoge-
     neous field (Apollo
     Rocket)

Figure 2. Charge per unit length on a spheroid

485

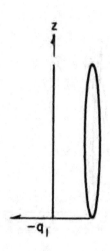

e) a negative charged spheroid

f) an uncharged spheroid
   in a homogeneous field

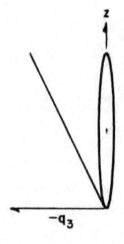

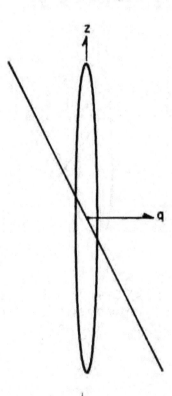

g) a negative charged spheroid in
   a homogeneous field (FFAR)

h) an uncharged spheroid of twice the length as in
    in figure 2g in a homogeneous field (FFAR)

Figure 2. Charge per unit length on a spheroid (con't)

Furthermore, we use the relation

$$\tanh^{-1} \frac{c'}{a'} = \frac{1}{2} \ln \frac{1 + \frac{c'}{a'}}{1 - \frac{c'}{a'}} = \frac{1}{2} \ln \frac{1 + \sqrt{1-\rho/a'}}{1 - \sqrt{1-\rho/a'}} \quad . \tag{30}$$

Substituting (29) and (3) into (28), yields

$$E_T' = -F \frac{a' \sqrt{1-\rho/a'}^3}{\rho \left(\frac{1}{2} \ln \frac{1 + \sqrt{1-\rho/a'}}{1 - \sqrt{1-\rho/a'}} - \sqrt{1 - \rho/a'} \right)} \quad . \tag{31}$$

For the FFAR $\rho/a' = 0.03$ and for the Apollo rocket $\rho/a' = 0.001$. Therefore, with an error of about 1.5 percent or less, we may neglect ρ/a' with respect to 1 and obtain a good approximation given by

$$E_T' = -F \frac{a'}{\rho \left(\frac{1}{2} \ln \frac{4a'}{\rho} - 1 \right)} \quad . \tag{32}$$

We now calculate the field at the tip of the charged spheroid E_T in the homogeneous field with a charge distribution as shown in figure 2c, and compare this field with (32). The first step will be to determine what charge will move the point of zero charge density to $z = a-a'$. From the condition $q_1 + q_2 = 0$ for $z = a-a'$, we obtain from (23) and (27)

$$\frac{Q}{2a} - \frac{2\pi\varepsilon F c(a-a')}{a \left(\tanh^{-1} \frac{c}{a} - \frac{c}{a} \right)} = 0$$

or

$$\frac{Q}{4\pi\varepsilon} = \frac{Fc(a-a')}{\tanh^{-1} \frac{c}{a} - \frac{c}{a}} \quad . \tag{33}$$

The field at the tip is obtained by superposition of the field of the uncharged spheroid in the homogeneous field given by (14) and that of the charged spheroid given by (8). Furthermore, we replace the factor $Q/4\pi\epsilon$ in (8) by the expression given in (33),

$$E_T = -F \frac{c^3}{ab\left(\tanh^{-1}\frac{c}{a} - \frac{c}{a}\right)} + \frac{Fc(a-a')}{\tanh^{-1}\frac{c}{a} - \frac{c}{a}} \frac{1}{b^2} . \qquad (34)$$

Transforming b and c into ρ and a, and neglecting ρ/a with respect to 1 as before, reduces (34) to

$$E_T = -F \frac{a'}{\rho\left(\frac{1}{2}\ln\frac{4a}{\rho} - 1\right)} . \qquad (35)$$

This is almost the same equation as (32). The only difference is that a' in the argument of the logrithim in (32) is replaced by a in (35). From (32) and (35), we obtain the relation

$$E_T' = \frac{\frac{1}{2}\ln\frac{4a}{\rho} - 1}{\frac{1}{2}\ln\frac{4a'}{\rho} - 1} E_T . \qquad (36)$$

We will use (36) to check our previous statement that the length of the flame tail has only a minor influence on the field at the tip. If we set $\rho = 0.1$ m, a' = 100 m, and a = 200 m, E_T' would be the field at the tip of the Apollo rocket with the length of the flame tail of 100 m, and E_T would be the field at the tip for a flame tail of 300 m length. By inserting the numerical values in (36), we obtain $E_T' = 1.11\ E_T$. This shows, indeed, that the length of the flame tail is unimportant to the field at the tip. For a longer tail, the field at the tip is slightly weakened. This result is contrary to the general belief which assumes that the longer the flame tail, the stronger the field at the tip of the rocket. The difference in the results is caused by enforcing the charge density zero at the exhaust pipes of the rocket.

It should be mentioned that the net charge Q, which is necessary to shift the point of zero charge density to a given location, is not identical or connected with the charge produced by the exhaust. The field generated by this charge must still be superimposed on the field given by (32) or (35).

To apply (36) to the case of the folding-fin aircraft rocket, we set ρ = 3.5 cm, a' = 120 cm, and a = 60 cm. From (36) follows E_T' = 0.72 E_T. In this case, the supplemented uncharged spheroid has a weaker tip field than the charged spheroid. If the 28 percent error should become important, the more accurate formula (35) should be used for calculating the field at the tip. Since a' = 2a, with a being equal to the half length of the rocket, (35) can be written as

$$E_T = - F \frac{2a}{\rho(\frac{1}{2} \ln \frac{4a}{\rho} - 1)} .$$ (37)

1.4 Simulation of a Large Rocket By a Small Rocket With a Trailing Wire.

We know from section 1.2 that the field concentration factor, $\eta = E_T/-F$, for the Apollo rocket is about 300. The question to be answered here is the following: Is it possible to produce the same field concentration factor with a small rocket where a trailing wire will supply the missing length. For a first check, we may use (37) with the radius of curvature ρ = 0.1 m, and the length of the wire, L = 2a = 100 m. The field concentration factor is then η = E_T/-F = 357. This result shows that is should be possible to simulate the Apollo rocket by a small rocket with a trailing wire. If the field concentration factor or, consequently, the breakdown field at the tip of the rocket, would be the only criterion for triggering lightning discharges, we could reduce the length L of the trailing wire in the same proportion as the radius of curvature ρ so that the ratio L/ρ remains the same.

The length of the folding-fin aircraft rocket is 1.20 m = 1200 mm. If we use this rocket without trailing wire, the radius of curvature should be about 1 mm to produce the field concentration factor of 357. However, such a sharp point would not trigger a lightning discharge but would, instead, go into corona discharge in a cloud field of about 10 kV/m. Such a rocket could not be used to discharge a cloud by triggering lightning discharges for a safe passage of the Apollo rocket. However, it can be used to simulate the Apollo rocket for checking a cloud prior to the Apollo launch with regard to the presence or absence of dangerous electric fields. A field-measuring payload for the FFAR has been developed and tested under this contract. The design and test results will be discussed in a separate report. It is sufficient to state here that the sensitivity level can be lowered by sharpening the corona point at the tip of the rocket to respond to fields as weak as 2-3 kV/m. Therefore, it would be no technical problem to adjust the sensitivity to simulate the Apollo or any other specified rocket.

From the experimental results discussed in the next section, we will learn that lightnings have been triggered with the FFAR topped with spheres of 3.5-cm radius. For

such a radius of curvature, the length of the trailing wire
must be 35 m to produce a simulation of the same lightning
triggering capacity as the Apollo rocket. The reduction in
length of the trailing wire may become important if it is
found to be technically difficult to pull wires of 100 m in
length by a rocket. A spheroid of 35 m length and a radius
of curvature at the tip of 3.5 cm was used to represent the
rocket with the trailing wire. This would mean that the
small axis is 0.78 m or that the spheroid is 1.56 m wide.
The diameter of the wire, however, will only be a fraction
of a millimeter. This means that the charge on the wire
cannot spread out over as large a surface as on the spheroidal model and, consequently, most of the charge will
concentrate on the body of the rocket. This would result
in a higher charge concentration and a higher field at the
tip of the rocket. The field concentration factor calculated
with the spheroidal model is, therefore, a minimum value.
To calculate the maximum value, we neglect the wire completely, match the spheroid to the rocket alone and treat
it as a body charged to the potential difference with respect
to its environment of ϕ = FL, where F = external field and
L = length of wire. This follows from the fact that the
rocket and the wire are tied to the potential value of the
external field at the lower end of the wire by the corona
discharge at this point. The field at the tip of the rocket
is then given by (9):

$$E_T = \frac{Q}{4\pi\epsilon} \frac{1}{a\rho} \quad ,$$

and the charge on the rocket follows from (6) if we set
ϕ = FL and u = a:

$$- FL = \frac{Q}{4\pi\epsilon} \frac{1}{2c} \ln \frac{a + c}{a - c} \quad .$$

With the approximation used before,

$$\frac{c}{a} = 1 - \frac{\rho}{a} \approx 1 - \frac{1}{2}\frac{\rho}{a} \pm \ldots$$

and neglecting ρ/a with respect to 1, we obtain

$$E_T = \frac{-2FL}{\rho \ln \frac{4a}{\rho}} \quad . \tag{38}$$

For L = 35 m, ρ = 3.5 cm, and a = 60 cm, we obtain from (38), a field-concentration factor η of

$$\eta = \frac{E_T}{-F} = 473 .$$

The maximum field-concentration factor is about 25 percent higher than the minimum field-concentration factor. The thinner the wire, the closer the real value will be to the maximum value. Thus, using a thin wire of about 1 mm radius, the length of 35 m can be reduced by 25 percent, which then results in a length of 26 m to simulate the Apollo rocket.

2. EXPERIMENTAL TESTS

The pilot tests reported in this section have to be viewed with an open mind. Some of the results are given to discussion and may find a different interpretation in the future. These experiments are to be considered as an attempt to work out an operational procedure, as an instrument-testing method and as a gathering of first results which may prove or disprove the feasibility of the approach. No serious discrepancies with the theoretical predictions have appeared, therefore, the approach seems to be sound. However, some improvement of the experimental design is indicated.

Fifteen rockets were fired into thunderstorms at the
Langmuir Observatory of the Institute of Mining and Technology
at Socorro, New Mexico to prove the feasibility of triggering
lightning discharges with the folding-fin aircraft rocket
Mighty Mouse. The dummy warheads of the rockets were topped
off with spheres of different diameters to inhibit the formation of corona discharge. If a discharge from a sphere occurs
it will be in the form of a spark discharge, which has a better
chance of growing into a lightning discharge than a corona
discharge. The size of the sphere is referred to in table 2
as S(small), M(medium), and L(large), indicating a diameter
of 5, 7, and 10 cm, respectively.

At the launch site, two field recorders were installed
to record the stationary field of the thunderstorm with a
field mill, and the field of the lightning stroke with a
capacitive loaded wire antenna. The NASA-6 airplane was
equipped with two field mills recording all three components
of the electric field underneath the cloud. The following
test procedure was followed whenever possible. The airplane
would check the field underneath clouds in the target area and
transmit the field values to the launch site by voice communication. When the field value surpassed 100 kV/m at a given
place, the airplane was ordered out of the target area and
rockets fired into the peak field area. After the firing,
the airplane was recalled to check the effect of the rocket-
triggered lightning on the cloud fields. Due to communication difficulties, the airplane did not participate in the
test on August 8 and the first test on August 13. The
results of the test are given in table 2.

On August 8, three rockets were fired, topped with medium,
small, and medium spheres. The first rocket fired at 17 h,
15 min, 38 sec, triggered a ground discharge. The second
small-sphere rocket and the third medium-sphere rocket fired

at 17 h, 19 min, 9 sec, and 17 h, 31 min, 12 sec, respectively, did not trigger lightning discharges. The launch azimuth and elevation was the same for all rockets, therefore, they all followed the same path. This seems to indicate that the cloud was discharged by the first rocket and did not recover to provide high enough fields for the triggering of further lightnings, or that the small-sphere rocket does not provide enough initial energy to trigger a lightning discharge. It is also possible that the electrified part of the cloud moved out of the target area before the third medium-sphere rocket was fired.

The next storm occurred on August 13, and medium- and large-sphere rockets were fired into the cloud. Both rockets triggered lightning discharges. The electric ground field generated by the first lightning of 12.5 kV/m was considerably larger than the electric field of the second lightning of only 3.5 kV/m even though the second rocket was equipped with the large sphere.

Figure 3 shows the lightning-field record on the upper trace and the cloud-field record on the lower trace. Time increases from left to right and seconds are marked off by the lower margin-pen. The numbers above the margin-pen trace indicate minutes. The time of the rocket firing is indicated by an arrow with the letters R.F. in the upper trace and the time of "thunder heard" by an arrow marked D. In the lower trace, the step-like field increases due to lightnings triggered by a medium-sphere rocket or large-sphere rocket are indicated by the letters L.M. or L.L.

During the next more active storm on the same day, rockets were fired into the same part of the storm until the last, namely the fourth one, did not trigger a lightning stroke (see table 2 and fig. 3b). Between the last and before-last firing, there was a time period of 2 min 7 sec.

Table 2. Experimental Results of the Soccorro Tests

1	2	3	4	5	6	7	8	9	10	
8.8.70 17 h 15.38	M	54	17	5.6	4	-25	-	-	-	
17 h 19.09	S	-	-	-	-	-25	-	-	-	
17 h 31.12	M	-	-	-	-	-25	-	-	-	
13.8.70 15 h 42.51	M	53	16	5.3	12.5	- 7.5	196	080	-	
15 h 44.52	L	59	21	6.9	3.5	- 6	196	080	-	
18 h 55.04	L	45	11	3.6	20	- 9	309	079	-	
18 h 56.49	M	27	9	3.0	10	-11	309	079	150	before
18 h 58.35	L	38	11	3.6	12	-11	309	079	-	
19 h 00.42	M	-	-	-	-	-11	309	079	50	after
14.8.70 15 h 26 46	M	61.0	-	-	1	- 1	325	080	60	before
15 h 27.14	L	59.5	-	-	5.5	- 1.2	325	080	-	
15 h 27.43	M	60.5	-	-	0.5	- 1	325	080	40	after
15 h 29.04	M	-	-	-	-	- 1	305	080	-	
15 h 29.36	L	-	-	-	-	- 1	305	080	-	
15 h 30.04	M	-	-	-	-	- 1	305	080	-	

Column

1. Date & Time of rocket firing.

2. The size of the sphere S, M, or L.

3. Time period Δt between rocket firing and triggered lightning in seconds.

4. Time period δt between lightning and thunder in seconds.

5. Distance of lightning from launch site calculated from δt with the assumption that the sound velocity is 330 m/sec. Distance in kilometers.

Column

6. The electric field at the launch site of the lightning discharge, ΔE in kV/m.

7. The thunderstorm field at the launch site before the lightning discharge, E in kV/m ground.

8. The azimuth of the launcher degrees referenced to north.

9. The elevation of the launcher degrees referenced to horizontal.

10. The vertical component of the electric field F in kV/m as measured by the airplane before and after the firing.

495

This means that this part of the storm did stay discharged for more than 2 min and that the average recovery time of the ground field of about 40 sec did not reflect the recovery of the field in the cloud. The field measurements of the airplane show similar results. The field underneath the cloud before the firing was about 150 kV/m and after the firing, about 50 kV/m. It took the airplane more than 2 min after the last firing to reach the location of the previous maximum field, which indicates that even 4 min after the last triggered lightning, the field at the cloudbase did not recover to its previous value of 150 kV/m. It is interesting to note that the field at the ground (column 7) remained at about the same value of -11 kV/m during and after the firing, not counting the field changes and their recovery curves caused by the triggered lightning discharges.

It may be noted from table 2, column 3, that the time difference between launch and lightning discharge for the first three rockets was 45, 27, and 38 sec, respectively. The apogee of the rocket occurred about 35 sec after launch, which would indicate that the second rocket triggered the lightning discharge on its upward pass. The discharges were hidden by the cloud and could not be seen from the ground. This leaves some doubt if the recorded stroke, which occurred 27 sec after the second rocket, was cuased by the rocket or if it was a natural discharge produced by the storm. In favor of the triggered lightning discharge is the fact that all three lightning discharges occurred at approximately he same distance, namely, 3.0 to 3.6 km from the launch site.

To check the assumption that lightning strokes were really triggered by the rockets, the following test was carried out on August 14. The rocket firing was delayed until the storm was in its dissipating stage. The natural lightning frequency was then about one discharge every 3 min.

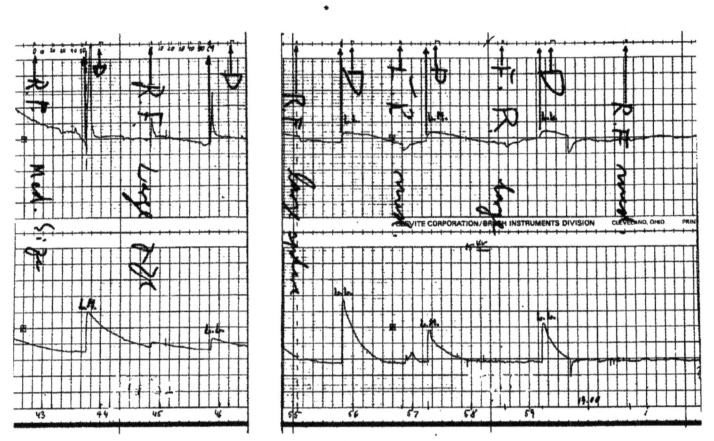

Figure 3. Field records of triggered lighning discharges on August 13, 1970.

a) first storm   b) second storm

It can be assumed that in the dissipating state, the altitude level at which the fields are still strong enough to produce lightning discharges, is confined to a small and limited region. Rockets fired in a fast sequence should trigger discharges at the same time interval after launch, if at all. There would be a good chance that in the time period of about 2 min between the launch of the first rocket and the impact of the last one, no natural discharges would occur if firing started right after a natural one. Two series of three rockets each were fired, the first at the spot where the airplane located the peak field of about 60 kV/m with an azimuth setting of 325°, and the second about 1 km to the west of the first one with an azimuth setting of 305°. The first three shots triggered charges at 61.0, 59.5, and 60.5 sec after launch. The close agreement of the time periods leaves practically no doubt that the lightning discharges were caused by the rockets. Note that the ground fields caused by the rockets with medium-sized spheres are extremely weak, as could be expected. The second series, deflected from the peak field area, triggered no lightning. No attempt will be made to strain the data with a statistical evaluation because the number of firings is too small and the feasibility of the approach is evident. Future tests should be carried out with the corona rocket checking the field before and after triggered lightning discharges and the triggering rocket should be equipped with a transmitter which stops transmitting if a discharge is triggered. The future experiments should include the use of larger rockets and rockets and/or dropsondes with trailing wires capable of triggering lightning discharges. The identification of triggered discharges would be much easier under such conditions.

There is one aspect in the problem of triggering lightning discharges which is virtually unexplored and that is the transition from corona to spark and from spark to lightning discharge. Laboratory experiments of Hermstein (1960) have shown that the radius of curvature of the electrode should be larger than about 2 cm to insure that the form of the discharge is a spark and not a corona discharge. (Hermstein's papers have been published in German, however, his results have been discussed in English by Loeb [1965]). It was for this reason that the smallest spheres used as payloads for the lightning triggering rocket had 2.5-cm radius. The question, still unanswered, is under what conditions does a spark grow into a lightning discharge?

Furthermore, applying laboratory results of the conditions for corona or spark discharges to the initial stages of a lightning discharge is by no means a safe procedure. It is most probable that the lightning discharge starts as a corona discharge on a precipitation particle. In the laboratory, the corona discharge, once established, is a very stable phenomena and requires a substantial field increase to be converted into a spark discharge. However, the lightning discharge manages this transition seemingly without difficulties. The major difference between the laboratory corona or spark discharge and the lightning discharge is that the corona and spark discharges are generated between two electrodes charged by a power supply capable of maintaining the voltage difference between the electrodes, at least at the beginning of the discharge process. The lightning discharge starts in free air and has to draw its power from the cloud field without the help of metallic electrodes. Voltage differences and distances which occur in lightning discharges cannot be simulated in the laboratory. In the next section we will discuss a lightning tower which shall make it possible to study the lightning discharge under more natural conditions.

3. THE LIGHTNING TOWER

It is a well-known fact that lightning strikes a high tower more often than any other spot on flat terrain. Therefore, towers have been and still are used for lightning study. Nevertheless, such a study is a very tedious enterprise because a tower is on the average not struck more than a few times per year. However, the towers used for lightning research are not well adapted for frequent lightning strikes. It will be the objective of this section to calculate the design criteria for a lightning tower which is specially suitable to study the problem of triggered lightning discharge. It would be desirable that the tower ignite lightning strikes, or at least long sparks, at ground field of about 3 kV/m, that means practically under every shower and thundercloud. If this can be achieved, such a tower could be used to discharge the lower part of the cloud and to protect a sensitive area against lightning strikes.

Since it is assumed that the lightning will start from the tower, the point of contact with the earth is known and fixed, and a good ground can be provided. Because of the expected frequent lightning strikes, the tower should be fenced to keep people out of the danger zone and work at the tower can only be done in fair weather. Measurements can be made from a remote place or from a well-screened room at the base of the tower.

The same criteria which led to the design of the lightning triggering rocket applies to the lightning tower. The field concentration factor of the tower has to be so high that for a given external field, the breakdown value is reached or surpassed at the tip of the tower. Furthermore, the radius of curvature of the tip should not be less than 2.5 cm. If the tower is 100 m high, the value $a/\rho = 10,000$ cm/ 2.5 cm = 4.10^3. For this value, we obtain from (32), the field concentration factor $\eta \doteq E_a/-F_0 = 1.27 \times 10^3$. This would mean that for a ground field of about 3 kV/m, the breakdown value of 3000 kV/m at the top of the tower would be reached. Sparking and maybe lightning would be triggered whenever the electric field at the ground increases to 3 kV/m or more. A ground field of 3 kV/m occurs under virtually every thunder shower. Therefore, there should be sufficient occasions during a thunderstorm season to study the transient condition from a spark to a lightning discharge at such a tower. The validity of the simulation of a tower by a spheroidal boss with a major semi-axis of 100 m and a radius of curvature of 2.5 cm at the top may be questioned because the small axis of such a spheroidal boss is only 1.58 m. This would mean that the base diameter of the simulated tower is only 3 m. However, it has been mentioned before that the small axis of the spheroid has only a minor influence on the field at the top. Therefore, we could increase the base of the tower by a factor of 10 or more if necessary without significantly changing the field at the top as long as the radius of curvature at the top and the height of the tower is maintained. There is no difficulty in providing the right radius of curvature. We top the tower with a pipe of several meters length and 5-cm diameter, which is closed at the end by a semi-spherical cap. This will give the highest point of the tower the desired radius of curvature.

A more serious influence on the field at the top may be introduced by the lattice structure of the tower. The capacity of a frame-type construction is certainly much less than that of the solid surface of the substitute spheroid. Following a similar line of reasoning as in the problem of the rocket with a trailing wire, we may assume that the top of the tower will be enclosed by sheet metal forming a body equivalent to a sphere with a radius a = 2 m. Disregarding the influence of the supporting tower, the sphere will have a potential difference ϕ_o = Fh against its environment. The lightning triggering sphere of the radius ρ = 2.5 cm, mounted on a thin rod, is at the distance d above the sphere surface. The little sphere would have the potential difference $\phi = \phi_o \frac{a}{a+d}$ with respect to the environment and the field at the sphere surface would be $E = \frac{\phi}{\rho} = \phi_o \frac{a}{\rho(a+d)}$. With ϕ_o = F·h, we obtain the field concentration factor

$$\eta = \frac{E}{F} = \frac{h}{\rho} \frac{a}{a+d} \quad . \tag{39}$$

The factor, h/ρ, with h = 100 m, ρ = 2.5 cm, which, in our example is 4×10^3, can be considered as the basic field concentration factor. It is the factor one would obtain by placing a grounded little sphere of radius ρ = 2.5 cm at an altitude of 100 m above ground in a homogeneous field F. This factor is modified by the presence of the larger sphere by a second factor a/a+d. This second factor is smaller than 1 but will approach 1 if a>>d. This means that either a should be very large or d, from the surface of the large sphere to the center of the small sphere, should be as small as possible. It should be large compared to the radius ρ to keep the mutual influence between the two spheres negligible. This is an assumption made implicitly by the derivation of (39). As an example, we may choose ρ = 2.5 cm, d = 50 cm, and a = 200 cm. In this case, the second factor becomes a/a+d = 0.8 and the complete field concentration factor for a 100 m high tower, $\eta = 4 \times 10^3 \times 0.8 = 3.2 \times 10^3$. This is about 2.5 times as much as the field concentration factor obtained for the spheroid. The main reason for the difference is that the spheroid has a solid surface down to the ground, whereas, the ground support of the large sphere is completely ignored. The broken surface of lattice structure of the real tower would result in a field concentration factor η, which lies in between the two extreme values, $1.27 \times 10^3 < \eta < 3.2 \times 10^3$.

From these two calculations, we may extract the general rules for the construction of a lightning tower. The lattice construction of the tower is acceptable from the electrostatic point of view. However, the top of the tower should carry a body with a solid surface. This body does not have to be a sphere, but may be constructed as a drum-like cylindrical box with well-rounded corners of about 2 m radius and 4 m height. The reason for the box is to provide a large capacity for the accumulation of electrical charge (or energy) and to generate a concentrated electric field at the top of the tower, which does not decay too fast with distance. On the top of the box, a little sphere is mounted on a slim rod which acts as a primer for triggering a lightning discharge. Provision should be made for an easy exchange of different kinds of primer electrodes. The radius of the little sphere may be varied from 2.5 cm to 10 cm, and the length of the supporting rod from 2.5 cm to 100 cm. The goal of the research using the lightning tower is to provide the missing information concerning the conditions under which a spark discharge grows into a lightning discharge. A second problem which can be studied under such natural conditions is the stability of corona discharge. If the spherical primer electrode is replaced by a corona needle, the triggering of lightning discharges should be inhibited at least until the electric field surpasses a certain threshold value. If this threshold value is much higher than the field necessary to trigger a lightning discharge with the spherical primer electrode, this investigation may lead to a simple protective device for the Apollo rocket against triggered lightning discharges.

4. RECOMMENDED RESEARCH AND DEVELOPMENT

In the first phase of this project, the feasibility of a project to trigger lightning strokes by small rockets with the purpose of discharging electrified clouds has been demonstrated by the test series at the Langmuir Observatory in Socorro, New Mexico, in a qualitative way. These tests should be expanded in several directions as, for instance, a larger number of firings, the use of different rockets, the use of different trigger mechanisms including dropsondes with or without trailing wires, improved monitoring systems, extensive research in the field and charge development of thunder and shower clouds, etc.

The overall aim will be to arrive at quantitative data to simulate the lightning triggering capability of larger space vehicles such as the Apollo rocket and, using the simulation, to investigate to what extent and for what time period an electrified cloud can be discharged for safe penetration of the specified space vehicle.

The different facets of this program will be discussed in more detail below. However, it shall be emphasized that a plan that is too detailed, or fixed adherence to a present program often turns out to be either impossible to carry out, or adverse to efficient progress, as long as the project is in its research stage. A certain flexibility must be maintained which permits emphasis to be shifted between different program points up to a point where some items may be completely abandoned and new research areas introduced as dictated by the results of intermediate steps. Such changes must be anticipated and accepted as long as the final goal is kept clearly in mind. Close coordination between APCL and KSC will be maintained if the need for re-programming becomes apparent.

4.1 FFAR Firings at Socorro and KSC

About 30 to 50 FFAR should be fired at Socorro and at
KSC. The purpose is to confirm the results of the 1970 Socorro
tests and to collect more data especially with regard to the
number of triggered lightning discharges necessary to discharge
the treated part of the cloud and to determine the recovery
time for different types of clouds. These tests are conducted
with improved monitoring equipment, such as field-measuring
rockets, the three station lightning locating networks and
thunder-recording instruments. The cloud-field and lightning-
field recorder as well as the NASA-6 airplane will be used in
the same manner as in the Socorro tests. The guideline for
one test run is as follows.

The NASA-6 airplane will locate the area underneath the
cloud where the electric field surpasses 60 to 100 kV/m.
Five rockets will be fired into this area in the following
sequence. First, one field-measuring rocket is fired to
determine the electric field strength along the rocket path.
This is followed by three or more lightning triggering rockets
until lightning is no longer being triggered. Finally, one
field-measuring rocket is fired into the same area to determine the discharging effects of the triggered lightnings.

The NASA-6 airplane will then check the field decay
under the discharged area. This sequence is repeated when
a new area at high electric fields is located by the NASA-6
airplane. An alternative follow-up procedure would be to
continue with the firing of field-measuring rockets spaced
at about 3 min intervals to determine the recovery time of
the thunderstorm. It should be kept in mind that the
thunderstorm may move in a 10 min period to bring an undis-
charged part of the storm into the previous target area.

At KSC, the tests will be conducted with the more
elaborate procedure that the airplane path is followed by

radar, that the coordinates of the airplane at the location of the peak field, as determined by the radar, are used to calculate the azimuth and elevation settings of the rocket launch and that the rocket path is also recorded by radar. This will add considerably to the quality of the obtained data and will facilitate the transition from the research to the operational phase.

4.2 Use of Larger Rockets

The FFAR has an apogee of about 16,000 ft. This means it will reach the lower part of a thunder cloud, where the ground discharges are generated, but is unable to penetrate the higher parts of the cloud to trigger cloud discharges. Therefore, it is necessary to conduct tests with rockets, which can reach altitudes of about 30,000 ft, to discharge the two maximum field areas in the cloud. The higher and farther reaching rockets also make it possible to use the larger target area of KSC more effectively. This is a very desirable advantage because the frequency of thunderstorms in the target area of KSC is much less than that in the Socorro rocket range. The payloads and test procedure will be essentially the same as that with the FFARs.

4.3 Dropsondes

It is proposed that dropsondes be used with trailing wires, which are dropped from an airplane to ignite lightning discharges in the weaker fields of shower clouds as an intermediate step to the use of rockets with trailing wires. Rockets with trailing wires of about 300 ft in length to simulate the Apollo rocket are not available at present and it may take some time to develop them. However, it is not a technical problem to develop a capsule for a target practice bomb with a roll of wire, 300 ft in length inside, which is released at the drop. It is anticipated that the higher field concentration factor at the tip of the target bomb caused by the long wire will initiate a discharge in weaker fields of shower clouds. It should also be possible by the trailing wire method, to trigger lightning discharges in a thunderstorm which is already partially discharged by small rockets, and decrease the remaining field still further. It shall be the purpose of these dropsonde tests to determine how far such discharging processes can be carried.

The procedure of the test has to be different from the rocket test due to the fact that the airplane needs time to turn around and drop a second trigger bomb into the same place. On the other hand, a fast-moving airplane permits a series of drops in rapid succession, so that the bombs are dispersed along the flight path and will discharge a larger part of the storm than can be discharged by series of rockets which follow the same path. Another advantage of the airplane is that it may hunt for a developing storm over a wide range of the ocean and it is not forced to wait until a storm moves or develops in the target area. In this case, however, the monitoring system is limited to the instrumentation carried by the NASA-6 airplane and to the field-measuring dropsondes carried by the bombing airplane. The receiving and recording instruments for the dropsonde would then have to be installed in the NASA-6.

4.4. Determination of the Electric Charge of the Apollo Rocket in Flight

There may be a substantial electric charge on the Apollo rocket produced by the motors, which will support or impede the triggering of a lightning discharge in a strong enough electric field. It is, therefore, important to measure the polarity and the amount of the rocket charge. Because of the

overriding electric effect of the steam clouds produced during
lift-off, the charge on the rocket cannot be determined by
ground measurements. The best means to measure the rocket
charge is the NASA-6 airplane equipped with field mills
which should pass the rocket in flight at the closest distance
possible. It is suggested that this measurement be carried
out during the launch of Apollo 15 and a flight plan to this
effect will be submitted.

5. REFERENCES

Brook, M., C.R. Holmes, and C.B. Moore (1970), Analysis of Apollo 12 lightning incident, NASA-01540.

Evans, W.H. (1969), Electric fields and conductivity in thunderstorms, J. Geophys. Res. <u>74</u>, No. 4, 439-948.

Freier, G.D. (1968), The electrical structure of thunderstorms, Proc. Internatl. Conf. Cloud Physics, Toronto, Canada.

Hermstein, W. (1960a), "Die Stromfaden-Entladung und ihr Ubergang in das Glimmen:, Archiv für Elektrotechnik XLV Band Heft 3, 209-224.

Hermstein, W. (1960b), "Die Entwicklung der postiven Vorentladungen in Luft zum Durchschlag:, Archiv fur Elektrotechnik XLV Bank, Heft 4, 279-288.

Kasemir, H.W. (1969), Lightning hazard to rockets during Launch I, ESSA Tech. Rept. ERL 143-APCL 11.

Kasemir, H.W. (1971), Calibration of atmospheric electric field meters and the determination of form factors at Kennedy Space Center, NOAA/APCL Tech. Memo 11.

Loeb, L.B. (1965), Electrical Coronas (Univ. of Calif. Press, Berkeley and Los Angeles, Calif.).

Ruhnke, Lothar H. (1971), A rocketborne instrument to measure electric fields inside electrified clouds, NOAA Tech. Rept. (in press).

From #rd Conference on Weather Modification of AMC, June 26-29, 1972, Rapid City, SD, pp. 237-238.

THEORY AND EXPERIMENTS TO THE PROBLEM OF ROCKET TRIGGERED LIGHTNING DISCHARGES

Heinz W. Kasemir

Atmospheric Physics and Chemistry Laboratory
Environmental Research Laboratories
National Oceanic and Atmospheric Administration
Boulder, Colorado

1. INTRODUCTION

The incentive for this research came from the lightning incident during the launch of Apollo 12. It was evident that even electrified shower clouds, which do not generate lightning discharges by themselves, present a danger to the electronic systems of larger rockets. In the case of Apollo 12, the rocket was not hit accidentally by a natural lightning generated by the cloud, but the lightning was initiated by the rocket itself. The high field concentration at the Q-ball on the tip of Apollo 12 caused breakdown in the comparatively weak cloud field. This suggested a method discharging electrified clouds by artificial lightning discharges, triggered by inexpensive small rockets shortly before the launch of a large rocket or space vehicle. Such research is now being conducted by APCL under contract from NASA's Kennedy Space Center. The 5-year project is in its second year. A brief summary of the results obtained during this period is given here. A more detailed description of the problem areas, and of he different phases of the project, is given in the project reports (Kasemir, 1969, 1970a, b; Ruhnke, 1971).

The research program is divided into theory and experiments, and the experimental section is further subdivided into instrument development and field tests.

2. THEORY

For the theoretical calculation, the rocket is represented by a conductive spheroid and the cloud field is assumed to be a homogeneous field, at least for the length of the rocket. With these assumptions, the field concentration factor c at the tip of the rocket with the radius of curvature R and the rocket length a is given by

$$c = \frac{E_T}{F} = \frac{a/R}{0.5\ln\dfrac{4a}{R} - 1} \tag{1}$$

where E_T is the field at the tip and F is the homogeneous cloud field.

A remarkable feature of equation (1) is that the field concentration factor c depends only on the ratio of a/R, but not at the individual magnitude of either a or R. This means that the Apollo rocket, with $a = 100$ m, $R = 0.1$ m, and $a/R = 1000$, has the same field concentration factor as the much smaller folding fin aircraft rocket Mighty Mouse, with $a = 1.2$ m if the radius of. curvature R is also kept proportionally smaller, in this case $R = 1.2$ mm. The field concentration factor for both of these rockets is then $c = 345$. A cloud field of 8.7 kV/m would produce the breakdown field of 3000 kV/m at the tip. At the small Mighty Mouse rocket with the sharp tip, the breakdown field will produce a stable continuous corona discharge, whereas at the rounded Q-ball of the Apollo rocket the breakdown field will produce a spark discharge which may grow into a lightning discharge.

A corona discharge payload for the Mighty Mouse rocket, which measures the electric field along the rocket path, has been designed (Ruhnke, 1971), as we will see later under instruments. Using the right ratio a/R, the Mighty Mouse can be adjusted to respond with corona discharge to the same external field to which any other chosen larger rocket would respond with a triggered lightning discharge. In this way, a simple small indicator rocket can be designed to check electrified clouds for content of a dangerous electric field, just before launch of a larger rocket. To preserve the lightning triggering capability of the large rocket, the radius of curvature at the tip of the smaller rocket should not be less than 2.5 cm. According to laboratory tests, this is the smallest radius of curvature which will reliably produce spark discharges. Trigger payloads with spheres of 3.5 cm radius at the tip have been designed for the Mighty Mouse. However, due to the smaller ratio $a/R = 34.4$, and the smaller field concentration factor $c = 24$, the external field necessary to generate lightning discharge is now 125 kV/m. For a simulation of the Apollo-Saturn V by a smaller rocket with $R = 3.5$ cm, the rocket length has to be $a = 35$m to give the ratio $a/R = 1000$. It is hoped that a trailing wire of 34 m length supplement the length of the Mighty Mouse to furnish a device of the same lightning triggering capability as the Apollo-Saturn V.

3   EXPERIMENTS

3.1   Instruments.

The field measuring payload for the Mighty Mouse rocket is described in detail by Ruhnke (1971). The corona current is used to measure the electric field. A high ohm resistor linearizes the relation between field and corona current. A simple glow lamp circuit converts the corona current into short pulses of audio frequency, which in turn modulate a 400 MHz transmitter.

The lightning triggering payload was recently also equipped with a transmitter, which is modulated by a constant audio frequency pulse of 1000 Hz. If a lightning is triggered, the transmitter is destroyed and stops transmission. The transmitter serves the purpose of identifying the triggered from the natural lightning discharge.

The NASA 6 airplane is equipped with two cylindrical field mills described by Kasemir (1969). The three components of the electric field are recorded with these instruments as the airplane flies underneath the clouds.

A skilled observer can guide the airplane to the area of maximum field by reading the field records. The purpose of the airplane is twofold. First, to pinpoint the maximum field area into which the measuring and triggering rockets are fired, and second, to determine if the field is high enough that lightning can be triggered on the cloud field.

At the ground, the cloud field is recorded with a field mill, and the lightning field with a horizontal wire antenna. The antenna gives a time record of all lightning strokes, but does not discriminate between natural and triggered discharges. This record is especially useful for statistical evaluation (Magaziner, 1972). This method is greatly improved by the lightning plotter, which records the position of the lightning strokes (Ruhnke, 1972). Natural lightning discharges, which occur outside the target area, can be separated from the triggered discharge and the one or two strokes – if any – which may occur inside the target area during the rocket flight.

The last ground equipment is the rocket payload receiver. The measuring payload records the field along the rocket path before and after a series of triggering payloads. This furnishes clear evidence of the extent that the cloud has been discharged by the triggered lightning discharges. The cessation of the trigger payload transmission with a simultaneous lightning field record of the wire antenna gives unmistakable proof that a lightning has been triggered.

3.2   Field Tests

The field tests were carried out with Mighty Mouse rockets without trailing wire. They were intended primarily as equipment tests and as a means to develop test procedures. However, they also yielded the first confirmation that lightings can be triggered under circumstances predicted by theory. The Socorro 1970 test is discussed in detail by Kasemir (1970b). Nine lightings were triggered out of 15 firings. Here the cases of not-triggered lightings are just as informative as the cases of triggered lightings. Firings were usually continued until the last one or two rockets did not trigger. It came as a surprise that three lightings could be triggered by rockets fired in 30 sec intervals into the same part of the cloud. The exact time interval (61.0; 59.5; 60.5 sec) between launch and lightning does not leave much doubt that all three lightings were triggered by the rockets. A statistical evaluation gave the probability of 0.98 for triggered rockers and of.0.02 for a natural occurrence.

The electric field measured by the airplane underneath the cloud before and after the firing was in one case 150 and 50 kV/m, and in a second case 60 and 40 kV/m. The theory predicts triggered lightings in fields of 125 kV/m or higher. This agrees very well with the first case. In the second case it leads to the rather obvious conclusion that the field inside the cloud is higher by at least a factor of 2 than underneath the cloud. The field measuring rocket – which was not available at the first Socorro test – will give the exact ratio. The second conclusion which can be drawn from the field values is that the field underneath the cloud is reduced by the triggered lightning discharges to 50 kV/m or less. This is confirmed by the fact that lightings were not triggered by the Mighty Mouse if the rocket was red into fields less than 60 kV/m.

4   REFERENCES

Kasemir, H. W., 1964: The cylindrical field mill. ECOM, U.S. Army Tech. Report 2526.

Kasemir, H. W., 1969: Lightning hazard launch. Part I. ESSA Tech. Report L 143-APCL 11.

Kasemir, H. W., 1970a: Lightning hazard to rockets during launch, Part II. ESSA Tech. report ERL 144-APCL 12.

Kasemir, H. W., 1970b: Basic theory and pilot experiments to the problem of triggered lightning discharges by rockets, NOAA, APCL, April 1971, Contract CC-88025-1970.

Magaziner, E. L., 1972: Evaluation of Mighty Mouse trigger rocket flights, Progress report II, Contract CC-88025 NASA Kennedy Space Center.

Ruhnke, L. H., 1971, A rocket-borne instrument to measure electric fields inside electrified clouds. NOAA Tech. Report ERL 206, APCL 20.

Ruhnke, L. H., 1972: Evaluation of an automatic lightning positioning system, NOAA Tech. Report, in preparation.

Lightning Suppression by Chaff Seeding and Triggered Lightning

HEINZ W. KASEMIR

LIGHTNING SUPPRESSION BY CHAFF SEEDING

The Physical Concept

The purpose of seeding a thunderstorm with chaff is to inhibit lightning discharges. The physical concept of this method is to increase the conductivity of the cloud by ionizing the air by corona discharge on the chaff fibers so that the electric field is kept below the value necessary to ignite lightning. The principal idea of chaff seeding can be easily demonstrated by a laboratory experiment. A metallic sphere of about 0.5-

m diameter is placed over a grounded plate with an air gap of 10 to 20 cm between sphere and ground. If the sphere is charged to several 100 kV—let's say, by a Van de Graff generator—sparks will flash over between the sphere and the grounded plate. If we now bring into the air gap one chaff fiber of about 5 cm length, sparkover will stop immediately. The chaff fiber is attached to the end of a long thin Teflon rod, that is, it is well insulated and doesn't touch either the sphere or the plate. If we remove the chaff fiber from the air gap the sparks will flash over again. In this experiment the charged sphere represents the thunderstorm, the plate is the earth's surface, and the sparks are cloud-to-ground lightning. The chaff fiber shows the effect of chaff seeding. The effect can readily be explained. Corona discharge at the two ends of the fiber produce a large number of ions which flow in a wide stream from the upper end of the fiber to the sphere and from the lower end to the plate. This ion flow increases the conductivity of the air between sphere and plate and the resulting current—if not actually shorting the sphere to ground—is such a load on the Van de Graff generator that

the voltage at the sphere drops below the flashover voltage.

This experiment shows us that there are three important conditions that must be fulfilled if chaff seeding of a thunderstorm is to be effective in suppressing lightning discharges. These conditions are (1) that the volume of air in a storm or between the storm and ground can be made conductive by the corona discharge on the chaff fibers; (2) that the electric field necessary to produce lightning is higher than that to produce corona discharge; and (3) that the current induced by corona discharge will load down the thunderstorm generator so that the electric field remains below the lightning igniting value.

Let us construct an electrical circuit diagram which is equivalent to a thunderstorm (see Figure 17.7.) On the left side of Figure 17.7, the generator symbols G_1 and G_2 represent the positive and negative charge generation in the top and the base of a storm which would result in a charge distribution in the cloud, as shown on the right side of the figure. G_1 and G_2 are assumed to be constant current generators related to the microphysical processes in the cloud

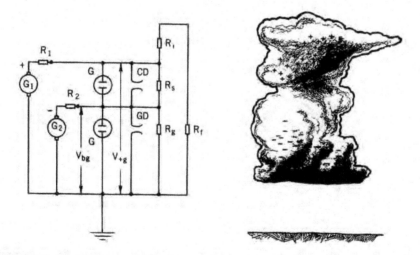

Figure 17.7 Left, circuit diagram of a thunderstorm. Right, charge distribution in a thunderstorm.

LIGHTNING MODIFICATION

which is steadily producing positive and negative ions. The resistors R_1 and R_2 associated with these generators control the supply current, that is, the charge production of the cloud. The potential from cloud bottom to ground V_{bg} due to G_2 and the potential from cloud top to ground V_{tg} due to G_1 both vary with the strength of the current source and also with the resistances between the respective terminal and ground. Three different means are provided in the circuit diagram for the dissipation of charge. These are, from left to right in Figure 17.7, the two glow lamps G, the two spark gaps CD and GD, the ohmic resistors R_s, and the series R_i, R_f, R_g. The current flow through the ohmic resistors represents the conduction current of a thunderstorm. R_s is the shunt resistor between the hot terminals of the two generators; it is determined by the conductivity of the column of air between the upper positive and the lower negative charge inside the storm. R_g is the resistance of the air column between the base of the cloud and the ground, and R_i the resistance of the air column between the top of the storm and the ionosphere. This branch of the circuit is closed by R_f which represents the resistance between the ionosphere and the earth in the fair weather areas. With the exception of R_f, the resistors have comparatively high values (on the order of hundreds of megohms) so that the charge produced by the generators cannot leak away very fast. As a consequence, the voltage at the terminals builds up to high values until a spark is ignited through the spark gaps CD or GD. These spark gaps picture the lightning discharges in a real thunderstorm, CD representing the cloud discharge and GD the ground discharge. The glow lamps of the circuit diagram have no natural equivalent in the thunderstorm. They represent the artificially introduced corona discharge of the chaff fibers. If the voltage across the glow

lamps reaches their breakdown value, the lamp will ignite and keep the voltage very effectively at this value. Any further increase in the current output of the generators will be shunted through the glow lamps. If the glow lamps have a lower ignition voltage than the spark gaps, no flashover at the spark gaps will occur.

It is generally assumed that the average thunderstorm produces a current output of about 3 A, which is equally divided between the three load circuits. Between the two main charge centers inside the storm through the shunt resistor R_s, 1 A flows as conduction current. From a technical point of view, this current may be considered a leakage current across faulty insulation between the poles of the generator. It is the purpose of chaff seeding to increase this current by increasing the conductivity between the generator poles so much that the voltage between the poles is reduced below the lightning igniting voltage. It will be the topic of the next section to discuss in detail how the conductivity is increased by the corona discharge on the chaff fibers.

Another part of the current output flows from the positive pole in the top of the storm to the ionosphere through the resistor R_i, spreads out in the ionosphere over the whole globe, flows down to earth through the resistor R_f, and returns to the negative pole of the thunderstorm generator through the resistor R_g. This is the contribution of the storm to the atmospheric electric fair weather field. The current flow in this branch — the global circuit — is assumed to be on the order of 1 A.

The last part of the thunderstorm charge production is dissipated in lightning discharges. If we assume that the average thunderstorm produces one lightning discharge every 30 sec and that 30 C is discharged by each lightning, 1-A continuous-supply current is required to feed the lightning activity. More severe storms may generate lightning discharges every 5 or 10 sec. With the same amount of coulomb discharged, the lightning supply current would be on the order of 3 to 6 A. As has been mentioned, the thunderstorm is in a technical sense, a current generator with a given current output; therefore, if lightning activity is to be suppressed, the lightning supply current of 1 to 6 A has to be channeled through another circuit, which in our case is the two glow lamps in parallel with the resistors R_s and R_g. To absorb this current without increase in the terminal voltage these resistors have to be decreased to $\frac{1}{2}$ to $\frac{1}{5}$ of their natural value by chaff seeding. For proper operation of the circuit the ignition voltage of the glow lamp should be less than the flashover voltage of the spark gap, and the glow lamp should be capable of absorbing the spark gap supply current without a substantial increase in the generator voltage.

Applied to the thunderstorm problem this means that the field to start corona discharge on the chaff fibers should be lower than the lightning-igniting field and that the conductivity of the chaff seeded volume should be increased to such an extent that the lightning supply current can be absorbed without a significant increase of the field. The first condition is met easily because the corona onset field of a chaff fiber 10 to 100 μ thick and 5 to 10 cm long is of the order of 30 kV/m, whereas the lightning-igniting field is about 500 kV/m, that is, more than 10 times as large. The fulfillment of the second condition is not so easily asserted and will be discussed in the next section.

The Effect of Corona Discharge at Chaff Fibers on the Conductivity of the Cloud

The conductivity of the air stems from the ionization of air molecules by cosmic rays and radioactive emanation of the ground. The ion production is given by the ionization constant q, which is of the order of 5 to 20×10^6 ion pairs/sec m^3. One single chaff fiber with a corona current of 1 μA will produce 6×10^{12} ion pairs/sec; that is, it would be equivalent to the natural ion production of a volume of 3×10^5 m^3 or of a cube of 68-m length, width, and height. If we require that the ion production in the cube should be increased three times, then the chaff fiber with 1-μA corona current would provide the required number of ion pairs for a cube with a side length of about 46 m. A cloud volume of $4 \times 4 \times 4$ km^3 = 64×10^9 m^3 contains 65×10^4 such cubes. This means that about half a million chaff fibers would be enough to increase the ion production in the cloud by a factor of 3. Implicit in this estimate is the assumption that a field of about 70 kV/m exists to produce the 1-μA corona current and that the chaff fibers are evenly distributed throughout the volume.

In estimating the increase of conductivity by an increase of ion production we have to take into account the loss by recombination and attachment. As the negative ion stream released by the chaff fiber is moving upward and the positive ion stream is moving downward, ions of opposite polarity do not meet; consequently, there is no ion loss by recombination of small ions. Only if the upward moving negative ion stream meets the downward moving positive ion stream produced by another chaff fiber floating at a higher level will there be recombination of small ions. However, this recombination or intermixing will be beneficial for a continuous current flow, as will be discussed later.

A heavy loss of small ions will occur in the first few seconds by attachment to cloud droplets until the droplets are charged to capacity and reject further attachment of ions of the same sign. An area of negative space charge will form above the chaff fiber and one of positive space charge below it. This will have two

LIGHTNING MODIFICATION

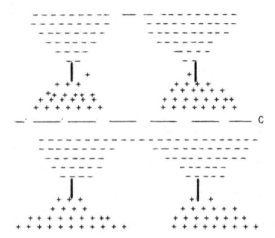

Figure 17.8 Positive and negative space charge distribution of an array of chaff fibers.

effects. First the presence of a negative space charge in the negative ion stream will deflect the ions to both sides until the hitherto uncharged cloud droplets at the sides become also negatively charged. This will result in a kind of trumpet-shaped space charge, as depicted in Figure 17.8. These negative and positive space charges forming above and below the single chaff fiber will generate an electric field, opposed to the existing field, which produces the corona discharge. If the space charge can accumulate unchecked or their opposing field is not cancelled by other means, then the corona discharge will be quenched by its own space-charge generation.

However, this result is applicable only to a single chaff fiber or to the uppermost or lowest layer of the seeded area. If we have two layers of chaff fibers, as in Figure 17.8, the fields of the inner positive and negative space charge layers will approximately cancel at the locus of the chaff fiber. The extrapolation to more than two layers of chaff fibers is evident. Such a volume filled with positive and negative space charge pockets resembles very strongly that of a dielectric material polarized by an external field. The local fields of the internal dipoles cancel each other by the sheer number of dipoles, and

the reduction of the primary field inside the dielectric can be interpreted as the effect of the surface charge at the boundary of the dielectric material.

In the case of chaff seeding for the purpose of lightning suppression, we require in addition to the field reducing effect that an enhanced current flow is carried through the seeded area. This would necessitate that at the contact area—marked by a dashed line C in Figure 17.8—between the positive space charge of the upper chaff fiber and the negative space charge of the lower chaff fiber a strong recombination between the opposite charged cloud droplets takes place, or that this contact area is penetrated by a sufficient number of positive small ions coming down from the upper chaff fiber and of negative small ions coming up from the lower chaff fiber. Even if the small ions recombine rapidly with cloud droplets or small ions of opposite polarity they reduce the space-charge pockets and stimulate a continuous corona discharge. For a continuous current flow through the seeded area it is necessary only that each chaff fiber carry the current through its own region of influence.

A numerical analysis to determine the size of the region of influence is extremely difficult. Probably the best way to solve this problem is to carry out measurements in a large cloud chamber equipped with a plate condenser, which is capable of generating fields in the order of 100 kV/m; however, there are three effects which tend to reduce the little pockets of space charge, thereby helping to prevent the quenching of the corona discharge, and to support the current flow. The effects are turbulent mixing, recombination of positively and negatively charged cloud droplets, and washout by precipitation. The enhanced coagulation of oppositely charged cloud droplets may even lead to precipitation enhancement. This would apply also to raindrops falling through the seeded area. The raindrop picks up a number of highly charged cloud droplets passing through, say, the upper negative space charge pocket. It becomes negatively charged itself. Entering the positive space charge pocket below, the raindrop now coagulates with the positive charged cloud droplets, loses its negative charge, and becomes positively charged. This process continues alternately until the raindrop leaves the seeded area. Besides reducing the space-charge pockets this also increases the growth rate of the raindrop.

We assume that an enhanced current flow can be carried through the seeded area. Given a sufficient overlapping of the region of influence of the chaff fibers, the large number of small ions liberated by corona discharge, the recombination of charged cloud droplets, the effect of turbulent mixing, and the washout of space-charge pockets by precipitation are lumped together in one material parameter, namely, the conductivity inside the seeded area, with the resulting effect that the conductivity inside the seeded area is increased by corona discharge. Accepting the assumption that the conductivity inside the seeded area is increased by chaff seeding, the problem of the field concentration at the surface of a conductive body in an electric field

comes immediately to mind. If we also generate by chaff seeding regions of field concentration, this method of lightning suppression may inadvertently turn out to be a method of lightning triggering. It is therefore quite important to have quantitative answers to the following questions: (1) how does the field concentration at the boundary depend on the conductivity ratio inside to outside of the seeded area; (2) how does it depend on the geometrical shape of the seeded area; and (3) how much is it changed if we are dealing with the inhomogeneous field of a thunderstorm generated by space charges instead of the general assumption of a homogeneous field.

The solution of the third problem is more difficult and has been worked out and discussed.[18] The solution to problems 1 and 2 is given in the next section.

The Field Concentration at the Boundary of the Chaff Seeded Area

A body of a given shape with the constant conductivity λ_2 is embedded into an environment of the constant conductivity λ_1. The whole system is exposed to a homogeneous electric field F_1 in the direction of the z axis of a Cartesian coordinate system x, y, and z. Find the field F_2 inside the body and the maximum field F_m at the boundary of the body. The body shall be represented either by an infinitely long elliptical cylinder or by a spheroid. The solution shall be given for the cylinder for the following cases: (ac) the small axis of the elliptical cross-section is in the z direction, (bc) both axes are equal, in which case the elliptical cylinder becomes a circular cylinder, and (cc) the large axis of the elliptical cross-section is in the z direction; and for the spheroid, for (as) a flat disc, with the short axis in the z direction, (bs) a sphere, and (cs) a prolonged spheroid, with the long axis in the z direction.

LIGHTNING MODIFICATION

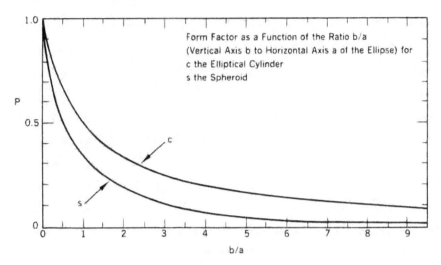

Figure 17.9 Form factor for ρ as a function of the ratio b/a.

The cases ac to bc may be encountered if the chaff is dispersed continuously from an airplane and the cases as to cs if the chaff is dispersed discontinuously in little packages from an airplane, dropsonde, or rocket.

The potential function of cases ac to cc can be obtained from *Static and Dynamic Electricity* by Smyth[37] if the dielectric problem is transformed into a current-flow problem by substituting the conductivities λ_1 and λ_2 for the two dielectric constants ξ_1 and ξ_2. The solution to problems as to cs is given by Kasemir.[13]

From these potential functions a very simple equation has been obtained for the maximum field concentration F_m at the boundary of the seeded area, as well as the attenuated field F_2 inside the seeded area. These equations are valid for all the problems stated earlier and are

$$\frac{F_2}{F_1} = \frac{\lambda_1/\lambda_2}{\lambda_1/\lambda_2 + (1 - \lambda_1/\lambda_2)\rho} \quad (17.1)$$

and

$$\frac{F_m}{F_1} = \frac{1}{\lambda_1/\lambda_2 + (1 - \lambda_1/\lambda_2)\rho}. \quad (17.2)$$

From Equations 17.1 and 17.2 also follows

$$F_2 = \frac{\lambda_1 F_m}{\lambda_2}, \quad (17.3)$$

where F_1 is the original field outside the seeded area, F_2 is the field inside the seeded area, F_m is the maximum field at the boundary, λ_1 is the conductivity outside the seeded area, λ_2 is the conductivity inside the seeded area, and ρ is the form factor, which depends only on the geometrical shape of the seeded area (Figure 17.9). With the horizontal half-axis a and the vertical half-axis b of either the elliptical cylinder or the spheroid, ρ is given for the elliptical cylinder by

$$\rho_c = \frac{a}{a + b} \quad (17.4)$$

and for the spheroid by

$$\rho_s = \frac{a^2}{a^2 - b^2}\left(1 - \frac{b}{\sqrt{a^2 - b^2}}\right.$$

$$\left. \cdot \tan^{-1}\frac{\sqrt{a^2 - b^2}}{b}\right). \quad (17.5)$$

For $a \to \infty$ the elliptical cylinder, as well as the spheroid, degenerate into a hori-

519

zontal infinite layer. Note that ρ_c and ρ_s approach 1. For $\rho = 1$ it follows from Equation 17.1

$$\frac{F_2}{F_1} = \frac{\lambda_1}{\lambda_2}. \qquad (17.6)$$

As F_1 and λ_1 are given parameters the field F_2—according to Equation 17.6—is inversely proportional to λ_2. This is a well-known result and follows from the assumption of a continuous current flow. If $i_1 = \lambda_1 F_1$ is the current density outside and $i_2 = \lambda_2 F_2$ inside the seeded area we obtain from Equation 17.6

$$i_2 = i_1. \qquad (17.7)$$

If we bring a conductor into a homogeneous field there is usually a field concentration at the boundary. However, in this case inserting $\rho = 1$ into Equation 17.2 we obtain

$$F_m = F_1. \qquad (17.8)$$

The maximum field F_m is equal to the original field F_1 outside the seeded area; that is, there is no field augmentation at the boundary.

If we go to the other extreme and let $b \to \infty$ then the elliptical cylinder deteriorates into a vertical infinite layer and the spheroid into an infinitely long circular cylinder. In this case ρ_s and ρ_c approach zero. From Equations 17.1 and 17.2, we obtain

$$F_2 = F_1 \qquad (17.9)$$

and

$$\frac{F_m}{F_1} = \frac{\lambda_2}{\lambda_1}. \qquad (17.10)$$

These are somewhat unexpected results. Equation 17.9 says that the field inside the seeded area is the same as outside the seeded area no matter how much we increase the conductivity inside the seeded area. This means that continuous seeding from a dropsonde, where we may generate a body of seeded area like a

prolonged spheroid, would have almost no field-reducing effect. Furthermore, we learn from Equation 17.10 that the field concentration on the ends of the prolonged spheroid is proportional to the inside-outside conductivity ratio of the seeded area. Both effects are favorable for generating lightnings and adverse to lightning suppression; therefore, the manner in which a cloud is seeded will have some effect on the outcome, that is, if lightning is suppressed or prematurely triggered.

Equation 17.2 answers completely questions 1 and 2 stated at the beginning of this section. The maximum field concentration F_m/F_1 at the boundary of the seeded area and the field attenuation F_2/F_1 inside the seeded area is shown in Figure 17.10 as a function of the ratio b/a (vertical to horizontal dimension) for the cylindrical as well as the spheroidal case.

To discuss the influence of the conductivity λ_2 of the seeded area on the field concentration we go first to the extreme case that $\lambda_2 = \infty$. This would be equivalent to the case of a metallic conductor in an electrostatic field. According to Equation 17.2 the field concentration factor is $F_m/F_1 = 1/\rho$. This means that the reciprocal of the form factor is identical with the electrostatic field concentration factor. For the sphere, for instance, $\rho = \frac{1}{3}$ and $F_m/F_1 = 3$. If λ_2 is not infinite, but a finite multiple of λ_1, the field concentration factor decreases. In the case of the sphere and for $\lambda_2 = 2\lambda_1$ it drops to 1.5. Finally, for $\lambda_2 = \lambda_1$ the field concentration factor is 1, as it should be. Question 2, about the influence of the geometrical shape on the field concentration factor, is even easier to answer because the maximum "electrostatic" field concentration factor for all shapes can be obtained from $1/\rho$, as shown in Figure 17.9. In general the smaller the ratio b/a (vertical to hori-

LIGHTNING MODIFICATION

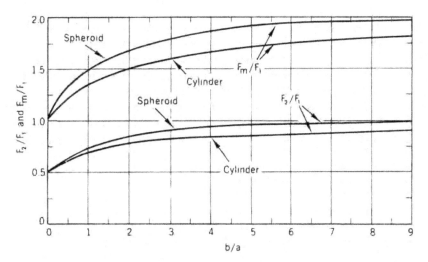

Figure 17.10 Field concentration F_m/F_1 at the boundary of the seeded area. Field attenuation F_2/F_1 inside the seeded area.

zontal dimension of the seeded area), the smaller the field concentration factor, with the cylindrical shape more favorable than the spheroidal shape. The electrostatic field concentration factor of the horizontal circular cylinder is 2 as compared to the sphere, where it is 3. Assuming again an inside conductivity $\lambda_2 = 2\lambda_1$, the value is reduced to 1.33 as compared to the sphere value of 1.5. This means that line seeding from an airplane flying horizontally through or below the cloud will produce the most desirable shape of the seeded area.

Using as a reference a 1973 paper by Kasemir,[19] we can answer question 3—how much is the field concentration factor changed if we are dealing with the inhomogeneous field of a thunderstorm instead of a generally assumed homogeneous field—as follows. The seeded area has an equalizing effect on the field distribution of a thunderstorm. It will decrease the maximum field value between the two main charge centers and will increase the outside field at the upper and lower boundary of the seeded area, analogous to the maximum field

concentration in a homogeneous field. However, this value is lower than in the case of the homogeneous field and if the seeded area is large enough to reach the poles of the thunderstorm generator, it is even lower than the maximum field of the not-seeded storm.[18]

Field Tests with Chaff Seeding

Field tests of chaff seeding underneath thunderstorms have been carried out at Flagstaff, Arizona in 1965 and 1966. For these tests a C-47 airplane was equipped with the following instruments: (*a*) two field mills measuring the three components of the electric field; (*b*) two chaff dispensers; (*c*) one corona discharge indicator; and (*d*) sensors for different meteorological and airplane parameters, including a strip chart and tape recorder.

Each field mill records two components of the atmospheric electric field and automatically eliminates the influence of the airplane charge on the measurement.[18]

The chaff dispenser contains the chaff not as fibers cut to a certain length and pressed a million apiece into little packages, but as a long strand of conductive fibers wound up on a reel. Ten reels constitute one dispenser unit housed in one wing tank. During operation the 10 strands are forced out through 10 guide holes at great speed, and before leaving the tank completely are chopped by a helical chopper into fibers of a preset length. This design has several features that are crucial for lightning suppression. (1) The chaff is emitted continuously. Bird nesting, that is, bunching together in clumps of several hundred or thousand fibers, is completely eliminated. (2) The chaff is distributed more evenly behind the airplane because a continuous stream of fibers and not individual packages emerges from the airplane. (3) It is possible to experiment with different lengths of fibers, the length depending on the speed of the chopper, which is easily adjustable.

The corona discharge indicator measures the electromagnetic emission of the chaff fibers as soon as they emerge from the chaff dispenser. This range should be limited to 50 to 100 m and the indication selective to corona discharge on the chaff only; that is, corona dis-

charge on the airplane itself should not be indicated. The last point is difficult to establish and needs a more detailed study. Otherwise the instrument seems to be working properly. One essential point has already been confirmed namely, that corona discharge occurs on chaff fibers if the electric field in the atmosphere surpasses a threshold value of about 25kV/m, which is in agreement with the laboratory tests.

The airplane would fly below developing thunderstorms or shower clouds and hunt for areas with electric fields of about 30 kV/m or more. If such an area was found and the field pattern had been established by several passes through this area, chaff would be ejected on two to four test runs. The electromagnetic emission of the corona discharge and the electric field were recorded during consequent passes back and forth through the seeded area until either corona discharge or the strong electric field disappeared. Figures 17.11 and 17.12 are typical examples of such flight records.

Figure 17.11 shows the corona discharge on the upper trace and the vertical field component on the lower trace. Seeding has been marked on the record

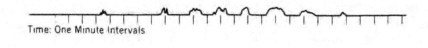

Time: One Minute Intervals

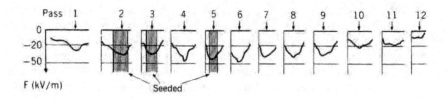

Corona Discharge Generated by Chaff Seeding
2 August 1966

Figure 17.11 Corona discharge generated by chaff seeding.

LIGHTNING MODIFICATION

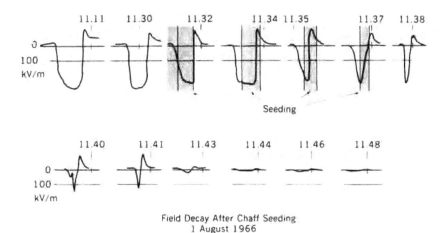

Field Decay After Chaff Seeding
1 August 1966

Figure 17.12   Field decay after seeding.

as shaded areas. Eleven passes have been made below the storm. On the second, third, and fifth passes, chaff was dispensed. On the second and fourth passes, the trace of corona discharge emission was small and irregular. At the third pass the plane missed the previous seeded area completely, and no corona discharge was recorded. But it seems that the chaff fibers spread out very rapidly and after 5 to 10 min the whole area was solidly filled with corona discharge emissions until the fields dropped below 20 kV/m.

Figure 17.12 shows the decay of a strong electric field after chaff seeding. It may be pointed out that between the first and the second pass about 20 min elapsed because the area was lost and could not be relocated any earlier. This proved to be fortunate because it shows that during this time the field remained at its high value of about 300 kV/m. After the area was found again chaff seeding began at the third and following passes. The decay of the field could be recognized 3 min after chaff seeding started. Ten minutes thereafter the field had completely collapsed.

These experiments have established

that corona discharge is generated if chaff fibers are dispersed in the electric field of thunderstorms exceeding values of 30 kV/m.

It seems highly probable that the decay of strong electric fields is caused or accelerated by corona current produced by the chaff fibers.

There are a number of open questions which must be answered before chaff seeding becomes an effective tool for lightning suppression. The most important task is to prove experimentally that the effect of chaff seeding on the conductivity inside the cloud is as stated above. This implies that chaff can be dispersed in a reasonable time through a large volume of the cloud. If we visualize the thin line of chaff ejected from the airplane on one seeding run, the fast field decay shown in Figure 17.12 is indeed remarkable.

TRIGGERED LIGHTNING

The two lightning strikes to Apollo 12 shortly after launch in 1968 prompted more systematic research into the problem of rocket triggered lightning discharges. This investigation may even lead to the possibility of temporarily discharging electrified clouds by triggered lightning. The pioneer work of this research was done by M. M. Newman[26-28] with J. R. Stahman and J. D. Robb,[29] who successfully triggered lightning discharges with rockets fired into a thundercloud from a ship. The rockets were connected to the ship by a trailing wire. This research has been continued by this author with L. R. Ruhnke and others.[15-17, 19, 35]

It is not realized in general that the airborne vehicles themselves cause the lightning strikes if they enter regions with high electric fields. It is very seldom that they are hit accidentally by natural lightning produced by the storm. The probability of lightning striking an airplane during the flight through the cloud is much too small to account for the number of hits. Furthermore, Apollo 12 was struck twice in a cloud which did not produce natural lightning. There are several other arguments in favor of the trigger hypothesis; however, the most convincing one is this research itself. If lightning is caused by the airborne vehicle, it should be possible to trigger lightning by test rockets under the right conditions any time. If lightning strikes to airborne vehicles are accidental hits by natural lightning when about 1000 to 10,000 rockets are fired into a storm, only one may be struck by natural lightning. The experimental results, which are discussed later, show quite clearly that the trigger hypothesis is the correct one.

From calculations given in detail[17] it can be deduced (1) that the cloud field, which caused the lightning strikes during the Apollo 12 launch, was only on the order of 10 kV/m, or (2) that the small "folding fin aircraft rocket Mighty Mouse" will trigger a lightning if fired into a cloud field of about 100 kV/m. Our next step then is to carry out an experimental test to prove or disprove the theory.

Experiments

The theory predicts that lightning can be triggered even with a small inexpensive rocket such as the 2.75 folding fin aircraft rocket Mighty Mouse (Figure 17.13). This is somewhat surprising because it was always thought that large rockets or rockets with long trailing wires are necessary to trigger lightning discharges. The price to pay for the cheap triggering method is the necessity of comparatively high fields on the order of 100 kV/m. Such fields occur only in thunderstorms, and even then are restricted to small areas and to certain time periods in the life cycle of the storm. This imposes two problems on the experimental procedure. First, how to locate the high field areas, and second, how to distinguish a triggered from a natural lightning. The first problem was solved by using an airplane equipped with a field recording system[18] to monitor electric field underneath thunderstorms in the rocket range. If the field reached the theoretically determined threshold value, rockets were fired into this area of the cloud. To distinguish a triggered from a natural lightning several criteria were applied. The distance of the lightning from the rocket launcher as determined by the time interval between lightning field change and thunder arrival should be inside the rocket range. Lightnings triggered by consecutively fired rockets should have the same distance from the launcher and/or occur at approximately the same time interval after launch or both. In addition to these criteria a statistical analysis has been carried out to determine the probability of

LIGHTNING MODIFICATION

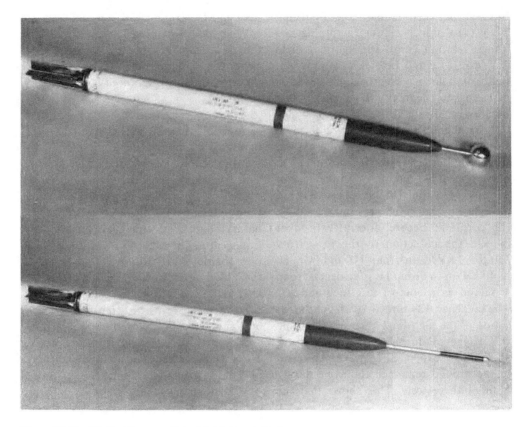

Figure 17.13   Mighty Mouse rocket with trigger payload.

the occurrence of a natural lightning at the time of the triggered lightning.

Field tests to trigger lightning discharges have been carried out at the Langmuir Observatory of the Institute of Mining and Technology at Socorro, New Mexico, with trigger rockets (Mighty Mouse, Figure 17.13) topped with metal spheres of 5, 7, or 10 cm diameter, referred to in Table 17.1 as S (small), M (medium), and L (large); a field mill and a wire antenna at the ground to record the cloud and the lightning field; and an airplane equipped with cylindrical field mills[18] (Figure 17.14) to record the three field components in the air. The airplane checked the field underneath clouds in the target area and transmitted the field

values to the launch site. When the field surpassed 100 kV/m—or at the last firing, 60 kV/m—the airplane left the target area and rockets were fired into the peak field region. After the firing the airplane was recalled to check the discharging effect of the triggered lightning on the cloud field. The distance of the lightning strike from the ground station was determined by the lightning-thunder time interval assuming the speed of sound to be 330 m/sec.

Table 17.1 shows the result of 15 rockets fired into storms at Socorro with the following parameters listed in the columns from left to right: (1) date and time; (2) diameter of the trigger sphere, L = 10 cm, M = 7 cm, S = 5 cm; (3) time interval between launch and lightning (sec); (4) distance between lightning and ground station (km); (5) lightning field (kV/m); (6) cloud field (kV/m); (7) azimuth of launcher (degree); (8) elevation of launcher (degree); (9) airplane field before and after launch (kV/m); and (10) test number. The rockets were fired in five small groups marked I to V, of two, three, or four rockets each aimed to answer specified questions. In the case of groups III and IV, where the field underneath the cloud was measured by the airplane before and after rocket launch, the reduction of the field from 150 to 50 kV/m in group III and from 60 to 40 kV/m in group IV is evident. Group V served as a control test to group IV. The launcher was turned 20° from the location of the high-field area. No lightning was triggered during this misalignment, as could be expected. However, the field at the ground remained the same during both of these tests. This shows that the field at the ground did not indicate localized high-field areas in the cloud and therefore cannot be used to determine either the lightning potential of a cloud or the area in which lightning is most likely to occur.

Another remarkable feature of test group IV is that the three triggered lightning discharges occurred practically the same time after the respective launch. This indicates that the lightnings were caused by the rocket penetrating a high-field area because an occurrence of natural lightning with such an accurate timing after three consecutive launches is improbable. Furthermore, it shows that the cloud area was not completely discharged by the first lightning but that enough charge remained to produce a second and third lightning discharge. However, the third lightning is rather weak, as can be recognized by the small lightning field at the ground (column 5).

We close this section with a brief

Figure 17.14 Airplane with top and nose field mill.

Table 17.1　Triggered Lightning Data from Field Test at Socorro, August 1970

Date, Time	L = 10 cm M = 7 cm S = 5 cm	Time Interval (sec)	Distance Lightn. Station (km)	Lightning Field Ground (kV/m)	Cloud Field Ground (kV/m)	Launch Azimuth (deg)	Launch Evelation (deg)	Cloud Field Airplane (kV/m)	Group
Aug. 8									
17 h 15.38	M	54	5.6	4	−25	—	—	—	
17 h 19.09	S	—	—	—	−25	—	—	—	I
17 h 31.12	M	—	—	—	−25	—	—	—	
Aug. 13									
15 h 42.51	M	53	5.3	12.5	−7.5	196	080	—	
15 h 44.52	L	59	6.9	3.5	−6	196	080	—	II
Aug. 13									
18 h 55.04	L	45	3.6	20	−9	309	079	Before 150	
18 h 56.49	M	27	3.0	10	−11	309	079	—	
18 h 58.35	L	38	3.6	12	−11	309	079	—	III
19 h 00.42	M	—	—	—	−11	309	079	After 50	
Aug. 14									
15 h 26.46	M	61.0	—	1	−1	325	080	Before 60	
15 h 27.14	L	59.5	—	5.5	−1.2	325	080	—	IV
15 h 27.43	M	60.5	—	0.5	−1	325	080	After 40	
Aug. 14									
15 h 29.04	M	—	—	—	−1	305	080	—	
15 h 29.36	L	—	—	—	−1	305	080	—	V
15 h 30.04	M	—	—	—	−1	305	080	—	

LIGHTNING MODIFICATION

assessment of the research aimed at discharging electrified clouds by triggered lightning discharges.

The theory given in my 1971 paper[17] seems to be broad enough to cover the essential physical facts of the triggered lightning problem, and its predictions agree with the experiments carried out so far. The instruments are adequate to measure the necessary parameters. The field tests must be considered to be in the preliminary stage.[17, 35] There is little doubt that lightning can be triggered even by such a small rocket as the folding fin aircraft rocket Mighty Mouse. Further tests are necessary to determine the electric parameters of the triggered lightning and its effect on the cloud charge. This leads into the problem of discharging the cloud, which is presently of a more speculative nature. Problems to be solved here, for instance, are the charge distribution in the cloud during its life cycle, the amount of charge destroyed by the individual lightning, and the regenerative power of the charge-producing mechanism in the cloud. The preliminary field test at Socorro indicates that the cloud field can be reduced by triggered lightning in a small area of the cloud to such an extent that further lightnings are not ignited by subsequent rockets, but the region discharged and the time it stays discharged are still open questions.

REFERENCES

1. Aufdermaur, A. N., and D. A. Johnson, Charge separation due to riming in an electric field, *Quart. J. Royal Meteorol. Soc.* **98**, 369–382, 1972.

2. Barrows, J. S., Weather modification and the prevention of lightning-caused forest fires, *Human Dimensions of Weather Modification*, W. R. Sewell, Ed., Dept. Geog. Research Paper 105, Univ. Chicago, Ill., 1966, pp. 169–182.

3. Baughman, R. G., and D. M. Fuquay, Hail and lightning occurrence in mountain thunderstorms, *J. Appl. Meteorol.* **9**(4), 657–660, 1970.

4. Chalmers, J. A., *Atmospheric Electricity*, 2nd ed., Pergamon Press, New York, 1967, 515 pp.

5. Colgate, S. A., Differential charge transport in thunderstorm clouds, *J. Geophys. Res.* **77**, 4511–4517, 1972.

6. Dawson, G. A., and D. G. Duff, Initiation of cloud-to-ground lightning strokes, *J. Geophys. Res.* **75**, 5858–5867, 1970.

7. Drake, J. C., Electrification accompanying the melting of ice particles, *Quart. J. Royal Meteorol. Soc.* **94**, 176–191, 1968.

8. Fuquay, D. M., Weather modification and forest fires, *Ground Level Climatology*, American Association for the Advancement of Science, 309–325, 1967.

9. Fuquay, D. M., and R. G. Baughman, Project Skyfire—Lightning Research, Final Report to National Science Foundation, Grant No. GP-2617, 1969, 59 pp.

10. Fuquay, D. M., A. R. Taylor, R. G. Hawe, and C. W. Schmid, Jr., Lightning discharges that caused forest fires, *J. Geophys. Res.* **77**(12) 2156–2158, 1972.

11. Gish, O. H., and G. R. Wait, Thunderstorms and the earth's general electrification, *J. Geophys. Res.* **55**, 473–484, 1950.

12. Israel, H. Bemerkung zum Energieumsatz im Gewitter, *Geofis. Pura Appl.* **24**, 3–11, 1953.

13. Kasemir, H. W., Zur Strömungstheorie des luftelektrischen Feldes II, *Arch. Meteorol. Geophys. Bioklimatol.* **5**(1), 56–70, 1952.

14. Kasemir, H. W., The thundercloud, in *Proceedings of the Third International Conference on Atmospheric Space Electricity*, S. Coroniti, Ed., Elsevier, New York, 1963.

15. Kasemir, H. W., Lightning hazard to rockets during launch I, *ESSA Tech. Rept., ERL 143-APCL 11*, 1969.

16. Kasemir, H. W. Lightning hazard to rockets during launch II, *ESSA Tech. Rept., ERL 144-APCL 12*, 1970.

17. Kasemir, H. W., Basic theory and pilot experiments to the problem of triggering lightning discharges by rockets, *NOAA Tech. Memo. ERL-APCL-12*, 1971.

18. Kasemir, H. W., The cylindrical field mill, *Met. Rundschau* **25**(2); 33–38, 1972.

19. Kasemir, H. W., Lightning suppression by

REFERENCES

chaff seeding. NOAA Tech. Rept. *ERL 284-APCL 30*, 1973, 32 pp.

20. Latham, Jr., and B. J. Mason, Generation of electric charge associated with the formation of soft hail in thunderclouds, *Proc. Roy. Soc. A*, **260**, 537–549, 1961.

21. Latham, J., and A. H. Miller, The role of ice specimen geometry and impact velocity in the Reynolds-Brook theory of thunderstorm electrification, *J. Atmos. Sci.* **22**, 505–508, 1965.

22. Latham, J., and C. D. Stow, Airborne studies of the electrical properties of large convective clouds, *Quart. J. Royal Meteorol. Soc.* **95**, 486–500, 1969.

23. Loeb, L. B., The mechanisms of stepped and dart leaders in cloud-to-ground lightning strokes, *J. Geophys. Res.* **71**, 4711–4721, 1966.

24. MacCready, P. B. Jr., and R. G. Baughman, The glaciation of an AgI-seeded cumulus cloud, *J. Appl. Meteorol.* 7(1), 132–135, 1968.

25. Mason, B. J., *The Physics of Clouds*, 2nd ed., Oxford Univ. Press, London, 1971, 671 pp.

26. Newman, M. M., Use of triggered lightning to study the discharge process in the channel, *Problems of Atmospheric and Space Electricity*, American Elsevier, New York, 1965, pp. 482–490.

27. Newman, M. M., Triggered lightning stroke at very close range, *J. Geophys. Res.* **72**(18), 4761–4764, 1969.

28. Newman, M. M., Lightning discharge simulation and triggered lightning, *Planetary Electrodynamics*, Vol. 2, Gordon and Breach, 1969, pp. 213–219.

29. Newman, M. M., J. R. Stahmann, and J. D. Robb, Experimental study of triggered natural lightning discharges, Final Report DS-67-3, Lightning and Transients Research Institute, Minneapolis, Minn., 1967.

30. Phillips, B. B., Charge distribution in a quasi-static thundercloud model, *Monthly Weather Rev.* **95**, 847–853, 1967.

31. Pierce, E. T., Thunder and lightning, *Shell Aviation News*, 9–13, 1958.

32. Pierce, E. T., Latitudinal variation of lightning parameters, *J. Appl. Meteorol.* **9**, 194–195, 1970.

33. Prupacher, H. R., E. H. Steinberger, and T. L. Wang, On the electrical effects which accompany the spontaneous growth of ice in supercooled, aqueous solutions, *Planetary Electrodynamics*, Vol. 1, S. Coroniti and J. Hughes, Eds., Gordon and Breach, 1969, pp. 283–306.

34. Reynolds, S. E., M. Brook, and M. F. Gourley, Thunderstorm charge separation, *J. Meteorol.* **14**, 426–436, 1957.

35. Ruhnke, L. R., A rocket-borne instrument to measure electric fields inside electrified clouds, *NOAA Tech. Rept. ERL 206-APCL 20*, 1971.

36. Schonland, B. F. J., *The Flight of Thunderbolts*, 2nd ed., Clarendon Press, Oxford, 1964, 182 pp.

37. Smyth, W. R., *Static and Dynamic Electricity*, McGraw-Hill, New York, 1950.

38. Stergis, C. G., G. C. Rein, and T. Kangas, Electric field measurements above thunderstorms, *J. Atmos. Terr. Phys.* **11**, 83–90, 1957.

39. Stow, C. D., On the prevention of lightning, *Bull. Am. Meteorol. Soc.* **50**(7), 514–520, 1969.

40. Taylor, A. R., Lightning effects on the forest complex, *Proc. Ann. Tall Timbers Fire Ecology Conf.* **9**, 127–150, 1969.

41. Uman, M. A., *Lightning*, McGraw-Hill, New York, 1969, 264 pp.

42. Uman, M. A., *Understanding Lightning*, Bek Technical Publications, Inc., Carnegie, Pa., 1971, 166 pp.

43. Viemeister, P. E., *The Lightning Book*, Doubleday, Garden City, N. Y., 1961, 316 pp.

44. Vonnegut, B., Some facts and speculations concerning the origin and role of thunderstorm electricity, *Meteorol. Monogr.* **5**, 224–241, 1963.

45. Workman, E. J., and S. E. Reynolds, Electrical phenomena occurring during the freezing of dilute aqueous solutions and their possible relationship to thunderstorm electricity, *Phys. Rev.* **78**, 254–259, 1950.

Lightning Suppression by Chaff Seeding at the Base of Thunderstorms

H. W. Kasemir, F. J. Holitza, W. E. Cobb, and W. D. Rust

Atmospheric Physics and Chemistry Laboratory, NOAA, Boulder, Colorado 80302

An aircraft equipped with cylindrical electric field mills and a chaff dispenser was used to release large numbers of 10-cm-long chaff fibers at the base of thunderclouds. The electric fields and lightning discharges were monitored during repeated flight passes at cloud base. Twenty-eight thunderstorms that met certain qualifications were seeded with chaff or left unseeded, the unseeded storms becoming 'control' cases. An analysis of the data from 10 thunderstorms seeded with chaff and 18 unseeded control storms shows that seeding thunderstorms with chaff reduced the number of observed lightnings to about one third or less of those observed in the control storms. A statistical evaluation of the occurrence of lightning discharges in the time interval before seeding and in the corresponding time interval of the control storms revealed that there was no significant difference between seeded and control storms. This indicates that both categories of storms belong to the same population. If the same test is applied to the time interval after seeding and the corresponding interval of the control storms, we find that there is a significant difference between seeded and control storms after seeding. Therefore the lightning reduction by chaff seeding is not due to chance.

Introduction

It has been proposed that lightning might be suppressed by seeding electrified clouds with chaff fibers [*Weickmann,* 1963; *Kasemir and Weickmann,* 1965]. In high-electric fields, ions produced near the tips of the chaff fibers by corona discharge can be expected to increase the local electrical conductivity within the cloud and to reduce the electric field to a value below that necessary to initiate lightning [*Kasemir,* 1973]. The feasibility of this lightning suppression concept can be tested by examining to what extent the lightning activity is modified by the release of large numbers of chaff fibers at the base of thunderstorms.

A preliminary step in this program was the investigation of the effect of chaff seeding on the electric field at cloud base [*Holitza and Kasemir,* 1974]. It was decided to study first the influence of chaff seeding on the electric field rather than on the lightning activity for two reasons: (1) the electric field is the prime cause of lightning, and thus any limitation of or decay in the field should reduce the lightning activity of a storm, and (2) since the electric field underneath a thunderstorm is usually high enough to produce corona discharge on dispersed chaff fibers but not sufficiently high to initiate lightning discharges, seeding at cloud base allowed us to study field decay at the place of seeding without the increased dangers associated with storm penetration.

The results of the 1972 field decay experiment [*Holitza and Kasemir,* 1974] show that the field at the base of the seeded storms decayed on the average about 5 times more rapidly than the field beneath storms that were not seeded. Thus, with respect to the electric field at least, chaff seeding produces the expected result. The experiment was continued during 1973, and we will discuss here the effects of chaff seeding at cloud base on the lightning activity of the thunderstorms that were investigated in 1972 and 1973.

Instrumentation

An aircraft was equipped with cylindrical field mills and a chaff dispenser. The field mills [*Kasemir,* 1972] were mounted on the nose and the top of the aircraft as depicted in Figure 1, and all three components of the electric field were recorded. The chaff dispenser holds a 10-kg roll of chaff rope that is 3.6

km long and consists of 3000 strands of 25-μm-diameter conducting fibers. The dispenser cuts the chaff into 10-cm lengths and will provide about 10^8 individual fibers at a rate of approximately 10^7 fibers/min.

Operational Procedure

The operational procedure developed for the field decay experiment of 1972, which is described in detail by *Holitza and Kasemir* [1974] and is summarized here, was also followed in 1973. As the aircraft proceeded to an area of developing cumulus congestus clouds, a particular cloud was selected for investigation. The initial choice was usually a precipitating cumulus that showed vigorous growth. Frequently, we would observe lightning flashes from the developing cloud as we approached it. The electric field was monitored during repeated flight passes that were made no lower than 100 m below the cloud base through or adjacent to the precipitation. Cloud base was typically about 4 km but ranged from 4 to 5 km. Rain, snow, and graupel were often encountered and occasionally hail.

If the field reached or surpassed 50 kV/m, the cloud would be seeded on the next pass provided the field maintained or surpassed the 50-kV/m level. During a seeding pass, chaff was released near cloud base in areas of clear and cloudy air, precipitation, and both updrafts and downdrafts. There was no attempt to seed any preferential location relative to the cloud, but only the presence of an electric field of at least 50 kV/m was used to determine where we seeded. Since the fall velocity of the chaff is quite small, we assume that the chaff moves under both electrostatic [*Holitza and Kasemir,* 1974] and aerodynamic forces but believe that the aerodynamic forces are generally dominant.

The time when the chaff dispenser started to operate is marked on the field record and is designated as the start of the first seeding run. In the case of a control storm the corresponding time is also marked on the record and is designated as the point of seedability of the control storm. Seeded storms were seeded on two consecutive passes. Since the average time of seeding for each pass was about 1 min, approximately 2×10^7 chaff fibers (2 kg) were released below each seeded storm. This is 20 times more chaff than the 10^6 chaff fibers that are indicated from theoretical considerations to be necessary for lightning suppression [*Kasemir,* 1968]. For

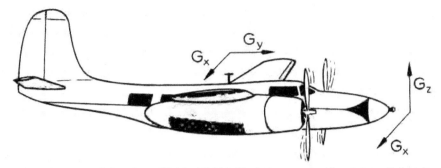

Fig. 1. A sketch of an aircraft showing cylindrical field mills that are mounted at points of aircraft symmetry. Also shown is the coordinate system used to define the three components of the electric field, given here in terms of potential gradients G. i.e., G points toward positive charge.

both seeded and control storms we continued to monitor the electric field at the base of the storm until it dissipated, or we were forced to leave the area for logistic or safety reasons.

In a recent report, *Imyanitov et al.* [1971] gave a critical review of lightning suppression by chaff seeding as reported by *Kasemir* [1968]. In this review they concluded that 100 tons of chaff are necessary to seed one storm. This conclusion, however, is based on three incorrect assumptions: (1) 5000 chaff fibers have a mass of 1 kg instead of the correct value of about 0.5 g, (2) the corona current produced by chaff should compensate for a 100-A conduction current when in fact it is only necessary to compensate for the 1 A required to sustain the lightning activity of the storm, and (3) ion losses due to recombination and attachment to cloud droplets should be accounted for, which *Kasemir* [1968] has done. We would like to emphasize that the chaff seeding reported in this paper was done with about 2 kg per storm and not 100 tons.

The choice of 50 kV/m as the minimum threshold field for chaff seeding was somewhat arbitrary. Such a threshold must be high enough to insure ion production at the chaff fibers by corona, should be low enough so that it occurs at the base of a reasonable number of thunderstorms, and should usually occur when lightning activity is already well established.

It was our intention by alternating between seeded and control storms to collect about the same number of each type of storm. This method worked very well in the first year and in the first half of the second year. At that time we had data from seven seeded and seven control storms. During the second half of the second year our chaff dispenser broke down twice. This prevented us from seeding as we desired, and therefore we were able to seed only three of the last 14 storms investigated. As a result, we ended up with a total of 10 seeded and 18 control storms.

RESULTS

For the statistical evaluation discussed later it is important first that a large enough number of seeded and control storms is available to give the statistical tests sufficient reliability and second that the data of all analyzed storms are complete at least for the important period of time after seeding for which the lightning suppression effect is to be determined. All storms that are complete in this time period and that meet the seedability criteria defined above are included in the data analysis. To obtain an almost complete set of data for the time after seeding, the data records have been truncated at 25 min after the point of seedability, although many storms were monitored for a longer time period. An equally long time interval of −25 min before the point of seedability has been chosen for the start of data evaluation.

The occurrence of a lightning discharge was determined from the field records by the characteristic discontinuity of the field caused by a lightning flash. There was no attempt to distinguish intracloud lightning from cloud to ground lightning. An electric field change of at least 2 kV/m was used as a threshold value, and lightning discharges that produced less than a 2-kV/m field change were disregarded. The reason for selecting any threshold level is to eliminate lightnings from nearby storms. From measurements at the ground we can expect a field change of about 2 kV/m from an 'average' lightning flash that occurs at a distance of about 7 km from the recording station [*Israel*, 1961; *Brook et al.*, 1962]. The diameter of our investigated storms was often about 10–16 km, so that an effective lightning reception range of 7 km would approximately equal the storm radius. It is possible, however, that the field change due to lightning at the cloud base is greater than that at the ground, which would result in a larger reception area. Therefore the same field records were also evaluated by using a larger field change threshold level of 4 kV/m. With this higher threshold, some of the weaker cloud discharges of the investigated storms were probably excluded.

The storms were divided into two time periods with a zero reference time set at 5 min after the start of seeding for the seeded storms and 5 min after the point of seedability for the control storms. We could have chosen as our zero time the point of seedability at the beginning of the first seeding pass. It is obvious, however, that any lightning suppression effect due to chaff seeding cannot be expected to occur when the first chaff fibers are being released from the airplane. It is necessary to allow time for the ions that are liberated by corona discharge to reach the cloud and to penetrate into it before they can have any effect on the lightning activity of the storm. Therefore the zero time that divides the first part, when there should be no influence of chaff seeding, and the second part, when an effect of chaff seeding may be expected, has been set at 5 min after the beginning of the first seeding run.

In comparison with the division of the data by the seedability point into 25 min before and 25 min after seedability, the zero time divides the data into a time period of 30 min before to 20 min after zero time. The chosen time interval will not bias the results in favor of a suppression effect, since almost no lightnings were observed during the interval of +20 to +30 min after seeding in the seeded storms, while there were many lightnings recorded during the same interval in the control storms.

The entire period of −30 to +20 min was subdivided into 5-min intervals, and the number of lightnings in each interval in each storm were determined from the field mill records. The number of lightnings per 5 min are defined as lightning density and

TABLE 1. Lightning Densities (per 5 min) of the Control Storms for the 2-kV/m Threshold

5-min Interval	1	2	3	4	5	6	7	8	9	10	11	12	13	14	15	16	17	18	Row Average
									Storm Number										
						Before Zero Time													
−6	3	6	12	1	2		0			1				2	0				3.0
−5	14	2	10	0	8		2			2	2			2	0	0			3.8
−4	3	11	7	2	5	0	4		0	1	3	2	0	3	1	7	0	0	2.9
−3	8	7	11	1	2	0	7		1	2	1	2	1	4	0	3	4	0	3.2
−2	10	1	4	2	3	3	8		0	2	2	7	0	7	1	4	0	0	3.2
−1	8	7	10	2	1	1	6	0	0	4	0	0	1	2	0	7	6	4	3.3
Average	7.7	5.7	9.0	1.3	3.5	1.0	4.5	0.	0.3	2.0	1.6	2.8	0.5	3.3	0.3	4.2	2.5	1.0	3.2
						After Zero Time													
1	15	1	12	2	1	0	1	1	0	4	1	0	4	5	0	15	2	3	3.7
2	12	2	17	2	0	0	3	10	0	3	0	1	3	4	1	21	3	3	4.7
3	20	2	10	3	1	0	4	3	2	3	2	3	2	3	0	0	6	1	3.6
4	10	2	17	4	0	0	2	3	3		0	0	1	0	0	7	4	1	3.2
Average	14.3	1.8	14.0	2.8	0.5	0	2.5	4.3	1.3	3.3	0.8	1.0	2.5	3.0	0.3	10.8	3.8	2.0	3.8

are designated as a_{ij} for the control storm and b_{ij} for the seeded storm. The index i indicates a particular 5-min time period and ranges from −6 to 4 (i.e., from −30 to 20 min); the index j identifies the individual storms and ranges from 1 to 18 for the control storms and from 1 to 10 for the seeded storms.

In Tables 1 and 2 the lightning densities of the control and seeded storms for the 2-kV/m threshold are given; Tables 3 and 4 contain the densities for the 4-kV/m threshold. If no lightning was recorded during a time interval, it is indicated by a zero. If no data were obtained due to late arrival or early departure of the aircraft, it is indicated by a no entry. These are omitted from the statistical evaluation and from the calculation of the average values. Each table is divided into two sections, the upper one containing the lightning densities before the zero time and the lower one the densities after the zero time.

The entries in the right-hand column of each table are the storm-averaged lightning densities A_i^m or B_i^m during each time interval i averaged over all storms:

$$A_i^m = (N_i)^{-1} \sum_i a_{ij} \qquad B_i^m = (N_i)^{-1} \sum_i b_{ij}$$

where A refers to control storms, B refers to seeded storms, the superscript $m = 1$ refers to times before zero and $m = 2$ refers to times after zero, and N_j is the number of lightning densities that were averaged. In the last row of each section of the tables is the interval-averaged lightning densities A_j^m or B_j^m averaged over all time intervals before and after the zero time of each storm j:

$$A_j^m = (N_i)^{-1} \sum_i a_{ij} \qquad B_j^m = (N_i)^{-1} \sum_i b_{ij}$$

where N_i is the number of lightning densities that are averaged. The last entry of the last row of each section is the average of all lightning densities A^m or B^m in that section:

$$A^m = (N)^{-1} \sum_{i,j} a_{ij} \qquad B^m = (N)^{-1} \sum_{i,j} b_{ij}$$

where N is the number of lightning densities in that section.

DATA ANALYSIS AND DISCUSSION

Shown in Figures 2 and 3 are plots of the interval-averaged lightning densities A_i^1 and A_i^2 of the control storms and B_i^1 and B_i^2 of the seeded storms for the 2-kV/m (Figure 2) and the

4-kV/m (Figure 3) field change threshold. The approximate seeding interval $i = -1$ is shown as a shaded area. It may be seen in these figures that before zero time the interval-averaged lightning density of the control storms is sometimes higher than that of the seeded storms and vice versa; however, after zero time the interval-averaged lightning density of the seeded storms was always less than that of the control storms and exhibited a continuous decrease to zero. This steady decay of the interval-averaged lightning density is similar to the field decay discussed by *Holitza and Kasemir* [1974]. Both are the results of a continuous neutralization of the effect of the lower cloud charge by the ion cloud liberated by the corona discharge from the chaff fibers.

We also compare the lightning activity of control and seeded storms by using the overall averages of the lightning densities A^1, A^2, B^1, and B^2 found in the tables. The important comparison is between the control and seeded storms after time zero. We define a lightning reduction ratio R^2 as B^2/A^2, where B^2 is the overall average lightning density of the seeded storms and A^2 is that of the control storms. From the results that use the 2-kV/m threshold (Tables 1 and 2) we obtain an

TABLE 2. Lightning Densities (per 5 min) of the Seeded Storms for the 2-kV/m Threshold

5-min Interval	1	2	3	4	5	6	7	8	9	10	Row Average
					Storm Number						
				Before Zero Time							
−6	4	1	3	0	0		6		11		3.6
−5	1	0	3	0	2		1	0	3	0	1.1
−4	1	0	5	1	2		9	1	11	0	3.3
−3	1	0	4	4	3	1	6	0	6	0	2.5
−2	1	0	0	2	3	2	2	2	3	0	1.5
−1	1	0	1	6	2	2	8	2	6	1	2.9
Average	1.5	0.2	2.7	2.2	2.0	1.7	5.3	1.0	6.7	0.2	2.4
				After Zero Time							
1	0	0	2	4	5	1	3	3	3	0	2.1
2	0	0	1	5	1	0	4	0	3	0	1.4
3	0	0	1	2	1	0	0	0	2	0	0.6
4	0	0	0	0	0	1	0	0	1	0	0.1
Average	0	0	1.0	2.8	1.8	0.3	1.8	0.8	2.3	0	1.1

TABLE 3. Lightning Densities (per 5 min) of the Control Storms for the 4-kV/m Threshold

5-min Interval	1	2	3	4	5	6	7	8	9	10	11	12	13	14	15	16	17	18	Row Average
								Storm Number											
							Before Zero Time												
−6	2	5	6	0	2		0			1				2	0				2.0
−5	13	0	4	0	6		2		2	1				2	0	0			2.7
−4	2	7	6	1	2	0	4		0	1	3	1	0	1	1	5	0	0	2.0
−3	5	5	6	0	2	0	4		1	1	1	0	1	0	0	1	3	0	1.8
−2	8	1	1	1	3	1	4		0	1	2	4	0	5	1	2	0	0	2.0
−1	6	4	8	1	1	1	4	0	0	2	0	0	0	2	0	5	3	1	2.1
Average	6.0	3.7	5.2	0.5	2.7	0.5	3.0	0	0.3	1.3	1.4	1.3	0.3	2.0	0.3	2.6	1.5	0.3	2.1
							After Zero Time												
1	11	0	10	1	0	0	0	0	0	3	0	0	4	5	0	9	2	1	2.6
2	10	2	14	0	0	0	3	4	0	1	0	1	3	3	0	19	2	2	3.6
3	18	2	10	2	1	0	4	1	2	2	2	2	2	3	0	0	5	0	3.1
4	6	2	16	3	0	0	2	1	3		0	0	1	0	0	6	3	0	2.5
Average	11.3	1.5	12.5	1.5	0.3	0	2.3	1.5	1.3	2.0	0.5	0.8	2.5	2.8	0	8.5	3.0	0.8	2.9

$R^2 = 0.20$. A similar calculation that uses the results with the 4-kV/m threshold (Tables 3 and 4) yields an $R^2 = 0.21$.

If we calculate the same ratio for the lightning activity of seeded to control storms before zero time, we obtain for both the 2-kV/m and the 4-kV/m threshold level, $R^1 = B^1/A^1 = 0.77$. This means that the seeded storms had a lower average lightning production than the control storms did. It seems to us that it is necessary to raise the lightning activity of the seeded storms to that of the control storms before we determine the lightning reduction ratio. We define this weighted lightning reduction ratio as $R = R^2/R^1$; $R = 0.38$ for the 2-kV/m threshold level, and $R = 0.27$ for the 4-kV/m threshold level. The results of this evaluation suggest to us, as did the plots of the average lightning densities, that the occurrence of lightning was substantially reduced in those storms seeded with chaff.

STATISTICAL ANALYSIS AND DISCUSSION

In addition to the evaluations above, we performed a statistical analysis of the lightning densities in order to obtain a measure of the likelihood that our results are due to chance.

The statistical test used to compare the seeded and control storms was the nonparametric Wilcoxon two-sample rank test, which makes no assumptions about the distribution of the data [*Brownlee*, 1965]. This analysis was applied to test the null hypothesis H_0 that there is no difference between the population of lightning densities of the seeded and control storms. We will reject H_0 if the significance level S is less than 0.05 and will accept H_0 if S is greater than 0.05.

We apply this test first to the lightning densities a_{ij} and b_{ij} for each 5-min interval i. This gives us the significance level S_i for each individual time interval i, which implies that there is ($S_i \le 0.05$) or that there is not ($S_i > 0.05$) a significant difference between lightning densities of seeded and control storms for each specified i. Analogous to Figures 2 and 3, we have plotted S_i versus i in Figure 4. We see from Figure 4 that for any time interval before zero time, S_i exceeds 0.1. For these intervals we accept H_0 and conclude that there is no significant difference between the population of lightning densities of seeded and control storms. This result for time intervals before seeding began also indicates that there was no bias in the selection of storms. For the time interval $i = 1$, i.e., 5 min after

TABLE 4. Lightning Densities (per 5 min) of the Seeded Storms for the 4-kV/m Threshold

5-min Interval	1	2	3	4	5	6	7	8	9	10	Row Average
				Storm Number							
				Before Zero Time							
−6	3	1	1	0	0		4		5		2.0
−5	1	0	2	0	2		1	0	3	0	1.0
−4	1	0	0	1	2		6	0	7	0	1.9
−3	0	0	2	1	3	0	6	0	4	0	1.6
−2	0	0	0	1	3	2	2	2	3	0	1.3
−1	1	0	0	4	2	2	6	2	3	1	2.1
Average	1.0	0.2	0.8	1.2	2.0	1.3	4.2	0.8	4.2	0.2	1.6
				After Zero Time							
1	0	0	0	3	5	1	2	2	2	0	1.5
2	0	0	1	4	1	0	0	0	1	0	0.7
3	0	0	1	1	1	0	0	0	0	0	0.3
4	0	0	0	0	0	0	0	0	0	0	0.0
Average	0	0	0.5	2.0	1.8	0.3	0.5	0.5	0.8	0	0.6

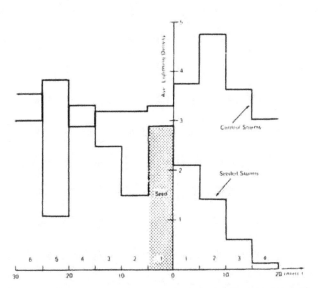

Fig. 2. The average lightning density for each time interval i, using a 2-kV/m field change threshold, for 10 seeded and 18 control storms during 1972 and 1973.

533

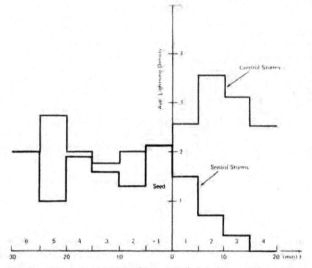

Fig. 3. The average lightning density for each time interval i, using a 4-kV/m field change threshold, for 10 seeded and 18 control storms during 1972 and 1973.

A_j^2 and B_j^2 after seeding. The significance level S^1 before seeding was 0.32 for the 2-kV/m threshold level and 0.35 for the 4-kV/m threshold level. This result again implies that there is no significant difference between the overall lightning activity of the seeded and control storms before seeding. The significance level S^2 after seeding is 0.01 for both the 2-kV/m and the 4-kV/m threshold level. This suggests to us that there is a significant difference between the lightning activity of the seeded and control storms after zero time. Therefore the reduction in lightning activity as indicated by the lightning reduction ratios which are calculated above is not due to chance.

CONCLUDING REMARKS

In the paper by *Holitza and Kasemir* [1974] it was argued that a significant lightning suppression effect will not result from chaff seeding underneath the cloud because ions would presumably be trapped on cloud particles and could not penetrate deeply into the cloud; thus the field inside the cloud would not be sufficiently reduced. The present findings make it necessary to review this assumption, since chaff seeding at the cloud base apparently does suppress the production of lightning. We therefore assume that (1) at least an adequate number of ions penetrate far enough into the cloud to reduce the high fields, (2) ions trapped on cloud particles may produce an adequately strong and an oppositely directed field that reduces the existing field within the cloud, or (3) a combination of these or other effects is present and reduces the electric field to a level below that necessary for lightning production.

We need to study the processes that govern the movement of the ions, which are created by corona discharge, within the storm and their interaction with cloud particles. In addition, to find an even more effective technique for lightning suppression, the location and the time of the origin of lightning, as well as the processes that generate breakdown fields within thunderstorms, need to be determined. For instance, the altitude and the temperature ranges in which cloud and ground discharges originate are not sufficiently well known and require further investigation.

seeding, the significance level S_i is still 0.4–0.5. In the plots of the average lightning densities (Figures 2 and 3) A_i and B_i, the difference between the seeded and control storms for the time interval $i = 1$ is already apparent, but our statistical analysis indicates that such a difference has no statistical significance. In the next interval, $i = 2$, S drops to about 0.05; in the remaining intervals, $i = 3$ and 4, it decreases further to 0.002 or less, and H_0 must be rejected.

We may summarize these results in the following way: (1) the lightning densities of the seeded and control storms before time zero belong to the same population, since there is no significant difference between them, and (2) a significant suppression of lightning starts approximately 10 min after seeding begins.

The same Wilcoxon two-sample rank test was also applied to the overall lightning densities A_j^1 and B_j^1 before seeding and

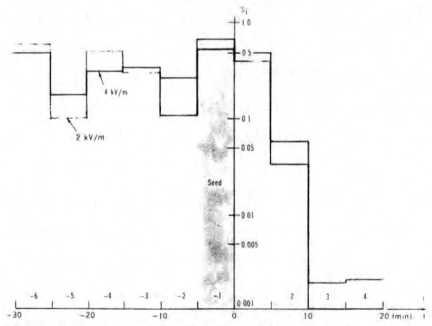

Fig. 4. The significance levels S_i that are obtained by a statistical comparison of the lightning densities of the seeded and control storms for each time interval i, using the 2-kV/m and 4-kV/m field change thresholds.

SUMMARY

The lightning activity of 10 thunderstorms that were seeded with chaff and 18 thunderstorms that were not seeded has been analyzed with the following results:

1. Seeding thunderstorms with chaff reduced the number of observed lightnings to about one third or less of those observed in control storms.

2. A statistical analysis that uses the nonparametric Wilcoxon two-sample rank test shows that there was no significant difference between the lightning activity of the seeded and control storms before seeding was started.

3. The Wilcoxon test also indicates that a significant suppression of lightning apparently began about 10 min after the beginning of seeding with chaff at the cloud base.

4. The lightning suppression effect was even more dominant as time progressed from 10 min after seeding. This is in contrast to the continued or even increased occurrence of lightning from the control storms.

Acknowledgments. We wish to express our gratitude to P. V. Tryon of the Statistical Engineering Laboratory of the National Bureau of Standards for his critical review and helpful suggestions and to B. R. Caldwell and F. Gould for their help in the field program. Thanks are also extended to B. Brodie, W. Powell, D. Pruess, L. Ray, and J. Hart, who piloted and maintained the aircraft. The cooperation of T. Floria and the controllers of the U.S. Federal Aviation Administration, who provided the air traffic control necessary for our research, is gratefully acknowledged. One of us (W.D.R.) participated in this study with support from the National Research Council under a National Research Council–NOAA Resident Research Associateship.

REFERENCES

Brook, M., N. Kitagara, and E. J. Workman, Quantitative study of strokes and continuous currents in lightning discharges to ground, *J. Geophys. Res.*, 67, 649–659, 1962.

Brownlee, K. A., *Statistical Theory and Methodology in Science and Engineering*, 2nd ed., pp. 251–256, John Wiley, New York, 1965.

Holitza, F. J., and H. W. Kasemir, Accelerated decay of thunderstorm electric fields by chaff seeding, *J. Geophys. Res.*, 79, 425–429, 1974.

Imyanitov, I. M., Ye. V. Chubarina, and Ya. M. Shvarts, *Elektrichestro Oblakov*, Gidrometrologicheskoye Press, Leningrad, 1971. (English translation, *NASA TT F718*, 1972.)

Israel, H., *Atmosphärische Elektrizität*, vol. 2, p. 141, Akademische, Leipzig, East Germany, 1961. (English translation, *Atmospheric Electricity*, vol. 2, p. 420, Program for Scientific Translations, Jerusalem, 1973.)

Kasemir, H. W., Lightning suppression by chaff seeding, final report, pp. 1–49, U.S. Army Electron. Command, Fort Monmouth, N. J., June 1968.

Kasemir, H. W., The cylindrical field mill, *Meteorol. Rundsch.*, 25, 33–38, 1972.

Kasemir, H. W., Lightning suppression by chaff seeding, *Tech. Rep. ERL 284-APCL*, p. 30, NOAA, Boulder, Colo., 1973.

Kasemir, H. W., and H. K. Weickmann, Modification of the electric field of thunderstorms, paper presented at International Conference on Cloud Physics, Tokyo, Japan, 1965.

Weickmann, H. K., A realistic appraisal of weather control, *J. Appl. Math. Phys.*, 14, 528–543, 1963.

(Received March 3, 1975;
revised October 31, 1975;
accepted November 13, 1975.)

Triggered Lightning

DON W. CLIFFORD AND HEINZ W. KASEMIR

Abstract—Although the triggering of lightning to ground stations using wire-trailing rockets has become somewhat routine, it has not yet been proven that free-flying aircraft trigger lightning. However, there is substantial evidence that aircraft are somehow involved in the initiation of the strikes which they experience. This paper reviews the evidence and arguments for aircraft-triggered lightning. The triggering mechanisms which have been suggested are discussed in light of ground-based triggering experience and recent flight data related to weather conditions and discharge parameters associated with strikes to aircraft. Some conclusions are drawn about the probable characteristics of aircraft-triggered lightning and the implications of triggering to aircraft safety.

Key Words—Triggered lightning, triggering mechanisms, weather conditions, discharge parameters, aircraft strikes, aircraft safety.

Manuscript received January 18, 1982.

D. W. Clifford is with McDonnell Aircraft Co., McDonnell Douglas Corp., St. Louis, MO 63166. (314) 232-5033.

H. W. Kasemir is with the Colorado Scienfitic Research Corp., Berthoud, CO 80513.

I. INTRODUCTION

FOLLOWING THE DISCUSSION of natural lightning processes in the preceding paper by Uman and Krider, this paper will examine the possible triggering of lightning as a result of introducing an aircraft or other aerospace vehicle into the atmospheric electrical system. The question of aircraft triggering is more than a scientific curiosity. If aircraft do indeed initiate the strikes which they experience, several important questions are raised about the traditional philosophy of lightning protection for aircraft. It is an important issue to the aerospace community because the design and test specifications currently in use can impact significantly the cost and performance of new-technology aircraft. They also impact the cost of the ground test facilities used to simulate the lightning hazard. Lightning design and test specifications are derived from ground-based measurements of natural cloud-to-ground lightning.

Very few attempts have been made to measure the parameters of lightning strikes to aircraft in flight until very recently. However, government-sponsored flight-research programs are now beginning to generate data which can be used to help clarify the atmospheric electrical hazards for aircraft and, possibly, provide more information about the role of the aircraft in initiating the lightning process.

The purpose of this paper, therefore, is to review the evidence and arguments for aircraft-triggered lightning, drawing upon the ground-based triggering experience for insights into the triggering process. The triggering arguments which have been advanced will be reviewed and compared with in-flight lightning-strike reports to deduce the flight and weather conditions most likely to produce triggered lightning. Based upon the conclusions drawn about the triggering process, the probable characteristics of aircraft-triggered lightning are discussed in light of their effects on aircraft safety.

II. GROUND-BASED TRIGGERING

Natural cloud-to-ground lightning is usually initiated by a step-leader that lowers negative charge from the cloud toward the earth. However, in many reported strikes to towers or high structures, the leader is observed to travel upward from the top of the structure into the clouds. These upward-going leaders are characterized by upward branching of the channel into the cloud. Since the leader is initiated from the structure, the lightning is said to be triggered by the structure. That is, the lightning would not have occurred if the structure were not present.

Probably the first documented occurrence of lightning triggered by a tall tower was reported by McEachron [1]. Photographs taken with a Boys camera of lightning strikes to the Empire State Building in New York showed that, for 53 lightning discharges, the ratio of triggered (upwards-moving leader) to natural (downwards-moving leader) lightning strikes was 16 to one. From extensive lightning research on Mount San Salvador in Switzerland, Berger in 1977 reported that, of 303 lightning discharges to the 70-m-high tower on the top of the mountain (314 m above sea level), 245 showed upward-moving and 50 downward-moving first-leader strokes [2]. Here, the ratio between triggered and natural lightning strikes is five to one.

Pierce [3] summarized data from several sources on the incidence of lightning flashes to tall structures and deduced that structures less than 150-m high do not trigger lightning, although they may be struck frequently by naturally occurring lightning. However, with increasing height there is a very rapid increase in the proportion of flashes initiated by the structure; for a structure height of 200 m, it is about 50 percent, and at 400 m it exceeds 80 percent.

In the case of a tall structure, the electrostatic-field configuration at its top is controlled by the general ambient field E_a, the height H of the structure, and its shape. Since the ambient field changes with distance h above the ground, Pierce calculated a quantity which he called the potential discontinuity, $V_d = \int_0^H E_a(h)dh$. V_d approximates the potential difference between the tip of the structure and the ambient atmosphere. V_d must exceed a certain threshold before a leader can be initiated; whether a streamer becomes a self-propagating leader depends on a critical value of E_a being exceeded. Based upon his observations, Pierce concludes that V_d must be around 10^6 V to initiate a lightning flash, leading to a requirement for ambient fields of a few to several kilovolts/meter, depending on the height of the structure.

III. ROCKET-TRIGGERED LIGHTNING

The first scientists to conduct the daring experiment of triggering lightning discharges intentionally for close-up study and measurements (distance of their observation room from the point of the lightning was about 20 m) were Newman and co-workers [4]. They used a small metal ship as the observation and instrument platform, and fired a small rocket, with a trailing wire connected to the ship, toward an overhead thunderstorm when the potential gradient was high enough to produce strong corona current (about 30 μA) from a pointed conductor on the mast. The electric field was established to be 15 to 20 kV/m. The wire length was usually about 100 m when the triggered lightning strike (upward-moving leader discharge) was initiated.

Since those early experiments, a French research group at Saint Privat d'Alliers has developed the rocket-triggering technique over land [5]. During the six-year period from 1973 to 1978 inclusive, over 100 wire-trailing rockets were fired from an instrumented launch tower, resulting in approximately 75 triggered-lightning strikes to the ground. During the summer of 1981, the French group cooperated with the U.S. Air Force in triggering approximately 40 strikes to a ground station near the Langmuir Laboratory in New Mexico [6]. A wide array of electrical and optical instruments used during these tests provided an excellent opportunity to study the triggered flashes in detail, resulting in several advances in the understanding of lightning phenomena.

In relation to the triggering parameters, the electrostatic field at the ground must be of the order of 10 kV/m with a negative polarity (in the cloud) and somewhat higher for positive cloud charges, to ensure a high probability of triggering [6]. The height reached by the wire varied from 50 to 530 m for 48 triggered strokes. In half of those cases, the height was between 50 and 200 m. Multiple subsequent strokes were observed only when triggering occurred before the wire reached 250 m.

Based upon these values, the minimum Pierce voltage discontinuity would be of the order of 500 kV at the time of triggering. However, it is likely that the average value is significantly higher, not only because of the higher altitudes reached by some rockets before triggering occurs, but because the electric-field gradient near the ground may be significantly lower than at higher altitudes. Standler and Winn [7] have shown that the field at the ground is limited by space charge produced by corona currents from grass, trees, and other sharp objects on the ground. Using balloon-borne field meters, they found that the magnitude of the field a hundred meters above the ground is several times larger than at the ground. In one case, the field 300 m above the ground was six times that at the ground. Therefore, an average value for the actual triggering-voltage discontinuity of the French rockets may be well into the megavolt range. The level in each instance apparently depends upon the charge density in the cloud and how near the cloud was to discharging spontaneously. For the purpose of this paper, however, the minimum triggering levels are of most interest and they will be used in relating the rocket-triggering results to the aircraft-triggering case.

As in the case of tall structures, the lightning discharges to the rockets are of the upward-moving leader variety, indicating that the flash is initiated by the rocket. It would seem that the voltage discontinuity required to trigger a flash by a rapidly moving rocket should be somewhat less than for a stationary structure. According to Brook et al., the reason is that corona discharge from a structure forms a cloud of space charge around the tip, shielding it from the field of the cloud. If the relative velocity between air and point is greater than the ion drift velocity (about 100 m/s), the screening space charge cannot develop, and the enhanced field strength penetrates in full force to the emitting point. The chances of initiating a lightning discharge should then be increased [8].

IV. NONGROUNDED VEHICLE TRIGGERING

The most famous incident of nongrounded rocket triggering of lightning was the Apollo 12 launch. During the early portion of the liftoff into a nonthunderous cloud, the rocket was struck twice by lightning; the first strike occurred 36 s after launch while the rocket was at an altitude of 2 km. The vehicle was struck again about 15 s later at an altitude of 4.2 km, 650 m above the freezing level. The event has been analyzed and described by Pierce [3], M. Brook et al. [8], Kasemir [9], and Godfrey et al. [10].

In a situation similar to the Apollo 12, but on a smaller scale, the French group routinely triggered lightning to the ground in New Mexico by firing a small rocket trailing only a 100-m length of conducting wire [6]. The triggered lightning passed through the rocket and wire with the lower end of the wire at a height of a few hundred meters above the ground. For this type of triggering, the field at the ground had to be slightly above the field level required for grounded rocket-triggering (15-18 kV/m).

For ungrounded objects, several workers [8], [9], [11] have calculated the field enhancement produced by placing a rounded conducting body in an ambient electric field. The distortion of a uniform electric field produced by the introduction of an extended prolate spheroid is illustrated in Fig. 1. The strength of the enhanced field is given by

$$E = E_a \frac{c^3}{a(a^2 - c^2)(\tanh^{-1} c/a - c/a)} \quad (1)$$

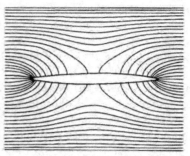

Fig. 1. Uniform electric field distorted by a conducting ellipsoid [8].

where a is the semimajor axis, b is the semiminor axis, and $c = (a^2 - b^2)^{1/2}$ is the eccentricity of the spheroid. The enhancement factor K at the tip is simply the ratio of the enhanced field to the ambient field

$$K = E/E_a = \frac{c^3}{a(a^2 - c^2)(\tanh^{-1} c/a - c/a)} \quad (2)$$

Values of K are given in Table I for various ratios of a to b. Kasemir and Perkins [12] used the facilities of the National Oceanic and Atmospheric Agency (NOAA) in Colorado to check the validity of the field-enhancement calculations. They placed a half spheroid made of conductive material on the ground plate of a large parallel-plate capacitor. The plates were square, 5 m on a side, and were separated by about two meters or less. The dimensions of the half spheroid were $a = 1$ m, $b = 0.25$ m. The enhancement factor calculated by (2) is $K = 13.26$. The air breakdown gradient adjusted to the 1600-m altitude of the Boulder NOAA laboratories was $G_b = 2.6$ MV/m. Therefore, the theoretically predicted plate-capacitor gradient needed to produce a spark discharge from the spheroid was 196 kV/m. The measured breakdown value was 201 kV/m, a deviation from the theoretical value of about two percent. The estimated accuracy of the field measurement was 5 percent. Consequently, the enhancement factors calculated by (2) were borne out in this case.

In calculating the enhancement factor for an actual vehicle in an ambient field, it is necessary to account for r, the radius of curvature at the tip. The tip gradient depends on a good fit of r much more than on a good fit of b or c. For a prolate spheroid, the maximum curvature at the tip is related to the ratio of a/b by $a/r = a^2/b^2$, where r is the radius at the tip. For prolate spheroid field enhancement values incorporating the tip radius, the plot of K versus a/r from Brook et al. in Fig. 2 may be used. For an actual vehicle, the radius used in the a/r term should be the radius of the appropriate vehicle extremity. Sharp points should be discounted, however, because the field decays rapidly from such a point. A few millimeters away, its presence is no longer a factor.

For the case of the Apollo 12 rocket, the tip of the rocket was capped with a 10-cm-diameter hemisphere. Depending then upon the manner of accounting for the exhaust plume, a value of K is obtained between 300 and 500. Assuming a plume length of 200 m, an ambient field strength of 2.5 kV/m in the direction of the plume would produce a triggering field in air of 2.4×10^6 V/m at the altitude of the strikes.

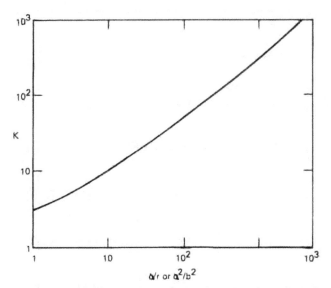

Fig. 2. Electric-field-enhancement factor versus the ratio of the height a of the half ellipsoid to the radius r of curvature of the tip [8].

TABLE I
FIELD-ENHANCEMENT FACTORS [11].

a/b	2	5	10	20	50	100
K	5.8	16	49	148	694	2330

When a very thin conductor such as a long wire is considered, (1) can be seen to include $E = E_a \cdot a$. This value is consistent with Pierce's assumption that, for an airborne, pointed conductor in electrical equilibrium with its atmospheric environment, the voltage discontinuity at the tip is approximately $V_d = E_a(l/2)$, where l is the electrical length of the conductor. Since voltage is being considered rather than the electric field, a factor of one half the length is required to reference the voltage at the end of the conductor to the midpoint of the conductor where the potential must be zero. Applying this criterion to the wire-trailing rockets in New Mexico, the 15-kV/m field and the 100-m wire length yield a triggering value of $V_d = 750$ kV, a little higher than the grounded case.

A similar, yet somewhat different, triggering configuration was reported by Clifford [13]. An analysis of naval-aircraft lightning-strike data revealed that an inordinately large number of lightning strikes were recorded to aircraft towing gunnery targets at the end of long steel cables. In these cases, the reports of outside observers are available because the incidents often involved a chase plane flying in formation with the tow aircraft at the time of the event. The steel cables were extended as far as 5 to 6 km (15 000 to 18 000 ft) behind and below the aircraft. At least ten such lightning-strike incidents have been reported in the last few years.

In every case where a chase-pilot report was available, unusual electrical activity was reported around the tow aircraft. The reports included descriptions of electrical halos, St. Elmo's fire, large electrical arcs, sparking on the fuselage and other terms descriptive of very-high-potential electrical effects. There were no measurements of ambient electric fields associated with these incidents. There was usually mention of some active weather in the vicinity. The tow cable case is interesting and worthwhile to use for demonstration of the vector calculation of V_d. It can be assumed that the gunnery target altitude is lower than the aircraft and that the length of cable forms some angle with the ambient field lines. With reference to Pierce's expression for the voltage discontinuity for the airborne case, we should use $V_d = 1/2 \int_0^L \bar{E}_a \cdot dl$.

$\bar{E}_a$ is the cloud-field vector and dl is the length-element vector along the trailing wire. In effect then, only the component of the wire length which is aligned with the electric field vector will contribute to the enhancement of the field.

For the tow-target case, if a triggering value V_d of 500 kV were required and a field value of only 10 kV/m were present, then the cable length would have to have a component in the direction of the field at least 100 m long. For the likely case of a vertical electric field, the droop in a 5-km-long tow cable would have to be in excess of 100 m, which is easily the case.

There may be many other factors involved in the consideration of lightning triggered by long conductors. However, the intent of this discussion is simply to illustrate that lightning is initiated by high structures, rockets with wires attached to the ground, and even by rockets or aircraft with trailing conductors (exhaust plume or wire) not connected to ground. In fact, Kasemir [9] reported cases where lightning was apparently triggered by small rockets with no trailing conductor. At any rate, it is clear that lightning can be triggered by long conductors injected into an ambient electric field and that the size of the conductor and the strength of the ambient field are the major factors in producing a triggered discharge. Voltage discontinuities as low as 500 to 750 kV may be adequate to trigger lightning from airborne systems under optimum conditions.

V. AIRCRAFT-TRIGGERED LIGHTNING

Few data exist, other than circumstantial evidence and much speculation, to prove or disprove whether aircraft actually trigger lightning. However, the circumstantial evidence is growing and the idea of aircraft triggering lightning is fairly well accepted, although no definite triggering mechanism has been identified. In the late 60's, Fitzgerald [14], [15], Vonnegut [16], Petterson and Wood [17], Cobb and Holitza [18], as well as Pierce [3], and Brook et al. [8], addressed the question of aircraft-triggered lightning. More recently, Nanevicz discussed the possibility of a triggered strike to an instrumented Lear jet [19], and Clifford [13] reopened the subject of triggered lightning based on a review of in-flight strike statistics and pilot reports of "static discharges."

Fitzgerald's studies included an extensive flight program in the "Rough Rider" series where an F-100 aircraft was flown through Florida thunderstorms. Fifty-five strikes to the aircraft were experienced during 205 storm penetrations as electric fields and radar cross sections were measured around and above the storm cloud by other aircraft. Most of the strikes to the aircraft occurred in, or in the edge of, an intense negative-space-charge concentration at the flight level as determined by the on-board horizontal-field sensor. Fitzgerald

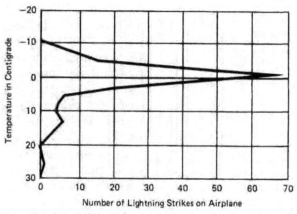

Fig. 3.   Lightning strikes to aircraft versus temperature in clouds [25].

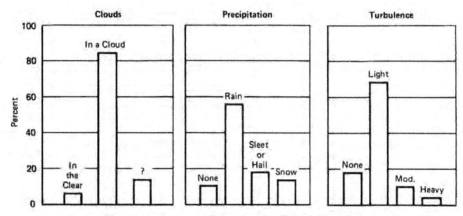

Fig. 4.   Environmental conditions at time of strike [22].

emphasized that the vertical electric-field component usually indicated that the aircraft was passing through a transition zone where positive over negative charge concentrations transitioned to negative over positive. It was thought that aircraft interception or near approach to highly charged precipitation streamers, as indicated by on-board field mills, was an important factor in triggering lightning strikes through the aircraft to charge-cell concentrations. The highest strike rate was experienced at an altitude of 30 000 feet and −40°C temperature.

The strikes to the F-100 normally occurred in moderately high field-strength regions. Fields near 390 kV/m were measured, but 69 percent of the strikes occurred when the field was less than 130 kV/m. Fitzgerald noted that different storms produced different lightning-strike frequencies. In two exceptional storms, strikes to the aircraft occurred on four out of five passes through the storms. These storms were in an early dissipating stage of their life cycle.

The 1976 strike reported by Nanevicz occurred just outside a cloud where the ambient field was about 68 kV/m. Using Pierce's ground-based triggering criterion, Nanevicz calculated a voltage discontinuity of 924 kV at the aircraft, concluding that the strike was consistent with Pierce's 10^6 V requirement for triggering. Using the airborne criterion, however, a voltage discontinuity of 431 kV (one-half the ground-based value) would be obtained. This value is close to the minimum rocket-triggered data. Using the same ap-

proach, similar calculations for the F-100 strikes reported by Fitzgerald yield voltage discontinuities well within the 500- to 750-kV range for most of the strikes. Note that, in these two cases, the aircraft were seeking out charge pockets in active thunderstorms. The same circumstances prevailed during a strike to a B-25 reported by Gunn in 1948 [20].

Under much different circumstances, Cobb and Holitza reported on three strikes to a DC-6 in 1967. Fairly low fields were measured at the time of these discharges. In one case, the vertical field was nil and the axial horizontal field was only 4 kV/m. (Could the low field readings have indicated that the aircraft was in the center of a charge concentration?) In the other cases, the maximum ambient field components were 16 and 27 kV/m (vertical), whereas the vertical dimension of the DC-6 was 9 m. The resulting voltage discontinuity is less than half of the minimum value established for triggering.

Each of the three strikes reported to the DC-6 was experienced near the freezing level within a cloud, and in each case both rain and graupel were present. In contrast to the F-100 strikes which occurred during penetrations of active storm cells at altitudes high above the freezing level, the DC-6 conditions are much more representative of the conditions existing at the time of most lightning strikes to aircraft during normal operations.

In a review of in-flight strike-statistic reports from several souces [21],–[24], Clifford [13] concluded that over 80 percent of lightning strikes to aircraft occur in conditions identical to those described by Cobb and Holitza at the time of the DC-6 strikes. The aircraft are almost always inside clouds, near the freezing level, in active mixed-phase precipitation, when the strikes occur. Fig. 3 is taken from a Soviet report [25] showing the sharp concentration of strikes to aircraft at the freezing level. Fig. 4 summarizes other environmental conditions at the time of strikes to aircraft in the summary by Fisher and Plumer [22].

Further insight into the aircraft-triggering process (if it indeed exists) is provided by a review of commercial-pilot reports by Harrison [21] and from Air Force and Navy lightning-strike-incident reports provided by the respective service Safety Centers and summarized by Clifford [13]. These reports indicate that pilots generally agree that strikes occur near the freezing level inside precipitating clouds, and therefore they spend as little time as possible in those regions.

The pilots characterize the discharges which occur in these circumstances as "static discharges," rather than lightning. They say that this type of discharge is often characterized by preliminary displays of corona or St. Elmo's fire and by a build-up of radio interference (static). The build-up intensifies for up to a few seconds before it is terminated by a loud, bright discharge. The intense electrical manifestations described by the pilots under these conditions are difficult to reconcile with the maximum voltage levels thought to be attained during normal triboelectric charging.

The pilots report that some strikes do occur suddenly and without warning. They believe that the sudden discharges are more severe and are more likely to be a natural-lightning discharge. However, the sudden type is rare and many pilots say they have never experienced any strike other than the slow build-up type.

One final piece of data from the in-flight lightning-strike reports, that is important to the triggering issue, is that the aircraft are almost never reported to be flying in electrically active storm clouds at the time they are struck. That is (as in the case of Apollo 12), no lightning was observed in the clouds other than the strike to the aircraft. While some readers may tend to be dubious about these data, it is quite reasonable to assume that the reports are accurate since both commercial and military pilots have a healthy respect for active thunderstorms and give them a wide berth. A thorough evaluation of the synoptic conditions and local weather by Harrison [21], and documented radar observations by Trunov [23], confirmed that active thunderstorms were not generally present at the time of the strikes to aircraft reported in their respective studies. Sometimes there was thunderstorm activity in the general vicinity (within a range of 10–50 km), but not always. Pilots generally will not penetrate a cloud if lightning activity is observed in the cloud or if the radar shows more than light (Level 2) precipitation. Therefore, the reports that strikes often occur in nonstormy clouds are believable.

It is clear in these cases that although the clouds in which the aircraft were flying were not active thunderclouds, they were still electrified. It is not always easy for a pilot to know whether there is lightning in a cloud. In the daytime, lightning discharges are hard to see and they may be hidden in the cloud or by a rain curtain. They must be very close by (say 100 to 500 m) for thunder to be heard, so unless the aircraft is instrumented with a Stormscope or field mills, it is quite possible for lightning to exist in the cloud without the pilot's knowledge. It is clear that moderately high charge concentrations can continue to exist in decaying thunderstrom clouds that have ceased to produce natural lightning. Therefore, it is possible for high fields to exist in clouds that are not electrically active. Nonetheless, it is probable that aircraft do trigger lightning in clouds that have produced no natural lightning, and it is possible that these are the majority of lightning strikes to aircraft.

VI. LIGHTNING TRIGGERING CONCEPTS

A. Enhanced-Field Breakdown Processes

In a sense, all lightning is triggered lightning, since some mechanism must initiate the discharge. As discussed in Uman and Krider's paper in this issue, in the case of natural cloud-to-ground lightning, it is believed that the process begins with a preliminary breakdown phase, possibly involving streamers connecting charge pockets in the cloud. The charged region then produces a step-leader which propagates through the uncharged air to ground. Intracloud processes have not been studied as thoroughly, but it is still thought that the breakdown is initiated with a similar preliminary breakdown phase, followed by a leader from one pocket of charge propagating to the oppositely charged region. Dawson and Winn[26] and Phelps [27] showed analytically and experimentally that streamers, once initiated, can propagate in low-ambient-field regions. The ambient field level required is a function of the initial energy of the streamer and its polarity.

In the case of lightning triggered by the presence of an aircraft, it is possible that the initiation process is different from that assumed for natural lightning. A qualitative description of a different triggering concept has been given previously by Kasemir [9], [28]. Basically, the concept involves the generation of corona and streamers from a physical object in a high-field region. The streamers develop into propagating leaders in both directions, following the field lines and feeding upon the field. Kasemir's discussion leads to the following conclusions.

1) The maximum gradients occur between the charge concentrations, not in the center of them. Indeed, in the center of the charge region, the gradient is small or zero. This fact follows from basic electrostatics which show that the field in a sphere of uniform charge distribution builds linearly with r from zero at the center (where the fields from all surrounding charges cancel) to a maximum at the surface of the sphere.

2) Since triggered-lightning discharges will originate in regions of the thundercloud that contain the strongest gradients, the most likely places for lightning initiation are the regions between two oppositely charged space-charge volumes, and the most unlikely places are in the center of the space-charge volumes. In most thunderstorm models, there are two locations with maximum gradients: between the upper positive and the middle negative, and between the middle negative

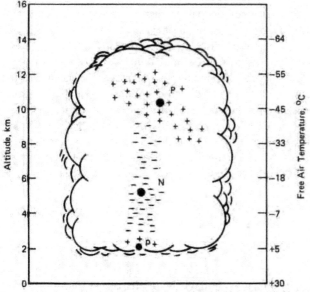

Fig. 5. Probable distribution of the thundercloud charges, P, N, and p for a South African thundercloud according to Malan (1952, 1963).

Fig. 6. Shuttle orbiter model in corona in a parallel-plane gap [12].

and the lower positive space-charge regions. This latter region coincides approximately with the freezing level, according to the conventional model of charge distribution in a thundercloud, as shown in Fig. 5. These maximum-gradient regions are where triggered discharges can be most easily initiated, according to Kasemir. It is interesting that, in the Apollo 12 incident, two discharges occurred to the rocket: one at an altitude of 2 km and the second at 4.2 km. The first strike apparently struck the ground, presumably discharging a pocket of charge in the lower part of the cloud. There was no observation of a ground flash in the second case, indicating an intracloud flash, discharging two pockets of charge at higher altitudes.

An experiment conducted by Kasemir and Perkins [12] in the NOAA facility described earlier supports Kasemir's triggering scenario. A model of the space-shuttle orbiter was suspended in the dc field of the parallel-plate gap. Fig. 6 shows the corona discharge on the orbiter in a vertical gradient of 150 kV/m. The upper plate of the capacitor was charged positive, and the luminous column of ionized air connecting the upper plate and the corona discharge points on the tip of the tail fin carried negative charge upward. Since the orbiter was insulated from the ground plate by three Teflon insulators, the negative corona discharge could not be sustained by negative charge residing originally on the

model. The loss of negative charge had to be balanced by a loss of the same amount of charge by positive corona located on the under part of the fuselage. In Fig. 6, a number of positive corona points can be observed stretched along the lower part of the fuselage. The spread of the points is due to the fact that the belly of the orbiter is relatively flat and has no sharp points on which the positive corona discharge could concentrate.

The production of corona from both sides of the model is by charge separation (induction charging) in a conductor exposed to an external electric field. The net charge on the model is near zero, and the energy for the corona discharge process is solely furnished by the potential energy of the external field.

Kasemir [28], Griffith and Phelps [29], and Dawson and Duff [30] discuss the possibility that the triggering process discussed above may, in fact, be the dominant mechanism in the initiation of natural lightning. Snowflakes or other precipitation particles are proposed as the nuclei which launch streamers and trigger the process. Because of their small size, the ambient fields required to trigger would be about 3×10^5 V/m; near the maximum values actually measured in clouds. For aircraft, the field required for triggering should be much smaller. The values of V_d discussed earlier of 500 to 750 kV would infer ambient fields as low as a few tens of kV/m for aircraft triggering, depending on the size of the aircraft.

Although the minimum field levels predicted for triggering seem to be consistent with some flight data, it is a fact that aircraft are not always struck, even when field levels well in excess of the minimum field required for triggering are present. The NASA F-106 research aircraft has frequently measured fields in excess of 100 kV/m without being struck [31]. In fact, research aircraft seeking lightning often have difficulty getting struck. Although Fitzgerald had fairly good success with 55 strikes in 205 penetrations, he measured field values in excess of 300 kV/m with no lightning strike. Other investigators have been frustrated by the lack of strikes encountered. Newman remarked in his paper [4] that "during 133 penetrations of thunderstorms, in about ten of which a 200-m trailing-wire antenna was used, only two lightning discharges were encountered with peak currents of about 7 kA. The small number of strikes, even in areas of extensive lightning activity, indicates that the presence of the aircraft is not a large factor in diverting to itself a natural lightning discharge." Newman goes on to speculate that "the electric field in which the aircraft is flying may, at times, be large enough for the aircraft to give rise to propagating streamers by which a discharge can be triggered."

Shaeffer [32] postulates that the aircraft must be adjacent to, or in close proximity to, a source of charge (presumably the cloud charge center) before leaders can propagate. He indicates that the enhanced field falls off so sharply with distance from the aircraft that leaders cannot propagate if the aircraft is far removed from the charge center.

B. Aircraft-Wake-Enhancement Factors

Pierce noted in his summary of triggered-lightning incidents that the only deviations from the general pattern of voltage discontinuity values were the aircraft cases. The deduced values of voltage discontinuity for those cases were as low as 50 kV, whereas for the other cases (tall structures, rocket trailing wire, Apollo 12), the values were always nearer 10^6 V. Pierce reasons, therefore, that other factors may be required to initiate an aircraft discharge, such as a high charge on the aircraft from precipitation charging.

Vonnegut, in his paper [16], included some discussion of possible mechanisms for producing ionization or charge in the wake of the aircraft. He believed that an aircraft charged to a high potential by triboelectric processes had a higher probability of triggering a strike.

Muhleison [11] and Shaeffer [32] have both examined and rejected the possibility of an ionized engine exhaust enhancing the ambient field. The electron concentrations beyond the luminous region of the exhaust are many orders of magnitude too low (10^5 versus 10^{12} p/cm^3 in a rocket exhaust or lightning streamer). However, Pierce [3] reasons that the ionic trail, although not of itself a good conductor, is nevertheless much more conducting than the undisturbed atmosphere, and a lightning leader encountering such a trail could be led toward the aircraft. Pierce's reasoning requires an existing leader, however, whereas the field conditions being considered are not yet capable of initiating such a leader.

The fact that a large majority of the reported strikes to aircraft occur around the freezing level in precipitation within clouds led Clifford [13] to postulate two possible mechanisms for producing the intense fields required to produce triggering and the electrical disturbances reported by pilots in nonstormy clouds. The first is an extraordinarily high rate of triboelectric charging produced in highly charged precipitation regions near the freezing level. The second is the production of an ionized wake produced by the passage of a highly charged aircraft through mixed-phase precipitation where supercooled water droplets and ice particles are present. Clifford thinks that the charge-exchange process in the wake can be so intense as to produce an ionized channel which acts as an electrical extension of the aircraft, greatly enhancing the ambient electric field between two charge centers. This field enhancement would lead to triggering in the same manner as discussed previously.

C. Summary of Triggering Concepts

In summarizing the triggering arguments, two types of conditions are considered. In the first case, the aircraft flies into a high electric field (greater than 50 kV/m) in proximity to a charged region. The enhancement of the field by the conducting vehicle produces fields at the extremities of the aircraft which are sufficiently high to launch streamers. The streamers develop into a leader which draws energy from the ambient field, and which eventually discharges and neutralizes the charge centers which gave rise to the initial ambient field. This type of strike may be experienced outside or inside a cloud, at any altitude where proximity to charge centers is close enough to experience the necessary field level. This type of discharge may be abrupt and fairly severe.

The second argument involves aircraft flying in mixed-phase precipitation inside nonstormy clouds. Several authors feel that the ambient fields are too low in this case to allow normal triggering and, therefore, some additional enhance-

ment mechanism must be postulated. They suggest various processes, usually involving the charge on the aircraft produced by heavy triboelectric charging. This type of discharge is characterized by a build-up of static charge manifestations for a few seconds before the discharge. Such build-ups are described in the large majority of strikes reported.

VII. CHARACTERISTICS OF TRIGGERED LIGHTNING

No measurement has been made of a lightning discharge known to have been triggered by aircraft. Therefore, triggered-lightning parameters must be based upon the characteristics of ground-based triggered lightning and upon the supposed mechanisms of an intracloud lightning discharge. Those parameters can then be compared with measurements and observed physical effects of discharges to aircraft.

A pronounced characteristic of ground-triggered lightning is the universal presence of a continuing-current component [4], [5]. Many times, *only* a continuing-current discharge is experienced, but at other times hybrid flashes (continuous current plus discrete strokes) are observed. During the early French rocket-triggering experiments, of 62 triggered flashes, 23 were low-amplitude continuous currents and 39 were hybrid flashes. It is notable that, in order to experience a hybrid flash with high-amplitude, multiple, discrete strokes, the rocket had to trigger the flash before it reached an altitude of 250 m. If the flash triggered at higher rocket altitudes, the resulting discharge often exhibited only a continuing-current component.

That such continuing-current discharges are to be expected from triggered lightning is explained by Brook *et al.* [8] who point out that discrete (pulse-type) lightning discharges must originate from regions of relatively high charge density. The linear dimensions of the charged volumes generating lightning discrete strokes in thunderclouds appear to be of the order of 300 to 500 m. Continuing-current discharges, however, originate in volumes 30 to 50 times greater. On the other hand, the ratio of charges transferred in continuing currents is two to six times greater than in discrete strokes. Brook *et al.* argue that, if the charged volumes are assumed to have about the same potential, the capacity of the volume drained by the continuing current is approximately three times as large, consistent with the greater amount of charge stored. However, the discrete-current maximums are 50 times greater than the continuing currents. It is, therefore, reasonable to assume that the effective impedance of the current source is approximately proportional to the volume in which it is stored. It is this consideration which leads to the belief that triggered lightning probably involves volumes of charge whose density is considerably below that which would lead to a natural discharge, and would, therefore, exhibit a current consistent with the notion of a high internal impedance. This notion is also consistent with the fact that the high-altitude rocket-triggering produced only continuing-current discharges. More field enhancement is required to produce a discharge from a diffuse charge region.

Normal cloud-to-ground lightnings, where tall structures or rockets are not involved, are initiated by stepped-leaders from the clouds to ground. Once the leader contacts ground, the channel is rapidly discharged. Consequently, measurements of natural cloud-to-ground flashes made on the ground are always characterized by one or more discrete current pulses of high amplitude with an occasional continuing current following after one or more of the subsequent return-stroke pulses.

A natural intracloud discharge consists basically of a probing leader that attempts to bridge the gap between charge centers in a thundercloud. The leader advances fairly steadily at a velocity of about 10^4 m/s, supplied by a comparatively continuous current from the source. Occasionally, the leader encounters a localized concentration of opposite charge; the encounter generates a current pulse known as a K-change along the channel pre-ionized by the leader. However, there is never an encounter with a charged surface of extent and conductivity comparable with that of earth. Consequently, no very rapid neutralization of charge can occur. The K-change current surges attain peaks (1 to 2 kA) an order of magnitude less than those of return strokes. Cianos and Pierce [33] model a severe intracloud discharge as a series of ten K-change pulses of 2 kA superimposed on a continuing current of 125 A, terminated by a 20-kA discharge (a severe K-change). Uman and Krider discuss intracloud discharges in more detail in the first paper of this issue.

Since most discharges to aircraft occur in clouds at altitudes around the freezing level (usually 3 to 5 km), the discharge parameters experienced by those aircraft must be much different than those generally quoted for cloud-to-ground lightning. If a ground stroke is involved, the aircraft would have to be in the upper regions of the channel, leading to much-reduced magnitudes of peak current and dI/dt. The stroke is more likely to be an intracloud discharge, simply because there are more cloud discharges than cloud-to-ground discharges. Whether or not the aircraft triggers the discharge, the discharge waveform at the aircraft will probably look much like a natural intracloud strike, i.e., a low-amplitude continuing current, with, in some cases, discrete current pulses superimposed. At the freezing level, this would be true even for a cloud-to-ground stroke. The current pulses could be expected to have amplitudes of a few to several kiloamperes and relatively slow rise times.

The "Rough Rider" measurements discussed earlier bear out the intracloud scenario. The pulses measured were typically a few hundred to a few thousand amperes amplitude. More recently, the French government has sponsored a flight research program using a large Transall C-160 aircraft [34]. The aircraft registered a large number of strikes during the 1978 campaign, characterized generally as occurring in precipitating clouds between 0° and −5°C in mixed ice and rain. Thirteen strikes to the aircraft were measured, and the resulting waveforms were classified into three categories. The three are depicted in Fig. 7. All strikes in each category exhibited a continuing-current component.

The first group (six of 13 measured) is characterized by a continuing current with a series of discrete pulses spaced about 100-μs apart on the continuing-current base. The pulses are typically a few kiloamperes and look like K-recoil pulses, or possibly step-leader pulses. The second category (three of 13) has one large-amplitude (20 70 kA) pulse superimposed midway on the continuing-current base. The third category

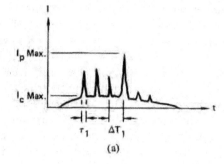

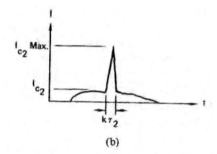

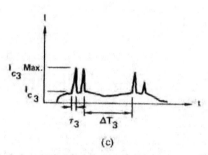

Fig. 7. In-flight lightning-strike waveforms measured on French Transall aircraft [34]. (a) Lightning current type 1 (six). $I_p \leqslant 4.5$ kA. Number of pulses: 3 to 12. 8 μs $< \tau_1 <$ 200 μs. 50 μs $< \Delta T_1 <$ 200 μs. I_{c_1} max $<$ 2 kA. (b) Lightning current type 2 (three). $I_p \leqslant$ 70 kA. 80 μs $< \tau_2 <$ 200 μs. $I_{c_2} \leqslant$ 500 A. (c) Lightning current type 3 (four). $I_p \leqslant$ 50 kA. Number of pulses: 1 to 8. 20 μs $< \tau_3 <$ 200 μs. 50 μs $< \Delta T_3 <$ 14 ms.

(four of 13) was characterized by groups of high-amplitude impulses occurring at intervals spaced up to 14-ms apart on the continuing-current base.

The fact that all of the French aircraft measurements exhibit a continuing-current base which builds up slowly implies that the discharges are triggered by the aircraft. They may be intracloud flashes or the aircraft may be near the origin of a cloud-to-ground strike. If the aircraft simply intercepted a propagating leader, the continuous current could jump up abruptly, exciting resonances on the aircraft. The peak-current amplitudes and rise times are consistent with Little's calculations [35] for return strokes at altitude using a lumped-transmission-line model. Since continuing currents characterize both intracloud and ground-triggered leaders, it is impossible to say which type of flash the French data represent.

A few strikes have been recorded to the NASA F-106 research aircraft [31]. Some were recorded at the freezing level in clouds, but others were recorded at much-higher altitudes. The instrumentation package [36] was designed to measure the fastest changes and, therefore, did not record continuing currents. The peak currents deduced from the derivative measurements were usually a few hundred amperes, and were very fast (fractions of a microsecond total duration). These measurements may be of preliminary streamers which occur before the main leader is launched [37]. One higher-current discharge (about 14 kA), measured on the F-106 late in the 1981 series [38], resembled the second category of strikes in Fig. 7.

In some cases at least, continuing-current discharges appear to occur to aircraft with no evidence of any high-energy, discrete discharge (as in a natural cloud-to-ground discharge). Typical of such incidents were discharges experienced by the NASA LRC F-106 and the USAF/NOAA C-130 research aircraft [39] while flying in precipitation near the freezing level. An inspection of strike effects on both aircraft by Clifford [37] revealed a series of light pit marks, typical of a swept stroke, trailing from the front of the aircraft back along the forward fuselage. The first pit mark of each series was no more pronounced than the others and there was no evidence of other attachment points forward of the first pit mark. The light pit marks indicate a low-amplitude continuing current, probably on the order of tens to hundreds of amperes, sweeping back along the vehicle. Such patterns seem to be the rule, rather than the exception, for many aircraft strikes.

Summarizing the expected parameters for a triggered strike, the discharge will be characterized by a continuing current of a few hundred up to one or two thousand amperes, persisting for tens to hundreds of milliseconds. Superimposed on the continuing current may be several discrete pulses of moderate to low amplitude (2 to 50 kA) with rise times of a few to many microseconds. Infrequently, a single high-amplitude discharge (reminiscent of a positive stroke to ground) may be experienced with higher amplitudes, but with relatively slow rise-times (tens of microseconds).

Although the triggered-discharge characteristics may not seem to be severe, there is evidence that triggered intracloud lightning can cause serious damage. For example, an Air Force C-130E crashed in Charleston, SC in November 1978 as a result of a verified lightning strike [40]. The aircraft was flying in the classical strike conditions, i.e., at the freezing level in a precipitating nonthunderous cloud, at the time of the strike. The lightning attached at the leading edge of the wing and swept back over the fuel-tank region in discrete steps. At one point, the dwell time was sufficient for the continuing current to burn a hole through the wing skin, detonating the fuel vapor and blowing off the outer wing section. The wing section was recovered from the crash scene and clearly shows the swept-stroke markings and the puncture hole.

Based on the French flight data discussed earlier, fairly-high-magnitude discrete pulses may be superimposed on the continuing current base of a triggered strike. However, the observed rates of rise were not high. The maximum dI/dt measured was 26 kA/μs, although for most (12 of 13), the average was closer to 2.5 kA/μs. It should be noted, however, that Weidman and Krider [41] have measured intracloud processes with very fast (submicrosecond) field transition times. These low amplitude but very fast field changes and the moderate amplitude current and voltage changes associated

with a triggered strike would constitute a serious threat to avionics systems, even if more severe peak current levels were not experienced.

VIII. SUMMARY AND CONCLUSION

A close inspection of reports on in-service lightning strikes to aircraft reveals that most aircraft strikes (probably over 80 percent) occur in nonstormy clouds, i.e., in clouds where no lightning was previously detected. The other 20 percent may occur inside or outside clouds in high-cross-field regions. Most of the strikes occur near the freezing level in mixed-phase precipitation, where ambient field levels are not high. Since no lightning was produced until the aircraft entered the cloud, there is a high probability that the presence of the aircraft initiates the discharge in some way.

Ground-triggered discharges from structures and wire-trailing rockets were examined to find the level of field-enhancement necessary to trigger a discharge when charge concentrations are overhead. Pierce's calculations suggest that ambient field levels had to be high enough to generate voltage discontinuities of 10^6 V at the tip of a fixed conductor, and 0.5 to 0.7 $\times$ 10^6 V for a rapidly moving ungrounded conductor. Electric-field records from at least one research aircraft struck at the freezing level in precipitating clouds indicated that the ambient fields were much too low to produce such levels. However, other strikes to an F-100, a B-25 and a Lear jet occurred when the ambient fields were high enough that enhancement by the aircraft could generate potential discontinuities meeting the airborne triggering criterion. However, in these instances the aircraft were flying in, or around, active thunderclouds in strong-cross-field regions and were not flying in the more common strike conditions inside the cloud.

Two different triggering concepts were discussed in this paper. The first is a direct application of the ground-triggering criteria for field enhancement. This concept assumes that, if a high enough ambient field is present (>50 kV/m), streamers from the aircraft can be produced which may grow into a leader, connecting charge centers and resulting in a lightning flash. Data were discussed which indicated that triggering might be enhanced if the aircraft were in proximity to a charge center.

Triggered-lightning discharges inside nonthunderous clouds have led to suggestions that additional field-enhancement mechanisms are needed to initiate a discharge. Some possible enhancement mechanisms were examined, including abnormally high static-charging conditions and ionized-wake enhancement.

The projected characteristics of triggered discharges were found to be comparable to natural intracloud discharges as described in the literature, and as measured in the various flight-research programs. The discharges are characterized by a continuing current base, which may or may not have discrete moderate-level current pulses superimposed. A distinguishing feature of triggered discharges may be the gradual build-up of the continuing current.

Triggered lightning probably does not pose as serious a hazard to aircraft as natural cloud-to-ground discharges near the ground (probably a rare occurrence). Very-high peak currents and high values of dI/dt are not present. However, the continuing currents are capable of causing direct structural damage (i.e., burn-through of wing skins) and the discrete current pulses can be of damaging magnitudes. Indirect (induced) effects on avionics equipment can be significant if systems are not properly designed for lightning protection.

An important question is whether aircraft ever intercept natural cloud-to-ground lightning at low altitudes, or whether all strikes are triggered. There are undocumented reports of aircraft being struck at low altitudes during landing and takeoff, and any flash to ground can produce subsequent strokes with very fast rise times, even though the initial stroke may be slow or even a dc continuing current. Consequently, the impact of these conclusions, though somewhat speculative, is that the great majority of strikes to aircraft will never approach the worst-case severe model parameters specified for aircraft design and test. However, a few strikes might, and the matter of whether or not to continue to design all aircraft for that low probability threat should be considered by the aerospace lightning community.

REFERENCES

[1] K. B. McEachron, "Lightning to the empire state building," *J. Franklin Institute*, vol. 227, pp. 149–217, 1939.

[2] K. Berger, "The earth flash," *Lightning*, vol. 1. ed. Golde, New York: Academic press, 1977.

[3] E. T. Pierce, "Triggered lightning and some unexpected lightning hazards," 138th Annual Meeting of the Amer. Asso. for the Advancement of Science, Philadelphia, Dec. 1971.

[4] M. M. Newman, J. R. Stahmann, J. D. Robb, E. A. Lewis, S. G. Martin and S. V. Zinn, "Triggered lightning strokes at very close range," *J. Geophys. Res.*, vol. 72, p. 4761, 1967.

[5] R. P. Fieux, C. H. Gary, B. P. Hutzler, A. R. Eybert-Berard, P. L. Hubert, A. C. Meesters, P. H. Perroud, J. H. Hamelin, and J. M. Person, "Research on aritificially-triggered lightning in France," *IEEE Trans. on Power Apparatus and systems*, vol. PAS-97, No. 3, pp. 725–733, May-June 1978.

[6] P. Hubert, "Triggered lightning at Langmuir Laboratory during TRIP-81," Centre d'Etude Nucleaires de Saclay Report No. Dph/EP/81-66, 23 Nov. 1981.

[7] R. B. Standler and W. P. Winn, "Effects of coronae on electric fields beneath thunderstorms," *Quarterly J. Royal Meteorological Society*, vol. 105, no. 443, pp. 285–302, Jan. 1979.

[8] M. Brook, C. R. Holmes, and C. B. Moore, "Lightning and rockets: some implications of the Apollo 12 lightning event," *Naval Research Reviews*, Apr. 1970.

[9] H. W. Kasemir, "Basic theory and pilot experiments to the problem of triggering lightning discharges by rockets," NOAA Tech. memo. ERL APCL 12, 1971.

[10] R. Godfrey, E. R. Mathews, and J. A. McDivitt, "Analysis of Apollo 12 lightning incident," NASA (MSC) Rep. 01540, Feb. 1970.

[11] R. P. Mulheisen, "Phenomenology of lightning/aircraft interaction," AGARD Lecture Series No. 110, Paper No. 3, June 1980.

[12] H. W. Kasemir and F. Perkins, "Lightning trigger field of the orbiter," NASA, KSC Contract CC 69694A, Final Rep., 1978.

[13] D. W. Clifford, "Another look at aircraft-triggered lightning," FAA/NASA/FIT Symposium on Lightning Technology, Hampton, VA, Apr. 1980.

[14] D. R. Fitzgerald, "Probable aircraft triggering of lightning in certain thunderstorms," *Monthly Weather Review*, vol. 95, p. 835, 1967.

[15] D. R. Fitzgerald, "Aircraft and rocket triggered natural lightning

discharges," Lightning and Static Electricity Conference, AFAL rep. TR-68-290, Part II, Dec. 1968.

[16] B. Vonnegut, "Electrical behavior of an aircraft in a thunderstorm," Federal Aviation Admin., Rep. FAA-ADS-36, Feb. 1965.

[17] B. J. Petterson and W. R. Wood, "Measurements of lightning strikes to aircraft," Final Report to Fed. Aviation Admin. under Contracts FA65WAI-94 and FA66NF-AP-12, Jan. 1968.

[18] W. E. Cobb and F. H. Holitza, "A note on lightning strikes to aircraft," *Monthly Weather Review*, vol. 96, no. 11, Nov. 1968.

[19] J. E. Nanevicz, R. C. Adamo, and R. T. Bly, "Airborne measurement of electromagnetic environment near thunderstorm cells (TRIP-76)," Stanford Res. Inst. Final Rep., Air Force (FDL) Contract NAS9-15101, Mar. 1977.

[20] R. Gunn, "Electric field intensity inside of natural clouds," *J. Appl. Phys.*, vol. 19, pp. 481–484, 1948.

[21] H. T. Harrison, "UAL turbojet experience with electrical discharges," UAL Meteorology Circular no. 57, Jan. 1967.

[22] F. A. Fisher and J. A. Plumer, "Lightning protection of aircraft," NASA Ref. Public. 1008, ch. 3, Oct. 1977.

[23] O. K. Trunov, "Conditions of lightning strike on air transports and certain general lightning protection requirements," 1975 Lightning and Static Electricity Conference, Culham Laboratory, England, Apr. 1975.

[24] B. L. Perry, "Lightning and static hazards relative to airworthiness," 1970 Lightning and Static Electricity Conference, SAE Rep. P-35, Dec. 1970.

[25] I. M. Imyanitov, "Aircraft electrification in clouds and precipitation," USAF Foreign Tech. Div. Rep. FTD-HC-23-544-70, Apr. 1971. Published originally in Elektrizatsiya Samoletov v Oblakakh i Osadkakh, pp. 1-211, 1970.

[26] G. A. Dawson and W. P. Winn, "A model for streamer propagation," Zeitschrift fur Physik, vol. 183, pp. 159–171, 1965.

[27] C. T. Phelps, "Field enhanced propagation of corona streamers," *J. Geophys. Res.*, vol. 76, no. 24, 20 Aug. 1971.

[28] H. W. Kasemir, "A contribution to the electrostatic theory of a lightning discharge," *J. Geophys. Res.*, vol. 65, no. 7, pp. 1873–1878, 1960.

[29] R. F. Griffiths and C. T. Phelps, "A model for lightning initiation arising form positive corona streamer development," *J. Geophys. Res.*, vol. 81, no. 21, July 1976.

[30] G. A. Dawson and D. G. Duff, "Initiation of cloud-to-ground lightning strokes," *J. Geophys. Res.*, vol. 75, no. 30, 20 Oct.

[31] F. L. Pitts and M. E. Thomas, "1980 direct strike lightning data," NASA Tech. Memo. 81946, Feb. 1981.

[32] J. F. Shaeffer, "Aircraft initiation of lightning," 1972 Lightning and Static Electricity Conference, USAF Rep. AFAL-TR—72-325, Dec. 1972.

[33] N. Cianos and E. T. Pierce, "A ground-lightning environment for engineering usage," Stanford Res. Inst. Tech. Rep. no. 1, Proj. 1834, Aug. 1972.

[34] Centre D/Essais Aeronautique De Toulouse, "Measure Des Characteristiques De La Foudre En Altitude," Essais No. 76/650000P.4 et Final, July 1979.

[35] P. F. Little, "Transmission line representation of a lightning return stroke," *J. Phys. D: Appl. Phys.*, vol. 11, pp. 1893–1910, 1978.

[36] F. L. Pitts, M. E. Thomas, R. E. Campbell, R. M. Thomas, K. P. Zaepfel, "In-flight lightning characteristics measuring system," FAA/FIT Workshop on Ground and Lightning Technology, Rep. FAA-RD-79-6, Mar. 1979.

[37] D. W. Clifford, "Characteristics of lightning strikes to aircraft," Intl. Aerospace Conference on Lightning and Static Electricity, Oxford, England, Mar. 23–25, 1982.

[38] F. L. Pitts, NASA Langley Research Center, personal communication, Nov. 1981.

[39] R. K. Baum, "Airborne lightning characteristics," FAA/NASA/FIT Sympos. on Lightning Tech., Hampton, VA, pp. 22–24 Apr.

[40] 1980. D. W. Clifford, "Aircraft mishap experience from atmospheric electricity hazards," NATO AGARD Lecture Series No. 110, paper No. 2, June 1980.

[41] C. D. Weidman and E. P. Krider, "The radiation field waveforms produced by intracloud lightning discharge processes," *J. Geophys. Res.*, vol. 84, no. C6, pp. 3159–3164, 1979.

ELECTROSTATIC MODEL OF LIGHTNING FLASHES TRIGGERED FROM THE GROUND

H. W. Kasemir
Colorado Scinetific Research Corporation
Berthoud, Colorado

Abstract: The electrostatic model of a lightning discharge is based on mathematics and theoretical physics. The purpose of this paper is to give the mathematical part of this theory. Equations are deduced, which link the gradient measurements at the ground whith the properties of the triggered lightning flash and the gradient and charge distribution of the thunder storm. These equations are required for the analysis of the data obtained during the Kennedy Space Center triggered lightning experiments.

I. Introduction

The electrostatic lightning theory is here applied to lightning triggered from the ground the provide the mathematical and theoretical background for an evaluation of the data obtained from the Kennedy Space Center triggered lightning experiments. These experiments are especially important and valuable because they combine the features of a laboratory experiment with the object of study being a real lightning discharge. Furthermore, the data from the extensive field mill network and from video cameras have been made available to the scientific community by Mr. William Jafferis of Kennedy Space Center. The results of the analysis of these data will be presented in another paper.

The concept of the electrostatic lightning theory and its application to different problems of the lighning discharge have been published by the author and presented at several scientific conferences. Reference [1], [2], [3], [4], [5], [6]. However, increasing requests that the - often rather brief - treatment of the mathematical side of the theory should be given in more detail, was another reason for addressing the mathematical formalism in this paper in a more extended manner.

The lightning channel is represented by a slim spheroid. Therefore the spheroidal coordinate system is the obvious choice. It is used however here in an unusual manner in so far as both the long axis "a" of the spheroid and the focal distance "c" are treated as variables. This is necessary to keep the spheroid representing the channel slim as the lightning grows. In the textbooks the focal distance "c" is assumed to be constant and the spheroidal coordinates are normalized to it, so that "c" itself appears in the equations only as a number "1", i.e., it is not treated as a variable. Since this normalization procedure is carried over to the Legendre's polynominals of first and second kind, which are the key functions used in the calculation, these functions as well as all other textbook equations in spheroidal coordinates have to be freed from the restricctive normalization and the "c" has to be restored to give these functions and equations the full flexibility.

Paragraph II and III deal with the modeling of a growing lightning channel, the different coordinate systems and related problems, and the Legendres's polynominals. Recurrence equations of higher order polynominals, differentials, and integrals are included with regard to computer programming of more complex models. In these two paragraphs the normalization problem and its pitfalls have been discussed in detail.

In paragraph IV the equations and functions of paragraph II and III are used to construct the cloud potential function of a simple cloud model in spheroidal coordinates. The close relationship between the charge distribution in a thundercloud and the properties of a lightning discharge is a new and attractive feature of the electrostatic lightning theory. One may say that combined with the laws of theoretical physics the potential function of the thunderstorm determines uniquely the lightning discharge.

Paragraph V shows how the harmonic solutions in spheroidal coordinates of Laplace's equation, which are the Legendre's polynominals of first and second kind, can be used to construct the lightning potential function in a general way and then adapt it by the boundary conditions to any cloud model where the cloud potential function has been expanded into a series of Legendre's polynominals of first kind. Once this process is understood the total calculation is reduced to solving a limited number of simple algebraic equations for one unknown in each. This process is shown in detail for the cloud potential function given in paragraph IV. A superposition, i.e., adding together, of the cloud and the lightning potential functions in a resulting potential function Φ solves our problem. From this potential function the lightning parameters such as tip gradient, net charge of and charge distribution on the channel, gradient at the ground, etc., can be easily calculated, and the resulting equations are listed in paragraph V and VI.

Useful textbooks are listed in Reference [7], [8], [9], [10]. Once the restoration of the focal length "c" is done, these textbooks give certainly a more extensive and complete treatment of the whole field of electrostatic theory. It may also be mentioned that the electrostatic lightning theory is only a first step and shall be extended into an electrodynamic theory and finally include plasma physics to deal with the interior phenomena of the lightning channel.

II Model and coordinate system

The lightning triggered from the ground is mathematically represented by the upper half of a spheroid, which grows vertical upwards from the horizontal ground plane. Fig.1. The lightning channel and the ground are electrical conductive and have both zero potential. The coordinates used in the calculation are the spheroidal coordinates u,v.ϕ shown in Fig.2. Figure 2 is a cross-section through the centerline in the vertical plain ϕ = const. Since this is a rotational symmetric problem the potential and gradient functions do not depend on ϕ. The surfaces u and v = const. are obtained by rotation around the centerline. The slim spheroid in the center of the set of confocal spheroids u represents the lightning channel and is singled out by the value u = a. The coordinate u as well as the special value a represent here the whole surface of the spheroid, but the numerical value of u or a is identical with the value of the long axis of the spheroid. The two short axis b, b can be obtained from the well known relation $b^2 = a^2 - c^2$, where c is the focal distance on the long axis from the centerpoint. See Fig. 2.

At this point we have to discuss a deviation from the ordinary use of spheroidal coordinates. The focal distance (eccentricity) is usually kept constant and u and v are normalized to c. The arguments of the two Legendre functions Pn and Qn - which will be discussed in the next section - are v/c and u/c. New, usually Greek letters, are then introduced in the literature as arguments and the normalized c replaced by 1. In this way one of the important parameters namely c disappears from the equations.

It is quite obvious from Fig. 2 that with the growth of the lightning channel from a to 2a the base of lightning widens from b to almost 2a, i.e. our lightning model of a slim spheroid deteriorates to a half sphere. This is the consequence of the constant c now hidden in the equations by the normalization.

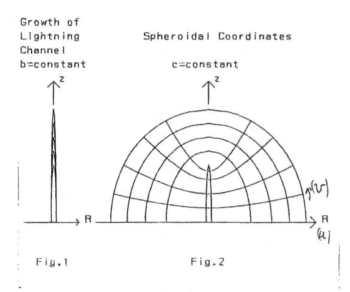

Growth of
Lightning
Channel
b=constant

Spheroidal Coordinates

c=constant

Fig.1   Fig.2

To achieve a growth as shown in Fig. 1. we have to keep b constant not c. To achieve this we have to restore first c in the normalized textbook equations and then give it freedom to grow in the same manner as a, to maintain the relation $c^2 = a^2 - b^2$. This results in a spheroidal coordinate system which continuously adjusts to the growth of the channel and keeps the channel slim. The mathematical equations including the Legendre functions remain the same as long as we restore c and treat it as a variable parameter.

The variability of c has still another aspect which is of some importance. The better known and more often used cylindrical and spherical coordinate systems are special cases of the more general spheroidal coordinates. It is not difficult to show that the spheroidal coordinates convert to cylindrical for c→∞ and to spherical coordinates for c→ o. Intuitively we can deduce this result from Fig.1 and 2. In Fig.1 we imagine that the lightning channel grows to a great length. The greater part is already outside the picture frame and what can still be seen from the lightning channel is of cylindrical shape. Therefore cylindrical coordinates could be used in approximation at distances small compared to length of channel and below the tip. On the other hand we see from Fig. 2 that in distances larger than twice the length of the spheroids convert approximately into spheres. Therefore spherical coordinates could be used for large distances from the channel.

We will briefly list now the three coordinate systems: cylindrical, spherical, and spheroidal with the relation to each other and to the cartesian coordinates x,y,z.

1. Cylindrical coordinates R, φ, z [m, ɭ, m]

$$R = \sqrt{x^2 + y^2} \quad ; \quad \phi = \text{atan}(y/x) \quad ; \quad z = z \quad (2.1)$$

2. Spherical coordinates r, φ, ϑ [m, ɭ, ɭ]

$$r = \sqrt{R^2 + z^2} \quad ; \quad \phi = \phi \quad ; \quad \cos(\vartheta) = z/r \quad (2.2)$$

3. Spheroidal coordinates u, φ, v [m, ɭ, m]
Relation to cylindrical coordinates

$$z = uv/c \quad ; \quad \phi = \phi \quad ; \quad R = \sqrt{(u^2 - c^2)(1 - v^2/c^2)} \quad (2.3)$$

Relation to spherical coordinates

$$r = \sqrt{u^2 + v^2 - c^2} \quad ; \quad \phi = \phi \quad ; \quad \cos(\vartheta) = \frac{uv/c}{\sqrt{u^2 + v^2 - c^2}} \quad (2.4)$$

In equation (2.3) and (2.4) c is restored. As an illustration to the spheroidal coordinates of fixed and variable c we will discuss briefly how Fig.1 (variable c) and Fig.2 (fixed c) are plotted. The plotter accepts an x value for the horizontal axis and a y value for the vertical axis. Therefore the conversion of spheroidal to cylindrical coordinates with x=R and y=z given in (2.3) would be the right choice to plot the spheroidal coordinate grid. If we now use the normalized value u/c of the literature we have to divide the first and last equation of (2.3) by c. This automatically normalizes z and R also to c. If c is fixed as in Fig. 2 this is still acceptable since we are still free to add another scale factor of our choice. To plot the confocal spheroids (ellipses) give u as a parameter in successively increasing steps (outer loop) and let v/c run from -1 to +1 in small steps for instance 0.01 (inner loop). Use the conversion equation (2.3) to obtain the plots coordinates R,z. To plot the hyperboloids exchange the role of u and v. Note that the numerical value which identifies the whole surface of a given hyperboloid is the same as the point of its apex on the z axis. However if c is variable as in Fig.1 this procedure would not work. Here we would like to plot the lightning channel u=a for different a but for constant b. The first root of R in (2.3) is then $a^2 - c^2 = b^2$ and $c^2 = a^2 - b^2$. This substitution in (2.3) leads then to the equation which has been used to draw Fig.1. This is one example to show how we can change a static spheroidal coordinate system into a time dependent one capable of representing the growth of the channel without drastically deforming the channel. It is only neccessary to define a and c as functions of time with the restriction that always

$$c^2(t) = a^2(t) - b^2.$$

Curved or curved linear coordinate systems like the spheroidal or the spherical and cylindrical coordinates require for the definition of the line elements certain factors, which are in orthogonal systems the three diagonal elements of the metric tensor. They are usually called the h-parameters. In a three dimensional coordinate system $x1$, $x2$, $x3$ with h-parameters h1, h2, h3 the line elements ds1,ds2,ds3 are defined by

$$ds1=h1dx1 \; ; \; ds2=h2dx2 \;, \; ds3=h3dx3 \quad (2.5)$$

The surface element dS1 on the surface $x1 = $ constant, dS2 on $x2$ and dS3 on $x3 = $ constant are given by

$$dS1=ds2ds3; \; dS2=ds3ds1, \; dS3=ds1ds2 \quad (2.6)$$

and the volume element dV is given by

$$dV=ds1ds2ds3 \quad (2.7)$$

The line elements are used in all vector operators such as grad, div, divgrad, ~~const~~ etc., the surface elements are required to calculate the surface or in our case the charge per unit length or the net charge on the lightning channel, and the volume element to obtain the charge of a given cloud volume or the inside the lightning channel if this should become necessary.

Table 1 lists the h-parameters for the cartesian, cylindrical, spherical, and spheroidal coordinate systems. The first column contains the coordinates in right handed order of the 4 systems in row 1 to 4. Column 2-4 contains the h-parameters h1, h2, h3 where h1 refers to the first listed coordinate, h2 to the second, and h3 to the third coordinate. In the following sections we will replace the identifying numerical index by the coordinate. For instance in spheroidal coordinates h1=hu, h2=hφ, h3=hv. In paragraph V we have to calculate the lightning gradient components LGu and LGv from the lightning potential function LP in spheroidal coordinates. We will then use the two equations

$$LGu = \frac{\partial LP}{\partial su} = \frac{1}{hu}\frac{\partial LP}{\partial u} \; ; \; LGv = \frac{\partial LP}{\partial sv} = \frac{1}{hv}\frac{\partial LP}{\partial v} \quad (2.8)$$

Furthermore we will need for the calculation of the charge per unit length and the net charge of the lightning channel the surface element dSu. From (2.5) and (2.6) replacing again the indicees 1,2,3 by u,φ,v we obtain

$$dSu=hφdφhvdv \quad (2.9)$$

Because of the rotational symmetry all functions and h-parameters are independent of φ. A simple integration of dφ from o to 2π turn the surface element dSu into the ring zone element dO.

$$dO = 2\pi \, hφ \, hv \, dv = 2\pi b \, (u^2-v^2)^{1/2} \, dv/c \quad (2.10)$$

III Legendre (1751-1833) functions of first and second kind.

Legendre polynominals of first kind $Pn(cos\vartheta)$ are fairly well known for their part in the harmonic solutions of Laplace's differential equation in spherical coordinates. A rotational symmetric potential function $F(r,\vartheta)$, which fulfills Laplaces's differential equation, can be expanded in these harmonics with arbitrary factors An

$$F(r,\vartheta) = \sum An \, Pn(cos\vartheta)/r^{n+1} \quad (3.1)$$

The first term for n=o is the potential function of a point charge (monopole), the second term that of a dipole and so on.

Legendre's polinominals of the second kind $Qn(u/c)$ are comparatively unknown. They play the part of $1/r^{n+1}$ in (3.1) in the harmonic solutions of Laplace's differential equations spheroidal coordinates. With the argument v/c replacing the cosϑ in (3.1) we can expand the potential function LP (u,v) of a lightning flash represented by a spheroid in a sum of products Bn Pn Qn.

$$LP(u,v) = \sum Bn \, Pn(v/c) \, Qn(u/c) \quad (3.2)$$

In analogy to the interpretation of the individual terms in (3.1) as potential functions of a monopole, dipole, etc., we can

	h1	h2	h3
x, y, z	1	1	1
R, ϕ, z	1	R	1
r, ϕ, ϑ	1	$r \sin(\vartheta)$	r
u, ϕ, v	$\dfrac{\sqrt{u^2-v^2}}{\sqrt{u^2-c^2}} = hu$	$\sqrt{(u^2-c^2)(1-v^2/c^2)} = R = h\phi$	$\dfrac{\sqrt{u^2-v^2}}{c\sqrt{1-v^2/c^2}} = hv$

Table 1

h-parameters of the different coordinates systems

interpret the individual terms of (3.2) as the potential functions of a line charge with constant charge per unit length, of a line charge with bipolar charge per unit length, and so on. This analogy is also evident if we let the lightning channel shrink to length zero. The constant charge distribution is squeezed into a point charge, the bipolar charge distribution into a dipole charge, etc. Mathematically this can easily be seen if we let c go to zero. Since v is always smaller than or equal to c it also goes to zero. It is now evident from the relation between spherical and spheroidal coordinates in (2.4) that in the first equation for v and c going to zero u goes to r and from the third equation that v/c goes to $\cos\vartheta$. Therefore $Pn(v/c)$ goes to $Pn(\cos\vartheta)$. That $Qn(u/c)$ goes to $1/r^{n+1}$ can be shown by expanding the ln-term in Qn into a power series. However this is somewhat more difficult and beyond the scope of this paper.

In the later calculation we will require polynominals up to the order of n=4 or 5. We will write down here a collection of equations with c already restored, which are useful for calculation and computer programming. Following Reference [10] (German and English text) the argument is x, i.e. for Pn x=v/c for Qn x=u/c. If the equation is equally valid for Pn and Qn the common name Kn (K=Kugelfunktionen) is used.

$$P_0 = 1 \; ; \qquad Q_0 = \frac{1}{2} \ln \frac{u/c + 1}{u/c - 1}$$

$$P_1 = v/c \; ; \qquad Q_1 = \frac{u}{c} Q_0 - 1 \qquad (3.3)$$

The recurrence equation, valid for both Pn and Qn, allows us to obtain the value for all functions with n>1

$$K_{n+1} = [(2n+1)xK_n - nK_{n-1}]/(n+1) \qquad (3.4)$$

If we for instance need the value of P3(x) we calculate first from P0 and P1 the value of P2 and then from P2 and P1 the value of P3 using the recurrence equation twice.

The Recurrence equation for the differentials Pn' and Qn' valid for Pn and Qn. We use here and later on the abreviation

$$K_n' = dK_n/dx \qquad (3.5)$$

$$K_n' = n(xK_n - K_{n-1})/(x^2 - 1) \qquad (3.6)$$

Here the differentiation of Kn is reduced to an expression containing Kn and Kn-1, where Kn-1 would have to be calculated anyway to obtain Kn. However here are some hidden traps which have to be watched. The differentiation later on is not to the normalized u/c or v/c but to u or v direct. The best way to handle this problem is to use the equation (3.6) anyway to obtain the Qn' or Pn' and follow it by a factor $d(u/c)/du=1/c$ or $d(v/c)/dv=1/c$, i.e. supplement (3.6) by the factor 1/c to obtain differentiation to u or v direct.

A similar situation exists by using textbook intergral equations. We will later need to integrate Legendre's polinominals of the first kind over the upper half of the spher-

oid from v=o to v=c to obtain the net charge of the triggered lightning. The requires the solution of integrals of the type

$$I = \int Pn(v/c)dv = c \int Pn(x)dx \qquad (3.7)$$

Reference [9] page 136 (5) gives the equation

$$\int Pn(x) \, dx = \frac{x \, Pn(x) - Pn-1(x)}{n + 1} \qquad (3.8)$$

When no boundaries are given an arbitrary constant should be added to fulfill the boundary condition set by the physical application. If we integrate let's say from v=o to a given v our integral solution with x=v/c would be

$$I = c \frac{xPn(x) - Pn-1(x) + Pn-1(o)}{n + 1} \qquad (3.9)$$

The cloud potential function CP in IV is given in a brief power series of z^n, with n=1..4. We have first to convert the z^n into spheroidal coordinates. From the first equation in (2.3)

$$z^n = u^n (v/c)^n \qquad (3.10)$$

To determine the arbitrary constants Bn in (3.2) we have to express the expression $(v/c)^n$ in Legendre's polynominals of first kind. We can use here an equation given by Reference [6] Volume I, page 439, equation (11) With v/c=x we have

$$x^n = \frac{n!}{1 * 3 * 5 * * * (2n-1)} [Pn(x) + (2n-3)\frac{1}{2}P(n-2) + \qquad (3.11)$$

$$(2n-7)\frac{2n-1}{2*4}P(n-4)(x)..]$$

This series breaks off if the last term has the polynominal P0 or P1, i.e. the written part of equation (3.11) is good up to n=4.

The last topic of this paragraph is to give a brief outline of the scheme to solve the problem of a lightning flash growing in the electric field of a thunderstorm.

We assume here that the potential function CP of the thunderstorm is given and that the strongest gradients are in the vertical direction i.e. along the z axis of the coordinate system. Lightning flashes will originate at the point of maximum gradient and grow to both sides up and down in the direction of the gradient vector. The zero point of the spheroidal coordinate system will be placed at the place of maximum gradient with the long axis in z direction. In case of the ground triggered lightning the zero is at the ground and the direction of the gradient and channel growth is upwards. We will furthermore assume that the cloud potential function is given as a brief power series of z in the form

$$CP = \sum An \, z^n \qquad (3.12)$$

If CP is not in this form, any function can usually be expanded into a power series of z. However in this case it may be more convenient to expand CP immediately into a series of Legendre's polynominals of first kind. In case of the expansion into z^n we convert z into spheroidal coordinates. With (2.3)

$$z^n = u^n (v/c)^n \qquad (3.13)$$

and with (3.11) we convert $(v/c)^n = x^n$ into the finite series of Pn. Collecting all the factors of the same Pn into a function Fn(u) we obtain the potential function CP in the form

$$CP = \sum Fn(u) Pn \qquad (3.14)$$

According to (3.2) the lightning potential function LP is given by

$$LP = \sum Bn Qn(u) Pn \qquad (3.2)$$

and the final potential function Φ is given by a superposition of (3.2) and (3.14)

$$\Phi = \sum [Fn(u) + Bn Qn(u)] Pn \qquad (3.15)$$

The last step is now to determine the, as yet, unknown constants Bn from the condition that at the conductive spheroid u=a the potential is constant, let's say C. In case of a cloud discharge or a leader stroke C is the cloud potential at the place of the lightning origin. In our case of a ground triggered lightning flash C=0. This gives in our case

$$Bn = -Fn(a)/Qn(a) \qquad (3.16)$$
or in the cloud discharge case

$$Bn = [C - Fn(a)]/Qn(a) \qquad (3.16a)$$

The whole purpose of the expansion of the cloud potential function CP is the determination of the constants Bn. However, for the calculation of the gradient, net charge, or charge per unit length, etc., it is often easier to keep the cloud potential function expressed in spheroidal coordinates.

IV Cloud potential function CP

The cloud potential function CP plays a more important part in the electrostatic lightning theory than is usually assumed or expected. It reflects the electric energy distribution in a thunderstorm and determines the way in which a developing lightning discharge can obtain energy from this reservoir and also the way in which this cannot be done. The cloud potential function remains still unknown since there are no practical ways to measure it and it is asessable only through a theoretical link. Therefore we assume here first a reasonable cloud model where the potential function can be calculated from a given charge distribution in the cloud. Then we calculate with the scheme given in the previous paragraph the development of a ground triggered lightning flash and see if we can spot any characteristic features on the lightning charge or the gradient at the ground. We may later be able to reverse the process and determine the cloud potential along the lightning channel from the ground gradient measurements.

It was predicted previously by the author Reference [1] that the charge distribution along the channel should be approximately proportional to the cloud potential

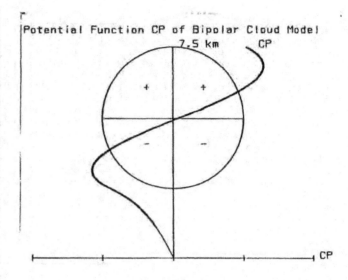

Fig. 3

function but with a reversed sign (mirror effect) and that the lightning should stop growing, if the difference between its own potential and that of cloud at the tip of the lightning channel approaches the same value. We will see that this prediction is confirmed by the present calculation.

Fig. 3 shows the cloud potential function in the center of the bipolar cloud model the author used for the energy calculation of a cloud discharge. Reference [6]. The sphere representing the cloud has in its upper half positive and in its lower half negative space charge. The vertical axis shows the altitude in Km and the horizontal axis the cloud potential CP. The wavy line marked CP displays the cloud potential function versus altitude. It starts from zero at the ground, takes a big negative swing with increasing altitude to a maximum value shortly after entering the cloud, which has its base at 2.5 km. Then it drops to zero at the center of the sphere at 5 Km altitude and continues with a positive swing through the positive space charge region reaching its positive maximum at about 6.8 Km altitude. This type of double swing is characteristic for any bipolar charge distribution in a cloud. However, the point of interest here is the zero crossing of the potential function at about 5 Km altitude at the dividing line between positive and negative space charge. A lightning strike triggered at the ground can grow no higher than to this point since it meets here its own potential zero and the gradient at the tip will go to zero. Even before this value is reached the tip gradient will drop below break down value and no ionization of the air before the tip can occur.

For our numerical calculation we will use a somewhat simpler mathematical representation of the cloud potential function as shown in Fig.3. However it has the same characteristic features as the CP shown in Fig. 3, namely it starts from zero at the ground, reaches its negative maximum at about 2.7 Km and crosses the zero line at 5Km altitude. The new cloud potential function is given by the simple mathematical expression

$$CP=A1z-A3z^3 \tag{4.1}$$

By differentiating to z we obtain the cloud gradient CG.

$$CG=A1-3A3z^2 \tag{4.2}$$

For z=o the gradient at the ground is

$$CG=A1 \tag{4.2a}$$

A1 is the gradient at the ground before the lightning flash is triggered and can be obtained from the field mill records. A3 can be expressed by A1 and h, where h is the altitude of the zero crossing of CP inside the cloud. We have from (4.1)

$$o=A1h-A3h^3; \text{ or } A3=A1/h^2 \tag{4.1a}$$

It is somewhat amazing that with equation (4.1) and the input of the two parameters A1 and h the triggered lightning is determined uniquely from start to end. What follows now is only the application of the equations given in paragraph II and III.

Conversion of (4.1) to spheroidal coordinates by (2.3)

$$CP=A1uv/c-A3(uv/c)^3 \tag{4.3}$$

Expansion of $(v/c)^n$ into Legendre's polinominals of first kind Pn by (3.11)

$$v/c=P1; \quad (v/c)^3=2P3/5 + 3P1/5 \tag{4.4}$$

Inserting the expression of (4.4) in (4.3) and sorting to Pn

$$CP=(A1u-3A3^3/5)P1-2A3u^3P3/5 \tag{4.5}$$

The factors of P1 and P3 in (4.5) determine the Fn(u) functions.

$$F1(u)=A1u-3A3u^3/5 \; ; \quad F3(u)=-2A3u^3/5 \tag{4.6}$$

and by differentiation to u

$$F1(u)'=A1-9aA3u^2/5 \; ; \quad F3(u)'=-6a3u^2/5 \tag{4.7}$$

From (3.16) and (4.6) we obtain the B1 and B3 for the lightning potential function

$$B1=-F1(a)/Q1(a)=-(A1a-3A3a^3/5)/Q1(a)$$
$$B3=-F3(a)/Q3(a)=+2A3a^3/5Q3(a) \tag{4.8}$$

The lightning potential function is then given by

$$LP=B1(a)Q1(u)P1+B3(a)Q3(u)P3 \tag{4.9}$$

The complete potential function Φ is obtained by a superposition of CP and LP. From (4.5), (4.6), and (4.9) we obtain the complete potential function in the form

$$\Phi=F1(u)P1+F3(u)P3+B1(a)Q1(u)P1+ \\ B3(a)Q3(u)P3 \tag{4.10}$$

The first two terms on the right side represent the cloud potential function CP and the last two terms the lightning potential function LP.

V Ground Triggered Lightning

For the following calculations we collect again the terms with P1 and P3 of (4.10) and define

$$G1(u)=F1(u)+B1(a)Q1(u)$$
$$G1'(u)=F1'(u)+B1(a)Q1'(u)/c$$
$$G3(u)=F3(u)+B3(a)Q3(u) \tag{5.1}$$
$$G3'(u)=F3'(u)+B3(a)Q3'(u)/c$$

and obtain from (4.10) with (5.1)

$$\Phi=G1(u)P1+G3(u)P3 \tag{5.2}$$

The gradient G_u in u direction is with (2.8)

$$G_u=[G1(u)'P1+G3(u)'P3]/hu \tag{5.3}$$

and the gradient G_v in v direction

$$G_v=[G1(u)P1'/c+G3(u)P3'/c]/hv \tag{5.4}$$

To obtain the gradient GT at the tip of the spheroid we set in (5.3) u=a and v=c. From this follows

$$P1=1 \; ; \; P3=1 \; ; \; hu=1 \text{ and}$$
$$GT=G1(a)'+G3(a)' \tag{5.5}$$

The charge density q at the surface of the spheroid is given by

$$q=-\epsilon G_u(a) \tag{5.6}$$

The charge of a ring zone element qr is from (5.3) with (2.10)

$$qr=qdO=-2\pi\epsilon b^2[G1'(a)P1+G3'(a)P3]dv/c \tag{5.7}$$

If we integrate qr from v/c=0 to v/c=1 we obtain the net charge Q of the spheroid. The integral of P1 yields the factor 1/2 and the integral of P3 the factor -1/8.

$$Q=-2\pi\epsilon b^2[G1'(a)/2-G3'(a)/8] \tag{5.8}$$

To obtain the charge per unit length Q1 we replace in P1 and P3 the argument v/c by z/a, and the differential dv/c by dz/a. This conversion follows for u=a from the relation (2.3) z=av/c. We divide then (5.7) by dz and obtain

$$qr/dz=Q1=-2\pi\epsilon b^2[G1'(a)P1(z/a)+ \\ G3'(a)P3(z/a)]/a \tag{5.9}$$

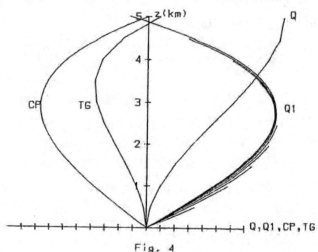

GROUND TRIGERED LIGHTNING
Charge per unit length Q1   Tip Gradient TG
Cloud Potential CP   Net charge Q

Fig. 4

Fig. 4 shows the cloud potential function CP and the lightning parameters tip gradient TG, net charge Q and charge per unit length Q1 (horizontal axis) versus channel length "a" or - in case of CP - altitude "z". For the tip gradient TG and the net charge Q there is only one value for each value of "a". Connecting these values for successively increasing values of "a" by straight lines results in the drawn curves. The situation for the charge distribution Q1 along the channel is different. For each "a" we have not only one point but a whole line full of points to draw. Therefore for each individual "a" a line is plotted from the ground to the tip of "a" representing the charge distribution on the channel for the specific length "a". Since the charge distribution on the next length "a1" is not exactly the same as on "a" each successive Q1 is slightly different from the previous one. This gives the whole Q1 line the feathered appearance. But even so, the Q1 line is a good mirror image on the z axis of the CP line as predicted. If we are able to reconstruct the charge per unit length on the lightning channel from the gradient measurements at the ground, which should not be difficult, we can obtain the cloud potential function versus altitude and by differentiating to z also the cloud gradient along this line.

The parameter which has a direct influence on the ground gradient is Q. From Q will follow by differentiation Q1.

The curve of the tip gradient TG shown in Fig. 4 also has a shape similar to the cloud potential function CP. The maximum is at a somewhat higher altitude of 3.5 Km instead of 3 Km of the CP curve. This may be due to the fact that the spheroid tip gets sharper as the length "a" of the channel increases. The radius of curvature at the tip decreases from 5cm for a=500m to 5mm for a=500m. Since the tip gradient is inverse proportional to the radius of curvature this effect alone will account for a ten times higher tip gradient at higher altitudes. This is one feature of the spheroidal model

which does not fit the natural lightning channel and has to be watched if quantitative results are used to predict experimental measurements. In spite of this shortcoming the tip gradient goes through zero close to the zero crossing of the cloud potential function CP. Since the lightning channel will stop growing even if the tip gradient falls below the breakdown value the zero cloud potential is an effective barrier which a ground connected lightning discharge including the return stroke cannot penetrate.

VI The gradient at the ground, Go

The plane z=o is given in spheroidal coordinates by v=o. Since $P1(o)=o$ and $P3(o)=o$ the u component G_u in (5.3) is also zero. The G_v component in (5.4) is simplified since for v=o hv=u/c, $dP1/dv=1/c$, and $dP3/dv=-3/2c$. Then

$$Go=G_v=[G1(u)-3G3(u)/2]/u \qquad (6.1)$$

or with (4.6), (4.8), (5.1)

$$Go=A1[1-aQ1(u)/uQ1(a)]+$$

$$A1\frac{3}{5}(a/h)^2[aQ1(u)/uQ1(a)-aQ3(u)/uQ3(a)] \qquad (6.2)$$

It is easy to see in (6.2) that at the foot point of triggered lightning where u=a the ground gradient Go=o. This is somewhat surprising, however we know this effect very well in a different situation, namely as the cone of protection against lightning strikes at the foot point of a high tower. The cone of protection is usually decribed as a cone with its apex at the tip of the tower and a circular base at the ground with a radius about equal to the height of the tower. Humans and buildings inside the circular base are supposed to be protected against lightning strikes. The word is - as in many cases - a misnomer, since the only physical explanation is, that in this area a down comming leader stroke following the field lines will hit the tower and not the ground. To picture this effect we would have to turn the cone upside down with its apex at the foot point of the tower and its base at the height of the tower. The inside of this cone then describes the area where the vertical field lines and the leader stroke following the lines will bend towards and hit the tower instead of going to the ground. This cone would better be described as cone of attraction. With respect to the ground triggered lightning research it would be an interesting experiment to record the gradient by -let's say - three field mills in 100, 500, and 1000 m distance from the trigger site to check if the gradient increases with increasing distance from the channel foot point as predicted by (6.1) or (6.2).

The importance of this research will become immedeatly clear if we realize that the ground triggered lightning is in essence the same as the lightning flash triggered by an aircraft or rocket. The loweer half of the spheroid, wich in our case is the mirror image of the upper half at the ground, is in the case of an aircraft the other half of the aircraft triggered lightning growing in the direction opposite to the first half. The aircraft itself replaces the trigger site at the ground. The two halves of the aircraft triggered lightning strike screen the aircraft from the high electric fields in the same manner as the ground triggered lightning screens the trigger site. Reference [5]. The mathematical treatment is in both cases the same. The consequenz of this result in respect to the immunisation of electronic equippment agaeinst triggered lichtning interference or damage is that research sould be directed to protection and screening against high currents and magnetic fields rather than against high voltage and induced charges. Laboratory tests with long sparks produced by pulsed high voltage generators seem to go in the wrong direction. The ground triggered lightning provides a much more realistic source of energy output of an aircraft triggered lightning than a spark, and will also give the information whether this energy is concentrated mainly in the electric or magnetic field and whether in the high or low frquency band. It is rather obvious that a spark of micro- or milliseconds duration is a poor simulation of the long continuous current of about one second duration, which is one of the charcteristic features of a triggered lightning discharge.

To plot the ground gradient as a function of the distance "R" for a given channel length "a" or as a function of "a" for a given field mill site with the distance "R" we have to express "u" by "R" and "c", and "c" by "a" and "b". For v=o these equations are

$$u=(R^2+c^2)^{1/2} \; ; \quad c=(a^2-b^2)^{1/2} \qquad (6.3)$$

Graphs and a more detailed discussion of Go and its relation with the lightning and cloud parameters will be the topic of another paper dealing with the evaluation of the KSC field mill measurements.

References

1. H.W.Kasemir, "Qualitative Uebersicht ueber Potential -, Feld -, und Ladungsverhaltnisse bei einer Blitzentladung in der Gewitterwolke," in "Das Gewitter" by Hans Israel, Akad. Verlags. Ges. Geest and Portig K.-G, Leipzig, (German) 1950
2. H.W.Kasemir, "A Contribution to the Electrostatic Theory of a Lightning Discharge," JGR,65,#7,1873-1878, 1960
3. D.W.Clifford, H.W.Kasemir,"Triggered Lightning", IEEE, Vol. EMC-24,No.2,137,1982
4. H.W.Kasemir, "Static Discharge and Triggered Lightning," Lightning and Static Electricity Conference, 24-1, 1983
5. H.W.Kasemir, "Theoretical and Experimental Determination of Field, Charge, and Current on an Aircraft Hit by Natural or Triggered Lightning," Lightning and Static Electricity Conference, 2-2, 1984
6. H.W.Kasemir, "Energy Problems of the Lightning Discharge," VII International Conference on Atmospheric Electricity, Albany, New York A.M.S. 1984
7. Frank-v.Mises:"Differential Gleichun-gen Der Physik I ", Vieweg and Sohn, Braunschweig (German) 1930
8. J.A.Stratton, "Electromagnetic Theory," McGraw-Hill Book Co., New York, 1941
9. W.R.Smythe, "Static and Dynamic Electricity," McGraw-Hill Book Co,. New York, 1950 .
10. Jahnke-Emde, "Tables of Higher Functions," B.G.Teubner Verlagsgesellschaft (German and English) Leipzig, 1948.

8B.4

ANALYSIS OF CLOUD, GROUND, AND TRIGGERED LIGHTNING
FLASHES FROM GRADIENTS MEASURED AT THE GROUND

Heinz W. Kasemir
Colorado Scientific Research Corporation
1604 S. County Road 15, Berthoud, CO 80513, U.S.A.

ABSTRACT

From electrostatic theory it follows that bipolar charge arrangements such as cloud flashes, stepped and dart leaders of ground flashes have bipolar ground gradients with polarity reversal at a certain distance, that monopolar charge arrangements such as return strokes and ground triggered lightning have monopolar gradients without polarity reversal. Gradient records of ground triggered lightning flashes revealed that an initial long lasting rise is a unique feature of these gradients. Using these and other already known characteristic features it seems possible to identify each lightning type from its ground gradient signature. Special emphasis is placed on the identification of tower as well as rocket triggered lightning from a single station. The results of the evaluation of gradient records obtained at Kennedy Space Center during the Rocket Triggered Lightning Project 1986 and 1987 are discussed.

1.   INTRODUCTION

The gradient data of natural and triggered lightning recorded at Kennedy Space Center during the Rocket Triggered Lightning Project (RTLP) 1986 and 1987 have been evaluated from different viewpoints. This paper deals with an investigation to determine differences and similarities between the gradients of triggered and natural lightning strikes and will provide physical explanations for different features in the gradient records.

Four gradient recording stations have been placed in the KSC area in distances of about 0.5, 2, 10, and 20 km from the trigger site. The instruments had a linear frequency response from 0.1 Hz to 1 MHz. The sample time of the analog to digital converter was set at 50 or 100 microseconds. The data have been recorded on floppy or hard discs.

In the first section the theoretical concept of a cloud, ground, and triggered lightning is explained and illustrated in the nine pictures of Fig. 1. The second section deals with similarities and differences between the three lightning types. The third section gives the results of an evaluation of the ground gradients of natural ground discharges on ground triggered lightning obtained at Kennedy Space Center. A key factor is used to identify from the records of a single station the two types of lightning discharges. In section four conclusions and results are summarized.

2.   GROUND GRADIENTS OF A CLOUD, GROUND AND TRIGGERED LIGHTNING DISCHARGE

Figure 1 is composed of nine individual pictures arranged in three rows and three columns. The pictures are identified in a matrix form by two digits, and will be referenced for instance as Pic. 21 for the picture in row 2, column 1. The theoretical sketches in the upper two rows are all composed in the same manner. The thunderstorm is represented by a sphere above a horizontal ground line, the lightning channel by a vertical line, and the theoretical ground gradients by curved lines ascending from or descending toward the ground line. Inserted in the ground line are two little square boxes, one right underneath the channel and one at the right end of the line. These represent two recording stations: one close to the nadir point of the channel, one far away from it. The corresponding gradient lines stop right overhead or beneath these boxes and start always to the left of them. The ground line serves in this case as the time axis for a display of the past gradient as function of time. In the first row the channel has grown to about half of its final length and in the second row to its full length for a cloud discharge (Pic.21) and up to the first return stroke for a ground discharge (Pic.22). In case of the triggered lightning (Pic.23) it may end here or it may continue to grow spreading out horizontally and produce a few weak return strokes which would correspond to the second, third and so on, return strokes of natural lightning flash.

The charge distribution on the lightning channel is indicated by + or − signs placed close to the channel. The space charge of the cloud is indicated only by two + signs at the inner surface of the upper half of the circle and by two − signs at the lower half. These signs mean that the whole upper half of the sphere is filled with positive and the lower half with negative space charge. We are dealing here with a bipolar thunderstorm model. The charge distribution on the channel of a leader and of a cloud discharge is bipolar, i.e. the charge in the upper and lower half of the channel is of opposite polarity, and the channel grows in both directions. Radar pictures by Mazur, et al. [1] show that the growth of leader and cloud discharges is indeed bidirectional. The charge distribution on the channel of a ground discharge after each return stroke and at the completion of the discharge is unipolar. All bipolar charge arrangements such as cloud flashes, stepped and dart leader have bipolar ground gradients meaning that at some distance from the center-point of the bipole the gradient undergoes a polarity reversal. All monopolar charge arrangements such as return strokes and in essence ground triggered lightning flashes have monopolar

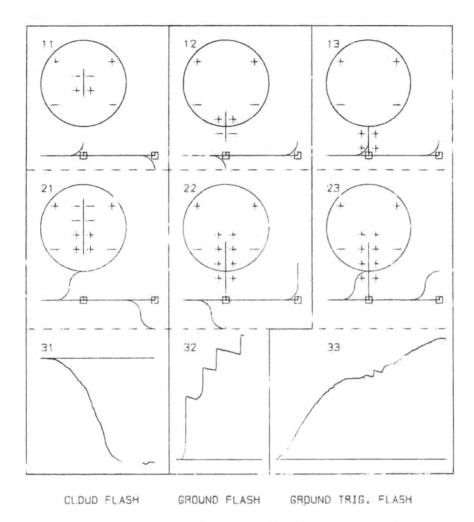

CLOUD FLASH   GROUND FLASH   GROUND TRIG. FLASH

Fig.1, Row 1 and 2 theoretical sketches, Row 3 recorded gradients

Pic.11 Cloud flash, initial phase, channel charge, gradient
Pic.21 Cloud flash, final phase, channel charge, gradient
Pic.31 Cloud flash, gradient record at far away station
Pic.12 Ground flash, initial phase, channel charge, gradient
Pic.22 Ground flash, final phase, channel charge, gradient
Pic.32 Ground flash, gradient record at far away station
Pic.13 Trig. flash, initial phase, channel charge, gradient
Pic.23 Trig. flash, final phase, channel charge, gradient
Pic.33 Triggered flash, gradient record at far away station

ground gradients meaning the polarity of the gradient remains the same at any distance. (See Pics in row 1 and row 2). We will use this later for identification of the triggered lightning gradient.

These generalized statements are based on the electrostatic lightning theory (2.3) where the cloud discharge and leader channel have a bipolar charge distribution and therefore a polarity reversal of the ground gradient at some distance from the channel. The more commonly accepted model is that of the unipolar cloud discharge and leader stroke. Charge of one sign is collected from the positive or from the negative charged volume of the cloud funneled into the lightning channel and transferred to the oppositely charged part of the cloud in case of a cloud discharge or to ground in case of a ground discharge.

This model is based on assumptions which are difficult to explain. A reasonable suggestion for a cloud charge collecting mechanism is still outstanding. Why charges of one sign concentrate themselves into a channel is a mystery since it is generally known that charges of the same sign repel each other and spread as far apart as possible. However, what concerns us here is the claim that a monopolar leader can cause a polarity reversal of the ground gradient which from an electrostatic point of view is not possible. To recognize this we only have to picture in our mind the field lines emerging from a point charge about a conductive ground plane. These field lines leave the point charge in radially outward directions, bend around in increasing arcs to end on the plane all in the same direction namely vertically downwards. Here is no polarity reversal in the field at the ground. We may move the charge around, increase or decrease its magnitude,

557

this will change the magnitude of the ground field, but nowhere its polarity. This will remain true if we extend the picture to any arrangement of charges of the same sign as for instance the negatively charged leader emerging from the negatively charged part of the cloud. This is a simple rearrangement of a constant amount of negative charge and can cause an amplitude change in the ground field but not a polarity reversal.

Since the leader field is usually separated from the cloud field, we separate the picture into two parts. The first one is that of a slightly decreasing cloud charge. This decreases the ground gradient in all distances from the cloud but doesn't produce a polarity reversal. The same is true for the reappearance of the lost cloud charge in the leader channel. This causes an increase of the ground field but not a polarity reversal either taken by itself or in superposition with the cloud field. However, the concept of a monopolar leader or cloud discharge with the attributes of the polarity reversal of a bipolar charge distribution in the channel is still the generally accepted model today. How can this be explained?

The most clear cut explanation of how this is done is given in Uman's books "Lightning" page 53 ... [4,5]. He first calculates the ground field of a line charge with constant charge per unit length (equation 3.8) as a model for the monopolar leader. The field of this leader model by itself doesn't have the polarity reversal in the ground field, as it should be and as can easily be confirmed from the equation. Then he calculates "the field change at the ground due to the decrease in the source (cloud) charge" in equation (3.9). But instead of subtracting it from the field of the cloud charge, as would be the obvious thing to do, he changes the polarity of this field and adds it to the channel field in equation (3.10). In essence, he creates from the decrease of the cloud charge a fictitious charge of opposite polarity, takes it out of the cloud field and adds it to the channel field. In this way the monopolar leader acquires the characteristics of a bipolar leader. The conceptual error here is that the decrease of the cloud charge is not equal to the creation of charge of opposite polarity.

It is here not the place to go into further discussion of the monopolar leader and cloud discharge model versus the bipolar model, but we may briefly state that the bipolar charge distribution is fundamentally different from the monopolar charge distribution and that the general statements and conclusions given in this paper cannot be explained by the monopolar concept.

3.   DIFFERENCES AND SIMILARITIES IN THE GROUND GRADIENTS OF CLOUD, GROUND, AND TRIGGERED LIGHTNING FLASHES

Pic.11 shows a cloud discharge at about half of its final length. It starts in the middle of the sphere. The upper half of the channel carries negative induction charges because it grows into the positive space

charge of the cloud and the lower half of the channel carries positive induction charges growing into the negative part of the cloud. The gradient recorder underneath the channel is affected more by the lower positive charge than by the higher negative charge and is positive. The far away recorder is more affected by the upper negative than by the lower positive charge on the channel and the gradient goes negative. This trend continues until the cloud flash stops, as can be seen in Pic.21. Pic.31 shows a typical record of the ground gradient of a cloud discharge far away (more than 10 km) from the storm.

The first leader of a ground discharge originating close to the bottom of the storm has a bipolar charge distribution but of opposite polarity to that of the cloud discharge, as shown in Pic. 12. This would also be true for the dart leaders of the following return strokes. However, the charge on the channel after the first return stroke is all positive and the gradient change on close by and far away stations is in the same direction (Pic.22). The recorded gradient shown in Pic.32 is that of a far away ground flash with four return strokes. Pic.22 shows the development of the flash only up to and including the first return stroke.

The case of the ground triggered lightning is very simple. The channel carries only positive charge, or if we have positive cloud charge overhead only negative charge. The gradient is the same polarity at all stations from close by to far away (Pic.13 and 23), and the charge distribution on channel is similar to that of a first return stroke of a ground discharge. A ground triggered lightning may or may not have in the second half of its lifetime a few weak return strokes. If they are present, Pic.33, they would correspond to the 2., 3., ... return strokes of a natural ground discharge.

A lightning triggered by a rocket not connected to ground or by airplane [Mazur, 1] is bipolar and bidirectional. It would produce the ground gradients of a cloud discharge, if it is triggered in the upper part of the cloud, or that of a ground discharge, if it is triggered in the lower part of the cloud and reaches the ground. This may make it difficult to differentiate between ground gradients caused by natural and air triggered lightning flashes.

From Pic.31, 32 and 33, it is obvious that there are remarkable differences in the gradient signatures of cloud, ground, and triggered lightning flashes beside the absence or presence of polarity changes. In the following discussion we use the abbreviations: Cloud flash = C, Ground flash = G, triggered lightning without return strokes = T, Triggered lightning with return strokes = TR. The gradient shown in Pic.33 is of a TR.

From the eleven triggered lightning gradients recorded at KSC three are of T with an average duration (lifetime) of 0.22 sec. and eight are of TR with an average duration of 0.84 sec. In comparison, the values given by M. A. Uman [4] for natural lightning are as follows: The average duration of C is

0.25 sec. and of G 0.2 sec. Therefore it is not possible to differentiate between a T gradient and a C gradient by the duration. For positive identification we have to use here the polarity reversal of the C gradient with distance which the T gradient has not. (See Fic. 21 and 23). T gradients cannot be confused with G gradients because they lack the return stroke signature. TR gradients cannot be confused with C gradients because in this case the C gradient lacks the return stroke signature. For the separation of TR gradients from G gradients the duration is already a good indicator since TR gradients last about 4 times as long as average G gradients. However, in this case, we have even more conclusive differences. The initial rise of the TR gradient before the first return stroke lasts about 0.4 sec. The stepped leader of the first return stroke in a G gradient has a duration of about 0.02 sec. [Uman,4]. See also Pic.32 compared to

Pic.33. Here we have a difference of durations of a factor 20. The physical reason for this large difference is that the stepped leader starts in a high field of 300 kV/m and consequently has a much higher velocity than the ground triggered lightning which starts in a low field of about 5 kV/m and with a relative low velocity. Furthermore, the ratio of the amplitude of the first return stroke and the stepped leader in the G gradient, (Pic.32) is 2.14 and the ratio of the amplitude of the first return stroke to that of the initial rise in the TR gradient (Pic.33) is 0.014. The difference between these two amplitude ratios is a factor of 150. The important fact here is that from the gradient records of one station alone, a ground triggered lightning flash containing return strokes can be identified and separated from the gradients of natural lightning discharges.

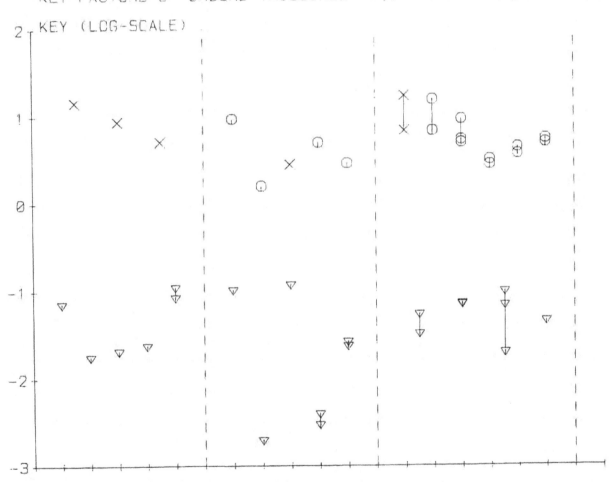

Fig.2

Key factors for ground triggered and natural cloud to ground discharges
   Vertical axis: Key factors log-scale 0.001 to 100
Symbol of Ground Discharges = ▽
Symbol of Rocket Triggered Discharges = o
Symbol of Tower Triggered Discharges = x
   Horizontal Axis: Events (Time)
Left Section Forenoon 28 August 1986
Middle Section Afternoon 18 August 1986
Right Section 14-16 July 1987

4. KEY FACTORS USED TO DISTINGUISH BETWEEN A GROUND TRIGGERED AND A NATURAL CLOUD TO GROUND LIGHTNING

The suggestion advanced in the previous section, that it may be possible to use the characteristic differences between the ground triggered and the natural cloud to ground discharge to separate these two types of discharge is here carried one step further for the case that the ground triggered lightning has at least one return stroke. We define the two key factors KT for ground triggered lightning and KN for natural cloud to ground discharges. For KT we use the following parameter, maximum amplitude AT and duration DT of the initial rise of the gradient, and the amplitude of the first return stroke RT. The corresponding parameters of the ground discharge are AN and DN for amplitude and duration of the stepped leader and RN for the amplitude of the first return stroke. We obtain then

$$KT = AT*DT/RT; \quad KN = AN*DN/RN \qquad (1)$$

From Fig. 1, Pic.33, we see that AT and DT are large compared to RT. Therefore, we would expect here a large number for KT of the triggered lightning. On the other hand it is apparent from Fig. 1, Pic. 32 that AN and DN are smaller than RN so we could expect a small number for KN of the cloud to ground discharge. Fig. 2 shows the key factors of ground triggered lightning. The symbol o is for rocket triggered discharges and the symbol x for lightning triggered from an as yet unidentified source. There is a striking similarity between the gradients of x and rocket triggered discharges indicating that x is triggered from the ground. Since it is well known under the name of "upward going leader" that high towers can initiate lightning discharges in strong thunderstorm fields and that there are many high towers in the Kennedy Space Center and U. S. Air Force Cape Canaveral area, it is a reasonable conclusion that the x gradients have been produced by tower triggered lightning. The key factors with the symbol ∇ are calculated from gradients of ground triggered lightning.

The vertical axis gives the values of the key factors in logarithmic scale from -3 to 2 also for values ranging from 0.001 to 100.

The horizontal axis gives the sequence of events, i.e., roughly the time. The vertical dashed lines separate the plot into three sections. In the left section the gradient records were obtained on the forenoon of 28 August 1986 and the records of the middle section in the afternoon of the same day. The key factors shown in the right section were calculated from the records obtained on 14-16 July 1987. The symbols connected by vertical solid lines are from the same lightning flash recorded at two or three stations. The four overlapping symbols in the right section are from the two stations 0.5 or 2 km distant from the trigger site. The other two stations were 10 or 20 km distant from the trigger site. The scatter of these factors points to the possibility that the key factor may be distance dependent.

In spite of the scattering of the key factors, there is still a quite noticeable separation between ground triggered and natural lightning by about a factor 10. The lowest rocket triggered value is 1.55 and the highest value of cloud to ground discharges is 0.121. If we calculate the average value of all rocket and tower triggered lightning and of all natural lightning, the respective values are 6.09 and 0.047. The average values are separated by a factor 130.

It is certainly possible that there are ground discharges with extremely long stepped leaders which could bring the key factor of natural ground discharges into the domain of the key factor of ground triggered lightning. The author has seen such leaders in Oklahoma which run along the cloud base for 10 km or more before they make contact with ground. Such cases can only be dealt with when gradient records are available. The first possibility which comes to mind is that all stepped leaders which show polarity reversal in the gradient are coming from the cloud and if followed by a return stroke indicate a natural ground discharge. The other possibility is to lift the restriction of single station evaluation and fall back on simultaneous records of two or more stations at least temporarily for the study phase.

In conclusion we may state that the key factor method to separate ground triggered from natural cloud to ground discharges using gradient records of a single station is not a fool proof final prescription but a working hypothesis restricted as yet to ground triggered lightning with at least one return stroke to help in future research.

5. RESULTS AND CONCLUSIONS

1. Bipolar charge arrangements such as cloud flashes or stepped and dart leader of ground discharges produce bipolar ground gradients having a polarity reversal of the gradient with increasing distance from the flash.

2. Monopolar charge arrangements such as return strokes and ground triggered discharges have monopolar ground gradients with the same polarity at any distance from the flash.

3. Ground triggered lightning with or without return strokes are characterized by the long lasting and continuous increase of the gradient of about 0.3 to 0.4 sec. duration. This initial rise of the gradient is a unique feature of the ground triggered lightning. It is monopolar and differs therefore in essence from a cloud discharge or the leaders of a ground discharge. It has the same charge distribution on the channel as a return stroke but differs vastly in duration.

4. The gradients of a triggered lightning flash without return strokes can be separated from cloud discharge gradients by the lack of gradient reversal with distance and from ground discharge gradients by the lack of return strokes.

The gradients of a ground triggered lightning flash with return strokes can be separated from cloud discharge gradients by the presence of return stroke signatures and from ground discharge gradients by the presence of the long initial rise of the gradient.

5. A key factor has been calculated for the ground discharge from the ratio of the leader amplitude to the first return stroke amplitude and multiplied by the duration of the leader. Correspondingly the key factor of the ground triggered lightning is calculated from the ratio of the amplitudes of the initial rise to that of the first return stroke multiplied by the duration of the initial rise. These key factors differ in size by at least a factor 10, the average values differ by a factor 130 in the evaluation of measured gradients obtained at Kennedy Space Center. These key factors will help to separate natural from ground triggered lightning discharges.

ACKNOWLEDGMENT

The author is grateful to the Naval Research Laboratory and to Kennedy Space Center of the National Aeronautic and Space Administration for support to obtain the data used in this paper.

REFERENCES

1. V.Mazur, B.D.Fisher and J.C.Gerlach, "Lightning Strikes to an Airplane in a Thunderstorm", J.Aircraft, 21, 1984, 607-611.

2. H.W.Kasemir, "A Contribution to the Electrostatic Theory of a Lightning Discharge", JGR, 65, #7, 1873-1878, 1960.

3. H.W.Kasemir, "Static Discharge And Triggered Lightning" International Aerospace and Ground Conference on Lightning and Static Electricity, Fort Worth, Texas, 24-1 to 24-11, 1983.

4. M.A.Uman, "Lightning", McGraw-Hill Book Company, 1969

5. M.A.Uman, "Lightning", Dover Publications, Inc. 1984.

Reprinted from Preprint Volume: VII International Conference on Atmospheric Electricity, June 3-8, 1984, Albany, N.Y. Published by the American Meteorological Society, Boston, Mass.

7.18

ENERGY PROBLEMS OF THE LIGHTNING DISCHARGE

Heinz W. Kasemir

Colorado Scientific Research Corporation
Berthoud, Colorado 80513

1. INTRODUCTION

The energy for a lightning discharge is provided by the electric field of the thunderstorm. Therefore it is necessary first to develop a mathematical analytic thunderstorm model before the lightning problems can be attacked. The potential function and the field and charge distribution of the thunderstorm model should be given in reasonably simple mathematical equations. This model will be developed in section 2. It has the features of a bipolar thunderstorm in the geometrical form of a sphere but can be extended to other charge distributions or shapes.

In section 3 the lightning is placed into the thunderstorm field at the point of maximum field value which in this model is at the center of the sphere. The lightning is represented by a slim conductive spheroid which starts at a natural (precipitation particle) or man made (airplane or rocket) nucleus and growth along the strongest field line upwards and downwards until it reaches the cloud boundary. All charges necessary to form the induction charges on the otherwise uncharged conductor are produced inside the channel by ionization. Since there is always an equal amount of positive and negative charge produced by the ionization process the net charge of the cloud discharge is and remains zero. There is no cloud charge collection process by a widespread filament network or any other mechanism causing a flow of charge from the environment into the lightning channel. This is the fundamental difference between the lightning model presented here and the generally accepted lightning model of today.

The physical concept of what may be called the electrostatic lightning theory was already introduced by the author (Kasemir 1950) about 34 years ago. This paper gives the mathematical treatment of this concept for the cloud discharge. Due to space limitation the mathematics of the ground discharge will be given in another publication.

A certain emphasis is placed on energy calculations for two reasons. First it will be shown at the end of this paper that the introduction of a conductive but uncharged lightning channel into the thunderstorm field will lower the energy content of the system, meaning that this is a process nature will follow and support. The second and more important reason is that energy is the link between the electric and plasma physics processes in the lightning channel. Without the incorporation of these processes into a lightning theory time dependent problems of the lightning discharge cannot be solved. The determination of the energy available for ionization, heating the channel, electric radiation etc., may help the incorporation of plasma physics in the lightning theory.

2. THE THUNDERSTORM MODEL

We assume for our model rotational symmetry and will alternately use the three rotational symmetric coordinate systems of spheroidal u, v, ϕ, cylindrical z, R, ϕ, and spherical r, θ, ϕ coordinates all with the same zero point in the center of a sphere of radius K. The sphere K represents our thunderstorm and is filled with positive charge in the upper and negative charge in the lower half of the sphere. This implies that the charge density q is given by a power series of odd powers of z.

$$q = a_1 z + a_3 z^3 + \cdots \qquad (1)$$

The author has developed a simple method to obtain the potential function of a charge distribution inside a sphere which is proportional to z^n. Take the spherical harmonics

$$\Phi = A \, r^{n+2} P_{n+2} \qquad (2)$$

and omit the first term in the Legendre's polynominal P_{n+2} which is indicated by using the symbol $[P]_{n+2}$. The potential function

$$\Phi = A \, r^{n+2} [P]_{n+2} \qquad (3)$$

is then a solution of Poisson's equation, where the space charge density is proportional to z^n. The method works for all n, $0 < n < \infty$.

The development or the proof of this method is beyond this paper. However, the reader should apply the Laplacian operator to (3) and convince himself of the correctness of the solution. We apply it here to the first term $n=1$ of (1) to obtain the potential function of our simple thunderstorm model. This spherical harmonic for $n=1+2=3$ is:

$$\Phi = Ar^3 P_3 = Ar^3 (5\cos^3\theta - 3\cos\theta)/2.$$

Omit the $\cos^3\theta$ term and absorb the 3/2 factor in A.

$$\Phi = -Ar^3\cos\theta \qquad (4)$$

$$q = -\varepsilon\Delta\Phi = \varepsilon A \, 10 r\cos\theta = \varepsilon A \, 10 z \qquad (5)$$

The potential function Φ obtained by the amputation method leads indeed to a charge distribution q which is proportional to z. To fulfill the boundary condition we add to (4) the harmonic solution $ACr\cos\theta$ with the arbitrary constant C, which matches the $\cos\theta$ dependence of (4). This term will not add to the charge distribution q since $\Delta(r\cos\theta) = 0$. The outside potential function Φ_o is also chosen to have the same $\cos\theta$ dependence, $\Delta\Phi_o = 0$, and an arbitrary constant AB. Therefore it will be the outside spherical harmonic of a dipole.

$$\Phi_o = AB\cos\theta/r^2 \qquad (6)$$

whereas the inside potential function is

$$\Phi_i = A \, (Cr-r^3) \cos\theta \qquad (7)$$

The two arbitrary constants B and C are determined by the boundary conditions that at the sphere surface $r=K$ the potential and the gradient should be continuous. A discontinuity in the potential function indicates a dipole layer, and a break in the gradient indicates a surface charge layer caused by an abrupt change of conductivity of the air at the cloud boundary. The existence of a dipole layer is never postulated nor measured, that of a surface charge layer is theoretically known as the screening charge layer but experimentally not well established. If it exists, it will affect only the outside potential function Φ_o by an additional factor k given by the ratio of outside to inside conductivity. With the assumption of potential and gradient continuity at the boundary $r=K$ we obtain

$$C = 5K^2/3 \quad , \quad B = 2K^5/3. \qquad (8)$$

With these values substituted in (6) and (7) we arrive at the final form of the inside and outside potential function of the thunderstorm model.

$$\Phi = A \, (5K^2 r/3 - r^3) \cos\theta \qquad (9)$$

$$\Phi_o = 2AK^5\cos\theta/3r^2 \qquad (10)$$

The other thunderstorm parameters like net charge of the upper or lower half, gradient or field distribution, maximum values of gradient or potential difference between the poles etc., can be obtained by easy differentiation or integration from equation (9). It will be sufficient here to give the final equations. As a kind of brief feasibility check we will calculate the numerical values of these parameters. However our main purpose here is to obtain the equation of the potential function along the z axis and an expression for the electric energy content of a thunderstorm. For comparison and discussion of the spread and manner in which the parameters were obtained and reported in the literature the reader is referred to the three most recent textbooks on lightning and thunderstorms, Uman 1969, Golde 1977, and Volland 1982.

Equation (9) requires the input of the two arbitrary constants A and K. We set $K=3km$ which gives the charged volume in the storm a horizontal and vertical diameter of 6km. For the determination of A we use for the value of the maximum gradient G_m at the center of the sphere and set $G_m=300kV/m$. From (9) we obtain

$$G_m = 5AK^2/3 \quad ,$$

$$A = 3G_m/5K^2 = 0.02 \; V/m^3 \qquad (11)$$

$$AK^3 = 540 \; MV$$

The numerical values of the thunderstorm parameter of this model based on the assumption $K = 3km$, $G_m = 300 \; kV/m$ are listed in column 2 of Table 1. Column 1 gives the parameter symbol and the equation derived from equation (9). Q is the net charge of the upper half of the sphere. The lower half contains the same amount of charge but of negative polarity. U is the electric energy of the model calculated according to the well known electrostatic energy equation; see for instance Stratton 1941. The index m in our q, Φ, and G indicates that these are maximum values. The third column gives maximum and minimum values of the parameters reported in the literature. These are generally known values and are not referenced here to individual authors. The reference list would need a page for itself. See however the above cited textbooks.

563

Table 1

Parameter Equation	Numerical Value	Literature
$q_m = 10 \, \varepsilon AK$	$3.5 nC/m^3$	
$\Phi_m = 0.83 AK^3$	$446 MV$	$100 - 1000$ MV
$G_m = 5AK^2/3$	$300 kV/m$	$300 - 500$ kV/m
$Q = 5\pi\varepsilon AK^4/2$	$125 C$	$25 - 300$ C
$U = 0.5 \int q\Phi_i dV$	$3.1 \; 10^{10}$ Ws	
$= 80\pi\varepsilon K(AK^3)^2/63$	$= 8600$ KWh	

The numerical values given in Table 1 are well in the range of values reported in the literature. The remarkable feature is that these values are based only on two input parameters namely K and G_m. This shows that the model - albeit a simple one - is in itself consistent. Furthermore, since this is an analytical model the interrelation of those parameters is clearly expressed in the equations and is of special interest. For instance, it is remarkable that the electric energy is proportional to K^7.

Figure 1 shows the dependence of Φ, G, and q on altitude z in the vertical center line of the sphere. The parameters are normalized to their maximum values given in Table 1.

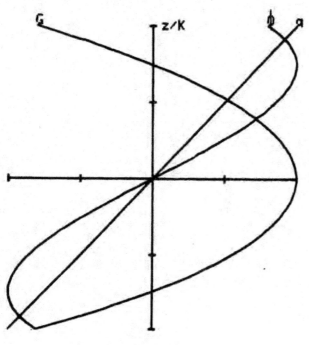

Fig. 1

Potential function Φ, Gradient G and Spacecharge density q as a function of altitude z.

3. THE CLOUD DISCHARGE.

In the following calculation we represent the lightning channel by an uncharged, conductive, slim spheroid with the long axis a in z direction, short axis b, and focal distance (excentricity) c. Since the cloud discharge starts at the maximum gradient at the center point of the sphere z=0 and growth vertically upwards and downwards following always the strongest gradient, which is along the z axis, we use as our primary potential function Φ_1 the function Φ_i given in (9) but restricted to the close environment of the z axis. This allows us to treat Φ_1 as a function of z only. To check the accuracy of this approximation we convert (9) into cylindrical coordinates.

$$\Phi_i = A \; [5K^2 z(1-3R^2/5K^2)/3 - z^3] \qquad (12)$$

R is the radial distance from the z axis. The core of the lightning channel has a radius of about R = 3cm and it is for this distance that we require the potential function to be a function of z only, at least to a good approximation. For R = 3cm and K = 3km the second term in the round parenthesis is 6×10^{-11} and can certainly be neglected against 1. Expressed in a different way we may state that inside a cylinder around the z axis with a radius of R = 400m neglect of the R dependence in equation (12) causes an error of less than 1%. Therefore we write Φ_1 as

$$\Phi_1 = A \; (5K^2 z/3 - z^3) \qquad (13)$$

We define the spheroidal coordinates u, v, ϕ in reference to the cylindrical coordinates by

$$z = uv/c, \; R = [(u^2-c^2)(c^2-v^2)]^{1/2}/c, \; \phi=\phi \quad (14)$$

Since $-1 < v/c < 1$ we use x=v/c as the argument of Legendre's polynomials P_n of the first kind. Then with the expansion of x^n into P_n (Frank-Mises 1930)

$$z = uP_1 \; , \; z^3 = u^3(2P_3/5 + 3P_1/5) \qquad (15)$$

we write (13) in spheroidal coordinates sorting to P_n

$$\Phi_1 = A(f_1 P_1 - f_3 P_3) \qquad (16)$$

$$f_1 = 5K^2 u/3 - 3u^3/5 \; , \; f_3 = 2u^3/5 \qquad (16a)$$

f_1 and f_3 are functions of u and the P_1 and P_3 are functions of x=v/c. The argument is not indicated in the following equations. The differentiation of f to u is indicated by a prime on the function f, $df_1/du = f_1'$. If u assumes the value "a", which determins the spheroid representing the lightning discharge, then $f_1(a)$, $f_3(a)$, and $f'_1(a)$, and $f'_3(a)$ denote the special values of these functions for u = a.

564

The solution of Laplace's differential equation in spheroidal coordinates is given by the product of Legendre's polynomials of the first kind P_n and the second kind Q_n. See for instance Smythe 1950. Selecting harmonics with P_1, and P_3 we write our secondary potential function, Φ_2 which describes the reaction of the spheroid as

$$\Phi_2 = A\,(-B_1 Q_1 P_1 + B_3 Q_3 P_3) \qquad (17)$$

and the final potential function Φ as

$$\Phi = \Phi_1 + \Phi_2 = A[(f_1 - B_1 Q_1)P_1 - (f_3 - B_3 Q_3)P_3] \qquad (18)$$

The two arbitrary constants B_1, and B_3 are determined by the boundary condition $\Phi = 0$ for $u = a$.

$$B_1 = f_1(a)/Q_1(a) \;,\; B_3 = f_3(a)/Q_3(a) \qquad (19)$$

Defining two new functions $F_1(u)$, $F_3(u)$ by

$$F_1 = f_1 - B_1 Q_1 = f_1(a)[f_1/f_1(a) - Q_1/Q_1(a)]$$
$$F_3 = f_3 - B_3 Q_3 = f_3(a)[f_3/f_3(a) - Q_3/Q_3(a)] \qquad (20)$$

(18) assumes the simple form

$$\Phi = A(F_1 P_1 - F_3 P_3) \qquad (21)$$

Before we continue with the calculation of the lightning parameters we would like to make a comment on dimensions, identification, and the Legendre's polynomials. The spheroidal coordinates u, v and the focal distance c have the dimension [m]. This gives the functions f_1, f_3, F_1, and F_3 the dimension $[m^3]$. The constant A determined in (11) has the dimension $[V/m^3]$. Since the argument of P_1 and P_3 is $x = v/c$ [m/m = 1] the P_1, P_3 used in (21) are dimensionless functions and the dimension of Φ is the dimension of AF_1 or AF_2, also [V] as it should be.

To identify the spheroid representing the lightning channel we set $u=a$. This implies that we know also either the focal distance c or the small axis b, since the spheroid cannot be defined by the long axis alone. This problem is hidden in the literature by normalizing u and c to c, for instance $y = u/c$ and $c/c = 1$. Change x and y to the corresponding greek letters ξ and η to obtain the normalized spheroidal coordinates used by Smythe. The Legendre's polynomials of the second kind Q_n are almost everywhere in the literature given in normalized form. To make them useful for this calculation whatever symbol is used for the argument has to be replaced by u/c. Using equation (8) on page 145 of Smythe 1950 we obtain with the proper substitution

$$Q_0 = 0.5\,\ln[(u+c)/(u-c)] \;,\; Q_1 = uQ_0/c - 1$$
$$Q_3 = 0.5(5u^3/c^3 - 3u/c)Q_0 - 5u^2/2c^2 + 2/3 \qquad (22)$$

and from equations (6), (7), (8) on page 168 of Smythe 1950

$$h_u = [(u^2-v^2)/(u^2-c^2)]^{1/2}$$
$$h_v = [(u^2-v^2)/(c^2-v^2)]^{1/2}$$
$$h_\phi = R = [(u^2-c^2)(c^2-v^2)]^{1/2}/c \qquad (23)$$

the three components of the metric tensor.

It is not difficult to calculate now the gradient in u direction G_u, the surface charge density $s = -\varepsilon G_u(a)$ on the spheroid $u=a$ and the net induction charge S on the upper (or lower) half of the spheroid.

$$G_u = A(F'_1 P_1 - F'_3 P_3)/h_u \qquad (24)$$
$$s = -\varepsilon G_u(a)$$
$$\quad = -\varepsilon A(F'_1(a)P_1 - F'_3(a)P_3)/h_u(a) \qquad (25)$$
$$S = -\varepsilon A\pi b^2(F'_1(a) + F'_3(a)/4) \qquad (26)$$

To calculate the energy the lightning obtains from the thunderstorm we use equation (70) from page 118 of Stratton 1941

$$U - U' = 0.5[\int_{V_0}\varepsilon E^2 dV + \int_{V1}\varepsilon(E-E')^2 dV] \qquad (27)$$

Equation (27) is the mathematical expression of the theorem that the energy U of a system of a fixed set of charges (thunderstorm) is lowered by the introduction of an uncharged conductor (lightning). The prime on U and E in (27) indicates that this is the energy U' and field E' of the system after the conductor has been introduced. DU = U-U' is the energy the lightning can obtain from the cloud field and convert into the energy consuming processes of the lightning channel, mainly ionization, heat, and radiation.

The minimum energy required for these processes determines the start and the stop of of a lightning discharge. However equation (27) can also be applied to the decremental growth of the lightning channel--lets say from one step to the next of a stepped leader.

We have to solve here the two integrals of (27)

$$I_1 = \int_{V_0}\varepsilon E^2 dV \;,\; I_2 = \int_{V1}\varepsilon(E-E')^2 dV \qquad (28)$$

The first integral I_1 has to be extended over the inside of the spheroid with the volume V_0. Since after the conductive spheroid is introduced the field E' inside the conductor is zero, E' doesn't appear in the first integral. E is the gradient of Φ_1 given in (13). Since it is a function of z only the integration in cylindrical coordinates is not difficult and we obtain

$$I_1 = \varepsilon A^2 V_0 K^4(25 - 6a^2/K^2 + 27a^4/7K^4) \qquad (29)$$

In the second integral E is again the gradient of Φ_1 and E' the gradient of $\Phi = \Phi_1 + \Phi_2$. Therefore $E-E' = -\nabla\Phi_1 + \nabla(\Phi_1 + \Phi_2) = \nabla\Phi_2$. The integral has to be extended over V_1 which is the volume outside the spheroid extended to infinity. Since the argument is the square of $\nabla\Phi_2$, with Φ_2 given by (17), we convert the volume integral into a surface integral using Green's first theorem. The surface extends over the surface of the spheroid and the surface over a large sphere with the radius going to infinity. However since the product of Φ_2 and its normal gradient vanishes with $1/r^5$ the surface integral over the large sphere is zero. It is not difficult to solve the integral over the surface of the spheroid and we obtain

$$I_2 = 4\pi\varepsilon b^2 A^2(B_3^2 Q_3 Q'_3/7 + B_1^2 Q_1 Q'_1/3) \qquad (30)$$

The energy gain DU of the lightning discharge is then given by

$$DU = (I_1 + I_2)/2 \qquad (31)$$

Comparison with numerical values given in the literature is of limited value since these values are based on a different concept. It may be sufficient to mention here that the net charge of the cloud discharge of our model is zero but the induction charges on the upper and lower part of the channel are $Q = \mp 6C$, for a cloud discharge that penetrates to the rim of the sphere. The energy gain of such a discharge is about $DU = 500$ KWh which is about 6% of the electric energy of the cloud (8600 KWh).

REFERENCES

Frank-v.Mises: "Differential Gleichungen Der Physik", Vieweg & Sohn, Braunschweig 1930

Golde, R.H.: "Lightning I, II", ed. R.H. Golde, Academic Press, New York 1977

Kasemir, H.W.:"Qualitative Übersicht über Potential-,Feld-, und Ladungs Verhältnisse bei einer Blitzentladung in der Gewitterwolke." Das Gewitter, ed. H. Israel, Geest & Portig, Leipzig 1950

Smythe, W.R.: "Static and Dynamic Electricity", McGraw-Hill, New York 1950

Stratton, J.A.: "Electromagnetic Theory" McGraw-Hill, New York 1941

Uman, M.A.: "Lightning" McGraw-Hill, New York 1969

Volland, H.: "Handbook of Atmospherics" CRC Press, Boca Raton, Florida 1982

7.19

BREAKDOWN WAVES IN RETURN-STROKE AND LEADER-STEP CHANNELS

Lee W. Parker

Lee W. Parker, Inc.
Concord, Massachusetts

Heinz W. Kasemir

Colorado Scientific Research Corporation
Berthoud, Colorado

1. INTRODUCTION

The mechanisms of the stepped leader and return stroke have not been satisfactorily explained. This paper is a follow-on to the new lightning theory proposed by Kasemir (this Conference). The new theory, based on an electrostatic model, quantifies how energy becomes available from the thunderstorm electrostatic field in the process of producing lightning. What is needed is a link between the electrostatic model and the plasma physics of the lightning channel. We are concerned with the conversion of the energy available into ionization, channel heating, electromagnetic radiation, etc. Our purpose is to raise questions and stimulate further discussions regarding the physics of the problem.

One experimental observable connected with the problem of energy conversion is the high apparent velocity of the return-stroke wave, of the order of 10^8 m/s, one-third the velocity of light. Similar velocities are exhibited by the waves in leader steps (typical lengths of order 100 m traversed in typical times of order one μs). (It should be noted that the velocities are those of moving bright spots, observed by optical means (Uman, 1969; Berger, 1977; Idone and Orville, 1982, Orville and Idone, 1982). The optical phenomenon may not necessarily be identical with the electrical phenomenon.) There have been many attempts to explain this high velocity (Cravath and Loeb, 1935; Loeb, 1965; Winn, 1967; Turcotte and Ong, 1968; Uman, 1969; Albright and Tidman, 1972; Kline and Siambis, 1972; Klingbeil et al, 1972; Fowler, 1974, 1976; Suzuki, 1977; Little, 1978; Gallimberti, 1979; Strawe, 1979; Lin et al, 1980; Hemmati, 1983; Fernsler, 1984; and other references cited in these, to sample a small fraction of the enormous literature available on the subject).

The wave-velocity results of most of the theories appear to be expressible in terms of the electron drift velocity and Townsend's ionization coefficient, and (in many cases) the "preionization" density in the channel as well. The predicted wave velocities are generally much lower than our "target value" of 10^8 m/s.

Other types of theories are also available, such as those using transmission-line models and circuit theory (Little, 1978; Strawe, 1979). These have no difficulty predicting wave velocities close to that of light. Circuit parameters can be adjusted to fit the observed data, but their physical significance is not clear. Lin et al (1980) have assumed current models in the channel that reproduce their observed electric and magnetic field waveforms.

2. BREAKDOWN-WAVE MODEL

The question arises: Can we understand the wave-front velocity on another basis such as energy conversion?

Assume the following, admittedly very crude, picture of a breakdown wave in a lightning return-stroke channel. At the instant that the leader reaches the ground the channel is a poor conductor but has a high potential ΔV (of order 10^8 V) corresponding to its position of origin in the cloud. (We are ignoring potential drops, junctions with upward connecting leaders, etc.) A breakdown wave (alternatively referred to in the literature as an "ionization potential wave") of zero potential moves up the channel from the ground with velocity v_W, and there is a relatively narrow wave-front or transition region separating the high-potential channel region upstream (ahead of the front) from the zero-potential channel region downstream (behind the front). Across the wave-front region there is a "ramp" function of potential variation, from zero to ΔV. The ramp (which may be identified with the bright spot) occupies an interval Δx of space (the thickness of the wave front), within which the field has an average value E_W.

As the wave passes any fixed position x, the gas at x is subjected to a time-varying electric field in the form of a pulse, of width $\Delta x/v_W$ and of average value E_W. During the time interval $\Delta x/v_W$, electrical energy is converted into ionization, channel heating, electromagnetic radiation, etc. Thus, the electric field energy is decreasing with time. During this same time interval, internal energy (ionization and heating) is increasing with time. In our simple model the wave speed adjusts itself so that the rate of decrease of electrical energy is equal to the rate of increase of internal energy (ionization, heating). (The decay of E_W is not entirely associated with conversion of electrostatic energy into internal energy. However, this approximation will afford insight into the physics.) We will next discuss the physics involved and will isolate some of the key parameters.

3. ENERGY CONVERSION AND WAVE VELOCITY

We now make certain assumptions leading to expressions for the rate of decrease of electrical energy and the rate of increase of internal energy. For the electrical energy per unit volume we take $\varepsilon E_w^2/2$, where ε is the permittivity and E_w is the average field. For the time interval we take $\Delta x/v_w$, which may be re-expressed (in terms of the wave potential ΔV) as $\Delta V/(E_w v_w)$. Hence the rate of decay of electrical energy per unit volume is assumed to be given by:

(d/dt)(electrical energy per unit volume)

$$= (\varepsilon E_w^3/2)v_w/\Delta V \qquad (1)$$

For the rate of increase of internal energy per unit volume we take $nkT/\Delta t$, where n = electron concentration, T = electron temperature, k = Boltzmann's constant, and Δt is a time delay for breakdown (identified with "formative" and/or "statistical" time lag in gas-discharge terminology) (Raether, 1964; Morgan, 1978). Equating the two rates and solving for v_w yields:

$$v_w = K\Delta V \qquad (2)$$

where:

$$K = (2nkT)/(\varepsilon E_w^3 \Delta t) \qquad (3)$$

Note that the wave speed v_w is proportional to the wave potential ΔV, through a coefficient K. The proportionality relationship is supported by experiments that will be discussed later.

Let the parameters be assigned the following provisional values:

$n = 10^{23}/m^3$

$kT = 1\ eV = 1.6 \times 10^{-19}\ J$
  $(nkT = 16000\ J/m^3)$

$\varepsilon = 0.884 \times 10^{-11}\ F/m$

$E_w = 3 \times 10^6\ V/m$
  $(= E_{BD},\ nominal\ breakdown\ field)$

$\Delta t = 10^{-7}\ s$

These values of n and kT have been suggested in the literature as typical for sparks and lightning channels (Uman, 1969; Barreto et al, 1977). The value of Δt is extremely variable (Morgan, 1978), but 10^{-7} s is a typical number. The resulting K is about 1300 m/V-s. This is three orders of magnitude too high for the return stroke, where the proportionality constant (K_L) is of the order of unity ($v_w = 10^8$ m/s with $\Delta V = 10^8$ V). The value of K can be brought down to unity by reasonable adjustment of the parameters. For example, Δt can increase by three orders of magnitude (allowed by the literature), or alternatively E_w can increase by one order of magnitude.

4. POSSIBLE CONNECTION WITH EXPERIMENT

An experiment of interest is that of a meter-long "guided discharge", in a gap between two electrodes, in which a breakdown "potential" wave initiated by a voltage pulse propagates along a channel preionized by a laser beam (Greig et al,

1980; Fernsler, 1984). The wave propagates across the gap with a velocity that is found to be proportional to the voltage pulse amplitude, as in Eq. (2). The measured coefficient of proportionality K is 50 m/V-s (i.e., between our theoretically estimated value, 1300, and the lightning value, unity). The experimentally-determined coefficient is an increasing function of the "preionization" density. An analysis of the proportionality coefficient K in terms of the preionization density has been performed by Fernsler (1984), involving avalanche theory and Townsend's ionization coefficient. This "first" wave, which bears some resemblance to a lightning leader process, is followed upon completing its traversal by a "return-stroke" discharge wave. The latter wave travels at much higher velocity (unmeasured, but estimated at 10^8 m/s or greater - J. R. Greig, personal communication).

Similar types of experiments have been performed earlier. An historical perspective is given by Loeb (1965), beginning with J. J. Thomson's 1893 experiments. Loeb describes results obtained in 1936 by Snoddy, Dieterich and Beams that are similar to those of Greig et al (1980), including wave velocity proportional to the applied potential, with a similar value for K. An interesting review of previous related theory and experiments is given by Winn (1967), including his own experiments, and their relevance to lightning.

Although there are some important differences between the conditions of this type of experiment and those of lightning (dimension scale too small, no leader steps and pauses, preionization channel not forged by the leader, "return stroke" draws current limited only by the power supply), there are also some important similarities, and this type of experiment would seem to be potentially valuable for investigating the plasma physics and energy-conversion processes in lightning.

We have thus begun to apply plasma physics to the problem of energy conversion in the lightning channel, a very important step in unraveling the nature of the phenomenon. Other attempts have been made, but much remains to be done.

5. REFERENCES

Albright, N.W. and D.A. Tidman, 1972: Ionizing potential waves and high-voltage breakdown streamers. Phys. Fluids, 15, 86-90.

Barreto, E., H. Jurenka and S.I. Reynolds, 1977: The formation of small sparks. J. Appl. Phys., 48, 4510-4520.

Berger, K., 1977: The earth flash. In Lightning, Vol. 1. R.H. Golde, Ed., pp. 119-190.

Cravath, A.M. and L.B. Loeb, 1935: The mechanism of the high velocity of propagation of lightning discharges. Physics (now J. Appl. Phys.), 6, 125-127.

Fernsler, R.F., 1984: General model of streamer propagation. Phys. Fluids (to be published).

Fowler, R.G., 1974: Nonlinear electron acoustic waves, Part I. Adv. Electronics Electron Physics, Vol. 35, Academic Press. Also, 1976: Nonlinear electron acoustic waves, Part II. Adv. Electronics Electron Physics, Vol. 41, Academic Press.

Gallimberti, I., 1979: The mechanism of the long spark formation. J. de Physique, 40, C7-193 to C7-250.

Greig, J.R., R. Pechacek, M. Raleigh, I.M. Vitkovitsky, R. Fernsler and J. Halle, 1980: Interaction of laser-induced ionization with electric fields. AIAA-80-1380 (AIAA 13th Fluid and Plasma Dynamics Conf., Snowmass, CO).

Hemmati, M., 1983: The exact solutions of the electron fluid dynamical equations, Ph.D. Thesis, U. Oklahoma.

Idone, V.P. and R.E. Orville, 1982: Lightning return stroke velocities in the Thunderstorm Research International Program (TRIP). J. Geophys. Res., 87, 4903-4916.

Kline, L.E. and J.G. Siambis, 1972: Computer simulation of electrical breakdown in gases. Phys. Rev. A, 5, 794-805.

Klingbeil, R., D.A. Tidman and R.F. Fernsler, 1972: Ionizing gas breakdown waves in strong electric fields. Phys. Fluids, 15, 1969-1973.

Lin, Y.T., M.A. Uman and R.B. Standler, 1980: Lightning return stroke models. J. Geophys. Res., 85, 1571-1583.

Little, P.F., 1978: Transmission line representation of a lightning return stroke. J. Phys. D.: Appl. Phys., 11, 1893-1910.

Loeb, L.B., 1965: Ionizing waves of potential gradient. Science, 148, 1417-1426.

Meek, J.M. and J.D. Craggs, 1953: Electrical Breakdown of Gases, Oxford Clarendon Press, London.

Morgan, C.G., 1978: Irradiation and time lags. In Electrical Breakdown of Gases, J.M. Meek and J.D. Craggs, Eds., John Wiley & Sons, New York. pp. 655-688.

Orville, R.E. and V.P. Idone, 1982: Lightning leader characteristics in the Thunderstorm Research International Program (TRIP). J. Geophys. Res., 87, 11177-11192.

Raether, H., 1964: Electron Avalanches and Breakdown in Gases, Butterworths, Washington.

Strawe, D.F., 1979: Non-linear modeling of lightning return strokes. FAA/FIT Workshop on Grounding and Lightning Technology, Melbourne, Florida. Report FAA-RD-79-6, U.S. D.O.T./FAA, 9-15.

Suzuki, T., 1977: Propagation of ionizing waves in glow discharge. J. Appl. Phys., 48, 5001-5007.

Turcotte, D.L. and R.S.B. Ong, 1968: The structure and propagation of ionizing wave fronts. J. Plasma Phys., 2, 145-155.

Uman, M.A., 1969: Lightning, McGraw-Hill, New York.

Winn, W.P., 1967: Ionizing space-charge waves in gases. J. Appl. Phys., 38, 783-790.

ELECTROSTATIC FIELDS OF GROUND TRIGGERED LIGHTNING

Lothar H. Ruhnke
Naval Research Laboratory, Washington D.C. 20375, USA

Heinz W. Kasemir
Colorado Scientific Research Corp., Berthaud CO 80513, USA

ABSTRACT: Electric fields of lightning triggered by wire rockets have been measured at several observation sites in Florida in 1986 and 1987 at distances of a few kilometers from the trigger location. Characteristic wave forms are described and show a continous rise in electric fields for 200 milliseconds followed by a period of high variability and ending with several return stroke signals. Electric fields are related to currents, charges and velocities of the positive leader of the triggered lightning. The relationship is tested on a physical model of triggered lightning suggested earlier by Kasemir.

INTRODUCTION

Lightning can be artificially triggered from the ground or from aircraft if environmental electric fields are high enough. A well developed technology exists to trigger lightning using wire-grounded rockets. The techniques have been described by Laroche et al. (1985), Horii (1982) and Uman (1987). Research on lighting has long been hampered by uncertainty of occurrence and therefore from lack of close and timely observations. The laboratory spark, although offering precise timing and close-up observations cannot well simulate features which occur over long distances like the stepping of leaders, the transition from streamer to leader or the branching and tortuosity of the natural lightning channel. Although triggered lightning is not in all respect equal to naturally occurring lightning, the artificially triggered lightning can provide the experimenter and theoretician with a new tool to explore lightning processes. Because of the possibility of close-up observations and the exact timing, the triggered lightning is an opportune subject of lightning and long spark research. At Kennedy Space Center a facility exists to trigger lightning at will for scientific purposes. Direct current measurements in the lightning channel, a network of surface electric field meters, and optical observations of the triggered lightning are available besides good local weather observations. Our goal is to use this facility to relate electrical measurements to a physical model of the triggered lightning.

1988 Proceedings Int. Conf. Atmospheric Electricity

MEASUREMENTS

Three observation stations located 0.5 km, 2.2 km and 9.6 km south of the trigger site were used to measure the electrostatic field at the surface. A 'slow antenna' with a high-pass characteristic and a time constant of 20 second was used to sense the electric field at 100 microsecond intervals for up to three seconds during each trigger event. After A/D conversion the data was stored with time signals for later analysis. In the summer of 1986 two triggered lightning events were recorded and in 1987 eight. All showed similar characteristics. The trigger # 103 of July 12, 1987 is typical for all. Its signal at the closest station is shown in Fig. 1.

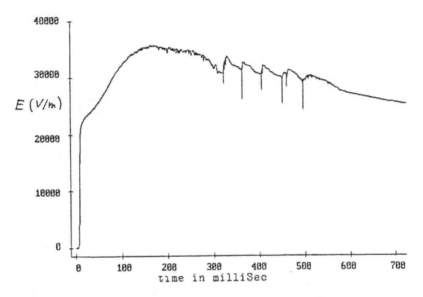

Fig. 1 Triggered Lightning # 103 July 12, 1987 17:27 EDT

For the first 10 milliseconds there is a sharp but continuous rise in the electric field. Its amplitude decreases with the third power of the distance at the ground indicating a location at low altitude of the related charges. For the next 200 milliseconds the rise in electric field is much slower, yet seems continuous at the sampling rate of 1 per 100 microsecond. This is followed by a period of many negative impulses (negative = negative charges overhead) for about 100 milliseconds. After that a series of returnstroke signals is most often observed, although at times no return strokes were associated with the trigger event. Laroche et al. (1982) observed similar waveforms of triggered lightning from a nine station fieldmill network but with less time resolution (5 millisecond sampling rate). Assuming a leader propagation velocity of about

1988 Proceedings Int. Conf. Atmospheric Electricity

$3*10^5$ m/sec., one can assume that the leader has entered the cloud within 10 milliseconds. For an interpretation of the waveforms one should first look at the easiest part, namely the first 10 milliseconds where one can assume that optical data will be available to support deductions arrived at from electrical measurements.

THEORY AND MODEL

For constructing a physical model of the triggered lightning one can make a few reasonable assumptions. During the vertical growth of the streamer/leader process one can assume that the potential of the leader is the same as the ground. From optical data it is reasonable to assume in first approximation a vertical line for the leader channel. This is of course no longer valid when the leader enters the cloud. Laroche et al. (1982) have shown strong horizontal components of the lightning channel inside clouds from electric field analysis. Charges on the channel will develop by influence of the ambient electric field. Assuming a constant diameter of the leader channel one can relate the charge per unit length on the channel directly to the potential distribution of the ambient electric field. Kasemir (1986) calculated the charge per unit length on a vertical leader for the case of a triggered lightning with the ambient electric field produced by a negative space charge overhead. He found that during the development of the vertical channel the charge on any part of the channel does not vary more than 10 percent during growth, and that any additional growth produces mainly an additional charge at the tip of the channel (Fig. 2). This then leads to the geometry of Fig. 3 and the relationship shown in equation (1) between the charge change dQ on the channel tip and the electric field change dE produced at the observation point at distance D from the base of the triggered lightning.

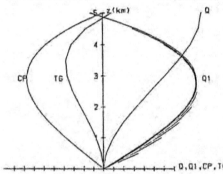

Fig. 2 Calculated Charge Distribution along Channel

$$dE = 1/(2*\pi\varepsilon) * dQ * h /(h^2 + D^2)^{(3/2)} \quad (1)$$

Measurement of dE_1 and dE_2 at two stations with their ratio R and different distances D_1 and D_2 from the trigger site will then give the channel height h using equations (1) and (2).

1988 Proceedings Int. Conf. Atmospheric Electricity

572

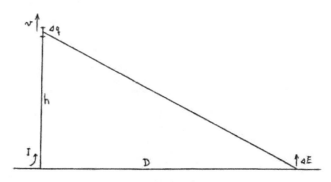

Fig. 3 Model Geometry

$$R = (dE_1/dE_2)^{(2/3)} \qquad\qquad (2)$$

$$h^2 = (D_2{}^2 - R * D_1{}^2)/(R - 1) \qquad (3)$$

Once h(t) is known one can by straight forward computation determine the current waveform i(t) at the base of the leader channel, the velocity of the upward propagating channel v(t) and the charge distribution q(h) along the channel (equations (4),(5) and (6)).

$$i(t) = dE/dt * 2\pi\varepsilon *(h^2 + D^2)^{(3/2)}/ h \qquad (4)$$

$$v(t) = dh/dt \qquad\qquad (5)$$

$$q(h) = i(t)/v(t) \qquad\qquad (6)$$

If the channel diameter would be known, one could also determine the ambient potential distribution before the lightning event from the known charge distribution along the channel.

GENERAL RESULTS

The leader channel starts usually several hundred meters above ground with a vertical velocity of about 2 to 3 10^5 m/sec. The vertical velocity is fairly constant or sometimes increasing with height, similar to the results of Kito et al. (1985). Fairly large variations in leader tip propagation, however, may occur from one strike to the next. Also sudden changes in propagation velocity are fairly common (Fig. 4). From 3 km altitude the vertical development is no longer dominant and excessive branching or horizontal propagation makes a two station analysis inappropiate. Current peaks occur about 7 to 10 millisec after initiation with magnitudes of 100 to several thousand amperes in agreement with direct measurements of Laroche et

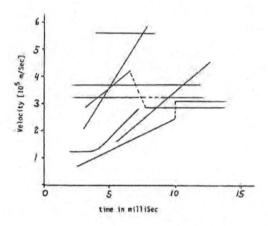

Fig. 4   Leader Propagation Velocity for
8 Triggered Lightning

al.(1985). The example in Fig. 5 shows a fairly strong peak
signal. The fact that a continuous current of several
hundred amperes flows for several hundred milliseconds
means that the positive leader propagates inside the cloud
much farther than to the center of the negative
space charge and for a much longer time than can be
explained by our simple model.
   The charge per unit length on the channel shows an
expected peak at 2 to 3 km altitude and decreases as the
channel apparently propagates toward the center of a
negative space charge (Fig.6). From the data of the charge
on the channel it can be inferred that the ambient

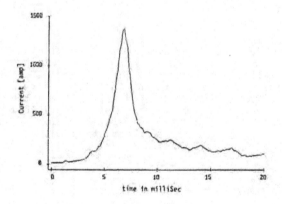

Fig. 5   Base Current of Lightning # 103

1988 Proceedings Int. Conf. Atmospheric Electricity

574

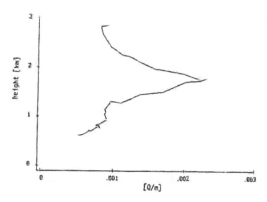

Fig.6 Charge Distribution on Leader Channel # 103

electrostatic field increases until the channel enters a
space charge region, then reverses polarity and decreases
in magnitude until the tip of the channel reaches ambient
potentials close to ground potential. Analysis of the first
10 milliseconds of a positive leader process of a triggered
lightning by electric field data from several stations
seems to be a reasonable complement to direct current and
optical measurements.

ACKNOWLEDGEMENT
    The assistence of Mike Gomez, Wolfram Kasemir and Dr.
Vlad Mazur for operating some of the field stations is
gratefully acknowledged.

REFERENCES

Horii,K., 1982: Experiment of Artificial Lightning
    Triggered with Rocket. Mem.Fac.Eng., Nagoya Univ., Japan,
    34, pp.77-112
Kasemir,H.W., 1986: Model of Lightning Flashes triggered
    from the Ground. International Aerospace and Ground Conf.
    on Lightning and Static Electricity., Dayton,Ohio. p.39-I
Kito,Y.,K.Horii,Y.Higashiyama,and K.Nakamura, 1985: Optical
    Aspects of Winter Lightning Discharge Triggered by
    Rocket-Wire Technique in Hokuriku District of Japan.
    J.Geophys.Res.,90, pp.6147-6157
Laroche,P.,A.Delannoy, and P.Metzger, 1982: Neutralized
    Electrical Charges Location during Triggered Lightning
    Flashes. International Aerospace and Ground Conference on
    Lightning and Static Electricity., Oxford, England
Laroche,P.,A.Eybert-Bérard, and L.Barret, 1985: Triggered
    Lightning Flash Characterization. 10th International
    Aerospace and Ground Conference on Lightning and Static
    Electricity,. Paris, pp.231-239
Uman,M., 1987: The Lightning Discharge. Academic Press.,New
    York, pp.205-230

1988 Proceedings Int. Conf. Atmospheric Electricity

575

SECTION 5. MEASUREMENT TECHNIQUES

1. An Apparatus for Simultaneous Registration of Potential Gradient and Air-Earth Current (Description and First Results), H. W. Kasemir, J. Atm. and Terr. Physics, Vol. 2,pp.32-37, 1951, Pergamon Press Ltd., London, 1951.

2. Measurement of the Air-Earth Current Density, H. W. Kasemir, Proc.First Intentional Conference on Atmospheric Electricity, AFCRC-TR-55-206, U.S. Dept. of Commerce, Office of Technical Services, Washington, D.C., 1954.

3. Antenna Problems of Measurements of the Air-Earth Current, H.W. Kasemir and L. H. Ruhnke, Proc. Second Int. Conf. on Atmospheric Electricity, Recent Advances in Atmospheric Electricity, Pergamon Press, New York, 1958.

4. A Radiosonde for Measuring the Air-Earth Current Density, H.W. Kasemir, USASRDL Technical Report 2125, Fort Monmouth, New Jersey, 1960.

5. Field Component Meter, H.-W. Kasemir, Tellus, Vol. 3, November, 1951.

6. The Cylindrical Field Mill, H. W. Kasemir, Meteorologische Rundschau, 25. Jahrgang, 2. Heft, 1972.

7. Electric Field Measurements from Airplanes, H. W. Kasemir, Proc. Fourth Symposium on Meteorological Observations and Instrumentation, Denver, CO, published by American Meteorological Society, Boston, April, 1978.

8. Evaluation of Mighty Mouse Trigger Rocket Flights, H. W. Kasemir, Progress Report on Kennedy Space Center Contract No. CC-88025, Part D., 1971.

9. Airborne Warning Systems for Natural and Aircraft-Initiated Lightning, L. W. Parker and H. W. Kasemir, IEEE Transactions on Electromagnetic Compatibility, VOL. EMC-24, No.2, May 1982.

10. Theoretical and Experimental Determination of Field, Charge, and Current on an Aircraft hit by Natural or Triggered Lightning, H. W. Kasemir, International Aerospace and Ground Conference on Lightning and Static Electricity, NASA, 1984.

11. Ranging and Azimuthal Problems of an Airborne Crossed Loop Used as a Single-Station Lightning Locator, L. W. Parker and H. W. Kasemir, Proc. 10th International Aerospace and Ground conference on Lightning and Static Electricity. Paris, France, 1985.

12. Predicted Aircraft Field Concentration Factors and their Relation to Triggered Lightning, L. W. Parker, H. W. Kasemir, Proc. International Aerospace and Ground Conference on Lightning and Static Electricity, Orlando, FL, 1984.

Journal of Atmospheric and Terrestrial Physics, 1951, Vol. 2, pp. 32 to 37. Pergamon Press Ltd., London

An apparatus
for simultaneous registration of potential gradient and air-earth current
(Description and first results)

H.-W. Kasemir

Luftelektrische Forschungsstelle Buchau a. F. des Observatoriums Friedrichshafen,
Landeswetterdienst Württemberg-Hohenzollern

(Received 1 November 1950)

ABSTRACT

An apparatus for the simultaneous continuous recording of the atmospheric electric potential gradient and the air-earth current is described. The CR constants are chosen in such a way that minor changes of the atmospheric electric elements in question are suppressed and the diurnal variations can readily be seen from the records. Modern technical means such as six-channel recorders, electrometer tubes, and low loss cables make it possible to register a great range of potential gradient and air-earth current. Moreover the apparatus is foolproof and convenient as to dimensions and weight so that it can be used on mountain observatories and expeditions without difficulty. Preliminary results obtained on mountain peaks are described, which will help in the interpretation of such records.

I. Introduction

During recent investigations on atmospheric electricity it has been established that, when recording only the potential gradient, it is very difficult to obtain proper knowledge on the atmospheric electric state of the atmosphere [1]. Recording of further elements, e.g. air-earth current or conductivity, involves many difficulties and only a small number of observers has been able to overcome them. Besides, when a second element was observed, usually the conductivity was recorded and it was only at Kew where the air-earth current was registered for a longer period. It is possible to calculate the air-earth current i from the potential gradient E and the conductivity λ from the equation $i = \lambda \times E$, as long as Ohms Law holds. It is possible that this condition is not fulfilled during windy weather periods, i.e. when strong convectional currents prevail.

Moreover, the air-earth current seems to be the primary and therefore the most informative of the three atmospheric electric elements, since during periods of the stationary state it depends very little on the variations of the local conductivity, but mainly on the variations of the atmospheric electric generator; that is, the world wide thunderstorm activity. Therefore, the air-earth current is the most important of the atmospheric electric elements and it is worthwhile to try its continuous registration.

Following the ideas outlined above H. Israël asked the author to develop an instrument for the simultaneous registration of potential gradient and air-earth current. Modern technical means make it possible to construct an apparatus that fulfills the conditions set by atmospheric electric investigators and is simple as to construction and operation. Also, its weight is so low that it invites scientists to take it along for work on expeditions and high mountain peaks. In the following

An apparatus for simultaneous registration of potential gradient and air-earth current

the instrument and first tests of its performance are described. Figure 1 shows the apparatus together with a six-channel recorder and Figure 2 shows the circuit diagram.

II. Registration of Potential Gradient

The potential gradient is picked up by a radioactive collector which does not work directly on the recorder as it does in the case of the electrostatic Benndorf-recorder but it controls the grid of an electrometer tube the plate current of which is registered by means of a two- or six-

Fig. 1. Amplifier and six-channel recorder for the registration of air-earth current and potential gradient.

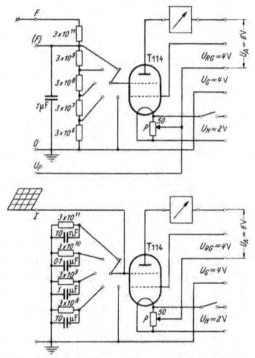

Fig. 2. Circuit diagram, above potential gradient, below air-earth current.

channel recorder. The scale of the recorder comprises a range of 0·1 mA. Its rugged construction makes the instrument convenient for use on expeditions where rough treatment cannot be avoided. Besides, up to six elements can be recorded on one strip of paper simultaneously so that it is possible to record not only potential gradient and air-earth current but in addition also meteorological elements such as temperature, humidity, sunshine, wind etc., a fact that facilitates studying of correlations.

It can be inferred from the circuit diagram (Figure 2) that the electrometer tube works as a simple one-stage d.c. amplifier. Its control grid receives the voltage from the radioactive collector via a voltage divider. One pole of the recorder is connected to the plate of the tube and the other one to the battery. Some points ought to be mentioned which are not always taken into proper consideration. The input resistance of the instrument should be higher than the resistance between air and collector surface, so that the input resistance does not load the collector.

3 JATP. Vol. 2

580

To prevent an overload of the tube the large voltage of the collector has to be reduced in such a way, that the control grid does not receive more than 1 v. This is done by means of a dropping resistor of $3 \times 10^{11} \, \Omega$. The grid leak resistor is constructed as a voltage divider which allows the sensitivity to be changed within a wide range. With this arrangement it is possible to cover practically all potential gradient variations under fine weather conditions as well as during periods of extremely high potential gradients (showers, hail storms, thunderstorms) (compare K. BURKHART [2]).

For getting the diurnal variations, hourly means of the BENNDORF registration are read and again reproduced on a far smaller time scale. Since the diurnal variations are of utmost importance it seems to be justified to construct the intrument in such a way that it records the diurnal variations automatically in a convenient

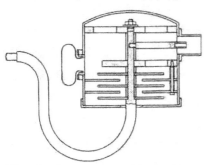

Fig. 3. Section of outside insulator.

manner. The short fluctuations are of no interest and can be filtered away by increasing the time constant; that is, by connecting a capacitance of low loss across the grid leak. A time constant of 1 hr can easily be achieved because of the large resistance of the grid leak. This procedure also allows the length of the paper transport to be reduced down to 24 cm per day or even less without losing essential details.

To make readings linear over the whole scale it is essential to let the tube work on the straight part of the characteristic. To this end one can use mechanical or electrical suppression of the zero point of the recorder. Bridge or compensation circuits would give the same results but are avoided, since additional batteries and heavier battery drain would make the apparatus less convenient for use on expeditions.

The potentiometer P (Figure 2) permits the selection of the grid bias and, as a consequence of this, of the plate current, and allows the zero reading of potential gradient to be placed on any desired value of the recorder scale.

The most critical point of the arrangement is the maintenance of the high insulation resistance which must be at least a hundred times larger than the grid leak and dropping resistor together. Low loss cable as is used for coaxial lines in high frequency technique can be employed to connect the collector to the input jack of the amplifier. The insulation of the cable is not influenced by humidity and it was found that a length of 60 ft did not cause any trouble. Thus the attention of the observer has only to be concentrated on the connections. In a room where the relative humidity can always be kept well below 100 % difficulties arise only seldom. Quite different are the conditions at the outside insulator that supports the collector, a section of which is shown in Figure 3. It is constructed on the lines of an air insulated condenser. Its upper room is dried by means of calcium chloride which is kept in the glass vessel to the left of the insulator housing. The outside air has to pass the long way between the condenser plates before it is able to touch the insulating parts. The calcium chloride keeps the relative humidity below 100 % and has only seldom to be replaced, even during periods of wet weather. On mountain

An apparatus for simultaneous registration of potential gradient and air-earth current

peaks the observer has only to watch that snow and hoar frost crystals do not short-circuit the collector. At sea level spider's webs and small insects often spoil the insulation. A special protective kind of glue smeared on the insulator guards against insects. The spider's webs have to be removed in the evening before they become wet and lose their insulation.

III. Registration of Air-Earth Current

The difficulties in recording air-earth current do not arise only because of its small quantity (about 3×10^{-12} A/m^2 at sea level and during fine weather). Every receiver, usually a metal plate or net, exposed to it, not only receives the air earth-current but also the influence charges caused by the fluctuations of the potential gradient which give rise to currents 100 times larger than the air earth-current itself. On the direct current caused by the air-earth current is superposed an a.c. of variable frequency. The a.c. can be filtered away by means of a filter of convenient time constant, say 1 hr, which is also desirable from another point of view outlined above. It has been mentioned before that the range of the recorder is 0·1 mA. A plate current change of 0·05 mA (that is half the recorder range) of the electrometer tube can be caused by changing the grid bias by about 1 v, $i.e.$ the grid leak resistor has to be so large that a current of 3×10^{-12} A causes a voltage drop of 1 v. It is possible to read currents down to 3×10^{-13} A from the registrations.

By means of a switch the grid leak resistance can be changed so as to keep the voltage variations of the control grid within 1 v, and to adapt the sensitivity of the amplifier to any occurring air-earth current. With regard to insulation and placing the electrical zero reading, the principles outlined in the previous chapter are valid.

The plate or net that picks up the air-earth current should always be at ground potential when placed level with the earth's surface [3]. The voltage drop across the grid leak resistor causes a potential difference between net and ground that varies with the air-earth current between 0 and 1 v. One could try to lift the receiving plate in such a way that its average potential is equal to that of the surrounding air. Under normal conditions the potential gradient is about 1 v/cm, $i.e.$, the receiving plane has to be lifted 0·5 cm. Such an adjustment is difficult, and besides experiments show that small differences in height and potential do not influence the results appreciably. The situation is quite different when the air-earth current becomes so large ($e.g.$ during showers) that the receiving plate acquires a potential difference of 100 v or more above ground. This occurs when the observer does not switch down the sensitivity in time. If this happens the receiver catches more radioactive particles than usual (receiving plate on ground potential). When the air-earth current returns to normal values the radioactive material on the net causes a higher conductivity and in consequence of this a higher air-earth—current than the normal one for a period of about 2 hr; $i.e.$, until the radioactive material has disappeared. This fact has to be taken into consideration when taking readings after periods of bad weather during which the sensitivity has not been switched down.

IV. Registrations on the Nebelhorn, on the Zugspitze and at Buchau

A first test of the instrument was made in September 1949 on the Nebelhorn (7333 ft). This test as well as that on the Zugspitze in March and April 1950 (9711 ft), was made to discover technical defects but the results yielded during this work are already of some general interest and give some hints as to what we can expect from the planned mountain programme [4]. The apparatus was housed in a primitive canvas shelter on the western ridge close to the summit of the Nebelhorn. After ten days of heavy rain during which the amplifier failed to work properly due to poor insulation (the instrument was then in its first stage of development) a

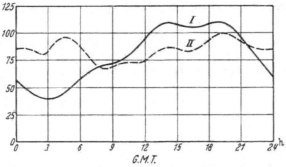

Fig. 4. Curve I: Diurnal variation of the air-earth current on the Nebelhorn. Curve II: Diurnal variation of the thunderstorm activity (F. J. W. WIPPLE and F. J. SCRASE).

period of five days of Foehn followed. The air-earth current was found to be about ten times higher than at sea level. The potential gradient, which was extremely disturbed by the sharp ridge and therefore expected to be very high, amounted only to slightly below 100 v/m. The daily variations in air-earth current were steady throughout these five days and resemble very much the world time variations of potential gradient found at sea (Figure 4) except for the maximum at 4 hr GMT which is, rather surprisingly, missing in the registrations of potential gradient. The similarity to the world time variations indicates the absence of serious convection. Moreover, at greater heights one has to expect the stationary state to be reached in a shorter period of time than at sea level, due to the higher conductivity. Although these facts are quite striking it is not advisable to derive further conclusions from this small number of observations.

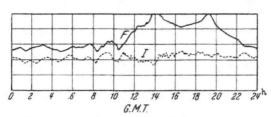

Fig. 5. Potential gradient (F) and air-earth current (I) of a typical day on the Zugspitze.

The registrations on the Zugspitze were started in the beginning of March 1950 in a period of high atmospheric pressure. P. LAUTNER [5] had previously carried out observations of potential gradient and air-earth current for three years on this mountain. The results differ considerably from those obtained on the Nebelhorn since no Foehn occurred during the observations. Figure 5 shows the variations of a typical day. The air-earth current does not have any apparent diurnal variation. The potential gradient has two maxima in the afternoon, namely one at 14 hr and the other one at 19 hr GMT. This result is well in accordance with the results yielded by LAUTNER on the Zugspitze and by other observers on high stations. The single period which is found on undisturbed days is most probably caused by sun radiation, which gives rise to convection at noon; that is, it lowers the conductivity and therefore lets the potential gradient rise. The air-earth current is not seriously affected by convection and therefore promises to throw more light on any world time variation than any other element. During the night and the morning hours potential gradient and air-earth current run fairly parallel. This fact demonstrates the absence of convection. For statistical investigations the material yielded on the Zugspitze is too poor. The potential

An apparatus for simultaneous registration of potential gradient and air-earth current

gradient variations observed at Buchau during the same time differ very much from those found on the Zugspitze due to local influences. In air-earth current the differences are smaller. As soon as more material is available statistical investigations will be started which should reveal important facts on the mechanism of the atmospheric electric field.

The registrations on the Zugspitze as well as those at Buchau seem to demonstrate that during undisturbed days air-earth current and potential gradient run parallel, especially during the night and the early hours of the morning. The parallelism is better on the Zugspitze than at Buchau and is upset as soon as convection starts. Since the material on which this conclusions is based is still poor it should be considered with caution; but it seems to show that the method proposed by H. ISRAËL (*loc.cit.* p. 32), *i.e.* to study convection by means of measuring atmospheric electric elements, promises some success.

Thanks are due to Dr. HANS ISRAËL for much helpful discussion and encouragement.

REFERENCES

[1] ISRAËL, H.; Die Tagesvariationen des elektrischen Widerstandes der Atmosphäre. Ann. Géophys. 1949 **5** 196–210. ISRAËL, H.; Probleme und Aufgaben der Luftelektrizität. Ber. dtsch. Wetterd. d. US-Zone 1950 (No. 12) 75–77. [2] BURKHART, K.; Luftelektrische Feldmessungen mit Elektrometerröhren. Z. Met. 1947 **1** 212–213. [3] WILSON, C. T. R.; On the measurement of the atmospheric electric potential gradient and the air-earth current. Proc. Roy. Soc. A 1908 **80** 537–547. [4] ISRAËL, H. see Nr. 1. [5] LAUTNER, P.; Deutsches Meteorologisches Jahrbuch, Bayern, 1926–28, 1929–30.

Printed by Universitätsdruckerei H. Stürtz AG., Würzburg

MEASUREMENT OF THE AIR-EARTH CURRENT DENSITY

H. W. Kasemir

Deutscher Wetterdienst
Meteorologisches Observatorium
Aachen, Germany

Abstract--There is no doubt today that the simultaneous registration of the atmospheric-electric field F, the air-earth current i, and the conductivity is necessary for the evaluation of atmospheric electric measurements. In this field the registration of the air-earth current i involves specific difficulties, not only with regard to the smallness of this current, but also because the conduction current is superposed on the large current of influence charges. This troubling influence current can be eliminated completely as may be shown theoretically and experimentally. From the Maxwell equation the following equation for the current J (through the measuring resistance) is deduced

$$J = (\theta/T)\,i + (1/T)[1 - \theta/T]e^{-t/T}\int i\, e^{t/T}dt$$

where t is the time, θ, the time constant of the atmospheric electric field at the measuring place, and T, the time constant of the input of the apparatus. The air-earth current i can be given as a function of time in any form. From this equation follows $J = i$, for $\theta = T$, that is, the current J flowing through the measuring resistance is identical with the air-earth current. The troubling influence currents, which are given by the second term of the equation, are compensated to zero. By a more electrotechnical substitute circuit diagram, the meaning of the condition $\theta = T$ is demonstrated. From the slow diurnal variations up to the quick variations of lightning flashes the recording of the air-earth current is independent of the frequency and correct in the amplitude. The registration apparatus is simple and the theory is clear so that the air-earth current is now accessible for the atmospheric electric technique of measurement.

For a long time it has been pointed out that simultaneous measurement of all of the three elements of atmospheric electricity is important. For the most part the potential gradient and the conductivity are recorded and the air-earth current is calculated from these elements. Normally, the expression 'air-earth current' is used for the conduction current whose carriers are ions moving in the direction of the electric field. This, however, should not be confused with the convection current which is carried by precipitation particles moving against the electric field and with Maxwell's dielectric current, commonly called 'influence current.' In the following discussion, we shall consider only the conduction and influence currents, but not the convection current.

A measurement of the air-earth current, that is, the conduction current, is very difficult since it is strongly disturbed by the influence current. In the following I will explain a method which yields a clean separation of the conduction current from the influence current. The air-earth current is usually measured by means of a plate exposed to the open air. This plate picks up the air-earth current which then flows through a calibrated resistance to the Earth and the voltage drop across the resistance is measured. With this method, the influence current is simultaneously being picked up. This effect is due to the existence of a capacity between the plate and the air. There exists also a certain capacity between the plate and the ground. Thus, a capacity lies parallel to the resistance of the air column above the plate as well as across the calibrated high resistance to the ground below the plate. We arrive at the following circuit diagram shown in Figure 1. The following is an explanation of the symbols used:

r = the resistance of the unit volume of the air

ϵ = the capacity of the unit volume of the air

R = the high ohm resistance to the ground

C = the capacity parallel to this resistance.

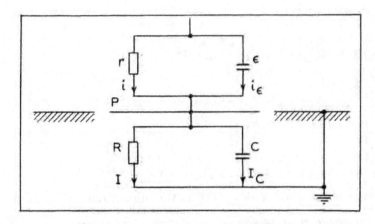

Fig. 1--Circuit for measurement of
air-earth current

Normally C is only a few $\mu\mu F$ but it may increase to much higher values up to 100,000 $\mu\mu F$ by long cables or by the addition of commercial condensers. We will see how the accurate measurement of the air-earth current depends on the proper magnitude of the condenser C. ϵ represents the influence effect of the field and is identical with the dielectric constant. This dielectric constant is interpreted in our diagram as the specific capacity of the unit volume of air, as will be shown later. The input of the recording device (electrometer or the grid of an electrometer tube) is connected with the plate P.

Now we will calculate the currents which flow through each of the resistances. The idea behind these calculations is to find a circuit which allows only the air-earth current but not the dielectric current to proceed through the resistance r, the plate, and the resistance R. The dielectric current, on the other hand, should go through the condenser ϵ, the plate, and the condenser C. At first, the calculation will be given in general terms by solving the differential equation which is derived from Maxwell's equations. Later on, we will see that the same result can be obtained simply through application of Ohm's law using the circuit diagram. Here we start with the Maxwell equation

$$\partial\theta/\partial t + i = c = \operatorname{curl} H \dots\dots\dots\dots\dots (1)$$

The first term means the Maxwell dielectric current, i is the conduction current; consequently, c, the sum of both of these terms is the complete Maxwell current. Curl H is the rotation of the magnetic field. We take the divergence of (1) and obtain

$$\operatorname{div}(\partial\theta/\partial t + i) = \operatorname{div} c = 0 \dots\dots\dots\dots (2)$$

divergence c = 0 indicates that for the unit volume considered the outflowing current c_2 must be equal to the inflowing current c_1. Thus, we arrive at the simple equation

$$c_1 = c_2 \dots\dots\dots\dots\dots\dots\dots\dots (3)$$

which means, mathematically expressed, the integration of (2).

This result will now be applied to our plate which is exposed to the air-earth current. The dielectric current $F\,\partial\theta/\partial t$ and the conduction current Fi flow through the plate with a cross section F. Flowing away from the plate are the currents I through resistance R and I_C through the capacity C. Consequently we get

$$F(\partial\theta/\partial t + i) = I + I_C \dots\dots\dots\dots\dots (4)$$

We are now interested in the current I through R as a function of the air-earth current i. Therefore, we have to try to eliminate from (4) the expressions $\partial\theta/\partial t$ and I_C. For the first expression this can be done by means of the well-known equations

$$\theta = \epsilon E \qquad \text{and} \qquad \lambda E = i \dots\dots\dots\dots (5)$$

In this equation ϵ means the dielectric constant, λ the conductivity, and E the electric field. Through differentiation with respect to time t, we get the expression

$$\partial\theta/\partial t = (\epsilon/\lambda)\, \partial i/\partial t \quad\dots\dots\dots\dots\dots\dots (6)$$

which gives the desired relation between the dielectric and the conduction current. To eliminate I_C we use the fact that the resistance R and the capacity C are affected by the same potential drop, indicated by the letter U. R, C, and U are interrelated through the simple equations $U = IR$ and $U = Q/C$. Q is the charge of the capacity C. The variation of Q with respect to time, dQ/dt, is the current I_C and we obtain a relation between the current I_C through the capacity C, and the current I through the resistance R.

$$I_C = dQ/dt = C\, dU/dt = CR\, dI/dt \quad\dots\dots\dots\dots\dots (7)$$

By means of (4), (6), and (7) we obtain the differential equation

$$CR\, dI/dt + I = [(\epsilon/\lambda)\, \partial i/\partial t + i]\, F \quad\dots\dots\dots\dots\dots (8)$$

The expression $T = CR$ is the time constant of our input circuit. The expression $\theta = \epsilon/\lambda$ is the well-known time constant of air. Eq. (8) can be solved without difficulties and we obtain

$$I = F\left[Ae^{-t/T} + (\theta/T)\, i + (1/T)(1 - \theta/T)\, e^{-t/T}\int_0^t ie^{t'/T}\,dt'\right] \quad\dots\dots\dots\dots (9)$$

At first sight, (9) appears to be rather complicated; we have to consider, however, that this equation is of general validity, since any function can be introduced for the air-earth current i. Even without specifying the function for i, important conclusions can be drawn from this equation. The first term of the right side with the integration constant A determines the transient response. It will decay with increasing time and therefore it does not have to be discussed any further. The equation, thus, assumes the simplified form

$$I = F\left[(\theta/T)\, i + (1/T)(1 - \theta/T)\, e^{-t/T}\int_0^t ie^{t'/T}\,dt'\right] \quad\dots\dots\dots\dots (10)$$

The last term of the right side of (10) will be zero for the case $\theta = T$; that is, it will be zero if the time constant of the input circuit is equal to the time constant of the air. Then we obtain from (10)

$$I = i\, F \quad\dots\dots\dots\dots\dots\dots\dots (11)$$

In this state the entire air-earth current picked up by the plate will go through the measuring resistance, whereas, all of the influence currents are deviated to ground through shunted capacity C. For example, we have to shunt the high ohm resistance of 10^{11} ohm to a parallel coupling of a capacity of about 9000 $\mu\mu$F in the case where the time constant of the air is about 15 minutes or 900 seconds. This shunting capacity is not, as heretofore, to be interpreted as damping capacity which more or less neutralizes the influence charges, but it has to be interpreted as balancing capacity by which we match the input of the measuring instrument to the atmospheric electrical circuit. This capacity is necessary to the correct balance by which we obtain not only an approximate but an exact registration of the air-earth current for all frequencies from zero to infinity. Variation from the fastest changes of the air-earth current, such as occurring during lightning discharges, down to the slowest diurnal variations is not only theoretically possible but has been practically carried out at the Meteorological Observatory at Aachen for quite some time.

If we choose to make the input capacity C smaller than is necessary for balancing, so a part of the influence current flows through the high resistance the circuit registers the well-known violent disturbances. If we choose to make the input capacity C larger than is necessary for balancing, a part of the air-earth current will be needed to charge the capacity. In this case the capacity C acts as a damping condenser and smooths out the fluctuations of the air-earth current. Both cases are of practical importance.

The first case, where the input capacity is too small, mismatching is suitable for recording the short-period current variations. This is again a measure for the exchange in the atmosphere which is normally difficult to record by other measuring devices.

The second case, where the input capacity is too large, mismatching is suitable for recording long-period current variations. For instance, in the investigation of the correlation between single weather periods and air-earth currents these long period current variations often occur.

We utilize (10) for investigation of mismatching by different frequencies of air-earth currents. The integral in (10) can be easily solved if we introduce the air-earth currents as sinusoidal alternating current.

$$i = i_0 e^{j\omega t} \quad \dots \dots \dots \dots \dots \dots \dots \dots \dots \quad (12)$$

j = root of minus one, signifies the imaginary unit, and ω signifies the angular frequency. Substituting (12) in (10) we obtain the following expression

$$I = [(1 + j\omega\theta)/(1 + j\omega T)] \, iF \dots \dots \dots \dots \dots \dots \quad (13)$$

If $\omega\theta$ and ωT are small in comparison with unity, that is, the duration of the current variation is great in comparison with θ and T, we can neglect the terms $j\omega\theta$ and $j\omega T$ compared to unity. We obtain again the single equation $I = i \, F$. If, however, the duration period of variation is small then we obtain approximately $I = (\theta/T) \, i \, F$.

If the ratio θ/T is larger or smaller than unity, then the measured current I will be larger or smaller than the air-earth current.

In closing I would like to show that (13) can, in a simple manner, be derived from the circuit diagram with the help of Ohm's law. Hereby, the main point is that we are able to interpret the dielectric constant ϵ as the capacity of the unit volume. We can do this with the same correctness as we interpret the conductivity λ as specific conductivity of the unit volume of air. We can show that best by a comparison of the definition equation. The conductance L of a square with the cross section F and the height h is given by the expression $L = \lambda F/h$, if λ is the specific conductivity. The capacity C of the same square is given by the well-known equation $C = \epsilon F/h$.

From these equations we obtain the specific conductivity as $\lambda = L \, h/F$ and the specific capacity as $\epsilon = C \, h/F$. The analogy of both formulas speaks for itself. ϵ has the dimension of a specific capacity, namely farad/m. This is an analog to the dimension of the specific conductivity mho/m. Since ϵ fulfills in addition all calculation rules which are derived for calculating capacities in electrical circuits, we can therefore interpret ϵ as capacity in our circuit diagram. Besides the presently used symbols we introduce for the current through the capacity ϵ the symbol i_ϵ. The current i_ϵ is identical to the Maxwell dielectric current or influence current density. In the manner we derived (4) we can now in a similar manner derive from the circuit diagram the following equation (Kirchhoffs law)

$$(i + i_\epsilon) \, F = I + I_C \dots \dots \dots \dots \dots \dots \dots \dots \dots \quad (14)$$

Now we have to eliminate from (14) the expression i_ϵ and I_C. Since the potential drop across r and ϵ must be equal, we obtain on the basis of the Ohm law $ir = (1/j\omega\epsilon) \, i_\epsilon$. In the same manner we obtain for the input circuit $IR = (1/j\omega C) \, I_C$. And from this get

$$i_\epsilon = ij\omega r\epsilon \quad \text{and} \quad I_C = Ij\omega RC \dots \dots \dots \dots \dots \dots \quad (15)$$

Taking into consideration that $r\epsilon = \epsilon/\lambda = \theta$ and $RC = T$, so we derive from (14) and (15)

$$iF \, (1 + j\omega\theta) = I \, (1 + j\omega T)$$

or also

$$I = (1 + j\omega\theta)/(1 + j\omega T) \; i \; F \ldots\ldots\ldots\ldots\ldots\ldots\ldots (13)$$

We thus obtain the same result as we have derived above by the integration of Maxwell's equation.

This method of matching the input of our measuring instruments with the atmospheric-electric circuit by means of shunting an accurate capacity is simple to apply from a technical viewpoint. As mentioned before, this method has proven its value by the registration of air-earth current at the Meteorological Observatory in Aachen. This method has worked excellently in registering the air-earth current during lightning discharges, thunder clouds and fair weather. I hope that this method will help other atmospheric electricity research stations to overcome the difficulties encountered in registering the air-earth current.

ANTENNA PROBLEMS OF MEASUREMENT OF THE AIR–EARTH CURRENT

HEINZ W. KASEMIR and LOTHAR H. RUHNKE
Meteorological Division, U.S. Army Signal Engineering Laboratories,
Fort Monmouth, New Jersey

1. INTRODUCTION

THE theory of atmospheric electricity evolved from the three dimensional potential theory. A charge distribution was assumed and the goal was to determine the potential function of the field distribution in space and on the ground by certain boundary conditions such as the conducting earth or the conducting ionosphere. The measuring technique was concerned mostly with field or charge measurements. With the detection of the conductivity of the air, two more elements were introduced, namely the conductivity of the air and the conduction current or air–earth current. The electrostatic theory gave way to an electrodynamic theory. The conception of the current source or generator was introduced and the field pattern was supplemented by the lines of current flow. Although, in the field of electrical engineering, we have developed the concept of the generator as an active two-terminal with internal resistance, open-circuit voltage, and short-circuit current, this concept has not been worked out in the three dimensional electrodynamic theory of atmospheric electricity. In the measuring technique the three dimensional current flow from the air is linked by our antenna with the one dimensional current flow in our measuring apparatus. These facts led to an investigation of the possibility of translating the fully developed conception of the one dimensional generator to the three dimensional generator. This paper will show how advantageous the translation was in solving problems of the air–earth current measurement.

2. THE ANTENNA FOR FIELD OR CURRENT MEASUREMENTS AS AN ACTIVE TWO-TERMINAL

To get any information of the atmospheric electric field or the electric current in the atmosphere, an antenna is used that picks up electric energy (Fig. 1a). The antenna can be understood as a generator or an active two-terminal. It is well known, as long as Ohm's law holds, that an active two-terminal is completely described by 3 characteristic elements, namely the open-circuit voltage ϕ, the short-circuit current I, and the internal resistance r, which may be complex*. The instrument is represented by the load resistor R, which may also be a complex quantity.

For an active two-terminal, two equivalent circuit diagrams are used. The first one interprets the generator as a voltage source with the internal resistor r in series (Fig. 1b), and the second one as a current source with the internal resistor r in parallel (Fig. 1c). Each of these circuit diagrams completely describes the generator so that, in principle, only one is necessary. But we can use either one of them to represent best the physical nature of the generator.

---

* Complex quantities are indicated by use of a bold italic character, for instance r.

HEINZ W. KASEMIR AND LOTHAR H. RUHNKE

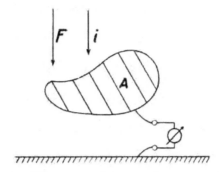

Fig. 1a—The antenna A as an active two-terminal.

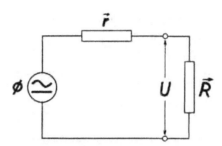

Fig. 1b—The active two-terminal as a voltage source.

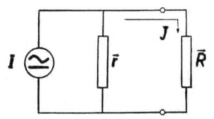

Fig. 1c—The active two-terminal as a current source.

If, the instrument measures the voltage drop U across the resistor R, we see from the circuit diagram Fig. 1b that

$$U = \frac{R}{r+R}\phi = \frac{R/r}{1+R/r}\phi. \tag{1}$$

If, on the other hand, the instrument measures the current J through the resistor R, it follows from the circuit diagram 1c that

$$\frac{J}{R} = \frac{I-J}{r} \text{ or } J = \frac{1}{1+R/r}I. \tag{2}$$

The fraction R/r is in general a complex quantity. But in case the complex resistors R and r have the same phase angle, the fraction becomes a real number. This case is

Antenna Problems of Measurement of the Air–Earth Current

defined as phase-matching and we will call the fraction R/r the matching factor a. From (1) and (2) follows then

$$U = \frac{a}{1+a}\phi, \text{ and} \tag{3}$$

$$J = \frac{1}{1+a}I. \tag{4}$$

In electrical engineering, matching usually implies that not only the phase angle but also the absolute values of R and r are the same. In this case the matching factor would be 1. In our application of the circuit diagram to atmospheric electric problems only the phase-matching is of importance. If a is a real number, we can see from Eqs. (3) and (4), that U and ϕ as well as J and I are in phase. Since this is valid for all frequencies, it holds also for any arbitrary time function.

Applied to the measurement of the air–earth current, it means that the measuring meter follows the variation of the air–earth current without delay. The matching condition for a plate antenna was discussed at the last Wentworth Conference[1]. There it was shown that the reactance between the plate and the air was given by an ohmic resistor and a capacitor in parallel. We will see here that this is true for any kind of antenna form and that the phase-matching condition time constant $T = RC$ of the apparatus impedance equal to the time constant of the air $\theta = \varepsilon/\lambda$ is a general condition for all antenna forms. Furthermore we will show how the internal resistor r can be measured or calculated, and under what conditions a field or an air–earth current measurement is achieved.

3. THE DERIVATION OF THE TWO-TERMINAL CHARACTERISTICS FROM ANTENNA DATA

(a) General phase-matching condition

Every antenna exposed to the atmospheric electric field receives not only the conduction-current density but also the displacement-current density. Let i be the conduction current density and $\partial \varepsilon F/\partial t = \varepsilon \partial i/\lambda \partial t$ the displacement current density. An integration of these current densities over the antenna surface S gives the total current I, which the antenna picks up from the air.

$$I = \int_S \left(i_0 + \frac{\varepsilon \partial i_0}{\lambda \partial t}\right) \, dS. \tag{5}$$

The subscript $_0$ indicates the value of i on the surface. We may write i_0 in the form

$$i_0 = i_{(S)} g_{(t)}, \tag{6}$$

where $i_{(S)}$ is a function of the surface coordinates only and $g_{(t)}$ is a pure time-function. Furthermore in (5) since the integration is taken over the surface, the time function can be written before the integral,

$$I = \left(g + \frac{\varepsilon}{\lambda}\frac{\partial g}{\delta t}\right)\int_S i_{(S)} \, dS. \tag{7}$$

The total current I will flow from the antenna through the resistor R and the capacity C to the ground. If I_R is the current through the resistor R and I_C the current through capacitor C we have

$$I = I_R + I_C. \tag{8}$$

Between the voltage drop U across R the charge Q of the capacitor C and the current I_C and I_R, the following relations hold.

$$I_C = dQ/dt = C\,dU/dt = CR\,dI_R/dt. \tag{9}$$

From (8) and (9) follows

$$I = I_R + CR\,dI_R/dt. \tag{10}$$

The product CR is the time constant T of the input of our measuring apparatus and the term ε/λ the time constant θ of the air. If we substitute these expressions and combine the Eqs. (7) and (10) we arrive at the differential equation for I_R.

$$\frac{dI_R}{dt} + \frac{I_R}{T} - \left(g + \theta\frac{dg}{dt}\right)\frac{\int i_{(S)}\,dS}{T} = 0. \tag{11}$$

The equation can be solved without difficulties. The solution is

$$I_R = I_a e^{-t/T} + \frac{e^{-t/T}}{T}\int_S i_{(S)}\,dS\int_0^t \left(g + \theta\frac{dg}{dt}\right)e^{t/T}\,dt. \tag{12}$$

The term $I_a e^{-t/T}$ describes the decay of a charge Q in case the capacitor C is charged at the time $t = 0$ by any other source than the conduction current. Here we consider the conduction current only, and set $I_a = 0$. Furthermore we simplify in (12) the time integral by partial integration and obtain the formula

$$I_R = \int_S i_{(S)}\,dS\left[\frac{\theta}{T}g - \frac{\theta}{T}g_0 e^{-t/T} + \left(1 - \frac{\theta}{T}\right)\frac{e^{-t/T}}{T}\int_0^t g e^{t/T}\,dt\right]. \tag{13}$$

g_0 is the value of g for $t = 0$ and the term $(\theta/T)g_0 e^{-t/T}$ is a transient, which will decay with the time constant T. From the remaining two terms in the brackets, the last one will be zero, if $\theta = T$, and we obtain the simple formula

$$I_R = g\int_S i_{(S)}\,dS. \tag{14}$$

Here the current I_R through our measuring resistor R is proportional to the time function g of the air–earth current. $\int_S i_{(S)}\,dS$ is a scale factor, which will be discussed later. If θ is not equal to T, the last term in Eq. (13) is not zero, and I_R is not proportional to g. We call $\theta = T$ the phase-matching condition. It means that I_R is in phase with the air–earth current density for any given time function g. The matching condition depends only on the time constant θ of the air and the time constant T of the apparatus

Antenna Problems of Measurement of the Air–Earth Current

impedance. It is independent of the geometrical form of the antenna which is implicated only in the surface integral. It is the same for the plate, the sphere, the vertical wire, the horizontal wire of any length and height or any other antenna form. There is only one limitation. The conductivity λ of the air should remain constant with time and be constant in the neighborhood of the antenna. The second condition is in most practical cases fulfilled, but the first one is not. The conductivity of the air varies in time with the change of locale and weather conditions. It is increased after rain, decreased in fog and clouds or by dust or heavy pollution in the air. So in practice the time constant of the recording apparatus is set on a fixed value matching the time constant of the air for an average conductivity value. By increasing conductivity a kind of damping effect will be noticed in the records for fast variations of the air–earth current. By decreasing conductivity an exaggeration of faster variations will occur. For diurnal variations of the air–earth current, these effects do not introduce a perceptible error, but for shorter periods in the range of minutes the deviation may be serious.

Usually the conductivity of the air is calculated from the sum of positive and negative ions. But for charging or discharging the antenna only one class of ions is effective according to the sign of the field. Therefore by calculating the time constant of the air from the dielectric constant and the conductivity by a positive field, only the positive conductivity should be taken into account, and when considering the conductivity by a negative field, only the negative conductivity should be taken into account. So the average time constant of the air of about 10 min will be doubled to 20 min. The same reasoning leads to the fact that the antenna picks up only the current supported by the ions of one sign. But all the problems concerning the electrode effect or measuring the full current with a plate near the ground will not be discussed in this paper.

Since phase-matching can be achieved by a load resistor, which consists of an ohmic resistor R and a capacitor C in parallel, we conclude from the circuit diagram (Fig. 1b), that the internal resistor r of the generator is constructed in a similar way, namely of an ohmic resistor r and a capacitor c in parallel (Fig. 2). Hereby r is the resistance between the antenna and the surrounding air and c is the capacity of the antenna. Since the relation holds $rc = \varepsilon/\lambda = \theta$, i.e.

$$r = \frac{\theta}{c} \tag{15}$$

we can determine the internal resistor r by measuring the capacity of the antenna and the time constant θ of the air. This very useful relation will be discussed later in more detail.

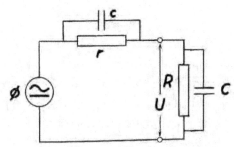

Fig. 2—The circuit diagram of the antenna for phase-matching.

HEINZ W. KASEMIR AND LOTHAR H. RUHNKE

(b) The open-circuit voltage as a product of the electric field and the effective height of the antenna

To determine the open-circuit voltage ϕ we insulate the antenna from the ground and measure the antenna voltage against the ground. The antenna will obtain this voltage by exposure to the atmospheric electric field F. ϕ will be proportional to the field strength F and a factor h, which has a dimension of a length.

$$\phi = Fh. \tag{16}$$

For an antenna of symmetrical form far enough from the ground so that the mirror effect can be neglected, h is the height measured from the ground to the midpoint of the antenna (Fig. 3a). We will call h the effective height of the antenna according to the same expression in the wave propagation theory. We do not get a similar simple situation in the usual arrangement, a plate leveled off with the ground, especially if we consider the pit in the ground necessary for mounting the insulators (Fig. 3b).

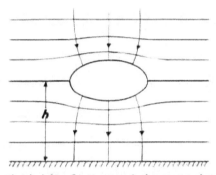

Fig. 3a—Effective height of a symmetrical antenna above flat ground.

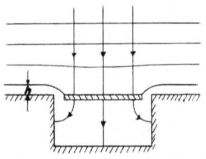

Fig. 3b—Effective height of a plate antenna leveled off with the ground.

Insulated from the ground, the plate will be part of an equipotential surface very near the ground indicating that the open circuit voltage is low. We will see later that this is not advantageous.

(c) The short-circuit current as a product of the air–earth current density and the effective area of the antenna

If the antenna is shortened to the ground it will pick up the greatest amount of current I from the air. We call this the short-circuit current. I is, of course, proportional to the air–earth current density i and to a factor of the dimension of an area. This area depends on the geometric form of the antenna only. We will call this factor the effective area M. The name shall point to the fact that the antenna draws a current from the air which would otherwise flow in an area of M m² of the flat ground.

$$I = iM. \tag{17}$$

This effective area is for the air–earth current measurement, the analog as the effective height is for the field measurement. We will see in the next chapter how it can be determined from the internal resistor or the capacity of the antenna.

(d) The internal resistance r

As can easily be seen from the circuit diagram Fig. 1b, the internal resistance r can be determined from the open circuit voltage $\phi = Fh$ and the short circuit current $I = iM$. Since r is the ohmic part of the complex resistor r, it is understood that I is the short-circuit conduction current, which flows through r, if the matching condition is fulfilled. For an even better understanding we may state, for the following, that the generator is a direct-current generator, and substitute the complex resistors by their real part. If we now set $R = 0$, it follows from Ohm's law, that

$$r = \frac{\phi}{I} = \frac{Fh}{iM}. \tag{18}$$

Since $i = \lambda F$, we obtain from (18)

$$r = \frac{h}{\lambda M}. \tag{19}$$

We see that r is determined by the two characteristic geometrical factors of the antenna, namely, the effective height and the effective area and the conductivity of the air. According to the relation $rc = \varepsilon/\lambda$, we obtain from (19)

$$M = \frac{hc}{\epsilon}. \tag{20}$$

Since no restrictions on the antenna forms were made by deriving Eq. (20) it is possible to calculate M for any antenna form, if the height h can be determined. As mentioned before for antennas of symmetrical forms the height h is the distance between ground and the midpoint of the antenna. For asymmetric antennas we have to measure the open circuit voltage ϕ and the atmospheric electric field F separately and calculate the effective height h by the equation

$$h = \frac{\phi}{F}. \tag{21}$$

The capacity of commonly used antennas range from 50 $\mu\mu$F to several hundred $\mu\mu$F and can easily be measured with any commercial capacity meter. Also the capacity of a great variety of antenna forms has been calculated from the geometrical data.

Heinz W. Kasemir and Lothar H. Ruhnke

(e) The load resistor R

The discussion of the influence of the load resistor becomes easy and clear by the use of the circuit diagram 1b. We may differentiate 3 cases:

1. R comparable with r.
2. $r \ll R$.
3. $R \ll r$.

In case 1 we read from the circuit diagram considering that

$$r = \frac{\epsilon}{\lambda c} \text{ and } \phi = Fh = ih/\lambda$$

$$U = \frac{R}{r+R}\phi = \frac{Rh}{\epsilon/\lambda c + R}F = \frac{Rh}{\lambda(\epsilon/\lambda c + R)}i. \tag{22}$$

The voltage U across the input terminals of the instrument depends, besides the atmospheric electric field or the air–earth current density, on the conductivity of the air λ. Therefore this condition is not useful for a pure field or a pure air–earth current measurement.

In case 2 with $r \ll R$, we obtain

$$U = Fh \text{ or } I = U/R = Fh/R,$$

a true field measurement. Since for practical antenna forms r is in the range of 10^{11}–$10^{12}\,\Omega$, the condition $r \ll R$ indicates for R the order of 10^{13}–$10^{14}\,\Omega$. This high impedance can be achieved by an electrostatic voltmeter. But it would be most difficult to maintain the required insulation of 10^{15}–$10^{16}\,\Omega$ for a longer period of time. To avoid this difficulty and to enable the use of an electrometer tube circuit, it is common to lower the internal resistance r to the range of 10^{9}–$10^{10}\,\Omega$ by a radioactive probe. In this case a grid resistor R in the range of 10^{11}–$10^{12}\,\Omega$ is applicable.

In case 3 where $R \ll r$ we obtain a true air–earth current measurement.

$$U = \frac{R}{r}\phi = R\frac{hc}{\epsilon}i, \text{ or } I = U/R = \frac{hc}{\epsilon}i = Mi.$$

The voltage U or the current I across the input terminals of the meter is proportional to the air–earth current density i. Here it is always possible to keep the measuring resistor R well below r. But since U drops down with the ratio R/r, the resistor R should not be smaller than necessary. For $R/r = 1/100$ an open circuit voltage of 50 V is required to shield a control voltage of $U = 0\cdot5$ V across the input terminals. Such an open circuit voltage can be achieved by the use of a wire antenna 1 m above the ground but not so easy by a plate antenna leveled off with the ground. Here the pit for housing the insulators should be at least $\frac{1}{2}$ m deep or the plate should be raised $\frac{1}{2}$ m above the ground.

4. ERRORS INTRODUCED BY ADDITIONAL GENERATORS

An error can be introduced in the measurement by additional generators. Two examples will be given here. The source of error can be conceived as a generator with its

Antenna Problems of Measurement of the Air–Earth Current

own characteristic elements. The antenna generator and the error generator can be connected in two different ways according to the physical nature of the error source. A parallel connection or a connection in series is possible.

(a) The Volta potential as a voltage generator

A Volta potential U_v exists between the antenna and the surrounding air. The internal resistance of this Volta potential generator is the same as it is defined for the antenna generator. Both voltages are connected in series as shown in Fig. 4.

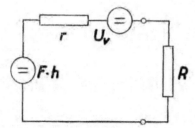

Fig. 4—The Volta potential generator in series with the antenna generator.

The current impressed by the Volta potential flows through the resistors R and r exactly as the antenna current. The ratio of these currents will be the same as the ratio of the generating voltages. That means if the open circuit voltage is large compared to the Volta potential, then the influence of the Volta potential will be negligible. In practice the Volta potential has the magnitude of about 1 V. There is no difficulty to obtain an open circuit voltage of 100 V, if the antenna is 1 m or more above the ground. But the Volta potential effect becomes noticeable as the distance of the antenna from the ground comes in the centimeter range. So by current measurements the plate antenna should have a clearance of about $\frac{1}{2}$ m from the ground. This condition is usually more or less fulfilled by the pit dug into the ground for housing the insulators.

(b) The piezo-electric effect as a current generator

The influence of charges generated by the piezo-electric effect of the cables, which connect the antenna with the input terminals of the apparatus, is far more serious. Here the generator is not connected in series with the antenna generator as in the case of the Volta potential but in parallel. Also it is evident from the physical point of view to consider the piezo-electric generator as a current source. So we obtain the diagram shown in Fig. 5, whereby the internal resistance of the error source is included in r.

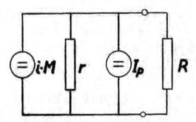

Fig. 5—The piezo-electric effect as a current generator in parallel to the antenna generator.

11

HEINZ W. KASEMIR AND LOTHAR H. RUHNKE

The piezo-electric current I_μ is directly added to the short-circuit current iM picked up by the antenna. Even the change of stress on the insulator of the cable caused by rapid temperature changes of a few degrees will cause a current large enough to compete with the air–earth current density. To keep the influence of this piezo-electric current negligible the effective area of the antenna should be as large as possible and the cables should be buried in the ground, where the temperature changes are slow. Another way to eliminate the cable effect is proposed by ISRAEL and DOLEZALEK[2] by the use of an electrometer amplifier with balanced input. Here the antenna is connected through a cable with the input 1, where an identical cable without antenna is connected to the input 2 of the amplifier. In this way the currents generated by the cables will cancel out.

5. THE CONVECTION CURRENT

The last point we will discuss is the influence of the convection current. It can be conceived as a current generator caused by charges moved by other than electrical forces such as gravity, wind or eddy diffusion. The generator is connected in parallel to our active two-terminal of the air–earth current antenna. Figure 6 shows the equivalent diagram. iM_1 is the conduction current picked up by the antenna, r its internal resistance, i_c is the convection current density, M_2 the effective area of the antenna for i_c and r_c the internal resistance of the convection current generator.

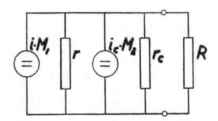

Fig. 6—The convection current generator and the conduction current generator in parallel.

We may define r_c as a resistance caused by uncharged particles moved by other than electric forces. These uncharged particles will pick up a small amount of the charge of the antenna by contact and carry it away. The water-drop collector acts in a similar way. Since for current measurement R has to be small compared to r as well as r_c, we can neglect the effect of r_c. So the influence of the convection current depends on the current density i_c and the effective area M_2.

As already pointed out by one of us[3] the wire antenna is the best antenna form to minimize the influence of the convection current. The effective area M_1 of the air–earth current for a horizontal wire is approximately Lh, where L is the length of the antenna and h the height above ground. The effective area M_2 for the convection current is only the projected area of the wire normal to the direction of the convection current. The largest possible area is Ld, where d is the diameter of the wire. The ratio of these areas is $M_1/M_2 = h/d$. For a wire antenna of the diameter of 1 mm and the height of 1 m, this ratio becomes 1000 : 1. This means the convection current has to be 1000 times larger than the conduction current to have the same effect on the antenna. If we use two wire antennas of the same length but of different heights, these will have the same effective area for the convection current but a different effective area for the conduction

Antenna Problems of Measurement of the Air–Earth Current

current. Feeding both of these antennas in an amplifier with a balanced input, the convection currents will cancel each other, and we will record the difference between the conduction currents $iL(h_1 - h_2)$. But even this difference will be proportional to the air–earth current density. In this way we will get a record of the pure conduction current. The same principle applies to a convection current measurement. We use 2 antennas with different effective areas for the convection current, for example a plate and a wire antenna, but with the same effective area for the conduction current. Here the conduction currents will cancel out and we achieve a pure convection current measurement.

REFERENCES

[1] H. W. KASEMIR. *Measurement of the air–earth current density:* Proceedings of the conference on atmospheric electricity. Geophysical Research Paper No. 42, Air Force Cambridge Research Center, Bedford, Massachusetts, 1955.

[2] H. ISRAËL and H. DOLEZALEK. Zur Methodik luftelektrischer Messungen IV: Störspannungen in luftelektrischen Messfühlern und ihre Verthütung, Gerlands Beiträge zur Geophysik, Leipzig 66, 129, 1957.

[3] H. W. KASEMIR. Zur Strömungstheorie des luftelektrischen Feldes III. Der Austauschgenerator: Arch. Met., Wien A 9, 357, 1956.

DISCUSSION

H. Dolezalek—I feel sorry that there is almost no time to discuss the measuring methods applied in the investigations of the papers we have heard. For example, the old discussion on the advantages and disadvantages of the field mills and the radioactive collector to be found in some European journals some decades ago could be taken up again right now. In this time I only intend to comment on Dr. Kasemir's paper, not by using his electrotechnical expressions of the active two terminal and the inner resistance of the antenna, but by applying a more simple picture. We may discuss whether air–earth current density should be measured at the ground or at the height of some meters, but if we intend to approach the surface conditions the well-known collecting plate should be used. In this case I think it goes almost without saying that the resistance between the plate and the ground, represented by the variable resistivity of the air in between, must be great with respect to the constant input resistance of the amplifier. A number of means can be applied to obtain that condition, but it would take too long to discuss these possibilities here.

15 June 1960

USASRDL Technical Report 2125

A RADIOSONDE FOR MEASURING THE AIR-EARTH CURRENT DENSITY

Heinz Kasemir

DA Task Nr. 3A99-07-001-05

U. S. ARMY SIGNAL RESEARCH AND DEVELOPMENT LABORATORY

FORT MONMOUTH, NEW JERSEY

601

ABSTRACT

An instrument is described which, in connection with an ordinary meteorological radiosonde, measures the air-earth current density as a function of altitude. The instrument is provided with three separate channels of high input impedance which make it possible to measure three atmospheric electric elements during the same flight. A detailed analysis of the measuring technique is given, and the results of 40 test flights are discussed.

CONTENTS

FIGURES

A RADIOSONDE FOR MEASURING THE AIR-EARTH CURRENT DENSITY

INTRODUCTION

The basic elements in atmospheric electricity are the electric field, the conductivity of the air, and the air-earth current density. At the ground, methods of measurement have been developed for each of them, but only field and conductivity measurements have been carried out from an airborne carrier.[1-9] As it becomes more and more obvious that all three elements have to be recorded to check Ohm's law, and in this way obtain complete and reliable information, there is a pressing need for a direct method of air-earth current measurement from an airborne carrier. The instrument described herein is specially designed for use with a radiosonde balloon, but the same principle would also apply to measurement from an airplane.

DISCUSSION

Description of the Apparatus

The current radiosonde consists of the balloon, the antenna system, the electrometer box, and Radiosonde AN/AMT-4A. The signal is received at the ground by a Rawin Set AN/GMD-1A and recorded by Radiosonde Recorder AN/TMQ-5A. As this system includes a direction-finding antenna, the angle of azimuth and elevation is recorded simultaneously during the flight so that the path of the balloon can be calculated from the height and the direction. The height of the balloon can be approximated from the air pressure measured by the radiosonde. In a standard meteorological radiosonde, the pressure-sensitive aneroid capsule also serves as a switch which, in turn, connects the thermistor, the humidity element, and two reference signals to the transmitter of the radiosonde.

To measure the atmospheric electric elements, the thermistor, the humidity element, and the reference signals are replaced by an electrometer tube circuit. The pressure switch is used to operate a set of relays in the input circuit of the electrometer box so that three different channels are connected in turn to the grid of the electrometer tube.

These channels can be used to measure three different atmospheric electric elements. Only current measurements are dealt with in this report; the three available channels are used to measure the current of the upper and lower antenna and to control the zero point of the electrometer tube.

Figure 1 shows the arrangement of the balloon, the antenna system, and the radiosonde. The antenna system consists of two vertical wires (I and II) of about 7 m length, one connected from the balloon to the radiosonde and the other hanging down from the radiosonde. This latter wire is held taut in flight by an inverted parachute at the lower end of the wire. Both antennas are insulated from the balloon and the parachute, respectively, by a 10-m-long nylon cord and are connected to the inputs 1 and 2 of the electrometer box. The impedances of inputs 1 and 2 are equal and consist of a high-ohm resistor with a capacitor in parallel. The capacitor serves to separate the displacement current from the conduction current. This problem, known as the "matching condition,"[10] will be discussed in detail later in this report.

In the normal fair-weather field, positive charge is collected by the upper antenna I and passed through the resistor of input 1. Negative charge is collected by the lower antenna II and passed through the resistor of input 2. These charges neutralize each other at the connection of the two input resistors. (This point is denoted by the ground symbol in Figure 1). The voltage drop across the resistor caused by the flow of positive or negative charge is measured by the radiosonde and is proportional to the positive or negative part of the air-earth current. If for some reason the number or the mobility of positive and negative ions are not equal, then the positive part of the air-earth current i_+ and the negative part i_- are not the

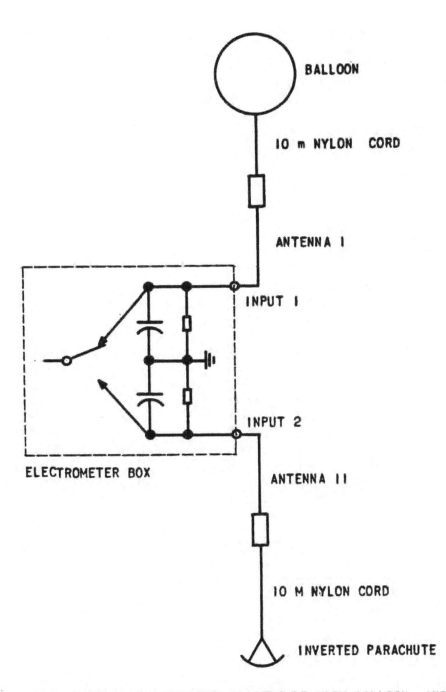

FIG. I   SCHEMATIC OF THE CURRENT RADIOSONDE WITH BALLOON AND ANTENNA SYSTEM

same. The upper and lower antenna would not collect the same amount of positive and negative charges, and the whole antenna system would accumulate charge of one sign. This would cause the rejection of a part of the larger current and the attraction of an additional part of the smaller current until in a steady-state condition both antenna currents are equal and proportional to $\frac{i_+ + i_-}{2}$.

The proportionality factor between the measured current I and the air-earth current density $i = i_+ + i_-$ depends on the geometric form of the antenna system and will be calculated in the next section.

The electrometer circuit (Fig. 2) was designed by Dr. J. Praglin of the Keithly Instrument Company. The first relay 1 in the input circuit connects, alternately, antennas I and II to the grid of the electrometer tube. After five operations of relay 1, the grid is disconnected from the antenna circuits by relay 2 and grounded for the zero check. The resistivity of the transistor is controlled by the electrometer tube EV-5886 and, in turn, the transistor controls the frequency of the blocking oscillator of the radiosonde. Full deflection of the radiosonde recorder is obtained by ±10V at the input of the electrometer box. With a 10^{12}-ohm resistor in the input circuit, this would correspond to a sensitivity of $\pm 10 \cdot 10^{-12}$A full scale.

In the application as an air-earth current instrument, a lower input resistor of 10^{11} ohm or even $5 \cdot 10^{10}$ ohm can be used, as outlined in the next section.

Effective Area of the Antenna System

The measurement of the air-earth current density with the radiosonde brings up four main problems. First, the determination of the model factor of the antenna system — here called the effective area; second, the shunting effect of the high conductivity in higher altitudes; third, the influence of initial charges; and fourth, the matching condition. Each problem is dealt with separately. Here, the effective area is calculated.

The air-earth current forms a homogeneous field of current flow of the current density i. From this the antenna system will receive a certain amount of current I. The quantity of I is proportional to the air-earth current density i. The proportionality factor M between i and I, $I = M \cdot i$, has the dimension of an area, and its value depends on the antenna form only. Therefore M is called the effective area of the antenna,[11] and it is calculated by representing the upper and lower antenna with the upper and lower half of a prolonged spheroid. The problem may then be formulated as follows: Given is a homogeneous field of current flow of the current density i. Inserted herein is a prolonged conducting spheroid with the long axis, called "a," parallel to the lines of current flow. Both small axes are equal and called "b." The eccentricity of the spheroid is c. The current flows into the upper half, and the lower half of the spheroid will be calculated.

The calculation is best worked out in elliptical coordinates u, v, and ϕ, but, because of their unfamiliarity the end results are converted to cylindrical coordinates z, r, and ϕ. The relationship between these two coordinate systems is given by the equations

$$\frac{z^2}{u^2} + \frac{r^2}{u^2 - c^2} = 1$$

$$\phi = \phi \tag{1}$$

$$\frac{z^2}{v^2} - \frac{r^2}{c^2 - v^2} = 1$$

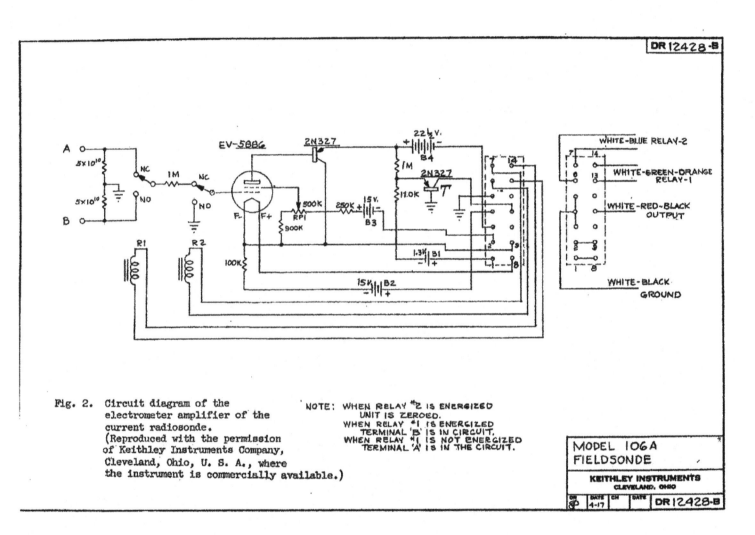

Fig. 2. Circuit diagram of the
electrometer amplifier of the
current radiosonde.
(Reproduced with the permission
of Keithley Instruments Company,
Cleveland, Ohio, U. S. A., where
the instrument is commercially available.)

NOTE: WHEN RELAY #2 IS ENERGIZED
UNIT IS ZEROED.
WHEN RELAY #1 IS ENERGIZED
TERMINAL 'B' IS IN CIRCUIT.
WHEN RELAY #1 IS NOT ENERGIZED
TERMINAL 'A' IS IN THE CIRCUIT.

The spheroid is given by setting u = a.

The potential function ϕ of a prolate spheroid inserted in a homogeneous field F is given by the equation

$$\phi = F u\, P_1 \left(1 - \frac{a}{u} \frac{Q_1}{Q_{1a}}\right) \tag{2}$$

Differentiation of (2) with respect to u and then setting u = a gives the field strength E at the surface of the spheroid.

$$E = -\frac{F P_1}{U_a Q_{1a}} \left(Q_{1a} - \frac{a}{c} Q_{1a}\right) = -\frac{F P_1}{U_a Q_{1a}} \frac{c^2}{a^2 - c^2} \tag{3}$$

According to Ohm's law, the relationship holds.

$$\lambda E = j \text{ and } \lambda F = i. \tag{4}$$

If j and i are substituted for E and F in equation (3), the current-density j at the surface of the spheroid is obtained.

$$j = -i \frac{P_1}{U_a Q_{1a}} \frac{c^2}{a^2 - c^2} \cdot \tag{5}$$

The integration with respect to the angle ϕ from 0 to 2π and along the long axis from v_1 to v_2 gives the current j_{12} flowing into a ring zone of the spheroid. The integral

$$j_{12} = \int_{v_1}^{v_2} j \, dS \text{ can be solved, and}$$

$$j_{12} = -i \frac{\pi}{Q_{1a}} (v_2 + v_1)(v_2 - v_1) \text{ is obtained.} \tag{6}$$

Now substitute the cylindrical for the elliptical coordinates.
From equation (1), with u = a

$$z = \frac{av}{c} \text{ or } v = \frac{c}{a} z \text{ is obtained.} \tag{7}$$

By inserting (7) into (6),

$$j_{12} = -i \frac{\pi}{Q_{1a}} \frac{c^2}{a^2} (z_2 + z_1)(z_2 - z_1) \tag{8}$$

To obtain the current I_u flowing into the upper half of the spheroid, set $z_2 = a$ and $z_1 = 0$.

$$I_u = -\frac{\pi c^2}{Q_{1a}} i \tag{9}$$

608

The current I_o flowing into the lower half is given by equation (8) for $z_2 = 0$ and $z_1 = -a$.

$$I_o = \frac{\pi c^2}{Q_{1a}} i \tag{10}$$

These currents I_u and I_o are of the same amount but of opposite polarity as it must be in a steady-state condition. They will cancel each other inside the spheroid.

The factor

$$M = \frac{\pi c^2}{Q_{1a}} \tag{11}$$

in equations (8) and (10) is the effective area of the antenna system to be calculated. As the wire antenna is long compared to its radius, the eccentricity c can be replaced by the axis a without a perceptible error. With $Q_{1a} = \frac{a}{2c} \ln \frac{a+c}{a-c} - 1 = \frac{a}{c} \ln \frac{a+c}{b} - 1 = \ln \frac{2a}{b} - 1$, the simple formula for the effective area is obtained.

$$M = \frac{\pi a^2}{\ln \frac{2a}{b} - 1} \tag{12}$$

The denominator of (12) changes very little for a wide variation of the length and the thickness of the antenna wire. The factor $\pi / \ln \frac{2a}{b} - 1$ has the approximate value of $\frac{1}{3}$, so the approximate formula $M = \frac{a^2}{3}$. $\tag{13}$

may be used.
In the range of M from 10 to 150 m^2 or a from 1.6 to 20 m, the error of (13) changes from -3 percent to +5 percent, being zero for M = 50 m^2. Figure 3 shows the effective area M as a function of the length "a" of the antenna. The diameter of the antenna wire is assumed to be 1.5 mm. The solid curve is calculated by the exact formula (12), while the dashed line shows the approximation according to formula (13). The length of the antenna can be chosen so that a current of the right range is delivered to the electrometer box, and an even calibration factor results. For instance, an antenna length of 5.21, 7.55, 12.22, or 17.60 m corresponds to an effective area of 10, 20, 50, or 100 m^2, respectively. For an average air-earth current density of $1 \cdot 10^{-12} \frac{A}{m^2}$, these antennas will deliver a current of 10^{-11}, $2 \cdot 10^{-11}$, $5 \cdot 10^{-11}$, or 10^{-10}A to the electrometer. For the instrument described here, an antenna length of 7.55 m has been used for the test flights in Greenland where the average air-earth current density is about $3 \cdot 10^{-12} \frac{A}{m^2}$. An antenna length of 12.22 m is best suited for the test flights in New Jersey, where the air-earth current density is lower.

Since the air-earth current density remains in the same range at all altitudes, it is not necessary to incorporate a range-switching device which is desirable for the field — or conductivity-radiosondes. For altitudes higher than 30 km, the resistance of the air between the two antennas becomes lower and lower and drops down to the value of the input resistor of the electrometer box, and below. In this case the air-earth current measurement changes to a field measurement. This problem will be discussed in the next section.

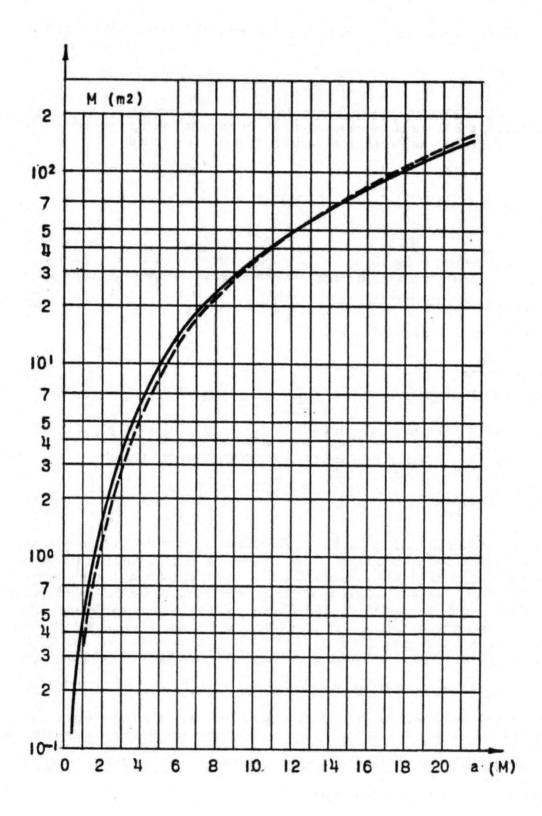

FIG. 3   THE EFFECTIVE AREA OF THE ANTENNA SYSTEM AS
A FUNCTION OF THE ANTENNA LENGTH

Internal Resistance of the Antenna System

As discussed in detail in reference 11, each antenna for measuring the air-earth current density can be represented as an active two-terminal determined by the open circuit voltage ϕ, the short-circuit current I and the internal resistance r. The influence of the internal resistance r on the measurement of the air-earth current density can be very easily recognized by a substitute circuit diagram, Fig. 4. Seen from the terminals 1 and 2 of the electrometer box, the antenna system imbedded in the field-of-current flow of the air-earth current can be represented by a current source I with the internal resistor r. Connected to this current source at the terminals 1 and 2 is the electrometer box with its input resistor R. In this case, R consists of the two $5 \cdot 10^{10}$-ohm resistors in series (Fig. 2). The current flowing through r shall be I_r and that flowing through the input resistor R of the instrument shall be I_R. It can be seen immediately from Fig. 4 that the two following relations hold.

$$I = I_r + I_R \tag{14}$$

$$r\,I_r = R\,I_R \tag{15}$$

I_r is eliminated from equations (14) and (15) and

$$I = \left(1 + \frac{R}{r} \right) I_R \tag{16}$$

is obtained.

If R is very small compared to r, the fraction R/r can be neglected.

$$I = I_R$$

This is the maximum current the antenna system is able to draw from the surrounding air. It is calculated in the preceding section that

$$I = M \cdot i \tag{17}$$

with the effective area M given by (12) or approximately by (13). In technical terminology, this current is called the short circuit current of the generator. If on the other hand R is very large compared to r, the whole current would be forced through the resistor r, and the open circuit voltage ϕ of the generator would be measured at the terminals 1 and 2.

$$\phi = I \cdot r = M \cdot i \cdot r \tag{18}$$

In this case the open circuit voltage would be the potential difference between the two antennas exposed to the atmospheric electric field F, but insolated from each other $(R \to \infty)$. If their mutual capacity is neglected, each antenna would assume the potential value of the undistorted field at their midpoints. As the distance of these midpoints is a, the potential difference, i.e., the open-circuit voltage, is given by the equation

$$\phi = F \cdot a = \frac{i}{\lambda} a \tag{19}$$

If equations (18) and (19) are combined and the value for M from equations (12) or (13) is introduced, the internal resistance r can be determined.

$$r = \frac{a}{\lambda M} = \frac{\ln \dfrac{2a}{b} - 1}{\lambda \pi a} \tag{20}$$

or approximately

$$r = \frac{3}{\lambda a} \tag{21}$$

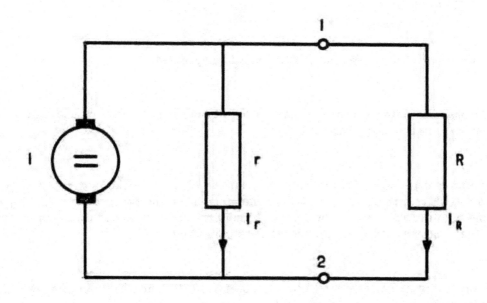

FIG. 4 SUBSTITUTE CIRCUIT DIAGRAM FOR AIR-EARTH CURRENT MEASUREMENT

The internal resistance, proportional to $1/\lambda$, decreases with an increase of λ. For a rough estimate it may be assumed that λ increases from its ground value $\lambda_0 = 2 \cdot 10^{-14} \dfrac{1}{ohm \cdot m}$ by a factor 10 for each 10 km altitude. With $a = 10$ m, obtained from equation (21), the values of r at 0, 10, 20, or 30 km altitude are $1 \cdot 5 \cdot 10^{13}$, $1 \cdot 5 \cdot 10^{12}$, $1 \cdot 5 \cdot 10^{11}$, or $1 \cdot 5 \cdot 10^{10}$ ohm, respectively. Compared with the input resistor $R = 10^{11} \Omega$, it is seen that up to 10-km altitude the condition for a true current measurement, R small compared to r, is fulfilled. At 20-km altitude, both resistors are of the same order; while at 30-km altitude, r is small compared to R. In the last case no longer would air-earth current density be measured; the atmospheric electric field would be measured instead. If the conductivity as a function of altitude were known, the recorded current I_R for higher altitudes could be corrected according to equation (16) to give the true air-earth density.

On the other hand, if the air-earth current density were constant in higher altitudes, as predicted by theory, the conductivity could be determined in addition to the air-earth current density. The data available at the present time show a constant current density for an average of several soundings. However, the individual flights may show a decrease, increase, or constancy with altitude. This makes it desirable to have a conductivity measurement made simultaneously with the current measurement.

There may also be another reason for the variation of the current density. One sounding will last for one and one-half hours and the current density may change during this time, feigning a variation with altitude.

One last point is made about the ratio R/r as a function of the antenna length a. If $U = I_R \cdot R$ is the input voltage necessary for the desired deflection at the recorder for the average air-earth current density i, from the relationship $I_R = M \cdot i$ is obtained,

$$R = \frac{U}{M \cdot i} \tag{22}$$

or with equation (13),

$$R = \frac{3U}{ia^2} \tag{23}$$

The internal resistor r is given by equation (21) $r = \dfrac{3}{\lambda a}$, and for the ratio R/r, equation

$$\frac{R}{r} = \frac{3U}{i} \frac{\lambda}{} \frac{1}{a} \tag{24}$$

is obtained. It can be seen that the fraction R/r is proportional to $1/a$. Therefore the condition $R/r \ll 1$ is better fulfilled for longer antennas than for shorter ones.

Influence of the Initial Charges at the Start

When the balloon and upper and lower antennas are released at the start of a flight, they are grounded either by direct contact with the ground or through the body of the operator. The influence charge of the antenna system connected to the ground is different from that in free flight. The assimilation of the charge distribution at the start to that at free flight may generate a transient of the recorded current density. It is the purpose of this section to determine the transient caused by these initial charges.

For the calculation, start from the equation (8) of the current j_{12} flowing into a ring zone of the spheroid of the length $z_2 - z_1$. If (8) is divided by $z_2 - z_1$, and this length kept a constant unit, a formula is obtained for the current per unit length j.

$$\underline{j} = -i \frac{\pi}{Q_{1a}} \frac{c^2}{a^2} (z_2 + z_1) \tag{22}$$

It is convenient to introduce a new variable, $z = \dfrac{z_2 + z_1}{2}$, which is the distance of the midpoint of the unit length from the midpoint of the spheroid $z = 0$. For a thin, long spheroid, $c = a$ may be assumed, with a negligible error and, as a good approximation, set $\dfrac{\pi}{Q_{1a}} = \dfrac{1}{3}$ as was done in the section, "Effective Area of the Antenna System."

The simple equation is then derived:

$$\underline{j} = -i \frac{2}{3} z. \tag{23}$$

The current flowing into the unit length of the spheroid increases linearly with the distance from the midpoint, but is independent of the cross section of the unit length. This means that the representation of the antenna system by a prolonged spheroid is valid even though the spheroid tapers off at the ends, whereas the wire antenna has a constant cross section over the whole length.

From the equation

$$E = \frac{q}{\epsilon} = \frac{j}{\lambda}$$

the relationship

$$q = \frac{\epsilon}{\lambda} j$$

is obtained between the charge and current density at the surface. An integration over a certain part of the surface would not alter the proportionality factor ϵ/λ. Therefore the following equations are also true:

$$\underline{q} = \frac{\epsilon}{\lambda} \underline{j}; \quad Q_u = \frac{\epsilon}{\lambda} I_u; \quad Q_o = \frac{\epsilon}{\lambda} I_o. \tag{25}$$

Substituting (23) in the first of equation of (25),

$$\underline{q} = -i \frac{2}{3} \frac{\epsilon}{\lambda} z \tag{26}$$

is obtained, which gives the distribution of the charge per unit length on the antenna system. Figures 5a, b, and c show different cases of charge distribution as they may occur during the radiosonde flight. The line from $-a$ to $+a$ of the vertical z axis represents the antenna system. At the midpoint of this line the input resistor R and the capacity C are indicated by their symbols. The horizontal axis gives the values of the charge q per unit length or, with a different scale factor, the values of the current j per unit length. Figure 5a represents the state during flight after the initial charges have disappeared. The area of the shaded triangle shows the net charges or the net currents of the upper and lower antenna. For the unit current density, the shaded triangle also represents the effective area of the antenna.

Figure 5b shows the charge distribution at the start when the full length of the antenna is played out, but the end of the lower antenna is still connected with the ground. To construct

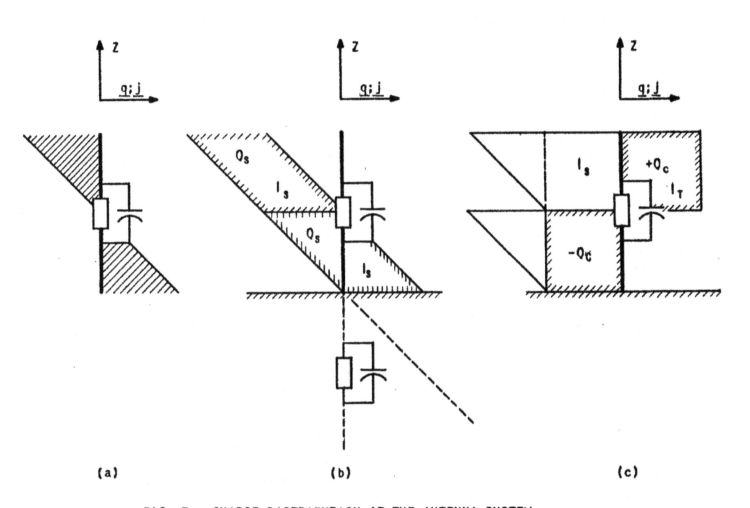

FIG. 5   CHARGE DISTRIBUTION AT THE ANTENNA SYSTEM
(a) DURING FREE FLIGHT

this picture, consider the mirror effect of the ground and supplement the antenna system by its image. In Fig. 5b the image is indicated by a vertical dashed line and the ground by the horizontal slanted line. Compared with Fig. 5a, the effective area of the upper antenna is three times as much as it is in free flight. This means that the current radiosonde connected to the ground measures three times the current measured in free flight. The two-fold surplus current will be called I_S.

As soon as the earth connection is broken and the balloon starts rising, the influence of the ground will fade away and the charge distribution of Fig. 5b will change to that of Fig. 5a. It can be seen from these figures that at the start the upper and the lower antennas have the same surplus of negative charge (indicated by the parallelogram with the shaded boundary in Fig. 5b) as compared to the charge in free flight. From this fact the conclusion can be drawn that the initial surplus charge is equally distributed over both antennas and that no charge will pass through the measuring resistor R as the antennas are discharged by the conduction current of the air. Applied to the radiosonde record, this would mean that immediately after the start of the flight the recorded current I_R would drop from its three-fold ground value to its true value in free flight and would not be affected by the decay of the initial charges. Figure 6a shows a trace of I_R versus time as it may be recorded at the start of the balloon flight.

The situation is different if the balloon and antenna are played out and released immediately. The input resistor R prevents the flow of the full amount of negative influence charge to the upper antenna. Then the missing part of the negative influence charge is taken out of the capacitor C, which remains positively charged. The rectangular boxes with the shaded boundary in Fig. 5c show the maximum amount of charge which may be stored in the capacitor. The capacitor will discharge through the resistor R, causing a transient current I_t of the same amount as the surplus current I_S, but opposite in sign. The transient current will decay with the time constant $R \cdot C$.

If the balloon with the antenna system is played out quickly, but kept anchored with the end of the lower antenna grounded until the transient current is dissipated, a trace at the recorder would be obtained, as shown in Fig. 6b. But if the balloon and antenna are released as soon as the antenna system is played out, a trace results, as shown in Fig. 6c. Here, the recorder would make a swing to the negative side and then return with the time constant $C \cdot R$ to its true positive value of the air-earth current in free flight.

It should be mentioned that the amplitude of the negative deflection cannot be greater than the amplitude of the positive air-earth current. If the negative deflection is greater, it is caused by mismatching, which will be discussed in the next section of this report. A trace as shown in Fig. 6d will result if there is a certain time lapse between the release and the start of the radiosonde, but not enough for the transient current to die out completely.

The best results would be obtained if the balloon is kept captive, with the lower antenna grounded until the transient current has disappeared. But in practice, such a procedure is not always possible. During windy weather the captive balloon will bounce up and down so that, beside the undesirable jerks on the antenna insulators, there is a good chance that the radiosonde will be smashed against the ground. So it is necessary to disregard the first part of the record until the transient current has decayed, or to measure the air-earth current density separately with a ground instrument and determine from this measurement the amplitude of the transient current.

Matching Problem

It is pointed out in reference 10 and 11 that each antenna system for measuring the air-earth current density will pick up both the conduction and the displacement current. Since the air-earth current is defined as the conduction current, the problem of how to separate and measure only the conduction current arises. If the conductivity of the air is constant or its variations are small and

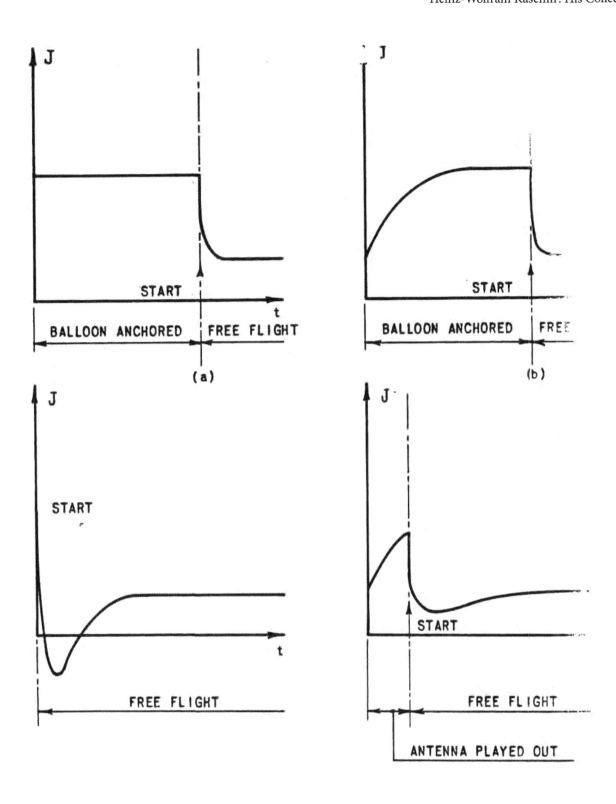

FIG. 6   RECORD OF THE CURRENT DURING THE START:
   (a) BALLOON HELD   ANCHORED ... GROUND BEFORE THE FLIGHT
   (b) ANTENNA PLAYED OUT AND BALLOON KEPT ANCHORED FOR A TI
       LONGER THAN THE TIME-CONSTANT OF THE INSTRUMENT.
   (c) ANTENNA PLAYED OUT AND BALLOON RELEASED IMMEDIATELY.
   (d) ANTENNA PLAYED OUT AND BALLOON KEPT ANCHORED FOR A
       SHORTER THAN THE TIME-CONSTANT OF THE INSTRUMENT

slow, this can be done by matching the time constant $T = R \cdot C$ of the measuring instrument to the time constant $\theta = \epsilon/\lambda$ of the surrounding air. Then the displacement current would flow through the capacitor C and only the conduction current will flow through the measuring resistor R. [10] If the conductivity λ and therefore also θ change by an appreciable amount, either T must be adjusted continuously to θ or, for a fixed T, serious mismatching must be expected. It is the purpose of this section to discuss the effect of mismatching.

If M is the effective area of the antenna, D the dielectric displacement, and i the conduction current density, then the complete current c flowing into the antenna is given by the equation

$$c = M\left(\frac{dD}{dt} + i\right) \tag{27}$$

Using the basic relationship $D = \epsilon E$ and $E = \frac{i}{\lambda}$ gives

$$D = \frac{\epsilon}{\lambda} i = \theta \cdot i \tag{28}$$

where $\theta = \frac{\epsilon}{\lambda}$ is the time constant of the air. λ and therefore θ are here functions of time. In this respect the treatment of the matching problem in this report is more general than that in previous publications. [10] [11]

Substituting equation (28) in (27) gives

$$c = M\left(\frac{d\theta i}{dt} + i\right) . \tag{29}$$

The complete current c is forced through the input impedance of the electrometer amplifier, which consists of a resistor R and a capacitor c in parallel. If Q is the charge of the capacitor, then $\frac{dQ}{dt} = I_c$ is the current flowing through the capacitor. I is the current flowing through the resistor R. The sum of I_c and I is again the complete current c, so

$$c = I_c + I = \frac{dQ}{dt} + I. \tag{30}$$

Analogous to equation (28), the following relations hold between Q, I, and the voltage U across the resistor or capacitor.

$$Q = UC; \quad U = IR$$

and from there

$$Q = RCI = TI. \tag{31}$$

$RC = T$ is the time constant of the input circuit, which is considered to be also a function of time. Combining equations (31), (30), and (29) gives

$$\frac{d(T \cdot I)}{dt} + I = M\left(\frac{d\theta i}{dt} + i\right) \tag{32}$$

The solution of this differential equation is

$$I = \frac{e^{-\int \frac{dt}{T}}}{T} \left\{ A + M \int_0^t \left[\frac{d}{dt}(\theta i) + i \right] e^{\int \frac{dt}{T}} dt \right\} \tag{33}$$

By partial integration of the term

$$\int i\, e^{\int \frac{dt}{T}}\, dt = i\, T\, e^{\int \frac{dt}{T}} - \int \frac{d\,(Ti)}{dt}\, e^{\int \frac{dt}{T}}\, dt$$

from equation (33) is obtained

$$I = \frac{A}{T} e^{-\int \frac{dt}{T}} + Mi + \frac{M}{T} e^{-\int \frac{dt}{T}} \int_0^t \left[\frac{d}{dt}(\theta - T)\, i \right] e^{\int \frac{dt}{T}}\, dt \tag{34}$$

The first term of the right side represents the decay of the initial charges. As this problem is dealt with in the previous sections, in general, set here $A = 0$. The second term is the true measurement of the air-earth current density. The error introduced by mismatching is shown by the third term. It depends on the change of the current density and the change of the conductivity with time. For $\theta = T$, the matching condition, the error becomes zero even if θ and T are not constant, but functions of time. If θ is smaller than T, the error term becomes negative and tries to delay the change of the recorded current I. This generates a kind of damping effect. If θ is larger than T, each change of the current density i will be exaggerated by the error term, so that a kind of over-shooting effect results. This damping or over-shooting effect tends to distort the record of the current radiosonde. For a few simple functions of i or θ, these distortions will be discussed in greater detail. T is set to be constant to approach the conditions of the real radiosonde. Even if it were theoretically possible by a simultaneous measurement of λ and an automatic matching mechanism to keep the matching condition and to achieve a true current density record, such an equipment would be too bulky and heavy for a radiosonde. With T = constant, equation (34) simplifies to

$$I = \frac{A}{T} e^{-\frac{t}{T}} + Mi + \frac{M}{T} e^{-\frac{t}{T}} \int_0^t \left[\frac{d}{dt}(\theta - T)\, i \right] e^{\frac{t}{T}}\, dt \tag{35}$$

The integral of the error term of (35) can be changed to a summation by partial integration. With the notation

$$\frac{d^n}{dt^n} \left[(\theta - T)\, i \right] = \left[(\theta - T)\, i \right]^{(n)} \tag{36}$$

$$I = \frac{A}{T} e^{-\frac{t}{T}} + Mi + M \sum_0^\infty (-1)^n T^n \left\{ \left[(\theta - T) i \right]^{(n+1)} - e^{-\frac{t}{T}} \left[(\theta - T) i \right]^{(n+1)}_{t=0} \right\} \tag{37}$$

is obtained.

This formula will be used to compute I for the following four sets of conditions:

    a. $A = 0;\ i = i_0 = const;\ \theta = \theta_0\, e^{\beta t}$

    b. $A = 0;\ i = i_0\,(1 + at);\ \theta = \theta_0 = const$

    c. $A = 0;\ i = i_0 = const;\ \theta = \theta_0\,(1 + \beta t)$

    d. $A = 0;\ i = i_0\,(1 + at);\ \theta = \theta_0\,(1 + \beta t)$

$$\tag{38}$$

The recorded current I is normalized by dividing by Mi_o, and with the notation $J = \dfrac{I}{Mi_o}$ the following equations are obtained for the four sets of conditions:

a. $\quad J = 1 + \dfrac{\beta\theta_o}{1 + \beta_T}\left(e^{\beta t} - e^{\frac{t}{T}}\right)$

b. $\quad J = 1 + a(\theta_o - T)\left(1 - e^{-\frac{t}{T}}\right) + at$ (39)

c. $\quad J = 1 + \beta\theta_o\left(1 - e^{\frac{t}{T}}\right)$

d. $\quad J = 1 + a\left[\left(1 + \dfrac{\beta}{a}\right)\theta_o - (1 + 2\beta\theta_o)T\right]\left(1 - e^{-\frac{t}{T}}\right) + a(1 + 2\beta\theta_o)t$

The case 38-a, 39-a will be encountered above the austausch layer, where the conductivity increases and the time constant θ decreases as an e-function. Fig. 7a shows a graph of equation (39)a with θ_o = 1200 sec, 600 sec, or 300 sec, respectively. T is always chosen to be 100 sec and β in this case is -10^{-3} 1/sec. This would be equivalent to an increase of λ by a factor of 10 every 8 km of height and to a balloon rise with a velocity of 3.5 m/sec. The current density is assumed to be constant. It is seen from Fig. 7a that the recorded current drops off in the first 250 sec from its ground value by about 90, 50, or 25 percent and returns then slowly to its original value. The 90, 50, or 25 percent would correspond to conductivity values of λ_o = 1.5, 3, or $6 \cdot 10^{-14} \dfrac{1}{\text{ohm} \cdot \text{m}}$. ($\lambda_o$ is here the sum of the positive and negative conductivity.)

The case of (38) b, c, and d is illustrated in Fig. 7b, c, and d. These conditions may be met inside the austausch layer. To plot the graph, the following assumptions have been made: The current density drops from its ground value i_o to $i_o/2$ at the top of the austaush layer in cases b and d and it is constant in case c. The time-constant θ drops in case c and d from the ground value θ_o to $\theta_o/2$ at the top of the austausch layer. This would be equivalent to an increase of λ from λ_o to $2\lambda_o$. In case b, θ is supposed to be constant. θ_o is chosen to be 1200 sec and Γ = 100 sec. The ascending velocity of the balloon is 3.5 m/sec. The thickness of the austausch layer is 1000 or 3000 m. The dashed line in all figures shows the normalized air-earth current $\dfrac{Mi}{Mi_o}$ and the solid line the normalized recorded current $J = \dfrac{I}{Mi_o}$. In Fig. 7b is seen the over-shooting effect of mismatching. The true air-earth current decreases to only the half of its ground value, but the recorded current is driven to negative side. This negative sweep at the beginning of the flight is an inherent feature of all fair-weather soundings at the Oakhurst tower area. It may very well indicate that the air-earth current density decreases with height inside the austausch layer. Similar results have been reported by Wigand, Rossmann, Hatakeyama, and Kraakevik[7,12,13,14] who have computed the air-earth current density from simultaneous records of the atmospheric electric field and the conductivity. These findings would point to the existence of an austausch generator.[15] A negative sweep of the recorded current would also be produced by an increase of the conductivity, even though the air-earth current remains constant. This can be seen from Fig. 7c. If the flight time of the balloon inside the austausch layer is long compared to the time constant T of the radiosonde, the negative displacement of the recorded current against the true air-earth current becomes a constant value in cases 7-b and c. According to equation (39) b and c, this final displacement for the case in Fig. 7b is $a(\theta_o - T)$ and for the case in Fig. 7c is $\beta\theta_o$. It is interesting to note

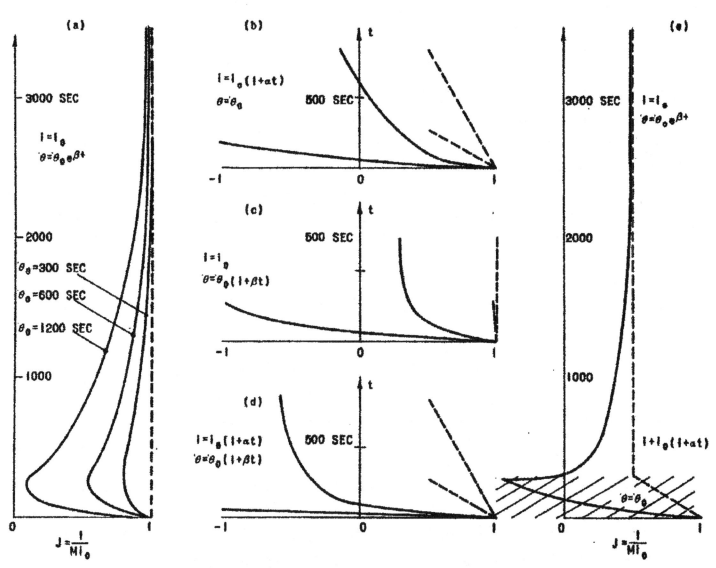

FIG. 7   INFLUENCE OF MISMATCHING ON THE RECORDED CURRENT J FOR THE FOLLOWING CASES:
(a) $I = I_0 = $ CONST; $\theta = \theta_0 e^{\beta t}$ (b) $I = I_0 (I + t)$; $\theta = \theta_0 = $ CONST.   (c) $I = I_0 = $ CONST; $\theta = \theta_0 (I + \beta +)$

(d) $I = I_0 (I + t)$; $\theta = \theta_0 (I + \beta t)$

(e) $I = I_0 (I + t)$; $\theta = \theta_0 = $ CONST FOR $0 \le t \le 286$ SEC
    $I = I_0 = $ CONST; $\theta = \theta_0 e^{\beta t}$ FOR $286 \le t \le 4000$ SEC

that in the case of Fig. 7b this displacement is proportional to the mismatching $(\theta_0 - T)$, but in the case of Fig. 7c it is not.

Figure 7d shows a record where both the current density and the time-constant of the air decrease with height. Here the negative sweep is even more pronounced than in Figs. 7b or c. It is learned from equation (39)d that beside the final constant displacement the rate of decrease of the recorded current is also different from the rate of decrease of the air-earth current.

Fig. 7e shows a combination of Figs. 7a and 7b. Here the assumption is made that inside the austausch layer, which has a thickness of 1000 m, the conductivity is constant $(\lambda = 3.10^{-14} \frac{1}{\text{ohm} \cdot \text{m}}$; $\theta = 600$ sec) and the current drops from its ground value to half of it at the top of the austausch layer. Above it the current density is constant, and the conductivity increases according to an e-function. The dashed line of Fig. 7e shows again the true current density, and the solid line shows the recorded current. Here is shown the sweep to the negative side inside the austausch layer and the slow return to the true value above it.

This analysis may be summarized as follows: If the conductivity and therefore the time-constant of the air as a function of height are known, or a reasonable assumption can be made, the recorded current of the radiosonde can be evaluated. The unusual feature of the negative sweep of the current record by a positive air-earth current can be explained. But in general, the record is a function of two variables — the air-earth current and the conductivity. Therefore the current record would have to be supplemented by a conductivity record to allow an unambiguous evaluation. It will be seen in the next section how much can be learned from air-earth current records as they are.

Test Flights

This discussion deals with 40 test flights. Twenty-five of them were made at the Oakhurst-tower area in New Jersey, and 15 were made at Thule in Greenland. The experimental data do not parallel the development of the theory because the flights were carried out prior to the theoretical analysis. Also, the author realizes that more records are needed to support some of the points of the theory. These flights are the only ones available now, and it will be the task of the future to fill in the gaps.

Figures 8a, b, and c show three flights at the Oakhurst-tower area with three different time-constants of T = 2000, 500, and 100 sec, respectively. The horizontal axis gives the recorded current density in units of $10^{-12} \frac{A}{m^2}$. The vertical axis gives the time rather than the altitude to show more clearly the time variation of the current density for the different degrees of mismatching.

The ascending velocity of the balloon is about $3.5 \frac{m}{sec}$ so that in each 100 seconds the balloon gains 350 m in altitude. The flights were made on clear, sunny days during the months of July and August. Figure 8a shows the sounding with a time-constant T = 2000 sec. This value is somewhat greater than the time-constant of the air at ground level. The current density starts at the ground with the comparatively high value of $3.10^{-12} \frac{A}{m^2}$ and decreases very little in the first 500 seconds of the flight. In this time the balloon would rise to an altitude of about 1.75 km. It is safe to assume that during this time the balloon was inside the austausch layer. From the height of 1.75 km to 2.20 km the current density drops rapidly from $2.8 \cdot 10^{-12}$ to

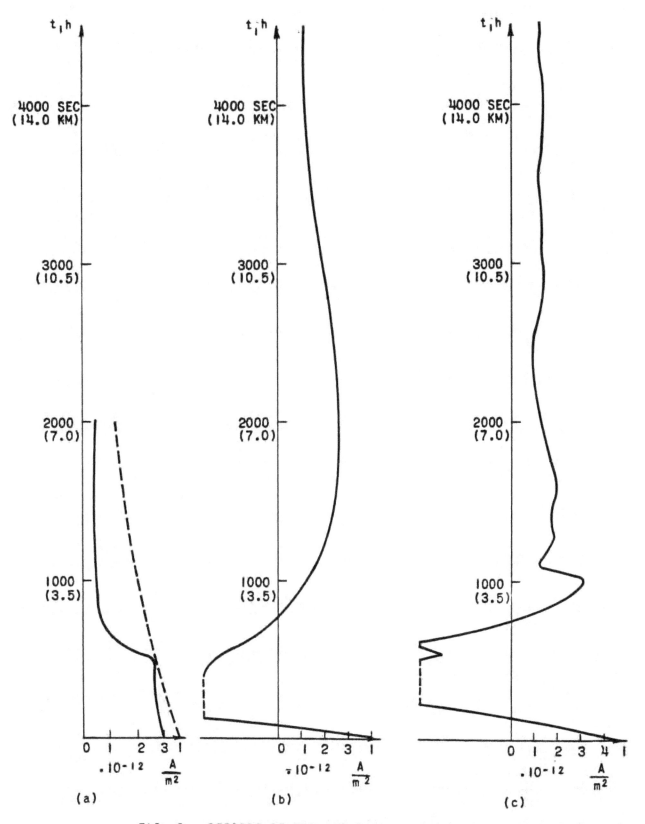

FIG. 8   RECORDS OF THE AIR-EARTH CURRENT AT THE OAKHURST TOWER:
(a) WITH AN INSTRUMENT TIME CONSTANT T=2000 SEC
(b) WITH AN INSTRUMENT TIME CONSTANT T= 500 SEC
(c) WITH AN INSTRUMENT TIME CONSTANT T= 100 SEC

$0.7 \cdot 10^{-12} \dfrac{A}{m^2}$. From there on the current density continues to decrease slowly until it reaches

the value of $0.4 \cdot 10^{-12} \dfrac{A}{m^2}$ and remains there to the end of the flight at about 6.5 km height.

It is learned from this flight that the current density above the austausch layer is fairly constant. This result is inherent in all flights. Therefore it may be expected that the recorded current above the austausch region is not, or very little, affected by mismatching. Inside the austausch layer the current density is much larger than above it — in this flight by a factor of 7 to 8. This points to the fact that there exists another generator which is equal to or more powerful than the world-wide thunderstorm generator. It is also interesting to see that the decrease of the current density at 2-km height is much faster than the decay of initial charges would be. (To facilitate the comparison, the decay curve with the time-constant of 2000 seconds is shown as the dashed line in Fig. 8a.) This indicates that the conductivity at this height does not differ very much from the ground value, so the matching condition is approximately fulfilled.

Figures 8b and c show the records of flights where the time-constant was 500 and 100 seconds, respectively. Here there is the condition $T << \theta$ and a large exaggeration of any change of the current density with time should be expected. If the current density decreases inside the austausch layer, a record similar to the theoretical curve of Figs. 7b or 7e should be obtained. The similarity between the theoretical and the practical result is close enough to suggest that the assumption of the theory, namely, the decrease of the current density, is also the reason for the negative sweep in the records. This sweep is much larger than the recorder range, so the pen stays for a while on the negative boundary (dashed portion of the curves in Figs. 8b and c). The decrease of the current density may also be concluded from the fact that the ground value of the current density is much larger than its value at high altitudes. But, as no effort had been made at that time to determine the part which the initial charges may add to the ground value, a quantitative evaluation of the ratio of ground value to high-altitude value is not possible.

During the International Geophysical Year, from 12 to 21 December 1958, two flights each day at 1100 and 2300 hours local time (1700 and 500 Greenwich Mean Time) were carried out. The time-constant of all these flights was 100 seconds. All records, with the exception of flight No. 16 on a rainy day, show the negative sweep inside the austausch layer and a fairly constant current density above. This can be seen from Fig. 9 which is an average curve of all these flights (except flight No. 16).

Some of the individual soundings show much more detail in the variation of the current density than the flight in Fig. 8c. One example is shown in Fig. 10a, which was a flight on an overcast day. The negative sweep is interrupted by two positive deflections (marked p, p in Fig. 10a) which indicate that the decrease of the current density has been interrupted by two short increases. As soon as the balloon entered the stratus cloud, a positive deflection, marked "c," was noticed on the record. A very similar deflection is recorded in higher altitudes, so it may be assumed that the stratus consisted of two distinct layers. Figure 10b shows the records of the one flight during rain. This curve shows large positive and negative deflections up to 1500 seconds flight time (about 5-km altitude), and it is the only flight where the recorder pen was driven to the positive boundary for some time. This illustrates the well-known relationship between meteorological conditions and atmospheric electric elements.

The same relationship is demonstrated by the records at Thule, Greenland. The exchange of air masses is reduced to almost zero and therefore the austausch generator effect is comparatively small. The sweep to the negative side occurs only occasionally and is not very pronounced.

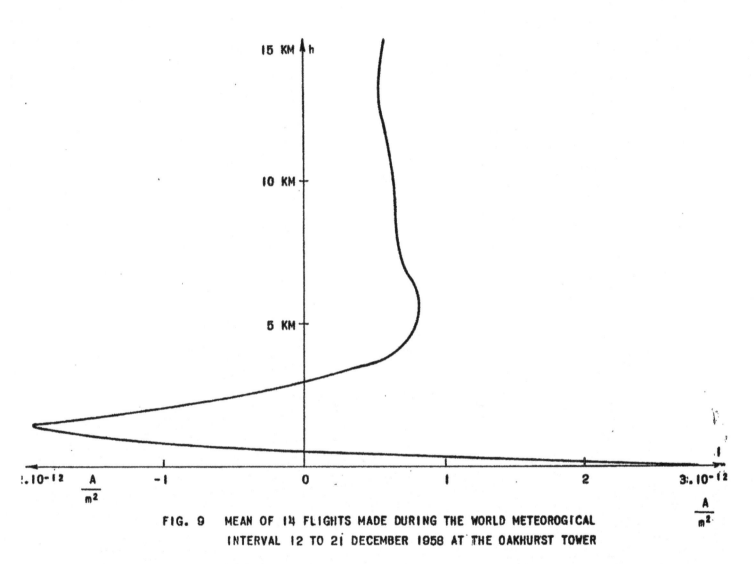

FIG. 9   MEAN OF 14 FLIGHTS MADE DURING THE WORLD METEOROGICAL
INTERVAL 12 TO 21 DECEMBER 1958 AT THE OAKHURST TOWER

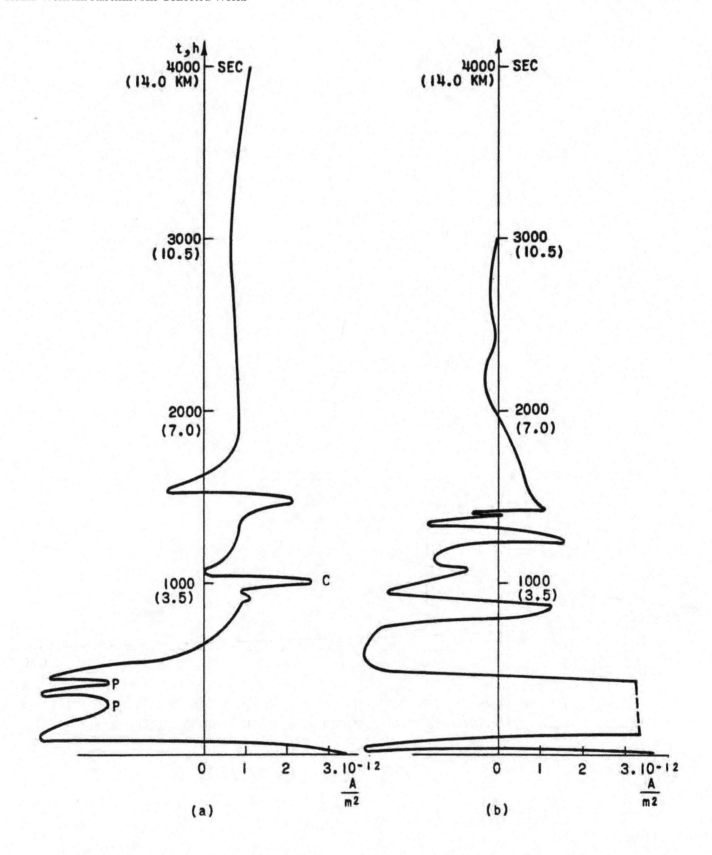

FIG. 10   TWO AIR-EARTH CURRENT RECORDS OF RADIOSONDE FLIGHTS ON:
(a) AN OVERCAST DAY
(b) A RAINY DAY

The mean curve of seven flights during fine weather on 24, 25, and 26 August 1958 is shown in Fig. 11. Here the negative sweep is not recognizable. After the decay of the initial charges, the recorded current density (solid line in Fig. 11) decreases very slowly with increasing altitude. This is the shunting effect of increasing conductivity as outlined in the section on "Internal Resistance of the Antenna System." If the conductivity as a function of height were known, formula (16) with 20) or (21) could be used to calculate the correct air-earth current density. As no simultaneous conductivity records are available, the conductivity measurement over Greenland of Kraakevik[16] have been used to correct the current radiosonde records; the result is shown by the dashed line of Fig. 11. It is seen that after the correction the current density is practically constant with height.

The second significant difference between the Greenland and the Oakhurst-tower records is that the current density over Greenland is about three times as large as the current density in higher altitudes over the Oakhurst-tower area. (Again it should be mentioned that the instrument records only half of the current, so the full air-earth current density over the Oakhurst-tower area is approximately $1 \cdot 10^{-12} \frac{A}{m^2}$ while that over Greenland is $3.2 \cdot 10^{-12} \frac{A}{m^2}$.) From this fact it may be concluded that the columnar resistance at the Oakhurst-tower is about three times as large as the columnar resistance at Greenland. If the Oakhurst-tower measurements are representative for medium and equatorial latitudes and the Thule measurements for arctic and antarctic latitudes, it would follow that a strong current flow passes through the ionosphere from the equatorial thunderstorm belt to the poles. It would be a worth-while task for a network of air-earth current radiosondes to prove this hypothesis.

Figure 11 is a typical example for about half of the flights at Greenland. There are also flights where the recorded current density slowly increases or decreases with altitude, which indicate some kind of generator effect. One flight during fog (Fig. 12) shows the large negative sweep similar to the records at the Oakhurst-tower area during fine weather. The conductivity in Greenland during fine weather is about 4 to 5 times greater than at the Oakhurst tower. The fog reduces the conductivity to about one-third of its fine-weather value, so the conductivity value during fog in Greenland is comparable to that during fair weather at the Oakhurst tower. The matching conditions of these two cases would be about the same. But this is hardly enough to explain the similarity of the records. It would seem that the electrification process is controlled by the same or similar physical forces in order to form this similar pattern of the recorded current.

CONCLUSIONS

The current recorded by the air-earth current radiosonde is a function of the air-earth current density and the conductivity. To evaluate the records it would be desirable to have the conductivity recorded simultaneously.

Test flights show that above the austausch layer the air-earth current density is fairly constant and the record seems to be not seriously affected by the increase of the conductivity with height. The theoretical analysis yields that this is the case, especially if the absolute value of the conductivity is high.

Inside the austausch layer the recorded current shows a sweep to the negative side. This can be theoretically explained by mismatching in case the air-earth current density decreases inside the austausch layer. A similar negative sweep would also be recorded if the air-earth current density is constant, but the conductivity increases. Probably both the decrease of the current density and the increase of the conductivity occur simultaneously. Complementation of the air-earth current radiosonde by a conductivity radiosonde would be necessary to separate these two effects.

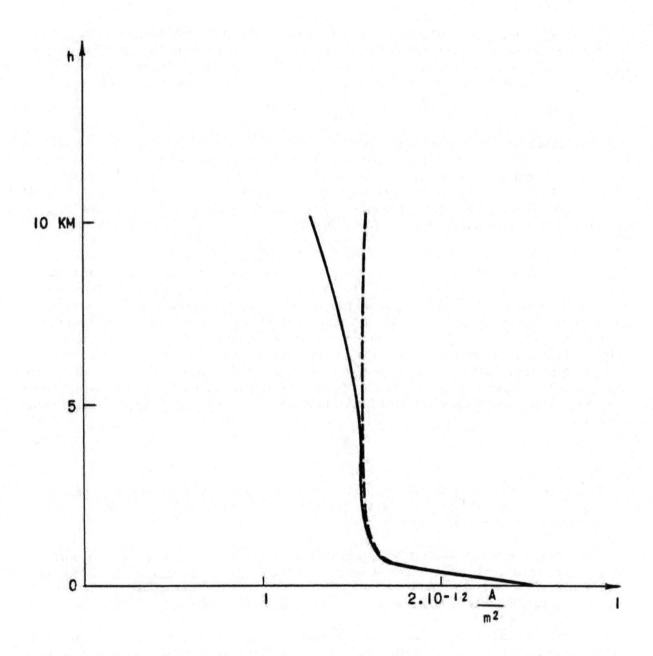

FIG. II   MEAN OF 7 FAIR-WEATHER FLIGHTS MADE AT THULE, GREENLAND

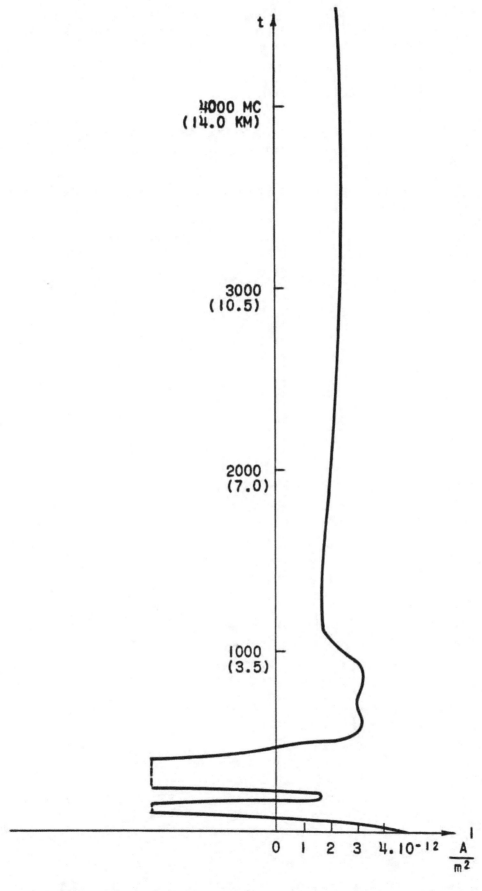

FIG. 12   FLIGHT RECORD DURING FOG AT GREENLAND

At Thule, Greenland, the soundings show that the air-earth current density is approximately constant from the ground-level up. The absolute value of the current density is about 3 to 4 times as large as the current density above the austausch layer at the Oakhurst tower in New Jersey.

REFERENCES

1. Koenigsfeld and Piranx, Ch., "Un nouvel èlectromètre pour la mesure des charges electrostatique par systeme electronique, son application à la mesure du potentiel atmosphèrique et a la radiosonde." Mém. I.R.M. XLV, 1950.

2. Koenigsfeld, L., "La mesure de la conductibilité par radiosonde," Inst. Roy. Met. Belg. Publ. Ser. B., No. 18, 1955.

3. Lugeon, J. and Bohnenblust, M., "Radiosondage du gradient potential et de la conductibilité électrique de l'air," Am. de la station central swisse de Meteorologie, 1956.

4. Venkiteshwaran, S. P., Dhar, N. C., and Huddar, B. B., "On the Measurement of the Electrical Potential Gradient in the Upper Air over Poona by Radiosonde," Proc. of the Indian Acad. of Sciences, XXXVII, 1953, S.260.

5. Ronike, G., "Über eine Radiosonde zur Messung des Potential Gefalles in der Freien Atmosphaere," Gerl. Beitr. z Physik 66, Heft 4, 1957, S.313.

6. Rein, C. G., Stergis, C. G., and Kangas, T., "An Airborne Electric Fieldmeter," IRE Transaction of Instrumentation, 1 - 6, No. 3, 1957.

7. Hatakeyama, H., Kobayashi, J., Kitaoka, T. and Uchikawa, K., "A Radiosonde Instrument for Measurement of Atmospheric Electricity and Its Flight Results," *Recent Advances in Atmospheric Electricity,* Pergannon Press, New York, 1958.

8. Jones, O. U., Maddever, R. S., Sanders, J. H., "Radiosonde Measurement of Vertical Electric Field and Conductivity in the Lower Atmosphere," *Recent Advances in Atmospheric Electricity,* Pergannon Press, New York, 1958.

9. Coroniti, S. C., Nazarck, A., Stergis, C. G., and Kotas, D. E., Seymour, D. W., and Werme, I. K., "Balloone-borne Conductivity Meter," AFCRC-TR-54-206, U. S. Dept. of Commerce, Office of Technical Services, Washington 25, D. C., 1954.

10. Kasemir, H. W., "Measurement of the Air-Earth Current Density," Proceedings on the Conference on Atmospheric Electricity, AFCRC-TR-55-222, U. S. Dept. of Commerce, Office of Technical Services, Washington 25, D. C., 1955.

11. Kasemir, H. W., and Ruhnke, L. H., "Antenna Problems of Measurements of the Air-Earth Current," *Recent Advances in Atmospheric Electricity,* Pergannon Press, New York, 1958.

12. Wigand, A., "Der Verticale Leitungstrom in der Atmosphaere und die Erhaltung des Elektrostatischen Erdfeldes," Phys. 2.22, 623, 1921.

13. Rossmann, F., "Luftelectrische Messungen Mittels Segelflugzeugen," Ber. d. Dtsch. Wetterd., U. S. Zone, Nr. 15, 1950.

14. Kraakewik, J. H., "Electrical Conduction and Convection Currents in the Troposphere," *Recent Advances in Atmospheric Electricity,* Pergannon Press, New York, 1958.

15. Kasemir, H. W., "Zur Stroemungs Theorie des Luft Elektrischen Feldes III. Der Austausch Generator." Arch. Met. Geophys. Biokl. Ser. A., 357-370, 1956.

16. Kraakewik, J. H., "The Electrical Conductivity and Current Density in the Troposphere," Doctor thesis at the University of Maryland, 1957.

APPENDIX
Symbols Used

i = air-earth current density $\left(\dfrac{A}{m^2}\right)$

λ = conductivity of the air $\left(\dfrac{1}{ohm \cdot m}\right)$

F = atmospheric electric field $\left(\dfrac{v}{m}\right)$

ϵ = dielectric constant $\left(\dfrac{F}{m}\right)$

j = current density at the surface of the spheroid $\left(\dfrac{Amp}{m^2}\right)$

E = field strength at the surface of the spheroid $\left(\dfrac{v}{m}\right)$

q = charge density at the surface of the spheroid $\left(\dfrac{coulomb}{m^2}\right)$

$\underline{j}$ = current per unit length at the surface of the spheroid $\left(\dfrac{Amp}{m}\right)$

$\underline{q}$ = charge per unit length at the surface of the spheroid $\left(\dfrac{coulomb}{m}\right)$

$I_u \; I_o$ = current flowing into the upper or lower half of the spheroid (Amp)

$Q_u \; Q_o$ = charge of the upper or lower half of the spheroid (coulomb)

$P_1 = \dfrac{v}{c}$ = spherical function of the first kind and first order

$Q_1 = \dfrac{u}{2c} \ln \dfrac{u+c}{u-c} - 1$ = spherical function of the second kind and first order

$U = \left(\dfrac{u^2 - v^2}{u^2 - c^2}\right)^{\frac{1}{2}}$

The subscript a of the functions Q_1 and U indicate that $u = a$

The superscript prime of the function Q_1 indicates the differentiation with respect to u/c.

$\dfrac{dQ_1}{d\,u/c} = Q_1'$.

$dS = \dfrac{2\pi}{c}\left(a^2 - v^2\right)^{\frac{1}{2}}\left(a^2 - c^2\right)^{\frac{1}{2}} dv$ = surface differential in elliptical coordinates.

The Field Component Meter

A Device for Measuring the Three Components of the Atmospheric Electrical Field and the Own Charge of an Airplane for Airplane Ascents

By H.-W. KASEMIR,

Atmospheric Electrical Research Center Buchau a. F. of the Observatory Friedrichshafen

(Manuscript received 1 October 1951)

Abstract

Recording the atmosperic-electric potential gradient from an aeroplane it is very useful to get all the three field-components. This is the case especially in investigations in greater altitudes and in thunderclouds. In this paper an apparatus is described working on the principle of the 'Feldmühle' (field strength meter) and giving simultaneous recording of the atmospheric-electric potential gradient separated in the three rectangular components of the field. Furthermore a method is pointed out to separate the field of the own charge of the aeroplane and to record this one separately. This enables the 'Feldkomponentenmühle' (field component meter) to be used in motor driven aeroplanes.

I. Introduction

All previously conducted measurements of the atmospheric electrical field in the free atmosphere started by the free balloon ascents to the modern measurements in gliders or engine-propelled aircraft (I, 2, 3, 4, 5, 9) are limited to the measurement or registration of the vertical field components. To the extent that one advances to altitudes of a few kilometers above the surface of the earth in weather that is undisturbed by atmospheric electricity, one can expect that the atmospheric electrical field is vertically oriented, and thus, the horizontal components are missing. For analyses of clouds and, in particular, of storm fields, however, aside from the vertical components, considerable horizontal components also emerge. In this manner, one is compelled to make certain assumptions about the geometric distribution of cloud charges, e.g. when registering the vertical components in the storm field, in order to even be able to assess these measurements.

Therefore, Simpson and Robinson (3) assume a charge distribution in overlapping spheres, while E. Wall (6) prefers the charge arrangement in layers in order to have easily visible theoretical potential and field ratios for the assessment of the measurements. Based on the meteorological relationships in the storm cloud together with the processes of forming electricity, H. Wichmann (7) derives another charge distribution, which can no longer be compiled into such easy geometric forms. Thus, a certain degree of insecurity remains with the assessment of the measurements, which would be able to be eliminated through the simultaneous registration of all three field components.

The horizontal components of the atmospheric electrical field achieve a significant meaning as well in the case of flights in fair weather if one reaches altitudes higher than the summits of storm clouds with the registering device.

THE FIELD COMPONENT METER

In this case, the electrical current must expand toward the sides and a horizontal field component must therefore dominate as well. This lateral balance will not initially occur in the ionosphere (8), but rather in lower laying atmospheric layers so that one may expect, e.g. the occurrence of horizontal field components in the space between the altitudes of 15 and 50 km.

For these and other reasons, the development of a measuring device for airplane ascents began as early as 1944 by the author, which simultaneously registers the three components of the atmospheric electrical field. In addition, this instrument was intended to provide the possibility of separating the field originating through the own charge of an airplane from the atmospheric electrical field and uniquely registering it. We know that engine-propelled aircraft are charged by the exhaust gases to voltages of 5-10000 V, such that the own field of the aircraft covers the weak atmospheric electrical fields at that altitude.

Unfortunately, the practical testing of the system in an airplane could no longer be conducted. Nonetheless, if this device is supposed to be described at this point, it will be done for this reason because, even today, practical flight testing does not seem certain for the foreseeable future in Germany.

2. Theoretical Considerations regarding Field Component Meter

The measuring device was built according to the principle of the automatically rotating Wilson plane, which is known under the name "field measurement machine" or especially as "field meter" in area of atmospheric electricity. In conjunction with the modern amplification technology, this measuring procedure has proven itself excellently for registering atmospheric electrical fields in the aircraft due to its insensitivity to mechanical stresses and due to the low insulation requirement.

The device ultimately measures the charge influenced by the outer field on a section of the aircraft's surface or the field strengths proportionate to this charge density.

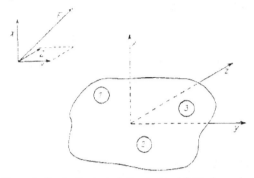

Figure I. Conductive body in the electrical field of a random direction.

To understand the measurement of the three components, a brief theoretical consideration should be established.

We will apply a randomly shaped conductive body into a randomly directed electrical field. We then dismantle this field with the field strength F according to our Cartesian coordinate system x, y, z into its components X, Z, Z. (Figure I.)

Furthermore, we will cut 3 random surface sections 1, 2, 3 from the body, which should however stay connected with the remaining surface of the body by a wire allowing for conducting. A charge q will be influenced by every component of the outer field on every surface section, which is proportionate to the respectively observed field components. We will identify the charges associated to the individual surface sections through the indexes 1, 2, 3 and subdivide the net charge in every surface segment into the parts that are induced by the field components X, Y, Z. We will identify these partial charges through further indexes X, Y, Z. Thus, it combines the net charge q1 on the surface section 1 from the partial charges q_{1x}, q_{1y}, q_{1z}. The same applies accordingly for the surface sections 2 and 3. Therefore, we obtain the following equation system.

$$q_1 = q_{1X} - q_{1Y} + q_{1Z}$$
$$q_2 = q_{2X} - q_{2Y} + q_{2Z} \qquad (1)$$
$$q_3 = q_{3X} + q_{3Y} - q_{3Z}.$$

The partial charges q_{1x}, q_{1y},........ are now proportionate to the field components X, Y, Z.

Will apply these with the aid of the proportionality factor a_1, b_1, c_1, ... and obtain from equation (I.)

$$q_1 = a_1 X - b_1 Y - c_1 Z$$
$$q_2 = a_2 X - b_2 Y - c_2 Z \qquad (2)$$
$$q_3 = a_3 X - b_3 Y - c_3 Z.$$

Because a_1, b_1, c_1, ... represent constant factors, which we can define and measure, e.g. as a partial capacity coefficient for the respective field direction, we may solve equation (2) according to the field components X, Y, Z and obtain with the new constants A_1, B_1, C_1,..........

$$X = A_1 q_1 - B_1 q_2 - C_1 q_3$$
$$Y = A_2 q_1 - B_2 q_2 - C_2 q_3 \qquad (3)$$
$$Z = A_3 q_1 - B_3 q_2 - C_3 q_3.$$

We can achieve the constants A_1, B_1, C_1,..........according to a known calculation from a_1, b_1, c_1, ... However, we would like to go over this equation because these constants are appropriately achieved through calibration with the practical implementation of the device.

As a result, we maintain that it is theoretically possible to determine the components of the outer field according to equation 3 through the measurement of the surface charge at three various points of the surface of the body, which is located in a randomly directed homogenous field.

Figure 2. Photograph of the field component meter.

In addition, the practical implementation demonstrates that is does not cause difficulties, to allow the device to automatically implement the necessary mathematical operation.

3. Practical Implementation of the Field Component Meter

The measurement of the surface charge or the field strength should be presumed as known. We will now turn to the technical construction of the field component meter. Figure 2 shows a photograph of such a device. The field meter comprises a short cylinder, the head of which is divided into 8 separately insulated segments. In this connection, 4 segments are allocated to the cylinder cover while the upper piece of the cylinder casing is also subdivided into 4 symmetrical partial sections. A somewhat larger cylindrical blade, in which notches have been made in such a way that during rotation respectively 2 opposing segments are released and shielded in the same rhythm, rotates over these segments. Every pair of segments of the cylinder casing is connected to each other through a resistor, which is connected with the surface of the aircraft through a center tap. We would like to clarify the mode of action of this arrangement in the schematic illustration of Figure 3a – 3d.

4 segments 1, 3 and 2, 4 of the cylinder casing are recorded on average. They are influenced by a horizontal field, the direction of which is provided parallel to the connection line of the segments 1, 3 and which stands vertical to the axis of the cylinder. From Figure 3b, we can see that segment 1 only carries a negative influence charge and segment 3 only a positive one. According to Figure 3a, in contrast, segments 2 and 4 carry a negative charge on their left half and a positive charge on their right half.

Now, we would like to follow up on the charge flow between segments 1 and 3 with a rotation of the blade. As a starting point, we will choose the position of the blade depicted

634

THE FIELD COMPONENT METER

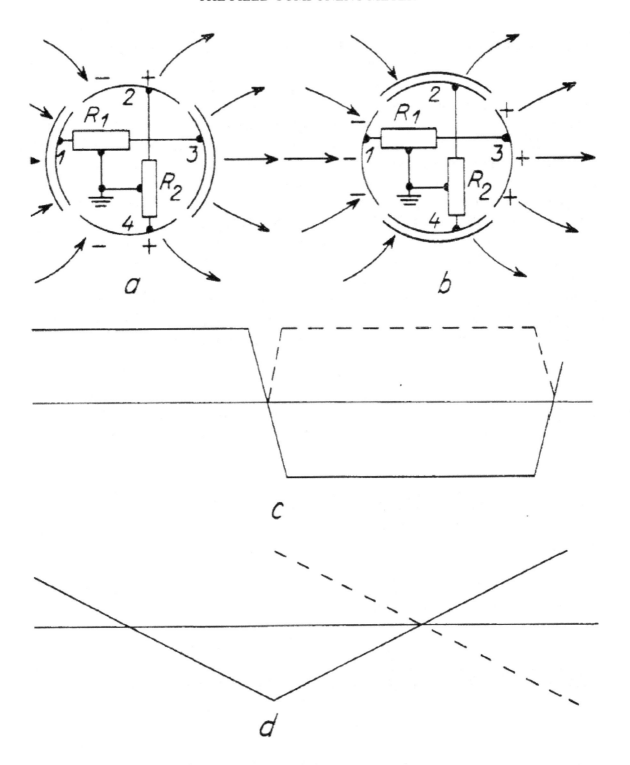

Figure 3a and 3b. Casing segments under the influence of a horizontal field.

Figure 3c. Voltage delivered by segments 1 and 3.

Figure 3d. Voltage delivered by segments 1 and 3.

635

in Figure 3a, in which segments 1 and 3 are completely shielded by the outer field. If the blade turns clockwise to the position 3b, then electrons flow over the resistor R_1 from segment 3 to segment 1 under the influence of the outer field. If the blade continues to turn until it reaches the position 3a, the influence charges over the resistor R_1 will balance, thus, electrons will flow from segment 1 to segment 3. Upon turning further, this process repeats itself anew. Beyond that, this continual charge flow through the resistor R_1 produces a voltage, the course of which is schematically represented in Figure 3c. This alternating voltage is then strengthened as usual through an amplifier and fed to the recorder through a force-controlled commutator. The amplitude of this voltage is proportionate to the incident field strength. The force-controlled commutator also ensures that the sign of the field is not lost.

Now we will observe the effect of the field on segments 2 and 4. If we also start from the position 3a, upon turning the blade in the position 3b, in the first half-time, the negative charges of segment 2 will balance with the positive charges of segment 4, while in the second half-time, the positive charges of segment 2 will balance with the negative charges of 4. If the blade is turned further to position 3a, the same charge flow will result, merely with the opposite sign. The curve 3d shows approximately the resulting voltage curve over the resistor R_2. Compared to the voltage curve over the resistor R_1, this is pushed in the phase to the extent that the resulting direct voltage will equal c through a

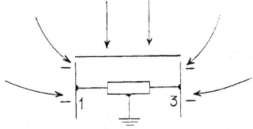

Figure 4. Influence of the segment pair 1 and 3 through a vertically aligned incident field.

compulsory rectification synchronous with the first electrical circuit. (The rectified voltage is marked with hashes in Figure 3c and 3d.)

Now if turn the field direction in such a manner that it coincides with the connection line of segments 2 and 4, then they will logically provide the voltage, that we just derived for segments 1 and 3, while the latter provides the rectified voltage of 0. Segments 1 and 3 as well as segments 2 and 4 respond through the device of the force-controlled rectification only to the field component, which is directed parallel to the connection line of the respective segments.

It is still necessary to convince ourselves that these segment pairs are also insensitive in relation to the field component, which is incident in the direction of the cylinder axis. For this purpose, we will observe Figure 4. Both segments 1 and 3 are represented in elevation in this figure.

Through the alignment of the incident field in the direction of the axis, the same sized negative charge influences both segments. This charge flows to segment 1 via the left half resistor and to segment 3 via the right half resistor. Because the current direction is opposite in both halves of the resistor, this compensates the voltage drop over the entire resistor to 0. The same applies for segment pair 2 and 4.

The requirement for this compensation is the symmetry of the field line diagram. If one mounts the field meter, e.g. on the top of the aircraft, the symmetry is ensured for the segment pair pointing in the direction of the wing of the aircraft when the meter is attached over the middle line of the aircraft. For the segment pair pointing toward the head and tail end, the electrically neutral line would have to be determined first. If this line was not able to be used for mounting due to any reason, it is possible to force the entire compensation through a grounding tap on the resistor.

In conclusion, we can ascertain that the segment pairs of the cylinder casing each give only one rectified voltage for respectively one component of an outer field.

THE FIELD COMPONENT METER

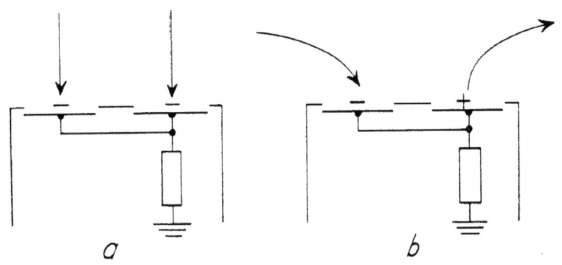

Figure 5a. Cover segment under the influence of a vertically aligned incident field.
Figure 5b. Cover segment under the influence of a horizontally aligned incident field.

Versus the respectively other two field components, the segment pair is insensitive, partially due to the force-controlled rectification, and partially due to a respective grounding of the spanning resistor.

The ratios for the segments in the cylinder cover are somewhat easier to understand. In this case, it is sufficient if just one of the two segment pairs is used. This is short-circuited according to Figure 5a and connected to the aircraft with a resistor.

The effect for incident fields aligning parallel to the cylinder axis is the same as with the normal single-component meter. In the case of incident fields aligning vertically to the cylinder axis, the effect is cancelled either already within the same segment or through the short-circuiting of both segments as can be readily seen in Figure 5b. If one does not place value on the fact that one obtains a separate input resistance for every component channel of the amplifier, one could forgo the segments in the cylinder cover altogether because incident fields aligning parallel to the cylinder axis can be registered with a switch according to Figure 6 through the cylinder casing segment.

In this instance, the influence charge flows over the applied resistance R_3 to segments 1 and 3. The voltage for the incident field component aligned vertically is relieved

through the resistor R_3. The cross resistance remains without effect in this process. The similarity of this switch to the implementations according to Figure 5a and 5b is so evident that an extensive discussion would only signify a repetition following the implementations made above.

With the previous implementations, it has been shown that in the case of one of the given designs of the field component meter respectively one segment pair always responds to only one field component, while it is not affected by the other two components.

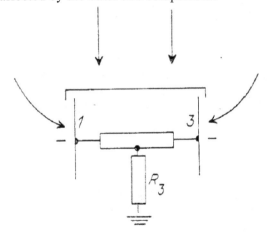

Figure 6. Registration of the vertical fields with the aid of the cylinder casing segments.

637

4. The Registration of the Own Charge of the Aircraft

A complication arises with the field registration in the aircraft by the fact that engine-propelled aircraft are charged through the exhaust gases so much that their own fields cover the weak atmospheric electrical fields at altitude. One can naturally bypass this difficulty by using a glider, from which one can remove a potentially present own charge with the aid of a compensator attached in the neutral line. On the other hand, precisely with measurements in rain and storm clouds, even the glider will assume an own charge of such an extent through the reception of charged precipitation, which cannot be completely eliminated with a compensator.

In the case of the component meter, the following method serves as a separation of these own charge fields. For this, a switch according to Figure 6 is used for the cylinder casing segments and a switch according to Figure 5a is used for the cylinder cover segments. We will identify the incident components aligning parallel to the axis of the meter with Fx and the field strength originating from the own charge with E. Partial capacity coefficients other than those for the outer field apply for the own charge field in the case of the individual segments. We will identify the former with b_1, b_2 and the latter with a_1, a_2.

With the switch according to Figure 7, we can now separate the own charge field from the outer field.

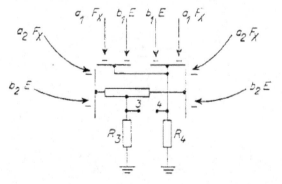

Figure 7. Switch for separating the effect of the own charge field from that of the outer atmospheric electrical field.

The voltage resulting from the resistor R_4 is provided through

$$U_4 = a_1 F_X - b_1 E, \quad (4)$$

while the following applies for the voltage resulting from R_3

$$U_3 = a_2 F_X - b_2 E. \quad (5)$$

Through the appropriate selection of resistor R_3 and R_4, we can achieve that $a_1 = a_2$. Then we obtain the input voltage U_e for the amplifier channel of the own charge by switching the resistors R_3 and R_4 against each other at the points 3 and 4 because it is

$$U_e = U_3 - U_4 = (b_1 - b_2) E. \quad (6)$$

Thus, only U^e is proportionate to the own charge field strength E. We will use the same switch diagram for the other combination of cylinder casing and cover segments. In this case, however, we will synchronize the resistors equivalent to R_3 and R_4 in such a manner that in the equations (4) and (5) b_1 will equal b_2. Then the own charge share will drop out and we will obtain

$$U_X = U_3 - U_4 = (a_1 - a_2) F_X. \quad (7)$$

Thus, U_N is a voltage that is only proportionate to the incident field component F_x aligning parallel to the axis of the meter.

A disturbance of the display of the horizontal field components by the own charge field strength will not occur if the casing segments are switched according to Figure 3a or 7. The same applies for the own charge field, which was presented for the vertical field component based on Figure 4. The effect of the own charge field is removed by the center tap of the cross resistance.

A practical difficulty could arise from the fact that the values for a_1 and a_2 as well as for b_1 and b_2 will be nearly the same size. Thus, we will obtain small values for U_e and U_N through the forming of the difference in parentheses from equation 6 and 7. Due to the state of the art in amplifier technology, amplifying even small alternating voltages is not difficult. However, it will be more difficult to achieve the precise balancing of the resistors. In addition, disturbance voltage

THE FIELD COMPONENT METER

of the meter originating from the contact voltage and that of the more or less sufficient grounding of the rotating blade must be kept small.

If it should turn out that this route can only be taken with great effort, there is always the possibility of using a simple auxiliary meter. This can then be mounted on an appropriate spot on the bottom of the aircraft. In this connection, the respective coefficients a_1, a_2 and b_1, b_2 obtain different signs, thus, strengthen each other. Otherwise, for the construction of the component meter, logically the same experiences apply that were gathered when constructing the simple field meter, about which however we will no longer go into at this point.

Summary

For the registration of the atmospheric electrical field in an aircraft, it would be extremely preferable to include all three field components. This is especially true with analyses at high altitudes and in storm fields.

In this work, an apparatus is described, which functions according to the field meter principle and allows the simultaneous registration of the components of the atmospheric electrical field analyzed according to Cartesian coordinates. Furthermore, a method is shown for using this device to separate the field of the own charge of the aircraft from the atmospheric electrical field and uniquely register the results for itself. As a result, it is possible to also use the field component meter in an engine-propelled aircraft.

LITERATURE

(1) EVERLING, E., u. A. WIGAND, 1921: Spannungsgefülle und vertikaler Leitungsstrom in der freien Atmosphäre nach Messungen bei Hochfahrten im Freiballon. An. d. Phys. 4, 66, 261 — 288.

(2) SIMPSON G. C., F. J. SCRASE, 1937: The Distribution of Electricity in Thunderclouds 1. Proc. R. Soc. A. 161, 309—352.

(3) SIMPSON G. C., G. D. ROBINSON, 1941: The Distribution of Electricity in Thunderclouds II. Proc. R. Soc. A. 177, 281—329, 1941.

(4) KASEMIR, H.-W., I. v. WITTICH, 1943: Feld- und Eigenladungsregistrierungen bei Flugzeugaufstiegen. ZWB Forschungsbericht Nr. 1886,

(5) ROSS GUNN, 1948: Electric Field Intensity Inside Natural Clouds. J. Applied Physics 19, 481—484.

(6) WALL, E.: Das Gewitter I. Wetter u. Klima 1 2, 7—21, 1948.
— Das Gewitter II. Wetter u. Klima 3 4, 65—74, 1948.
— Das Gewitter III. Wetter u. Klima 7 8, 193—204, 1948.
— Das Gewitter IV. Wetter u. Klima 11 12, 321—326, 1948.

(7) Wichmann, W.: Grundprobleme der Physik des Gewitters. Wolfenbüttler Verlagsanstalt. Wolfenbüttel u. Hannover 1948.

(8) ISRAEL, H., u. H.-W. KASEMIR, 1949: In welcher Höhe geht der weltweite luftelektrische Ausgleich vor sich. Ann. d. Geophysique 5, 313—324.

(9) R. LECOLAZET, 1948: Étude Expérimentale de l'état éleqtrique des cumulus de beau temps, interprétation des résultates. Ann. de Géoph., 4. 81 — 95.

Sonderdruck aus
„Meteorologische Rundschau", 25. Jahrgang, 2. Heft, 1972
Seite 33—38

The Cylindrical Field Mill*

H. W. Kasemir

Boulder, Colorado

Summary. A field-mill-type instrument is described which is especially suited to recording the atmospheric electric field from an airplane. The instrument measures the two components of the electric fields vertical to its axis and rejects the component in the direction of the axis. This feature can be used to eliminate the influence of the airplane charge on the measurement of the atmospheric electric field so that weak fields down to the order of 2 volts per meter can be recorded. Furthermore, each cylindrical field mill delivers two components of the electric field so that only two field mills of the cylindrical type are required to obtain all 3 components of the atmospheric electric field. Problems involving the correct places for mounting the field mills on the plane and the determination of the calibration factor are discussed. A brief analysis of the field records is given for the simple case of an electric charge of one sign in different positions to the airplane path. Ways are pointed out to either avoid strong electric fields during the flight or locate and penetrate them.

Zusammenfassung. Es wird eine Feldmühle beschrieben, die zur Messung des elektrischen Felds der Atmosphäre vom Flugzeug aus entwickelt wurde. Das Gerät mißt die beiden Komponenten des elektrischen Felds senkrecht zu seiner Achse und berücksichtigt nicht die Komponente in Richtung der Achsen. Diese Tatsache kann benutzt werden, um den Einfluß der Ladung des Flugzeugs auf das elektrische Feld der Atmosphäre auszuschalten, und damit können schwache Felder bis herab zu 2 V/m registriert werden. Da weiterhin jede zylindrische Feldmühle zwei Komponenten des elektrischen Felds mißt, können mit zwei Feldmühlen alle 3 Komponenten in der Atmosphäre erfaßt werden. Es werden die Fragen diskutiert, die sich aus der Anbringung der Feldmühle am Flugzeug und aus der Eichung ergeben. Eine kurze Analyse von Feldmessungen für den einfachen Fall einer elektrischen Ladung eines Vorzeichens an verschiedenen Punkten des Flugwegs wird gegeben. Möglichkeiten werden angegeben, um entweder starke Felder während eines Flugs zu vermeiden oder sie zu orten und zu durchfliegen.

1. Introduction

In general the field mill is a device that measures a surface charge density q. The surface charge density is proportional to the electric field E on that surface. With ε being the dielectric constant, the following relation holds

$$E = q/\varepsilon . \qquad (1)$$

As ε is constant in most cases, in space and time, the quantity measured by the field mill can be calibrated in terms of the electric field as well as in terms of the surface charge density. A field calibration is usually preferred. The working principle of the common type field mill is as follows: A certain area of the surface of the instrument is periodically shielded and exposed to the electric field by a rotating shutter (this type will be called, in the following, the shutter field mill). The surface charge flowing from the ground to the surface in the exposed position, and back to the ground in the shielded position, is measured either as current flow or as voltage drop over a resistor or a capacitor connecting the sensitive area with the ground. The purpose of this set-up is to change a constant electric field into an alter-

nating current or voltage, which can be easily amplified and recorded. The energy for the movement of the surface charge is supplied by the motor, which drives the rotating shutter. No energy is bled from the field to be measured, so the measurement is truly electrostatic.

The difference in the working mechanism of the cylindrical field mill compared to the shutter field mill is that the sensitive area is not shielded and exposed to the electric field by a shutter, but is moved periodically between two positions in which the surface charge induced by the electric field is different (Kasemir [2, 3]; Matthias [1]). The difference in the surface charges, which is also proportional to the influencing electric field, is then amplified and recorded. The cylindrical field mill has a number of advantages over the shutter field mill which will be discussed in detail in this paper. Furthermore, it has the peculiar feature of being capable of measuring two components of an electric field vector and rejecting the third. This makes the cylindrical field mill especially suited for field measurements from an airplane because it can be constructed and mounted in such a way that the influence of the airplane charge is automatically cancelled.

2. Theory

The sensor areas of the cylindrical field mill are the two halves of a cylinder mantle as shown in Figs. 1a and b. Figs. 2a and b are photographs of the finished instrument in which the two segments can be easily recognized

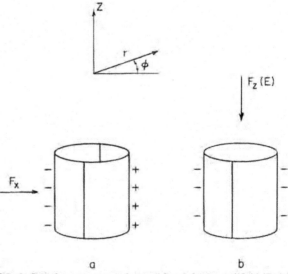

Fig. 1. Cylinder segments influenced by: a horizontal field F_x (a) and a vertical field F_z, or the field E of the airplane charge (b)

* H. Israël zum Gedenken gewidmet.

3 Meteorol. Rdsch., 25. Jahrg.

Fig. 2a. The cylindrical field mill

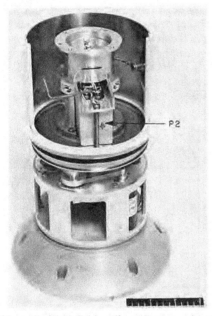

Fig. 2b. The cylindrical field mill, with top and one cylinder segment removed ($P2$ = operational amplifier)

on the end piece is calculated. As this calculation is rather tedious, and in practice the field mill is calibrated in an artificial field of known strength, the simple calculation using the infinite cylinder is sufficient for our purposes here.

The potential function Φ of an infinite cylinder placed in a homogeneous field F is given by the equation

$$\Phi = - F r \cos\varphi \, (1 - a^2/r^2) \, . \qquad (2)$$

Here, cylindrical coordinates z, r, φ are used, with the axis of the cylinder being the z axis of the coordinate system (Fig. 1a). By differentiation with respect to r and then setting $r = a$, the electric field and, with equation (1), the charge density q on the cylinder surface are obtained

$$q = 2 \, \varepsilon \, F \cos\varphi \, . \qquad (3)$$

Integration along the z axis over a length L and between angles φ_1 and φ_2 yields the charge Q of the so defined area of the cylinder mantle.

$$Q = - 2 \, a \, L \, \varepsilon \, F \, (\sin\varphi_2 - \sin\varphi_1) \, . \qquad (4)$$

If $\varphi_2 = \varphi_1 + \pi$, the charge of a half segment of the cylinder mantle is obtained.

$$Q = - 2 \, \varepsilon \, F \, 2 \, a \, L \, \sin\varphi_1 \, . \qquad (5)$$

This, then, would give the influence charge of one of the sensitive areas of the field mill, whereby φ_1 is the angle of the cut of the cylinder with respect to the direction of the field. If the cylinder rotates with a circular frequency ω, φ_1 increases by ωt, where t is the time, and from equation (5) follows

$$Q = - 2 \, \varepsilon \, F \, 2 \, a \, L \, \sin\,(\omega \, t + \varphi_1) \, . \qquad (6)$$

Differentiation with respect to time t yields the current I caused by the flow of the charge Q.

$$I = - 2 \, \varepsilon \, F \, 2 \, a \, L \, \omega \cos\,(\omega \, t + \varphi_1) \, . \qquad (7)$$

This current is used to drive an operational amplifier with the feedback resistor R (Fig. 3). The resulting output voltage U_o of the amplifier is then given by

$$U_o = - IR = 2 \, \varepsilon \, F \, 2 \, a \, L \, \omega \, R \cos\,(\omega \, t + \varphi_1) \, . \qquad (8)$$

The rotating cylinder head of the instrument has a radius of $a = 5$ cm and a length $L = 10$ cm. The feedback resistor is 22 MΩ. The amplitude of the output voltage U_o for an electric field of 100 volts per meter is in this case $U_o = 77$ millivolts.

in the head of the cylinder. The head rotates at 30 revolutions per second. Exposed to a horizontal electric field F_x (Fig. 1), negative and potive influence charges will flow to the two cylinder segments alternating in sign with the rotation of the head. The current flow caused by the influence charges is amplified, rectified, and recorded. The calculation will show that the amplitude of this alternating current is proportional to the electric field F.

To calculate the charge density on the surface, the field mill may be substituted by an infinitely long cylinder of radius a, exposed to a homogenous electric field F vertical to its axis. However, in reality, the field mill is only a short piece of this infinitely long cylinder, and a better approximation can be achieved if at least half of the infinite cylinder is cut away and the charge density

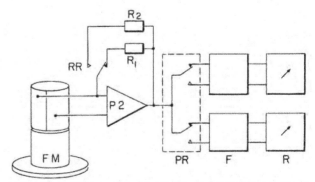

Fig. 3. Block diagram of the electric circuit of the two-component cylindrical field mill. FM cylindrical field mill, $P2$ operational amplifier, RR range relays, R_1, R_2 feedback resistors, PR phase-sensitive rectifier, F filter, R recorder

641

The amplitude of the output voltage U_o is proportional to the absolute value of the field vector of an electric field in a plane vertical to the axis of the field mill. The direction of the field vector with reference to the field mill coordinates can be obtained by recording the phase shift between U_o and a fixed voltage U of an auxiliary generator mounted on the axis of the field mill. This means that the absolute value and the direction of the field vector are the natural outputs of the cylindrical field mill. Using two phase-sensitive rectifiers and filters, the output voltage U_o of the amplifier can be split up into two parts, one of which is proportional to $F_x = F\cos\varphi_1$, and the other one to $F_y = F\sin\varphi_1$, as will be shown in the next section. In this way the two components of the electric field vertical to the axis of the field mill can be obtained.

To calculate the influence charge of the field component F_z parallel to the axis of the cylinder, the field mill cannot be represented by a cylinder of infinite length. Half of the cylinder has to be cut away and the open end closed by a circular disc.

An equal amount of field lines will end on each sensitive area, and a proportional amount of influence charge is thus induced (Fig. 1b). But the difference to the prior case of the horizontal field is that the amount of influence charge due to the axial field remains constant even if the cylinder head is rotating. This follows immediately from the rotational symmetry of the instrument. Therefore, the differentiation with time of the constant charge yields zero output current. This is the reason the cylindrical field mill will not respond to the component of the electric field in the direction of the cylinder axis. The same is true if the field mill should carry a charge of its own as would occur if attached to an airplane charged by the exhaust fumes. An equal amount of charge will flow to the sensitive areas, but this charge will not be modulated by the rotation of the field mill and therefore will not contribute to the current of the horizontal field. Only sudden field changes will generate short pulses of the output voltage. These transient currents would be suppressed by the filter network of the phase-sensitive rectifier.

A certain precaution must be exercised in choosing the spot on which to mount the field mill on the plane. The criterion for a good place is one where the charge density at the sensor areas of the cylindrical field mill, which results from the airplane charge, does not change by the rotation of the cylinder head. This can be checked by measuring the capacity of the sensor at different positions of the cylinder head. The best places are those where the rotational symmetry is greatest, for instance, at the nose or in the middle of the fuselage of the plane.

3. Instrument

In general, the field mill consists of four parts: 1. the field mill proper with the sensitive areas and the drive mechanism, which either rotates a periodically screening shield above the sensitive areas in the case of a shutter field mill or which rotates the cylinder head with its two sensitive segments as is the case of the cylindrical field mill; 2. the amplifier, which amplifies the electric signal, delivered by the field mill, to a level necessary to drive a recorder; 3. phase-sensitive rectifier, which rectifies the ac output of the amplifier to dc (in the case of the cylindrical field mill, there are two phase-sensitive rectifiers 90 degrees out of phase to each other, which produce

3*

outputs proportional to the x and y components of the electric field), and 4. one- or two-channel recorder.

The compactness of the amplifier P_2 allows it to be housed inside the cylinder head (Fig. 2b). This has two advantages: 1. The leads from the sensitive areas to the input of the amplifier can be kept extremely short. Excessive stray or cable capacity is so avoided and therefore the capacity load is very small. 2. Slip rings have to be used to transduct the signal from the rotating head to the fixed base of the field mill. If this can be done at the output of the amplifier instead of at the input, the noise introduced by the slip rings becomes negligible. Recently a rotating capacitor has been used to transmit the signal from the receiving segment in the field mill head to the input terminal of the amplifier, which in this case is located in the base of the field mill. This has all the advantages of short leads and low stray capacity and avoids signal transmission through slip rings.

The disadvantage of placing the amplifier in the cylinder head is that range switching becomes more difficult. The large dynamic range and the extremely low drift and noise level of the amplifier permit range switching at the output for about 10^2. However, the atmospheric electric field may change by a factor of 10^3. To accommodate this wide range, a range relay RR is incorporated in the feedback circuit of the amplifier, which can be operated by remote control and which reduces the feedback resistor by a factor of 100 (Fig. 3). In this way the dynamic range can be used twice. The range switch of this instrument permits the following settings: full deflection of the recorder for 50, 150, 500, 1500, 5000, 15000, 50000, 150000, and 500000 volts per meter.

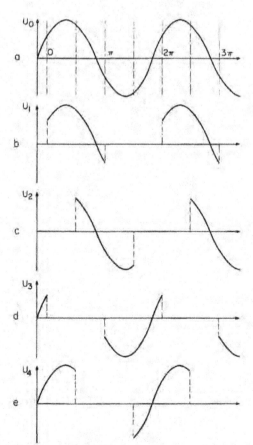

Fig. 4. Output voltage of (a) operational amplifier, (b) rectifier 1, (c) rectifier 2, (d) rectifier 3, (e) rectifier 4

Fig. 4a shows the sinusoidal output voltage U_o of the amplifier P_2. The vertical dashed lines indicate the make and break points of four phase sensitive rectifiers marked PR in Fig. 3. The output voltages U_1 to U_4 of these rectifiers are shown in Figs. 4b—e. These voltages are smoothed by the filter network F, paired and recorded on two strip chart recorders R (Fig. 3). The dc component U_{no} of the voltage U_n ($n = 1, 2, 3, 4$) can be calculated by the first term of a Fourier expansion.

$$U_{no} = 2\varepsilon F 2 a L \omega R \frac{1}{2\pi} \int_{c\pi}^{b\pi} \cos(\omega t + \varphi_1)\, d\omega t. \quad (9)$$

Usually the boundaries of the Fourier integral are 0 and 2π, but as evident from Fig. 4b—e, the functions $U_1 \ldots U_4$ are different from zero only at certain intervals. These intervals are given by the boundaries $c\pi$ and $b\pi$, whereby c and b can be expressed in the following manner:

$$c = \frac{n-1}{2}; \quad b = \frac{n+1}{2}; \quad n = 1, 2, 3, 4. \quad (10)$$

The solution of (9) with (10) yields

$$U_1 = AF \sin \varphi_1$$
$$U_2 = -AF \cos \varphi_1$$
$$U^3 = +AF \sin \varphi_1$$
$$U^4 = +AF \cos \varphi_2 \quad (11)$$

$$\text{when} \quad A = 2\varepsilon 2 a L \omega R/\pi$$

U_1 and U_3 are proportional to $F \sin \varphi_1 = F_y$ and therefore measure the y component of the electric field; while U_2 and U_4 proportional to $F \cos \varphi_1 = F_x$ measure the x component. As U_1 and U_3 as well as U_2 and U_4 have opposite signs, they are paired and fied into the imput terminals of a balanced potentiometer recorder. Thereby the signal strength is doubled, and any dc drift of the operational amplifier is cancelled. With the dimensions and the electrical circuit of the field mill discussed here, a field $F = 100$ volts per meter would produce ± 10 millivolts at the input terminals of the recorder. The recorder sensitivity is adjustable between 10 and 100 millivolt full deflection, which makes it possible to compensate for the different model factors of the different field components.

A few remarks are devoted to the noise voltage introduced by the contact potential. In the case of the shutter field mill, the screening shutter moves closely over the sensitive segments alternately shielding and exposing them to the external field. Any existing contact potential between the shutter and the segments is modulated in the same way as the external field and therefore cannot be cancelled out by a phase-sensitive rectifier. To keep the contact potential low, the shutter and the segments are usually gold or chrome plated. But the slightest scratch in the plating that exposes the substratum material will introduce the contact potential between the substratum material and the gold or chrome, and generate a high noise level. In the case of the cylindrical field mill, the screening shutter is absent. The two sensitive areas do not change their position with respect to each other, even if the cylinder head is rotating. Therefore, an existing contact potential between them is not modulated and cannot introduce an output signal. This

is the reason the noise level of the cylindrical field mill even without gold plating is lower at least by a factor 10 than that of a shutter field mill.

The sensitive areas of the described instrument are made from stainless steel. This brings a decisive advantage if the field mill is used in thunderstorm research. Hailstones impinging on the surface of the field mill with the airplane velocity of about 50 m per second would probably chop off the gold plating, but have no effect on a surface of solid stainless steel.

4. The Calibration Factor

If the field mill is placed in the middle of a sufficently large plate condenser with the field mill axis parallel to the equipotential lines, the applied field can be calculated from the distance of the plates and the voltage difference between them. In this way the basic calibration can be obtained. For airplane measurement, the field distortion factor of the plane itself has to be determined. In our case, one field mill was mounted on the nose and one centered on the lower fuselage (belly) of the plane (Fig. 5). These spots were chosen for their relatively good symmetry with respect to the main axis of the plane, insuring the cancellation of the field of the airplane charge. The nose field mill measures the horizontal and the vertical component F_x and F_z of the atmospheric electric field. The x axis is in the direction of the wings, and the z axis is in the direction upper to lower part (back to belly) of the plane. The fuselage field mill measures the x and y components in the horizontale plane, whereby the x component is again in the direction of the wings and the y component in the direction from the mose to tail of the plane.

Each of these components is affected by a different model factor. The following procedure for their determination was applied.

The atmospheric electric fair weather field is assumed to be vertical. This assumption was proved correct by most of the fair weather flights. The fair weather field was recorded by an atmospheric electric radiosonde, which had been calibrated in a plate condenser at the ground. The model factor M_z for the vertical component of the nose field mill had been obtained by having the airplane circle around the slowly rising radiosonde balloon and comparing the two recorded values. The model factor M_x of the nose field mill was then determined by banking the airplane to a 45-degree angle, which splits the vertical fair weather field into two equal components F_x and F_z with regard to the coordinate system x, y, z fixed to the main axis of the plane. M_x was then determined from M_z. The same method also yielded the model factor M_{x1} of the fuselage field mill. An attempt was made to obtain the remaining model factor M_y by diving, but the airplane could not be tilted down to a

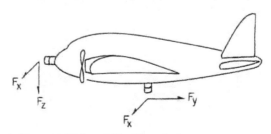

Fig. 5. Nose and fuselage field mill on airplane

45-degree angle. Especially, such a maneuver could not be performed near the ground, where the atmospheric electric field is strongest.

More successful was the method where the plane was flown near a thunderstorm at an altitude where large horizontal fields were present. The plane described a figure 8 where it crosses its previous path by a 90-degree angle. F_y of one path is F_x of the following path and vice versa, so that M_y can be determined from M_x. The figure 8 was run through several times to make sure that the field has not changed during the time necessary to perform the loop.

It should be mentioned that even if the model factors M_x, M_y, M_z are not equal, they are constant; and the phase-sensitive rectifier delivers each component with its correct model factor. This follows very clearly from the principle of superposition of electric fields.

5. The Interpretation of the Record of the Field Cylindrical Field Mill in Flight

The records of the three components of the electric field during actual flight were at first rather confusing and difficult to interpret until the basic pattern had been worked out. Fig. 6 gives 12 pictures of records as they appear if a single positive or negative space charge is approached and passed. The charge may be dead-center to the approaching plane up or down or right or left. For each of these cases a specific combination of the three component traces F_x, F_y, F_z will appear. These combinations are shown in Fig. 6a—l. On each small picture the airplane (seen from the back) is shown at the top with the charge marked by a plus or a minus sign according to its polarity. In the different pictures the charge is now placed in different positions with respect to the plane and the resulting traces of the F_x, F_y, and F_z recorder are shown underneath.

The first 6 pictures (Fig. 6a—f) deal with a positive charge, and the last 6 pictures (Fig. 6g—l) with a negative charge. The shape of the trace of F_y is the same for all pictures with a positive charge. F_y increases from zero to a positive maximum, then crosses the zero line and goes rapidly to a negative maximum after which

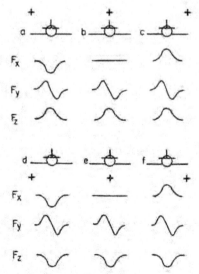

Fig. 6a—f. Records of the field components F_x, F_y, F_z by passing a positive charge (6a—f). The plus sign indicates the position of the charge with respect to the airplane

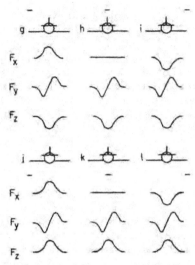

Fig. 6g—l. Records of the field components F_x, F_y, F_z by passing a negative charge (6g—l). The minus sign indicates the position of the charge with respect to the airplane

it decays to zero again. The zero crossing between the positive and the negative maximum indicates the point when the airplane passes through or sideways of the charge center.

The trace of F_y of a negative charge is the mirror image to that of a positive charge (Fig. 6g—l). It goes from zero to a negative maximum, crosses the zero line to a positive maximum, and decays to zero afterwards. Therefore, the sign of the first maximum of F_y identifies the polarity of the charge which the airplane approaches. It is pointed out that under static conditions it is not possible to determine if a positive deflection of F_y is caused by a positive charge in front or by a negative in back, but the movement of the airplane supplies the necessary information to resolve this ambiguity.

With the polarity of the charge determined from the F_y trace, it is easy to interpret the significance of the F_x and F_z traces. These are always simple excursions to one side, either positive or negative. If, for instance, the maximum of the F_z trace has the same polarity as the first maximum of the F_y trace, the charge is located above the airplane level. If the F_z trace remains zero during the excursion of the F_y trace, the charge center is at the same altitude as the plane. Similar simple rules can be worked out for the F_x component. If the F_x trace remains zero (Fig. 6b, e, h, k), the plane moves exactly above, below, or through the center of the space charge. If the maximum of F_x has the same polarity as the first maximum of F_y, the charge center is to the right of the plane; and in case of opposite polarity, it is to the left. These rules are valid for positive as well as negative charges.

It is possible to direct the airplane during the flight towards the charge center or away from it, whichever case is desirable. For instance, in thunderstorm research, where it is essential to locate the main charge centers, the field records may start out as those in Fig. 6c as the plane approaches the storm center. If F_y is still before its first positive maximum, the positive field increase of F_x, F_y, and F_z indicates that a positive charge center is located in front, above, and to the right of the plane. A right turn is then made until F_x becomes zero, as in Fig. 6b. The airplane then flies straight ahead, always keeping $F_x = 0$. At the same time, the plane should climb until $F_{\bar{z}} = 0$ and then level off. If the F_y trace has now passed its positive maximum and changed its polarity to the negative maximum, the zero crossing indicates that at this moment the plane is in the center of the positive charge.

In commercial air traffic it will be desirable to avoid the strong electric fields of thunderstorms because they are the birthplaces of lightning discharges. If in this case the field records indicate that a positive charge is above and to the right of the plane, the flight direction has to be changed to the left until again the F_x component is zero. During the left turn the F_y component passes through zero to its negative maximum, which indicates that the charge center is now to the rear of the plane. The flight is then continued in a straight direction until the field is sufficiently diminished.

Thunderstorms usually contain not only one but different charge centers of different polarities and in different altitudes. In this case the traces of the F_x, F_y, F_z components will be more complex than the simple forms of Fig. 6a—l. But any complex pattern can be analyzed or synthesized by a superposition of the basic patterns of Fig. 6.

It shall be mentioned that a similar analysis can be worked out for the field components in spherical coordinates. Such a representation would conform better to the vector character of the electric field given by vector amplitude and direction. However, the field distortion by the airplane, which results in different model factors for the different x, y, z components, would change the spherical to spheroidal coordinates. To correct this distortion by adjustment of individual amplifiers one would have to use first the representation in F_x, F_y, F_z components and after correction of the model factors calculate electronically the spherical components of the electric field.

References. Matthias, A.: Fortschritte in der Aufklärung der Gewittereinflüsse auf Leitungsanlagen. Elektrizitätswirtschaft **25**, 297—308 (1926). — Kasemir, H. W.: Die Zylinderfeldstärkenmühle mit kleinen Segmenten. Flugfunkforschungsinstitut Oberpfaffenhofen e.V., Außenstelle Gräfelfing. Beauftr. Hochfreque. Fragen. Ref. III A/1306 (1944); — Die Feldkomponentenmühle. Tellus **3**, 240—247 (1951).

Reprinted from Preprint Volume: Fourth Symposium on
Meteorological Observations and Instrumentation, Apr.
10-14, 1978, Denver, Colo. Published by the American
Meteorological Society, Boston, Mass.

ELECTRIC FIELD MEASUREMENTS FROM AIRPLANES

Heinz W. Kasemir

National Oceanic and Atmospheric Administration
Environmental Research Laboratories
Atmospheric Physics and Chemistry Laboratory
Boulder, Colorado

1.   INTRODUCTION

Electric field measurement may be carried out for two different purposes, namely when the electric field itself is the primary objective or when the electric field is used as an indicator for certain meteorological conditions. In the first category fall, for instance, research of the atmospheric electric global circuit, thunderstorm and tornado research, triggered lightning and advanced warning of high field areas, corona discharge that interferes with communication, or the use of the electric field for automatic level flight control of drones or model airplanes. In the second category, are problems relating electric parameters with the turbulence layer, air pollution, fog, and visibility. In cloud physics, high electric fields are good indicators of melting zones or the presence of certain hydrometeors as for instance graupel and hail.

If basic or applied research has to be carried on in the troposphere to about 12 km altitude the airplane equipped with electric sensors is probably the best platform to study electrically related problems. In this paper we will discuss two methods to measure the three components of the atmospheric electric field vector using either radioactive probes or field mills. For each system the discussion is subdivided into three parts, namely construction of the sensors, separation of the field components and the airplane charge, and calibration.

2.   RADIOACTIVE PROBE SYSTEM

Radioactive probe systems for airplane use have been described by Vonnegut et. al., 1961 and Markson, 1976. These systems are restricted to measuring the vertical component of the electric field vector and the main purpose is the investigation of the fair weather field. We will deal here with the problems of recording of all three components and the latter discussed cylindrical field mill system has been used extensively in thunderstorm research including storm penetrations. However, many points in sensor design and calibration procedure mentioned by Vonnegut and Markson are similar to the ones discussed here. For priority credit the reader is referred to these publications. One novel feature of Vonnegut's et. al. design is the capacitive coupling of the sensors to the electrometer amplifier. This eliminates the high ohmic voltage divider with its troublesome high ohm resistor and offers the opportunity to nullify the influence of the airplane charge before the signal is fed into the electrometer amplifier. Lane-Smith, 1977, discusses the dependence of the internal resistor of the radioactive probe on wind speed. Between 10 and 100 m/s wind speed the internal resistor of a 500 microcurie probe is of the order of 10^{10} ohm. This value is in agreement with the one quoted by Vonnegut et. al. Markson mentioned a 5% signal reduction by a 10^{12} ohm load resistance on the probe. This would indicate an internal resistance of $5 \cdot 10^{10}$ ohm of the probe. It is not mentioned if the signal reduction was determined at the ground. According to Lane-Smith the internal resistance may be lower by a factor 5 in a flight wind of 50 m/s.

The radioactive probe mounted on a metal rod in a distance of about .2 to .5 meter above the airplane surface measures the potential difference between the location of the probe and the airplane. Such a potential difference may be caused by a charge on the airplane and an external electric field. The field is then obtained by dividing the potential difference by the distance between probe and reference surface. Hereby it is assumed that the field is constant, i. e., homogeneous between probe and reference surface. At the flat surface of the earth this is a reasonable assumption, however, on an airplane a fairly constant field can be expected only at a few carefully selected locations.

2.1   Construction of Radioactive Probe Sensors

Points to keep in mind when constructing a radioactive probe device are the following:

2.1.1   The internal resistance of a radioactive probe of about 500 microcurie is of the order of 10^{10} ohm. To measure the open circuit voltage with 10% accuracy the input resistor of the amplifiers should be by a factor 10 higher, i.e., 10^{11} ohm. Some investigators have measured the short circuit current instead of the open circuit voltage. This reduces the impedance and insulation requirements substantially. However, this method should be thoroughly tested under flying conditions before it is applied.

2.1.2   If the probe were located 20 cm above

the airplane surface, it would pick up 10,000 volts or more in an external field of 50 kV/m. The high ohm input resistor has to be selected accordingly. Because of the strigent requirement on the linearity of the circuit the temperature as well as the voltage coefficient of the high ohm resistor has to be compensated for in the electrical circuit or taken into account in the calibration.

2.1.3   The insulation between probe and probe mount has to be 5.10^{12} ohm or more under all flying conditions. To avoid condensation of moisture on the exposed insulator surface, heating of the insulator may be required. Deep grooves in the insulator will add substantially to the insulating surface, Vonnegut et. al., 1961, Figure 1.

2.1.4   Mechanic stress on the insulator will produce substantial electrostatic charges (Piezo electric effect). Therefore, the supporting rod should follow the streamlines of the windflow with the mount in front of it as shown in Fig. 1a. A similar design was published by Vonnegut et. al. 1961 and was also used by Markson 1976.

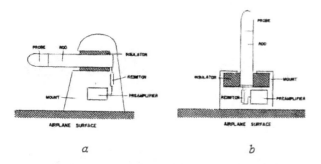

Figure 1.   a. Recommended construction of radioactive probe sensors.
b.   Not recommended construction of radioactive probe sensors.

A construction as shown in Fig. 1b where the rod crosses the stream lines is not recommended because the flight wind will exert strong pressure on the rod that is then transmitted to the insulator.

2.1.5   Mount, rod and probes should have as a unit a sturdy streamlined construction with well rounded corners to avoid corona discharge at higher external fields, Fig. 1a. Corona discharge may be produced on sharp corners in external fields as low as 30 kV/m and will limit the maximum field that can be measured. The short circuit method is especially sensitive to this limitation. Vonnegut et. al. have emphasized the limitation of corona discharge on accurate measurements of fields above 10 to 100,000 V/m. This may be true if the corona discharge starts at the probe mount or on sharp points in the immediate neighborhood of the probe. The corona current is in the order of 1 to maybe 100 microamperes and with an airplane speed of 100 m/s the charge released by corona is drawn out into a line charge of 10^{-8} to 10^{-6} C/m charge density. Such a line charge will generate a radial field of about 100 to 10,000 V/m in in 2 m distance. It depends on the distance of the sensor from corona producing points and

on the strength of the corona current if or if not the corona field is negligable or a measureable fraction of the external field. However, Vonnegut's point is well taken and a closer investigation is necessary if the radioactive probe system is used in thunderstorm research. In respect to the later described cylindrical field mill it shall be mentioned that the instrument has been calibrated in a plate condensor producing fields up to 300 kV/m. No corona has been observed and linearity between field and instrument output has not been impaired. Therefore the author is confident that the nose field mill far away from corona points on the airplane was not affected by corona fields. The redundant record of the F_X component by the nose and the top field mill insures the proper operation of both field mills as long as the two F_X records agree with each other.

2.1.6   The high ohm resistor and the first electrometer amplifier should be located in the probe mount (Fig. 1a) to avoid long cables leading to the cockpit. These cables have a noticable capacity to the grounded screen and produce microphonic noise because they are subjected to airplane vibrations. This will cause excessive damping and a small signal to noise ratio if the cables are located in the high ohmic input circuit.

2.2   Separation of field components and airplane charge

2.2.1   Placement of sensors on the airplane:

One essential factor in the separation of the field components and the airplane charge is the placement of the sensors on the airplane. In an ideal situation each sensor should be affected by only one field component. However, this can be realized only if the sensor carrier is of spherical shape. The best solution for an airplane would be to locate two sensors on the opposite sides of a flat surface, such as the wing or tail fin, that is normal to the direction of the field component to be measured. We define a cartesian coordinate system x, y, z in accordance with the main axes of the airplane so that the x, y, z axes have the positive directions left to right wing tip, tail to nose, and lower to upper part of the fuselage respectively. The x component could be measured by two sensors mounted on opposite points of the vertical tail fin, Figure 2a, and the z component by two sensors mounted on opposite sides of the left or right wing, Figure 2b.

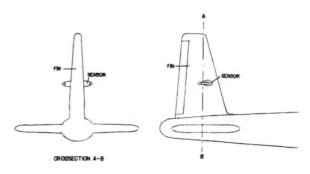

Figure 2a.   Placement of tail fin sensors.

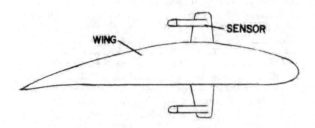

Figure 2b. Placement of wing sensors.

Such an arrangement has the advantage that the fin sensors will measure the same potential of the F_y and F_z component of an external field and of the field caused by a net charge on the plane, however the F_x component will produce voltages of the same amount but of opposite polarities on the two sensors. An analog subtractor circuit will then separate the x component from the influence of all other components and the airplane charge. A similar situation exists for the F_z component with regard to the two wing sensors. The influence of the F_x and F_y components of an external field and of the airplane charge is approximately eliminated by a subtractor circuit so that the output voltage of this circuit is proportional to the F_z component only.

The field concentration factor of the wing or the tail fin may be calculated with sufficient accuracy using as substitutes elliptical cylinders of appropriate width and thickness, Smythe, 1950, P. 94. For the ratio long to small axis of 1m to 20cm the field concentration E/F as a function of the distance D above the cylindrical surface is shown in Figure 3 by the solid line. F is hereby the homogeneous external field and E the field in the neighborhood of the cylinder. The field concentration is shown for the two extreme cases when the probe is located above the flat portion of the cylinder surface in direction of the small axis and in front of the curved portion in the direction of the long axis. Above the small axis the field is quite constant and the field concentration factor close to 1. The value at the surface is 1.2 and 60 cm above the surface 1.15.

The dotted line a little above the solid line gives the field as it would be determined by the radioactive probe method where the measured potential U is divided by the distance D. In figure 3 this field value is normalized to the external field F so that the dotted line shows the parameter U/DF. In 20 cm above the surface there is practically no difference between the actual field and that determined by the radioactive probe method. This offers the advantage that the F_z as well as the F_x component can be calibrated by applying a hard voltage to the rod. The 20% field concentration by the wing or the fin can be compensated by choosing the proper length of the rod or it has to be taken into account by a correction factor. In contrast to the practically homogeneous field above the flat portion of the wing the field E over the curved portion caused this time by an external horizontal field F is quite inhomogeneous. At the surface the field concentration factor E/F is 6 and drops in 20cm distance in front of the surface to about 1.9.

This is shown in Figure 3 by the solid line marked "Horizontal Field". The dotted line also marked "Horizontal Field" is again the field U/DF obtained from the potential U measured by the probe. However, here the value in 20 cm distances is U/DF = 2,83.

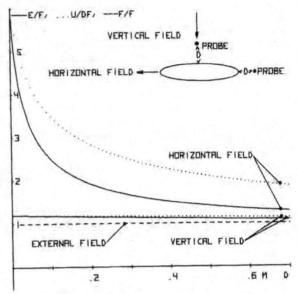

Figure 3. Field concentration caused by an elliptical cylinder exposed to a vertical field F in the direction of the short axis and to a horizontal field F in the direction of the long axis.

Dashed line --- external field F
Solid line ___ calculated field E/F as a function of distance D above the cylindrical surface
Dotted line ... Field U/DF as measured by radioactive sensors.

The main advantage using locations for the sensors where the field is approximately constant with distance from the surface is that this is true for the distorted external field as well as for the field generated by the airplane charge. Adjusting for the slight assymetry of the wing by increasing the amplification of the downward facing sensor—or in Vonnegut's case increasing the coupling capacitor—would result in the same output of the upper and lower sensor for the charge generated field and at the same time for the same absolute value but of opposite polarity of the two outputs due to the external field F_z. This effect will simplify the separation of the field components.

It may be briefly mentioned that the symmetry of the flat surface is required for easy calculation. In case of the wing, however, the slightly curved tear drop cross-section of an airfoil is based on a mathematical method (two dimensional conformal transformation) similar to that used in potential theory. Therefore it is not difficult to calculate the field around a wing model with an airfoil cross-section instead of using an elliptical cylinder. The transformation is given by Smythe, 1950, P. 94 and in more detail by Rothe, Ollendorff and Pohlhausen, 1931.

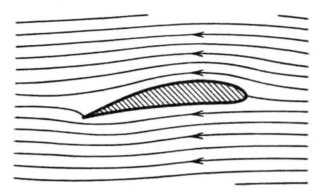

Figure 4. Equipotential lines of a homogeneous field around an airfoil after Rothe, Ollendorff, and Pohlhausen, 1931.

The equipotential lines around an airfoil exposed to vertical field are shown in Figure 4, (copy from the book of Rothe et.al.) Note that the distance between successive equipotential lines above and below the airfoil is approximately constant indicating a constant field, however the spacing above is slightly smaller than below, meaning that the field concentration factor is somewhat larger above than below the airfoil.

The whole discussion of sensor placement is based on the assumption that the plane is an equipotential surface, i.e. that the surface is conductive. Unfortunately the common custom to paint the plane with silicon based paint produces an excellent insulating surface. Accumulation of surface charge produced by friction of dust or precipitation particles may cause uncontrolled electric fields. Our remedy here was to render a large area around the sensor conductive by applying conductive tape or paint. This is a point that is usually neglected or not recognized.

2.2.2 Separation of field components and calibration:

The following calculation and procedure to separate and calibrate the field components is valid for radioactive sensors as well as shutter-type field mills. It has been shown by Kasemir, 1951, that with four sensors as a minimum all three field components and the airplane charge can be obtained. All four sensor outputs S_1, S_2, S_3, S_4 are functions of F_x, F_y, F_z. and Q. In general the field concentration factors $a_n \ldots d_n$ for each component and the charge will be different for each sensor. The symmetric placement of the sensors discussed above will make these factors equal for the wing and the tail sensor pair some of them with opposite signs. This simplifies the separating analog computer circuit given in a block diagram in Figure 5, and the separation procedure. The field concentration factors for the F_z and F_x component can be calculated for the fin and wing sensors as outlined above. The consistency of these calculations can be checked by 45° banking turns in a fair weather field, as will be discussed later. The field concentration factor for F_y is determined by a dive-climb operation of the aircraft and the influence of the airplane charge Q is eliminated by charging the airplane in flight

with a corona point driven by a high voltage supply operated from inside the plane. The separation and calibration procedure is carried out in a fair weather field where only the F_z is different from zero. The F_x and F_y components are produced artificially by banking and dive-climb operation of the aircraft. The charge of the aircraft is changed by corona emission. The separation of the field components and the charge is achieved by potentiometers P_n adjusted so that in turn the influence of the charge and the not desired components are eliminated from the output of the individual computer units using the proper flight operations.

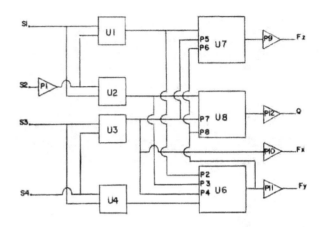

Figure 5. Block diagram of an analog computer circuit to separate the three components F_x, F_y, F_z of an external field and the airplane charge, Q.

As shown in Figure 5, the computer consists of 7 adder-subtractor units represented by square boxes with code numbers U_1 to U_4 and U_6 to U_8. The symbols U_n refer to the output voltage of the unit and to the corresponding term in the following calculation. The units U_6 to U_8 have weighted inputs adjustable by potentiometers P_2 to P_8. In the calculation the parameters P_n, n = 1 ... 12 identify the adjustable factors set by the potentiometers. The triangular symbols in Figure 5 represent operational amplifiers with adjustable gain. They are identified by their potentiometers P_1 and P_9 to P_{12}. S_1 and S_2 on the left side of Figure 5 are the output voltages of the upper and lower wing sensors preamplifiers and S_3 and S_4 those of the right and left fin sensor amplifiers. At the right side of Figure 5 the outputs of P_9 to P_{12} are proportional to F_z, F_x, and F_y and Q. These amplifiers are provided for setting convenient scale factors during the calibration procedure.

Because of the symmetric placement of the sensor mentioned above we have the following four equations for the four sensor outputs:

$$S_1 = a_1F_x + b_1F_y + c_1F_z + d_1U \qquad (1)$$

$$S_2 = a_2F_x + b_2F_y - c_1F_z + d_1U \qquad (2)$$

$$S_3 = a_3F_x - b_3F_y + c_3F_z + d_3U \qquad (3)$$

$$S_4 = -a_3F_x - b_3F_y + c_3F_z + d_3U \qquad (4)$$

We build the sum and differences of the wing and fin sensors with the adder-subtractor units U_1 to U_4.

$$U_1 = S_1 - S_2 = 2 c_1 F_z + (a_1 - a_2) F_x + (b_1 - b_2) F_y \qquad (5)$$

$$U_2 = S_1 + S_2 = (a_1 + a_2) F_x + (b_1 + b_2) F_y + 2d_1 Q \qquad (6)$$

$$U_3 = S_3 - S_4 = 2a_3 F_x \qquad (7)$$

$$U_4 = S_3 + S_4 = -2b_3 F_y + 2c_3 F_z + 2d_3 Q \qquad (8)$$

In level flight $F_x = F_y = 0$ and

$$U_1 = 2c_1 F_z; \quad U_2 = 2d_1 U; \quad U_3 = 2a_3 F_x; \quad U_4 = 2d_3 Q \qquad (9)$$

We check now the independence of U_1 and U_3 from Q by changing the charge of the airplane with the corona point. If U_1 shows to be influenced by the Q change, P_1 is adjusted so that the influence is eliminated. P_1 will balance the slight asymmetry of the airfoil profile and should agree with the calculation. U_3 should be zero and remain so during charge change assuming that probe strength, preamplifier gain, position, etc. of the fin sensors are identical. In equation (1) and (2) it is assumed that the balancing of the wing sensors is already accomplished.

We now clear U_4 from the influence of F_x F_z, and Q in adder-subtractor unit U_6. Here the following operation is performed:

$$U_6 = U_4 - P_2 U_1 - P_3 U_2 + P_4 U_3 \qquad (8)$$

In level flight $F_x = F_y = 0$ and U_6 reduces to

$$U_6 = 2c_3 F_z + 2d_3 U - P_2 2c_1 F_z - P_3 2d_1 Q \qquad (8a)$$

changing charge and adjusting P_3 until U_6 remains constant results in

$$U_6 = 2c_3 F_z - P_2 2C_1 F_z; \quad 2d_3 = P_3 2d_1 \qquad (8b)$$

We adjust P_2 until $U_6 = 0$. This eliminates F_z when

$$2c_3 = P_2 2c_1 \qquad (8c)$$

To eliminate F_x we use banking terms. Introducing the neglected F_x terms in U_6 we have

$$U_6 = -P_2 (a_1 - a_2) F_x - P_3 (a_1 + a_2) F_x + P_4 2a_3 F_x$$

During the banking term we adjust P_4 so that U_6 again is zero.

$$P_4 2a_3 = P_2(a_1 - a_2) + P_3 (a_1 + a_2) \qquad (8d)$$

U_6 will respond now only to the F_y component. Collecting the F_y terms from U_1, U_2 and U_4

$$U_6 = -[2b_3 + P_2(b_1 - b_2) + P_3(b_1 + b_2)]F_y = -AF_y \qquad (10)$$

With F_x and F_y separated we can clear U_1 and U_2 from their influence.

$$U_7 = U_1 - P_5 U_3 + P_6 U_6 = 2c_1 F_z + (a_1 - a_2) F_x + (b_1 - b_2) F_y - P_5 2a_3 F_x - P_6 A F_y \qquad (11)$$

$$U_8 = U_2 - P_7 U_3 + P_8 U_6 = 2d_1 Q + (a_1 + a_2) F_x + (b_1 + b_2) F_y - P_7 2a_3 F_x - P_8 A F_y \qquad (12)$$

We adjust P_5 and P_7 during banking turns and P_6 and P_8 during dive-climb operation until the influence of F_x and F_y on F_z and Q vanishes. Then it is

$$U_7 = 2c_1 F_z; \quad U_8 = 2d_1 Q \qquad (13)$$

$$a_1 - a_2 = P_5 2a_3; \quad b_1 - b_2 = P_6 A \\ a_1 + a_2 = P_7 2a_3; \quad b_1 + b_2 = P_8 A \qquad (14)$$

During the bank and dive-climb operation F_z will attenuate according to $\cos \alpha$, where α is the bank or dive-climb tilt angle of the plane. This should not be interpreted as an influence of the F_y and F_x component. The contribution of these components will have opposite polarities in left and right banks and dive and climb. Therefore adjustments should be made to the same attenuation of F_z using always the same tilt angle for left and right banks and dive and climb. It is quite possible that $a_1 = a_2$ and $b_1 = b_2$. This depends on the symmetry of the wing with regard to these components and the location of the wing probes. This situation exists if with broken connection to P_5 and P_6 the attenuation of F_z during left and right bank and during climb and dive is the same. In this case the adder-subtractor U_7 is unnecessary and the whole separation procedure simplifies accordingly.

2.2.3   The calibration procedure would start at the ground. Let's assume that we have calculated that the field concentration factor for the upper wing sensor is $c_1 = 1.2$. In a 100V/m external F_z field the upper sensor with the probe being 20cm above the wing surface would be at 24V potential. We apply 24V to the sensor and adjust P_9 so that our meter or recorder shows a convenient deflection. Since in flight the lower sensor will add the same voltage, the deflection of our meter in a 100 V/m F_z field would be twice as much, meaning that the present deflection will correspond to an external F_z field of 50 V/m. If the airfoil calculation showed that the field concentration below the wing is only $c_2 = 1.15$ we apply -23V to the lower sensor and adjust P_1 so that the lower sensor will cause the same deflection on the meter as the upper sensor with the 24V input voltage. If the field concentration factor a_3 of the F_x component for the wing sensors is calculated we set P_{10} so that for a 100 V/m external F_x field the same deflection occurs on the F_x meter as for a 100 V/m F_z field on the F_z meter. P_1, P_9, and P_{10} should not need to be readjusted in flight if the calculation is sufficiently representative and strength of probes, location of sensors, high ohmic voltage divider, preamplifier gain etc. for all sensors is sufficiently accurate. In flight we will have to execute the separation procedure, that is partly only a check on our ground setting, go through exact 45° banking turns to check the constancy of our absolute F_z and F_x ground calibration and in climb and dive operation set P_{11} so the $F_y = F_z \sin \alpha$, where α is the tilt angle of the plane and F_z is vertical field in level flight.

3. THE CYLINDRICAL FIELD MILL SYSTEM

Figure 6. Field mill with excentric ring for charge compensation.

Figure 7. Field mill with NRL charge compensator.

The cylindrical field mill (Figures 6 and 7) has been designed specifically for use on an airplane. The principal features of construction and operation have been described by Kasemir, 1951, and 1972. Therefore, it will suffice to mention briefly some important points.

3.1   The field mill measures the surface charge density, in the case of the shutter type field mill, on its sensitive plates or, in the case of the cylindrical field mill, on a cylindrical surface. The surface charge density is proportional to the field at the surface. The homogeneity of the field for a certain distance above the field mill is not required for easy calibration as it is for the radioactive probe system. If the field mill is attached to the airplane it forms a part of the airplane surface.

3.2   The input impedance of the field mill is in the order of 1 megohm as compared to 100,000 megohm of the radioactive probe. This makes it much easier to maintain the required insulation in all weather conditions.

3.3   The surface charge density is modulated by the rotating shutter of the shutter type field mill or by the rotating cylindrical head of the cylindrical field mill. This converts the d.c. into an a.c. input current or voltage. The amplification by a factor 10^5 can be achieved with a two stage amplifier. Therefore the required range of 1 V/m to 500 kV/m can be covered by calibrated range switches located in the two amplifier stages. This avoids the problem of

zero drift of a d.c. amplifier and of exceeding the dynamic range of the amplifier.

3.4   The limit of sensitivity is set mainly by contact potential and is about 5 V/m for the shutter type field mill and about .1 V/m for the cylindrical field mill. The better sensitivity of the cylindrical field mill is due to the fact that the modulation of the contact potential is avoided by the rotational symmetry of the cylindrical head. Based on the same principle the influence of the airplane charge is eliminated before the signal is amplified. Since this is an important point it will be discussed later in more detail.

3.5   The cylindrical field mill measures the field vector in a plane orthogonal to its axis. For instance in case of the nose field mill the axis is in the y direction and the field mill will measure the field vector in the x - z plane. By a phase sensitive rectifier the field vector can be split into the F_x and F_y component of the external field. The top field mill on the upper side of the fuselage has its axis in the z direction and consequently records the F_x and F_y component of the external field. Therefore, the nose and top field mills record all three components of the external field. Hereby, the F_x component is measured twice and gives an excellent continuous check on the proper performance of both field mills.

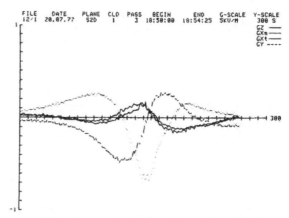

Figure 8. Gradient G_z, G_{xn}, G_{xt}, G_y in flight passing a thunderstorm.

Figure 8 shows a record of the three field components recorded at Kennedy Space Center by the airplane S2D of the Naval Research Laboratory during a path tangential to a thunderstorm. The two F_x components are shown by solid lines. The occasional small deviations between the two lines give a good idea of the overall accuracy. The apparent noise in the record shown in Figure 8 is not inherent in the field mills and subsequent amplifiers. It is caused by insufficient shock mount of the tape recorder. The author is not aware of a redundant record of a radioactive probe or a field mill system. However, redundant records are very important to check the proper operation of the system. It is worthwhile to mention that the two cylindrical field mills can replace six radioactive probe sensors or six shutter type field mills. In addition the airplane charge is automatically eliminated and an analog computer circuit to separate the different components is not required.

651

3.6   The elimination of the influence of the airplane charge on the field measurement is accomplished by a.) placement of the field mills and b.) by an excentric ring clamped on the base for the field mill as shown for the nose field mill in Figure 6. If the airplane were a perfect symmetric body, let's say a sphere, a cylindrical field mill mounted on this body would not record the field caused by a charge on the body. If the head of the mill with its two sensor plates rotates, the same amount of field lines will end, or better, emerge from the sensor plates regardless of their position during rotation. Consequently the surface charge on the plates in not modulated by always the same, and no a. c. signal is passed through the amplifier. If the body is not symmetric with respect to the rotational axis of the field mill as is obvious by the different slants of the nose cone in Figure 6, the asymmetry is compensated by the excentric lip pointing downward in Figure 6. The right position of the excentric ring is found by charging the airplane in flight and recording the produced charge vector. At the ground the ring is then placed in a new position and the new charge vector recorded. The vector difference shows the influence of the excentric ring and an improved position of the ring can be determined. These steps have to be repeated until a satisfactory compensation of the asymmetry of the airplane by the ring is accomplished for the nose and top field mills. The underlying principle here is essentially the same as nulling of a capacity bridge. It is also similar to the method of the adjustable capacitive coupler used by Vonnegut et. al., 1961. Charge compensation in the order of 100:1 has been accomplished with this method. Elimination of the influence of the airplane charge is essential, however, the described process is tedious, time consuming and costly in flight hours.

Very recently a charge compensator has been constructed by Wolfram Kasemir of the Naval Research Laboratory as shown in Figure 7. (Personal communication). Four rods extend through a screening bowl mounted on the base of the field mill. The rods are arrayed in pairs in the x and z plane of the nose mill. From each pair one rod can be extended farther out of the bowl or retracted into the bowl by a small motor controlled from the cabin. The airplane is now charged and the influence of the charge on the F_x and F_z component can be nullified by adjusting the extension of the corresponding movable rod. A similar device is mounted on the top field mill. Charge compensation down to the most sensitive range can be accomplished in 5 minutes.

3.7   Separation of the field components is done at the ground by placing a rubbed teflon rod underneath the nose field mill or in front of the top field mill. The charged teflon rod produces in the first case a strong vertical field in z direction and in the second case a strong horizontal field in y direction. The phase of the rectifiers is then adjusted so that the F_x meters have zero deflection and the F_z or F_y meters maximum deflection. The whole separation procedure of the field components can be done at the ground in a very short time.

3.8   The calibration of the cylindrical

field mill system follows a pattern similar to the calibration of the radioactive probe system. First the F_z component has to be calibrated. Since the nose field mill of the parked airplane is about 1.50 to 2 m above the ground the effect of the ground is only in the order of 10 to 20%. The nose field mill can be calibrated in fair weather at the ground against a calibrated ground field mill. By comparing the field record before and after touch down during landing the correction factor between the flying and grounded airplane is determined. This correction factor will be different for different airplanes. In the case of the NRL plane S2D the field measured by the flying airplane is about 7% less than that measured by the plane on the ground. Against this ground calibrated and corrected F_z component the F_x component is calibrated in flight by right and left 45° banking turns. The F_y component is calibrated against the F_z component during climbs and dives of a measured tilt angle.

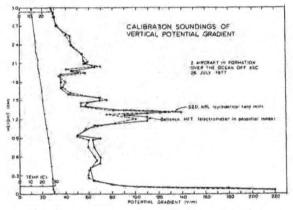

Figure 9. Comparison flight of Bellanca MIT with radioactive probes and S2D NRL with cylindrical field mills.

Figure 9 shows a comparison flight between the Bellanca of MIT equipped with radioactive sensors and the NRL plane S2D equipped with cylindrical field mills. The airplanes were flying side by side ascending from ground level to about 3 km altitude. The field decayed from a ground level value of 220 V/m to about 28 V/m at 3 km altitude but with sudden strong increases and decreases in small layers of 100 to 200 m thickness. The records of both planes display this fine structure with good agreement and the absolute calibration is practically identical. To the authors knowledge this is the first comparison flight between two airplanes equipped with different field measuring devices and the agreement is truly remarkable.

4.   REFERENCES

Kasemir, H. W.: "Die Feldkomponentenmühle", *Tellus, 3*, p. 240-247, 1951.

Kasemir, H. W.: "The Cylindrical Field Mill", *Met. Rundschau, 25, No. 2*, p. 33-38, 1972.

Lane-Smith, D. R.: "Review of instrumentation for atmospheric electricity", p. 189-201, *Electrical Processes in Atmospheres,* Ed. Hans Dolezalek, Reinold Reiter, Dietrich Steinkopf Verlag, Darmstadt, 1977.

Markson, R.: "Ionspheric Potential Variations Obtained from Aircraft Measurement of Potential Gradient." *J.G.R.*, *Vol. 81, No. 12*, p. 1980-1990, 1976.

Rothe, Ollendorff, Pohlhausen: "Funktionentheorie und ihre Anwendung in der Technik", Springer Verlag, Berlin, 1931.

Smythe, W. R.: "Static and Dynamic Electricity", McGraw-Hill Book Co., New York, 1950.

Vonnegut, B., C. B. Moore, and F. J. Mallahan: "Adjustable Potential-Gradient-Measuring Apparatus for Airplane Use", *J. G. R.*, *Vol. 66, No. 8*, p. 2393-2397, 1961

EVALUATION OF MIGHTY MOUSE

TRIGGER ROCKET FLIGHTS

CONTRACT NO. CC-88025
IMMEDIATE OBJECTIVES, PART D

PROGRESS REPORT

Project Leader: H. W. Kasemir
              NOAA - APCL R31
              Boulder, Colorado 80302

Principal Investigator: E. L. Magaziner
              NOAA - APCL R31
              Boulder, Colorado 80302

CONTENTS

LIST OF FIGURES

I.   INTRODUCTION

The purpose of the Mighty Mouse trigger rockets is to initiate lightning strokes when fired into a thunderstorm or into a cloud where high electric fields are present. The primary instrument which directs the rocket launches is the NASA-6 aircraft, equipped with the cylindrical field mills. A description of the role of NASA-6 in launch direction appears in another report on this contract: "The Role of the NASA-6 in the Mighty Mouse Experiments."

Several instruments are in use or development for the evaluation of the trigger rocket flights. These include a capacitively loaded antenna to record primarily the occurrence of lightning, a lightning plotting system to determine the positions of lightnings, and a transmitter borne by the triggering rocket which signals a lightning strike to the rocket. It is clear that when all three instruments are operational, the triggering of a lightning by a rocket can be unambigously determined. The occurrence of the lightning can be verified by the antenna. The lightning plotter can verify that the lightning position coincides with that of the rocket. The rocket borne transmitter signal can verify that the lightning stroke was not a near miss.

Data available at the present consists entirely of lightning occurrence recordings. Nevertheless, this data can be used to show that the rockets triggered lightnings. There remains uncertainty as to which and how many lightnings were rocket triggered.

II. ROCKET FIRINGS AT KSC, JUNE, 1971

Lightning occurrence recordings for periods during eighteen trigger rocket launches were made available by KSC from their June 1971 tests. Thirty six sixty second portions were extracted from the data, eighteen from T_0 to $T_0 + 60$ for each launch time T_0, and eighteen randomly chosen sixty second portions during which no rockets were in flight. The sixty second intervals from the first group were superimposed so that the launch times co-incide and the number of lightnings from $T - 1$ to $T + 1$ for each T from T_0 to $T_0 + 60$ were counted. The results of this appear in Figure 1, top, and will be referred to as the experimental group. The intervals from the second group were also superimposed obtaining the control group which appears in Figure 1, bottom.

The control group represents a natural, undisturbed record of lightning activity during the time of the experiments. The mean for the control case is 2.79 which means that, since there are eighteen cases, the mean number of lightnings in any two second interval is .156. The standard deviation for the control case is 1.23. Statistically, this means that the probability of choosing a two second interval with more than .33 lightnings is less than .01.

The experimental group appears quite different from the control group. The most outstanding features are the two peaks at $T_o + 28$ and $T_o + 37$. A statistic depending on the relative ranks of the control and experimental data points is used to estimate the significance of the apparent differences between the two groups. The null hypothesis is that the trigger rockets had no effect, specifically, that if f and g are the cumulative distribution functions for the control and experimental group respectively, then f = g. The result is that the null hypothesis can be rejected at the .008 significance level and hence that the rockets did indeed have an effect. In addition, the statistical test used is consistent with respect to the class of alternatives f > g and hence it can be significantly concluded that the experimental data is stochastically larger than the control data, that is, the effect of the rockets was to increase the number of lightnings.

The two peaks in the experimental group are highly significant. These peaks represent .500 and .333 lightnings in a two ·second interval, and as mentioned above, the probability of this occurring naturally is less than .01. In addition, the zero value separating the two peaks is 2.3 standard deviations from the mean and hence is significant on the .011 level. The physical interpretation is that the trigger rockets first initiate lightning strikes approximately twenty eight seconds after launch. The zero at thirty two seconds indicates an absence of strokes following the discharge by the rocket. The second triggering is at approximately thirty eight seconds after launch and is again followed by an absence of lightning strokes.

It is not possible to determine from the present data whether each rocket triggers two strokes, some rockets trigger two strokes or rockets tend to trigger strokes either at twenty eight or at thirty eight seconds after launch.

III. ROCKET FIRINGS AT SOCORRO IN 1970, 1971

The scarcity of firing opportunities at the Socorro experiments was discussed and illustrated at the September 21, 1971 meeting with KSC personnel. The only clearly successful test occurred on August 14, 1970. A description and evaluation of this experiment appears in a memorandum from Dr. Kasemir to Mr. Bailey dated August 19, 1970. On page 5 of the memorandum, a series of three shots which triggered lightnings at 61.0, 59.5 and 60.5 seconds after launch are cited. A statistical analysis of the lightning data recorded during the experiment was performed for this report. The results indicate that the probability of a natural occurrence of two lightning strokes after equal ($\pm$.5 seconds) time intervals from two consecutive rocket launches is only .06 and hence there is little doubt that the three strokes were rocket triggered.

IV. CONCLUSIONS

The results in this report were deduced from records of lightning occurrence only but already show that the trigger rockets perform as expected. A follow-up using all of the instrumentation described in the Introduction is now most desirable to obtain accurate data on critical parameters such as position of rocket in the cloud at time of trigger, length of time that cloud remains discharged, differences between triggered and natural lightnings, etc. It is this type of data that will lead to the final objectives of the program.

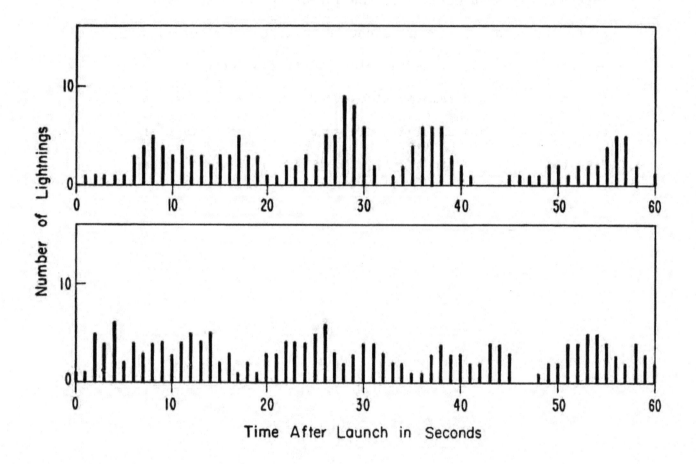

Figure 1. Top: The Experimental Group for the KSC Rocket Launches.
Bottom: The Control Group for the KSC Rocket Launches.

Airborne Warning Systems for Natural and Aircraft-Initiated Lightning

LEE W. PARKER, MEMBER, IEEE, AND HEINZ W. KASEMIR

Abstract—Possible airborne warning systems other than radar for avoidance of thunderstorms and lightning strikes to aircraft can include one or more of the various types of instruments surveyed here. One class of warning system considered is concerned with lightning strikes initiated ("triggered") by the aircraft itself when in high-field environments within electrified but "nonthundery" clouds. Since data show high correlations between strikes to aircraft and high-field environments, warnings can be provided by various types of field detectors considered. The other class of warning system is concerned with distant thunderstorm warnings provided by the many different types of RF sferics detectors considered, and by optical detectors. The sferics detectors sense electromagnetic fields radiated by lightning and by prelightning discharges. They include crossed-loop, multiple-loop, time-of-arrival, interferometer, pulse-height, spectral-amplitude, and *E*- and *H*-amplitude systems.

Manuscript received January 11, 1982.
L. W. Parker is with Lee W. Parker, Inc., 252 Lexington Road, Concord, MA 01742. (617) 369-5370.
H. W. Kasemir is with Colorado Scientific Research Corp., 1604 S. County Road 15, Route 1, Berthoud, CO 80513.

Direction-finding errors are discussed in detail, especially site errors (generally unrecognized) of crossed loops mounted on aircraft. Theoretical assessments of the latter are made using analytical models as well as a 3-dimensional computer model to simulate aircraft structures.

Key Words—Warning systems, lightning, natural, aircraft-initiated, RF detectors, optical detectors, errors

I. INTRODUCTION

THE PRESENT survey was undertaken because a need exists for reliable and inexpensive (light-weight) airborne lightning warning and avoidance systems. Two applications of warning systems would be in a) warnings of distant storms, enabling a pilot to avoid severe weather and possible encounters with hail, ice, and dangerous air motions, and in b) warnings of possible imminent lightning strikes to the aircraft in electrified clouds. Use of the familiar airborne weather radar is relatively straightforward with respect to a), but not with respect to b). One disadvantage, however, of airborne weather radar is its relatively high cost. Another is the difficulty sometimes encountered, due to attenuation-masking effects, in identifying thunderstorms hidden behind heavy showers. Airborne radar, the masking problem, and relations of precipitation echoes to lightning, are discussed in [1].

We deal here with two general classes of lightning warning systems, distant warnings and near-zone warnings. The distant-warning detectors can be electrical or optical in nature. By detecting the electromagnetic radiation emitted by lightning and predischarges ("atmospherics" or more briefly "sferics"), RF detectors can warn of the existence of electrical activity at a distance. The electrical activity is usually associated with severe weather. The connection is not well understood at present. A number of electrification mechanisms have been hypothesized. A recent survey of the status of theoretical electrification models has been made by Parker [2].

The paper is consequently divided into two parts. The first part is concerned with detectors providing near-range warnings. These are electrostatic-field detectors that can be useful for warning and avoidance of lightning strikes to the aircraft. They warn of the presence of high electric fields. High electric fields exist, not only in thunderstorms that may already contain lightning, but also in nonthundery clouds where their presence may cause the aircraft to inadvertently initiate or "trigger" a strike. It may be inferred from pilot reports that these triggered-lightning incidents outnumber by far the natural lightning strikes (Clifford and Kasemir, this issue).

We present in Section II some lightning-strike and electric-field statistics associated with aircraft, and discuss the physical background of lightning triggering by aircraft. Some original ideas, reported for the first time in [1] and highly relevant to the design of warning systems for triggered strikes, are discussed briefly.

As candidates for electric-field detection instumentation, we consider in Section III field mills, radioactive probes, and corona points. These are primarily near-range detectors. They have attractive characteristics and have been operated on (research) aircraft. Moreover, their capability may be extended by capacitive bypasses so as also to detect RF pulses from distant lightning.

The second part of the paper is concerned with detectors providing distant warnings. Sections IV, V, and VI deal with bearing detection (direction-finding), primarily by magnetic loops (IV), time-of-arrival (TOA) and interferometer VHF systems (V), and optical systems (VI), respectively. (A UHF system, the "electrograph", is included as well in Section V) Section VII treats range detection, primarily by single-station instruments. These include crossed loops, a 500-kHz pulse-height detector, an electric-field-amplitude detector, and an H/E amplitude ratio detector, among others.

Also discussed are the bearing errors of crossed loops, especially site errors associated with their location on the airplane. Since these can be serious even when the instrument is mounted symmetrially on a fuselage, we have developed a 3-D computer program for predicting these errors on realistic airplane geometries. Numerical solutions are reported for a T-39 airplane, previously used in test flights. (Solutions are consistent with results using simple analytic models.)

By a combination of theory and experiment, one may determine correction factors for site errors affecting a crossed loop, for any given airplane geometry. The results of this determination would suggest optimum locations for the placement of the loop on the airplane. The correction factors need be determined only once. If this is done, the site errors can be completely eliminated by suitable adjustment of the electronic amplification.

Section VIII consists of summary remarks on the detection/location systems for distant lightning and their bearing errors. The reader is referred to [1] for some tentative rankings of all systems considered as candidates for airborne use, according to availability, simplicity of construction, maintenance, calibration and interpretation of data, extent tested, potential accuracy, versatility, and insusceptibility to errors. The versatility category is of interest in that some instruments can have more than one function, e.g., both range and bearing detection, or both near-range and far-range detection. Many possible combinations of different types of instrumentation may be effective, e.g., field and optical detectors in addition to radar.

We believe that many of the detectors surveyed have virtues worthy of consideration for airborne warning systems. However, they require careful testing and evaluation.

II. LIGHTNING-STRIKE AND FIELD STATISTICS

A. Strike Statistics

In the last few years, a large amount of information on lightning strikes to aircraft has been collected and reported in the literature (for example, by Fitzgerald [3]–[5], Corn [6], DuBro [7], Clifford [8], Fisher and Plumer [9], Newman and Robb [10], and Harrison [11]). Airline pilot surveys show that over 80 percent of the lightning strikes to modern jet airliners occur within clouds, in precipitation and in turbulence. In over half of the strikes, "St. Elmo's fire" (corona) was observed before the strike. Most strikes occur near the freezing level, from $-5°C$ to $+5°C$, and generally while descending or ascending through about 3-km altitude.

Lightning is a hazard to both commercial and military aircraft. Ways and means have to be found to minimize the lightning danger to flight personnel and passengers, and the lightning damage to airplanes and electronic equipment.

(Incidentally, fly-by-wire and digital-avionics aircarft may be extremely vulnerable.) The development of accurate and inexpensive avoidance instruments is one approach to minimizing the hazard. If the electrical conditions leading to discharges involving the airplane can be determined, instruments and data systems can be developed to provide warnings and avoidance procedures to pilots. We are concerned here with such warning and avoidance systems. Although the electrical conditions of discharges are discussed in the paper by Clifford and Kasemir (this issue), a brief discussion of the physics involved is warranted here, oriented toward the instrumentation that would be required.

Pilot observations relevant to the warning and avoidance problem are summarized by Clifford [8] as follows.

Pilots generally agree that there are two distinct classes of lightning strikes to aircraft in flight. The first and most common variety usually occurs while flying in precipitation at temperatures near freezing. This type is preceded by a buildup of static noise in the communication gear, due to corona (visible at night). The buildup may continue for several seconds before the strike occurs.

The second variety occurs abruptly without warning. It is most likely to be encountered in or near ongoing thunderstorms, in contrast to the former variety which is often experienced in precipitation that has no connection with thunderstorms. Pilots tend to believe that the slow buildup type of discharge is not a true lightning strike but rather a discharge of excess charge ("P-static") built up on the aircarft by flight through the precipitation. This nonthunderstorm type greatly outnumbers the other. Both kinds can create a brilliant flash and a boom which can be heard throughout the airplane.

We believe that, in the "common" variety of discharge, the pilots are experiencing a strike initiated or "triggered" by the aircraft upon entering a region of high electric field. This view was proposed by us earlier [1]. Triggering of lightning by aircraft in high fields has also been discussed by Fitzgerald [3]-[5], Vonnegut [12], Pierce [13], [14], Shaeffer [15], and Kasemir and Perkins [16], among others. The rare variety of strike, that occurs without warning, is an accidental hit of the airplane by a natural lightning that originates somewhere else in an ongoing thunderstorm.

It is now also evident that large rockets can initiate strikes. After the Apollo 12 incident, where the rocket was struck by lightning twice, it was pointed out by one of us (HWK) that the Apollo spacecraft was not struck accidentally by two lightnings generated by a cloud that otherwise did not produce lightning discharges, but that the rocket triggered these lightnings by itself, upon penetrating high-electric-field regions in the cloud. This explanation was accepted since the alternative, namely, that the cloud waited just to countdown zero and then fired two lightning bolts at the rocket while not being capable of producing other lightning discharges before or after launch, appears highly unlikely.

The distinction between the natural and triggered strike is essential for the development of warning and avoidance instrumentation. We will discuss briefly some basic features of a lightning discharge that are relevant to the triggering problem.

The electric field in a thunderstorm is usually not high enough to reach the breakdown value of 2000 to 3000 kV/m (the lower values occuring at higher altitudes). However, a raindrop, being electrostatically similar to a conductive sphere, has a field concentration factor of 3 at the upper and lower points (at its "poles") in the direction of the thunderstorm field. Therefore, the external field necessary to cause breakdown is thereby reduced by a factor of 3. In addition to this effect, the raindrop will deform into a spheroidal shape under the influence of the field, and the field concentration factors at the poles or tips will increase accordingly. Values of 5 to 10 may be reached by this deformation. This brings the thunderstorm field required to produce breakdown at the tips down to values of 300 to 400 kV/m, values that have been measured in thunderstorms. If breakdown is reached at the tips of an elongated raindrop, the raindrop will go into corona discharge. In this stage, the raindrop is similar to an airplane in that corona discharge is observed by the pilots before a lightning is triggered. The corona discharge is a relatively stable discharge and, without a further energy input, the plane as well as the drop could remain in corona discharge until the external field drops below the breakdown value. Note that, although corona may be produced by high fields and precipitation charging (see below), not all corona develops into a major discharge.

With respect to triggering, the airplane has advantages over the raindrop, as follows. The field-concentration factors on wingtips, rudder, antenna masts, etc., are larger than on a raindrop. The capacity of the aircraft is larger than that of a raindrop, thus providing more energy-storage capability. The speed of the airplane is much larger than the fall velocity of a raindrop, leading to a much higher precipitation charging rate by the impact of other precipitation particles. These considerations explain why the airplane may still trigger a lightning discharge under conditions where the raindrop is not capable of doing so, i.e., in a nonthundery but highly electrified cloud. This explanation also fits with the pilots' observation that electric discharges triggered by airplanes in nonthundery clouds outnumber those occurring in thunderstorms.

With respect to triggering by precipitation charging, it is commonly argued, according to Clifford [8], that sufficient charge cannot be built up on an airplane to supply the energy required to produce a discharge that looks and sounds like lightning, so that, therefore, the pilots' explanation cannot be valid. It is our opinion (as discussed in more detail in [1]) that the role played by precipitation charging is as follows. The main electrical energy for the common type of lightning strike will be provided by the field of an electrified cloud. With respect to the main energy supply, there is no difference between a triggered and a natural lightning. However, the electric charge on the aircraft due to precipitation ("P-static"), which is especially strong in the melting zone of a cloud, may contribute to the triggering of a lightning discharge by the aircraft in cases where the field concentrations at the extremities of the aircraft are not sufficient to inititate a lightning discharge. Therefore, the role of precipitation charging is not to provide energy for a full-grown lightning, but to convert the corona discharge into long streamers which can then grow in the external cloud field into a proper lightning discharge.

All thunderstorm warning devices based on the electromagnetic waves emitted by lighning discharges (Sections IV-VII) are blind to the presence of electric fields too low to produce natural lightning but high enough to produce airplane-triggered lightning discharges. One needs instrumentation capable of detecting electric fields. Among the possible instruments are the field mill, the radioactive probe and the corona point, which will be discussed in the next section. (These instruments are versatile in that they have the dual capability of also being operated so as to detect sferics radiated by distant lightning.)

Fitzgerald [3]–[5] has analyzed data from instrumented aircraft penetrating thunderstorms during the 1965 Rough Rider program. His findings are summarized in [1] and in the paper by Clifford and Kasemir (this issue). Thunderstorm penetrations with instrumented aircraft that have been struck include those made by Cobb and Holitza [18], Nanevicz *et al.* [19], Musil and Prodan [20], and a French research aircarft [21]. Their observations further corroborate those considered in this section. Presently on-going penetration experiments with the intention of eliciting strikes to an instrumented aircraft are those of NASA [22], using an armored F-106. These experiments were continued through the summer of 1981 (F. L. Pitts, personal communication).

It is appropriate to end this discussion with the following quotations from [3]. In 1946, L. P. Harrison suggested [3] that "the field distortion or augmentation created by the presence of the aircraft may raise an initially high but subcritical potential gradient to the level where breakdown occurs at or near the aircraft. If conditions are suitable, the streamer could then continue to propagate between charge centers and a discharge would occur." Moreover, intense strikes are frequently associated with dissipating storms. Here we quote Fitzgerald [3].

"The data presented suggest that thunderstorms in their early stages of dissipation retain sufficiently large charge centers to account for one or more lightning discharges if a suitable means of initiating a streamer becomes available. It is likely that an aircraft entering a storm in this condition will act to 'trigger' a lightning discharge. These clouds may have little turbulence and no distinctive echo pattern on a typical Air Traffic Control radar. In normal IFR flight operations in regions with thunderstorms merged with showers and cloud decks, the routine radar avoidance of the presently most-active storm portions may readily lead to flight through a decaying storm and the possibility of an isolated lightning incident to the aircraft."

B. Corona and Triggering

In a set of relevant ground-based experiments, the effects of corona-producing points on the trigger-breakdown field of the Shuttle-Orbiter were investigated by Kasemir and Perkins [16]. They used a scale model of the spacecraft placed between the plates of a large-plate condenser. As part of the investigation, they also used a highly-polished spheroid to determine the trigger-breakdown field in the absence of, and in the presence of, corona-producing points on the spheroid. Three principal results are the following.

a) The field-enhancement factor of a spheroid can be calculated analytically, and the external breakdown field determined. An excellent agreement was found between the theoretically-calculated and the experimentally-measured breakdown field.

b) The trigger-breakdown field is about 33 percent less with corona points than without (see a)).

c) The percent reduction of the trigger-breakdown field (33 percent) due to the presence of corona points is only weakly dependent on the nature of the points (form, length, sharpness, etc.).

Important ramifications of these results regarding triggered lightning strikes to aircraft are that

a) not only is the likelihood of a triggered strike enhanced (33 percent lower trigger fields) by the presence of corona (in addition to altitude-dependent reduction of the trigger field), but that

b) all aircraft have so many sharp metallic protrusions acting as possible corona points (antennas, pitot tubes, landing gear, nuts, lightning arresters, exhaust nozzle rims, edges, corners, etc.) that they are essentially always in corona in strong fields.

Ordinarily the corona effect is not taken into account in the literature.

C. Field Statistics

We now turn to a discussion of the field distribution in a thundercloud (Figs. 1 and 2, previously unpublished). Fig. 1 shows a graph of the vertical gradient in a particular storm versus altitude or temperature. The horizontal axis gives the gradient from −200 kV/m to +250 kV/m. The vertical axis gives the altitude from about 3.8 to 6 km and has a temperature scale from +3° to −12°C. The data were collected by two airplanes equipped with field mills making repeated (26) storm penetrations at different altitudes during the NOAA lightning-suppression project (by Holitza and Kasemir [23], and Kasemir *et al.* [24]). The base of the cloud was at 4-km altitude. The peak of the negative gradient of −150 kV/m occurred at about 4.9-km altitude at a temperature of −2°C; the maximum positive gradient of 200 kV/m occurred at 5.5-km altitude or the −10°C temperature level. The lower negative peak gradient indicates the altitude level where ground discharges would originate and the upper positive gradient where intracloud discharges would originate. However, both gradient values were, in this case, too low to initiate lightning discharges.

It is of interest to note here that the thicknesses of the high-gradient layers are rather small, of the order of only 200 to 300 m.[1] This narrowness is even more apparent in Fig. 2. The vertical axis again gives the altitude; the horizontal axis represents time and is marked with the days from July 27 to August 21, 1974. On thunderstorm days, measurements were carried out of the type discussed in connection with Fig. 1. In Fig. 2, however, only altitudes of cloud top and base (marked by crosses), of maximum positive gradient (marked by circles), and maximum negative gradient (marked by triangles) are given. The numbers above the gradient symbols

---

[1] The horizontal dimension may, however, extend up to several kilometers.

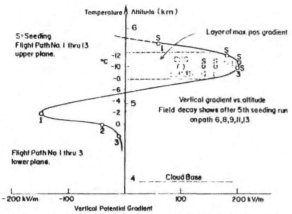

Fig. 1. Potential gradient versus altitude in thunderstorms.

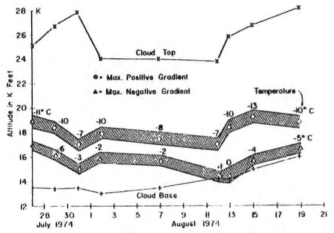

Fig. 2. High-gradient layers and associated temperatures in thunderstorms.

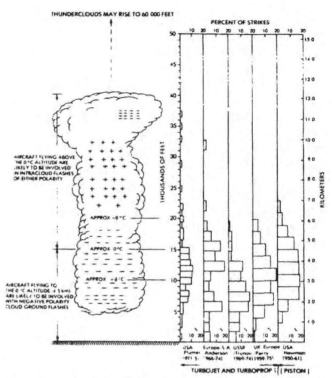

Fig. 3. Aircraft lightning-strike incidence versus altitude (after Fisher and Plumer, 1977 [9]).

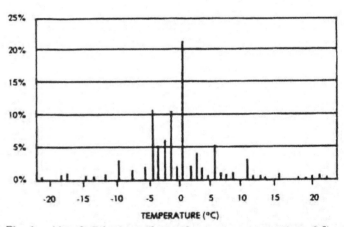

Fig. 4. Aircraft lightning-strike incidence versus temperature (after Fisher and Plumer, 1977 [9]).

indicate the temperature. The boundary lines of the shaded bands mark the 60-kV/m values enclosing the maximum positive gradient in the upper band, and the −60-kV/m gradient values enclosing the maximum negative gradient in the lower band. We see that maximum positive gradients have been located between the −7° to −13°C temperature levels, and maximum negative gradients between the +1° and −6°C levels. The thickness of the bands is about 1000 ft = 300 m. It is important to note that the temperatures as well as altitude levels of the maximum gradients are not fixed to an exact temperature or altitude but may change from day to day.

The thunderstorms were small-to-moderate heat storms occurring in Colorado and Wyoming, with ground level about 5 kft above sea level. A word of caution should be added regarding Figs. 1 and 2. The blank area in Fig. 2 between 20 and 28 kft altitude does not mean that this layer doesn't contain high gradients. It simply means that this area could not be investigated because it was above the ceiling of the aircraft. (For gradient data at higher altitudes, see Fitzgerald [4].) It should also be mentioned that it is not easy to locate the high-gradient areas. Without the records of all three components of the gradient provided by the field mills and by on-the-spot evaluation, it could not have been done at all. Even so, it is easily possible that some high-gradient area could have been missed, since they seem to occur in narrowly restricted regions. The importance of such data to avionics

does not have to be stressed. The pursuit of this type of research to include different storms (frontal storms, severe storms, storms over land with lower elevation, storms over sea, storms in different seasons, nonthundery clouds, etc.) is highly recommended.

A tentative conclusion can be drawn from Figs. 1 and 2. The thinness of the high-gradient areas suggests that a quick change in altitude should be made if an airplane finds itself in the state of continuously increasing corona discharge. The change should always be to *higher* altitudes even though there is a chance of penetrating the higher positive-gradient layer. A triggered lightning in the lower negative-gradient region may penetrate to the ground and involve the plane in a ground discharge. The lightning triggered in the upper high-gradient area will most likely be an intracloud discharge that doesn't have the destructive power of a ground discharge. (See natural lightning analysis by Kasemir [78].)

It is interesting to compare Figs. 1 and 2 with the figures "aircraft lightning-strike incidence versus altitude," and "lightning strikes to aircraft as a function of temperature," presented by Fisher and Plumer [9] and reproduced here as Figs. 3 and 4. Most of the lightning strikes in Fig. 3 occur in an altitude range of 5 to 18 kft and in Fig. 4 in a temperature range of $-10°$ to $+6°$C. The higher altitude and lower temperature boundary in Figs. 3 and 4 are in excellent agreement with the boundaries given in Figs. 1 and 2. However, the lower altitude (5 kft) and higher temperature ($+6°$C) boundary in Figs. 3 and 4 show a much lower altitude than the boundary (14 kft, $+1°$C) given in Fig. 2. This difference can be easily explained by the high ground level of 5600 feet above sea level in Colorado and Wyoming. Therefore, within reason, the agreement between the two sets of data has to be classified as good.

Some relationships of lightning activity (and strikes to aircraft) to radar echoes have been considered by Fitzgerald [25] (see also [1, sec. VI]).

III. FIELD MEASUREMENTS

We consider in this section 3 types of instruments for measuring electrostatic fields, all suitable for airborne use. We discuss first the field mill, then the radioactive probe, and finally the corona-point detector, all well-known instruments that have been used extensively for many years. A considerable amount of historical and technical data on these and many other instruments may be found in the comprehensive texts by Israël [27] and Chalmers [28].

A. Field Mills

The field mill is an electrostatic voltmeter, of rugged design but high sensitivity. This is an electrostatic induction type of instrument. A forerunner of the modern version was designed by C. T. R. Wilson [29]. There are many possible designs for such an instrument, e.g., planar-shutter field mill, cylindrical field mill, rotating wire, etc. In the same family of instruments as the field mill are the electrostatic flux-meter (test plate moving, not fixed), induction voltmeter and agrimeter (test plate grounded when exposed, connected to measuring instrument when shielded [28]). We will discuss two types, the planar-shutter field mill and the cylindrical field mill. Field mills are capable of measuring atmospheric electric fields over a wide range, from 1 V/m to 500 000 V/m. Since high electric fields are a characteristic feature of thunderstorms, the field mill is an ideal tool for thunderstorm research and is widely used for that purpose. Its capability of small and compact design makes it suitable for airborne application.

An extensive planar-shutter-type field-mill network at the ground has been in operation at Kennedy Space Center for thunderstorm warnings during the launch of important rockets, for instance the Apollo series, the Mariner program, and so on. Many airplanes have been equipped with field mills for thunderstorm warning, and for thunderstorm and lightning research projects sponsored by federal agencies. For instance, 8 airplanes equipped with field mills participated at the Apollo-Soyuz launch at Kennedy Space Center. Other examples include projects conducted by the U.S. Army in 1956–1958,

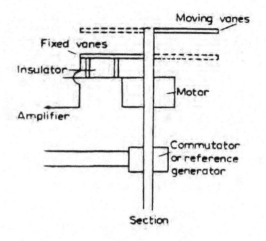

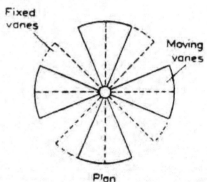

Fig. 5. Field-mill schematic, planar-shutter type (after Chalmers, 1967 [28]).

the U.S. Air Force Rough Rider lightning-research project, the U.S. Navy fog-research project, the NASA lightning-triggering project, the NOAA lightning-suppression project, and the TRIP projects 1976–1979. It should be mentioned that in the NASA lightning-triggering project (rockets fired to trigger lightning), in the NOAA lightning-suppression project (chaff seeding to dissipate the charge concentrations), and in the Air Force Rough Rider project (investigating probabilities of lightning strikes), the field-mill-equipped aircraft were used to locate the high-electric-field areas in the storm with the intention of *penetrating* and not avoiding these areas of maximum lightning probability.

Briefly, in a planar-shutter type of field mill, a rotating grounded electrode alternately shields a fixed insulated electrode from, and exposes it to, the electric field external to the instrument. A diagram of this type of field mill is shown in Fig. 5 (from Chalmers [28]). The ac voltage output of the fixed electrode is proportional to the strength of the field. The planar-shutter type field mill was designed to measure the atmospheric electric field at the ground (see Israel [27, vol. 2]). In this case, the field has only a vertical component. The sensitivity is about 5 V/m. The instrument has a frequency range, the upper limit of which is determined by the rotation rate of the shutter and the number of vane sectors, that is, the frequency with which the openings are covered and uncovered. Typically, for example, the upper frequency may be 10 Hz for a 30-Hz rotation frequency. With a capacitive bypass, however, the frequency range can be extended into the RF region and above (Smith [30]). This means that short pulses ("field changes") radiated by lightning discharges can be

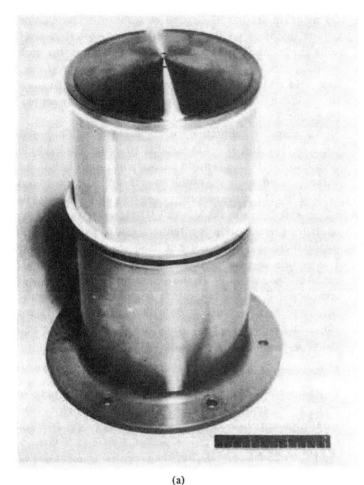

(a)

(b)

(c)

Fig. 6. Cylindrical field mill. (a) Upper left: Two halves, with upper half the rotating split-cylinder sensor head, the lower half the stationary base. White strips are Teflon insulators. (b) Upper right: Opened sensor head with one cylinder-segment removed. (c) Bottom: Mounting on airplane nose, with eccentric "humpring" charge compensator added. Dark line along center is a reflection of unknown origin.

detected. The limitation here is given by the design of the electronic amplifiers, and not by the sensor head (i.e., the field mill proper).

In general, the electric-field vector may be measured by using multiple field mills with different orientations to infer the three components of the field. At the ground, however, the electric-field vector is normal to the earth's surface. Therefore, one field mill suffices to obain the vector.

The extension of the ground-field-mill instrument to airborne operation involves a system of at least 4 planar-shutter-type field mills, mounted, if possible, at symmetric locations on the airplane surface, e.g., the two wingtips, or the top and bottom of the fuselage, or similar positions. With an analog or digital computer circuit, the four inputs can be analyzed, yielding the three components of the external field vector, separated from the field generated by the electrostatic charge on the airplane.

Airborne versions of the planar-shutter field mill have been designed and flown by Gunn [31], Kasemir [32], [33], Clark [34], and Fitzgerald [35], and more recently by C. R. Holmes at New Mexico Tech and by D. E. Olson at the University of Minnesota (personal communications).

The cylindrical field mill (Fig. 6) was specifically designed for airborne use by Kasemir [36], [37]. Its rotating electrodes alternately shield and expose one another, but neither is grounded. Its sensitivity is 1 V/m, which gives it a larger range of detection of thunderstorms than the planar-shutter type. The basic frequency range is also 0 to 10 Hz, with the above-mentioned extendability to RF frequencies via a capacitive bypass. The cylindrical field mill has an important advantage in being much less influenced by the airplane charge. This influence can be reduced significantly by use of a charge compensator or "humpring" [37] (Fig. 6).

On an aircraft, two cylindrical field mills are sufficient, one measuring two of the three components of the field vector, the other measuring the third component, plus one component already obtained by the first mill. This redundant measurement of one of the field components by both field mills has proven to be valuable and provides a continuous check on the proper operation of the field mills. Thus two cylindrical field mills are equivalent to five shutter-type mills.

The vertical cylindrical field mill is sensitive to noise produced by rain because one of the sensitive cylinders is exposed to impact by rain. If the shutter-type field mills are mounted in such a way that none of them is facing the flight direction they are protected from direct impact of rain, and they are comparatively noise free.

Compared with RF sferics sensors (discussed later, Section IV), the field mill is, in its static form, a short-range warning device capable of sensing fields and thunderstorms at distances up to about 10 mi. At this distance lightning discharge pulses are also clearly visible on the record. However, using the capacitive bypass or RF mode, the dc and ELF part of the signal is filtered out. This eliminates the capability of recording thunderstorm fields, but increases the signal-to-noise ratio of the lightning field. Therefore, lightning discharges can be detected at distances up to 100 mi. It should be mentioned that the use of the dc and RF modes of operation are not mutually exclusive, but can be used simultaneously in separate channels.

Besides providing lightning information, the field-measuring capability of a field mill has two distinct advantages. First, it warns of the possibility of triggered lightning in highly-electrified clouds that contain no natural lightning (see discussion in Section II). Second, it warns of the production of corona discharge on the airplane that may interfere severely with communication. The field-vector display in the cockpit would, in both cases, indicate the direction away from the high-field area. (Kasemir [36] discusses how the "language of the field mills" may be interpreted to yield the desired information.)

An absolute calibration of an airplane field-mill system requires time and effort at installation. Since the field mill is used mostly as a scientific instrument, its calibration, accuracy, response, etc., problems are fully recognized and dealt with. The same problems exist also for the calibration of RF sensors (Sections IV, V, and VII). However, in this case the necessity of calibration is often disregarded as being of minor importance, or sometimes not even recognized as existing This results in bearing and range errors discussed later. Because of the similarity of the problem, much can be learned regarding RF sensor calibration from field-mill calibrations. However, to date one has been content here with a coarse empirical calibration, and errors, for instance site errors discussed later, remain usually unrecognized.

B. Radioactive Probes

Radioactive probes (collectors) have long been used to measure the potential of a point in space with respect to the ground potential, even before the electron tube was invented (see Israël [27, vol. 2]). The potential difference divided by the height above ground gives the potential gradient, provided the gradient is constant. The probe operates by ionizing the surrounding air within a roughly spherical region, of radius about 5 cm at sea level and 10 cm at 10-km altitude. The ionized and slightly conductive air discharges or charges the probe until its potential is essentially equal to that of its near environment. Parachute dropsonde versions were used in 1928 and 1940, and a radiosonde in 1942 (Israël [27, vol. 2]).

The amount of radioactive material involved depends on the material used. Typically, 10 μCi of radium or 100-500 μCi of polonium may be used. Polonium is a pure alpha-emitter and is considered safe (used commercially to prevent static charging on phonograph records).

In principle, this is an extremely simple device. However, there are a number of problems that have to be considered in the construction of a field-measuring device using radioactive probes. Some of these are the following.

The probe must be well insulated from the ground (or airplane surface). The insulation requirement is severe: insulation resistance should not be less than 10^{13} to 10^{14} Ω. The slightest film of moisture due to humid air, condensation, etc., would short-circuit the probe to ground. (However, Teflon is a hydrophobic insulator, and with it the problem may not be too severe (R. Markson, personal communication).

The load placed on the radioactive probe must have an impedance of the order of at least 10^{12} Ω so that the probe can maintain its open-circuit voltage. In presently-available technology, this requirement dictates the use of field-effect transistors (FET's) as first-stage amplifiers. These amplifiers can handle comfortably input voltages in the range from millivolts to several tens of volts. However, they would be damaged by 50 V or 100 V at their input terminals. On the other hand, a radioactive probe located at 50 cm above ground would supply in normal fair weather of 100 V/m an input voltage of 50 V. In a moderate thunderstorm field of 10 kV/m, it would supply an input voltage of 5 kV.

Hence the FET amplifier requires a high-ohmic voltage divider to reduce the input voltage. However, high-ohmic resistors have not only a temperature coefficient but also a voltage coefficient. That is, their resistance value drops significantly for voltages higher than about 1000 V.

In addition, the radioactive probe itself saturates at fields on the order of 5 kV/m.

In spite of all these problems, the radioactive probe is used as a field meter since it also has some unique advantages. It and the corona detector (below) are to our knowledge the only field-measuring devices that have no moving parts. It is of light weight, and has negligible power requirements including the electronics. Its construction costs are very moderate. It is essentially a fair-weather instrument, but its capability for measuring small fields down to 1 V/m or less would make it useful for detection of thunderstorm fields at a large distance.

Radioactive probes have been operated successfully on aircraft outside clouds by Vonnegut *et al.* [38] and Markson [39]. In the only airborne comparison test known to us in which a radioactive probe system and a field-mill system were flown simultaneously on two separate airplanes flying side by side, the radioactive-probe data and field-mill data tracked each other extremely well [37].

C. Corona Detector

The relation between the current I flowing from a corona point and the electric field E producing the corona discharge is given, according to Whipple and Scrase [40], by

$$I = a(E^2 - E_0{}^2); E \geqslant E_0 \qquad (1)$$

where a is called the "corona constant" and has an empirical value in the range 10^{-13} to 10^{-16} A·m²/V². Corona discharge does not start until the field at the corona point exceeds the breakdown value in air. The ratio of the field at the point to the corona-producing field E is the field concentration factor. This factor is fairly constant for the individual point, but it depends on the height of the point above ground, usually given by the length of the rod carrying the point, and on the radius of curvature of the point. Therefore, large differences in the field concentration factor exist from point to point.

The field concentration factor determines the onset field E_0, which has a value of about 1 kV/m for a field concentration factor of 3000. If the external field E is smaller than the onset field, the corona current is zero (not negative). The onset field and corona constant for positive fields differ by about 10 percent from those for negative fields. Moreover, these parameters depend on the wind. This wind dependence in airborne applications has been extensively investigated by Chapman [41].

An intensive study of corona discharges has been made by Loeb and his co-workers [42]. Ground-based corona-current detection of high fields to warn of a possible first lightning strike is discussed in [83].

Balloon-borne corona points were used as early as 1937 and 1941 by Simpson and Scrase [43] and Simpson and Robinson [44] to determine the polarity of electric fields in thunderclouds. More recently, measurements of thunderstorm fields using corona points were made by Weber and Few [45] using balloon-borne ("coronasonde") payloads, and by L. Ruhnke [46] using Mighty Mouse rocket-borne payloads. At Ruhnke's suggestion, R. Markson (personal communication) complemented his airborne radioactive-probe system by an equivalent corona-point system, in order to extend the field-measuring range of the radioactive probes to higher field values. Markson's results showed that the two systems tracked one another well in their overlap range (R. Markson, personal communication).

The combination of these two devices (radioactive probes and corona points) deserves further investigation. Both systems are simple and lightweight. The corona points have the advantage of being operated in the short-circuit mode and therefore do not need high-ohmic insulation. They have the additional advantage of being rugged and fairly immune to rain or humidity. Furthermore, they would be more sensitive to corona discharge than the communication equipment of the airplane, and can be designed as a system to measure all three components of the electric-field vector.

D. Extension to Long-Distance (Sferics) Detection

As with field mills, radioactive probes and corona detectors are primarily short-range warning devices, but can also be extended to detect RF pulses from distant lightning. The extension again consists of using capacitive bypasses. The rod used to mount the radioactive probe or the corona point would serve as an antenna for lightning RF signals.

IV. RF SFERICS BEARING DETECTION (DIRECTION-FINDING) BY MAGNETIC LOOPS

The electromagnetic radiation emitted by a lightning discharge is usually known to radio or TV users as "atmospherics" or "sferics". It can be used in different ways to determine the direction of the source of the emission, i.e., the lightning discharge. We will consider in this section magnetic loops. TOA systems and interferometer systems will be discussed in a later section (Section V). We begin with a discussion of the theory of the crossed-loop system and its errors. Following this, we will consider several systems based on magnetic loops.

A. Crossed Magnetic Loops as Direction Finders

Low-frequency direction-finding by narrowband magnetic loops is a rather old and well-known technique (e.g., Watson Watt and Herd [47]). The theory of the magnetic crossed loop is as follows.

A time-varying magnetic field penetrating the area enclosed by a wire loop will induce in the wire a voltage and current proportional to the time rate-of-change of the magnetic field and to the area of the loop. If the normal vector of the loop surface and the magnetic-field vector are parallel, the induced voltage is a maximum. If the two vectors are orthogonal the induced voltage is zero. At an arbitrary angle α of incidence, the induced voltage depends on the cosine of the angle α. If there are two loops whose planes are at right angles to one another (crossed loops) the induced voltages in the two loops are proportional to $\cos \alpha$ for one loop and to $\sin \alpha$ for the other loop. The ratio of the two loop voltages is a function of the incidence angle α, but not of the magnetic field strength since this cancels out. From this ratio, therefore, the azimuthal direction of the lightning discharge is easily determined. Note that only two components of the magnetic vector can be detected with a crossed-loop antenna. Since the antenna is usually mounted on a horizontal surface (e.g., the top of the fuselage) the visibility of the crossed loop is limited to the horizontal plane. Kohl [48] has attempted to obtain all 3 components of the magnetic vector with his multiple-loop antenna. (See discussion below.) There remains, however, a $180°$ ambiguity in the two-loop system [1]. This can be resolved by using an omnidirectional electric-field antenna in conjunction with the crossed loops. However, electrical noise or uncontrolled phase shifts in the electronic circuits may make the system unreliable.

B. Errors in Bearing Detection (Direction-Finding) by Crossed Loops

In theoretical discussions, it is usually assumed that the electromagmetic field is linearly polarized, implying that the source current flows in a straight line. The lightning channel, however, is not a straight line, but generally follows a rather tortuous path with branches in all directions. Such a source current has many significant horizontal components. Consequently, the assumption of linear polarization can result in rather large errors in direction determination. This source of error will not be considered here. Another source of error not considered here is that due to the sky-wave from ionospheric reflection, but for our purposes the effect should be negligible [1]. The two principlal sources of errors are assumed to those due to 1) nonvertical channels, and 2) site errors.

1) Slanted-(Nonvertical)-Lightning-Channel Errors: Even if the current can be assumed to flow in a straight line, in general it is nonvertically oriented. If we are restricted to detection of two magnetic-field components (parallel to the ground plane), the "slant" lightning channel, if it is at any significant altitude, plus its image in the conducting ground plane, will cause the crossed loop to determine an erroneous direction. That is, the magnetic vector at the crossed loop, instead of being perpendicular to the line of sight will have another direction and a different source direction (perpendicular to the observed magnetic vector) will be inferred.

This effect has been analyzed by Kalakowsky and Lewis [49] and Uman et al. [50], but only with the crossed-loop detector at the ground. This situation results in simplifications. The more-general case where the detector is also at a nonnegligible altitude (i.e., airborne) above a conducting ground plane does not appear to have been considered in the literature. We present the following results of our analysis of the more general case (see details in [1]).

The error in direction due to a slanted lightning channel may be expressed as

$$\text{"misdirection" angle} = \arctan \frac{GP_y/P_z}{1 + GP_x/P_z} \qquad (2)$$

where P_x, P_y, and P_z are the components of the lightning dipole radiator (P_x horizontal in direction of observer, P_y horizontal and transverse to this, P_z vertical), and where G is a geometrical factor

$$G = (H + Qh)/D \qquad (3)$$

where D is the horizontal distance from the dipole to the detector, H is the altitude of the dipole, and h is the altitude of the aircraft, both above a conducting ground plane. Q is given by

$$Q = [(x_1{}^3 - x_2{}^3)\dot{P} + (x_1{}^2 - x_2{}^2)\ddot{P}/c]/$$
$$[(x_1{}^3 + x_2{}^3)\dot{P} + (x_1{}^2 + x_2{}^2)\ddot{P}/c] \qquad (4)$$

where $x_1 = [D^2 + (H - h)^2]^{-1/2}$, $x_2 = [D^2 + (H + h)^2]^{-1/2}$, and $\dot{P}$ and $\ddot{P}$ denote first and second time-derivatives of the moment P. In the case of harmonic radiation of wavenumber k, we may replace $\ddot{P}/c$ by $jk\dot{P}$, so that Q becomes a complex function of k, x_1, and x_2. Note that Q vanishes when the detector is on the ground plane ($h = 0$), and $G = H/D$ from (3) then leads to the simple result given by [49] and [50].

2) Site Error (Analytical Model): Another kind of error leading to errors in direction-finding by magnetic crossed loops is reradiation (or scattering) of the incident magnetic field by nearby conducting surfaces (site error). When the crossed loop is mounted on an aircraft conducting surface, this error can be important. We are unaware of any treatment of this effect for airborne application (although the work of Horner (e.g., [51]) is well known for treating site errors at the ground due to buried cables, hills, etc.). An extensive analysis of this effect for airborne application is given in [1], where we assumed the magnetostatic limit (wavelength $\gg$ aircraft dimensions) and solved Laplace's equation to obtain the perturbation field. For an arbitrary angle of incident direction, the longitudinal and transverse magnetic-field components are enhanced by different factors, giving rise to misdirection or site error. (This approach was suggested by C. E. Baum, personal communication. The analysis here is new.) This misdirection error depends on the geometry of the aircraft and the position where the detector is mounted. Some key results and sample applications follow.

Assume that the fuselage is modeled by a prolate spheroid and that the detector is centered on top. If the incident direction of the wave is θ_0 (where $\theta_0 = 0°$ and $90°$ if the direction of incidence is parallel and perpendicular, respectively, to the long axis), then it can be shown [1] that the "misdirection"

672

TABLE I
MISDIRECTION (BEARING ERROR) OF CROSSED LOOP ON
SPHEROIDAL FUSELAGE $\equiv \Delta\theta$, IN DEGREES

incident angle θ_0 (degrees)	needle limit t=0.0 R=0.5	t=0.1 R=0.5207	t=0.2 R=0.5591	t=0.5 R=0.7100	sphere limit t=1.0 R=1.0
0	0.	0.	0.	0.	0.
5	- 2.50	- 2.39	- 2.20	- 1.45	0.
10	- 4.96	- 4.75	- 4.37	- 2.86	0.
15	- 7.37	- 7.05	- 6.48	- 4.23	0.
20	- 9.69	- 9.26	- 8.50	- 5.51	0.
25	-11.88	-11.34	-10.39	- 6.68	0.
30	-13.90	-13.26	-12.11	- 7.71	0.
35	-15.70	-14.96	-13.62	- 8.57	0.
40	-17.24	-16.39	-14.87	- 9.22	0.
45	-18.43	-17.48	-15.79	- 9.63	0.
50	-19.21	-18.16	-16.33	- 9.76	0.
55	-19.47	-18.35	-16.40	- 9.60	0.
60	-19.11	-17.94	-15.93	- 9.12	0.
65	-18.00	-16.83	-14.83	- 8.30	0.
70	-16.05	-14.94	-13.07	- 7.14	0.
75	-13.19	-12.22	-10.61	- 5.68	0.
80	- 9.43	- 8.70	- 7.51	- 3.95	0.
85	- 4.92	- 4.53	- 3.90	- 2.02	0.
90	0.	0.	0.	0.	0.

($\Delta\theta = \theta - \theta_0$; θ = sensed angle; θ_0 = incident angle; t = aspect ratio = minor/major axis ratio; R = misdirection factor, (6)).

angle θ is given by the analytical expression

"misdirection" angle $\theta = \arctan(R \tan\theta_0)$ (5)

(error $\equiv \theta - \theta_0$)

where the factor R is a function of the aspect ratio t (ratio of minor to major axes),

$$R(t) = 0.5(1 - 2t^2 + N)/(1 - N) \qquad (6)$$

with N defined by

$$N(t) = \frac{t^2}{2(1-t^2)^{1/2}} \ln\left[\frac{1 + (1-t^2)^{1/2}}{1 - (1-t^2)^{1/2}}\right]. \qquad (7)$$

As the aspect ratio varies from the limit $t \to 0$ (an infinitely-long cylinder) to the opposite limit $t \to 1$ (sphere), the factor R varies from 0.5 to 1.0. The misdirection error, defined by $\Delta\theta = \theta - \theta_0$, is tabulated as a function of θ_0 in Table I, for values of $t = 0, 0.1, 0.2, 0.5,$ and 1.0. (The corresponding values of R are 0.5, 0.5207, 0.5591, 0.7100, and 1.0.) Note that the misdirection error is zero if θ_0 is $0°$ or $90°$, in all cases, and has a maximum magnitude for some θ_0.

Thus if the fuselage is modeled by an infinitely-long cyl-inder $(t = 0)$, Table I shows that the largest misdirection error (independent of cylinder radius) is $-19.5°$, occurring at an incident direction $\theta_0 = 55°$. For an aspect ratio $t = 0.1$ (approximating an F-106 Delta Dart), the largest error is $-18.4°$, at $\theta_0 = 55°$. For an aspect ratio $t = 0.2$ (more nearly characterizing a C-130 geometry which lies between 0.1 and 0.2), the largest error is $-16.4°$, at $\theta_0 = 55°$.

As t increases further (the fuselage becomes thicker compared with its length), the maximum error decreases toward zero, while the corresponding θ_0 moves slowly toward $45°$. The errors increase if the detector position is moved toward the nose or tail.

Also treated analytically in [1] is the case of a cylinder of elliptic cross section, which can model situations where the detector is on a wing or vertical tailfin. For these simpler cases, (5) is again valid, but with a different ratio factor R now defined by $1/(1 + t')$, where t' is the aspect ratio (ratio of vertical-to-horizontal axes). Thus for a circular cylinder, $t' = 1$ and $R = 0.5$, yielding results as above. In the middle of a flat wing, $t' \ll 1$, $R \cong 1$ and the error is small. On the top edge of a thin tailfin, $t' \gg 1$, $R \ll 1$, θ becomes zero, and the maximum error becomes the approximately constant value $-\theta_0$, except near $90°$.

3) Site Error (Numerical Model–T-39 Aircraft): For more-detailed realistic geometries, computer methods must be employed. Our 3-D computer code numerically solves the Laplace equation in integral form by a method of moments, subject to the boundary conditions of uniform field at infinity and zero normal gradient at the aircraft surface. The aircraft surface is approximated by a multifaceted surface where each facet is a small quadrilateral panel, as illustrated in Figs. 7 and 8. (The code was adapted from a fluid-flow code due to J. L. Hess.)

In light of data obtained in 1977 by the Air Force Flight Dynamics Laboratory in flight tests of Stormscope [52], [53] (see Section IV-C) we employed our computer to obtain a preliminary assessment of the possible influence of site errors. The aircraft is a T-39, with the instrument installed near the leading edge of (and on the underside of) the wing tip. The model portrayed in Figs. 7 and 8 shows the wing modeled reasonably realistically, while the fuselage, whose detailed structure should be unimportant in this case, is modeled crudely. (Detailed fuselage structure can be modeled more realistically, if necessary.)

The panels in Figs. 7 and 8 are labelled by letters *A–H*, denoting various sections, with Sections *A* and *B* on the fuselage, and Sections *C–G* on the wing and Section *H* on the wingtip. Each section has 12 panels, with 1–6 on the upper surface and 7–12 denoting image positions (not shown) on the under surface (with 7 under 6, 8 under 5, ⋯, and 12 under 1). The Stormscope instrument position is on Panel *G*-11 (under Panel *G*-2).

Some selected preliminary results are as follows, indicating panel location, misdirection, and angle of incidence at which this occurs. For each section (Sections *A–H*), we give the optimum panel location.

Sec. *A*-9: $+11°$ at $135°$ (bottom of fuselage at midwing)

B: (no good location, vertical plane)

C-3: $+16°$ at $30°$ (top of wing, behind leading edge)

C-10: $+18°$ at $5°$ (bottom of wing, behind leading edge)

D-2: $+8°$ at $15°$ (top of wing, adjacent to leading edge)

D-11: $+12°$ at $5°$ (bottom of wing, adjacent to leading edge)

E-11: $+5°$ at $15°$ (bottom of wing, adjacent to leading edge)

E-2: $+10°$ at $160°$ (top of wing, adjacent to leading edge)

F-5: $+3°$ at $110°$ (top of wing, ahead of trailing edge)

F-10: $+3°$ at $25°$ (bottom of wing, behind leading edge)

G-10: $+4°$ at $80°$ (adjacent to wingtip, bottom, behind leading edge)

G-3: $+6°$ at $115°$ (adjacent to wingtip, top, behind leading edge)

G-11: $-7°$ at $170°$ (Stormscope location, bottom, leading corner of wingtip)

G-2: $+10°$ at $95°$ (mirror of Stormscope location, top surface)

H: (no good location, vertical plane)

The foregoing represent optimum locations (where the misdirections are minimal). The misdirections are larger at other locations within a section.

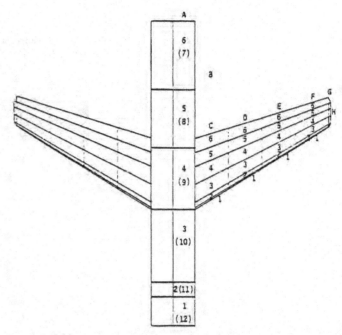

Fig. 7. T-39 computer simulation model, plan view, for crossed-loop site-error analysis. See text.

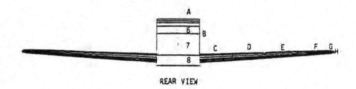

REAR VIEW

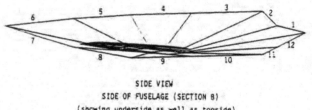

SIDE VIEW
SIDE OF FUSELAGE (SECTION B)
(showing underside as well as topside)

Fig. 8. T-39 computer simulation model, other views, for crossed-loop site-error analysis. See text.

The following conclusions may be drawn. The optimum fuselage location is underneath, at midwing position. The optimum locations on the wing (and in fact on the whole airplane) are near the wingtip and away from the fuselage, either on top and ahead of the trailing edge, or underneath and behind the leading edge. The actual Stormscope location used was a reasonable choice (in the absence of data on site errors) but could have been improved. The reported bearing discrepancies [53] are consistent with the computed misdirection near the wingtip, of the order of $10°$.

It should be mentioned that these figures apply to an incident field lying entirely in the horizontal plane. The presence of a vertical field component would be associated with a slanted lightning channel. In this way, the error due to slant would be coupled with site error. It is straightforward to include the vertical component in the calculations, i.e., to study the site errors associated with slanted channels, in any subsequent extension of the present work.

The overall agreement between the analytical calculations (e.g., using cylinder and ellipsoid models) and the numbers obtained from the computer model is satisfying. However, it is evident that more-detailed information can be obtained with the computer-model calculations. The analytical calculations are limited to simple analytical shapes, whereas with the computer model one can treat more realistic airplane shapes.

C. Stormscope Instrument

A modern commercial version of the crossed loop, the Stormscope instrument presently available for airborne application [54], uses a single crossed loop for the determination of the azimuthal direction (bearing) of lightning discharges with respect to the aircraft. The operating frequency band is centered on 50 kHz, that is, low frequency. It also indicates range (discussed in Section VII). Ranges of 40, 100, and 200 nmi may be selected. Lightning positions are displayed as dots on a CRT "scope". In the present version, the dots do not move so as to reflect the change in position of the aircraft. Such dots cannot show the spatial extent of the source. Also, the manufacturer claims that Stormscope rejects signals from horizontal discharges. (Hence some date from nonvertical discharges would be lost.) The manufacturer claims a range accuracy of ±10 percent, minimal errors from reradiation by nearby metal surfaces, and invulnerability to corona noise because of special signal-processing techniques.

The Stormscope is of considerable interest because of its compact geometry and relatively low cost. Its accuracy and reliability are still in the process of being assessed.

The range is indicated by the radial distance of the dot from the center of the display. According to the manufacturer, the range of the lightning discharge is obtained by computer evaluation of signal strength, time to peak value, decay time, spectral content, and comparison of electric- and magnetic-field amplitudes. The details of this evaluation could not be obtained at the present time. Therefore, the accuracy of the range determination is still an open question.

Stormscope Tests: In-flight tests of Stormscope were performed by the Air Force Flight Dynamics Laboratory in 1977 [52], [53], with funding by the FAA. The instrument was installed near the leading edge of (and on the underside of) the wing tip of a T-39B aircraft. This outboard location was chosen to reduce spurious responses because the instrument is very sensitive to electrical noise generated within the aircraft. (Note that this location was modeled in the computations of the preceding Section IV-B.)

The tests were performed near thunderstorms at Kennedy Space Center, in conjunction with on-board air-weather radar

(X-band), ground-weather radar, and the ground-based LDAR system for locating lightning discharge sources. (LDAR is an elaborate time-of-arrival system discussed later.)

Some key results of the Stormscope tests are:

1) Some reasonable correlations are obtained of Stormscope dots with weather-radar precipitation echoes, with some discrepancies.

2) Stormscope dots are much more dispersed both in range and azimuth than LDAR dots, although roughly in the same general area. (The angular dispersion is roughly consistent with the theoretically computed misdirection of the order of $10°$.)

3) Isolated Stormscope dots sometimes appear in the fair-weather region roughly $180°$ opposite the main activity. We believe these could be due to the $180°$ sensing errors discussed in [1].

4) With respect to range, Stormscope tends to indicate activity more distant than LDAR. Moreover, Stormscope sometimes shows a "radial spread" or "spoke" effect, a radial distortion of dots apparently due to range errors (see Section VII-B below).

In summary, Baum and Seymour [53] believe that, despite the apparent overspread of its dots (which in our opinion probably results from range and bearing errors), Stormscope is useful in providing a conservative warning of severe weather, i.e., larger avoidance areas than LDAR.

Spurious indications by Stormscope (high "false-alarm rates") have also been observed in ground-based tests for mining applications [55].

On the other hand, it should be noted that some private pilots have found Stormscope helpful in avoiding thunderstorms ([56], and R. Rozelle and R. Collins, personal communications, 1980).

D. Gated Wide-Band Direction-Finding System

A recently-developed crossed-loop detector system by Krider *et al.* [57], [58], called the Lightning Location and Protection (LLP) system, operates with gated wide-band (GWB) electronics. This system is presently ground based and is designed to locate exclusively cloud-to-ground return strokes, by triangulation using two or more well-separated stations. It rejects signals due to intracloud discharges, background noise, and return strokes lowering positive charge to ground. For an accepted signal, the bearing error due to slant in the lightning channel can be reduced, to under $1°$ according to NSSL tests by M. W. Maier (personal communication, 1980), by gating on the first few microseconds of the signal. This corresponds to radiation from the lowest 100 m of the return-stroke channel. The misdirection is minimized because the source and the detector are both essentially on the ground plane (see "Slanted Lightning Channel Errors" above). The LLP system uses 1 kHz to 1 MHz as its operating frequency band. It is commercially available and has recently seen widespread ground-based deployment by the USDI Bureau of Land Management to detect forest fires [58]. The LLP system can be used as close as 1 km from a return-stroke channel (M. A. Uman, personal communication). A similar system has been reported by Bent [59], using however 1 kHz to 100 MHz.

With respect to possible airborne application, a single-station version of the LLP system probably can be adapted to locate return strokes and display the data in a manner similar to Stormscope. The LLP system would, as well as Stormscope, be subject to site errors. These can be minimized by appropriate calibration.

E. Kohl's Multiple Loops

A system of multiple magnetic loops has been developed by Kohl [48] for determining the direction of sources of RF signals including sferics. It is, in essence, a removal of the two-component limitation of the crossed-loop antenna, and extension to a 3-component sensor by the addition of a third loop in the third direction. Airborne experiments have been sponsored by AFGL, and the system has been flown experi-mentally on a Canadian ASW patrol aircraft (D. A. Kohl, private communication, 1980). The system operates at 500 kHz and is capable of detecting a magnetic vector of arbitrary orientation, using appropriate electronics. It therefore has in principle a (vector detection) capability greater than that of a single pair of crossed loops. This implies for example that bearing errors due to slanted lightning channels (but not site errors) can be eliminated.

V. HF-VHF-UHF BEARING DETECTORS

In this section, we consider certain high-frequency systems for direction-finding and location of sferics sources. TOA systems, an interferometer system, and two other systems, are discussed.

A. TOA Systems

Two examples of time-of-arrival systems for detecting sferics from discharges in electrified clouds are a) LDAR (Lightning Detection and Ranging) [60], and b) Taylor's Lightning Mapping System [61]. The two systems operate similarly in that all the antennas in an array receive pulse signals radiated from the same discharge source located at a point in space at a given time. Using the differences in the times of arrival of the pulse at the individual antennas, an analysis may be performed yielding the coordinates of the source (angles of azimuth and elevation, plus range with a sufficient number of antennas). Both systems operate in the approximate frequency range 20-80 MHz. A small amount of radiation from return strokes (< 10 MHz) is detected. Both systems also depend on fast rise times of the signals to mark the times of arrival. Hence, weak signals and slow rises are rejected.

Two kinds of time resolution are involved:

1) the time resolution for a given pulse, which controls the accuracy of the angle determination, and
2) the time window for resolving individual pulses.

The time window for resolving individual pulses appears to be about 50 ns for Taylor's system, and about 100 μs for LDAR. The angular error ($\Delta\theta$) is related to the time resolution for a given pulse (Δt) approximately by

$$\Delta\theta° \sim 60 \frac{c\Delta t}{d} \tag{8}$$

where c is the speed of light and d is the baseline dimension. The use of different definitions of Δt may explain differences in estimates of accuracy among researchers. Equation (8) indicates that the LDAR system with its long baseline d, of the order of tens of kilometers, has a potentially high location accuracy. However, difficulties may arise because of the long baseline. If the time between emissions of separate VHF pulses in the clouds is of the order of, or less than, the pulse propaga-tion time (30–50 μs) between stations, the system may become confused. (A hyperbolic direction-finding system with high accuracy and much-larger baselines was developed earlier by Lewis et al. [62].)

The extremely long LDAR baseline dimension makes it obviously inapplicable as an airborne system. Nevertheless, this instrument has its value for testing airborne systems (e.g., Stormscope versus LDAR tests), and it can also potentially be useful for communicating lightning activity data to pilots from the ground (C. L. Lennon, personal communication, 1980).

As opposed to using arrays of antennas in the ground plane as in LDAR, Taylor's group consists of two pairs of antennas, one with a horizontal baseline, and the other with a vertical baseline. The horizontal-baseline pair is used for azimuth alone, and the vertical-baseline pair is used for eleva-tion alone. The baseline of this group is of the order of 10 m (as opposed to the 10-km scale of the LDAR group). Thus its output is azimuth and elevation (a direction line). (Taylor uses two such groups, separated on the order of 10 km, plus triangulation to determine range.)

The relatively small baseline of the Taylor group seems more applicable to airborne use than the LDAR. However, the reduction of the baseline to the order of 10 m or less (suitable for aircraft) can result in large angular errors. For example, if we use (8) with $d = 3.6$ m, and $\Delta t \sim (60$ MHz$)^{-1}$ $\sim 1.7 \times 10^{-8}$ s from bandwidth considerations, we obtain $\Delta\theta \sim 85°$. This result suggests that high precision may be difficult to obtain with TOA-type airborne systems. However, the prediction of errors is apparently not as straightforward as implied in the foregoing. In this connection, Taylor (private communication, 1980) states the following:

"Bandwidth considerations for obtaining Δt are totally irrelevant for Taylor's system since the difference in time of arrival at each pair of antennas is obtained from the initial rise of the pulse waveform. Laboratory tests showed Δt could be measured to <0.5 ns (no shorter interval was at-tempted). Field tests using the LDAR calibration-pulse trans-mitter atop the VAB (Vehicle Assembly Building) at KSC and using an airborne pulse transmitter on a NASA aircraft for Oklahoma calibrations indicated the azimuth and elevation angles were determined to $<0.5°$ error, thus showing that Δt was obtained in an operational mode to <0.4 ns. For $d = 1$ m in (8), we therefore have $\Delta\theta = 7.2°$. For $d = 3.6$ m, very reasonable for aircraft installation, we have $\Delta\theta = 2°$."

B. Interferometer System

This instrument has been used in radio-astronomy to locate accurately extraterrestrial sources of radio emission. It has been adapted to lightning location by Hayenga and Warwick (Warwick et al. [63]; Hayenga and Warwick [64]). In the present version, a 10-percent-relative-bandwidth receiver receives VHF radiation at 34.3 MHz emitted by breakdown processes occurring at the tip of a lightning channel. The frequency f is one of several adjustable parameters, to be defined later. The relative phases of the signals arriving at a pair of omni-directional antennas contain the desired information regarding the direction of arrival of the signals. The antennas are separated in the current version by a baseline d equal to twice the wavelength λ. (In the present version, $d = 17.4$ m $= 2\lambda$.) The accuracy of determination of the source direction depends on the accuracy with which the relative phase can be determined. The determination is simplified by mixing the outputs of the antennas with local-oscillator signals offset in frequency by an amount f_0 (typically 200 kHz) much lower than f. The signals are then multiplied with each other, producing an interference pattern with a sinusoidal modulation having the frequency f_0. The phase of the modulation, which can be determined accurately from successive zero-crossing times of the signal, is directly related to the relative phase of the original-frequency signals arriving at the two antennas.

The inherent high accuracy of source-direction determination by this method is limited in part, however, by the observation time τ (time of averaging) of a given train of waves. The method assumes that the train of waves is sufficiently long to produce an interference pattern, and that the radiation comes from a single source of small size during the time of observation. If the radiation comes from multiple sources, the direction determination may be in error. Hence the received radiation is averaged over a sufficiently short time interval to minimize the possibility of confusion with other sources. In the present design, this time interval τ is of the order of 1-2 μs. Thus pulses separated by 2 μs or more and their associated sources can be resolved, and pulses lasting much longer than 2 μs can be sampled every 2 μs to determine motion during the pulse.

From the above phase shift, one infers the polar angle of the source with respect to the baseline direction to determine one angle. Crossed baselines (two elements on each line) give the vector direction (both azimuth and elevation). Using two such groups widely separated can yield the source position by triangulation. The above technique was designed (by Warwick and Hayenga) for detecting the steps of a lightning stepped leader.

The accuracy of source-angle determination (azimuth and elevation) depends on four parameters of the system: wavelength λ, baseline d, bandwidth B, and integrating or averaging time τ. In degrees, the angle error may be expressed (aside from a trigonometric factor) as

$$\Delta\theta \sim \frac{90}{2\pi} \frac{\lambda}{d} \frac{1}{\sqrt{2B\tau}} \tag{9}$$

Hence, for 34.3 MHz and $d = 2\lambda = 17.4$ m, for $B = 4$ MHz (roughly 10 percent of the frequency), and for $\tau \sim 5/B \sim 10^{-6}$ s, we obtain the error

$$\Delta\theta \sim 2.3°.$$

In a possible application to airborne use, d would decrease. Let $d = 3.6$ m, as before. The frequency can be increased to 167 MHz to keep the same ratio of λ/d. The bandwidth B would increase to about 16.7 MHz, while the time τ can remain the same. Thus (9) would yield a smaller $\Delta\theta$ because of the larger value of $\sqrt{B}$, yielding an even higher accuracy (by a factor of about 2).

The interferometer and TOA systems both seem to have considerable freedom in the choice of the parameters. They seem readily adaptable to airborne use. The possibility of range determination in addition to azimuth and elevation should be looked into. With two stations, e.g., one on each wing of a large airplane, range may be determined by triangulation.

C. Electrograph and Crossed-Adcock Systems

A lightning warning ("Electrograph") system operating at UHF in the 900-MHz region has been proposed by E. A. Lewis and his co-workers. A directional parabolic antenna scans both azimuth and elevation to detect radio-noise pulses originating in small-scale electrical discharges (predischarges) preceding overt lightning flashes (Harvey and Lewis, [65]). In this sense, the system is similar to the lightning-mapping TOA and inter-ferometer systems.

A sferics system for lightning bearing detection in the broadcast band of frequencies (0.5 to 1.5 MHz) was developed by Stergis and Doyle [66]. Two Adcock antennas placed at 90° to one another were selected over crossed loops because of the Adcock lower sensitivity to polarization errors (due to down-coming sky-waves). Tests of crossed loops showed that 20°-30° bearing errors were not uncommon. The accuracy claimed is about 5°, with a potential reduction to 2°.

With respect to possible airborne applications, the Electrograph is being designed for this purpose (R. B. Harvey, personal communication). It is uncertain whether the Adcock system can be made sufficiently compact. Also, the Adcock system may be as susceptible as crossed loops to aircraft site errors. These questions need investigation.

VI. OPTICAL BEARING DETECTORS

Flash location by using the optical signals from lightning has some similarities to the sferics methods discussed above. For example, Kidder [67] has described a South African system employing several cameras to give bearings, and subsequent triangulation to locate the discharge [74]. Edgar and Turman have employed silicon-photodiode systems on a satellite for detection of lightning from space [68], [69].

An optical system for detection and recording of lightning on the ground, from aircraft and from space (the Orbiter) is under development by Vonnegut and his co-workers [70]-[72]. Their system consists of a photocell and a super-8 sound motion-picture camera, designed to be hand-held.

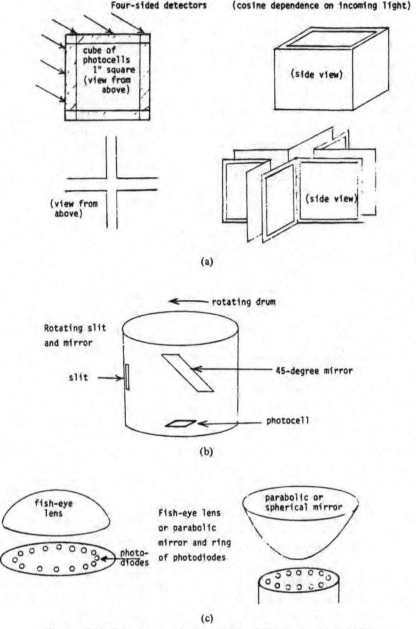

Fig. 9. Optical detectors. (a) Four-sided detectors. (b) Rotating slit and mirror. (c) Fish-eye lens or parabolic mirror, and ring of photodiodes.

Both the Vonnegut and the Edgar–Turman systems can detect lightning in clouds illuminated by direct sunlight, even when the lightning cannot be seen by the human eye. Daylight interference is also avoidable to some extent by use of a filter for the H-alpha line, which is much stronger in lightning than in daylight.

With respect to range detection, a limitation of an optical system is the line of sight. Hence, such a system would probably involve short-range triangulation.

With respect to bearing errors, in principle there should be none for vertical lightning channels. However, there is generally significant horizontal branching, and many lightning channels are approximately horizontal. These can represent broad sources, leading to large bearing errors. Another source-

broadening effect occurs even if the channel is vertical. Namely, the light emitted by the lightning can undergo nearly isotropic scattering with little attenuation within the same cloud or by nearby clouds, so that the light source as seen by the detector can be broad in extent. In addition, the attenuation may become serious if the line of sight passes through a large thickness of cloud material, e.g., parallel to a frontal system.

Some possible constructions of optical detectors are suggested by the sketches in Fig. 9. The sketches labelled (a) in the figure use the dependence of photocell output on the angle of incidence of the incoming light (i.e., a cosine law). The detectors on the 4 side-surfaces of the cube or on the 8 surfaces of the cross (E. P. Krider, personal communication) will have different outputs, in general dependent on the angle of the incident illumination. These outputs can be compared and analyzed to infer the correct direction.

The sketch labelled (b) is a rotating slit and mirror, using a single photocell. The sketches labelled (c) use a "fish-eye" lens or a parabolic mirror to direct incoming light to one of a large number of photodiodes arranged in a circle.

The dimensions of the above systems are very small, only of the order of an inch. Their cost is quite modest.

VII. RF-SFERICS RANGE DETECTION

A. Range by Multiple Stations

To obtain the range of a lightning discharge, two bearing-detecting (direction-finding) stations (sferics or optical) may be used with triangulation. This requires a) that the baseline of the system be of the same order as the range of the source (lightning discharge), and b) that the direction of the source be different from that of the baseline. Otherwise, small errors in reading the angle could result in large errors in the determination of the range. The TOA and interferometer systems may be exempt.

A remedy in the case of ground-based stations is to use 3 stations at the vertices of an equilateral triangle, with sides (baselines) on the order of tens of kilometers. This arrangement allows redundant verification of the position determination by each pair of stations. On an airplane, however, if one is restricted to a single station, the range would have to be determined by a different means. Single-station range detection has proved to be a difficult problem. A number of solutions have been suggested.

B. Stormscope Single-Station Range Detector

In our discussion of Stormscope (Section IV), we referred to its range indication. One would infer from the patent description [54] that the range is obtained from assuming that the detected magnetic-field strength H is inversely proportional to the distance r. If one is sufficiently far from a radiating dipole (so that the induction term is negligible compared with the radiation term—see (11) below) then $H = K/r$, where K is a constant proportional to the time-derivative of the current. If K were known and truly constant (all lightning discharges having the same di/dt), then the range would be simply given by K/H.

However, lightning discharges are highly variable and are distributed with respect to di/dt. This fact may contribute to the Stormscope "spoke" effect (Section IV-C above). Data on parameters of lightning discharges are difficult to obtain. For cloud-to-ground strokes, data have been obtained by Berger. Some of these are tabulated by Golde [73]. The data for negative strokes may be fitted roughly by a log-normal distribution with standard deviation about 5 dB. The corresponding figure given by Pierce [74] is 7 dB for VLF emissions. This standard deviation, corresponding to factors of about 2 and 1/2 with respect to the median, would thus apply also to the range determinations. Hence, variability alone would lead

to both underestimates and overestimates in range by factors of 2.

An additional problem with using magnetic-field strength is site error. Depending on the location of the instrument on the airplane, the apparent magnetic-field strength H can be enhanced by as much as a factor of 2. Under the $r = K/H$ assumption, this enhancement would lead to underestimates of r by a factor of 2. The site effect can also cause diminutions of H, leading to overestimates of r.

C. Kohl's 500 kHz Sferics Single-Station Range Detector

Following investigations by Norinder, Malan, and Horner of sferics radiation properties, Kohl ([75] and references cited therein) notes that, at 500 kHz, most of the radiation is not from the return stroke per se but rather from the breakdown processes associated with the lightning (i.e., predischarges). Using a simple omnidirectional electric antenna (wire), the pulse spectrum at 500 kHz is found to contain maximum pulse amplitudes essentially the same for all lightning bursts. Thus peak pulse measurements using spectrum analyzers can form the basis of a range determination when properly calibrated against radar echoes. The variation is monotonic with range and occurs from direct-path propagation losses (i.e., ground waves). The variation depends on the conductivity of the soil.

Empirically, one must detect pulses over several minutes to have a high probability of having detected the peak pulse(s). That is, large numbers of pulses (~500) must be detected. Also, large amounts of radar data are required to reduce the probable error. Kohl [75] reports a range-detection standard deviation of ±2.6 km over all data ranging from 2–274 km. In a recent private communication, he claims his range error to be under 10 percent in the range 25-200 miles. In addition, a rough fit to his data is given by amplitude proportional to $1/r^{1.1}$.

A difficulty of the calibration with respect to applicability to airborne use arises because a new calibration is initially required at any new location because of the changed ground-conductivity characteristics.

Regarding the existence of an invariant maximum pulse height, Kohl says that there is (at 500 kHz) "an apparent maximum limit of the radiation energy at that frequency. The nature of the sources at the instant that the maximum occurs is unknown, but it arises during the complicated electrical breakdown processes associated with lightning. This limit appears to be independent of major variations in return strokes." We believe that the breakdown processes referred to by Kohl are associated with the leader.

D. Spectral-Amplitude Ratio and Group-Time-Delay Difference

There is a possibility of using the low-frequency propagation characteristics of the earth-ionosphere waveguide [76]. This would use spectral-amplitude ratios (SAR) based on two VLF frequencies, or the group-delay-time difference (GDD) based on three VLF frequencies, with sharply-tuned receivers, for sferics range determination. Further details are given in [1] and [76].

It may be possible to adapt this method to airborne use (H. Volland, personal communication, 1980). However, the method seems to require large amounts of statistical data, which is time-consuming. Hence, adaptation would require careful consideration.

E. Electric-Field Single-Station Range Detection

Other possible range-detection methods can make use of the dependence on distance of electric-field amplitude E or the ratio of magnetic-to-electric-field amplitudes H/E. In order to clarify the relationship of field amplitude to source strength and distance, we model the source by an oscillating dipole and consider the equations for the electric and magnetic fields:

$$E(r, t) = M(t)/4\pi e r^3 + dM(t)/dt/4\pi e c r^2$$
$$+ d^2M(t)/dt^2/4\pi e c^2 r \qquad (10)$$

and

$$H(r, t) = \mu dM(t)/dt/4\pi r^2 + \mu d^2M(t)/dt^2/4\pi c r. \qquad (11)$$

here we denote

$E(r, t), H(r, t)$ = electric and magnetic field amplitudes
$M(t)$ = dipole moment (charge × length)
r = distance of receiving station from centerpoint of dipole
c = velocity of light
ϵ = electric permittivity of air
μ = magnetic permeability of air.

These equations (which can be found in equivalent form in many textbooks (e.g., [77]) are based on the following assumptions: The radiating source is a dipole of length l with its axis vertical. Its length and orientation are constant in time, and the current is constant along its length. The distance r is large compared with l. (Lightning discharges do not generally fulfill all of these requirements.)

The equations are valid for a receiving station at the ground where the electric field has only one component, perpendicular to the ground plane, while the magnetic field has one (azimuthal) component parallel to the ground plane. Therefore, it is possible to express E and H as scalar functions. For an observer in space above the ground plane (e.g., on an airplane) $\vec{E}$ and $\vec{H}$ are three-dimensional vectors, and (10) and (11) must be supplemented each by two more equations for the other two vector components.

In (10), the $1/r^3$ term is designated as the "electrostatic-field" term, while in (10) and (11) the $1/r^2$ and $1/r$ terms are designated as the "induction-field" and "radiation-field" terms, respectively. The $1/r^3$ term dominates in the "near zone," while the $1/r$ terms dominate in the "far zone."

The values of the scalars (or vectors) E and H depend not only on range r, but also on the time-derivatives of the dipole moment, M, dM/dt, and d^2M/dt^2. If the electric (or

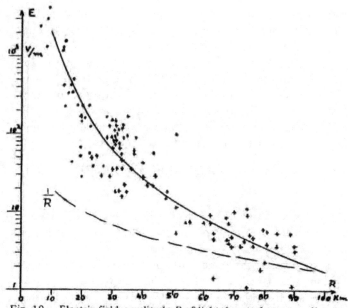

Fig. 10. Electric-field amplitude E of lightning strokes versus distance R. Dots: New Mexico data. Crosses: Alabama data.

magnetic)-field amplitude is to be used as an indicator of range, it must be assumed that M and its time-derivatives are the same—or approximately the same—for all lightning discharges detected. This is not the case, however, and variability among different lightning discharges can amount to an order of magnitude or more. A nearby weak discharge may thus produce a field strength of magnitude similar to that of a distant strong discharge.

Therefore, a dispersion or spread may be expected in data representing field amplitude versus distance. The data in Fig. 10, representing electric-field amplitude versus distance (Kasemir [78]), illustrate such a spread. The electric field and distance are given by the ordinate and abscissa, respectively. The field was measured with a capacitive-loaded antenna with a time constant of 5 s. The distance was obtained by triangulation from visual observations at two stations separated by a baseline 23.5 km in length.

We may read this figure in two ways. One way is to determine the spread in field values at a given distance (e.g., 30 V/m to 230 V/m at 20 km, and 3.2 V/m to 10 V/m at 70 km). We may, on the other hand, determine the spread in distance values at a given field value, e.g., 15 km to 23 km at 200 V/m, and 66 km to 90 km at 4 V/m. The first set of data is more nearly representative of the "near zone" given by the first term on the right-hand side of (10), and has a range spread of ±21 percent about its central value of 19 km. The second set of data is more nearly representative of the "far zone" given by the third term on the right-hand side of (10), and has a range spread of ±19 percent about its central value of 78 km. It appears from these and other samples that the relative error is about ±20 percent of the range, independent of the range itself. This value gives a rough idea of the error to be expected if the electric-field amplitude is used for range detection. The solid curve in Fig. 10 corresponds to a $1/r^3$ variation. For comparison, an inverse-distance law (dashed curve) is also shown, normalized to the solid curve at 100 km and 2 V/m. It is clear from these results that the $1/r^3$ curve fits the data quite well, almost to 100 km. However, the dashed curve indicates that a $1/r$ variation better represents the data beyond about 80 km. It is suggested that a fit to the data may be more effective, as well as physically more nearly correct, using 3 terms ($1/r^3$, $1/r^2$ and $1/r$) corresponding to (10).

Range detection based on the $1/r^3$ assumption has also been investigated by Ruhnke [79]. Both his results and data of the same type obtained earlier by Pierce [80] appear to show a considerably greater spread than that indicated by Fig. 10.

F. Single-Station Range Detection by Flash Counters

A related problem in error determination arises in the measurement of the lightning density (flashes per square kilometer) by lightning flash counters [81], [83]. That is, one records the discharges within a defined radius from the instrument. A wide-band low-frequency receiver is used, responding to sferics electric fields having amplitudes greater than a certain threshold level. The problem is then to determine the distance that corresponds to this threshold value.

It was discovered that the accuracy of the radius determination was poor due to the great variability of the source strength. However, by suitable statistical treatment of large amounts of data, Horner [81] established an effective range by comparisons between counter records and observations using visual and aural techniques.

In addition, there is a strong influence on the received signal strength by the differences in wave-propagation characteristics during day and night. This effect is usually ignored by reports on testing of range indicators based on the amplitude of lightning-generated electromagnetic pulses. Tests for airborne applications of these techniques should include both day and night data.

G. Single-Station Range Detection by H/E Ratio

To eliminate the influence of the source strength (represented in the "dipole model" of (10) and (11) by the dipole moment and its derivatives), Ruhnke [82] suggests measurement of the ratio of the magnetic to electric field amplitudes H/E in a sharply-tuned detector. At wavelength λ in the frequency domain, and in the "near zone" where r is much smaller than $\lambda/2\pi$, according to the dipole model, H/E increases linearly with r because of the $1/r^3$ dependence of E and the $1/r^2$ dependence of H. In the vicinity of $r = \lambda/2\pi$, or equivalently for frequency ν in the vicinity of $\nu(\text{kHz}) = 50/r$ (km), the ratio H/E has a peak value. For larger values of r, H/E drops off and in the "far zone" (where r is much larger than $\lambda/2\pi$) becomes constant since E and H are both proportional to $1/r$. Ruhnke suggests that frequencies lower than $50/r$ (km) be used to maintain variability of H/E with distance. Thus 1 kHz should be used for distances up to 50 km, and lower frequencies for greater distances. For small distances, the finite length of the lightning channel must be included in the theory. Assuming a simplified return-stroke model, at 1 kHz, Ruhnke shows that, beyond about 3 km, the

H/E variation with distance is similar to that of H/E given by the dipole model. In addition, Ruhnke [82] performed experiments at Kennedy Space Center which showed that, at 1 kHz, H/E depends on distance for at least 30 mi. Ruhnke suggests that experimental verifications be made of such a detector by correlations with other lightning-location methods, such as crossed loops or all-sky cameras. It would seem worthwhile to pursue this approach. Still needed, however, is research on questions similar to those affecting other types of detectors such as crossed loops: site errors, errors due to horizontal lightning channels, errors due to inhomogeneous ground conductivity along the propagation path, intrinsic lightning-source variability, and day-night differences.

VIII. BEARING ERRORS AND SUMMARY REMARKS

The magnetic crossed loop is a well-known direction-finder. It and the multiple-loop system are obvious candidates for adaptation as airborne bearing detectors. They have, in fact, already been operated on aircraft, while the other bearing detectors mentioned have only been operated at the ground.

The commercially available narrow-band version of the crossed loop called Stormscope (Section IV-C) is apparently simple and low in cost. However, field tests (both airborne and ground-based) made to date raise questions regarding its accuracy and reliability. Moreover, details of its data-processing operations are not available. Therefore, more extensive tests should be carried out. In this connection, our analysis (in Section IV-B here and in greater detail in [1]) of errors of crossed loops as bearing and range detectors should be helpful in recognizing errors and in suggesting improvements.

Another commercially available crossed-loop system (LLP, for Lightning Location and Protection) employs gated wide-band (GWB) electronics to detect the waveform. (Section IV-D.) It is presently ground-based and gates on the initial part of return-stroke signals of cloud-to-ground discharges. It has been deployed for lightning-caused forest-fire detection.

A third system of this type employs more than two loops (multiple loops). (Section IV-E.)

The narrow-band crossed loop also has a 180° ambiguity [1]. This can be removed by using an electric field antenna in conjunction with the loop. However, electrical noise or uncontrolled phase shifts in the electronic circuits may make the system unreliable. A GWB return-stroke system may not have this need.

Magnetic loops sense direction from the orientation of the incident magnetic-field vector. They are subject to errors associated with distortions of the magnetic vector that can be serious. In [1], we analyzed in detail two of the most important types of distortion error ("misdirection"), both of which change the apparent source direction. The results of these analyses are summarized in Section IV-B.

One of these errors is caused by "slant" or inclination in the (nonvertical) lightning-radiation channel. Another is caused by eddy currents induced in the aircraft skin that produce distortions in the local net magnetic vector (site errors). The channel-slant error can be large when the channel is inclined at a large angle from the vertical. This type of error can be minimized a) at the ground in the case of return strokes by employing GWB electronics, and b) presumably also by using multiple loops for more general cases.

In the case of the site error, even for radiation from vertical lightning channels the induced skin currents can cause a crossed-loop bearing error (Section IV-B and [1]) of about $20°$ even when the instrument is symmetrically located on a long fuselage. This may seem contrary to expectation, but is easily shown based on simple analytical models. The error can be even larger if the instrument is mounted near the nose or tail, or near a wing-edge (depending on location), or a vertical-fin edge. (The associated distortion of magnetic-field amplitude as well as direction would also cause errors in range detection by systems relating magnetic amplitude to distance.) Site errors are also predicted theoretically for a T-39 airplane, the aircraft previously used in an in-flight evaluation of Stormscope. For this prediction, we have used a 3-D computer code. The results are consistent with the flight results for the particular mounting location used, and general agreement is found between the results from the simple analytical (ellipsoid, ((5)-(7)) models [1] and those from the computer model. Obviously, more detailed information associated with complex airplane structures can be obtained from the computer model.

The multiple loops, TOA and interferometer may be less susceptible than the crossed loop to bearing errors due to lightning slant. However, they may be equally susceptible to site errors, which should be investigated. Estimates are presented of the accuracy of the TOA and interferometer systems ((8) and (9) in Section V). If these instruments prove to have high accuracy and insusceptibility to errors with respect to bearing, multiple-station versions of these systems mounted on aircraft could be considered for lightning location by triangulation.

Optical systems as bearing detectors are considered in Section VI. Several possible constructions are suggested. In principle, an optical detector has, of course, no site error and would appear to have a minimal bearing error for a vertical lightning channel, but cloud illuminations by reflection/scattering of the light over extended regions (e.g., "heat" or "sheet" lightning) would spread the apparent source and make the direction correspondingly uncertain. An optical system is inexpensive and simple, however, and can be used, for example, in conjunction with one of the range detectors.

Section VII deals with single-station range detection. As a single-station range detector, the crossed loop would be expected to be inaccurate in its present state of development which apparently assumes a $1/r$ dependence of the magnetic amplitude. Inaccuracies would be due in part to a) lightning source variability, and in part to b) site error (skin currents). With respect to single-station range detection, four possible systems are considered besides the crossed loops, namely, an electric amplitude detector, an H/E amplitude-ratio detector, a 500-kHz pulse-height-analysis detector, and low-frequency dispersion characteristics of the earth–ionosphere waveguide. Dipole-source model equations (10) and (11) are used in discussing the behavior of the field amplitudes.

ACKNOWLEDGMENT

This survey was supported by the Air Force under Contract F19628-79-C-0161. The comments and suggestions of many researchers cited are appreciated. We also wish to thank E. G. Holeman for his expert assistance in the 3-D computer modeling.

REFERENCES

[1] L. W. Parker, and H. W. Kasemir, "Airborne lightning warning systems: A survey," AFGL-TR-80-0226, 1980.

[2] L. W. Parker, "Thundercloud electrification models in atmospheric electricity and meteorology," Lee W. Parker, Inc. Rep. NASA CR-161441 (1980). (Also, presented at the 6th Int. Conf. Atmos. Elec., Manchester, England, 1980).

[3] D. R. Fitzgerald, "Probable aircraft 'triggering' of lightning in certain thunderstorms," Mon. Wea. Rev., vol. 95, pp. 835–842, 1967.

[4] ——, "USAF flight lightning research," Proc. Lightning Static Elec. Conf., 1968. AFAL Report AFAL-TR-68-290 Part II (1969).

[5] ——, "Aircraft and rocket triggered natural lightning discharges," 1970 Lightning Static Elec. Conf., AFAL Rep., 1970.

[6] P. B. Corn, "Lightning as a hazard to aviation," Amer. Meteor. Soc. 11th Conf. Severe Local Storms, Kansas City, MO, Oct. 2–5, 1979.

[7] G. A. DuBro, Ed. Agard Lecture Series No. 110, Atmospheric Electricity-Aircraft Interaction, 1980.

[8] D. W. Clifford, "Aircraft mishap experience from atmospheric electricity hazards," Agard Lecture Series No. 110, Atmospheric Electricity-Aircraft Interaction, 1980.

[9] F. A. Fisher, and J. A. Plumer, "Lightning protection of aircraft," NASA Ref. Publ. 1008 Oct. 1977.

[10] M. M. Newman, and J. D. Robb, "Protection for aircraft," Lightning, vol. 2, R. H. Golde, Ed. New York: Academic Press, 1977, pp. 659–696.

[11] H. T. Harrison, "UAL Turbojet experience with electrical discharges," UAL Meterological Circular no. 57, United Airlines, Chicago, IL, Jan. 1965.

[12] B. Vonnegut, "Electrical behavior of an airplane in a thunderstorm," Arthur D. Little, Inc. Rep. FAA-ADS-36, 1965.

[13] E. T. Pierce, "Triggered lightning and some unsuspected lightning hazards," presented at Amer. Assoc. Adv. Sci. 138th Annual Meeting, Philadelphia, PA, 1971.

[14] E. T. Pierce, "Triggered lightning and its application to rockets and aircraft," 1972 Lightning and Static Elec. Conf., AFAL-TR-72-325, 1972.

[15] J. F. Shaeffer, "Aircraft initiation of lightning," presented at 1972 Lightning and Static Elec. Conf., AFAL-TR-72-325, 1972.

[16] H. W. Kasemir, and F. Perkins, "Lightning trigger field of the Orbiter," KSC Contract CC 69694A Final Rep., NOAA, Oct. 1978.

[17] M. A. Uman, Lightning. New York: McGraw-Hill, 1969.

[18] W. E. Cobb, and F. J. Holitza, "A note on lightning strikes to aircraft," Mon. Wea. Rev., vol. 96, pp. 807–808, 1968.

[19] J. E. Nanevicz, R. T. Bly, Jr., and R. C. Adamo, "Airborne measurement of electromagnetic environment near thunderstorm cells. (TRIP-1976)," Tech. Rep. AFFDL-TR-77-62, Aug. 1977.

[20] D. J. Musil, and J. Prodan, "Direct effects of lightning on an aircraft during intentional penetrations of thunderstorms," in Lightning Tech., NASA Conf. Publ. 2128, FAA-RD-80-30, 1980, pp. 363–370.

[21] Centre d'Essais Aeronautique de Toulouse, "Mesure des characteristiques de la foudre en altitude," Essais No. 76/650000P.4 et Finale, July 1979.

[22] F. L. Pitts, R. M. Thomas, K. P. Zaepfel, M. E. Thomas, and R. E. Campbell, "In-flight lightning characteristics measurement system," FAA/FIT Workshop on Grounding and Lightning Technology, Melbourne, FL, Rep. FAA-RD-79-6, 1979, pp. 105–111. See also F. L. Pitts, and M. E. Thomas, Initial direct strike lightning data, NASA Tech. Memo. 81867 1980.

[23] F. J. Holitza, and H. W. Kasemir, "Accelerated decay of thunderstorm electric fields by chaff seeding," J. Geophys. Res., vol. 79, pp. 425–429, 1974.

[24] H. W. Kasemir, F. J. Holitza, W. E. Cobb, and W. D. Rust, "Lightning suppression by chaff seeding at the base of thunderstorms," J. Geophys. Res., vol. 81, pp. 1965–1970, 1976.

[25] D. R. Fitzgerald, "Some relationships of lightning to radar echoes," in Hq. AWS Aerospace Sciences Rev., Rep. AWSRP

105-2, 78-4, Dec. 1978.

[26] M. A. Uman, "Spark simulation of natural lightning," 1972 Lightning Static Elec. Conf., AFAL-TR-72-325 1972.

[27] H. Israël, *Atmospheric Electricity*, (Transl. from German, Israel Program for Scientific Translations), Nat. Tech. Inf. Service. No. TT-67-51394/1, vol. 1 (1971), and TT-67-51394/2, vol. 2 (1973).

[28] J. A. Chalmers, *Atmospheric Electricity*. New York: Pergamon Press, 1967.

[29] C. T. R. Wilson, "Investigations on lightning discharges and on the electric field of thunderstorms," *Phil. Trans.* vol. A221, pp. 73–115, 1920.

[30] L. G. Smith, "Electric field meter with extended frequency range," *Rev. Sci. Instrum.*, vol. 25, pp. 510–513, 1954.

[31] R. Gunn, (Basic Field Mill patent, first installed in dirigible U.S.S. Los Angeles, 1930), U. S. Patent no. 1,919,215, June 25, 1933.

[32] H. W. Kasemir, cited in footnote by H. Lueder, "Electric recording of approaching thunderstorms and the fine structures of the atmospheric electric thunderstorm field" (transl. from German), *Meteor. Z.*, vol. 60, pp. 340–351, 1943.

[33] H. W. Kasemir, "The field-component meter . . .," *Tellus*, vol. 3, pp. 240–247, 1951. (In German, translation available.)

[34] J. F. Clark, "Airborne measurement of atmospheric potential gradient," *J. Geophys. Res.*, vol. 62, pp. 617–628, 1957.

[35] D. R. Fitgerald, "Measurement techniques in clouds," *Problems of Atmospheric and Space Electricity*, S. C. Coroniti, Ed. sterdam: Elsevier, 1965, pp. 199–214.

[36] H. W. Kasemir, "The cylindrical field mill" (citing earlier work by Matthias (1926) and Kasemir (1944 and 1951)), Meteorologische Rundschau, vol. 25, pp. 33–38, 1972.

[37] H. W. Kasemir, "Electric field measurements from airplanes," Amer. Meteorol. Soc. Fourth Symp. Meteorol. Observations Instr., Denver, CO, Apr. 10–14, 1978.

[38] B. Vonnegut, C. B. Moore and F. J. Mallahan, "Adjustable potential gradient-measuring apparatus for airplane use," *J. Geophys. Res.*, vol. 66, pp. 2393–2397, 1961.

[39] R. Markson, "Ionospheric potential variations obtained from aircraft measurements of potential gradient," *J. Geophys. Res.*, vol. 81, pp. 1980–1990, 1976.

[40] F. J. W. Whipple, and F. J. Scrase, "Point discharge in the electric field of the earth," *Geophys. Mem. Lond.*, vol. 68, pp. 1–20, 1936.

[41] S. Chapman, "Corona-point discharge in wind and application to thunderclouds," *Recent Advances in Atmospheric Electricity*, L. G. Smith, Ed. New York: Pergamon Press, 1959, pp. 277–288. See also S. Chapman, "The magnitude of corona-point discharge current," *J. Atmos. Sci.*, vol. 34, pp. 1801–1809, 1977.

[42] L. B. Loeb, *Electrical Coronas*. Berkeley, CA: U. Calif. Press, 1965.

[43] G. C. Simpson, and F. J. Scrase, "The distribution of electricity in thunderclouds," *Proc. Roy. Soc. London, England*, vol. A161, pp. 309–352, 1937.

[44] G. C. Simpson, and G. D. Robinson, "The distribution of electricity in thunderclouds II," *Proc. Roy. Soc. London, England*, vol. A177, pp. 281–329, 1941.

[45] M. E. Weber, and A. A. Few, "A balloon-borne instrument to induce corona currents as a measure of electric fields in thunderclouds," *Geophys. Res. Lett.*, vol. 5, pp. 253–256, 1978.

[46] L. H. Ruhnke, "A rocket-borne instrument to measure electric fields inside electrified clouds," NOAA Tech. Rep. ERL 206-APCL 20, May 1971.

[47] R. A. Watson Watt, and J. F. Herd, "An instantaneous direct-reading radio-goniometer," *J. Instn. Elec. Engrs.*, vol. 64, pp. 611–622, 1926.

[48] D. A. Kohl, Direction Finder, U.S. Patent no. 3,242,495, Mar. 22, 1966.

[49] C. B. Kalakowsky, and E. A. Lewis, VLF sferics of very large virtual source strength, AFCRL-66-629, 1966.

[50] M. A. Uman, Y. T. Lin, and E. P. Krider, "Errors in magnetic direction finding due to nonvertical lightning channels," *Radio Sci.*, vol. 15, pp. 35–39, 1980.

[51] F. Horner, "The accuracy of the location of sources of atmospherics by radio direction-finding," in *Proc. Instn. Elec. Engrs.*, Pt. III, vol. 101, pp. 383–390, 1954.

[52] T. J. Seymour, and R. K. Baum, "Evaluation of the Ryan Stormscope as a severe weather avoidance system for aircraft—preliminary report," FAA/FIT Workshop on Grounding and Lightning Technology, Melbourne, FL, Rep. FAA-RD-79-6, 1979, pp. 29–35.

[53] R. K. Baum, and T. J. Seymour, "In-flight evaluation of a severe weather avoidance system for aircraft," AFFDL Report AFWAL-TR-80-3022, May 1980.

[54] P. A. Ryan, and N. Spitzer, Stormscope, U.S. Patent no. 4,023,408, May 17, 1977.

[55] R. L. Johnson, D. E. Janota, and J. E. Hay, "A study of the comparative performance of six lightning warning systems," in Lightning Tech., NASA Conf. Publ. 2128, FAA-RD-80-30, 1980, pp. 187–203.

[56] R. Rozelle, "Weather avoidance, an alternative to radar," *The AOPA Pilot*, Nov. 1979, pp. 95–105.

[57] E. P. Krider, R. C. Noggle, and M. A. Uman, "A gated wideband magnetic direction-finder for lightning return strokes," *J. Appl. Meteor.*, vol. 15, pp. 302–306, 1976.

[58] E. P. Krider, R. C. Noggle, A. E. Pifer, and D. L. Vance, "Lightning direction-finding systems for forest fire detection," *Bull. Amer. Meteor. Soc.*, vol. 61, pp. 980–986, 1980.

[59] R. B. Bent, "A new approach to lightning positioning and tracking," FAA/FIT Workshop Grounding Lightning Tech., Melbourne, FL. Rep. FAA-RD-79-6 supplement 1A, 1979.

[60] H. A. Poehler, and C. L. Lennon, "Lightning detection and ranging system LDAR/System description and performance objectives," NASA Tech. Memo 74105, June 20, 1979.

[61] W. L. Taylor, "A VHF technique for space-time mapping of lightning discharge processes," *J. Geophys. Res.*, vol. 83, pp. 3575–3583, 1978.

[62] E. A. Lewis, R. B. Harvey, and J. E. Rasmussen, Hyperbolic direction finding with sferics of transatlantic origin, AFCRL Report AFCRL-62-178, 1962.

[63] J. W. Warwick, C. O. Hayenga, and J. W. Brosnahan, "Interferometric directions of lightning sources at 34 MHz," *J. Geophys. Res.*, vol. 84, pp. 2457–2468, 1979.

[64] C. O. Hayenga, and J. W. Warwick, "Two-dimensional interferometric positions of VHF lightning sources," *J. Geophys. Res.*, vol. 86, pp. 7451–7462, 1981.

[65] R. B. Harvey, and E. A. Lewis, "Radio mapping of 250 to 925 MHz noise sources in clouds," AFCRL Rep. AFCRL-72-0078, Jan. 1972.

[66] C. G. Stergis, and J. W. Doyle, "Location of near lightning discharges," in *Recent Advances in Atmospheric Electricity*, L. G. Smith, Ed. New York: Pergamon Press, 1959, pp. 589–597.

[67] R. E. Kidder, "The location of lightning flashes at ranges less than 100 km," *J. Atmos. Terr. Phys.*, vol. 35, pp. 283–290, 1973.

[68] B. C. Edgar, "State of technology in optical systems," in *Proc. Workshop Need Lightning Obs. Space*, NASA Rep. NASA CP-2095, L. S. Christensen, W. Frost, and W. W. Vaughan, eds. July 1979, pp. 81–87. See also B. N. Turman, "A review of satellite lightning experiments," same report, pp. 61–80.

[69] B. N. Turman, and B. C. Edgar, "Global lightning distributions at dawn and dusk," submitted to *J. Geophys. Res.* See also B. C. Edgar, and B. N. Turman, "Optical pulse characteristics of lightning as observed from space," submitted to *J. Geophys. Res.*

[70] R. F. Griffiths, and B. Vonnegut, "Tape recorder photocell instrument for detecting and recording lightning strokes," *Weather*, pp. 254–257, Aug. 1975.

[71] B. Vonnegut, and R. E. Passarelli, Jr., "Modified cine sound camera for photographing thunderstorms and recording lightning," *J. Appl. Meteor.*, vol. 17, pp. 1079–1081, 1978.

[72] B. Vonnegut, O. H. Vaughan, Jr., and M. Brook, Nighttime/daytime optical survey of lightning and convective phenomena experiment (NOSL), NASA TM-78261, (1980).

[73] R. H. Golde, "Lightning currents and related parameters," in *Lightning*, vol. 1, R. H. Golde, Ed. New York: Academic Press, 1977, pp. 309–350.

[74] E. T. Pierce, "Lightning warning and avoidance," in *Lightning*, vol. 2, R. H. Golde, ed. New York: Academic Press, 1977, pp. 497–519.

[75] D. A. Kohl, "A 500-kHz sferics range detector," *J. Appl. Meteor.*, vol. 8, pp. 610–617, 1969.

[76] J. Schäfer, H. Volland, P. Ingmann, A. J. Eriksson, and G. Heydt, "A network of automatic atmospherics analysators," in Lightning Tech., NASA Conf. Publ. 2128, FAA-RD-80-30 1980, pp. 215–225.

[77] E. T. Pierce, "Atmospherics and radio noise," in *Lightning*, vol. 1, R. H. Golde, Ed. New York: Academic Press, 1977, pp. 351–384.

[78] H. W. Kasemir, "Analysis of the electrostatic field of a lightning

stroke, "U.S. Army Electronics R and D Laboratory Tech. Rep. 2321, Nov. 1962; see also H. W. Kasemir, "Contribution to the electrostatic theory of lightning charge," *J. Geophys. Res.*, vol. 65, pp. 1873–1878, 1960.

[79] L. H. Ruhnke, "Distance to lightning strokes as determined from electrostatic field strength measurements," *J. Appl. Meteor.*, vol. 1, 544–547, 1962.

[80] E. T. Pierce, "Electrostatic field changes due to lightning," *Quart. J. Roy. Metor. Soc.*, vol. 81, pp. 211–228, 1955.

[81] F. Horner, "The design and use of instruments for counting local lightning flashes," *Proc. Instn. Elec. Engrs.*, vol. 107, Part B, pp. 321–330, 1960.

[82] L. H. Ruhnke, "Determining distance to lightning strokes from a single station," NOAA Technical Rep. ERL 195-APCL 16, Jan. 1971.

[83] C. B. Moore, P. R. Leavitt, B. Vonnegut, and E. A. Vrablik, "Some atmospheric electric instruments for use in Air Force operations, Arthur D. Little, Inc. Rep. AFCRL-62-233, 1962.

THEORETICAL AND EXPERIMENTAL DETERMINATION OF FIELD,CHARGE, AND CURRENT ON AN AIRCRAFT HIT BY NATURAL OR TRIGGERED LIGHTNING.

Heinz W. Kasemir

Colorado Scientific Research Corporation
Berthoud, Colorado 80513

1. INTRODUCTION

We discuss in this paper the difference in the flight records of electric parameters for the case that a.) the aircraft triggered the lightning discharge and b.) the aircraft was hit by a natural lightning. The electric parameters are the three components of the thunderstorm gradient GX, GY, GZ, the gradient GQ produced by the aircraft charge and the current I flowing through the aircraft. All parameters are functions of time and their time records are shown in a qualitative way in 24 pictures with a time window of about three seconds. About one second in the middle of the trace, marked by "L" and framed by two vertical lines, covers the events during the life time of the strike. There is space left before and after the lightning area to show the events of about one second before and after the lightning discharge.

The construction of these pictures is based on the electrostatic lightning theory, Kasemir(1) 1960. The application of this theory to the triggered lightning has been discussed by the author on the previous Conference on Lightning and Static Electricity Kasemir (2) 1983. Since flight records of the parameters specified above in the frequency range of 0 to about 1 kHz or more could not be found in the literature,the following discussion should be considered only as a theoretical prediction. We will devote first a few remarks to the problem what is, can, or should be measured so that the data can be physically interpreted.This would provide together with the theory a solid base for the study of the remaining unsolved problems.

2. PROBLEMS OF MEASUREMENTS AND DEFINITIONS

2.1 Frequency range.

Records of the thunderstorm gradient and the aircraft charge in the frequency range from 0 to about 10 kHz

appear best to identify the major components of a cloud and a ground discharge, and to connect them with the thunderstorm field. Field mills are used to measure electric fields from an aircraft. However their frequency range is from 0 to 10 Hz. This is not enough for the present purpose and the frequency range should be extended to at least 1kHz but better to 10 kHz. L.Smith (3) 1954 has suggested a method how this frequency extension can be achieved.

2.2 Field or Gradient Components.

Different sign conventions are used in physics and atmospheric electricity for the field E and the dielectric displacement D=εE. If φ is the potential function of the thunderstorm and if q is the surface charge density on the aircraft we have in

Physics	Atmospheric Electricity	
G = grad φ	G = grad φ	(1)
E = -G	E = G	(2)
D = εE=-εG	D = εE=εG	(3)
q = D=εE=-εG	q = -D=-εE=-εG	(4)

We see that the definition of G in (1) is the same in physics and atmospheric electricity and using G in (4) instead of E or D the definition of q is also the same in both fields. We will use here as much as possible the parameters G and q to avoid confusion introduced by the different sign convention in E and D. When, however, E and D is used, it will be in accordance with the physical sign convention.

The replacement of D by q has the additional advantage that at the metallic surface of the sensor or antenna the expression "surface charge density q" conveys a better physical picture of the phenomena than the expression "dielectric displacement D". q also unites all the different sensors or instruments appearing under different names such as field mill, D-dot sensor, fast or slow antenna, etc. All measure the surface charge density integrated over the exposed area in one way or another.

The antenna or sensor is in all cases a capacitive coupler between the source and the first stage amplifier. The attributes "differentiating, slow, fast, etc." are not the properties of the sensor or antenna but of the first stage amplifier. This makes it relatively easy to adapt the measurement-s to any desired frequency range or band width by changing or modifying the first stage amplifier. It will not be necessary to change the antenna or the sensor.

The surface charge density q on each sensor is composed of contributions from each of the parameters GX,GY,GZ,GQ. Each of these parameters has its own scalefactor a,b,c,d, also called field concentration factors.

$$q = aGX + bGY + cGZ + dGQ \qquad (5)$$

Therefore the output of one sensor is a composite of all parameters and an analysis of the data of one or even two or three sensors in term of the specified parameters is not possible without making restrictive assumptions, which have to be justified.

Rudolph and Perala (4) 1982 have given an extensive analysis of theoretical and experimental records of the D-dot sensor in the higher frequency range of about 5 kHz to 20 MHz. The parameter D-dot is generally better known under the name "Maxwell's Displacement Current". At the sensor it will convert into the complete Maxwell current, i.e. including the conduction current. There is an interesting parallelism in the discussion of the physical aspects during the approach of a leader to the aircraft given here later on and in the contract report of Rudolph and Perala. However the main emphasis in Rudolph and Perala's analysis is placed on the high frequency part of Maxwell's displacement current which will not be discussed in this paper.

Usually it requieres four sensors and an accurate determination of the sixteen scalefactors to be able to solve the four linear equations of type (5) and to separate the GX,GY,GZ,and GQ. The difficult part here is not the mathematical calculation but the determination of the scalefactors. However there is a good chance that this can be done in an effective way with a computer model of the aircraft.

3. THUNDERSTORM GRADIENTS GX,GY,GZ.

The Cartesian coordinate system x,y,z has its zero point at the center of the aircraft with positive x to the right, positive y forward, and positive z upward. The lightning will be triggered and grow in the direction of the maximum gradient and it is from this direction that the aircraft will be struck. It will simplify the discussion if we assume that the gradient is in the z direction (upper to lower fuselage) and that the x and y gradient components are zero. If the maximum gradient is in the y direction, then the strike will be triggered in or will hit from the y direction and correspondingly for the x direction. For these cases we have only to replace the label "GZ" in Fig. 1 by "GY" or "GX" to obtain the corresponding pictures for GY and GX. If the direction of the gradient has a x, y, and z component then we have to use vector addition to obtain the vector gradient.

Fig.1 contains 24 pictures identified by numbers 1 to 24 and arranged in 4 columns and 6 rows. The first three rows show the pictures for a cloud discharge the last three rows are for a grond discharge. In each set of three rows the upper row displays GZ, the middle row GQ, and the lower row I. The first two columns contain the picturs for a triggered lightning and the last two columns those of a natural lightning. Hereby is the first column of a set of two for an aircraft not charged by precipitation and the second column for the aircraft charged by precipitation. The picture with the number n is referred to as Pic.n.

Pic.1 shows a GZ record that can be expected in case a cloud discharge has been triggered and no precipitation charging of the aircraft has occurred. Before the discharge is triggered the gradient increases until the trigger value of the gradient is reached. As soon as the channel develops in the +z and -z direction the Gz component is more and more screened with the growth of the channel. In Pic.1 this is shown as a drop to zero, but the zero value should be taken only as an indication of a reduction of the GZ components and should leave room to explain features introduced by other effects. For instance, the definition of the scalefactors is based on a homogenuous external field. This assumption is not valid any more if a lightning channel is close or attached to the aircraft. We will now have mutual influence between the GZ and GQ channel. These and other interferences and deviations from the simple electrostatic model need further study. We cover this point with the general statement that all predicted curves inside the area "L" are open to corrections due

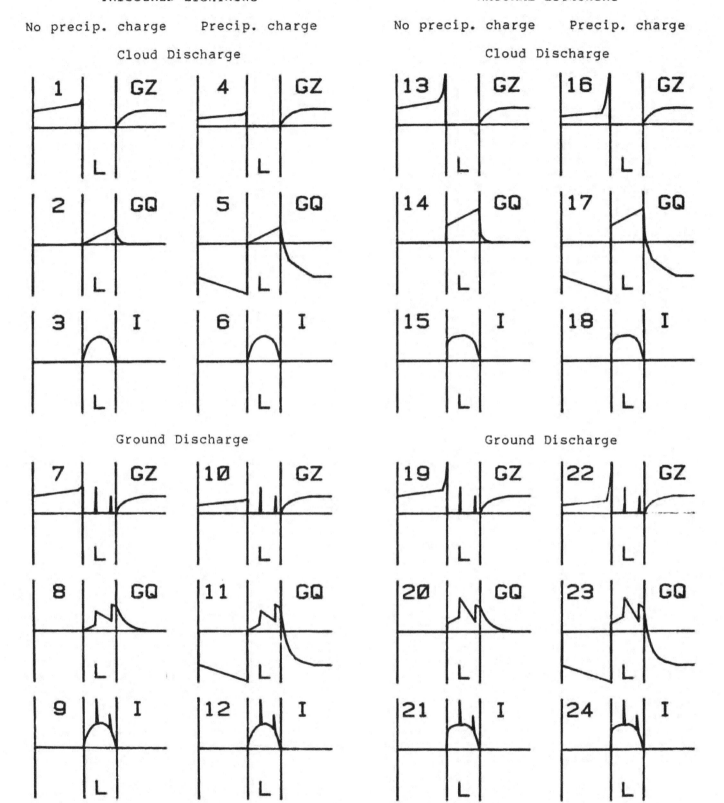

TRIGGERED LIGHTNING   NATURAL LIGHTNING

No precip. charge   Precip. charge   No precip. charge   Precip. charge

Cloud Discharge   Cloud Discharge

Ground Discharge   Ground Discharge

GZ Z-component of thunderstorm gradient
GQ aircraft charge gradient
 I lightning current flowing through aircraft

Fig. 1

687

to inputs outside the electrostatic assumptions.

We may however draw one general important conclusion from the screening effect of the channel. We see from (5) that with the assumption GX=GY=GZ=0 the q sensor output depends only on one parameter namely GQ. This provides a situation, where a physical interpretation of the data of only one sensor is possible. If more sensor outputs are available and all traces show the same polarity and the same time dependance, then we are on fairly good ground with our assumption, that we are dealing with the charge on the aircraft. The aircraft can be charged either positiv or negativ but not both at the same time. If in addition the scale factors are properly determined and we divide each sensor output by its scale factor and obtain so numerical agreement between all sensor outputs, this should furnish sufficient evidence that our assumption is correct.

If however two sensor outputs show a differnt sign or the numerical values are not equal, this will be an indication of the influence of other effects.

The reduction of the gradient (or field) by a high tower or pole at the foot of the tower is well known under the name "inside the cone of protection". This reduction can easily be measured even in a fair weather field. If we replace the ground plane by the mirror image of the tower at the ground, we have a fair replica of the situation of a cloud discharge triggered by aircraft. This replica would even be more complete if a lightning discharge is triggered by the tower. Less similar is the case when the tower or the aircraft is struck by a lightning discharge, as we will see later on.

Pic.13 shows the GZ trace of the aircraft struck by a cloud discharge. This trace is identical with that shown in Pic.1 with one important difference, namely the sharp gradient increase just before the lightning makes contact with the aircraft. The approaching lightning tip generates in front it a high field concentration of the thunderstorm field which will affect the gradient record even at a distance of a hundred or of several hundred meters. If we assume an average velocity of 100 to 1000 km/s of a stepped leader, Uman (5) 1969, then the rise of the gradient should begin in the order of milliseconds before contact is made.The gradient in the last millisecond would reach values above the breakdown value and probably exceed the dynamic range of the field mills. In any case, there would be an unmistakable

strong pulse just before contact of a lightning strike. Such a pulse will not be a feature of the GZ trace of a triggered discharge.

To complete the discussion of the GZ records of a cloud discharge we turn now to Pic.4 and 16 which show the influence of precipitation charge on the aircraft. The gradient GZ in Pic.4 before the lightning is triggered is low compared to Pic.1. This should indicate that part of the high gradient required at the aircraft surface to trigger the discharge is due to aircraft charge caused by precipitation. Therefore in Pic.5 corresponding to Pic.4 the gradient GQ is high, whereas in Pic.2, which corresponds to Pic. 1, the gradient GQ is zero.

In Pic.5 GQ is negative but in Pic.4 GZ is positive. This does not mean that GZ is reduced but that GZ produces on the aircraft induction charges of opposite polarity so that regardless of the polarity of GQ there is always one half of the aircraft surface where the gradient is increased. Then the breakdown value can be reached with the support of precipitation charging, whereas without precipitation charging the thunderstorm gradient alone may not be strong enough to produce the breakdown value.

It is common practice to take the breakdown field at the aircraft as an indicator that a lightning discharge will be triggered. This, however, would be an erroneous assumption. Reaching breakdown value on one part of the aircraft indicates only that corona discharge will start at this point. Therefore we may call this field or gradient value the corona on set field or gradient. There is still a gap in our knowledge under what conditions a corona discharge can be turned into a triggerd lightning discharge.

The last GZ record to discuss in this series is given in Pic.16 which shows the GZ trace of a charged aircraft struck by a cloud discharge. The charge on the aircraft will not be able either to attract nor to deflect the cloud discharge. The approaching lightning tip will still produce the sharp gradient rise starting a few milliseconds before contact.

Pic.7,10,19, and 22 show GZ records which can be expected if the aircraft either triggers (Pic.7 and 10) or is struck (Pic.19 and 22) by a ground discharge. The trace before the discharge is triggered or the aircraft is struck and the trace after the discharge is the same as that of the corresponding cloud discharge given in Pic.1,4,13, and 16 with the possible exception that the polarity may be reversed. There is again the sharp (negative) increase of GZ just before the discharge strikes the aircraft in Pic.19 and 22 which indicates that this was not a triggered lightning but a lightning strike.

In all Figures 7,10,19, and 22 there are two sharp spikes in the "L" area indicating the passing of two return strokes. There is a large voltage difference between the return stroke channel extending from the ground to the tip and having approximately ground potential and the remaining leader stroke extending from the tip to the end of the leader stroke in the cloud still having its original cloud potential of the order of -100 MV. This is indicated by the bright spot at the tip of the return stroke. If the tip length is about 100m, along the tip exist an average field of 1 MV/m. This strong field is necessary to produce ionization to raise the tip channel from the low conductivity of the leader channel to the high conductivity of the return stroke channel. If the velocity of the return stroke is of the order of 10000 to 100000 km/s (Uman (5) 1969) there is only a time period of 10 to 100ns to accomplish the ionization. In such a time period the tip would also pass through the aircraft and would then produce a gradient pulse of 1 MV/m going up and down in 10 to 100 ns. This pulse should be detectable by a D or D-dot sensor with a high frequency response even considering that the high conductivity of the aircraft will reduce its amplitude to a large degree. Such a fast pulse may exite resonance in the aircraft body. The current will rise in this time like a ramp function from the low level of the leader stroke to the higher level of the return stroke and decay at the end of the return stroke. There is a difference of 3 powers of 10 between the time periods of the current and the gradient pulse.

4. THE LIGHTNING CURRENT.

Pic.3 and 6 show the lightning current of a triggered cloud discharge without and with precipitation charging of the aircraft. The current input of precipitation charge is in the order of mA and will be negligible against the 10 to 100 A of a cloud discharge. Therefore both figures show the same trace. The current will rise from zero to a maximum and then drop off to zero again. Ripples from a stepping process or even K-changes, which would show up as little spikes in the current flow, are not reproduced in the pictures.

Pic.15 and 18 show the current through an aircraft struck by a cloud discharge. The I trace is almost the same as in the triggered lightning case (Pic.3 and 6) with one exception. The current starts here not from zero but jumps up to the value of the current already flowing in the lightning channel as soon as contact is made.

The same distinguishing feature can be seen in Pic.9, and 12 of the triggered ground discharge compared to Pic. 21, and 24 of the strike by a ground discharge. In the first case the current has a smooth rise from zero and in the second case the current jumps to the current value of the hitting lightning discharge at the time of contact.

THE SUDDEN INCREASE IN THE MEASURED PARAMETER, CURRENT AS WELL AS GRADIENT AT THE TIME OF CONTACT IS THE EARMARK OF A LIGHTNING STRIKE. WE WILL SEE LATER THAT THE SAME FEATURE CAN BE EASILY RECOGNIZED IN THE GQ TRACE. THE TRIGGERED LIGHTNING HAS NOT THIS CHARACTERISTIC PULSE IN THE GZ TRACE NOR THE STEP-LIKE INCREASE IN THE GQ AND I TRACE AT TRIGGER TIME.

The differences between the cloud and ground discharge traces are the spikes produced by the return stroke. We know this already from numerous field records on the ground. However, measured from an aircraft which is a part of the channel, the return stroke doesn't register in the GZ record as a step but as a spike, which should even be much shorter than the current spike, as discussed above. The current spike has another feature worth mentioning. It doesn't reverse the current flow of the leader stroke. The "return stroke" current doesn't return but flows in the same direction as the leader stroke current. The name "return stroke" fits only the optical phenomena for which it was originally coined.

Fig.2 shows two composite sketches of current traces measured from a French Transall aircraft during a research project of the Centre D'ESSAIS Aeronautique De Toulouse (6) 1979. The upper picture is based on three records and the lower picture on four. The maximum current peak of 70 kA in the upper and of 50 kA in the lower picture combined with the characteristic spikes of the right duration in the range of 20 to 200 microseconds identify these traces as ground discharge currents with one or four return strokes respectively. Since these are hand drawn graphs and it is not clear how accurate the beginning of the lightning current is represented, we can only say that there is a hint that the upper picture may be an example of a triggered discharge and the lower picture that of a lightning strike to the

689

aircraft. Unmistakable, however, is the fact that the return stroke current is of the same direction as the leader current.

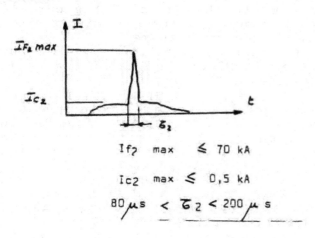

If2 max ≤ 70 kA

Ic2 max ≤ 0,5 kA

80 µs < δ_2 < 200 µs

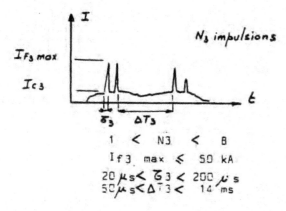

N_3 impulsions

1 < N3 < 8

If3 max ≤ 50 kA

20 µs < δ_3 < 200 µs

50 µs < ΔT_3 < 14 ms

Ground discharge trough Transall aircraft Upper Pic. 1ret.str.,lower Pic. 4ret.Str.

Fig.2 (Centre D'Essais Aero.(6) 1979)

5. THE GRADIENT GQ OF THE AIRCRAFT CHARGE

In case the aircraft is not connected to the lightning channel but charged by precipitation the charge density on a given point on the aircraft is not too difficult to calculate theoretically. This problem can be solved - as mentioned before - by using a computer model of the aircraft. However, if the aircraft has triggered a lightning or is struck by a lightning and is consequently a part of the lightning channel the problem to determine the charge density on a specified surface point by theoretical calculation requires a two step approach.

Our first problem here is to determine the charge density at the lightning channel itself and specifically at the place which will be occupied by the aircraft. Fig.3 (Kasemir (1) 1960) shows in Fig.3a the thunderstorm model of Simpson and Wilson. This is composed of three

spheres arranged vertically on top of each other. Each sphere is filled with space charge, alternating in sign, as shown in Fig.3a. The three spheres represent the main space charge volumes in the storm. The key to the determination of the charge distribution on a lightning channel is the potential function φ of the thunderstorm. The curve φ in Fig.3b,c,d shows the potential function as a function of altitude z (vertical axis) in the center line of the storm for x=y=0. The horizontal axis shows the values of φ or with a possible zero shift the values of the charge per unit length ql on the lightning channel. ql is approximately proportional to the charge density q on the lightning channel. We will assume later on, that this proportionality can be extended to the aircraft when it becomes a part of the lightning channel.

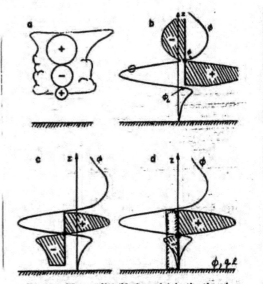

Fig. 3. Charge distribution: (a) in the thunderstorm model of Simpson Wilson; (b) on an intracloud stroke; (c) on a leader stroke of a cloud to ground discharge; (d) on the ground discharge after completion of the main stroke.

Potential function φ,horizontal axis
Charge per unit length ql,horizontal axis
Altitude z, vertical axis

Fig.3 (Kasemir (1) 1960)

In Fig.3b a cloud discharge is represented by a thick vertical line slightly to the right of the z axis. The point where it crosses the potential line is marked by a black circle. This can be the point where the cloud discharge originated or the point where the aircraft triggered the cloud discharge. The cloud discharge originating or the aircraft flying at this altitude would have assumed automatically the potential

690

φL of the storm at this point. Since the lightning channel as well as the airplane can be treated as conductors the whole lightning channel and all points on the airplane surface have the potential value of φL.

We define here the concept of the "bridged potential" φB (translation of the German word "überbrücktes Potential")

$$\phi B = \phi - \phi L \qquad (6)$$

φB and φ are functions of z whereas φL is a constant. φL provides here a new reference potential or zero shift for the potential function φB, however φB as defined in (6) is not in any physical sense a potential discontinuity. The relation with the charge per unit length ql and the charge density q is given by

$$ql = -kl\phi B, \quad q = -k\phi B \qquad (7)$$

The factors kl and k are based on the electrostatic lightning theory and can be calculated analytically. kl is a true constant, i.e. independent of z, for a homogeneous thunderstorm field and has the value of about 25 pF/m. It depends only on the long and small axis of a spheroid, which represents the lightning channel in the theoretical calculation. For the inhomogeneous field of a thunderstorm kl depends weakly on z. However for our purpose here the approximation by using the constant value also for an inhomogeneous thunderstorm field is sufficiently accurate.

The charge per unit length ql is shown in Fig.3b as the mirror image of the potential function, mirrored on the vertical line representing the lightning channel. The hatched area shows the negative net charge on the upper part and the positive net charge on the lower part of the channel. Since these net charges are equal in amount but of opposite polarity the net charge of the whole channel is zero. The reason for the rule "net charge zero for leader and cloud discharges" is that these discharges can not pick up charges from the cloud. This has been discussed in detail by Kasemir (2) 1983.

We interrupt here the discussion of Fig.3 to draw an important conclusion for Pic.2,5,14, and 16. The charge per unit length ql and consequently the charge density on the lightning channel and the aircraft is zero at the point of origin (marked by the black circle in Fig.3b.) Equations (6) and (7) lead to the same result since at this point φ=φL and according to equation (6) φB = 0.

If the lightning growth in such a manner that this condition remains during the lifetime of the discharge the airplane will not receive any charge from the lightning discharge. Add to this the fact that - as discussed above - the aircraft is also screened by the discharge from the thunderstorm field the aircraft experiences from a triggered cloud discharge only the weak current of few amperes. All field mills should record zero or very low values.

This result fits very well with the pilots' observation that the corona hiss in the communication gear stops immediately with the start of the triggered lightning - in the pilots' terminology this is called the "electrostatic discharge"- and that these electrostatic discharges cause no or only little damage to the aircraft.

Coming back to the topic of this paper to determine the characteristic features of a triggered and a strike by a lightning discharge we may conclude that a triggered lightning always starts with charge zero on the aircraft. This can be seen in Pic.2,5,8 and 11.The increase of the charge during the lifetime of the discharge will be explained in the next paragraph. The drop of the charge to zero of an aircraft charged previously by precipitation as shown in Pic.5 and 11 is only a guess. It is based on the assumption that any previous charge on the aircraft will be absorbed into the lightning channel and that the final charge distribution is dictated by the lightning discharge. With regard to the lightning-hit the aircraft will assume the charge of the channeltip at the time of contact.

Let's assume that in Fig.3b the aircraft is placed below the point of the lightning origin. The lightning channel coming down from above will have at its tip positive induction charge. As soon as contact is made the aircraft will receive positive charge from the channel. From then on the aircraft charge is determined by the channel charge at this altitude. If the aircraft was flying above the point of the lightning origin the events will be the same only that in this case the aircraft will be charged negative. Pic. 14,17,20 and 23 show the cases when the aircraft was charged positive by the lightning discharge. The modification introduced by a previous precipitation charge is the same as discussed for the case of the triggered lightning discharge, i.e. the previous aircraft charge will be absorbed by the lightning channel.

The explanation why the aircraft charge or the charge distribution on the lightning channel can change during the lifetime of the discharge. Physically this can be explained by the fact that the lightning channel will not grow with the same speed at the upper and the lower tip of the channel. There are several reasons for this: 1) Positive breakdown of the air ahead of the lower tip will not occur with the same speed as the negative breakdown ahead of the upper tip. 2) The field at the upper and the lower tip are not the same. 3) The extent of the region ahead of the tip where the breakdown value is reached or surpassed is not the same at the upper and lower tip. 4) The growth of the lower tip is stopped because the tip approaches an area which has the same potential value as the lightning channel. The field ahead of the tip would drop to zero and the ionization of the air for the growth of lightning channel would stop. The result of this uneven growth is that the lightning channel changes its potential ϕL and with it the charge distribution on the channel.

All this is based on the electrostatic lightning theory. It would take too much space to discuss it here in more detail. From equation (6) it is obvious that a change of ϕL will result in a change of ϕB, and from equation (7) that a change of ϕB will result in a change of ql and q. This possibility is indicated by the increase in GQ shown in all GQ picture of Fig.1 The general conclusion is that a change of GQ can be expected but not necessarily in the ramp like shape or polarity shown in the pictures.

Fig.3c. shows the charge distribution of the first leader stroke shortly before ground contact is made. The leader stroke may be triggered or may hit an aircraft flying through the lower part of or below the storm. With the exception of the reverse polarity the leader would have a similar effect on the GQ records as that explained for the case of a cloud discharge.

The last feature to be discussed is the signature of the return stroke. Fig.3c and d show the charge distribution on the channel shortly before and after the return stroke. This is another example of the influence of the lightning potential ϕL on the charge distribution on the channel with the additional variation that the net charge zero rule of the leader and cloud discharge is replaced by the condition $\phi L=0$ after ground contact is made. Equation (6) and (7) still hold but with $\phi L=0$.

$$\phi B = \phi \qquad (6a)$$

$$ql = -kl\phi; \quad q = -k\phi \qquad (7a)$$

The mirror rule for the construction of the new charge distribution on the lightning channel remains valid, but note that the lightning channel is displaced from the position on the left of the vertical axis z in Fig.3c to the position of the z axis itself in Fig.3d. This does not indicate a movement of the channel in space but a change in the potential value let's say from $\phi L=-70MV$ before ground contact to $\phi L = 0V$ after ground contact.

We see from Fig.3d that compared to Fig.3c the positive charge area has increased and the negative charge area has shrunk, indicating that the lightning channel has lost negative charge to the ground. This is in agreement with the fact stated quite often in the literatur that the ground discharge brings negative charge to earth. However the mechanism suggested here is quite different from that of discharging the leader to ground.

Along the whole channel the charge per unit length ql and with it the charge density q has increased due to the return stroke. This causes the sudden increase in the GQ traces in Pic.8,11 and Pic.20 and 23. This steplike increase is the same for the triggered and the natural ground discharge. After the first leader stroke there will be no difference between a triggered and a natural lightning strike and the general assumption that a triggered lightning (electrostatic discharge) is relative harmless cannot be extended to the triggered ground discharge.

6. SUMMARY AND CONCLUSIONS.

Assuming that the gradient components GX, GY, GZ and the gradient GQ caused by the aircraft charge can be separated and recorded in the frequency range of 0 to 10 kHz there are certain characteristic features in the records which will differentiate the triggered lightning from the strike of a natural lightning. These distinguishing features are centered at the time of contact by a lightning strike or at the time of triggering the lightning discharge.

The contact is marked by a sharp increase in the GX, GY, GZ record, starting a few ms before contact. The GQ record should also show a step-like increase at the time of contact.

In case of a triggered lightning the G-components should show only a smooth drop of the thunderstorm value to zero or a very low value and the GQ trace should show a smooth drop to zero if the aircraft was charged by precipitation previously. If the GQ trace was previouly zero, then it would remain zero at the time of triggering.

Since no records have been found in the frequency range and with the separation of parameters the characteristic features mentioned above should be considered as a theoretical prediction.

The physical rationale of these features have been discussed in detail in this paper. Suggestions have been made how and what to measure so that the data can be analysed with a physical point of view in mind. The deductions and conclusions are based on the electrostatic lightning theory which enables us to make not only qualitative but also quantitative predictions.

REFERENCES:
1. H.W. Kasemir,"A Contribution to the Electrostatic Theory of a Lightning Discharge," JGR,65,#7,1873-1878,   1960

2. H.W. Kasemir,"Static Discharge and Triggered Lightning," Lightning and Static Elelctricity Conference,24-1,1983

3. L.G. Smith,"An Electric Field Meter with extended Frequency Range," Rev.Sci.Instr. 25,510-513,   1954

4. T.Rudolph,R.Perala,"Interpretation Methology of In-Flight Data," Electro Magnetic Applications,Inc.   1982

5. M.A. Uman,"Lightning," McGraw Book Company   1979

6. Centre D'Essais Aeronautique De Toulouse "Measure Des Characteristiques De La Foudre En Altitude," Essais No. 76/650000,   1979.

RANGING AND AZIMUTHAL PROBLEMS OF AN AIRBORNE CROSSED LOOP USED AS A SINGLE-STATION LIGHTNING LOCATOR

L.W. Parker and H.W. Kasemir*

M-S 31, GTE Government Systems Corp., One Research Drive, Westborough, Massachusetts 01581, U.S.A.
*Colorado Scientific Research Corp., 1604 S. County Road 15, Berthoud, Colorado 80513, U.S.A.

Abstract - Data from recent flight tests, useful for assessing the lightning-location capability of a commercial airborne crossed loop mounted on a C-130 aircraft, are analyzed under a novel interpretation. The new interpretation identifies data from individual flashes. These and associated airborne radar cloud data lead to an improved assessment of the instrument's range and azimuth errors. The value of identifying individual discharges lies in the fact that the multiple return strokes of the same flash may be assured to originate in the same location. Therefore, if each dot on the display is associated uniquely with one return stroke, the radial distance and azimuth are the same for all return-stroke dots caused by the same lightning. The radial and azimuthal spreads can then be determined. The results obtained show (a) large radial spreads that severely deteriorate the instrument's ranging capability, (b) reasonably good azimuth capability for distant lightnings, and (c) deterioration of azimuth capability for nearby lightnings, but also for all single-stroke flashes. A novel method is also suggested as a hopeful possibility for overcoming the ranging difficulty. It should be investigated in future work.

1.0 INTRODUCTION

In our recent survey [1], we analyzed many lightning warning systems suitable for airborne use. The present paper is concerned with the accuracy of one of these systems, a single-station crossed loop used as an on-board lightning locator [1,2]. The data were obtained with Stormscope, a commercially-available instrument (operating in the 50-kHz region) that represents this type of detector. In flight tests performed by the Air Force in 1981 [3,4], the instrument was installed on a WC-130 Lockheed Hercules transport aircraft, operated by NOAA. The primary mission of the joint NOAA-Air Force program was to characterize the radiation emitted by lightning and its coupling to the aircraft. The aircraft was suitably instrumented with electric and magnetic field sensors for this purpose [5], and Stormscope was "piggy-backed" thereon.

To our knowledge, these data constitute the first available of their kind suitable for assessment of an instrument mounted on a large aircraft. Data obtained prior to 1981, including the Air Force's own tests[6] and experiences of operators of small private aircraft [7], have involved small or mid-sized aircraft. (New data involving a Convair 580 is anticipated from 1984 tests.)

The Stormscope data in conjunction with airborne radar data were acquired along a flight path toward and around an isolated storm located by the airborne radar. (An additional valuable feature is that the Stormscope display screen was manually cleared frequently during the flight.) These data, and our own analysis associating Stormscope dots with return strokes of individual lightning flashes, allowed an assessment to be made of the instrument's azimuth and ranging, and therefore also lightning-location, capabilities [3]. This association (suggested also in the preliminary study by J. Reazer [4]) represents an advance in that the analysis and statistical procedures are thus applicable to individual lightning discharges where the Stormscope dots are correlated in time, as opposed to the usual Stormscope display where clusters of dots are accumulated over many lightning discharges. The value of identifying individual discharges lies in the fact that the multiple return strokes of the same flash may be regarded as originating in the same place. Therefore, the radial distance and the azimuth should be the same for all return-stroke dots caused by the same lightning, and the radial or azimuthal spread can be determined.

In order to make clear why we can relate the individual Stormscope dots to lightning strokes, it will be helpful to review here the temporal phenomenology of a cloud-to-ground discharge. (This information may be found for example in Ref. 8.) The average lightning discharge (or flash) to ground is multiple, i.e., includes a series of several individual component strokes, i.e., return strokes. (The terms "flash" and "stroke" are attributable to Schonland.) During a moderate storm the average interval between flashes is 20 seconds. The overall duration of a flash is of the order of one second, and the duration of an individual component stroke is of the order of one millisecond. The multiplicity of a flash (number of strokes) has an average value between 3 and 4, but can vary from one to more than 20. The probability of any given number occurring decreases as the number increases. Each stroke consists of a high-current pulse, and is preceded by a low-current leader process. The mean time interval between strokes is about 30 ms or longer, and tends to increase with the number of strokes. Positive flashes, as opposed to negative ones, usually are single-stroke flashes.

Hence, the multiple strokes of a flash can easily be resolved in time if the resolution capability of the instrument is under a millisecond. We believe that the Stormscope instrument resolves the component return strokes of a single flash, and displays these strokes as individual dots, occurring within a time interval of one second. This interpretation leads to new types of results. The results to be presented are based on our recent report [3].

2.0 STORMSCOPE DATA ANALYSIS

The types of data obtained in the 1981 measurements that are relevant here are as follows:

a. Airborne video tape record of Stormscope displays, which provided the Stormscope data used here.

b. Timing data from an on-board time code generator (synchronized to the ground station time base).

c. Digitized airborne weather radar data, stored on flexible disk or magnetic tape. This radar data was valuable for the present purposes.

No independent "ground-truth" lightning location data were available. It should be mentioned that the data from the other on-board sensors with which the aircraft was heavily instrumented (B-dot, D-dot, I-dot) to measure submicrosecond-time-scale lightning-induced transients were tailored to a different investigation and were generally not suitable for the problem of interest here.

It should also be noted that the data-reduction effort, by Reazer and ourselves [3,4], was considerable, requiring manual timing, hand measurements, and copying of individual Stormscope dots from the video monitor screen onto drawings, from the video tape, frame by frame. (This laborious procedure can be avoided in the future by automatically digitizing the Stormscope data.) Using the same video tape and radar data as Reazer [4], we have extracted additional valuable data and reorganized all of the data to obtain new types of results, to be described below.

Figures 1 and 2 summarize the principal results obtained with respect to azimuth and ranging capability, as follows.

We organized the Stormscope dot data into about 50 individual cloud-to-ground flashes, with an average of about 4 dots (strokes) per flash, corresponding to a particular time period on 25 August 1981. During this time period, there was essentially a single storm cloud indicated by the radar. Combining radar data (digitized) with aircraft position and heading data at several positions allowed us to locate the ground position of the cloud and to reconstruct the aircraft flight path. Figure 1 shows the inferred cloud position and the flight path toward and around the cloud. The cloud is defined to have a circular shape with a typical diameter of 2.5 nmi. The flight path and the instantaneous Stormscope dot displays at various positions along the path allowed a test of how well Stormscope could track a well-defined lightning source. The numbers at positions along the path (from 1 to 41) not only label the positions, but also denote the "Flash Indices" [3] of the individual flashes occurring when the airplane was at those positions. The encircled indices (Flashes 1, 5, 7, 16, 20, 22) designate the flashes giving rise to the selected Stormscope displays that will be discussed below, together with the "rays" (azimuth lines) emanating from various points along the path. Figure 2 presents range data for the 41 flashes and shows how the range data compare with the true cloud position. We will discuss Fig. 2 in detail later.

2.1 Combined Cloud-Dot Displays

Figures 3-8 are diagrams of simulated compass-rose displays of Stormscope dot data with superimposed radar cloud data for the six selected individual flashes mentioned above. The cloud is represented diagrammatically by quadrilaterals whose vertices lie on ovals drawn by us around the densest portions of the radar clouds (not shown). The diagonals of the quadrilaterals correspond roughly to the major and minor axes of the ovals. Greater detail in defining the cloud shape is not warranted for the present purposes. The inference of the cloud position was not a trivial task in view of shortcomings in the available data (see Ref. 3). All six displays to be discussed have a maximum range of 50 nmi and a 360-degree azimuthal view. The cloud range and radial spreads of the dots can be estimated from the figures, but can also be read off directly from Figure 2.

Figure 3 (Flash 1, at 14:14:03) shows a single dot on the 264-degree radial, at a range of 43 nmi. The cloud (lightning source) on the other hand is near the 0-degree radial, at a range of about 30 nmi. This represents a poor azimuth detection which appears to be typical of single-dot (single-stroke) flashes.

Figure 4 (Flash 5, at 14:15:48) shows 7 dots on the 0-degree radial, with ranges distributed from 10 to 30 nmi. The cloud is near the 0-degree radial (straight ahead of the airplane) and at a range of 22 nmi. In this case, the azimuth detection is good. The radial spread is not good (as also indicated in Figure 2) and is mostly inward.

Figure 5 (Flash 7, at 14:16:24) shows 5 dots on the 0-degree radial, aligned with the cloud. The ranges are distributed from 18 to 45 nmi, with the cloud at 18 nmi. Thus, the azimuth detection is good, while the radial spread is poor and entirely outward.

As Figure 1 indicates, Flashes 1, 5, and 7 all occur while the airplane is heading directly toward the cloud.

Figure 6 (Flash 16, at 14:18:08) shows 3 dots, one on the 0-degree radial and two on the 12-degree radial, aligned with the cloud. The radial spread is from 21 to 50 nmi, with the cloud at 11 nmi. The azimuthal detection is still quite good, while the radial spread is poor, and entirely outward. At this time the airplane has begun its turn to circle around the cloud (Figure 1). The good azimuth defined by the two inner dots is also shown by the ray emanating from Point 16 in Figure 1.

Figure 7 (Flash 20, at 14:18:56) shows 6 dots, the outermost 4 of which are well aligned with the cloud (8 nmi) on the 25-degree radial. The radial spread is from 3 to 26 nmi, that is, inward and outward but predominantly outward. The azimuthal detection is still good.

Figure 8 (Flash 22, at 14:20:34) shows 13 dots, spread over 90 degrees in azimuth and over 3 to 26 nmi in range. The cloud is at 3 nmi, that is, nearby on the 75-degree radial. This occurs at the position of closest approach (Figure 1). This striking degradation in azimuth detection is associated with the closeness of the lightning, as will be discussed.

3.0 SITE ERRORS

Reference 3 [App. C; see also Refs. 1 and 2] presents an original 3-D computer model and numerical computations of the site error at the Stormscope location and at other locations on the C-130 aircraft. (These correction factors have to be determined only once, since they depend only on aircraft geometry. The azimuth errors due to real lightning sources cannot be calculated precise-

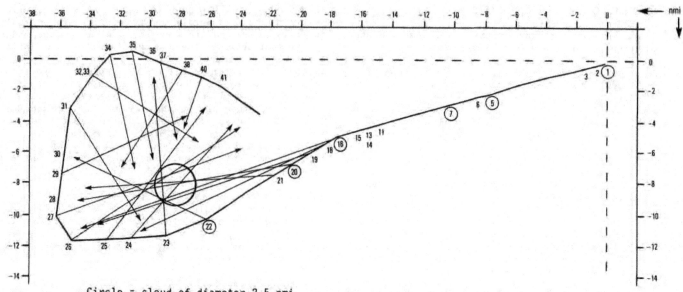

Circle = cloud of diameter 2.5 nmi.
Encircled numbers denote selected flashes discussed in text and in Figs. 3-8.
Arrows denote mean azimuths (see text).

FIGURE 1.   FLIGHT PATH AND STORMSCOPE AZIMUTHS AT VARIOUS POSITIONS WHERE INDIVIDUAL FLASHES OCCUR (25 August 1981).

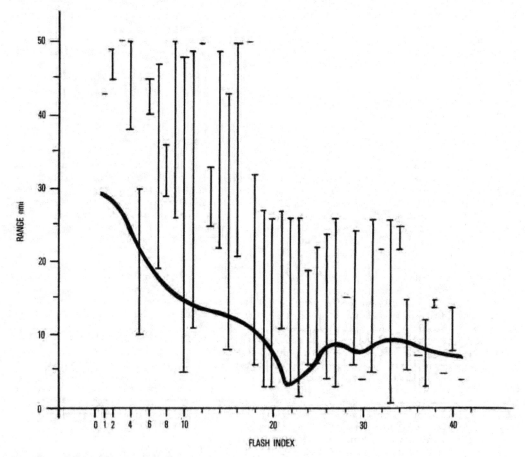

Heavy curve = cloud position
Single horizontal bar denotes range of a single-dot flash.
Vertical bars represent radial spread of dots caused by individual flashes.

FIGURE 2.   STORMSCOPE RANGE SPREADS versus FLASH INDEX and CLOUD POSITION ALONG FLIGHT PATH (25 August 1981).

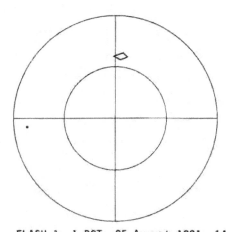

FIGURE 3.   FLASH 1. 1 DOT. 25 August 1981. 14:14:03.
Airplane at origin of coordinates.
Cloud quadrilateral: azimuth =   0°,
                       range =   30 nmi.
Stormscope single dot: azimuth = 264°,
                       range =   43 nmi.

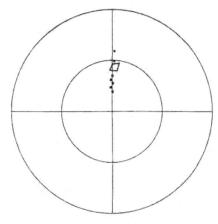

FIGURE 4.   FLASH 5. 7 DOTS. 25 August 1981. 14:15:48.
Airplane at origin of coordinates.
Cloud quadrilateral: azimuth =   0°,
                      range = 22 nmi.
Stormscope dots: azimuth =   0°,
                      average range = 20 nmi.
                      range spread = 20 nmi.

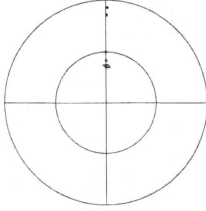

FIGURE 5.   FLASH 7. 5 DOTS. 25 August 1981. 14:16:24.
Airplane at origin of coordinates.
Cloud quadrilateral: azimuth =   0°,
                      range = 18 nmi.
Stormscope dots: azimuth =   0°,
                      average range = 32 nmi,
                      range spread = 27 nmi.

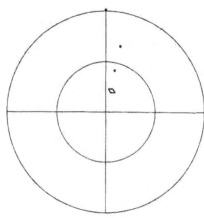

FIGURE 6.   FLASH 16. 3 DOTS. 25 August 1981. 14:18:08.
Airplane at origin of coordinates.
Cloud quadrilateral: azimuth = 15°,
                       range = 11 nmi.
Stormscope dots: azimuth = 0°-12°,
                       average range = 36 nmi,
                       range spread = 29 nmi.

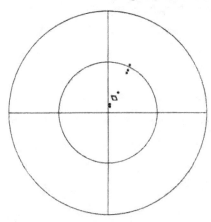

FIGURE 7.   FLASH 20. 6 DOTS. 25 August 1981. 14:18:56.
Airplane at origin of coordinates.
Cloud quadrilateral: azimuth = 25°,
                      range =   8 nmi.
Stormscope dots: azimuth = 10°-25°,
                      average range = 15 nmi,
                      range spread = 23 nmi.

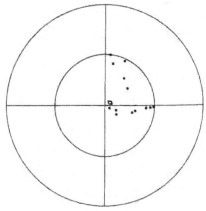

FIGURE 8.   FLASH 22. 13 DOTS. 25 August 1981. 14:20:34.
Airplane at origin of coordinates.
Cloud quadrilateral: azimuth = 75°,
                      range =   3 nmi.
Stormscope dots: azimuth = 6°-125°,
                      average range = 15 nmi,
                      range spread = 23 nmi.

697

ly, but can be determined by measurement.) An interesting result is that for the location chosen, under the "platypus" or "beaver tail," the site error is essentially identical to that of a crossed loop centered on a long cylinder. This problem has a known analytic solution. Hence, the site error calibration function for a crossed loop at the Stormscope location can be represented analytically, by

$$A_o = \text{arc tan } (2 \tan A),$$

where A denotes the apparent azimuth angle and A_o denotes the true azimuth angle.

This site-error formula predicts that there is no error in the forward and backward directions (zero and 180 degrees), and at 90 degrees. Also, the observed azimuth is more nearly in the forward (or backward) direction than the true azimuth. In the first four example flashes (Flashes 1, 5, 7, 16,) shown in Figs. 3-6 all azimuths are in the forward direction. Hence, there is no site error. There is also no significant site error in Flash 20, which occurs near the forward direction. In Flash 22, however, the extreme azimuthal scatter dominates the site errors. In other cases [3] the large errors in azimuth (greater than 20 degrees) when they occur cannot be explained as site errors.

4.0 DISCUSSION

In this section the Stormscope ("SS") data are analyzed to assess the capability of a crossed loop to track lightning activity, in particular cloud-to-ground flashes. First we consider the ranging capability, and next the azimuth capability. Then we will discuss the significance of the results.

4.1 Stormscope Ranging

Most of the data reveals severe radial spreading of the dots. To quantify this spreading effect, we consider the minimum and maximum ranges of the dots in the 41 flashes [3] associated with the aircraft trajectory in Fig. 1. (These minimum and maximum values can be obtained from the data given in Ref. 3.) Figure 2 shows a set of vertical "range-spread" bars defining the spread for each flash. Ten flashes consisting of only one dot/stroke (i.e., Flashes 1, 3, 12, 17, 28, 30, 32, 36, 39, 41) have no vertical range-spread bars, only a short horizontal "tick" mark representing the range of the single dot. Also included is the range of the radar cloud in each flash, denoted by the heavy curve.

For Flashes 1-18, the SS display range was 50 nmi. After Flash 18 the SS display range was switched to 25 nmi, which is the display range for the remaining Flashes 19-41. It is evident that in none of the flashes does the SS range reasonably approximate the cloud distance. Also, the maximum SS range in most cases overestimates greatly the cloud range. This could lead a pilot to infer that the electrical discharges are considerably further than they actually are. On the other hand, the inward spread may alarm the pilot unnecessarily. The inward spread could also account for the high "false alarm" rates experienced in the mining-operations warning tests of Ref. 9. It should be noted that there is a tendency for the radial spread to occupy the available display range. Thus, for Flashes 1-18, the maximum range tends to cluster around the 50 nmi limit of the display, while for Flashes 19-34, it tends to cluster around the 25 nmi limit. It is interesting that some improvement in ranging seems to occur in Flashes 35-41.

4.2 Stormscope Azimuths

Stormscope (SS) sensing of lightning flash azimuths should be reasonably good, in view of the well-established technology of magnetic crossed loops, in the absence of site errors, and when the radiator (lightning channel) is vertical and the magnetic vector of the electromagnetic field is perpendicular to the crossed-loop axis (that is, lies in the horizontal plane and is perpendicular to the line joining the radiator position and the sensor position)[2]. The following SS characteristics are evident:

a. In most cases, particularly at large distances, say 20-30 nmi, from the cloud (lightning source), with the aircraft headed generally toward the cloud, the SS dot azimuths track the cloud azimuth reasonably well. There should be no site error when the source is in the forward direction.

b. When the aircraft is near the cloud, say within 3-4 nmi, the SS dot azimuth sensing capability deteriorates. The azimuths either are wrong or fan out to the extent that they become indeterminate (e.g. Fig. 8). For example, at the aircraft flight path positions nearest to the cloud, the azimuths are fanned out over 90 degrees. Moreover, the multiplicity of dots increases, to as many as 17 per flash. The spread in azimuths cannot be explained as site errors.

c. Of the 10 single-dot flashes, 7 give erroneous azimuths. These are not explainable as site errors.

4.3 Continuous Triangulation on Aircraft Trajectory

Figure 1 shows the cloud and the aircraft flight path, with Stormscope (SS) azimuth lines emanating from various "flash positions" along the flight path. Each azimuth line represents an average of dot azimuths taken from the appropriate figure of Ref. 3. A single azimuth line has been selected here to represent each flash, whereas there are in many cases fans of azimuths.

The collection of azimuth lines generally and clearly point toward the cloud region. As the aircraft circles around the cloud, the SS azimuth lines generally point to the interior of the region. However, the distance to the source is still unknown due to the large radial spread. This figure suggests that an improved method of locating a lightning source using an airborne crossed loop is to accumulate successive azimuths (by an onboard computer) and compute their points of intersection continuously. The spread of ranges thus obtained by this "continuous triangulation" method should be much reduced, even though some remaining scatter is to be expected.

5.0 CONCLUSIONS

The principal results and conclusions, that should be valid for any crossed-loop lightning direction finder, are as follows.

Crossed-loop azimuths are reasonably good for distant lightnings, but deteriorate for nearby lightnings. The large errors in azimuth when they occur cannot be explained as site errors. The azimuthal spread is of the order of ± 3 degrees (for distant lightnings, beyond about 20 nmi), while the radial spread is of the order of 20 nmi. Thus, the radial spread of dots is excessive, and the ranges indicated by the dots are therefore fair at best, but generally poor to nonexistent. This is one of the main problems for single-station crossed-loop lightning locators.

The presence of site errors is suggested in relatively few cases. In most cases there was either no site error, or the azimuthal errors were excessive and not systematic (nearby lightning or single-stroke flashes).

A possible method is suggested for processing the azimuth data (of Stormscope or any other airborne azimuth detector) in such a way as to infer range while reducing the effective radial spread. This involves a technique of "continuous triangulation", whereby (using an on-board computer) one accumulates successive azimuths along the flight path and computes their points of intersection continuously. The system thus operates effectively as a two-station system, and the spread of ranges thereby obtained should be much reduced even though some remaining scatter is to be expected.

The outstanding feature of this research is as follows:

One can be sure that all the Stormscope dots that are generated by one lightning flash originate at the same location. This is always questionable if one tries to evaluate clusters.

5.1 Unanswered Questions

Several questions remain unanswered, but tentative explanations are offered, as follows:

a. Why does the azimuth capability deteriorate when the aircraft is near the lightning? Perhaps we are no longer in the "radiation zone" of the radiator. At 50 kHz the wavelength is 6 km or 3.2 nmi. Within this range we are in the induction zone of a point-dipole radiator. Moreover, since the length of a lightning channel is of the order of 6 km, the channel no longer appears as a point dipole at a range of 6 km. Also, the lightning channels may have horizontal components that can make significant contributions at close ranges. This effect (nonvertical lightning) would degrade the azimuth-detection capability of a crossed loop[2].

b. Why does the multiplicity of dots increase greatly near the lightning source? There may be, in addition to the strong pulse of a return stroke, many weak pulses radiated by the lightning channel, for instance, K-changes or electromagnetic reflections at the branch points of the channel. These weak pulses may be picked up at close range. Another possibility is the contribution of intracloud discharges to SS data; this cannot be clearly ruled out, particularly at close ranges. The manufacturer, however, appears to have designed the instrument to detect only cloud-to-ground discharges.

c. Why do single-dot flashes tend to give erroneous azimuths? We have no obvious explanation for this. There may be other storms not seen by the radar that could account for odd azimuths.

6.0 ACKNOWLEDGEMENT

We are grateful to C. Mashburn of Warner Robins Air Logistics Center for his encouragement and support, and to G. duBro, J. Reazer, P. Rustan, A. Serrano, and L. Walko of the Atmospheric Electricity Hazards Group at Air Force Wright Aeronautical Laboratories for providing the data and many helpful discussions. This work was sponsored by the U.S. Air Force under Contract F09603-83-C-1680.

7.0 REFERENCES

1. Parker, L. W. and H. W. Kasemir, Airborne warning systems for natural and aircraft-initiated lightning, IEEE Trans. Electromagnetic Compatibility EMC-24, 137-158 (1982).

2. Parker, L. W., Errors in lightning direction-finding by airborne crossed loops, International Aerospace and Ground Conference on Lightning and Static Electricity, Ft. Worth, DOT/FAA/CT-83/25, Paper #57 (1983).

3. Parker, L. W. and H. W. Kasemir, Assessment of a lightning detector as an aid in strike avoidance by a C-130 aircraft, Lee W. Parker, Inc. Final Report to Warner Robins ALC under Contract F09603-83-C-1680 (Sept. 1984).

4. Reazer, J., Data acquisition for evaluation of an airborne lightning detection system, Draft Report (Sept. 1982); Walko, L. C. and J. Reazer, Air Force Flight Dynamics Laboratory Report AFWAL-TR-83-3083 (Sept. 1983).

5. Rustan, P. L., B. P. Kuhlman, A. Serrano, J. Reazer and M. Risley, Airborne lightning characterization, Air Force Flight Dynamics Laboratory Report AFWAL-TR-83-3013 (Jan. 1983).

6. Baum, R. K. and T. J. Seymour, In-flight evaluation of a severe weather avoidance system for aircraft, AFFDL Report AFWAL-TR-80-3022 (1980).

7. Rozelle, R., Weather avoidance, an alternative to radar, The AOPA Pilot, November (1979), pp. 95-105. (Also, personal communication, 1980).

8. Uman, M., Lightning, Dover Publications, New York (1984). See also Uman, M., A review of natural lightning: experimental data and modeling, IEEE Trans. Electromagnetic Compatibility EMC-24, 79-112 (1982).

9. Johnson, R. L., D. E. Janota and J. E. Hay, An operational comparison of lightning warning systems, J. Appl. Met. 21, 703-707 (1982).

PREDICTED AIRCRAFT FIELD CONCENTRATION FACTORS AND THEIR RELATION TO TRIGGERED LIGHTNING *

Lee W. Parker
Lee W. Parker, Inc.
252 Lexington Road
Concord, Mass. 01742

and

Heinz W. Kasemir
Colorado Scientific Research Corporation
1604 S. County Road 15
Berthoud, CO 80513

ABSTRACT

A 3-D computer model (PESTAT code) has been developed for predicting geometric field concentration factors (FCF's) on the surfaces of an aircraft that could trigger lightning upon entering the field of a strongly-electrified cloud. The present computational approach has the practical advantage that it provides a quick survey of FCF's and therefore a simple assessment of the danger to the aircraft. The FCF's and the measured ambient field vector are key parameters that would be used by a warning system for triggered strikes.

Application to a C-130 aircraft geometry yields theoretical FCF's ranging up to about 180, depending on the direction of the cloud field relative to the aircraft. The relation of the predicted FCF's and the associated critical fields to corona onset and triggered strikes is discussed. The common method of assessing triggered-lightning danger using only FCF's and a nominal value for the breakdown field is questionable, because the criterion being used leads to the critical field for corona onset. As opposed to this, the lightning danger depends on a different criterion for the critical ambient field, namely, that field which turns corona discharge into a full-grown lightning discharge.

* International Aerospace and Ground Conference
on Lightning and Static Electricity
Orlando, Florida, 26-28 June 1984

LEE W. PARKER, INC.
252 Lexington Road
CONCORD, MASSACHUSETTS 01742
Phone 617 369-1490

1. INTRODUCTION

IT IS INCREASINGLY EVIDENT, from the collected data on aircraft lightning accidents, pilot reports, and theoretical considerations, that most of the lightning strikes to aircraft are triggered, i.e., initiated by the aircraft itself, as opposed to accidental hits by natural lightning (see Kasemir, this conference, and Refs. 1-3).* (These triggered strikes are usually not recognized as such, and are often called "electrostatic discharges.") Hence, in assessments of lightning danger to aircraft, the emphasis should be shifted from the study of conventional lightning warning devices, that are based on detecting the presence or absence of natural lightning, to the study of warning systems that can assess the danger of a triggered discharge.

A triggered strike can occur in highly electrified clouds that do not necessarily produce natural lightning. The probability of this type of strike becomes high when the aircraft enters a high-field region. This occurs because the electric field intensity at the aircraft extremities (wing-tips, nose, tail structure, etc.) is locally enhanced over its ambient value. The enhancement is due to the nature of the surface charge distribution on the aircraft, induced by the external field. It does not depend on the presence of any intrinsic charge on the aircraft, such as "P-static" charge (the net intrinsic charge can be zero). When the enhancement at one or more points on the surface is such that the breakdown-field value is exceeded, corona discharge begins and a strike may be initiated. The enhancement multiplication factor for local field intensity is called the "field concentration factor" (FCF).

It is desirable to have a capability for predicting field concentration factors (FCF's) for any given aircraft geometry. A computer code has been developed for this purpose (PESTAT, for "Parker ElectroSTATics"). Section 2 presents our approach, based on (a) the numerical representation of the aircraft by a collection of quadrilateral or triangular surface "elements," or "patches," and on (b) the solution of a set of simultaneous equations for the surface charge densities on the surface elements. The simultaneous equations represent a discrete approximation to an integral equation that is equivalent to the Laplace equation for the electrostatic field.

Section 3 presents some preliminary numerical solutions for a C-130 Hercules aircraft. The computed field concentration factors (FCF's) range up to about 200, depending on the direction of the external field, with the highest values occurring as expected at the wing-tips, nose and tail extremities. From the FCF's, we derive critical "trigger-field" profiles (projections of surface in 3-D space defining

*Numbers in parentheses designate References at end of paper.

critical ambient field intensity as a function of direction) such that breakdown occurs somewhere on the aircraft. In Sec. 4 we discuss these results and the relation between corona onset and triggered strikes. We also emphasize the gaps and inconsistencies in our knowledge, and the difficulty of relating electrostatic theory to triggered lightning. The theory must include plasma physics.

Appendices A and B provide auxiliary analytical data as follows. Appendix A presents the computer algorithm used by the PESTAT code to evaluate the potential due to the charge of a quadrilateral planar surface area. Appendix B provides and applies analytical formulas for the fields at and near the tips of ellipsoids, that can be used as simple analytical models to calculate field concentrations (for aircraft wings, for example).

The magnitude and direction of the ambient electric field, and the geometric field-concentration factors of the aircraft, are key parameters that would be used by a warning system for triggered strikes. The present computational approach has the practical advantage that it provides a quick survey of concentration factors, and therefore a simple assessment of the danger to the aircraft.

2. COMPUTER MODEL

The computer model of the aircraft is constructed of quadrilateral surface elements or patches (with occasional triangles representing degenerate quadrilaterals). (See Fig. 1.)

One may solve Laplace's equation for the electric potential V in 3 dimensions, namely,

$$\frac{\partial^2 V}{\partial x^2} + \frac{\partial^2 V}{\partial y^2} + \frac{\partial^2 V}{\partial z^2} = 0 \qquad (1)$$

From this solution one may obtain the electric field vector $\vec{E}$ at any point from the gradient of V. However, the solution depends on the boundary conditions. Far from the aircraft the electric field must approach the ambient value $\vec{E}_0$. At the (conducting) aircraft surface, the induced surface charges must distribute themselves so as to cause complete cancellation of the field in the interior. That is, the field must be purely normal at every surface point, and the surface must be equipotential.

Rather than solve Eq. (1) as a differential equation, say, by finite differences or finite elements (5,6), it is more convenient for our purposes to solve instead the equivalent integral-equation form of Laplace's equation. This means that the potential V at any "field point" (point of observation) can be written as an integral of surface charge density σ over the aircraft surface, plus the "external" potential V_{ext} (that would exist there in the absence of the aircraft) due to the ambient field:

$$V = \int \frac{\sigma da}{(4\pi\epsilon)R} \quad + \quad V_{ext} \qquad (2)$$

where ϵ is the permittivity, da is the element of surface area, and R is the distance between the field point and the surface point on da where σ is defined. We do not know σ in advance, but may solve for it by equating V (as given by Eq. (2), at every point on the surface) to the aircraft body potential V_b. Another condition is that the integral of σ itself gives the total (i.e. net) charge Q. Thus, σ satisfies the integral equation:

$$\int \frac{\sigma da}{R} \quad - \quad (4\pi\epsilon)V_b = -(4\pi\epsilon)V_{ext} \qquad (3)$$

subject to the condition:

$$\int \sigma da = Q \qquad (4)$$

In the problem of interest here, σ and V_b are an unknown function and constant, while V_{ext} and the total charge Q are considered as given. In the case where Q=0, the σ function is sometimes referred to as the "influence charge."

In our model these relations are approximated by dividing the aircraft surface into small quadrilateral patches with constant values of σ on the individual patches, and performing the required geometric integrations over all patches (Appendix A). There results a system of simultaneous equations for the values of the σ's, defined to be at the centroids of the patches. Thus, Eqs. (3) and (4) are replaced by the N+1 simultaneous algebraic equations:

$$\sum_{j}^{N} A_{ij}\sigma_j \quad - \quad (4\pi\epsilon)V_b = -(4\pi\epsilon)V_{i,ext} \qquad (5)$$

$$(\text{where } i = 1, 2, \ldots, N)$$

and

$$\sum_{j}^{N} (\text{Area})_j\sigma_j = Q \qquad (6)$$

where the subscript j is the index for the j-th patch and its σ, and runs from 1 to N where N is the total number of patches. The symbol $(\text{Area})_j$ denotes the area of the j-th patch, σ_j denotes the value of σ on the j-th patch, and A_{ij} denotes a geometrically-determined matrix element representing the potential at the i-th patch centroid due to unit charge density on the j-th patch. The symbol $V_{i,ext}$ denotes the value of the external potential at the i-th patch. The unknowns are comprised of the N values of σ_j plus the "body" potential V_b, i.e., N+1 unknowns. The method for evaluating the matrix elements (A_{ij}) is given in Appendix A.

When the voltage V_b is given and the charge Q is unknown, Q replaces V_b as an unknown so that one now solves for the σ's and Q. Equations (5) and (6) can easily be generalized to include multiple bodies with arbitrarily assigned voltages or charges.

Solving for the σ's and V_b (or σ's and Q) completes the computational task. The solution can be obtained numerically either by direct inversion or by relaxation techniques. Our computer code PESTAT implements these procedures (next section).

The procedure for evaluating a matrix element giving the potential at a "field point" P due to a patch arbitrarily located with unit charge density on it is given in Appendix A, using an original analysis that applies to any polygon. Our formulation was developed independently of the formulation for triangular patches published earlier by Rao et al (7). Of course, one can also assume simple point charges, as Shaeffer has done (8). Shaeffer's paper gives a good description of the method (he calls it a "moment method" approach) and shows how to include external fields. (Rao et al do not consider external fields.) However, the point-charge method gives poor results close to the surface, and hence, although simple, is not satisfactory for FCF calculations. We believe that the present method is superior to others available. Its power is demonstrated by results of the type discussed below.

VALIDATION OF PESTAT - An important result discussed in the next section is the field concentration factor (FCF) of about 200 produced at the wing-tips of a C-130 aircraft by a horizontal field aligned with the wing. To assess the accuracy of PESTAT in its prediction of this rather large factor, a series of numerical calculations were made for ellipsoids, whose field concentration factors can be calculated analytically (Appendix B). In particular, the large FCF of about 200 is also analytically predicted at the tip of an ellipsoid modeling the wing of a C-130 aircraft. Good agreement was obtained, provided a sufficiently large number of patches was used. The requirement became more severe with increasing eccentricities. For the C-130 FCF's to be discussed next (high eccentricities), the estimated error is about 10 percent (too high).

3. NUMERICAL FIELD CONCENTRATION FACTORS FOR A C-130 AIRCRAFT

A C-130 Hercules aircraft was modeled using of the order of 1000 quadrilateral patches. A general view of the computer model is shown in Fig. 1, showing this to be a reasonably realistic representation for obtaining preliminary solutions. We have omitted details such as engine pods and fuel tanks under the wing on the assumption that these relatively small structures, being somewhat distant from extremities such as wing-tips, nose, tail fins, etc., should make negligible contributions to the field concentration factors (FCF's) at the extremities. These additional small structures can be modeled, to any extent desired, but at a cost of more unknowns and greater computational expense.

Figure 2 shows 3 views of the computer model, projected onto the 3 principal planes of

702

the computational problem space:
  (a) the x-y plane
  (b) the y-z plane
  (c) the z-x plane

Figure 3 shows the FCF computed using the PESTAT code, plotted in polar coordinates of the x-y plane (horizontal plane = plane of Fig. 2a). The FCF varies from a maximum of about 180 in the y-direction (wing tip to wing tip) to a minimum of about 130 in the x-direction (nose-to-tail direction). The FCF profile has roughly an oval shape.

Figure 4 shows the FCF profile in the y-z plane (vertical plane containing the wing-tip-to-wing-tip line = plane of Fig. 2b). Here the FCF profile, which has a scalloped shape, varies from a maximum of about 180 in the y-direction (wing tip to wing tip) to a minimum of about 100 in the direction 30° from the vertical direction. The FCF is about 114 in the vertical direction.

Figure 5 shows the FCF profile in the z-x plane (vertical plane containing the nose-to-tail line = plane of Fig. 2c). The profile in this plane is not symmetric about the nose-to-tail or vertical directions. Instead, the profile is dumbbell-shaped, with its long axis at about 40° with respect to the x (nose-to-tail) direction. The maximum FCF is about 175. The short axis of the dumbbell represents a minimum FCF of about 60. In this direction the electric field is aimed towards the concavity of the aircraft profile where the tail meets the fuselage.

Since the numerical results probably overestimate the FCF's by about 10 percent, these results should be reduced by 10 percent. The reduction has not been applied in the following discussion. Additional numerical solutions were obtained for the case where tips of the ellipsoid modeling the C-130 wing were truncated. In the center of each flat end-face the FCF was down to about 40, increasing to about 200 at the edges.

CRITICAL FIELD - Next, we use the FCF profiles to define critical "trigger field" (E_{crit}) profiles in each of the 3 principal geometric planes (x-y, y-z, z-x). The critical field is obtained by dividing a nominal value of "breakdown" field, which we assume to be 3000 kV/m, by the FCF function in Figs. 3-5. This yields the trigger-field E_{crit} profiles shown in Figs. 6-8 for the 3 principal planes. The E_{crit} profiles can have curious forms, as shown particularly in Figs. 7 and 8. Of interest is the minimum value of E_{crit}, about 17 kV/m, which occurs along the wing-tip-to-wing-tip direction in the x-y plane (Fig. 6) and in the y-z plane (Fig. 7), and at about 40 degrees from the horizontal in the z-x plane (Fig. 8, in the direction from above the tail to below the nose). The z-x behavior is due to the influence of the tail-structure-fuselage interaction.

The significance of an E_{crit} profile is as follows. When the aircraft finds itself in an ambient (cloud) vector field having a given direction and magnitude, the surface field somewhere on the aircraft surface (e.g. at an extremity) exceeds the breakdown value 3000 kV/m

when the ambient field magnitude exceeds E_{crit} for the given direction. (The ambient field vector can be measured in flight using field mills. A field mill system can be designed to provide this type of data to the pilot on a continuous basis (1,9 for example).)

In the next section we discuss the physical significance of these E_{crit} profiles, and their relation to corona onset and triggered strikes.

4. TRIGGER FIELDS, CORONA AND STRIKES TO AIRCRAFT

The "trigger-field" profiles (E_{crit}) shown in Figs. 6-8 represent the critical ambient (electrified-cloud) fields that would induce breakdown at the aircraft extremities. These are therefore at least corona-onset fields, and represent a lower bound for the lightning-trigger field. However, in general the lightning-trigger fields are observed to be different from the fields required for the onset of corona discharge. There are gaps and inconsistencies in our knowledge that prevent our making a precise connection of corona-onset fields with those that would produce triggered strikes. Nonetheless, although our knowledge is incomplete, the results of this paper represent preliminary steps that must be taken along the path toward a complete understanding of the triggering phenomenon.

Triggered strikes to instrumented aircraft have been reported as occurring in a variety of conditions, certainly in high fields, up to 360 kV/m (10), but occasionally also in surprisingly low fields, down to 4 kV/m (11). If we were to interpret Figs. 6-8 as trigger-fields for strikes, they would imply that triggering can occur in a field as low as 17 kV/m when it is in the right direction. In addition, since the breakdown field is actually lower by a factor of about 3 (1000 kV/m) because of altitude, the predicted trigger field would be as low as 6 kV/m.

As examples of observed triggering and related corona phenomena, to which we ultimately may relate our predicted field concentration factors but which presently remain unexplained, we discuss next some pilot observations, and results of a relevant experiment.

PILOT EXPERIENCE - Pilot observations relevant to the triggering and related corona phenomenon are summarized by Clifford (3) as follows.

"Pilots generally agree that there are two distinct classes of lightning strikes to aircraft in flight. The first and most common variety usually occurs while flying in precipitation at temperatures near freezing. This type is preceded by a buildup of static noise in the communication gear, due to corona (visible at night). The buildup may continue for several seconds before the strike occurs.

The second variety occurs abruptly without warning. It is most likely to be encountered in or near ongoing thunderstorms, in contrast to the former variety which is often experienced in precipitation that has no connection with thunderstorms. Pilots tend to believe that the slow buildup type of discharge is not a true lightning strike but rather a discharge of excess charge ('P-static') built up on the aircraft by flight through the precipitation. This non-thunderstorm type greatly outnumbers the other. Both kinds can create a brilliant flash and a boom which can be heard throughout the airplane."

We believe that in the "common" variety of discharge the pilots are experiencing a strike initiated or triggered by the aircraft upon entering a region of high electric field. The charge on the aircraft is not important for the energy budget but may be important for the triggering process (see below). This view was proposed by us earlier (1). Triggering of lightning by aircraft in high fields has also been discussed by Fitzgerald (9,10), Vonnegut (12), Pierce (13,14), Shaeffer (15), and Kasemir and Perkins (16), among others. The rare variety of strike, that occurs without warning, is an accidental hit of the airplane by a natural lightning that originates somewhere else in an ongoing thunderstorm.

In our opinion (expressed earlier in (1,2)), the role played by precipitation charging is as follows. The main electrical energy for the common (triggered) type of lightning strike is provided by the field of the electrified cloud. However, the electric charge on the aircraft due to precipitation ("P-static"), which is especially strong in the melting zone of a cloud, may contribute to the triggering of a lightning discharge by the aircraft in cases where the field concentrations at the extremities of the aircraft are not quite sufficient to initiate a lightning discharge. Therefore, the role of precipitation charging is not to provide energy for a full-grown lightning, but merely to help convert the corona discharge into long streamers which can then grow in the external cloud field into a proper lightning discharge. Evidence for the probable importance of precipitation charging is provided by pilot reports and by data obtained by Fitzgerald (10). Its role is discussed by Clifford and Kasemir (2).

The fact that pilots observe corona discharge in advance of a triggered strike is not surprising. Due to its many sharp protrusions (antennas, pitot tubes, landing gear, nuts, rivets, lightning arresters, exhaust nozzle rims, edges, corners, etc.), an aircraft always goes into corona discharge when it enters a region of sufficiently high field. Corona begins at a point as soon as the field at the tip exceeds the breakdown value. This can occur for ambient fields as low as a few kV/m, depending on how small the tip radius of curvature is (Appendix B). The corona discharge is relatively stable, and, unless the ambient field increases beyond some as yet unknown critical value, the aircraft will remain in corona discharge until the external field drops below the critical value for corona onset. Note that although corona may be produced by high fields (and precipitation charging), not all corona develops into a major discharge. The reasons for this are not clear. What is perhaps surprising is that many experiments have been performed in attempts to

deliberately provoke triggered strikes (both with instrumented aircraft and with wire-trailing aircraft and rockets), while few have succeeded (17-19). (Franklin's famous kite experiment might also fit into this category.) In most cases, copious corona discharges have been manifested, but without accompanying strikes. The occurrence of a soft hiss in the communication gear indicating the onset of corona is well known by pilots. This phenomenon is common and can become severe enough to black out communication, but an accompanying lightning strike is relatively rare.

CORONA AND TRIGGERING EXPERIMENT - In a set of relevant ground-based experiments, the effects of corona-producing points on the trigger-breakdown field of the shuttle-orbiter were investigated by Kasemir and Perkins (16). They used a scale model of the spacecraft placed between the plates of a meter-sized large plate condenser. As part of the investigation they also used a highly-polished spheroid to determine the trigger-breakdown field in the absence of, and in the presence of, corona-producing points on the spheroid. Details are given in Refs. (2) and (16), but some of the principal results are the following:

(a) The observed trigger-breakdown field of the polished spheroid without corona-producing points was predicted accurately from the solution of Laplace's equation. Our predictions of FCF's for a C-130, shown in Figs. 6-8, are similar types of solutions.

(b) The trigger field was reduced (by 33 percent) due to the presence of corona points, and this reduction was only weakly dependent on the nature of the points (form, length, sharpness, etc.).

These results suggest that:

(a) Theoretically-calculated trigger-field profiles such as those in Figs. 6-8 might be valid lower-bound predictors for aircraft triggering of lightning.

(b) The likelihood of a triggered strike seems to be enhanced by the presence of corona (in addition to altitude-dependent reduction of the trigger field). Ordinarily the corona effect is not taken into account in the literature.

There is a caveat in extrapolating these experimental results to the triggering of lightning by aircraft. The gap between the plates was only about one meter. Hence the corona "plumes" or "filaments" could easily bridge the gap, and give one the impression that lightning may be triggered if the field exceeds breakdown values over a distance of the order of a meter. This may be a false impression caused by the unrealistically small scale of the experiment. In an actual triggered lightning discharge, the field may need to exceed breakdown over distances of the order of many meters. That is, the corona filaments or plumes may need to be many meters long before the external field can convert them into a full lightning discharge. The required distance is not presently known and represents a gap in our knowledge. This gap cannot be filled by electrostatic arguments alone. In Appendix B we show analytically. that, even for a large ellipsoid modeling a C-130 wing, it is difficult electrostatically to produce "super-breakdown" fields over off-tip distances of the order of several meters.

The theory must deal eventually with the conversion of energy from electrostatic into gas ionization and heating, and will involve plasma physics, but such a theory has not yet been developed (4,20).

APPENDIX A: EVALUATION OF MATRIX ELEMENTS FOR THE "PESTAT" CODE

Consider a planar quadrilateral "patch" as shown in Fig. 9, and a field point P at which the potential due to the charge on the patch is to be evaluated. We have chosen a coordinate system in which the patch lies in the x-y plane, and the field point P is on the z-axis, at a height z above the plane, as in the figure. (For arbitrarily-oriented patches, a suitable rotation is required to achieve the orientation of the figure.) Let the 4 vertices of the patch be labelled 1, 2, 3 and 4, as in the figure. Let the radial distances from P to each of the 4 vertices be labelled R_1, R_2, R_3 and R_4, as in the figure. The derivation is lengthy; only results are given.

The matrix element is the sum of 4 contributions, from each of the 4 line segments of the patch. Consider the segment from Vertex 1 to Vertex 2, called Segment 1-2. Let the equation of Segment 1-2 be given (in the x-y plane) by:

$$y = A x + B \qquad (A-1)$$

where A and B are determined by the x,y coordinates of Vertices 1 and 2. Let C and D be denoted by

$$C = (1 + A^2)^{\frac{1}{2}} \qquad (A-2)$$

$$D = Ay + x \qquad (A-3)$$

Then define F_1 and F_2, associated with Vertices 1 and 2, by:

$$F_j = x_j \ln (y_j + R_j) + (B/C) \ln (CR_j + D_j)$$

$$- z \tan^{-1}[(y_j D_j - A R_j^2)/z R_j] \qquad (A-4)$$

where j takes on the respective values 1 and 2. Then the contribution of Segment 1-2 is:

$$(\Delta F)_{12} = F_2 - F_1 \qquad (A-5)$$

Similar contributions are obtained from the remaining 3 sides, namely, $(\Delta F)_{23}$, $(\Delta F)_{34}$, and $(\Delta F)_{41}$, as we trace the patch in the clockwise direction. In the special case where the line

segment is vertical (A is infinite), its contribution vanishes (set $\Delta F=0$).

The potential at P due to the patch charge is then the sum of the ΔF's. The matrix element $A_{ij}(= \int da/R)$ is evaluated by choosing P to be the centroid of the i-th patch, and evaluating the contribution to the potential at P due to unit charge density on the j-th patch (the patch of the figure):

$$A_{ij} = (\Delta F)_{12} + (\Delta F)_{23} + (\Delta F)_{34} + (\Delta F)_{41} \qquad (A-6)$$

Note that this procedure represents a line-integral formulation. The algorithm is valid for any polygon.

APPENDIX B: FIELD ENHANCEMENTS NEAR TIPS OF ELLIPSOIDS

To provide analytic insight for field enhancements or "field concentration factors" (FCF's) near the tips of elongated bodies (such as wings, fuselages, corona wires, etc.), we consider a triaxial ellipsoid as an analytically tractable model. The ellipsoid is defined geometrically by its semi-axes a, b and c, listed in order of decreasing dimension (a is the longest, c is the shortest). The ellipsoid is located in an ambient field E_0 in the direction of its long axis. The net charge is zero for present purposes. (This simplifies the analysis, which can be extended to include net charge. A charged ellipsoid will be discussed later.) We will be interested in two ratios: (1) the ratio of the tip field (E_{tip}) to the ambient field (E_0), and (2) the ratio of the field off the tip (E) to E_{tip} or to E_0. The derivations are lengthy, and only the results are presented. The ratios may be expressed as follows:

At the tip of the ellipsoid in the direction of the long axis (x-direction), where x=a at the tip, we have:

$$\frac{E_{tip}}{E_0} = \frac{a^2}{bc} \frac{\sin^3\phi}{D(\phi,m)} \quad (= FCF) \qquad (B-1)$$

where:

$$\sin\phi = (1 - c^2/a^2)^{\frac{1}{2}} = \text{(focal radius)}/a \qquad (B-2)$$

$$m = (1 - b^2/a^2)/(1 - c^2/a^2) \qquad (B-3)$$

and $D(\phi,m)$ is defined in terms of the Jacobi elliptic integrals of the first and second kind, $F(\phi,m)$ and $E(\phi,m)$, as:

$$D(\phi,m) = [F(\phi,m) - E(\phi,m)]/m \qquad (B-4)$$

where

$$F(\phi,m) = \int_0^\phi \frac{d\theta}{(1 - m \sin^2\theta)^{\frac{1}{2}}} \qquad (B-5)$$

$$E(\phi,m) = \int_0^\phi d\theta(1 - m\sin^2\theta)^{\frac{1}{2}} \qquad (B-6)$$

(See M. Abramowitz and I. A. Stegun, "Handbook of Mathematical Functions," New York: Dover, 1965.)
Off the tip, where $x > a$, we have:

$$\frac{E}{E_{tip}} = T_1 + T_2 \text{ (off-tip field of ellipsoid)} \quad (B-7)$$

where

$$T_1 = \frac{E_o}{E_{tip}}\left[1 - \frac{D(\phi',m)}{D(\phi,m)}\right] \qquad (B-8)$$

with

$$\sin\phi' = (1 - c^2/a^2)^{\frac{1}{2}}/(x/a) \qquad (B-9)$$

and

$$T_2 = \frac{abc}{x(x^2 - a^2 + b^2)^{\frac{1}{2}}(x^2 - a^2 + c^2)^{\frac{1}{2}}} \qquad (B-10)$$

At the tip, where $x=a$, we have $\phi'=\phi$, $T_1=0$, and $T_2=1$. Off the tip, as x increases from a towards infinity, T_2 falls off monotonically from unity towards zero, while T_1 rises monotonically from zero to E_o/E_{tip}, that is, a small number compared with unity in the cases of interest here. Hence, T_1 may be approximately neglected over most of the off-tip range of x of interest here.

This completes the set of exact formulas for the tip and off-tip fields, or FCF's, for ellipsoids.

Next we present some important limiting cases of these equations.

SPHEROID - An important case is that of the prolate spheroid, where $b=c$. In this case we have $m=1$, and:

$$E(\phi,1) = \sin\phi = (1 - b^2/a^2)^{\frac{1}{2}}$$
$$= \text{(focal radius)}/a \qquad (B-11)$$

$$F(\phi,1) = \int_0^\phi d\theta \sec\theta$$
$$= \frac{1}{2}\ln\left(\frac{1 + \sin\phi}{1 - \sin\phi}\right) \qquad (B-12)$$

Then we have

$$D(\phi,1) = \frac{1}{2}\ln\left(\frac{1 + \sin\phi}{1 - \sin\phi}\right) - \sin\phi \qquad (B-13)$$

and E_{tip}/E_o becomes:

$$\frac{E_{tip}}{E_o} = \frac{a^2\sin^2\phi}{b^2 Q_1} \text{ (spheroid)} \qquad (B-14)$$

where

$$Q_1 = \frac{1}{2\sin\phi}\ln\left(\frac{1 + \sin\phi}{1 - \sin\phi}\right) - 1 \qquad (B-15)$$

For the off-tip field of the spheroid, we obtain:

$$\sin\phi = (1 - b^2/a^2)^{\frac{1}{2}} \qquad (B-16)$$

$$\sin\phi' = (1 - b^2/a^2)^{\frac{1}{2}}/(x/a) \qquad (B-17)$$

$$D(\phi,1) = Q_1/\sin\phi \qquad (B-18)$$

$$D(\phi',1) = Q_1'/\sin\phi' \qquad (B-19)$$

where

$$Q_1' = \frac{1}{2\sin\phi'}\ln\left(\frac{1 + \sin\phi'}{1 - \sin\phi'}\right) - 1 \qquad (B-20)$$

so that

$$T_1 = \frac{E_o}{E_{tip}}\left[1 - \frac{Q_1'}{Q_1}\frac{\sin\phi}{\sin\phi'}\right] \qquad (B-21)$$
$$\text{(spheroid)}$$

and

$$T_2 = \frac{ab^2}{x(x^2 - a^2 + b^2)} \text{ (spheroid)} \qquad (B-22)$$

and $E/E_{tip} = T_1 + T_2$, as given by Eq. (B-7).

SLENDER ELLIPSOIDS AND SPHEROIDS - Other cases of interest are those of slim ellipsoids and spheroids (b/a and c/a small compared with unity). It can be shown that the approximation obtained from Eqs. (B-1) through (B-6) for the tip field is:

$$\frac{E_{tip}}{E_o} \simeq \frac{a^2/bc}{\ln\left(\frac{4a}{b+c}\right) - 1} \text{ (= FCF for ellipsoid)} \quad (B-23)$$

Neglecting T_1, we have from Eqs. (B-7) through (B-10) for the off-tip field:

$$\frac{E}{E_{tip}} \simeq T_2$$

$$= \frac{abc}{x(x^2 - a^2 + b^2)^{\frac{1}{2}}(x^2 - a^2 + c^2)^{\frac{1}{2}}} \qquad (B-24)$$
$$\text{(ellipsoid)}$$

For the spheroid ($b=c$), introducing the useful definition of tip radius of curvature $r = b^2/a$, we have the approximations:

706

$$\frac{E_{tip}}{E_0} \cong \frac{a^2/b^2}{\ln(2a/b) - 1}$$
(spheroid)

$$= \frac{2a/r}{\ln(4a/r) - 2} \quad (= \text{FCF for spheroid}) \quad \text{(B-25)}$$
(spheroid)

and

$$\frac{E}{E_{tip}} \cong \frac{ab^2}{x(x^2 - a^2 + b^2)} = \frac{a^2 r}{x(x^2 - a^2 + ar)} \quad \text{(B-26)}$$
(spheroid) (spheroid)

FIELD OF A CORONA POINT - We now apply the formulas for a spheroid to a corona point. Consider a rod, of thickness 1.4 cm at its center, with a half-length $a = 0.5$ m, and with sharpened tips having radius of curvature $r = 10^{-4}$ m (0.1 mm). Then according to Eq. (B-25) the FCF at the tip is about 1300. Assuming 3000 kV/m is the breakdown field strength, breakdown is achieved at the tip when the ambient field $E_0 = 2.3$ kV/m. This value of E_0, which corresponds to the quantity we call E_{crit} in the text, is the observed order of magnitude for corona-point onset fields, about 2 kV/m (L. H. Ruhnke, personal communication). Off the tip, however, the field falls rapidly from its tip value. According to Eq. (B-26), E falls from E_{tip} to $E_{tip}/3$ in a distance Δx only about 10^{-4} m from the tip. This small distance is of the order of r, the tip radius of curvature. The fall-off with distance Δx occurs so rapidly that even for extremely (and unrealistically) large ambient fields, breakdown is exceeded only within minute distances from the tip. As an example, in order to have "super-breakdown" fields (off-tip fields exceeding 3000 kV/m) at all values of x between $x = 0.50$ m (at the tip) and $x = 0.55$ m (an interval of $\Delta x = 5$ cm off the tip), the external or ambient field must exceed the unrealistic value 2.7 MV/m!

TIP FIELD: ELLIPSOID MODEL OF C-130 WING - We next model the wing of a C-130 aircraft by a triaxial ellipsoid having semiaxes a, b and c = 20 m, 2.4 m, and 0.36 m. Using the exact formula (B-1) we compute the FCF at the tip, due to an ambient field E_0 directed along the long axis. We obtain FCF = E_{tip}/E_0 = 192. The ellipsoid is sufficiently slim, however, to allow the use of the approximate formula (B-23), which yields FCF = 196, a sleight overestimation by about 2 percent. The corresponding numerical calculation by the PESTAT code yields the field concentration factor FCF = 210, an overestimation by about 10 percent. (This error is not serious. It depends on the number of patches used, and can be reduced by sufficiently increasing the number of patches.)

OFF-TIP FIELD: ELLIPSOID MODEL OF C-130 WING - It is appropriate to use the approximation for E/E_{tip} given by Eq. (B-24) for the off-tip field. Here, just as for the corona point (above), we consider the off-tip distance, Δx, the off-tip FCF given by multiplying E/E_{tip} by 196, and E_{crit} obtained by dividing 3000 kV/m by FCF. Thus, we obtain the following tabulation:

Δx (m)	E/E_{tip}	FCF=E/E_0	E_0 (E=3000 kV/m)
0 m	1.00	196	15 kV/m
0.01	0.48	94	32
0.02	0.35	68	44
0.05	0.21	41	72
0.1	0.14	26	110
0.2	0.081	16	190
0.5	0.037	7.2	420
1.0	0.019	3.7	820
2.0	0.0090	1.8	1700

The first column gives the off-tip distance, the second shows the fall-off of the field relative to the tip value, the third gives the effective FCF for that position, and the fourth gives the ambient field required to produce breakdown at that position. The fall-off with distance is extremely rapid, the first factor of 2 occurring within a distance 0.01 m (approximately twice the smaller tip radius of curvature, 0.006 m); then the field falls off more slowly. From the fourth column we see that even for the largest observed ambient fields (of the order of 400 kV/m), the off-tip interval Δx in which "super-breakdown" fields occur is less than 0.5 m. To stretch the interval to 2 meters would require an unrealistic 1700-kV/m ambient field, a stringent requirement associated with the small tip radius of curvature. If we choose alternatively to increase the ellipsoid tip radius of curvature to fit the tip geometry of interest, this would reduce both the eccentricity of the ellipsoid model and the FCF at its tip. This exercise illustrates the difficulty, using electrostatic models for aircraft-size bodies, of achieving "super-breakdown" fields over off-tip intervals of several meters.

CHARGED ELLIPSOID - In estimating the effects of precipitation charge, it is useful to have available formulas for the tip and off-tip fields of a charged ellipsoid. It can be shown that the formulas are almost identical to the formulas for an uncharged ellipsoid in an ambient field E_0. For the tip field, we obtain a good approximation by replacing E_0 in Eq. (B-23) by an "effective ambient field" V_b/a, where V_b is the body potential. The off-tip field variation is well approximated by the right-hand side of Eq. (B-24) without the first factor, a/x. Thus, rate of field fall-off is about the same as for uncharged ellipsoids. If we assume a vehicle potential $V_b = 0.1$ MV on the model ellipsoid, the field at the tip is about 1000 kV/m, i.e., somewhat under breakdown magnitude.

707

ACKNOWLEDGEMENTS

We wish to express our appreciation to L. H. Ruhnke (NRL) for stimulating discussions of corona and triggering problems, to E. G. Holeman for valuable assistance in the development of our PESTAT code, and to A. W. Glisson (U. Miss.) for kindly sending a Fortran listing of his STATIC code (see Rao et al), which proved helpful in validating our PESTAT code.

REFERENCES

1. L.W. Parker and H.W. Kasemir, "Airborne Warning Systems for Natural and Aircraft-Initiated Lightning," IEEE Transactions on Electromagnetic Compatibility, Vol. EMC-24, No. 2, pp. 137-158, May 1982.

2. D.W. Clifford and H.W. Kasemir, "Triggered Lightning," IEEE Transactions on Electromagnetic Compatibility, Vol. EMC-24, No. 2, pp. 112-122, May 1982.

3. D.W. Clifford, "Aircraft Mishap Experience from Atmospheric Electricity Hazards," NATO AGARD Lecture Series No. 110, Paper No. 2, June 1980.

4. L.W. Parker and H.W. Kasemir, "Breakdown Waves in Lightning Return-Stroke and Leader-Step Channels," 7th International Conference on Atmospheric Electricity, State University of New York at Albany, NY, 4-8 June 1984.

5. L.W. Parker, "Calculation of Sheath and Wake Structure About a Pillbox-Shaped Spacecraft in a Flowing Plasma," in Proceedings of the Spacecraft Charging Technology Conference, C.P. Pike and R.R. Lovell, editors, AFGL-TR-77-0051, pp. 331-366, Feb. 1977.

6. O.C. Zienkiewicz, "The Finite Element Method in Engineering Science," New York: McGraw-Hill, 1971.

7. S.M. Rao, A.W. Glisson, D.R. Wilton and B.S. Vidula, "A Simple Numerical Solution Procedure for Statics Problems Involving Arbitrary-Shaped Surfaces," IEEE Transactions on Antennas and Propagation, Vol. AP-27, No. 5, pp. 604-608, Sept. 1979.

8. J.F. Shaeffer, "Electrostatic Field Solutions for Irregular Electrodes Described by Potential or Net Charge," McDonnell Aircraft Company Report MDC A 1997, Dec. 1972.

9. D.R. Fitzgerald, "Experimental Studies of Thunderstorm Electrification," Air Force Geophysics Laboratory Report AFGL-TR-76-0128, June 1976.

10. D.R. Fitzgerald, "Probable Aircraft Triggering of Lightning Discharges in Certain Thunderstorms," Monthly Weather Review, Vol. 95, No. 12, pp. 835-842, Dec. 1967.

11. W.E. Cobb and F.J. Holitza, "A Note on Lightning Strikes to Aircraft," Monthly Weather Review, Vol. 96, No. 11, pp. 807-808, 1968.

12. B. Vonnegut, "Electrical Behavior of an Airplane in a Thunderstorm," Arthur D. Little, Inc. Report FAA-ADS-36, Feb. 1965.

13. E.T. Pierce, "Triggered Lightning and Some Unsuspected Lightning Hazards," American Association for the Advancement of Science 138th Annual Meeting, Philadelphia, PA, 1971.

14. E.T. Pierce, "Triggered Lightning and Its Application to Rockets and Aircraft," 1972 Lightning and Static Electricity Conference, AFAL-TR-72-325, Dec. 1972.

15. J.F. Shaeffer, "Aircraft Initiation of Lightning," 1972 Lightning and Static Electricity Conference, AFAL-TR-72-325, Dec. 1972.

16. H.W. Kasemir and F. Perkins, "Lightning Trigger Field of the Orbiter," Kennedy Space Center Contract CC 69694A Final Report, NOAA, Oct. 1978.

17. M.M. Newman and J.D. Robb, "Protection for Aircraft," in "Lightning," Vol. 2, R.H. Golde, editor, pp. 659-696 (esp. p. 662 ff.), New York: Academic Press, 1977.

18. P. Laroche, A. Eybert-Berard, P. Richard, P. Hubert, G. Labaune, and L. Barret, "A Contribution to the Analysis of Triggered Lightning: First Results Obtained During the TRIP 82 Experiment," Paper No. 22 in 1983 International Aerospace and Ground Conference on Lightning and Static Electricity, DOT/FAA/CT-83/25(A), Oct. 1983.

19. B.D. Fisher and J.A. Plumer, "Lightning Attachment Patterns and Flight Conditions Experienced by the NASA F-106B Airplane," Paper No. 26 in 1983 International Aerospace and Ground Conference on Lightning and Static Electricity, DOT/FAA/CT-83/25(A), Oct. 1983.

20. E. Barreto, H. Jurenka and S.I. Reynolds, "The Formation of Small Sparks," Journal of Applied Physics, Vol. 48, pp. 4510-4520, 1977.

708

FIGURE CAPTIONS

Fig. 1 - General view of a C-130 aircraft computer model using quadrilateral patches

Fig. 2 - Three views of the computer model, projected onto (a) the x-y plane, (b) the y-z plane, and (c) the z-x plane (of the computational problem space)

Fig. 3 - Field concentration factor (FCF) profile in polar coordinates, in the x-y plane

Fig. 4 - Field concentration factor (FCF) profile in polar coordinates, in the y-z plane

Fig. 5 - Field concentration factor (FCF) profile in polar coordinates, in the z-x plane

Fig. 6 - Critical field profile (in kV/m) in polar coordinates, in the x-y plane, obtained by dividing 3000 kV/m by the FCF profile of Fig. 3 (see text)

Fig. 7 - Critical field profile (in kV/m) in polar coordinates, in the y-z plane, obtained by dividing 3000 kV/m by the FCF profile of Fig. 4 (see text)

Fig. 8 - Critical field profile (in kV/m) in polar coordinates, in the z-x plane, obtained by dividing 3000 kV/m by the FCF profile of Fig. 5 (see text)

Fig. 9 - Quadrilateral patch coordinate system for computation of potential at field point P

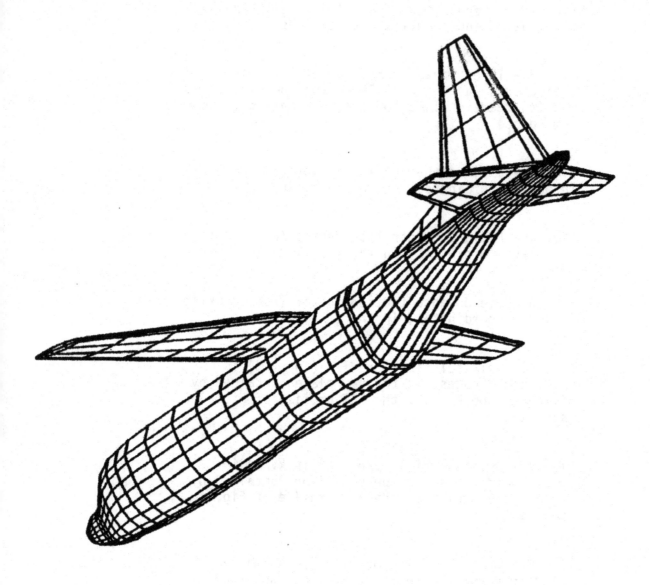

Fig. 1 - General view of a C-130 aircraft computer model using quadrilateral patches

(a)

(b)

(c)

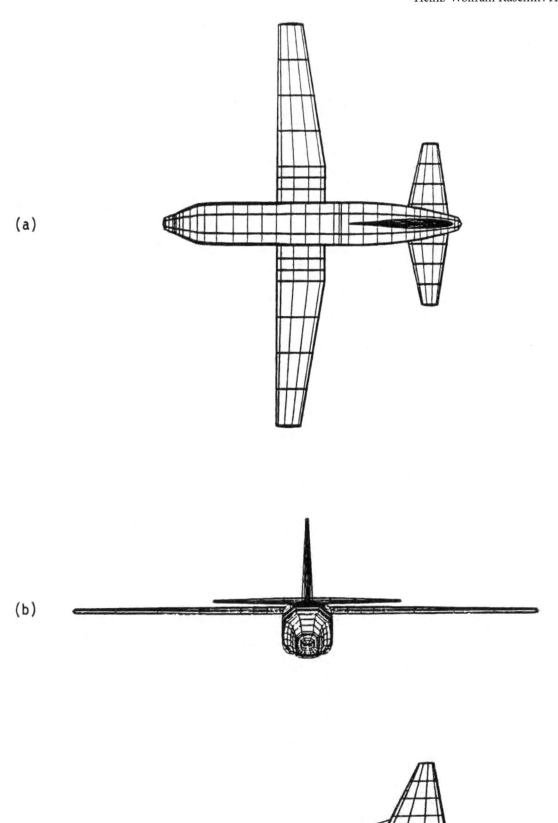

Fig. 2 - Three views of the computer model, projected onto:
(a) the x-y plane, (b) the y-z plane, and (c) the z-x plane
(of the computational problem space)

711

FIELD CONCENTRATION FACTOR

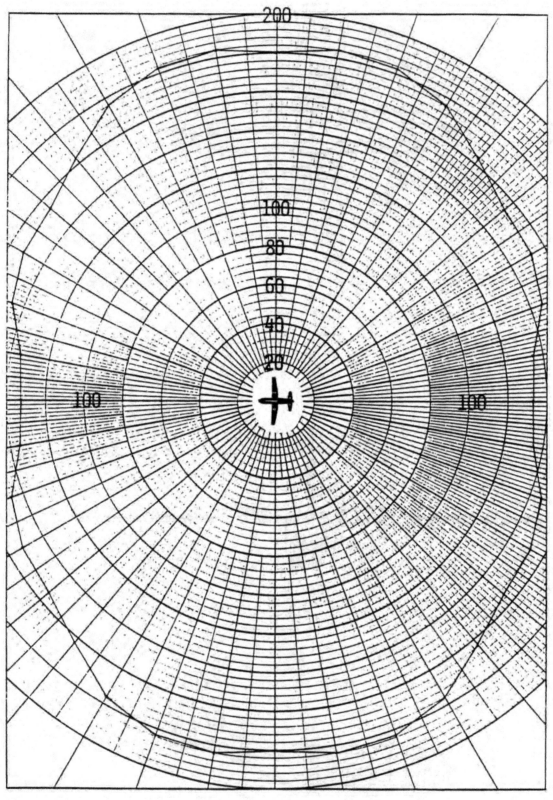

Fig. 3 - Field concentration factor (FCF) profile
in polar coordinates, in the x-y plane

FIELD CONCENTRATION FACTOR

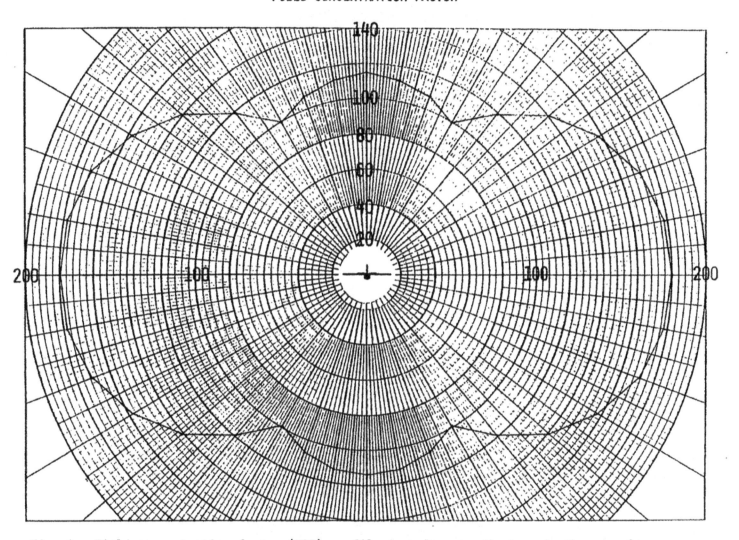

Fig. 4 - Field concentration factor (FCF) profile in polar coordinates, in the y-z plane

FIELD CONCENTRATION FACTOR

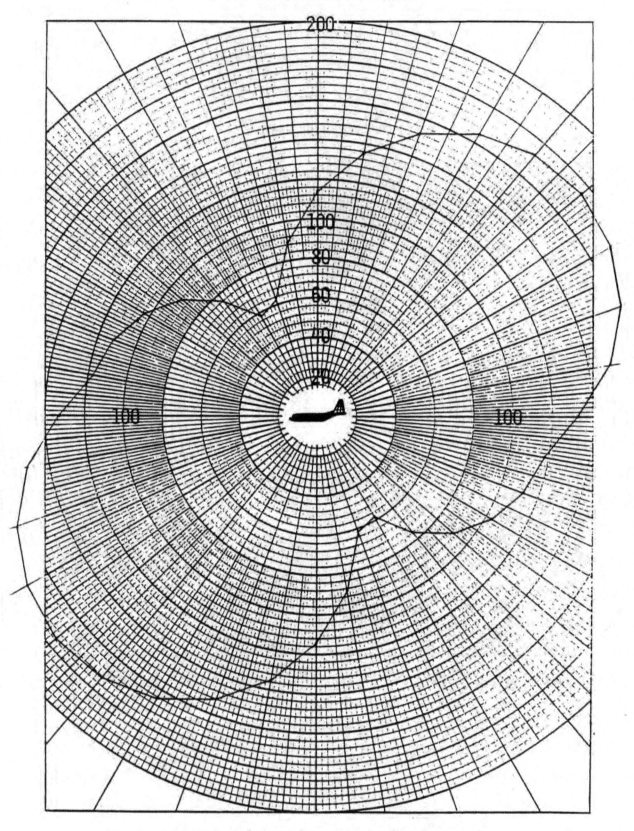

Fig. 5 - Field concentration factor (FCF) profile
in polar coordinates, in the z-x plane

CRITICAL FIELD

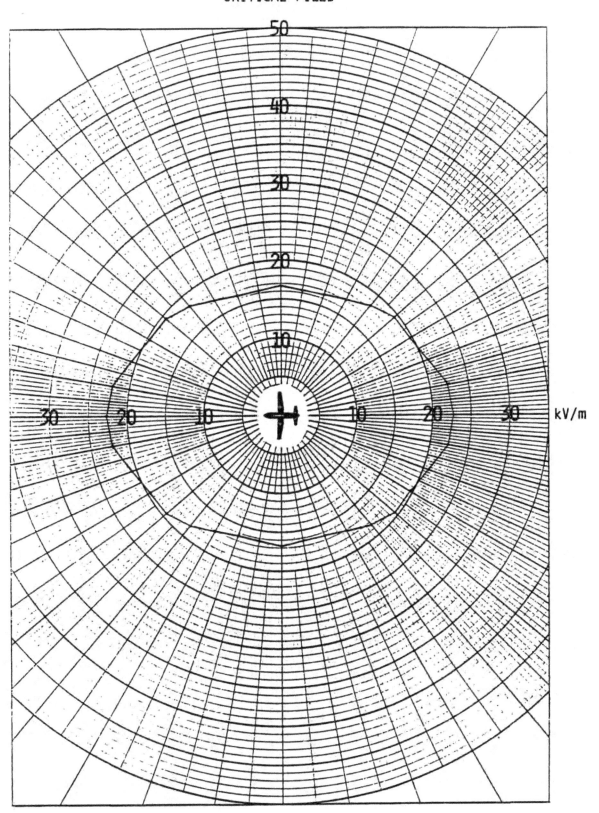

Fig. 6 - Critical field profile (in kV/m) in polar coordinates,
in the x-y plane, obtained by dividing 3000 kV/m
by the FCF profile of Fig. 3 (see text)

CRITICAL FIELD

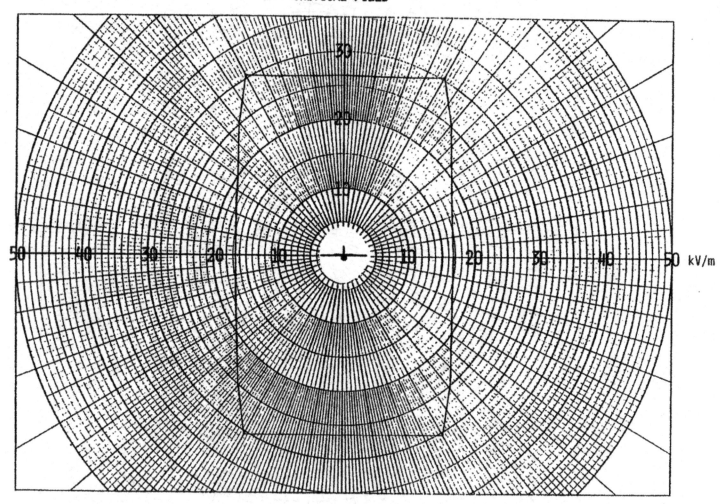

Fig. 7 - Critical field profile (in kV/m) in polar coordinates, in the y-z plane,
obtained by dividing 3000 kV/m by the FCF profile of Fig. 4 (see text)

CRITICAL FIELD

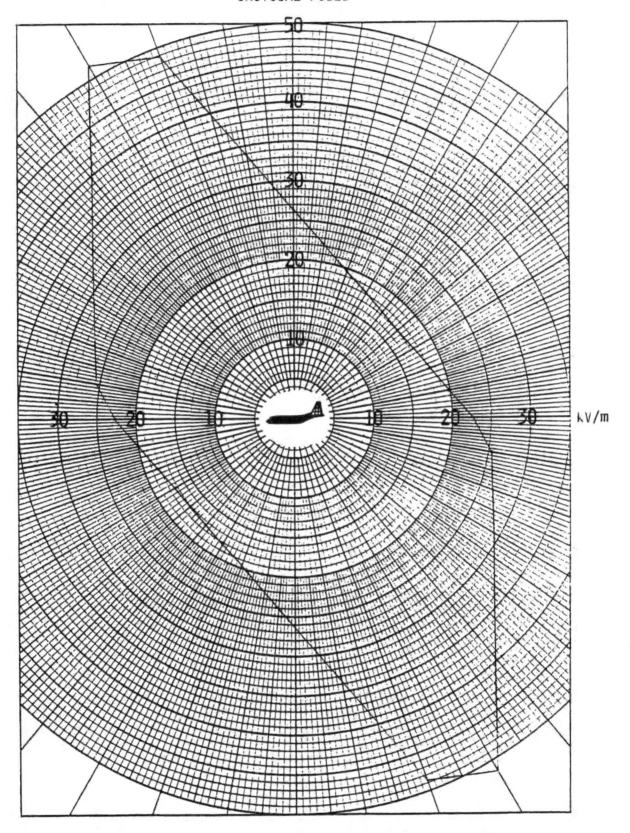

Fig. 8 - Critical field profile (in kV/m) in polar coordinates,
in the z-x plane, obtained by dividing 3000 kV/m
by the FCF profile of Fig. 5 (see text)

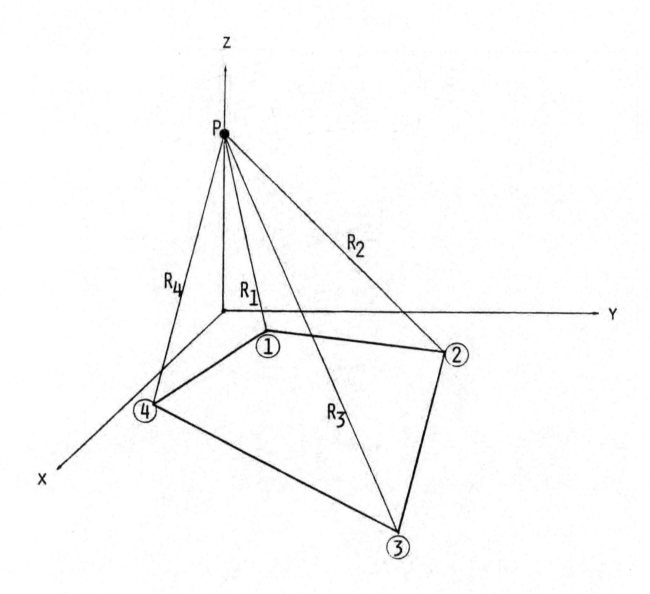

Fig. 9 - Quadrilateral patch coordinate system
for computation of potential
at field point P